ATOMIC MASSES OF SOME IMPORTANT ISOTOPES

1 atomic mass unit $= 1.66 \times 10^{-24}$ gm
$$= 931 \text{ MeV}$$

Except in the case of the neutron and the proton, the mass given refers to the neutral atom.

Name	Symbol	Atomic mass
Carbon 12	$_6\mathrm{C}^{12}$	12.000000
neutron	n	1.008665
proton	p	1.007272
Hydrogen	$_1\mathrm{H}^1$	1.007825
Deuterium	$_1\mathrm{H}^2$ or $_1\mathrm{D}^2$	2.014102
Tritium	$_1\mathrm{H}^3$ or $_1\mathrm{T}^3$	3.016049
Helium 3	$_2\mathrm{He}^3$	3.016029
Helium 4	$_2\mathrm{He}^4$	4.002603
Nitrogen 14	$_7\mathrm{N}^{14}$	14.003074
Oxygen 16	$_8\mathrm{O}^{16}$	15.994915
Uranium 235	$_{92}\mathrm{U}^{235}$	235.0439
Uranium 238	$_{92}\mathrm{U}^{238}$	238.0508

Physics

Physics

Third Edition

Kenneth R. Atkins *University of Pennsylvania*

John Wiley & Sons, Inc. *New York · London · Sydney · Toronto*

Library of Congress Cataloging in Publication Data:

Atkins, Kenneth Robert, 1920–
   Physics.

Includes index.
1. Physics.
QC21.2.A84   1975   530   75-11677
ISBN 0-471-03629-3

Printed in the United States of America

10 9 8 7 6 5 4 3 2 1

Cover Design by Angie Lee

Preface to the First Edition

This book is intended for use in a two-semester, elementary physics course which aims at a proper appreciation of the progress which has been made toward an understanding of the basic physical nature of the universe. It therefore emphasizes such fundamental concepts as conservation laws, symmetry principles, the nature of particles and fields, indeterminacy in quantum mechanics, the nature of space and time in the theory of relativity and the recent explosive developments in the study of fundamental particles. At the same time, it does not neglect such important traditional matters as classical dynamics, kinetic theory, magnetism and the interference and diffraction of waves. However, in order to limit the total amount of material and to avoid confusion of the main theme by the pursuit of side issues, it does omit secondary matters such as hydrodynamics, surface tension, thermoelectricity, geometrical optics and electronics. Although the subject matter is always fundamental, and often profound, an attempt has been made to avoid abstruseness and to adopt a hard-headed approach based upon ideas and phenomena with which the reader is already familiar.

Very little previous mathematical training is required. Elementary algebra and trigonometry are used, but are explained in a mathematical appendix. Calculus is not used. An early chapter is devoted to a careful introduction of vectors, which are then used extensively.

This book should therefore be particularly suited to a liberal arts course intended mainly for non-scientists. It is, in fact, based upon a course which was taught over a period of five years at the University of Pennsylvania to a mixed class containing students majoring in the humanities, premedical students, predental students, architects and a few professional scientists. Since for most of these students it was their terminal course in physics, it was felt that they should not be allowed to go away without some understanding of the significance of the great achievements of physics:

v

Newtonian mechanics, the theory of the electromagnetic field, relativity, quantum mechanics, and the nature of fundamental particles.

The broad approach of this book might also be appropriate in an introductory course for science majors and engineers, who often concentrate so heavily on the details that they never have the opportunity to view the subject as a whole. Such an introductory course might be taken in the freshman or sophomore year of college, but many students are capable of tackling it in high school. High school teachers might also find this book useful in a refresher course.

An unusual feature of this book is the space devoted to the theory of relativity, which occupies four of the thirty-three chapters. This could be justified on the grounds that relativity has made revolutionary contributions to our understanding of time and space and on the grounds that it is so profound a subject that it must be explained in careful detail if it is worth explaining at all. In my opinion, however, the overriding justification is that there is no topic in physics more likely to excite the enthusiasm of the students and to start them thinking and arguing.

The aim of a physics course should be much more than to train the students to solve problems. Nevertheless, questions and problems can help the student to understand the text and to know that he has understood it. The problems have been divided into three categories. Type A problems are very simple and require little more than the insertion of numbers into formulas presented in the text. Type B problems are sufficiently difficult to act as a guide to whether the student has really understood the text. Type C problems are much more difficult and are intended to stimulate and challenge the very best students. In addition to the problems, there are several questions at the end of each chapter. These questions do not ask the student to repeat parrotlike what he has read in the chapter, but are intended to stimulate further thought on the issues involved. Most of them are difficult and a few have no generally accepted answer (for example, question 2 of Chapter 25).

The controversial question of units has been dealt with in the following way. CGS and MKS units have both been used with almost equal emphasis in the chapters on dynamics and heat. The English gravitational system has been almost completely avoided, except that occasionally distances have been expressed in miles, speeds in miles per hour and weights in pounds or tons, when it was felt that this would give the reader a better impression of their magnitude. In general, however, it was felt that the well-known complications introduced by the English system more than offset any advantages arising from its use in everyday life. In the chapters on electromagnetism and atomic physics, the electrostatic CGS system has been given preference. In this system the elementary formulas which are the main concern of this book assume a simple form. Moreover, the velocity of light plays an important role in the formulas of electromagnetism, and this is clearly brought out by the CGS system, but often obscured by the use of the permittivity constant ϵ_0 and the permeability constant μ_0 in the rationalized MKS system. The practical system of electromagnetic units is introduced by quoting the numerical relationship between each practical unit and its CGS counterpart. This is necessary because most students are already familiar with volts, amperes, and watts. It also prepares the way for the indispensable electron volt. To solve problems, the student is encouraged to convert practical units immediately into CGS units and to convert electron volts into ergs.

The two right hand rules adopted in the chapters on electromagnetism are uncommon, although one of them is used in the PSSC text. They were chosen for the following reasons. The right hand is used in both cases and so there can be no confusion about when to use which hand. The thumb indicates the velocity of the charge or the direction of the electric current in both cases, and the direction in which the hand pushes always represents the unknown direction which is being sought. When the angles involved are not right angles, the author's experience is that the chosen rules are less likely than some other rules to lead to neurosis or strained muscles. Finally, if the student or the instructor has previously learned other rules, the chosen rules are sufficiently different to avoid confusion.

The student is encouraged to use powers of ten rather than special units or prefixes. If the wavelength of an x-ray is expressed as 5×10^{-8} cm and the wavelength of visible light as 5×10^{-5} cm, their relative magnitudes are immediately obvious. For non-scientists in particular, the common practice of expressing the x-ray wavelength as 5 Å and the light wavelength as 500 mμ seems to be highly undesirable.

In the diagrams, vectors representing different physical quantities have been distinguished from one another by the use of several different kinds of arrows, and throughout the book a particular vector quantity is always represented by the same kind of arrow. It is not necessary for the reader to memorize this vector code, but it may improve the clarity of the diagrams as he subconsciously comes to realize that, for example, a particular kind of black arrow always represents an electric field, whereas the same kind of arrow in brown always represents a magnetic field[*].

I have benefited greatly from discussions with many of my colleagues, and in particular Professors R. D. Amado, F. Ayzenberg-Selove, S. A. Bludman, H. Brody, M. Cohen, C. W. F. Everitt, K. Geller, J. Halpern, E. G. Muirhead, G. I. Opat, H. Primakoff, and W. Selove. I am grateful to all those who so willingly supplied material for the illustrations. They are acknowledged in the relevant captions, and the word "courtesy" used there is no exaggeration. Particular mention should be made of Dr. Brian Thompson for his beautiful photographs of interference and diffraction, one of which is used on the cover. I am especially grateful to my typist, Mrs. Lillian Faison, the artist, George Kelvin, and the staff of John Wiley & Sons, who not only tolerated but encouraged a somewhat exacting author. Finally, a book such as this, which is based upon a course of lectures, owes an indefinable debt to the questions and comments of many students.

Philadelphia
July, 1964

K. R. Atkins

[*] In the second and third editions the vector code has been modified and electric and magnetic fields are no longer represented by similar arrows in different colors.

Preface to the Third Edition

The reception given to the first and second editions indicates that the approach to the teaching of introductory physics embodied in this book meets the needs of a large group of physics teachers, and therefore no attempt has been made to change the essential character of the book for this third edition. There are certain changes of detail suggested by experience, and there is some new material of topical interest. For the benefit of those familiar with earlier editions, there are the following major changes and additions.

The discussion of conservation of momentum has been amplified by the inclusion of Section 8-3. Chapter 25 on general relativity has been extensively rewritten, and Section 25-2 contains a more thorough treatment of Mach's principle. Most of the new additions are concerned with astrophysics or cosmology in order to take advantage of the way in which the exciting new developments in these subjects illuminate some of the fundamental principles of physics. Specifically, Section 10-6 discusses the role played by rotation in the universe. Section 11-9 is concerned with the physical significance of the various temperatures encountered in different places throughout the universe. The discussion in Section 13-3 of the cosmological significance of the ratio of the strengths of electrical and gravitational forces has been brought up to date. Section 25-6 now contains a more thorough discussion of the gravitational red shift and its astrophysical applications, while Section 25-7 explains how this might have dramatic consequences in the case of "black holes." In Section 28-5 the Heisenberg uncertainty principle is applied to the question of the stability of white dwarf stars and neutron stars. Section 31-4 is a general discussion of available sources of energy with particular emphasis on tracing back these sources to their primary origin in the universe. A discussion of neutrons in neutron stars has been added to Section 32-1. In each case the astrophysical or cosmological topic is introduced in direct relationship to the physical principle that it illustrates.

Several new problems have been included at the end of each chapter. Some of these are related to the new astrophysical topics, but most of them are simple problems in the "A" category. In the earlier editions there was emphasis on difficult problems designed to stimulate the interest of the better students. These problems have been retained, but more consideration is now given to students who find it easier to learn the subject if they are first given simple straightforward problems that can be solved by a direct application of the new laws and equations introduced in the chapter.

Since the first edition of this book the MKS system of units has been formalized as the SI or International System and has received wide acceptance. This third edition contains an appendix explaining SI units and quoting the form taken by the equations of electromagnetism when these units are used. A simple set of rules is given for converting the SI equations into their corresponding CGS form.

Kenneth R. Atkins

Philadelphia
1975

Contents

Physics

The
Way
Ahead

1-1   Describing the World Around Us

Physics is an attempt to describe, in as fundamental and penetrating a way as possible, the nature and behavior of the world around us. Before dismissing this sentence as an abstract and profitless philosophical statement of the type that authors must use to get their books smoothly under way, the reader should accept the challenge, lift his, or her, eyes from the page, and look at the world around. As I do this myself, I am first conscious of a multitude of shapes and colors; a subtle interplay of patches of color on the wallpaper; an expanse of bright red carpet; through the window, the intricate shape of a green tree against a flat expanse of blue sky. On the wall is a reproduction of Picasso's "Boy with a Horse." On the bookshelf the most conspicuous item is a bulky *Complete Works of Shakespeare*. From the record player I hear the sounds of Schubert's Octet in F Major. Whatever the circumstances surrounding the reader, he will probably not have to look far to find things equally complicated and marvellous, and certainly the mind that he is using to consider these matters is an example of one of the most complicated and marvellous structures that can be contemplated.

Obviously, a complete description of the world around us must include the tree, the mind of man, the art of Picasso, the plays of Shakespeare, and the music of Schubert. Even more obviously, these things are not the primary concern of physics. However, if I lift my Picasso off its hook and then let it go, it falls to the ground. The same thing happens with the *Complete Works of Shakespeare* and the record of Schubert's Octet. Moreover, if all three objects are held at the same height and released at the same time, they all strike the ground at the same time (ignoring small, almost imperceptible differences in falling time mainly due to the resistance of the air). This is the sort of thing that is the concern of physics. Whatever is relevant to the aesthetic appeal of the Picasso is clearly irrelevant to the rate at which it falls to the ground, and we are led to make a very general, fundamental statement that: "*All unsupported bodies fall toward the center of the earth at the same rate.*" Of course we shall have to be more precise about the meaning of the phrase "at the same rate," and also we might be well advised to specify that the bodies are to fall through an evacuated space, in order to eliminate the effect of the air which, in an extreme case, makes a balloon rise. But when, following in the steps of Newton, we have refined the idea to the point where the statement becomes: "*Any two bodies attract one another with a force that is proportional to the product of their masses and inversely proportional to the square of the distance between them,*" then we have arrived at one of the most important and basic laws of physics, Newton's law of universal gravitation.

The patches of color that are an essential part of our immediate experience can be analyzed by means of an instrument known as a spectroscope, which is essentially a prism of glass. If white light, such as the sunlight now streaming in through my windows, falls upon this glass prism, it is spread out into a spectrum of the colors of the rainbow, as shown in figure 1-1. All the rays of light are bent by the prism, but the red light is bent least and the violet light most, the other colors coming in between, as shown in the diagram. The colors of this spectrum are the basic components of all visible light. If the spectroscope were arranged to accept only light from my red carpet, then only the red region of the spectrum would be illuminated; if the light came from the green tree, the spectrum would be illuminated mainly in the green region; and if the light came from the blue sky it would be the blue region that would be illuminated (see figure 1-2). However, if the

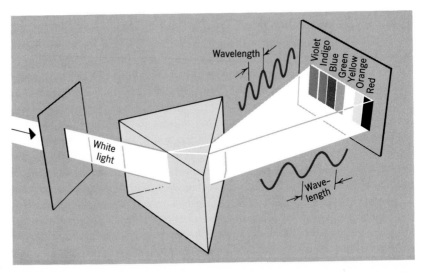

Figure 1-1 A glass prism resolves white light into a spectrum of basic colors. The different colors correspond to different wavelengths.

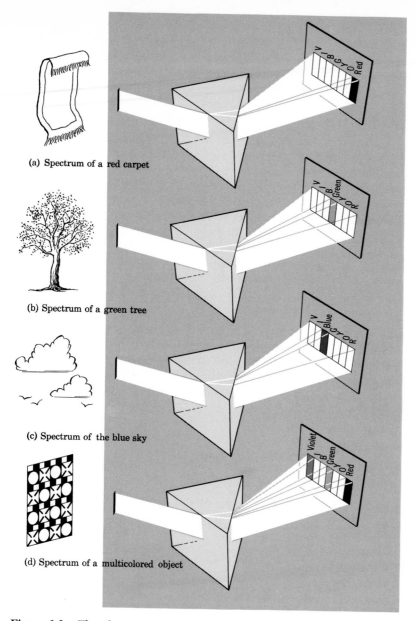

(a) Spectrum of a red carpet

(b) Spectrum of a green tree

(c) Spectrum of the blue sky

(d) Spectrum of a multicolored object

Figure 1-2 The glass prism analyzes various samples of light into basic color components.

spectroscope were made to accept light from my multicolored wallpaper, then several regions of the spectrum would be illuminated, as shown in figure 1-2d. In this way, any sample of visible light can be analyzed into some or all of the basic colors of the spectrum mixed in appropriate proportions.

We are again making the sort of analysis that is the concern of physics, but physics is not satisfied to stop at this stage. It must penetrate more deeply into the heart of the matter and ask, "What is the nature of light, and what is the difference between various basic colors?" This particular question led to several centuries of speculation, culminating in the early nineteenth century with a burst of ingenious experimental activity that provided the following answer: *"Light is a wave, similar to a wave on the surface of the ocean."* The quantity that characterizes the color is the **wavelength,** or the distance between adjacent crests of the wave. For red light this distance is about twice as great as for violet light (see figure 1-1).

There is much more to be said about the nature of light. An ocean wave requires the presence of the ocean, but a light wave can travel through empty space, as it does between the sun and the earth. The explanation of this, provided later in the nineteenth century, was that *"Light is a wave in an electromagnetic field."* Now the analysis has penetrated so deeply that it can no longer be expressed in familiar terms and the uninitiated reader cannot be expected to understand even vaguely what is meant. The explanation is in fact so lengthy that it will not be attempted until later in this book. Even so, it is a nineteenth century explanation, and today the physicist would rather say that *"Light consists of particles (or quanta or photons) whose behavior is governed by a wave in an electromagnetic field,"* all of which we shall attempt to explain later on. Having thus overreached ourselves to illustrate how physics is always penetrating more and more deeply into the mystery of the behavior of things, let us also emphasize again the other aspect of the matter. Whatever is true in these statements about light is equally true of the light from the red carpet and the light from the painting by Picasso. They are an attempt to find something common to all visual experiences.

1-2   Is it True?

It is generally believed that physics deals with indisputable facts and absolute truth. Actually, although the facts are usually indisputable, they are frequently incomplete, and the interpretation of the incomplete facts leads to theories that are only rough approximations to the truth. The situation can be fully appreciated only when one has an understanding of the whole of physics, but we shall attempt to illustrate it by reference to the atomic theory with which, it is safe to assume, the reader already has some familiarity.

In the nineteenth century, when this theory was progressing from triumph to triumph in its clarification of chemistry and the behavior of gases, its proponents might have expounded it in the form of the following postulates.

Postulate A. Matter cannot be divided indefinitely into smaller and smaller pieces: at a certain stage it will be found to consist of very small, submicroscopic entities called atoms, which cannot be divided any further.

Postulate B. Atoms cannot be created or destroyed.

Postulate C. All the atoms of a particular chemical element have identical properties; for example, they all have the same mass.

Postulate D. Atoms of different chemical elements have different properties; for example, they have different masses.

Postulate E. An atom of one chemical element cannot be changed into an atom of a different chemical element.

Now let us examine these postulates in the light of modern knowledge, pointing out the extent to which they are false, but also (to avoid generating too sceptical an attitude) pointing out the extent to which they are good, useful approximations to the true situation.

Postulate A. Atoms can be further subdivided into small fundamental particles known as electrons, protons, and neutrons. It is an open question whether these fundamental particles themselves can be subdivided still further. However, an atom is a stable, compact entity that can remain unchanged for long periods of time, and exceptional measures have to be taken to break it up into its constituent parts.

Postulate B. Modern physicists are aware of processes during which atoms are created out of energy, and reverse processes during which atoms are destroyed and changed into energy. However, these processes are very rare in everyday experience.

Postulate C. The atoms of a particular chemical element can occur in different forms, known as isotopes, with different masses. The element is determined by the number of electrons in the atom (which is the same as the number of protons), but the number of neutrons can then vary to give various isotopes. However, each element has only a small number of stable isotopes and the naturally occurring form of the element usually contains these stable isotopes in fixed proportions.

Postulate D. It is possible for an isotope of one element to have almost the same mass as an isotope of another element. The two atoms are then called isobars. However, isobars do have slightly different masses, and so this postulate is strictly true.

Postulate E. It is well known that nuclear physicists can readily convert an atom of one element into an atom of another element. This process is basic to the performance of a nuclear reactor and an atomic, or hydrogen, bomb. However, the process is not very common in everyday life and is certainly not relevant to ordinary chemical reactions.

In the course of this book we shall discuss several cases of well-established theories that have been found to be not quite true and have had to be modified to bring them a little closer to the truth. An outstanding example is Newtonian mechanics, which is quite adequate to describe the motion of bodies that are not moving extremely fast, and is therefore all that is needed to predict the path of a guided missile, but which breaks down when the bodies begin to move with velocities comparable with the velocity of light and has to be replaced by Einstein's special theory of relativity. It follows inescapably that our best modern physical theories are probably only rough approximations to the truth and will eventually have to be replaced by something better. For example, the special theory of relativity does not cope adequately with the subject of gravitation, and so Einstein was led to formulate the general theory of relativity. However, there are very few experimental tests of the general theory and it has not been accepted with the same degree of confidence as the special theory. Many physicists believe that in this particular field there will be some surprising developments during the next few decades.

Physics must therefore be regarded as an evolving subject approaching closer and closer to the truth but never quite attaining it.

1-3   Classical Physics and Modern Physics

During its evolution, physics has passed through two major phases, which are commonly referred to as classical physics and modern physics. Because these two phases differ so radically, in their attitude toward the nature of the universe and in their philosophical implications, their separation is a profitable one that we shall continually emphasize. In order to give the reader some perspective on the detailed explanations of the succeeding chapters, we shall now present a very brief summary of the evolution of physics.

Classical physics started in earnest in the seventeenth century when Galileo and Newton discovered the laws governing the motion of bodies. It came to regard the universe as a collection of isolated bodies separated by regions of empty, featureless space. The bodies exerted forces on one another in spite of the lack of any obvious direct connection through the intervening space. At first, the bodies discussed were the sun, the planets, the earth's moon, the moons of the other planets, bodies falling to the earth, cannon balls shot from cannons, and so on. The forces were initially gravitational forces, electric and magnetic forces, and certain incompletely understood forces such as the upward force that the top of a table exerts on a plate to prevent it from falling under gravity.

In the nineteenth century the idea that all matter is composed of **atoms** was fully accepted, and the universe then came to be regarded as a collection of isolated atoms moving through empty space and exerting forces on one another. This point of view seemed capable of providing a complete explanation of all the phenomena associated with heat, sound, electricity, magnetism, and the various properties of matter, such as elasticity, viscosity, and surface tension. However, the nature of electricity and the origin of the interatomic forces remained obscure until early in the twentieth century when it was discovered that atoms are themselves complicated structures built up out of **fundamental particles.** The most abundant and most important of these fundamental particles are the **electron,** which is a very light particle and the unit of negative electricity; the **proton,** which is a more massive particle than the electron and the unit of positive electricity; and the **neutron,** which has about the same mass as the proton, but no electric charge. With the discovery of these particles it became necessary to modify the picture of the universe and to think in terms of electrons, protons, and neutrons moving through empty space. This promised to give an even more complete and fundamental description of physical phenomena.

An important feature of this view of the universe was the implication that once the fundamental laws of physics had been formulated, it would be possible to describe completely the motion of the particles and therefore to predict the state of the universe at all future times. This view therefore encouraged the belief that everything that will happen in the future is determined by what has happened in the past and that our lives are governed by Inescapable Destiny. The two capital letters are used to give the phrase an ominous ring and the rest is left to the imagination of the reader.

This brief summary of the situation is an oversimplication, and not all classical physicists would have subscribed to all the above views, particularly the philosophical implications. Moreover, there is an alternative view of the universe that has received considerable attention. This view refuses to regard space as empty and featureless, and prefers to endow all points in space with physical properties. It is sometimes expressed in the extreme form that all of space is filled with a continuous, indivisible fluid frequently called the **ether.** This point of view received considerable support at the beginning of the nineteenth century when Young and Fresnel performed a series of decisive experiments to demonstrate that light is a wave, a wave that can travel through "empty space." At the same time Faraday performed some crucial experiments on the nature of electricity and magnetism. He came to the conclusion that all the space surrounding an electric charge must be visualized as a field of electric force somewhat similar to a continuous all-pervading liquid (the ether) in a state of strain, and that similarly, the region surrounding a bar magnet must be regarded as a field of magnetic force. This implied that every point in space, whether it is occupied by matter or not, is associated with two quantities that specify the nature of the electric and magnetic fields at that point.

This concept of an **electromagnetic field** proved very profitable and the difficult subject of electricity and magnetism yielded to theoretical treatment during the second half of the nineteenth century, when Maxwell expressed the fundamental equations of the subject, not in terms of the behavior of electrically charged particles, but in terms of the behavior of the electric and magnetic fields at all points in space. One consequence of these equations is that they predict the existence of waves that travel through the electromagnetic field. Maxwell was able to show that these waves have all the known properties of light. Light is therefore basically an electromagnetic phenomenon. As the nineteenth century neared its end, Hertz succeeded in generating electromagnetic waves of long wavelength, which we now call radio waves, and he found that their velocity was exactly equal to the velocity of light.

All this emphasis on the behavior of light raised the interesting question "How does the velocity of light depend upon the velocity of the source of the light and the velocity of the observer who sees the light"? The search for an answer to this apparently simple question produced **Einstein's theory of relativity** and drastic modifications in Newtonian mechanics, which had been the basis of classical physics. Einstein presented his theory in 1905. In 1901, Planck had taken the first step in the direction of quantum mechanics. The dawn of the twentieth century thus coincided almost exactly with the dawn of modern physics.

Planck's idea was that light is not emitted continuously but in little bundles of energy called **quanta.** Evidence soon accumulated that light, which had been so firmly established in the nineteenth century as a wave motion, frequently behaves like a stream of particles. To complete the dilemma, electrons, protons, atoms, and molecules, which behaved obviously like particles in many experiments, were found in other experiments to manifest unmistakable properties of wave motion. The answer to this riddle of the duality of particle behavior and wave behavior is contained in the elaborate mathematical formalism of **quantum mechanics.** At its heart is the famous **Heisenberg uncertainty principle,** which tells us that even if an entity like an electron is to be regarded as a particle, it is impossible to know both the exact position and the exact velocity of this particle. The role of the wave associated with the particle is to tell us the probabilities of the particle's being found in various places. Probability and chance then lie at the core of our description of the universe. It becomes impossible to predict precisely any future state of the universe; one can merely state the probabilities of various possible occurrences. Inescapable Destiny is replaced by Fickle Chance!

Modern physics includes both particles and fields. The number of types of fundamental particle have increased rapidly in recent years and it is relevant to ask if they are all fundamental or whether something simpler is hidden beneath. Some elegant relationships are beginning to reveal themselves among these fundamental particles, and the next giant step in the evolution of physics may not be far ahead.

1-4 The Limitations of This Book

The reader is now beginning to realize what this book is all about. It places the emphasis on "pure" physics rather than "applied" physics. Although applied physics is not part of our main purpose, we do not wish to underestimate the importance of applied physics in furthering the welfare of man and improving his control over his environment. Moreover, pure and applied physics have always been inseparable. To mention only one example, the discovery of the electron was a consequence of the simple desire to understand the inner working of nature, but it led inevitably to the development of electronic devices, radio communication, radar, electronic automation, and large computers.

Conversely, most of modern experimental physics, however "pure" its motivation, would be impossible without these electronic devices.

Even within the confines of what would normally be considered pure physics, this book will always place the emphasis on the fundamental principles and will rarely be able to do justice to the way in which these principles have been applied to describe the detailed behavior of matter. For example, we shall consider the structure of the atom and the principles of quantum mechanics which govern the atom's behavior, but it will not be possible to consider in detail all the elements in the periodic table and the way in which the electronic structure of each atom determines the physical and chemical properties of each element. Neither shall we attempt to describe how in recent years the fundamental principles have achieved enormous successes in the explanation of the detailed properties of solids. Again, it is not our intention to underestimate the interest or importance of these matters; but they are complicated, and to discuss them properly would distract us from our main purpose of describing the foundations on which the detailed explanations are built.

Modern physics will be given the attention it deserves, but has not always received, in a comprehensive survey of physics. This does not mean, however, that we shall sweep aside the errors of the past, and present only the latest fashion in physical theory. Modern physics is only a halfway house toward the ultimate truth and it should not be presented with an air of finality. Also, since physics is an evolutionary progress towards better and better understanding, modern physics is a structure built on top of classical physics, and it would not be possible to give a simple exposition of modern physics without first having discussed classical physics. We shall therefore adopt a type of historical approach, but not in the form of a narrative arranged in strict chronological order. Even when discussing the physics of the seventeenth century we shall assume a knowledge of the physics of the twentieth century. The policy of King George III toward his American Colonies takes on its full significance in the light of the events of the nineteenth and twentieth centuries. Similarly, Newtonian mechanics is revealed in its full power and elegance when allied with the atomic theory of the nineteenth century, and its limitations are realized in the light of the twentieth century developments embodied in the theories of relativity and quantum mechanics. The reader should therefore be cautioned that this pseudohistorical approach is aimed purely at clarity of exposition and may therefore create some false impressions of how physical theory did in fact evolve. The reader who values the fullness of his education would be well advised to read some of the many admirable books on the history of science.

Above all else is the hope that the reader will be left with an overwhelming feeling of the intriguing mystery underlying the nature of existence and the beauty and elegance of man's attempt to understand this mystery. Physical theory has an aesthetic appeal which bears comparison with the paintings of Picasso, the plays of Shakespeare, or the music of Schubert.

QUESTIONS

1. You no doubt start out with a preconceived attitude toward physics, based on a previous course you have taken or on what you have heard and read about the subject. Make a careful introspective analysis of your reasons for this attitude. Can you justify it convincingly to yourself, your classmates, and your instructor?
2. The following physical quantities will be very carefully discussed in this book: (*a*) acceleration, (*b*) force, (*c*) energy, (*d*) voltage. What is your present understanding of each of these concepts? How precisely can you define each one? Can you refine your ideas to the point of assigning numerical values to these quantities?

PROBLEMS

Before embarking upon the detailed discussion of the succeeding chapters, check that you are familiar with the necessary mathematical techniques. Read the Mathematical Appendix and attempt the problems given there.

PARTICLES IN MOTION

2

A Mathematical Description of the Universe

2-1 Bodies in Space

Scientists of the seventeenth century were particularly interested in astronomy, and it is therefore easy to understand why some of them regarded the universe as a collection of isolated bodies separated by vast regions of empty space. The appearance of the sky on a clear night illustrates this well (figure 2-1). The stars are points of light whose diameters are clearly very much smaller than the distances between them, and most of the space in the sky appears to be empty. The actual situation is even more impressive than it appears to be. The distance from the sun to the nearest star, Alpha Centauri, is about thirty million times the diameter of the sun! Newton was not aware of this fact, since the first measurement of the distance of a star was not made until the year 1838, but he was aware of the distances involved in the solar system.

Let us consider how the solar system would appear to an observer viewing it from outside. If we shrink the scale until the sun is about the size of a football placed in the center of a football field, the planets can then be represented roughly as follows: Mercury is a grain of sand at a distance of about 10 yards from the sun; Venus is an apple seed at a distance of about 20 yards from the sun; the earth is another apple seed at a distance of about 25 yards from the sun; Mars is a poppy seed at a distance of about 40 yards from the sun, just inside the field; Jupiter is a golf ball somewhere in the stands; Saturn is a table tennis ball in the

Figure 2-1 A portion of the night sky. "Isolated bodies separated by vast regions of empty space." (Mount Wilson and Palomar Observatories.)

parking lot outside; Uranus is a pea three blocks away; Neptune is a pea about half a mile away; and Pluto is a grain of sand about three quarters of a mile away. The moon would be represented by a grain of sand about $2\frac{1}{2}$ inches from the apple seed that represents the earth.

2-2   The Universe as a Collection of Particles

For some of the purposes of astronomy, then, the earth may be treated as a small particle in space—but we all know that it is much more complicated than that. What about the air, the rocks, the oceans, and the myriad forms of complicated living organisms that teem over the surface of the earth? We now know that any object, however complex, can be visualized as a sufficiently complicated structure built out of atoms, and that the atom itself is a structure composed of the three fundamental particles, electrons, protons, and neutrons. The protons and neutrons are packed tightly together into the nucleus, which occupies a region of diameter about 10^{-12} cm at the center of the atom. The electron has a diameter of about 10^{-13} cm and is about 10^{-8} cm from the nucleus. The atom is therefore about 10,000 times larger than the nucleus and about

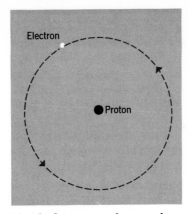 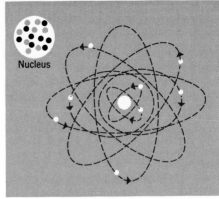

(a) A hydrogen atom has a nucleus consisting of a single proton and a single electron is revolving about this nucleus.

(b) In the oxygen atom the nucleus contains eight protons and eight neutrons tightly packed together. Eight electrons revolve about this nucleus.

Figure 2-2 The structure of the hydrogen and oxygen atoms.

100,000 times larger than the electron. The atom, like the solar system, is mainly empty space.

Figure 2-2 shows the structure of two atoms; the simplest atom, hydrogen; and a slightly more complicated atom, oxygen. By bringing together two atoms of hydrogen and one atom of oxygen, and modifying the orbits of some of the electrons, we obtain a molecule of water, as shown in figure 2-3. Using mainly atoms of carbon, hydrogen, oxygen, and occasionally atoms of other elements, we can build up complicated structures representing organic molecules and eventually, when the structures become extremely complex, living organisms. In this way one might hope to describe something as complex as a nerve cell in a human brain.

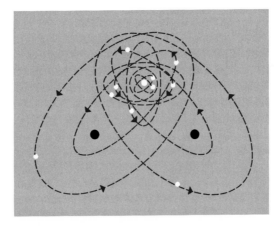

Figure 2-3 The construction of a water molecule.

Consider, then, the following suggestion for giving a complete description of the whole universe. Assume that the universe is composed entirely of small particles such as electrons, protons, and neutrons. At a given instant of time, specify exactly the position of each particle in the universe. This will give a complete description of the instantaneous appearance of any structure in the universe. For example, given the positions of all the particles inside a human brain, we would have a complete description of the state of that brain. Now repeat this procedure at all instants of time. Then, according to this viewpoint, everything that could be known about the universe would be known. For example, by describing in this way the changing pattern of particles in the human brain, we might hope to have a complete description of all the thoughts passing through that brain.

It is not possible! In the year 1927 it was finally shown not to be possible. For three centuries before that, it might well have been. The developments of the year 1927 (the Heisenberg uncertainty principle) will not be discussed until chapter 28. Meanwhile, it will serve our purpose well, and considerably simplify the exposition, to assume that the universe *is* a collection of particles which *can* be located precisely at points in space.

2-3 Fixing the Positions of the Particles

How shall we set about this program of specifying the exact position of each particle at each instant of time? Suppose that I toss a piece of chalk across the room and ask where it is half a second after leaving my hand. A physicist cannot be satisfied by a vague statement such as: "The chalk

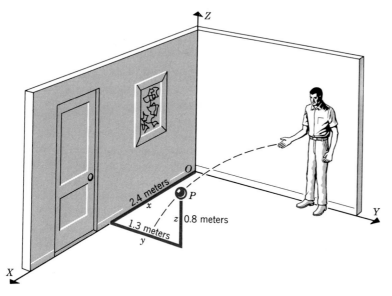

Figure 2-4 Cartesian coordinates.

A Mathematical Description of the Universe

is somewhere over there about midway between the painting and the door."
He must be more precise, and he achieves precision by making use of
numbers.

We can proceed in the following way. We can choose one corner of
the room as our reference point and call it the **origin** O (see figure 2-4).
We can then say that, to move from the origin O to the piece of chalk,
we must go 2.4 meters along the length of the room, then go 1.3 meters
across the width of the room, and finally rise 0.8 meter above the floor.
If you have had sufficient previous mathematical training, you may recognize that this is equivalent to the mathematical procedure of choosing the
three edges of the room, OX, OY, and OZ to be **Cartesian axes** (examine
figure 2-4 carefully). The point P at which the chalk is located is specified
by giving its three **Cartesian coordinates,** $x = 2.4$ meters, $y = 1.3$ meters,
and $z = 0.8$ meter.

Figure 2-5 shows how to find the position of the point P relative to the
Cartesian axes for several cases in which the Cartesian coordinates x, y,
and z have been given definite numerical values. It also illustrates the fact
that a coordinate may be negative. If the distance along the axis is measured
out by starting at O and moving in the direction of the arrow-head, then
the coordinate is positive; but if it is necessary to move from O in a direction
opposite to the arrow-head, then the coordinate is negative. This is best
understood by studying the figure carefully.

Notice that in order to specify the position of a particle completely it
is necessary to know *three* distances. This is what is meant by the statement that **space has three dimensions.**

To proceed with the description of the universe, the three Cartesian
coordinates are all that are needed to determine the position of the
particle at the instant of time when it is at P. Suppose that the Cartesian
coordinates of *all* the particles in the universe are also known at this
same instant of time. Then, according to the present viewpoint, this
knowledge provides a complete instantaneous description of the state of
the universe. We are assuming that we also know the nature of each
particle, whether it is an electron, proton, neutron, or one of the other
less common fundamental particles.

A complete description of the behavior of the universe requires that
the coordinates of all the particles be known at *all instants of time.* For
example, the particular particle we are discussing does not necessarily
remain at P but may move along the path shown in figure 2-6. Some
instant of time must be chosen as the zero of time. Suppose that t
seconds after this instant of zero time the particle is at P with coordinates
(x, y, z). t' seconds after the instant of zero time the particle will have
reached a different point P' with coordinates (x', y', z'). For every value
of the time there is a point on the path where the particle is located at
that time. Thus, an instantaneous description of the particle involves the
three space coordinates x, y, and z and also the time t. Time is therefore
a sort of *fourth dimension* and it is possible to talk about the "four-
dimensional space-time continuum." This expression is often used in the
theory of relativity, since in this theory there is a mathematical trick

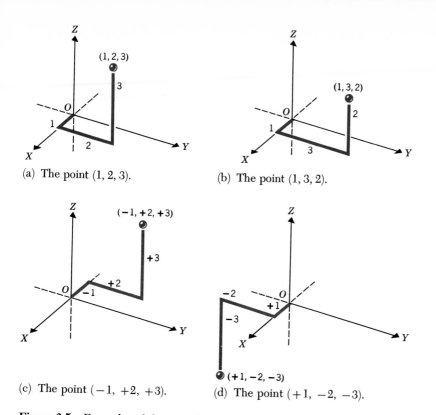

(a) The point (1, 2, 3).

(b) The point (1, 3, 2).

(c) The point (−1, +2, +3).

(d) The point (+1, −2, −3).

Figure 2-5 Examples of the use of Cartesian coordinates. It is assumed that the distances are measured in terms of some unit distance. For example, the point (1, 2, 3) might be more precisely written (1 cm, 2 cm, 3 cm).

whereby time can be incorporated into the equations in the same way as the coordinates of space. However, we should not be dazzled by the mathematics, and we should always remember that time is an essentially different thing from space. It would be better to talk about the "three-plus-one dimensional space-time continuum."

The essential thing is that, when we are given the four numbers

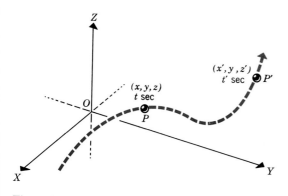

A Mathematical Description of the Universe

Figure 2-6 The path of a particle through space and time.

17

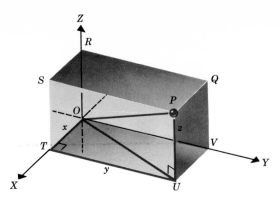

Figure 2-7 Illustrating example 2-3-1.

(x, y, z, t), they tell us that t seconds after the agreed zero of time the particle is at a point whose Cartesian coordinates are x, y, and z with respect to an agreed set of Cartesian axes. If this set of four numbers is given for all values of the time t, then the motion of the particle is completely described.

Example 2-3-1

Find the distance of the point (x, y, z) from the origin O.

Referring to figure 2-7, consider triangle OUP, in which the angle OUP is a right angle. From Pythagoras' theorem,

$$OP^2 = OU^2 + UP^2 \tag{2-1}$$

Consider triangle OTU in which the angle OTU is a right angle. Using Pythagoras' theorem again,

$$OU^2 = OT^2 + TU^2 \tag{2-2}$$

Substituting this value of OU^2 in the first equation

$$OP^2 = OT^2 + TU^2 + UP^2 \tag{2-3}$$

But $OT = x$, $TU = y$, $UP = z$, and so

$$OP^2 = x^2 + y^2 + z^2 \tag{2-4}$$

The distance of the point (x, y, z) from the origin O is therefore

$$OP = \sqrt{x^2 + y^2 + z^2} \tag{2-5}$$

2-4 Mass, Length, and Time

The above method of describing the universe involves two basic types of measurement—measurements of length and measurements of time. In addition the particles have certain intrinsic properties, the most important of which is mass. This section is concerned with the units of mass, length, and time.

To measure the length of the line on this page in centimeters, you would take a rule graduated in centimeters, lay it alongside the line, and adjust the zero mark on the scale to be opposite one end of the line. You would then notice that the other end was opposite the mark labeled 9 and you would then say that the length of the line is 9 centimeters. There are two aspects of this statement. One is the *number* 9, and the other is the *unit of length,* the centimeter. The full significance of the statement is that you are provided with an agreed unit of length called the centimeter and nine of these units laid end to end have the same length as the line. The laying end to end is done by the manufacturer of the rule who has taken pains to see that the distance between adjacent marks is as nearly as possible equal to this agreed unit of length. The unit is just as important as the number. If the centimeter had originally been chosen to be just one half as long as it now is, then the length of the line would have been 18 centimeters.

The unit of length used to be defined in a very practical way by reference to a certain platinum-iridium bar that is kept under very careful conditions at Sèvres in France. The distance between two scratches on the surface of this bar was *defined* as exactly 1 **meter,** or 100 **centimeters.** The "very careful conditions" are elaborately designed to ensure that the length of the bar does not change; for example, the temperature is kept constant to avoid thermal expansion or contraction of the metal. However, physicists have long been uneasy about the arbitrariness of this standard and the possibility that the distance between the scratches might vary slightly due to unknown factors. In 1960 it was finally decided by international agreement that the unit of length should be defined with reference to the wavelength of an orange line in the spectrum of atoms of a single, pure isotope of krypton (krypton 86). The wavelength is, of course, the distance between adjacent peaks of the wave (see figure 1-1). The meter is now defined to be exactly 1,650,763.73 times the wavelength of this particular krypton line. To the best of our knowledge, this is the present distance between the two scratches on the platinum-iridium bar.

In the same vault at Sèvres where the standard meter is kept there is a cylinder of platinum-iridium which, by international agreement, is taken to represent a mass of exactly 1 **kilogram** or 1000 **grams.** The physical nature of mass will be discussed in chapter 6.

The unit of time is the **second,** which of course is related to the rotation of the earth on its axis. The earth rotates once in 1 day, which is equivalent to $24 \times 60 \times 60 = 86,400$ sec. However, the procedure used by astronomers to define the length of the day is extremely complicated. The obvious definition of the length of the day might seem to be that it is the time between successive arrivals of the sun at its highest position in the sky. Unfortunately, because of the elliptical motion of the earth around the sun, this time varies from day to day throughout the year and it becomes necessary to take an appropriate average over the whole year. The resulting average is called the *mean solar day* and a better definition of 1 second is that it is 1/86,400 of a mean solar day. However, the motion of the earth is complicated by many factors, some known and some unknown, and the mean solar day varies from year to year. Therefore, in order to define the second with the minimum of ambiguity, it was decided by international agreement that a particular year, the year 1900, was exactly 31,556,925.9747 sec long.

The erratic behavior of the rotation of the earth makes it a very unsuitable basis for the definition of the unit of time, and so in 1967 a new international agreement based the definition of the second on the more fundamental, more reproducible behavior of an "atomic clock." In the type of atomic clock chosen, atoms of cesium 133 are made to vibrate by radio waves similar to those used in radar. The second is now defined as the time interval during which a cesium atom makes 9,192,631,770 such vibrations. Within the limits imposed by the accuracy of present measurements, it is the same as the second defined in the previous paragraph.

There are two major systems of units used in physics, the MKS system and the CGS system.

MKS System *or Meter-Kilogram-Second System*

The unit of length is the meter (m), and is 1,650,763.73 times the wavelength of the krypton-86 line. The unit of mass is the kilogram (kg), and it is the mass of the platinum-iridium cylinder kept at Sèvres. The unit of time is the second (sec), and it is the time interval during which a cesium atom vibrates 9,192,631,770 times in an atomic clock.

CGS System, *or Centimeter-Gram-Second System*

The unit of length is the centimeter (cm), and it is 1/100 of a meter or 16,507.6373 times the wavelength of the krypton-86 line. The unit of mass is the gram (gm) and it is 1/1000 of the mass of the platinum-iridium cylinder. The unit of time is again the second (sec).

Since the fundamental standards are the same in both systems, it might appear that there is no real difference between them. The point is that in the MKS system all lengths must be expressed in meters and all masses in kilograms, whereas in the CGS System all lengths must be in centimeters and all masses in grams. A velocity might be given as 3 meters per second in the MKS system, but in the CGS system it would be 300 centimeters per second. The density of water is 1 *gram* per cubic *centimeter* in the CGS system, but 1000 *kilograms* per cubic *meter* in the MKS system. The two systems must never be mixed by expressing masses in grams and lengths in meters, for example, as in the highly undesirable statement that the density of water is 1,000,000 grams per cubic meter. It is also inadvisable to depart from either system, as in the expression of a velocity as 3 millimeters per second. This should be either 0.3 centimeters per second or 3×10^{-3} meters per second. Expressing a speed as 36 kilometers per hour is quite acceptable when driving an automobile in France, but when computations in physics are concerned, it should be changed to 10 meters per second (MKS System) or 1000 centimeters per second (CGS System).

The Golden Rule, then, is

> Decide at the very beginning of a calculation whether you are going to use the MKS or the CGS System, and stay with this decision. If it is the MKS System, express all lengths in meters, all masses in kilograms, and all time intervals in seconds. If it is the CGS System, express all lengths in centimeters, all masses in grams, and all time intervals in seconds.

The measurement of many physical quantities involves only measurements of mass, length, and time, and this fact can be used to express the units of these quantities. For example, a velocity is a length divided by a time and may therefore be written

MKS System: 7 meters per second, 7 m/sec, or 7 m sec^{-1}.
CGS System: 700 centimeters per second, 700 cm/sec, or 700 cm sec^{-1}.

A density is a mass divided by a volume, and a volume is a length multiplied by a length multiplied by a length. The density of water may therefore be written

MKS System: 1000 kilograms per cubic meter, 1000 kg/m^3, or 1000 kg m^{-3}.
CGS System: 1 gram per cubic centimeter, 1 gm/cm^3, or 1 gm cm^{-3}.

As we shall see later, *linear momentum* is mass multiplied by velocity. Its units are therefore kg m sec^{-1}; or gm cm sec^{-1}. *Angular momentum* is linear momentum multiplied by a length, and its units are kg m^2 sec^{-1}; or gm cm^2 sec^{-1}.

There is a third system of units, the **English system** or **gravitational system,** in which the unit of length is the **foot,** and the unit of time is the **second.** Instead of defining a unit of *mass* it defines a unit of *force* called the **pound.** This system is extensively used in engineering and in practical affairs, but almost never by physicists. Its use in an introductory text on pure physics would lead only to unnecessary confusion, with no compensating gain, so it will be studiously avoided.

Example 2-4-1

If the density of water is 1.000 gm cm^{-3} in the CGS system, what is it in the MKS System?

1 cubic centimeter of water has a mass of 1 gram.
1 cubic meter = $(100)^3 = 10^6$ cubic centimeters.

Therefore 1 cubic meter of water has a mass of 10^6 grams.

10^6 grams = 10^3 kilograms.

1 cubic meter of water has a mass of 10^3 kilograms. Therefore, in the MKS system, the density of water is 10^3 kg m^{-3}.

1. What is your reaction to the suggestion that your thoughts can be completely described in terms of the changing positions of electrons, protons, and neutrons in your brain? Try to estimate the extent to which this reaction is based upon emotion and the extent to which you can support it by reference to observational facts and the use of logic.

2. Which of the following distances is comparable with the length of your index finger?
(a) 10^6 cm, (b) 10 m, (c) 0.1 m, (d) 10^{-3} cm, (e) 10^{-6} m.

3. Think of an object comparable in size with each of the following distances.
(a) 10^7 m, (b) 10^5 cm, (c) 10^3 cm, (d) 1 m, (e) 10^{-3} m, (f) 10^{-8} cm, (g) 10^{-12} cm.

4. In each case state whether the quantity is given in the MKS system of units, the CGS system of units, or neither. (a) An acceleration of 3.6 centimeters per second per second. (b) One pound of butter. (c) A five-year-old child. (d) The record for the 100-meter dash is 8.34 seconds. (e) An angular momentum of 10 gm cm^2 sec^{-1}. (f) A density of 20 grams per cubic meter.

5. The sine of an angle θ is defined as the ratio of the length of the opposite side to the length of the hypotenuse of a certain triangle. Does the value of sin θ depend upon the unit of length used?

6. Does the measure of an angle in radians depend upon the unit of length used?

7. Consider the ratio of the length of a particular day (January 1st, 1950, for example) to the total length of the particular year (1950) in which it occurred. Does this ratio depend upon the choice of the year 1900 in the definition of the second?

8. After a laborious calculation, you conclude that the distance from the center of the earth to the center of the moon is 3.8×10^{10} cm and that the diameter of the moon is 3.1×10^{10} cm. Relying only upon your knowledge of the appearance of the moon in the night sky, can you think of a simple reason why these results cannot both be correct?

9. If the standard meter were *defined* as the distance between two scratches on a certain platinum-iridium bar, would it be completely pointless to discuss the question of whether the length of this bar changes with time or environmental conditions?

10. Suppose that very accurate measurements suggested that the diameter of the earth is gradually shrinking. How could we be sure that this was not really due to a gradual increase in the length of the standard meter?

11. Imagine that a theorist has suggested that the wavelength of the orange line in the spectrum of krypton 86 is gradually increasing as the universe grows older. Defend him against the charge that this theory is meaningless. Suggest some other lengths with which the wavelength might be compared.

12. Discuss the following statement: "The principal requirement of the standard of length is that its variations in length shall be much less than the variations of the lengths of most other things."

13. Why does the departure of the earth's orbit from a perfect circle result in a variation of the length of the solar day throughout the year? Suggest reasons why the *mean* solar day might vary from year to year.

PROBLEMS

A

1. Draw diagrams similar to figure 2-5 for the points $(1, 1, 2)$, $(5, 1, 3)$, $(-1, -2, -3)$, $(-1, -1, +1)$, $(+1, +2, -3)$, and $(+1, -2, +3)$.
2. The top speed of a certain aircraft is 420 mph. An international commission decides to define a new mile which is three times as long as the present one. What is the top speed of the aircraft in "new-miles" per hour?
3. Convert the following quantities to the CGS system of units. You may use the table of conversion factors on the inside of the cover. (*a*) 5 miles, (*b*) 50 millimeters, (*c*) 30 m, (*d*) 5 kg, (*e*) 5×10^{-3} kg, (*f*) 18 months, (*g*) 6 minutes, (*h*) 1000 mph, (*i*) 5 kilometers per hour, (*j*) 0.2 mm/sec, (*k*) 15 m sec^{-1}.
4. Convert the quantities in the previous question to the MKS system of units.
5. From your knowledge of the number of seconds in a minute, the number of minutes in an hour, and so on, calculate the number of seconds in a year. Why does your answer differ from the number quoted in the text?
6. Approximately how many atoms placed side by side in a straight line touching one another would stretch across a lecture room 10 meters wide?
7. Calculate to the nearest second the time for a cesium atom in an atomic clock to make 2×10^{12} vibrations.

B

8. The speed limit on a certain expressway is 60 mph. An international commission decides that the hour shall be re-defined so that it takes the earth only 20 hours to rotate once on its axis. What should the new speed limit be?
9. About how many seconds are there in the lifetime of an average man? Express your answer in the form $n \times 10^m$, where n and m are integers.
10. A famous traveler took 80 days to go around the world. At the same average speed, how long would it take him to go to (*a*) the moon, (*b*) the sun, (*c*) Pluto (6×10^{14} cm away), (*d*) the nearest star, Alpha Centauri (4×10^{18} cm away)?
11. A rocket is able to achieve speeds in the vicinity of 30,000 mph. At this speed, how long would it take to encircle the earth at the equa-

tor? How long would it take to go to the celestial objects mentioned in the previous question?

12. A student calculates that the velocity which must be given to a rocket to enable it to escape completely from the earth is 721 cm sec^{-1}. Is this reasonable? Why or why not?

13. How far from the origin is the point (1 cm, 2 cm, 3 cm)?

C

14. Modify the proof given in example 2-3-1 to show that the distance OP is still $\sqrt{x^2 + y^2 + z^2}$ when x and y are positive, but z is negative.

15. The point P is (x, y, z) and the point P' is (x', y', z'). Show that the distance of P' from P is

$$PP' = \sqrt{(x' - x)^2 + (y' - y)^2 + (z' - z)^2}$$

In the first instance assume that x' is greater than x, y' is greater than y, and z' is greater than z. Then consider a case such as x' greater than x, y' greater than y, but z' less than z.

16. What is the distance between the point (1, 1, 2) and the point (2, 3, 4)? The unit of length is assumed to be the centimeter.

17. What is the distance between the points $(-1, +2, +1)$ and $(+1, +1, -2)$?

18. What is the distance between the points $(+a, +b, +c)$ and $(-a, -b, -c)$? First find the answer by using the formula of problem 12. Then draw a diagram that will help to convince you that your answer is right. (If you find this difficult to visualize, consider first a two dimensional diagram when $c = 0$).

19. What is the distance between the points $(+a, +b, +c)$ and $(+a, +b, -c)$? Again draw a diagram to elucidate the answer.

20. What is the distance between the points $(+a, +b, +c)$ and $(-a, -b, +c)$? Draw the diagram for this case.

21. The Empire State Building is 102 stories high. Working in centimeters, place a lower limit on the height of the building. Make this lower limit as large as you can, but be prepared to advance a completely convincing argument to prove that the height could not possibly be less than this.

22. Think of an object which has a mass of about (a) 1 gm, (b) 100 gm, (c) 1 kg, (d) 100 kg, (e) 10^6 gm, (f) 10^{16} gm. (Hint: Most substances have a density in the range between 1 gm/cm^3 and 10 gm/cm.^{3} Take an average density of 5 gm/cm^3 and estimate the volume of the object. Calculate the length of the side of a cube with the same volume. This will act as a guide to the approximate size of the object. Then, if you wish, you can look up the density and estimate more precisely the volume of the object you choose.)

23. Assuming that the oceans occupy about one half of the surface of the earth and have an average depth of the order of magnitude of $2\frac{1}{2}$ miles, calculate approximately what fraction of the mass of the earth is contained in the oceans.

Vectors

3-1 The Nature of a Vector

In a three-dimensional space there are many physical quantities that have direction as well as magnitude. Such quantities are called **vectors.** One of the simplest examples of a vector is the **displacement** of one point from another. The distance between Philadelphia and New York is 90 miles, but you cannot reach New York by traveling 90 miles in a southerly direction from Philadelphia. You must travel 90 miles in exactly the right direction, which is approximately northeast. The displacement of New York from Philadelphia is a vector, and it has both magnitude and direction. Its magnitude is 90 miles and its direction is northeast.

A second example of a vector is **velocity.** From a practical point of view, traveling due west at 60 miles per hour (mph) is very different from traveling due east at 60 mph. To describe completely the velocity of a body, one must give the direction in which it is moving in addition to the distance it travels in a given time. Strictly speaking the word "velocity" should always indicate a vector, whereas the word "speed" should be used to indicate the magnitude of the velocity, its direction being ignored. In this book we shall follow this convention whenever the distinction is important, but in common usage the word "velocity" is often used instead of "speed."

A third example of a vector is a force. Until we have defined force precisely, the reader may rely upon his intuitive notion of force as a push or pull. Clearly, if one is trying to push an automobile up a hill, the *direction* in which one pushes is as important as the strength one exerts.

A quantity that has magnitude but no direction is called a **scalar.** Examples of scalars are time, mass, speed, temperature, electric charge, volume, density, and energy.

3-2 Representation of a Vector

In print, the symbol for a vector is set in boldface type, whereas the symbol for a scalar is set in italics. The vector representing a velocity would therefore appear as v, whereas the corresponding speed would be v. In handwriting, this distinction is not possible and it is convenient to indicate a vector quantity by placing an arrow above it; for example $\vec{v}$.

The vector displacement from point P to point Q might be written $\overrightarrow{PQ}$, with an additional implication that the displacement is in the direction from P to Q, rather than from Q to P. The magnitude of the displacement, a scalar quantity, is written PQ.

A vector is represented pictorially by a line that is drawn in the same direction as the vector and has a length proportional to the magnitude of the vector on some agreed scale. An arrowhead indicates one of the two possible directions along the line. Figure 3-1a is a map of part of the northeastern United States drawn on a scale of 1 inch to 200 miles. Figure 3-1b shows how the displacement of New York from Philadelphia can be represented by an arrow drawn in the right direction with a length of 0.45 inches which, according to the scale of the map, represents the 90 miles between the two cities. A velocity of 60 mph due north could be represented by a line with an arrowhead pointing due north and a length of 6 inches, if it had been agreed that 1 inch was to represent 10 mph. A velocity of 40 mph due west would then be represented by a line 4 inches long pointing due west and therefore at right angles to the first line.

Multiplication of a vector by a scalar does not change the direction of the vector but multiplies its magnitude by the scalar in question. The vector $\mathbf{A}$ has magnitude A and the vector $c\mathbf{A}$ therefore has magnitude cA and is in the same direction as $\mathbf{A}$. Multiplication of a vector by a minus sign reverses its direction so that it points in exactly the opposite direction. Figure 3-2 illustrates these points.

3-3   Addition of Vectors

Figure 3-3 is another map of part of the northeastern United States. The displacement from Philadelphia to New York is represented by the vector $\overrightarrow{PQ}$. The displacement from New York to Albany is represented

PARTICLES IN MOTION

(a) The locations of Philadelphia and New York on a map.

(b) The vectorial representation of the displacement of New York from Philadelphia.

Figure 3-1

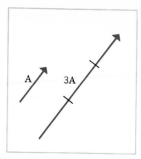

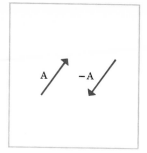

(a) Multiplication of a vector A by a scalar quantity of $c = 3$.

(b) The distinction between **A** and −**A**.

Figure 3-2

by the vector $\overrightarrow{QR}$. From the point of view of one's eventual location, a trip from Philadelphia to New York followed by a trip from New York to Albany has exactly the same end result as a trip directly from Philadelphia to Albany. The vector $\overrightarrow{PR}$ is therefore exactly equivalent to the sum of the vectors $\overrightarrow{PQ}$ and $\overrightarrow{QR}$. This is written

$$\overrightarrow{PR} = \overrightarrow{PQ} + \overrightarrow{QR} \tag{3-1}$$

Throughout the book statements or equations of particular importance will be enclosed between vertical colored bars. Notice that this is a vector equation and must be clearly distinguished from an ordinary algebraic equation involving scalar quantities, with which the reader may already be familiar. It is *not* true, for example, that

$$PR = PQ + QR \tag{3-2}$$

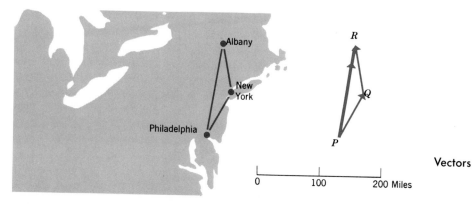

Vectors

Figure 3-3 Use of a map to illustrate the addition of vectors.

27

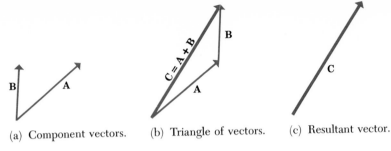

(a) Component vectors. (b) Triangle of vectors. (c) Resultant vector.

Figure 3-4 Addition of two vectors using the "triangle of vectors."

since one covers fewer miles in going directly from Philadelphia to Albany than if one went via New York.

The method used above to add the vectors $\overrightarrow{PQ}$ and $\overrightarrow{QR}$ and obtain the vector $\overrightarrow{PR}$ was to construct a triangle PQR. This method works for the addition of any two vectors, whether they are displacements, velocities, forces, or any other type of vector. The procedure is sometimes referred to as the **triangle of vectors** and is illustrated in a general case in figure 3-4. To add the vectors **A** and **B**, construct a triangle in which **A** and **B** are adjacent sides. The sum of **A** and **B** is then the third side **C** and one can write

$$\mathbf{C} = \mathbf{A} + \mathbf{B} \qquad (3\text{-}3)$$

Notice that the arrowheads on the vectors being added together, **A** and **B**, both proceed in the same direction around the triangle (counterclockwise in figure 3-4b), but that the arrowhead on the resultant vector **C** points in the opposite direction (clockwise in figure 3-4b). The effect of **C** is exactly the same as the combined effect of the vectors **A** and **B**. **C** is called the **resultant** of **A** and **B**, whereas **A** and **B** are the **components** of **C**. For all practical purposes, the vectors **A** and **B** can be replaced by the single vector **C**. Conversely, if one started with the single vector **C**, it would be permissible to remove it and replace it by the two vectors **A** and **B**, if this should prove convenient for any reason. Let us emphasize again that the addition of vectors is a very special procedure clearly distinct from the addition of the scalar magnitudes of the quantities. It is not the same thing as $A + B$ and

$$C = A + B \text{ is } not \text{ true} \qquad (3\text{-}4)$$

A similar procedure is used to add together more than two vectors as illustrated in figure 3-5. The vectors are joined together, following one after the other with tail joined to head. The resultant is then obtained by joining the tail of the first vector to the head of the last vector. This diagram is sometimes called the **polygon of vectors**. There is no reason why the vectors should all be in the same plane. It is easy to imagine a

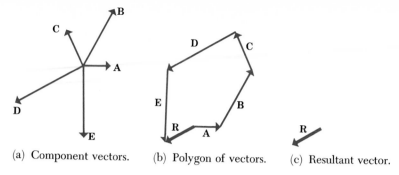

(a) Component vectors.　(b) Polygon of vectors.　(c) Resultant vector.

Figure 3-5　Addition of several vectors using the "polygon of vectors."

skew polygon with individual vectors jutting out in various directions in three-dimensional space. Notice that the arrowheads on the component vectors are all pointing in the same direction around the polygon, whereas the arrowhead on the resultant vector points in the opposite direction.

The procedure for subtracting the vector **B** from the vector **A** is very simple. Reverse the direction of **B** and then *add* it to **A**. This is shown in figure 3-6. The reader can easily convince himself that, if

$$A - B = D \tag{3-5}$$

then

$$A = B + D \tag{3-6}$$

Example 3-3-1

An aircraft is trying to fly due north at a speed of 100 m/sec but is subject to a cross-wind blowing from the west to the east at 50 m/sec. What is the actual velocity of the aircraft relative to the surface of the earth?

The aircraft flies northward at 100 m/sec relative to the air, but the air is moving eastward at 50 m/sec, so the resultant velocity of the aircraft is the vector sum of these two velocities. In the vector diagram of figure 3-7, 1 cm represents 20 m/sec. The northward velocity of the aircraft is therefore represented by a vector 5 cm long and the eastward velocity of the air by a vector 2.5 cm long. Drawing the triangle of vectors, PQR, the resultant $\overrightarrow{PR}$ is 5.59 cm long and the magnitude of the

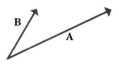

(a) The vectors **A** and **B**.　(b) Subtracting **B** from **A**.

Figure 3-6　Subtraction of vectors.

Vectors

29

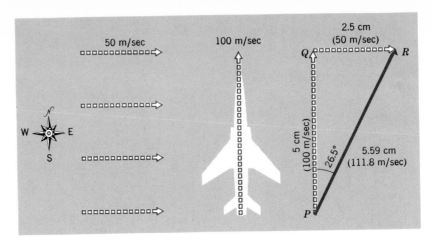

Figure 3-7 Illustrating example 3-3-1.

resultant velocity is therefore $5.59 \times 20 = 111.8$ m/sec. The angle QPR, which this resultant makes with the northerly direction, can be measured with a protractor and is found to be $26.5°$.

Example 3-3-2

A motor boat can move with a maximum speed of 10 m/sec, relative to the water. A river 400 m wide flowing at 5 m/sec must be crossed in the shortest possible time to reach a point on the other bank directly opposite the starting point. In which direction must the boat be pointed and how long will it take to cross?

If the boat were pointed directly at the opposite bank, then during the crossing it would drift downstream and it would not reach the other bank at a point directly opposite the starting point. It must therefore be pointed in a direction tilted in the upstream direction as shown in figure 3-8. As illustrated in the vector diagram PQR, the result of adding the velocity of the boat relative to the water to the velocity of the water must be a resultant velocity v pointing directly toward the opposite

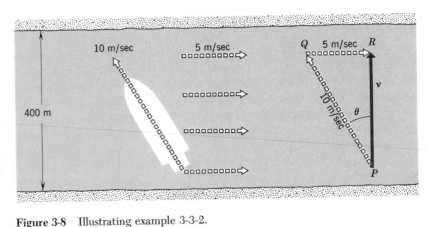

Figure 3-8 Illustrating example 3-3-2.

bank. We cannot draw this triangle of vectors immediately because we do not know the angle θ between the direction of motion and the direction straight across the stream. However, inspecting the triangle PQR and remembering that in trigonometry the sine of the angle θ is defined as

$$\sin \theta = \frac{QR}{PQ}$$

$$= \frac{5}{10} = 0.5 \tag{3-7}$$

we refer to tables of sines and find that the angle whose sine is 0.5 is 30°. The boat must therefore be pointed upstream at an angle of 30° from the direction perpendicular to the bank.

Applying Pythagoras' theorem to the triangle PQR

$$PQ^2 = QR^2 + PR^2 \tag{3-8}$$

or $\quad PR^2 = PQ^2 - QR^2 \tag{3-9}$

that is, $v^2 = 10^2 - 5^2 = 75 \tag{3-10}$

$$v = \sqrt{75} = 8.66 \text{ m/sec} \tag{3-11}$$

The boat therefore crosses the river at a speed of 8.66 m/sec. Since the distance across the river is 400 m, the time taken is $400/8.66 = 46.2$ seconds.

3-4   Resolving a Vector into Two Rectangular Components

Consider any vector **A**, as in figure 3-9*a*. With the starting point of the vector as origin O, draw any two axes OX and OY at right angles to one another. Let the angle between OX and the direction of **A** be θ, which is determined by our initial choice of the direction OX. From the end P of the vector **A** drop a perpendicular PQ to the axis OX (the angle PQO is then 90°). Now consider the triangle of vectors OPQ in figure 3-9*b*.

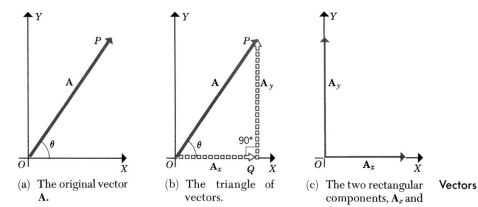

(a) The original vector **A**.

(b) The triangle of vectors.

(c) The two rectangular components, A_x and A_y.

Vectors

Figure 3-9   Resolving a vector into two rectangular components.

Clearly

$$\overrightarrow{OP} = \overrightarrow{OQ} + \overrightarrow{QP}$$

or $A = A_x + A_y$ (3-12)

The original vector of figure 3-9a may therefore be removed and replaced by the two vectors A_x and A_y of figure 3-9c. A_x and A_y are called the two **rectangular components** of **A**. A_x is the **x-component** and A_y is the **y-component**. (*Note carefully that the two rectangular components replace the original vector. Avoid the mistake made by many students who add the two components to the original vector, thus counting it twice over!*)

In the triangle OPQ the cosine of the angle θ is defined as

$$\cos \theta = \frac{OQ}{OP} = \frac{A_x}{A} \qquad\qquad (3\text{-}13)$$

Therefore

$$A_x = A \cos \theta \qquad\qquad (3\text{-}14)$$

Similarly the sine of the angle θ is defined as

$$\sin \theta = \frac{QP}{OP} = \frac{A_y}{A} \qquad\qquad (3\text{-}15)$$

Therefore

$$A_y = A \sin \theta \qquad\qquad (3\text{-}16)$$

Given the magnitude A of the original vector **A** and the angle θ it makes with the axis OX, the rectangular components are therefore found by using the two formulas

$$A_x = A \cos \theta \qquad\qquad (3\text{-}17)$$
$$A_y = A \sin \theta \qquad\qquad (3\text{-}18)$$
θ is the angle between the vector **A** and the x-axis

When the angle θ is given, tables are available for the quantities $\cos \theta$ and $\sin \theta$.

Example 3-4-1

An aircraft is climbing with a steady speed of 200 m/sec at an angle of 20° to the horizontal (see figure 3-10). What are the horizontal and vertical components of its velocity?

Horizontal component $= 200 \cos 20°$
Vertical component $= 200 \sin 20°$

Trigonometric tables tell us that

$\cos 20° = 0.9397$ and $\sin 20° = 0.3420$

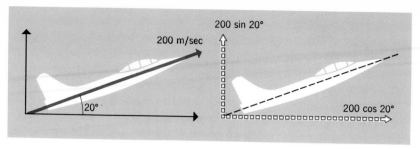

Figure 3-10 Illustrating example 3-4-1.

Therefore, horizontal component $= 200 \times 0.9397$
$$= 187.94 \text{ m/sec}$$

Vertical component $= 200 \times 0.3420$
$$= 68.40 \text{ m/sec}$$

Notice that the sum of 187.94 and 68.40 is *not* 200, but you can check that $(187.94)^2 + (68.40)^2 = (200)^2$. This sort of extra check on the results of a calculation is good policy, since it helps to reveal any theoretical or arithmetical errors that might otherwise go undetected.

3-5 Finding the Resultant Vector from Its Two Rectangular Components

Applying Pythagoras' Theorem to triangle OPQ of figure 3-9b,

$$OP^2 = OQ^2 + QP^2 \tag{3-19}$$
$$\text{or} \quad A^2 = A_x^2 + A_y^2 \tag{3-20}$$
$$A = \sqrt{A_x^2 + A_y^2} \tag{3-21}$$

The definition of the tangent of the angle θ is

$$\tan \theta = \frac{QP}{OQ} = \frac{A_y}{A_x} \tag{3-22}$$

This may also be written

$$\theta = \tan^{-1}\left(\frac{A_y}{A_x}\right) \tag{3-23}$$

which means that θ is the angle whose tangent has the value A_y/A_x. Given the number A_y/A_x, tables are available to find the angle whose tangent is equal to this number.

If we are given the two rectangular components $\mathbf{A}_x$ and $\mathbf{A}_y$ of the vector $\mathbf{A}$, the magnitude and direction of the vector itself may therefore be found using the two formulas

Vectors

$$A = \sqrt{A_x^2 + A_y^2} \tag{3-24}$$
$$\theta = \tan^{-1}(A_y/A_x) \tag{3-25}$$

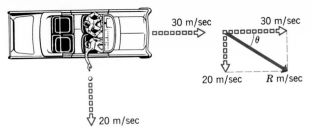

Figure 3-11 Illustrating example 3-5-1.

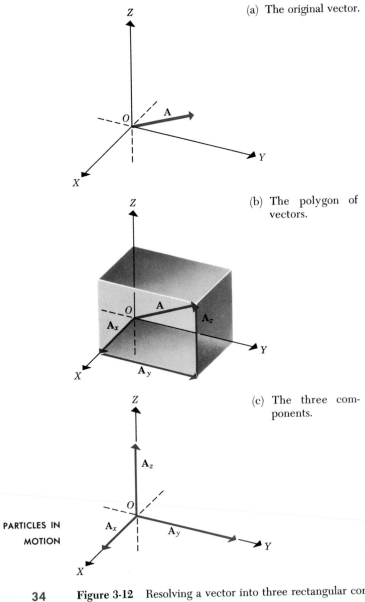

(a) The original vector.

(b) The polygon of vectors.

(c) The three components.

Figure 3-12 Resolving a vector into three rectangular components.

Example 3-5-1

A boy can throw a baseball horizontally with a speed of 20 m/sec. If he performs this feat in a convertible that is moving at 30 m/sec in a direction perpendicular to the direction in which he is throwing (see figure 3-11), what will be the actual speed and direction of motion of the baseball?

If the resultant velocity is R m/sec at an angle θ to the direction in which the convertible is moving, then

$$R^2 = (20)^2 + (30)^2 = 1300$$
$$R = \sqrt{1300} = 36.06 \text{ m/sec}$$

Also, $\tan \theta = \dfrac{20}{30} = 0.666$

From tables of tangents, $\theta = 33.69°$. Therefore, the ball has a speed of 36.06 m/sec in a direction at an angle of 33.69° to the direction in which the convertible is traveling.

3-6 Resolving a Vector into Three Rectangular Components

In three-dimensional space it is frequently necessary to resolve a vector into three rectangular components along the three Cartesian axes OX, OY, and OZ. This is illustrated in figure 3-12, which should by now be self-explanatory. As an exercise the reader should prove the following:

$$A = \sqrt{A_x{}^2 + A_y{}^2 + A_z{}^2} \tag{3-26}$$

The cosine of the angle between **A** and $\mathbf{A}_x$ is A_x/A.
The cosine of the angle between **A** and $\mathbf{A}_y$ is A_y/A.
The cosine of the angle between **A** and $\mathbf{A}_z$ is A_z/A.

QUESTIONS

1. A vector has zero magnitude. Is it necessary to specify its direction?
2. Can a vector have zero magnitude if one of its components is not zero?
3. Several vectors, which are not all in the same plane, add together to give a zero resultant. What is the minimum number of vectors which can satisfy this requirement?
4. Ten vectors add together to give a zero resultant. Is it possible that nine of these vectors are in the same plane but that the tenth is not in this plane?
5. Two racing cars are traveling around a circular track at steady speeds of 90 mph and 100 mph. The driver of the slower car has an elaborate piece of equipment which enables him to measure the velocity of the faster car *relative to himself*. Does this relative velocity ever become zero? Does it matter whether the cars are going round in the same or opposite directions? If the two cars were

going around at the same speed in opposite directions, would the relative velocity ever be zero?

PROBLEMS

A

1. The vector **v** is a velocity of 30 m/sec toward the northeast. What are the magnitudes and directions of the vectors (*a*) 5**v**, (*b*) −**v**, (*c*) −3**v**, (*d*) (3**v** − **v**), (*e*) (**v** + 2**v**), (*f*) (**v** − 2**v**).
2. Draw diagrams to find the magnitude and direction of the sum (**A** + **B**) of each pair of vectors in the figure.

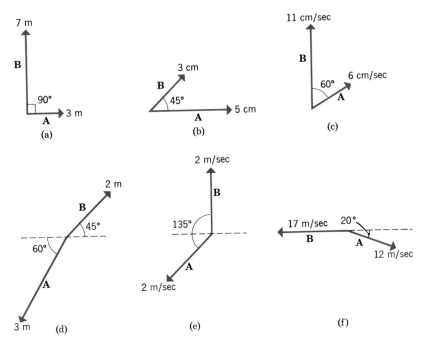

Problems 3-2 and 3-3

3. Find the vector (**A** − **B**) for each of the pairs in the figure.
4. Add together the three vectors shown in the diagram.

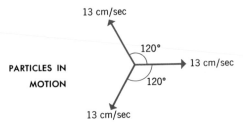

Problem 3-4

5. A hiker walks 5 miles in a straight line toward the west and then 5 miles in a straight line toward the north. What is his resultant displacement?

6. A hiker walks 10 miles in a straight line in a direction which is 60° north of west. How far must he then walk in a straight line toward the east in order to reach a destination due north of his starting point?

7. A ship is sailing at 30 knots toward the northeast. What is the easterly component of its velocity?

8. Draw diagrams to find the sum of the vectors in each part of the figure.

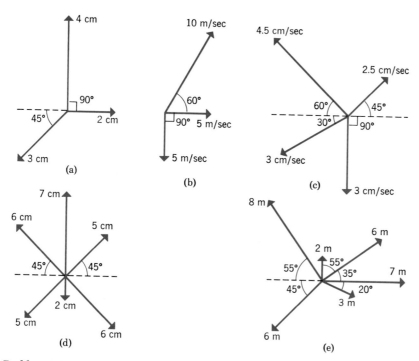

(a)

(b)

(c)

(d)

(e)

Problem 3-8

9. Repeat problem 8(e), but add the vectors together in a different sequence to see if this affects the answer.

10. A train is traveling at 50 mph in a direction 30° east of north. Resolve its velocity into two rectangular components due north and due east.

11. A missile is launched with a speed of 1200 m/sec at an angle of 60° to the horizontal. Resolve its velocity into horizontal and vertical components.

12. A plane takes off at an angle of 30° to the horizontal. The horizontal component of its velocity is known to be 200 mph. What is its actual speed? What is the vertical component of its velocity?

13. Find the magnitude and direction of the vectors whose components are (a) $A_x = +3$, $A_y = +3$ (b) $A_x = +3$, $A_y = -3$ (c) $A_x = +3$,

Vectors

37

$A_y = +4$ (d) $A_x = +5, A_y = +7$ (e) $A_x = -7, A_y = +2$ (f) $A_x = -3, A_y = -5$.

B

14. A mailman's round is made up as follows:
    (a) $\frac{1}{2}$ mile due east, (b) $\frac{1}{4}$ mile due north, (c) $\frac{3}{4}$ mile northwest, (d) $\frac{1}{2}$ mile due south, (e) 1 mile, southwest. When he has finished, what is his displacement relative to his starting point?

15. Raindrops are falling vertically at 20 mph. At what angle should you tilt your umbrella if you are walking at 10 mph?

16. A racing pigeon can fly with a speed of 30 m/sec. The wind is blowing from the west with a speed of 15 m/sec. In which direction must the bird fly in order to reach a destination northeast of his starting point? If the distance to be covered is 25 kilometers, how long will it take?

17. A balloon is rising with a vertical velocity of 2 m/sec. If, in addition, the wind is blowing with a velocity of 5 m/sec, in which direction is the balloon moving?

18. A bird is flying due north with a speed of 30 mph. A train is traveling due east with a speed of 30 mph. What is the apparent magnitude and direction of the velocity of the bird from the point of view of a passenger in the train?

19. A train has a velocity of 50 mph. If I walk at 10 mph from one side of the train to the point directly opposite on the other side, what are the magnitude and direction of my velocity relative to the tracks?

20. A river has a velocity of 2 m/sec. The bow of a rowboat is pointed directly toward the opposite bank and it is rowed with a speed of 5 m/sec relative to the water. If the river is 50 m wide, how long will it take to row across and how far downstream will the boat have drifted in this time? What is the angle between the bank and the direction of motion of the boat?

C

21. Referring to example 3-3-2, in which direction must the boat be pointed to reach the opposite bank in the shortest possible time, if the point at which it reaches this bank is unimportant? How long will it take to cross and how far downstream will it have drifted?

22. A rifle is fired from a jeep traveling with a speed of 10 m/sec. The target is stationary 100 m away in a direction at right angles to the velocity of the jeep. The muzzle velocity of the bullet is 200 m/sec. If the rifle is aimed directly at the center of the target, how far to one side will the bullet strike? How far to one side must it be aimed in order to hit the center?

23. When launching a satellite, it is possible to utilize the rotation of the earth to increase the velocity of launching. Suppose that the rocket is fired vertically at the equator with a speed of 8000 m/sec relative to the surface of the earth. What is the velocity of launching from

the point of view of an observer who is moving around the sun with the earth, but is not rotating like the earth?

24. A vector of magnitude 10 units is inclined at the same angle to each of the three axes, OX, OY, and OZ. What are the magnitudes of its three rectangular components?

25. A missile is fired toward the northwest with a speed of 5000 m/sec at 45° to the horizontal. Find the three rectangular components of its velocity if the X axis points toward the west, the Y axis toward the north and the Z axis vertically upward.

26. Two vectors **A** and **B** have rectangular components A_x, A_y, and B_x, B_y. Their resultant is $\mathbf{R} = \mathbf{A} + \mathbf{B}$. Prove that
   (a) The components of **R** are $(A_x + B_x)$ and $(A_y + B_y)$.
   (b) $R^2 = \sqrt{(A_x + B_x)^2 + (A_y + B_y)^2}$
   (c) If θ is the angle between **R** and the X axis

$$\tan \theta = \frac{A_y + B_y}{A_x + B_x}$$

   (d) Extend these results to the addition of any number of vectors, **A, B, C, D** and so on.
   (e) Extend the results to three dimensions.

27. The vectors **A** and **B** are perpendicular to one another and both lie in the XY plane. Prove that $A_x B_x + A_y B_y = 0$.

Vectors

4

Motion along a Straight Line

4-1   Velocity along a Straight Line

When the velocity of a body is constant, the velocity is very simply defined as the distance traveled divided by the time taken; but when the velocity changes with time, a more careful definition is required. This provides a good example of how physics must frequently concern itself with the precise nature of the quantities it is discussing.

Suppose that the motion is restricted to a straight line, and choose the X axis to coincide with this line (figure 4-1). Under these circumstances the instantaneous position of the particle is uniquely determined by its distance from the origin O, that is, its x coordinate. At the time t_1, let this coordinate be x_1. At a later time t_2, let the coordinate be x_2. Then, however complicated the motion during this time interval, the

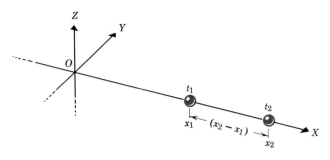

Figure 4-1   Motion along the X axis.

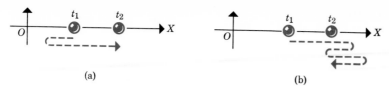

(a) (b)

Figure 4-2 At time t_1, the body is at x_1, and at time t_2 it is at x_2, but its motion during the intervening time interval might have been very complicated.

net distance traveled is $(x_2 - x_1)$ and the time taken is $(t_2 - t_1)$, so that it is possible to define an **average velocity** in the following way:

Average velocity during the time interval between t_1 and t_2

$$= \bar{v} = \frac{(x_2 - x_1)}{(t_2 - t_1)} \tag{4-1}$$

A bar above a symbol frequently implies that it is some sort of average.

However, the simplicity of this definition can be deceptive. Figure 4-2 illustrates two ways in which the body might have traveled between the two points, and in both cases the average velocity reveals very little about the true nature of the motion. What is really needed is a definition of the *instantaneous* velocity at each definite instant of time during the motion. Notice, for example, that in figure 4-2*a* the instantaneous velocity at time t_1 is to the left, whereas the average velocity between t_1 and t_2 is to the right. The way to avoid this difficulty is obvious: choose the instant of time t_2 sufficiently close to t_1 so that the motion has not had time to reverse itself during the interval $t_2 - t_1$, and then, at least, the body will always be traveling in the same direction. Even so, the magnitude of the velocity might change so much between t_1 and t_2 that the average velocity would bear no direct relationship to the instantaneous velocity at time t_1. Suppose, for example, that the body starts out very slowly and covers only $1/100$ of the distance $x_2 - x_1$ during the first half of the time interval $t_2 - t_1$, but then suddenly accelerates and covers the remaining $99/100$ during the second half of the time interval. After thinking about such examples, it becomes clear that the only way to avoid such difficulties is to make the time interval $t_2 - t_1$ so short that the velocity does not have time to change between t_1 and t_2. Since we cannot, in general, know how rapidly the velocity is changing, the only thing to do is to make $t_2 - t_1$ very small indeed. How small? The mathematical answer is that, although $t_2 - t_1$ must be larger than zero, it should be smaller than any number you choose to quote to me, however small. The practical answer is that $t_2 - t_1$ must be as small as possible, subject to the conditions that the small distance $x_2 - x_1$ and the small time interval $t_2 - t_1$ can be measured with sufficient accuracy for the purpose in hand, bearing in mind that when a quantity becomes very small it becomes increasingly more difficult to measure it.

(a) A millipede running along a scale.

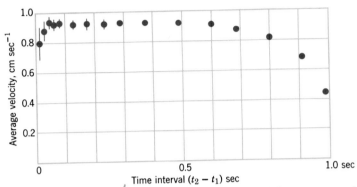

(b) The average velocity depends upon the length of the time interval.

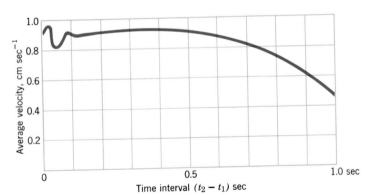

(c) The result of measuring the average velocity with perfect accuracy.

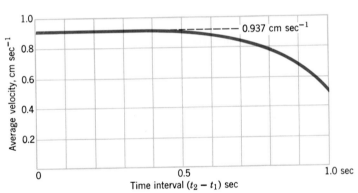

(d) The millipede wobbled as he passed the zero mark.

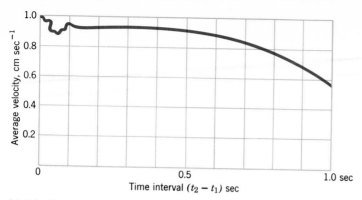

(e) The leg movements may produce an additional ripple. This ripple is less conspicuous for large time intervals because of the averaging procedure.

Figure 4-3 (*continued*)

To illustrate these points, consider the concrete example of a millipede running along a wooden rule graduated in centimeters (figure 4-3). Suppose that we have at our disposal a very elaborate timing mechanism that can measure time intervals with very high accuracy and that we set this mechanism going at the instant when the front end of the millipede passes the zero of the scale. We then observe the scale reading of his front end after various intervals of time. Suppose, however, that the scale readings are likely to be in error by about 0.001 cm. Table 4-1

Table 4-1

Time interval since zero mark was passed	Scale reading of front end of millipede	Average velocity $\frac{(x_2 - x_1)}{(t_2 - t_1)}$ cm sec^{-1}	
t_2 seconds	x_2 cm	$(x_1 = 0, t_1 = 0)$	
0.0100	0.008	0.8	
0.0250	0.022	0.88	
0.0500	0.048	0.96	
0.0750	0.070	0.933	
0.1000	0.094	0.940	
0.1500	0.140	0.933	
0.2000	0.188	0.940	
0.2500	0.235	0.940	
0.3000	0.280	0.933	Motion
0.4000	0.376	0.940	Along a
0.5000	0.467	0.934	Straight
0.6000	0.560	0.933	Line
0.7000	0.617	0.881	
0.8000	0.651	0.814	
0.9000	0.618	0.689	
1.0000	0.468	0.468	**43**

gives a set of readings that might be obtained under these circumstances. The first column gives the time t_2 sec at which the scale reading of the front end of the millipede was measured and found to be x_2 cm as given in the second column. The third column gives the calculated value of the average velocity between the instant at which the millipede passed the zero mark (when $t_1 = 0$ and $x_1 = 0$) and the later instant t_2 at which the scale reading was x_2.

Close scrutiny of the data reveals that, at some instant between 0.8 and 0.9 sec, the millipede began to walk backward. However, if the only observation of the millipede had been made after 1 sec, the observer would have been ignorant of this change of direction and he would have deduced an average velocity of 0.468 cm sec^{-1} to the right. But as one considers shorter and shorter time intervals, the average velocity rises until, for intervals in the range 0.5 to 0.1 seconds, it steadies down to a value in the vicinity of 0.94 cm sec^{-1}, slight fluctuations around this value being readily attributable to the 0.001 cm error in reading the scale. Notice, however, that when the time interval is less than 0.1 second and the total distance traveled $x_2 - x_1$ is only a few times larger than the error, then this error makes itself felt in the form of an erratic variation in the apparent value of the velocity. All of this is illustrated in figure 4-3b in which the average velocity is plotted against the time interval used. (Note that figure 4-3b is not a plot of the instantaneous velocity at various times t_2, but a plot of the average velocity between the instants t_1 and t_2. The *instantaneous* velocity at $t_2 = 1$ second, for example, is backward and therefore negative.) On the graph, each point is associated with a vertical line that represents the range of values inside which the "true" average velocity lies. The exact value of this "true" velocity is unknown because of the error in measuring $x_2 - x_1$.

A reasonable interpretation of these results would be that the millipede started out with a velocity of about 0.94 cm sec^{-1} and then slowed down and turned back. If the distances and times had been measured with perfect accuracy, the curve obtained might have looked like the one in figure 4-3c. We are interested in the initial velocity at the zero mark and, assuming perfect accuracy, this could obviously be obtained to any desired precision by taking a very small time interval $t_2 - t_1$, which would eliminate the effect of the continual slowing down of the millipede.

It is important to consider another possibility. With increased accuracy the curve might have the form shown in figure 4-3d. Perhaps, as the millipede passed the zero mark he was hit head-on by a puff of wind that made him wobble slightly, producing an up and down oscillation of the graph. There was no evidence of the wobble in the previous inaccurate measurements, because the wobble took place over a distance comparable with the smallest distance that could then be measured with any accuracy. Pressing the matter further and achieving still greater accuracy, one might discover that the curve is really as shown in figure 4-3e and the additional ripple is attributable to the jerky motion of the millipede's numerous legs.

This complicated discussion has been given to emphasize two points. Theoretically, the instantaneous velocity of a body is found by studying its motion over a very small time interval—the smaller the better. In practice, however, the motion has to be measured with instruments of limited accuracy and this automatically places a limit on how small the time interval can be. We must fully realize that our knowledge of the universe is limited by the crudity of our methods of observation, and, as these methods are improved and their accuracy increases, we should not be surprised to find that new and finer details are revealed.

The thoughtful student might object that in fact there is a "true" velocity of the body and that we can *imagine* a perfect measurement with instruments refined to the point of perfect accuracy, and we can then use the mathematical approach of making the time interval as small as is necessary to reveal all the fine details. This eminently reasonable argument is refuted by modern physics. As we shall explain in a later chapter, one of the fundamental features of quantum mechanics is that it will not allow us to decrease the size of the time interval without limit. If we attempt to do this, however perfect our measuring instruments, there comes a point where an accurate measurement of velocity becomes impossible. Fortunately, in the case of an everyday moving object such as a baseball this point is not reached until the time interval has the incredibly small value of 10^{-35} seconds. So we are not aware of this in everyday experience. For electrons in atoms, though, the point is crucial, as we shall discover later in this book.

At this stage let us be satisfied with the well-behaved nature of baseballs and the following definition of instantaneous velocity:

> **The instantaneous velocity** of a body is obtained by dividing the distance it travels by the time taken, with the time interval as short as it possibly can be without exceeding the limits of accuracy of the measuring equipment or running into the difficulties raised by quantum mechanics.

$$v = \frac{x_2 - x_1}{t_2 - t_1} \text{ when } t_2 - t_1 \text{ is sufficiently small} \qquad (4\text{-}2)$$

Velocity is a vector, and so far we have defined only its magnitude. Its direction is, of course, the direction of the straight line to which the motion was restricted, but there are two opposite ways of moving along this line. With the help of figure 4-4, it is easy to see that if the body moves to the right, then x_2 is greater than x_1, $x_2 - x_1$ is positive, and the velocity $(x_2 - x_1)/(t_2 - t_1)$ is positive. For example, suppose that, initially, $t_1 = 2$ sec and $x_1 = +3$ cm, and that, at a later time, $t_2 = 5$ sec and $x_2 = +8$ cm. Then the average velocity is $(8 - 3)/(5 - 2) = \frac{5}{3}$ cm sec^{-1}.

If the body moves to the left, x_2 is less than x_1, $x_2 - x_1$ is negative, and the velocity is negative. For example, suppose that, initially, $t_1 = 2$ sec and $x_1 = +3$ cm, and that, at a later time, $t_2 = 5$ sec and

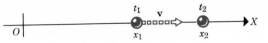

(a) A positive velocity to the right.

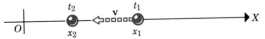

(b) A negative velocity to the left.

Figure 4-4

$x_2 = +1$ cm. Then the average velocity is $(1 - 3)/(5 - 2) = -\frac{2}{3}$ cm sec^{-1}.

The student who has not had much experience with algebra should note these arguments carefully. Points to the right of the origin have positive x coordinates; points to the left of the origin have negative x coordinates. Velocities to the right are positive; velocities to the left are negative.

4-2 Acceleration along a Straight Line

Velocity is the rate of change of distance with time. Acceleration is the rate of change of velocity with time. The procedure that has been given for defining instantaneous velocity can readily be adapted to define instantaneous acceleration. At successive instants of time t_1, t_2, t_3, t_4, let the x coordinates be x_1, x_2, x_3, x_4, as in figure 4-5a. If t_2 approaches very close to t_1, the instantaneous velocity at time t_1 is

$$v_1 = \frac{x_2 - x_1}{t_2 - t_1} \tag{4-3}$$

when $t_2 - t_1$ is sufficiently small.

(a) Four successive positions of the body.

(b) By making t_2 approach t_1 we obtain the velocity v_1 at t_1. By making t_4 approach t_3 we obtain the velocity v_3 at t_3.

(c) By making t_3 approach t_1 we can obtain the acceleration a at t_1.

Figure 4-5

Similarly, if t_4 approaches close to t_3, the instantaneous velocity at t_3 is

$$v_3 = \frac{x_4 - x_3}{t_4 - t_3} \qquad (4\text{-}4)$$

when $t_4 - t_3$ is sufficiently small.

The **average acceleration** during the time interval between t_1 and t_3 is defined as

$$\bar{a} = \frac{v_3 - v_1}{t_3 - t_1} \qquad (4\text{-}5)$$

Note that: Acceleration $= \dfrac{\text{Change in velocity}}{\text{Time taken}}$ $\qquad (4\text{-}6)$

However, the acceleration itself may change with time, and so the **instantaneous acceleration** at time t_1 is defined as

$$a = \frac{v_3 - v_1}{t_3 - t_1} \text{ when } t_3 - t_1 \text{ is sufficiently small} \qquad (4\text{-}7)$$

A velocity is a distance divided by a time, and its units are m sec^{-1} or cm sec^{-1}. An acceleration is a velocity divided by a time, or a distance twice divided by a time. Its units are therefore m sec^{-2} or cm sec^{-2}, which can also be written meters per second per second and centimeters per second per second or m/sec^2 and cm/sec^2.

Acceleration, like velocity, is a vector. The significance of this will be explained more fully in section 5-2. For motion in a straight line, the sign of the acceleration must be given careful thought. The sign of a is clearly the same as the sign of $v_3 - v_1$, since $t_3 - t_1$ is always positive. A positive acceleration is represented by an arrow pointing in the same direction as the positive direction of x, which is usually to the right. A negative acceleration is represented by an arrow pointing in the opposite direction. To make certain that you understand this, consider carefully the following statements, thinking in each case about the sign of $v_3 - v_1$. If the velocity is positive (to the right) and its magnitude is increasing, the acceleration is positive (to the right). If the velocity is positive (to the right) and its magnitude is decreasing, the acceleration is negative (to the left). If the velocity is negative (to the left) and its magnitude is increasing, then the acceleration is negative (to the left). If the velocity is negative (to the left) and its magnitude is decreasing, then the acceleration is positive (to the right).

4-3   Motion in a Straight Line with Constant Acceleration

The simplest case of accelerated motion in a straight line occurs when the acceleration is constant. Let the constant acceleration be a. Choose any convenient instant to be the zero of time and then choose the origin of the X axis to coincide with the position of the body at the zero of time

Figure 4-6 Motion in a straight line with constant acceleration. The origin of x is chosen to be the position of the body at the zero of time. The velocity at the origin is v_0.

(figure 4-6). Then $x = 0$ when $t = 0$. It is very important to note this special choice of origin, since the formulas we shall quote are not true unless the origin is chosen in this way. At the zero of time, when the body is at the origin of x, let its velocity be v_0. At time t, the x coordinate of the body is x, its velocity is v, and its acceleration is, of course, a. Since the acceleration remains constant, the average acceleration between times 0 and t is the same as the instantaneous acceleration at any time, a.

Therefore
$$\bar{a} = \frac{v - v_0}{t - 0} = a \tag{4-8}$$

that is,
$$\frac{v - v_0}{t} = a \tag{4-9}$$

Multiplying both sides by t

$$v - v_0 = at \tag{4-10}$$

Adding v_0 to each side

$$v = v_0 + at \tag{4-11}$$

The student who has only limited experience of algebraic equations may find it helpful to translate such an equation into words.

(Velocity at time t)

= (Velocity at zero time when the body is at the origin)

+ (Acceleration) × (Time elapsed since body was at the origin)   (4-12)

The above equation gives us the velocity v at any instant of time t. Let us now ask where the body is at this instant. At the origin, the velocity of the body is v_0. If it had maintained this velocity without acceleration, after a time t it would have been displaced through a distance $v_0 t$ (displacement = constant velocity × time). At the end of the time interval t the velocity is $v = v_0 + at$. If it had had this velocity all the time, it would have been displaced through a distance $vt = (v_0 + at)t = v_0 t + at^2$. The actual displacement obviously lies somewhere in between these two values. With the help of some slightly complicated mathematics, it can be shown that the displacement is exactly half way between the two values and is $v_0 t + \frac{1}{2}at^2$. This is the displacement from the origin and is therefore the x coordinate

$$x = v_0 t + \tfrac{1}{2}at^2 \tag{4-13}$$

If proper attention has been given to the sign of v_o and a, the sign of x is automatically correct. Note particularly that the equation gives the *position* of the particle at any instant of time, and this is not necessarily the same as the "total distance traveled." If I go on a journey from Philadelphia to New York, which is 90 miles away, and back again, there is some sense in saying that I have traveled a total distance of 180 miles, but at the end of the journey the change in my x coordinate would be zero. Consider the case when a is in the opposite direction to v_o with, for example, v_o positive and a negative. The body starts to move to the right of the origin but eventually the retardation reduces the velocity to zero and then reverses it. The body moves a certain distance to the right, turns around, and returns to the origin. After the reversal of the velocity, the distance x of the body from the origin is clearly smaller than the "total distance traveled."

In some problems it is necessary to calculate the velocity when the body is in a certain position and the time is unknown. The equation that makes this possible can be deduced from the previous equations. The procedure will illustrate how careful attention should be paid to quantities whose values are given and to the quantity that is to be calculated, whereas quantities which are not given and which do not need to be known should be eliminated from the equations. In the present instance we are given the initial velocity v_o (at $x = 0$, $t = 0$), the acceleration a and the final x coordinate x. The quantity to be calculated is the final velocity v. The time t is irrelevant to the problem and should therefore be eliminated from the equations. The equation

$$v = v_o + at \tag{4-14}$$

can be easily rearranged to give

$$t = \frac{v - v_o}{a} \tag{4-15}$$

Substituting this value of t in

$$x = v_o t + \tfrac{1}{2}at^2 \tag{4-16}$$

one obtains

$$x = v_o \frac{(v - v_o)}{a} + \tfrac{1}{2}a \frac{(v - v_o)^2}{a^2} \tag{4-17}$$

Multiplying out the brackets and canceling one a in the numerator and denominator of the last term,

$$x = \frac{(v_o v - v_o{}^2)}{a} + \frac{(v^2 - 2vv_o + v_o{}^2)}{2a}$$

$$= \frac{(2v_o v - 2v_o{}^2 + v^2 - 2vv_o + v_o{}^2)}{2a}$$

$$= \frac{(v^2 - v_o{}^2)}{2a} \tag{4-18}$$

Multiplying both sides by $2a$

$$2ax = v^2 - v_o^2 \tag{4-19}$$

Adding v_o^2 to both sides

$$v^2 = v_o^2 + 2ax \tag{4-20}$$

This is the desired equation, since the left hand side contains only the quantity v, which is to be calculated, and the right hand side contains only the quantities which are known (v_o, a, and x).

The three important equations will now be presented together. Remember that these equations are meaningless unless the exact significance of each symbol is understood. They are true only if the origin is chosen so that $x = 0$ when $t = 0$, and if v_o is the velocity at the origin. Also, they are applicable only if the acceleration is constant.

$$v = v_o + at \tag{4-21}$$
$$x = v_o t + \tfrac{1}{2}at^2 \tag{4-22}$$
$$v^2 = v_o^2 + 2ax \tag{4-23}$$

In any particular problem, the student will have to decide which of these three equations to use. To do this he should notice that each equation contains only four different kinds of symbol. Three quantities will be given in the problem and a fourth quantity will have to be calculated. He should therefore choose the equation that involves the three known quantities plus the quantity to be calculated, and nothing else. For example, suppose he is given the initial velocity v_o and the acceleration a and is asked to calculate the position x when the velocity has a certain value v. Clearly he should use equation 4-23, because this contains only the given quantities v_o, a, and v and the desired quantity x.

4-4 A Freely Falling Body

The most important example of motion in a straight line with constant acceleration is the case of a body falling freely in the earth's gravitational field. If the body falls in a vacuum, or if the conditions are such that the resistance of the air can be neglected, and if the total height through which the body falls is small compared with the earth's radius (which is about 6.37×10^8 cm), then, as the body falls its acceleration always points vertically downward and has the same value at all instants. Moreover, at a particular place on the earth's surface this acceleration is the same for all falling bodies, independently of their size, shape, or mass, and it is called the **acceleration due to gravity**. It is usually represented by the symbol g and its value is approximately 9.8 m sec^{-2} or 980 cm sec^{-2}. We shall elaborate these points in chapter 7.

The photograph of figure 4-7 shows the special case when the body is released from rest and $v_0 = 0$. Equations 4-21, 4-22, and 4-23 then take

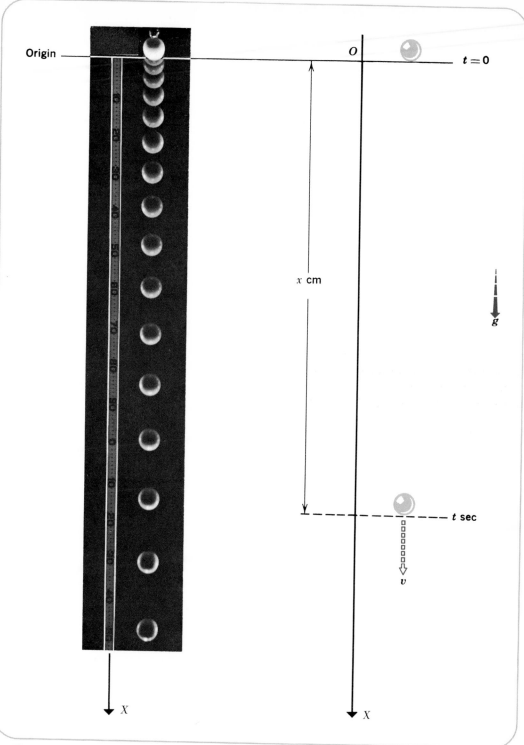

Figure 4-7 A ball falling from rest photographed at successive intervals of $\frac{1}{30}$th of a second. (From PSSC Physics, D.C. Heath and Company, Boston, 1965.)

51

the simplified form

$$v = gt \tag{4-24}$$

$$x = \tfrac{1}{2}gt^2 \tag{4-25}$$

$$v^2 = 2gx \tag{4-26}$$

In general, however, it is possible to consider cases when the body is initially projected vertically downward or vertically upward with an initial speed v_0. The equations then revert to the more general form

$$v = v_0 + gt \tag{4-27}$$

$$x = v_0 t + \tfrac{1}{2}gt^2 \tag{4-28}$$

$$v^2 = v_0{}^2 + 2gx \tag{4-29}$$

If the body is released from rest or projected vertically downward, the acceleration due to gravity g has the effect of continually increasing the downward speed by an amount g every second. The equations quoted above are easy to apply with all the quantities involved taken to be positive.

However, if the body is projected vertically *upward*, the *downward* acceleration g first reduces the upward velocity until the body is brought to rest at the highest point of its path. The downward acceleration then builds up a downward velocity and the body falls from its highest position with a gradually increasing speed. The equations can still cope with this situation, but it is very important to keep clearly in mind which quantities are positive and which are negative. This is illustrated by the following examples.

Example 4-4-1

A man standing on the roof of a building 30 m high throws a ball vertically downward with an initial velocity of 500 cm sec^{-1} as it leaves his hand (see figure 4-8). The acceleration due to gravity is 9.8 m sec^{-2}. (a) What is the velocity of the ball after it has been falling for 0.5 sec? (b) Where is the ball after 1.5 sec? (c) What is the velocity of the ball as it strikes the ground?

Decide to use the MKS system of units. Then the initial velocity must be expressed as 5 *meters* per second. Place the origin at the top of the building. Then $x = 0$ when $t = 0$. Let the positive direction of x be downward. The initial velocity is downward and therefore positive, so $v_0 = +5$ m sec^{-1}. The acceleration is downward and therefore positive, so $a = g = +9.8$ m sec^{-2}.

(a) In this part of the problem one is given v_0, g, and t, and must deduce v. The correct equation to use is therefore 4-27.

$$v = v_0 + gt$$
$$= (+5) + (+9.8)(+0.5)$$
$$= 5 + 4.9$$
$$= +9.9$$

After 0.5 sec the velocity is 9.9 m sec^{-1} downward.

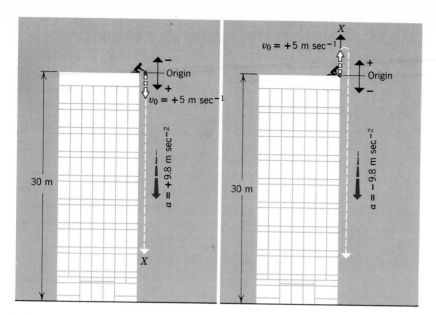

Figure 4-8 Illustration for example 4-4-1.

Figure 4-9 Illustration for example 4-4-2 (second method).

(b) In this part of the problem one is given v_o, g, and t, and must calculate x. The correct equation to use is therefore 4-28.

$$x = v_o t + \tfrac{1}{2}gt^2$$
$$= (+5)(+1.5) + \tfrac{1}{2}(+9.8)(+1.5)^2$$
$$= 7.5 + (4.9)(2.25)$$
$$= 7.5 + 11.025$$
$$= 18.525$$

After 1.5 sec the ball is 18.525 m below the roof or 11.475 m above the ground.

(c) When the ball strikes the ground $x = +30$ m. So one is given v_o, g, and x, and asked to calculate v. The correct equation is therefore 4-29.

$$v^2 = v_o{}^2 + 2gx$$
$$= (+5)^2 + 2(+9.8)(+30)$$
$$= 25 + 588$$
$$v^2 = 613$$
$$v = 24.76$$

When it strikes the ground, the ball has a velocity of 24.76 m sec^{-1}.

Example 4-4-2

Repeat the previous problem for the case when the ball is thrown *upward* with a velocity of 5 m sec^{-1}. Also, (d), what is the maximum height reached by the ball?

Motion
Along a
Straight
Line

The only difference is that the initial velocity is in the negative direction and so $v_o = -5$ m sec^{-1}. Otherwise the arguments for parts (a), (b), and (c) are identical.

(a) $v = v_o + gt$

$\qquad = -5 + (+9.8)(+0.5)$

$\qquad = -5 + 4.9$

$\qquad = -0.1$

After 0.5 sec the body is moving *upward* with a velocity of 0.1 m sec^{-1}.

(b) $x = v_o t + \frac{1}{2}gt^2$

$\qquad = (-5)(+1.5) + \frac{1}{2}(+9.8)(+1.5)^2$

$\qquad = -7.5 + 11.025$

$\qquad = +3.525$

After 1.5 sec the ball is 3.525 m *below* the roof.

(c) $v^2 = v_0{}^2 + 2gx$

$\qquad = (-5)^2 + 2(+9.8)(+30)$

$\qquad = +25 + 588$

$\qquad = 613$

$\qquad v = 24.76$

When it strikes the ground, the ball has a velocity of 24.76 m sec^{-1}, exactly the same as in the previous problem.

(d) The ball starts to move upward but the downward acceleration steadily reduces the magnitude of its velocity. When the ball reaches its highest position, the velocity has been reduced to zero and subsequently the velocity is reversed and the ball begins to move downward. In the highest position, therefore, it is known that $v = 0$. The known quantities are therefore v_o, g, and v, and the desired quantity is x. The equation to use is 4-29.

$v^2 = v_o{}^2 + 2gx$

$0^2 = (-5)^2 + 2(+9.8)x$

$0 = 25 + 19.6x$

$19.6x = -25$

$x = -1.277$

The negative sign indicates that the highest position is *above* the roof where the x coordinates are negative. The maximum height reached by the ball is therefore 1.277 m above the roof, or 31.277 m above the ground.

It is interesting to repeat this problem with the origin still at the roof, but the positive direction of x *upward* (see figure 4-9). Then the initial velocity is upward and positive ($v_o = +5$ m sec^{-1}), but the acceleration is downward and therefore negative ($a = g = -9.8$ m sec^{-2}). Proceeding as before, but with due regard to the sign of each quantity:

(a) $v = v_o + gt$

$\quad = (+5) + (-9.8)(0.5)$

$\quad = +5 - 4.9$

$\quad = +0.1$

After 0.5 sec, the velocity is 0.1 m sec^{-1} upward.

(b) $x = v_o t + \frac{1}{2}gt^2$

$\quad = (+5)(+1.5) + \frac{1}{2}(-9.8)(+1.5)^2$

$\quad = +7.5 - 11.025$

$\quad = -3.525$

The negative sign now indicates that the position of the ball is *below* the roof. Therefore, after 1.5 sec the ball is 3.525 m below the roof.

(c) When the ball strikes the ground, its x coordinate is -30 m.

$v^2 = v_o^2 + 2gx$

$\quad = (+5)^2 + 2(-9.8)(-30)$

$\quad = 25 + 588$

$\quad = 613$

$v = \sqrt{613} = \pm 24.76$

When the ball strikes the ground, its velocity is 24.76 m sec^{-1}. Since a square root can be positive or negative, this procedure does not tell us the direction of the velocity.

(d) $\qquad v^2 = v_o^2 + 2gx$

$\qquad 0 = (+5)^2 + 2(-9.8)x$

$\qquad 0 = 25 - 19.6x$

$\qquad 19.6x = 25$

$\qquad x = +1.277$

The positive sign now indicates that the highest position is 1.277 m *above* the roof. We therefore obtain exactly the same answers because obviously the behavior of the ball is independent of our choice for the direction of the x axis—but notice how important it is to consider carefully the sign of each quantity.

QUESTIONS

1. How might the data of table 4-1 be used to obtain a rough idea of the variation of instantaneous velocity with time during the first second of the motion?

2. If the instantaneous velocity of a body is zero, is its instantaneous acceleration necessarily zero? Give examples.

3. If a body moving along a straight line reverses the direction of its velocity twice, what can be said about the acceleration?

4. If a body moving along a straight line reverses the direction of its velocity only once, does this necessarily imply that the acceleration is always in the same direction? Explain your answer carefully.

5. Interpreting the word "speed" to mean an intrinsically positive quantity, at what point in the following motions does the speed have its smallest value? (*a*) A body released from rest. (*b*) A body thrown vertically downward. (*c*) A body thrown vertically upward.

6. Is it possible to have a straight line motion with *v* and *a* always in opposite directions? (Hint: Consider a rocket launched vertically with a very high velocity.)

7. Can you give a simple argument to refute a theory that suggests that the resistance of the air can be allowed for by assigning a smaller value to the acceleration due to gravity, *g*?

PROBLEMS

In all problems involving falling bodies, neglect air resistance.

In some problems speeds are given in miles per hour when it is easier to appreciate the magnitude of the velocity in this way. Some of these problems can be worked through without changing to CGS or MKS units.

A

1. A particle starts from the origin with a velocity of $+15$ cm/sec. Its acceleration is always zero. What is its *x* coordinate 3 sec later?

2. A particle starts from the origin with a velocity of -15 cm/sec. Its acceleration is always zero. What is its *x* coordinate 3 sec later?

3. A body falls vertically from rest. (*a*) What is its velocity after it has been falling for 3 sec? (*b*) How far has it then fallen?

4. A body falls vertically from rest. How long does it take to build up a speed of 200 cm sec^{-1}?

5. A body falls vertically from rest. How long does it take to build up a speed of 60 mph?

6. Through what height must a body fall from rest in order to attain a speed of 500 cm sec^{-1}?

7. A body is dropped from rest at the top of a building 40 m high. What is its speed when it hits the ground?

8. A ball is thrown vertically downward with an initial speed of 10 m/sec. What is its speed after it has been falling for 2 sec?

9. A ball is thrown vertically upward with a speed of 8 m/sec. What is its speed at its highest point?

10. A car traveling along a straight highway maintains a speed of 60 mph for 1 hour and then 30 mph for 2 hours. What was its average velocity over the 3-hour interval? What would its average velocity have been if it had turned round at the end of the first hour?

11. A car traveling along a straight highway maintains a speed of 60 mph for the first 20 miles and then 30 mph for the next 40 miles. What was its average velocity over the whole 60 mile run?

12. A rocket is launched vertically upward in a straight line. Immediately after the instant of launching, the burning of the fuel produces an upward acceleration of 4.5×10^3 cm/sec² and this acceleration is maintained at the same value until all the fuel is used up. If the rocket must achieve a final velocity of 8×10^5 cm/sec, how long must the fuel last?

13. A car traveling at 60 mph applies its brakes and comes to rest in 0.8 sec. Assuming the acceleration (or, if you prefer, deceleration) was constant during this time, calculate its value in the CGS system of units. How far did the car travel while its brakes were applied?

14. A body moving along a straight line has an initial velocity of $+6.3$ m/sec and maintains a steady acceleration of $+0.25$ m/sec². What is its velocity after 3 sec? How far has it traveled during this time?

15. A body moving along the X axis passes the origin with a velocity of $+3$ cm/sec. It maintains a steady acceleration of $+0.2$ cm/sec². What is its velocity when its x coordinate is $+40$ cm?

B

16. A body moving along the X axis with a constant acceleration passes the origin with a velocity of $+5$ cm/sec. 2 sec later its x coordinate is $+6$ cm. Calculate the magnitude of the acceleration and specify its direction.

17. A skater is gliding over the ice with a velocity of 2.6 m/sec. Another skater pushes him from behind and produces a steady acceleration of 1.3 m/sec² until his velocity has been raised to 18.2 m/sec. How long did the push last?

18. An electron is emitted from a hot metal wire with an unknown velocity. It then enters a device 4 cm long which accelerates it at a rate of 3×10^{16} cm/sec² in the same direction in which it is moving. At the far end of this 4 cm it enters another device which measures its velocity and finds it to be 7×10^8 cm/sec. With what velocity was the electron emitted from the hot wire?

19. On a certain straight highway the speed limit is 60 mph. A policeman times a car with a stop watch which may introduce an error of 0.5 sec and concludes that the speed of the car is 62 mph. What is the minimum distance in yards over which the car must be timed before the policeman can be quite certain that it was exceeding the speed limit?

20. A stone is dropped from rest at the top of a building 150 m high. How long will it take to reach the ground? With what velocity will it strike the ground?

21. An automobile crashes into a tree at 60 mph. From what height would it have to be dropped to hit the ground at the same speed? About how many stories would there be in a building this high?

22. When the automobile crashed into the tree at 60 mph, its front end

was brought to rest in a very short distance, but the body buckled and the driver traveled a further distance of 1 foot (30.5 cm). Assuming that his acceleration was constant, calculate its value and compare it with the acceleration due to gravity.

23. An engine traveling along a straight track passes a station with a velocity of 15 m/sec. It then maintains a steady forward acceleration of 0.02 m/sec² and reaches the next station 10 minutes later. What is the distance between the two stations?

24. If you were to throw a ball vertically upward, about how high do you think you could throw it? Deduce the maximum velocity with which you are capable of throwing the ball. Convert it into mph.

25. A ball is thrown vertically upward with a velocity of 7 m/sec. How high does it rise? How long does it take to reach its highest position? How long does it take to fall from its highest position back to the ground? What is its velocity upon returning to the ground?

26. A ball is thrown vertically downward from a balloon with a velocity of 800 cm/sec. What is its velocity 2.5 sec later? How far has it then fallen? What is its velocity after falling 12 m?

C

27. A ball is thrown vertically upward and just reaches the top of a building 100 m high. At the same instant that the first ball is thrown from the ground, a second ball is dropped from rest at the top of the building. At what height do the two balls pass one another?

28. A toy rocket is launched vertically upward with a steady acceleration of 13 m/sec² during the 7 sec that the fuel lasts. What height does it reach? How long a time elapses between the start of the launching and the return to the ground? What is its velocity when it returns to the ground?

29. The velocity with which a bullet emerges from the muzzle of a certain rifle is always the same. If the rifle is fired vertically upward, the bullet reaches a height of 150 m. From what height must it be fired vertically downward so that the bullet strikes the ground with twice its muzzle velocity?

30. A ball is dropped from rest at the top of a tall building. 2 sec later another ball is thrown downward from the same starting point with an initial velocity of 25 m sec⁻¹. For how long has the first ball been falling at the instant when the second ball passes it? How far have they then both fallen?

31. Two bodies are dropped simultaneously from rest at heights of 125 m and 35 m above the ground. What is their difference in height after they have been falling for (a) 0.5 sec; (b) 1.0 sec, (c) 2.0 sec, (d) 3.0 sec, (e) 4.0 sec, (f) 5.0 sec. When one of these bodies hits the ground, it does not bounce but sticks to the surface.

32. Repeat problem 31 with the assumption that when a body strikes the ground it rebounds with its speed halved.

33. A ball is thrown vertically upward with a velocity v_o. If the acceleration due to gravity is g, show that the time t_o which elapses be-

fore it returns to the hand of the thrower is $t_o = 2v_o/g$. A movie film of the motion is taken and is then run through the projector backward. Prove that the reversed motion of the ball is indistinguishable from its actual motion. (Hint: Compare the position and velocity at time t with the position and velocity at time $t_o - t$. Explain carefully why this comparison is relevant.)

34. A car is capable of a maximum acceleration a_1 and a maximum deceleration a_2 (both a_1 and a_2 are to be considered as intrinsically positive). What is the minimum time in which it can cover a distance d if it must start from rest and finish at rest?

35. Repeat the previous problem with a maximum acceleration a_2 and a maximum deceleration a_1. Compare the two answers, and discuss the situation in relation to the procedure introduced into problem 33, in which a movie was taken and then run backward through the projector. Consider the form of graphs showing the variation of velocity with time and the variation of distance with time in both cases.

Motion
Along a
Straight
Line

5

Motion along a Curved Path

5-1 Velocity along a Curved Path

When a body moves along a curved path in three-dimensional space, its instantaneous velocity requires a more elaborate definition, but the vectorial nature of the velocity then becomes apparent. Consider the motion of the body in figure 5-1. Suppose that at time t_1 the body is at the point P_1 and at a later time t_2 has moved on to the point P_2. During the time interval $t_2 - t_1$ the body has suffered a vectorial displacement $\overrightarrow{P_1P_2}$. The average velocity during this time interval is defined as:

$$\bar{\mathbf{v}} = \frac{\overrightarrow{P_1P_2}}{t_2 - t_1} \tag{5-1}$$

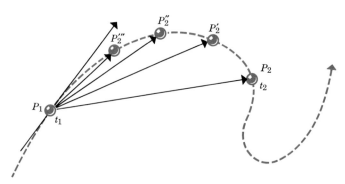

Figure 5-1 The velocity of a body moving along a curved path.

This is a vector. It is, in fact, the vector $\overrightarrow{P_1P_2}$ multiplied by the scalar quantity $1/(t_2 - t_1)$ and is clearly in the same direction as $\overrightarrow{P_1P_2}$.

To define the instantaneous velocity at time t_1, one must clearly make the time interval $t_2 - t_1$ as short as possible, subject to all the considerations discussed in connection with motion along a straight line.

Instantaneous velocity at P_1 is

$$\mathbf{v} = \frac{\overrightarrow{P_1P_2}}{(t_2 - t_1)} \text{ in the limit when } t_2 - t_1 \text{ is sufficiently small}$$

$$(5\text{-}2)$$

In figure 5-1, P_2, P_2', P_2'', P_3''' is a sequence of points obtained by making the time interval $t_2 - t_1$ shorter and shorter. It is apparent that the direction of $\overrightarrow{P_1P_2}$ approaches closer and closer to the direction of the tangent to the curved path at P_1.

The instantaneous velocity at any point on the curved path is in the direction of the tangent to the curve at that point.

5-2 Acceleration along a Curved Path

Referring to figure 5-2, suppose that at time t_1 the body is at P_1, and its instantaneous velocity is $\mathbf{v}_1$, and that at time t_2 the body is at P_2 and its instantaneous velocity is $\mathbf{v}_2$. The triangle of vectors accompanying this figure shows how to calculate the vector $(\mathbf{v}_2 - \mathbf{v}_1)$, which is the vectorial change in velocity between t_1 and t_2. Notice that the vector

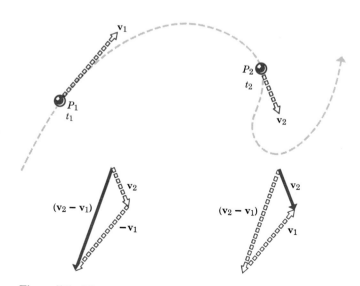

Motion
Along a
Curved
Path

Figure 5-2 The vectorial change in velocity between times t_1 and t_2 from which the average acceleration can be calculated.

61

$(\mathbf{v}_2 - \mathbf{v}_1)$ is the vector that must be added to $\mathbf{v}_1$ in order to change it into $\mathbf{v}_2$. It is therefore the additional velocity that the body has acquired during its motion from P_1 to P_2. Since acceleration is change in velocity divided by time, the definition of the average acceleration is

$$\bar{\mathbf{a}} = \frac{\mathbf{v}_2 - \mathbf{v}_1}{t_2 - t_1} \tag{5-3}$$

This is a vector and is in the same direction as $\mathbf{v}_2 - \mathbf{v}_1$.

Going through the usual procedure of making $t_2 - t_1$ as short as possible, one obtains the instantaneous acceleration

Instantaneous acceleration at time t_1 is

$$\mathbf{a} = \frac{(\mathbf{v}_2 - \mathbf{v}_1)}{(t_2 - t_1)} \text{ in the limit when } t_2 - t_1 \text{ is sufficiently small} \tag{5-4}$$

The direction of **a** depends upon the details of the motion and it is not necessarily along the tangent or perpendicular to the tangent.

5-3 Uniform Circular Motion

The simplest case of curved motion is motion in a circle with constant speed. Referring to figure 5-3, the radius of the circle is r, its center is O and the body is at the point P_o at the zero of time, $t = 0$. After t seconds the body has moved to P and has traced out a distance s along the perimeter of the circle. During this time the radius OP has moved through an angle θ (the Greek letter theta), which is most conveniently measured in **radians**. The measure of an angle in radians is the length of the arc which it cuts off the perimeter of a circle divided by the radius of that circle.

$$\theta \text{ (in radians)} = \frac{s}{r} \tag{5-5}$$

Notice that θ is the ratio of one length s to another length r. Its numerical value is therefore independent of the choice of the unit of length and, of course, the units of time and mass are irrelevant. We say that θ is a dimensionless quantity, or that the radian is dimensionless.

If the body makes 1 complete revolution, OP rotates through $360°$, but s is then the complete perimeter of the circle, $2\pi r$, and so the angle in radians is

$$\theta = \frac{s}{r}$$

$$= \frac{2\pi r}{r}$$

$$= 2\pi \text{ for 1 complete revolution} \tag{5-6}$$

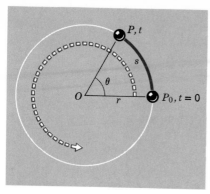

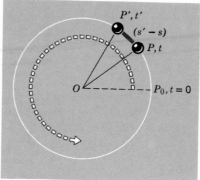

Figure 5-3 Motion in a circle.

Figure 5-4 Method of deducing the instantaneous velocity of the body.

So $\qquad 360° = 2\pi$ radians $\qquad\qquad$ (5-7)

It follows that $180° = \pi$ radians $\qquad\qquad$ (5-8)

$$90° = \frac{\pi}{2} \text{ radians} \qquad\qquad (5\text{-}9)$$

$$1 \text{ radian} = \frac{360°}{2\pi}$$

$$= 57.3° \text{ approximately} \qquad\qquad (5\text{-}10)$$

The rate at which the angle changes with time is a new kind of velocity. It is called the **angular velocity** and is represented by the symbol ω (the Greek letter omega). If the angle is θ_1 radians at time t_1 and θ_2 radians at a slightly later time t_2, then the definition of ω is:

Angular velocity,

$$\omega = \frac{\theta_2 - \theta_1}{t_2 - t_1} \qquad\qquad (5\text{-}11)$$

in the limit when $(t_2 - t_1)$ is sufficiently small.

An angular velocity is quite different from an ordinary velocity, v, which is sometimes called a linear velocity to emphasize the distinction. The units of v are m sec^{-1} in the MKS system or cm sec^{-1} in the CGS system. The units of ω are radians per second, which can be abbreviated to rad sec^{-1}. Since the radian is dimensionless, this is the same as sec^{-1}. (It pays to leave in the word radians, though, because this reminds us of what we are talking about). Notice, for example, that the numerical value of v is different in the MKS and CGS systems, but the numerical value of ω is the same in both systems.

Motion Along a Curved Path

63

The definition of uniform motion is that the angular velocity should be the same at all instants of time. This implies that the angle swept out by the radius OP is proportional to the time.

$$\theta = \omega t \tag{5-12}$$

The reader should verify for himself that this is consistent with the definition of ω in equation 5-11, and with the initial conditions $\theta = 0$ when $t = 0$. Since, from equation 5-5

$$
\begin{aligned}
s &= r\theta \\
&= r\omega t
\end{aligned}
\tag{5-13}
$$

$$s = vt \tag{5-14}$$

$$v = \omega r \tag{5-15}$$

v is the speed with which the body traces out the perimeter of the circle. We shall now show that it is the same thing as the magnitude of the tangential velocity. Figure 5-4 shows the body moving from P to P' during the time interval from t to t'. The magnitude of the average velocity over this time interval is

$$\text{Average velocity} = \frac{PP'}{(t' - t)} \tag{5-16}$$

To obtain the instantaneous velocity at P, we must make the time interval as short as possible and P' as close to P as possible. Now, it is possible to prove rigorously that, when P' approaches very close to P, the difference in length between the arc PP' and the chord PP' is negligibly small, and so

$$\text{Instantaneous velocity at } P = \frac{\text{arc } PP'}{(t' - t)} \tag{5-17}$$

The length of the arc P_0P is $s = vt$, and similarly the length of the arc P_0P' is $s' = vt'$. The length of the arc PP' is therefore

$$
\begin{aligned}
\text{arc } PP' &= s' - s \\
&= vt' - vt \\
&= v(t' - t)
\end{aligned}
\tag{5-18}
$$

The instantaneous velocity at P is therefore

$$
\begin{aligned}
\text{Instantaneous velocity at } P &= \frac{v(t' - t)}{(t' - t)} \\
&= v
\end{aligned}
\tag{5-19}
$$

As P' approaches P the direction of $\overrightarrow{PP'}$ coincides with the tangent to the circle at P, and so, in accordance with what has already been said in Section 5-1, the direction of the instantaneous velocity is along the

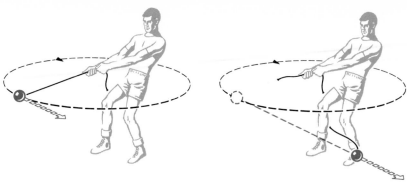

Figure 5-5

tangent to the circle and is at right angles to the radius *OP*. Imagine, for example, a ball on the end of a string being whirled around in a horizontal circle (figure 5-5). If the string suddenly snaps, the ball flies off along a straight line tangential to the circle.

5-4 Acceleration of a Body Moving Uniformly in a Circle

It is easy to fall into the error of believing that, since the body moves around the circle with constant speed, it has no acceleration. However, its velocity is a vector and, although the *magnitude* of this vector does not change, its *direction* is continually changing. In fact the direction of the vector rotates through 360° during every revolution of the body. A change in the direction of the velocity at constant magnitude is just as real an acceleration as a change in the magnitude. Figure 5-6 shows how

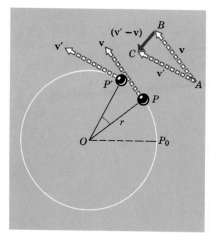

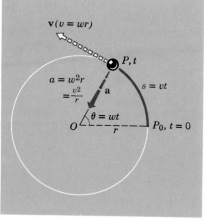

Motion Along a Curved Path

Figure 5-6 Method of deducing the instantaneous acceleration.

Figure 5-7 A summary of the important equations for uniform motion in a circle.

65

the direction of the velocity changes as the body moves from P to P'. The triangle of vectors ABC on the right of this figure shows the initial velocity $\mathbf{v}$ at P, the final velocity $\mathbf{v}'$ at P' and the vector $(\mathbf{v}' - \mathbf{v})$ which must be added to $\mathbf{v}$ in order to change it into $\mathbf{v}'$. This vector $(\mathbf{v}' - \mathbf{v})$ is the change in velocity between times t and t' and the average acceleration is $(\mathbf{v}' - \mathbf{v})/(t' - t)$. Calculating this quantity in the limit when $(t' - t)$ is very small and P' is close to P, we shall obtain the instantaneous acceleration at P.

The vector representing the velocity is always at right angles to the radius OP. OP and $\mathbf{v}$ therefore rotate together like two rigid rods firmly attached at right angles to one another. It should be apparent that when OP rotates through a certain angle, the vector $\mathbf{v}$ rotates through exactly the same angle. The angle POP' is therefore equal to the angle BAC. Also $OP = OP'$ and $AB = AC$, since the velocity does not change its magnitude as it rotates. POP' and BAC are therefore similar isosceles triangles and

$$\frac{BC}{AB} = \frac{PP'}{OP}$$

or

$$\frac{BC}{v} = \frac{PP'}{r} \tag{5-20}$$

We have already pointed out that when P' is very close to P the chord PP' is very nearly equal to the arc PP' and, according to equation 5-18, is

$$PP' = v(t' - t) \tag{5-21}$$

Therefore, $\dfrac{BC}{v} = \dfrac{v(t' - t)}{r}$

$$BC = \frac{v^2(t' - t)}{r} \tag{5-22}$$

The magnitude of the acceleration at P is

$$a = \frac{BC}{(t' - t)} \text{ when } (t' - t) \text{ is very small}$$

$$= \frac{v^2(t' - t)}{r(t' - t)}$$

$$a = \frac{v^2}{r} \tag{5-23}$$

Since, from equation 5-15, $v = \omega r$, this may also be written

$$a = \frac{(\omega r)^2}{r}$$

$$a = \omega^2 r \tag{5-24}$$

In the limit when P' is very close to P the angle CAB is very little larger than $0°$. Since angle ABC is equal to angle ACB, and the three angles of the triangle ABC must add up to $180°$,

Angle ABC = Angle ACB = $90°$ approximately $\qquad$ (5-25)

when P' is very close to P. The direction of the acceleration is the direction of $\overrightarrow{BC}$ and it is therefore perpendicular to **v**. Since **v** is along the tangent to the circle at P, the acceleration **a** must be along the radius. Notice that the acceleration points *toward* the center of the circle (see figure 5-7).

> When a body moves in a circle with constant angular velocity ω and constant speed v, it has an acceleration which always points along the radius toward the center of the circle and whose magnitude is
>
> $a = \omega^2 r$ $\qquad$ (5-26)
>
> or
>
> $a = \dfrac{v^2}{r}$ $\qquad$ (5-27)

Example 5-4-1

Ignoring the motion of the earth around the sun and the motion of the sun through space, calculate (a) the angular velocity, (b) the velocity, and (c) the acceleration of a body resting on the ground at the equator.

Because of the rotation of the earth the body at the equator moves in a circle whose radius is equal to the radius of the earth (figure 5-8).

Figure 5-8 Illustration for example 5-4-1.

Motion
Along a
Curved
Path

r = radius of earth

$\quad = 6.37 \times 10^6$ meters

We are going to use the MKS system of units. One revolution, which is 2π radians, takes 1 day or $24 \times 60 \times 60$ seconds.

(a) The angular velocity is therefore

$$\omega = \frac{2\pi}{24 \times 60 \times 60}$$

$$= 7.27 \times 10^{-5} \text{ radians per second}$$

(b) The velocity is

$$v = \omega r$$

$$= (7.27 \times 10^{-5}) \times (6.37 \times 10^6)$$

$$= 4.64 \times 10^2 \text{ m/sec}$$

Since 1 mph = 0.447 m/sec

$$v = \frac{4.64 \times 10^2}{0.447} \text{ mph}$$

$$= 1040 \text{ mph}$$

which is almost twice as fast as a commercial jet plane.

(c) The acceleration toward the center of the earth is

$$a = \frac{v^2}{r}$$

$$= \frac{(4.64 \times 10^2)^2}{6.38 \times 10^6}$$

$$= 3.37 \times 10^{-2} \text{ m/sec}^2$$

which is about $\frac{1}{3}\%$ of g, the acceleration due to gravity.

5-5 A Body Projected Horizontally

In the previous chapter we discussed the straight line motion of a body falling vertically downward or thrown vertically upward. We shall now extend the discussion by introducing a horizontal component of the velocity so that the path is curved.

Consider, for example, a ball thrown over the edge of a cliff with an initial velocity u_0 entirely in the horizontal direction, as in figure 5-9. The acceleration produced by gravity, g, is entirely in the vertical direction and cannot change the horizontal component of the velocity. This implies that a fielder starting from the foot of the cliff at the instant the ball is thrown and running outward with a constant horizontal velocity u_0 always remains directly underneath the ball.

Meanwhile, the ball continues to fall in the vertical direction in exactly the same way as if it had no horizontal velocity. Imagine a second ball dropped from rest at the top of the cliff at exactly the same instant that the first ball is thrown horizontally outward. The two balls fall

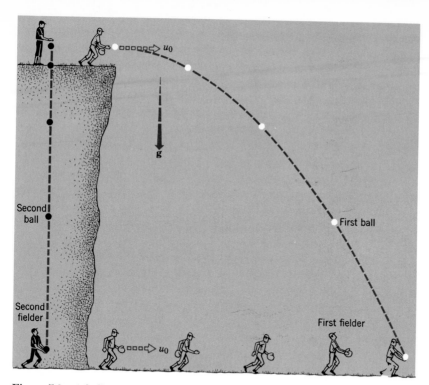

Figure 5-9 A ball projected horizontally over the edge of a cliff.

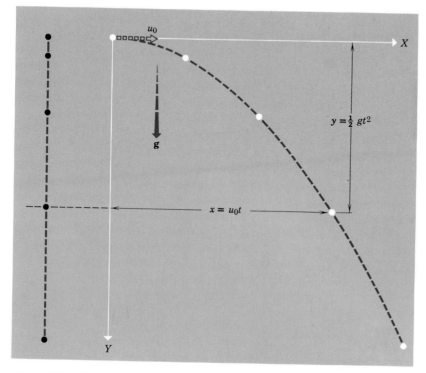

$$y = \tfrac{1}{2} gt^2$$

$$x = u_0 t$$

Motion
Along a
Curved
Path

Figure 5-10 The use of Cartesian coordinates to discuss figure 5-9.

vertically in exactly the same way and are always at the same height. A second fielder standing stationary at the foot of the cliff catches the second ball at exactly the same instant that the running fielder catches the first ball.

To tackle the problem mathematically, use Cartesian axes with the origin at the top of the cliff. Let the x axis be horizontal with its positive direction outward from the cliff, and let the y axis be vertical with its positive direction downward from the top of the cliff (figure 5-10). After a time t, the horizontal distance covered by the first ball is the same as the horizontal distance covered by the running fielder and is

$$x = u_0 t \tag{5-28}$$

Meanwhile, the vertical distance through which either ball has fallen is obtained by using equation 4-25 from Chapter 4. We must remember, though, that we are now considering motion in the y direction rather than the x direction, that the initial velocity v_0 in the vertical direction is zero, and that the acceleration a is $+g$, which is positive because it is downward and the positive y direction is also downward. Equation 4-25 therefore becomes

$$y = \tfrac{1}{2}gt^2 \tag{5-29}$$

To obtain y as a function of x, we can use equation 5-28 to obtain an expression for t,

$$t = \frac{x^2}{u_0} \tag{5-30}$$

and then insert this value of t into equation 5-29

$$y = \tfrac{1}{2}g\left(\frac{x}{u_0}\right)^2$$

or

$$y = \left(\frac{g}{2u_0{}^2}\right)x^2 \tag{5-31}$$

This is actually the equation of a parabola with its vertex at the top of the cliff.

Example 5-5-1

The cliff is 20 m high and the ball is thrown from the top with a horizontal velocity of 8.0 m sec^{-1}. (a) How long will it take to reach the horizontal plane at the foot of the cliff? (b) How far from the foot of the cliff will it strike the ground? (c) What will be its speed as it strikes the ground?

(a) Use the MKS system of units. We are given that $u_0 = 8.0$ m sec^{-1}, $g = +9.8$ m sec^{-2}, and $y = +20$ m when the ball strikes the ground. We wish to know the value of t when the ball strikes the ground, and so we choose equation 5-29 because it contains t and the known quantities y and g:

$$20 = \tfrac{1}{2} \times 9.8t^2$$

Whence

$$t = \sqrt{\frac{2 \times 20}{9.8}}$$

$$= 2.02 \text{ sec}$$

The ball strikes the ground 2.02 sec after it was thrown.

(b) Equation 5-28 tells us that after 2.02 sec the ball has traveled a horizontal distance

$$x = 8.0 \times 2.02$$

$$= 16.16 \text{ m}$$

The ball strikes the ground 16.2 m from the foot of the cliff.

(c) Apply equation 4-27 in Chapter 4 to the vertical motion. After 2.02 sec, the vertical component of the velocity is

$$v = 0 + (9.8 \times 2.02)$$

$$= 19.8 \text{ m sec}^{-1}$$

Since the horizontal component of the velocity is always $u_0 = 8 \text{ m sec}^{-1}$, the resultant velocity V is given by

$$V^2 = u_0{}^2 + v^2$$

$$= 8^2 + 19.8^2$$

$$= 64 + 392$$

$$V = \sqrt{456}$$

$$= 21.4 \text{ m sec}^{-1}$$

The speed of the ball as it strikes the ground is 21.4 m sec^{-1}.

5-6 A Body Projected in Any Direction

Now imagine a body projected upward with an initial speed V_0 at an angle θ to the horizontal, as in figure 5-11. As we shall shortly prove, its path is again a parabola. Let us calculate h, the maximum height to which it rises, and R, its range on a horizontal plane.

As shown in figure 5-12, take the origin at the point of projection, the x axis horizontal, and the y axis vertical with the positive y direction now *upward*. The initial velocity V_0 may be replaced by a horizontal x-component $V_0 \cos \theta$ and a vertical y-component $V_0 \sin \theta$ upward. As in the previous section, the horizontal component is unaffected by the acceleration due to gravity and remains constant. After a time t, the displacement in the horizontal x direction is

$$x = (V_0 \cos \theta)t \tag{5-32}$$

Apply equation 4-28 to the vertical motion in the y direction, with the initial velocity $v_0 = V_0 \sin \theta$, and the acceleration $a = -g$, which is nega-

Motion
Along a
Curved
Path

71

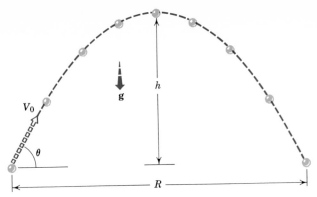

Figure 5-11 A body projected at an angle θ to the horizontal.

tive because g is downward whereas the positive y direction is upward. The analogue of equation 4-28 then gives us the height of the projectile at time t:

$$y = (V_0 \sin \theta)t - \tfrac{1}{2}gt^2 \tag{5-33}$$

Similarly, equation 4-27 gives us the vertical component of the velocity at any instant

$$V_y = V_0 \sin \theta - gt \tag{5-34}$$

The highest point is reached when V_y has been reduced to zero:

$$V_0 \sin \theta - gt = 0 \tag{5-35}$$

or

$$t = \frac{V_0 \sin \theta}{g} \tag{5-36}$$

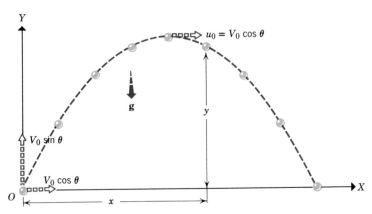

Figure 5-12 The use of Cartesian axes to discuss figure 5-11.

This value of t may be substituted in equation 5-33 to give the maximum height:

$$h = \left(V_0 \sin \theta\right)\left(\frac{V_0 \sin \theta}{g}\right) - \frac{g}{2}\left(\frac{V_0 \sin \theta}{g}\right)^2$$

$$h = \frac{V_0^2 \sin^2 \theta}{2g} \tag{5-37}$$

When the body is in its highest position, the vertical component of its velocity is zero and its resultant velocity is $V_0 \cos \theta$ horizontally. This is identical with the situation discussed in the previous section if u_0 is put equal to $V_0 \cos \theta$. The previous proof that the path is parabolic can therefore be carried over to the present situation.

To obtain the range R, use the fact that when the body strikes the ground again $y = 0$. At this instant of time, t_s, equation 5-33 becomes

$$(V_0 \sin \theta)t_s - \tfrac{1}{2}gt_s^2 = 0$$

$$(V_0 \sin \theta)t_s = \tfrac{1}{2}gt_s^2 \tag{5-38}$$

This equation is satisfied by $t_s = 0$, but this obviously corresponds to the instant of projection, when it is also true that $y = 0$. If t_s is not zero, then both sides of the equation can be divided by t_s:

$$V_0 \sin \theta = \tfrac{1}{2}gt_s$$

Whence

$$t_s = \frac{2V_0 \sin \theta}{g} \tag{5-39}$$

This is an interesting piece of information because it tells us how long the flight lasts. At the instant t_s, the x-coordinate is the range R, and so equation 5-32 implies that

$$R = V_0 \cos \theta \, \frac{2V_0 \sin \theta}{g}$$

$$R = \frac{2V_0^2 \sin \theta \cos \theta}{g} \tag{5-40}$$

Using the trigonometric relationship

$$\sin 2\theta = 2 \sin \theta \cos \theta$$

$$R = \frac{V_0^2 \sin 2\theta}{g} \tag{5-41}$$

Example 5-6-1

If the speed of projection V_0 is fixed, what must be the angle of projection θ to achieve the maximum range? What is then the maximum height?

Under these circumstances, the only variable in equation 5-41 is $\sin 2\theta$, which has a maximum value of 1 when $2\theta = 90°$ or $\theta = 45°$. The maximum range is therefore realized when the angle of projection is 45° and its value is

$$R_{max} = \frac{V_0^2}{g} \tag{5-42}$$

Since $\sin 45° = 1/\sqrt{2}$, the equation 5-37 tells us that the height reached is one quarter of the range

$$h = \frac{V_0^2}{4g} \tag{5-43}$$

QUESTIONS

1. If the acceleration of a body is constant in magnitude and direction, is its path necessarily a straight line? Give examples.
2. If the path of a body is a straight line, what can be said about (a) the nature of the acceleration **a**, (b) the relation between **v** and **a** at all instants.
3. If the acceleration of a body is always constant in magnitude, but not in direction, is the path necessarily either a straight line or a circle? Explain.
4. Is it possible for a body to move in a circle with a speed which varies in such a way that the acceleration points *outward* along a radius?
5. When a body is projected at an arbitrary angle θ, at what point in its path does the speed have a minimum value? Prove your answer rigorously.
6. If the initial speed V_0 is fixed, what must be the angle of projection θ to achieve a maximum height h? Prove your answer rigorously.

PROBLEMS

A

1. Express the following angles in radians (a) 30°, (b) 45°, (c) 60°, (d) 270°.
2. A string is just long enough to go around the perimeter of a square. It is then laid along an arc of the circle which passes through all four corners of the square. What is the measure in radians of the angle subtended by this arc at the center of the circle?
3. Engineers sometimes measure a rate of rotation in revolutions per minute or rpm. What is the angular velocity in rad sec^{-1} corresponding to 100 rpm?
4. A small ball is being whirled in a circle on the end of a string 15-cm long. What is the angle in radians swept through by the string while the ball moves along an arc of length 25 cm?
5. A particle moves with a constant speed of 7 cm sec^{-1} in a circle of radius 2 cm. What is its angular velocity?
6. A particle moving in a circle of radius r with a constant speed v takes

a time T to complete one revolution. Find the equation which (a) gives the angular velocity ω in terms of the time T, (b) gives the speed v in terms of r and T, (c) gives the radial acceleration a in terms of r and T.

7. A phonograph record with a diameter of 12 inches (30.5 cm) rotates at $33\frac{1}{3}$ rpm. What is the velocity of a point on its rim? What is the acceleration of this point?

8. A body moves uniformly in a circle of diameter 10 cm with an angular velocity of 3.5 radians/sec. What is its speed? What is its radial acceleration?

9. A body moves around a circle of radius 3 m with a constant speed of 30 cm/sec. What is its radial acceleration?

10. A cannon is mounted on a hill 150 m high, and its barrel points horizontally. It is trying to hit the gate of a city on the plain below, a horizontal distance of 350 m away. What should be the velocity of the cannon ball as it emerges from the muzzle?

B

11. The second hand of a wrist watch is 1.2 cm long from the center of the dial to the tip of the hand. Calculate the velocity and acceleration of the tip.

12. Calculate the velocity, angular velocity, and acceleration of the moon in its approximately circular orbit about the earth.

13. Calculate the velocity, angular velocity, and acceleration of the earth in its approximately circular orbit about the sun.

14. A simple theory of the hydrogen atom concludes that the electron moves around the proton with a velocity of 2.2×10^8 cm/sec in a circular orbit of radius 5.3×10^{-9} cm. Calculate the acceleration of the electron and compare it with the acceleration due to gravity, g.

15. A space station has the shape of a doughnut. The distance from the center of the hole to the center of the "dough" is 100 m. With what angular velocity must it be made to rotate so that a spaceman living in the "dough" experiences the same acceleration as on the surface of the earth?

16. A plate is knocked off a table with a horizontal velocity of 42 cm sec^{-1}. If the table is 76 cm high, what is the angle between the velocity of the plate and the horizontal at the instant just before it hits the floor?

17. If a rocket is fired at $60°$ to the horizontal with an initial speed of 500 mph, calculate its time of flight and range (in miles) on a horizontal plane.

C

18. A body moves with a constant angular velocity of 2.5 rad sec^{-1} in a circle of radius 15 cm contained in a vertical plane in the earth's gravitational field. What is its acceleration (a) in its highest position (b) in its lowest position? Explain carefully why there is, or is not, a difference between the two answers.

19. It is found that a rotating body breaks when some point in it

Motion
Along a
Curved
Path

reaches a maximum velocity. This velocity is approximately the speed of sound in the material and its value is typically 5×10^3 m/sec. To achieve a maximum radial acceleration, does it pay to rotate a small body or a large body? With a radius of about 1 cm, how many times larger than g is the maximum radial acceleration?

20. A tire has a diameter of 50 cm. What is the acceleration of the highest point on it when the car has a steady speed of 60 mph? (The point on the tire touching the road is known to be instantaneously at rest.)

21. The technique used in the sport of hammer throwing is to whirl the hammer around in a circle several times before releasing it. The hammer is about 4 feet long and the record throw is somewhat longer than 200 feet. Using common sense and guessing the values of some relevant quantities, calculate approximately the time taken for the last revolution just before the throw.

B

NEWTONIAN MECHANICS

6

The
Laws
of
Motion

6-1 Newton's First Law of Motion

Within the framework of classical physics, let us pursue our plan of visualizing the universe as a collection of particles in motion. In the two preceding chapters we devoted our attention to the description of this motion and we considered carefully such concepts as velocity and acceleration. We shall now turn to the basic question of why each particle moves along the particular path it does and how we might predict the details of its path. The underlying idea is that the details of the motion are determined by the interactions between the particle and the other particles in the universe. To use a language that is already familiar to the reader, but the precise significance of which will have to be described in detail, the other particles exert forces on the particle which push and pull it along the twists and turns of its path.

Before pursuing this point, a still more basic question must be answered. How would the particles behave if they did not interact with one another, if they were entirely oblivious of one another's existence? Would they all remain at rest, move in straight lines or circles, or oscillate backward and forward about fixed centers? The answer is not obvious and can only be found by carefully observing the universe and discovering which assumption best fits the facts. The answer is contained in Newton's first law of motion.

Newton's first law of motion:

In the absence of any interaction with the rest of the universe, a body would either remain at rest or move continually in the same straight line with a constant velocity. That is, the velocity, regarded as a vector, would remain constant; and rest would be the special case of zero velocity. Alternatively, one can say that the acceleration would be zero.

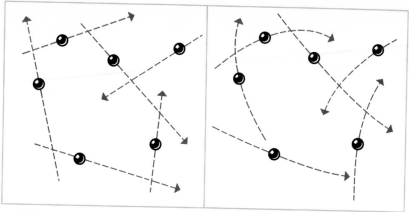

(a) If the particles did not interact they would move in straight lines with constant velocities.

(b) When they interact they accelerate one another and distort one another's paths.

Figure 6-1

6-2 Interactions and Determinism

In the absence of interactions, the particles in the universe would move in straight lines with constant velocities as in figure 6-1*a*. The role of the interactions is to change the velocities, to produce accelerations and hence complicated paths, as in figure 6-1*b*. A large part of classical physics consists of formulating the laws enabling calculation of the acceleration produced in one particle by a second particle. The procedure is somewhat as follows. Suppose that, at a fixed instant of time, we know the positions and velocities of all the particles, which are labeled 1, 2, 3, 4, 5, etc., as in figure 6-2. To calculate the acceleration of particle 1, let us consider first its interaction with particle 2. If we know

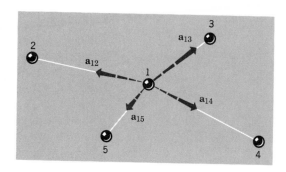

Figure 6-2   The acceleration of body 1 is the vector sum of contributions due to interactions with each of the other bodies in the universe.

the distance between 1 and 2, their velocities and certain properties of the two bodies such as their masses and electric charges, the fundamental laws enable us to calculate the acceleration a_{12} induced in 1 by its interaction with 2. Similarly we can calculate the acceleration a_{13} induced in 1 by its interaction with 3 and so on to obtain the contribution to the acceleration of 1 from all the other bodies in the universe. The vectorial sum $a_{12} + a_{13} + a_{14} + a_{15} + \cdots$ then gives us the total acceleration of body 1 at the fixed instant of time we are considering. The accelerations of all the other bodies at this same instant of time can be calculated in a similar way. It can be shown that this procedure enables us to describe the path of each particle in complete detail and hence calculate the positions, velocities, and accelerations of all the particles at all future instants of time and all past instants of time.

Determinism in classical physics

In classical physics, if the positions and velocities of all the particles are known at one instant of time, then the fundamental laws describing the interactions of the particles with one another can be used to calculate in precise detail the behavior of the particles at all past, present, and future times.

This implication of classical physics clearly has profound philosophical significance. It would seem to imply that the future of the universe is completely determined by its past and that there is no such thing as free will or human intervention. Suppose that I take the extreme point of view that my thoughts are merely a consequence of the positions and velocities of the electrons, protons, and neutrons in my brain. Then, if I know their positions and velocities at 12:30 p.m. and the positions and velocities of certain other particles in my immediate vicinity, with sufficient effort and the use of sufficiently complicated computers, I can presumably predict that at 12:55 p.m. the positions and velocities of these particles in my brain will be such as to cause me to make a decision to go out to lunch. Any feeling I have that this decision was made voluntarily is pure delusion. If, in the process of crossing the road to the lunch counter, I am struck by an automobile and killed, this could presumably have been predicted from the previous positions and velocities of the particles in the universe and was quite inescapable. These conclusions are very difficult to avoid if one accepts without question the basic approach of classical physics. Fortunately, modern quantum mechanics provides a loophole, as we shall discover in a later chapter, although it is not clear that it provides us with an alternative which is any more comforting.

The richness and variety of physical theory is partly due to the fact that the interaction between two particles has several aspects, which will now be enumerated.

A. Gravitational forces. These are always attractive and exist between any pair of particles in the universe.

B. Electromagnetic forces. These are present if the two particles have electric or magnetic properties.

C. Nuclear forces. These are the forces which hold the neutrons and protons together in the nucleus of an atom. They are important only when the two particles are very close together, within about 10^{-12} cm of one another.

D. Various other interactions are currently being studied at the frontier of physics and are not yet properly understood. These interactions are not conspicuous in everyday experience.

During our presentation of Newtonian mechanics we shall concentrate on gravitational and electromagnetic forces. The other forces are more complicated and will have to be carefully explained at the appropriate places later in the book.

6-3   Gravitation and Gravitational Mass

Imagine any two bodies 1 and 2 at a distance R apart, as in figure 6-3. They might be two fundamental particles, or the earth and the moon, or the earth and a falling body, or the sun and a planet, or two stars. If they are not electrically charged and not magnetized, there are no electromagnetic forces between them. If they are much farther apart than 10^{-12} cm, the nuclear forces can also be neglected. Then only their gravitational interaction need be considered. (The gravitational interaction is still present even when there are other types of force, but sometimes these other forces are much stronger and completely obscure the weak gravitational interaction.) In order to avoid mathematical complications of a non-essential nature, we shall assume that the distance between the bodies is very large compared with their size. As we have already explained in section 2-1, this requirement is frequently satisfied, especially in the solar system.

Gravitational phenomena, such as the motion of the planets and their moons and the behavior of bodies falling to the earth, can be explained

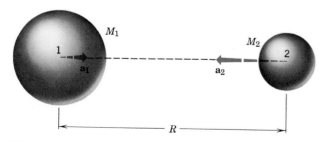

Figure 6-3   The gravitational interaction of two bodies. M_1 and M_2 are their gravitational masses.

in terms of the following assumption. The gravitational effect of body 2 on body 1 is such as to give body 1 an acceleration pointing directly toward body 2 of magnitude

$$a_1 = \frac{GM_2}{R^2} \tag{6-1}$$

Similarly, interchanging the roles of the two bodies, the gravitational effect of body 1 on body 2 is such as to give body 2 an acceleration pointing directly toward body 1 of magnitude

$$a_2 = \frac{GM_1}{R^2} \tag{6-2}$$

The first thing to notice about these formulas is that a_1 and a_2 vary inversely with the square of R. As the distance between the two bodies is increased, the gravitational accelerations they produce in one another decrease rapidly. G is a "universal constant of nature" called the **gravitational constant.** It should not be confused with the acceleration due to gravity g, which is not a universal constant but has a value that depends upon the accidental circumstance of being located at a particular spot near the earth's surface and even varies from place to place over the surface of the earth. The numerical value of G, which is assumed to be the same for any pair of bodies anywhere in the universe, is

$$G = 6.67 \times 10^{-11} \text{ in the MKS system} \tag{6-3}$$
$$= 6.67 \times 10^{-8} \text{ in the CGS system} \tag{6-4}$$

(As an exercise in units the reader might like to derive the value in the CGS system given the value in the MKS system.)

The quantities M_1 and M_2 are the **gravitational masses** of the bodies 1 and 2 and equations 6-1 and 6-2 serve to define these masses. This means that we can associate with any body an invariable number M which, when inserted into an equation such as 6-1, enables us to calculate the gravitational acceleration which the body produces in any other body. The procedure for determining the value of M will be described in a simplified form in the next paragraph. Gravitational mass is always positive. As we explained in section 2-4, the unit of mass, the standard kilogram, is arbitrarily defined as the mass of a particular platinum-iridium cylinder. This arbitrary choice determines the numerical value of the constant G. This can be illustrated by the following sequence of imaginary experiments. Let M_2 be the standard kilogram mass. Place a small mass M_1 at a measured distance R meters from it, and then measure the acceleration a_1 of M_1 due to M_2, say in some remote region of space where the effect of all other masses can be neglected. Then

$$a_1 = \frac{G \times 1}{R^2} \text{ if } M_2 = 1 \text{ kilogram} \tag{6-5}$$

$$G = a_1 R^2 \tag{6-6}$$

Since R and a_1 have both been measured, this experiment serves to measure G, which will be found to have the value given in equation 6-3. However, it has this particular value only because the standard kilogram is as large as it is.

Now take an unknown mass M, place a small mass m at a measured distance R away from it and measure the acceleration a of the mass m, all other masses in the universe being too far away to influence the result. Then

$$a = \frac{GM}{R^2} \tag{6-7}$$

$$M = \frac{aR^2}{G} \tag{6-8}$$

Since a and R have been measured and G has been determined by the first experiment, this second experiment serves to measure the unknown mass M. Notice that the measurement of a and R involves only basic measurements of length and time. We have used the MKS system in the above analysis. As an exercise, the student may wish to consider the slight modifications in the argument if the CGS system were used.

6-4 Electromagnetism and Inertial Mass

A careful examination of equations 6-1 and 6-2 will convince you that the gravitational mass of a body is a measure of its ability to produce an acceleration in *another* body. As long as we confine ourselves to gravitational interactions, there is no need to introduce any other definition of mass. Equations 6-1 and 6-2 are completely adequate to describe all purely gravitational phenomena, such as the behavior of the solar system. However, once we introduce electromagnetic interactions we are led to the concept of **inertial mass,** which rather seems to measure the ability of a body to resist the production of an acceleration in itself. Gravitational mass and inertial mass appear to be entirely different concepts and yet, to the best of our present knowledge, they are always exactly equal to one another for any body. We do not properly understand why this should be so.

Electromagnetism is a complicated subject that we shall discuss more fully later. For our present purpose, consider the particularly simple case of two electrically charged bodies 1 and 2 that are instantaneously at rest at a distance R apart, as in figure 6-4. We can associate with the bodies two numbers q_1 and q_2, which are called their **electric charges.** Unlike gravitational mass, electric charge can be either positive or negative. If the two charges are both positive or both negative they repel one another and accelerate directly away from one another. But

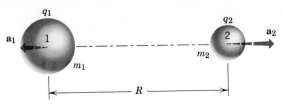

Figure 6-4 The electrostatic interaction of two bodies. q_1 and q_2 are their electrical charges, and m_1 and m_2 are their inertial masses.

if one is positive and the other negative they attract one another and accelerate directly toward one another. In addition to the gravitational acceleration, which is always present, the electrical interaction gives body 1 an acceleration

$$a_1 = \frac{1}{m_1}\frac{q_1 q_2}{R^2} \tag{6-9}$$

and body 2 an acceleration

$$a_2 = \frac{1}{m_2}\frac{q_1 q_2}{R^2} \tag{6-10}$$

m_1 and m_2 are the inertial masses of bodies 1 and 2 respectively. The equations take this particularly simple form only in the CGS system, but, since the complications introduced into the MKS formulation are not directly related to the physical behavior of the universe, we shall ignore them here. In the CGS system the units of charge are statcoulombs. Notice that electromagnetic accelerations also vary inversely with the square of the distance between the two bodies.

Notice that in equation 6-9 for the electromagnetic acceleration of body 1 the inertial mass of body 1 appears in the denominator, whereas in equation 6-1 for the gravitational acceleration of body 1 the gravitational mass of body 2 appears in the numerator. This is the reason why we say that the gravitational mass produces gravitational accelerations in other bodies, whereas the inertial mass resists the production of an electromagnetic acceleration in the body itself.

Equations 6-9 and 6-10 imply that

$$\frac{a_1}{a_2} = \frac{m_2}{m_1} \tag{6-11}$$

The accelerations are in inverse ratio to the masses. (This is true in the MKS system as well as the CGS system.) If we measure the two accelerations, we can deduce the ratio of the two inertial masses. If one of the bodies is the standard one kilogram cylinder of platinum-iridium, its inertial mass is defined to be unity in the MKS system and 1000 in the CGS system, and so the mass of the other body can be deduced. Comparing masses in pairs in this way, the inertial masses of all bodies can be determined.

The simple procedures that we have described for measuring gravitational and inertial masses are not, of course, used in practice. A complete description and analysis of the procedures actually used would be very lengthy and might, by virtue of its complexity, prove more confusing than helpful. However, when properly understood, these procedures are equivalent in principle to the direct procedures we have described. The standard kilogram is taken to be the MKS unit of both gravitational and inertial mass, but in the case of other bodies there seems to be no initial reason why the measured gravitational mass should turn out to be equal to the measured inertial mass. In fact, however, the gravitational mass has never been found to differ from the inertial mass by more than 1 part in 10^4, which is the limit of the experimental accuracy. It is conceivable that, if the experimental accuracy could be improved, a small difference might be found. Meanwhile, it is assumed that gravitational mass is exactly the same as inertial mass. Throughout the rest of this book we shall cease to make any distinction between them.

6-5   Force: Newton's Second Law

Newton's second law of motion is not a profound statement about the physical behavior of the universe, but rather a precise definition of the concept of **force**. If a body of mass m has an acceleration **a** (figure 6-5), then the force acting on it is *defined* as the product of its mass and its acceleration.

Newton's second law of motion

Force = Mass × Acceleration

$$F = ma \qquad (6\text{-}12)$$

Newtons = Kilograms × Meter sec^{-2}

Dynes   = Grams × Centimeter sec^{-2}

We are no longer making any distinction between gravitational and inertial mass and so the mass m is a uniquely defined quantity. Notice that equation 6-12 is a vector equation. Force is a vector and its direc-

$F = ma$

a

m

Figure 6-5   Illustrating the definition of force.

tion is the same as the direction of the acceleration **a.** Its magnitude is equal to the product of the scalar mass and the scalar magnitude of **a.**

It is not necessary to imagine force as though it were a little elf pushing the body along and exerting a strength which, for some magic reason, is always found to be equal to the product of mass and acceleration. Force is merely another name for the product of mass and acceleration. Its importance in physics is that this particular product turns out to be very useful.

The units of force are clearly kg m sec^{-2} or gm cm sec^{-2}, but the concept of force is so important that the units have special names in both systems. In the MKS system the unit of force is the **newton** and in the CGS system it is the **dyne.** If m is measured in kilograms and a in meters per sec per sec, then F is measured in newtons. If m is measured in grams and a in centimeters per sec per sec, then F is measured in dynes.

Example 6-5-1

Prove that

1 newton = 10^5 dyne

1 newton produces an acceleration of 1 m sec^{-2} in a mass of 1 kg. That is, 1 newton produces an acceleration of 100 cm sec^{-2} in a mass of 1000 gm.

But $F = ma$

If $m = 1000$ gm and $a = 100$ cm sec^{-2}

$F = 1000 \times 100$ dyne

$\quad = 10^5$ dyne

So 1 newton = 10^5 dyne

$$(6\text{-}13)$$

6-6 Newton's Third Law of Motion

One of the reasons why force is such a useful concept is that the forces two bodies exert on one another are frequently equal in magnitude but opposite in direction.

Newton's third law of motion

If body 1 exerts on body 2 a force $\mathbf{F}_{12}$, then body 2 exerts on body 1 a force $\mathbf{F}_{21}$ which is equal in magnitude but exactly opposite in direction.

$$(6\text{-}14)$$

$$\mathbf{F}_{12} = -\mathbf{F}_{21}$$

Let us first illustrate this with reference to gravitational forces. Figure 6-6 shows the now familiar case of two masses M_1 and M_2 at a distance R

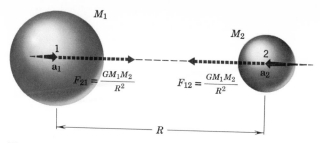

Figure 6-6 Gravitational forces between two bodies.

apart. The acceleration produced in 1 by the gravitational attraction of 2 is

$$a_1 = \frac{GM_2}{R^2}$$

(6-15)

The force exerted by 2 on 1 is therefore

$$F_{21} = M_1 a_1$$
$$= \frac{GM_1 M_2}{R^2}$$

(6-16)

Similarly, interchanging the roles of the two bodies, the acceleration produced in 2 by 1 is

$$a_2 = \frac{GM_1}{R^2}$$

(6-17)

The force exerted by 1 on 2 is therefore

$$F_{12} = M_2 a_2$$
$$= \frac{GM_1 M_2}{R^2}$$

(6-18)

The two forces obviously have the same magnitude, but since the force on either body points directly toward the other body, the two forces are in exactly opposite directions.

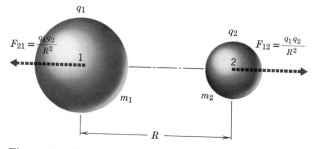

Figure 6-7 Electrostatic forces between two bodies.

$$F_{12} = -F_{21} \text{ for gravitational forces} \qquad (6\text{-}19)$$

Using the concept of force, Newton was able to express the law governing gravitational interactions in a particularly elegant and symmetrical form.

Newton's universal law of gravitation

Any two bodies in the universe have a gravitational attraction for one another. If their masses are M_1 and M_2 and their distance apart, R, is large compared with the size of either, then the force on either body points directly toward the other body and has a magnitude

$$F = \frac{GM_1M_2}{R^2} \qquad (6\text{-}20)$$

Now consider the electrostatic forces between two stationary electric charges. Figure 6-7 shows two positive charges q_1 and q_2 a distance R apart. The acceleration of 1 due to the electrostatic effect of 2 is

$$a_1 = \frac{1}{m_1} \frac{q_1q_2}{R^2} \qquad (6\text{-}21)$$

The electrostatic force exerted by 2 on 1 is therefore

$$
\begin{aligned}
F_{21} &= m_1a_1 \\
&= \frac{q_1q_2}{R^2}
\end{aligned}
\qquad (6\text{-}22)
$$

The acceleration of 2 due to the electrostatic effect of 1 is

$$a_2 = \frac{1}{m_2} \frac{q_1q_2}{R^2} \qquad (6\text{-}23)$$

The electrostatic force exerted by 1 on 2 is therefore

$$
\begin{aligned}
F_{12} &= m_2a_2 \\
&= \frac{q_1q_2}{R^2}
\end{aligned}
\qquad (6\text{-}24)
$$

So

$$F_{12} = -F_{21} \text{ for electrostatic forces} \qquad (6\text{-}25)$$

The two forces are again equal in magnitude but opposite in direction. The fundamental law of electrostatics, discovered by the French physicist Coulomb, can be stated in the following form.

Coulomb's law for electrostatic forces

If an electric charge q_1 is at a distance R from a charge q_2, the force on either charge due to the other has magnitude

$$F = \frac{q_1 q_2}{R^2} \tag{6-26}$$

If the charges are both positive or both negative, the forces are repulsive. If the two charges have opposite signs the forces are attractive.

As we shall explain more fully later, the simple form of equation 6-26 is valid only in the CGS system.

Notice particularly that the two forces referred to in Newton's third law act on different bodies. They never act on the same body.

Newton's third law is implicit in the nature of gravitational interactions as expressed by equations 6-1 and 6-2, and in the nature of electrostatic interactions as expressed by equations 6-9 and 6-10. Newton believed that his law was universally valid for all forces. In a later chapter, when we are discussing magnetic forces, we shall encounter a clear-cut case where the third law is not true.

6-7 Interatomic Forces

In everyday life many forces are encountered which seem to have no simple relation to either gravitational or electrostatic interactions. A few examples are: the pull of a muscle when we lift a weight; the tug of a tractor dragging a log along the ground; the upward force exerted by the top of a desk to support a book; the buoyancy of the air raising a balloon. These forces are now known to be due to the forces that atoms exert on one another and they are basically electromagnetic in character. The atom as a whole is electrically neutral because each negatively charged electron is balanced by a positively charged proton. However, as the electron moves around inside the atom, the electromagnetic force it exerts on an electron moving around inside a neighboring atom does not quite average out to zero. This cannot be completely understood without recourse to quantum mechanics. It is found that when the atoms are far apart they attract one another, but when they get too close they begin to repel one another. In a solid, the atoms are arranged in an orderly array at such a distance apart that, if they are pulled further apart and then released, they attract each other back into position, but if they are pushed closer together they repel each other back into position.

Interatomic forces between electrically neutral atoms always obey Newton's third law. The force exerted on atom 1 by atom 2 is equal in magnitude but opposite in direction to the force exerted on atom 2 by atom 1. When the small interatomic forces are added together for the

The Laws
of Motion

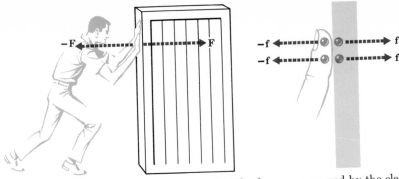

(a) A man attempting to push a crate illustrates Newton's third law.

(b) The forces are caused by the electromagnetic interaction between atoms in the crate and atoms in the man's hand.

Figure 6-8

whole body, the third law is found to be obeyed on a large scale and this is one important reason why the third law is so obviously valid in everyday experience. Those instances where the third law is violated are more subtle and do not have obvious consequences in everyday life.

To illustrate these points, consider a man trying to push a crate that is too heavy to move (figure 6-8). From a macroscopic point of view we would say that the man's hand is exerting a force $\mathbf{F}$ on the crate, while the crate is exerting an equal and opposite force $-\mathbf{F}$ on the man's hand. The origin of all this is that the atoms in the skin of the man's hand are in close proximity to certain atoms on the surface of the crate and these atoms exert repulsive electromagnetic forces on one another. The force exerted by an atom of the man's skin on a nearby atom of the crate is equal and opposite to the force exerted by the atom of the crate on the atom of the man's skin. When these forces are added together for all the atoms, the resultant force $\mathbf{F}$ on all the atoms of the crate is equal and opposite to the resultant force $-\mathbf{F}$ on all the atoms in the skin of the man's hand.

6-8 The Intuitive Concept of Mass

We all have direct experience of the nature of mass. It is much easier to push one's way through a light bamboo curtain than through a heavy revolving door. We can easily distinguish between being hit on the head by a table tennis ball and being hit on the head by a baseball, even though they have the same velocity. These experiences all involve interatomic forces. As the hand pushes on the revolving door, the atoms of the door exert forces on the atoms of the skin. These atoms in turn move closer to the atoms underneath them and exert forces which are thereby transmitted deeper and deeper into the hand. The forces are passed on from atom to atom until they reach the atoms of our muscles, which readjust themselves in some complicated way in order to produce the necessary forces. When the baseball strikes the head, atoms on the outside of the ball first push against atoms in the skin. The force is transmitted inward from atom to atom, displacing the atoms from their rightful positions and thereby producing a certain amount of damage. The physiological details are complicated, but the basic processes are interatomic and the forces are electromagnetic. The mass we experience intuitively is therefore the inertial mass of equations 6-9 and 6-10.

Mass is often defined as "the quantity of matter in a body." To be of value in physics this concept has to be made more precise and more quantitative. An obvious approach is to say that mass is a measure of the number of atoms in a body. This cannot cope with the fact that there are many different kinds of atoms with different masses. It is more unpleasant to be hit on the head by a lead brick than by a lump of charcoal containing the same number of atoms. An atom of lead is about seventeen times more massive than an atom of carbon.

The difference between various atoms is related to the number of protons, neutrons, and electrons they contain. Can we say that the mass of a body is a measure of the number of fundamental particles it contains? Unfortunately not, because fundamental particles do not all have the same mass. The neutron has a mass 1.0014 times as large as the mass of the proton, and the proton has a mass 1836.1 times the mass of the electron. A deuterium atom containing 3 particles (one proton, one neutron, and one electron) is not 3/2 times as massive as a hydrogen atom containing two particles (one proton and one electron). It is approximately twice as massive.

The basic problem is this: imagine a proton and an electron exerting electrostatic forces on one another. In figure 6-4, suppose that body 1 is an electron and body 2 is a proton. In equation 6-9, for the acceleration of the electron, we insert the mass of an electron. In equation 6-10, for the acceleration of the proton, we insert the mass of the proton, which is 1836.1 times larger. But what is the significance of this peculiar number 1836.1, which is not even an integer? This can justifiably be said to be the major unsolved problem of contemporary physics. It is complicated by the following facts which we shall discuss in more detail later in the book. According to the theory of relativity, the mass of a body is found to vary with its velocity. Mass and energy are similar quantities and can be converted into one another. The mass of an atom is not even the sum of the masses of the fundamental particles of which it is composed!

The major obstacle to a proper understanding of mass is probably our intuitive feeling that we know what it is before we start. Rather than attempting to justify this intuitive feeling, we would do better to realize how little we really understand the concept. At this stage, it is probably best to regard mass as a number which has to be inserted in equations such as equation 6-9 in order to get the right answer.

QUESTIONS

1. Argue as strongly as you can in favor of the theory that a body will come to rest unless a force is exerted on it to keep it in motion. Having established a good case, refute it.

2. What apparent evidence is there that a body left to itself in outer space moves in a circle around the earth? What observations are inconsistent with this point of view? When considering questions of this kind, try to distinguish between theories that you have always taken for granted and theories that you can support by observational evidence. For example, how do you know that the earth is rotating?

3. Taking into account the whole of your experience, do you believe that the future of the universe is completely predetermined?

4. If a body is instantaneously at rest, is the force on it necessarily zero?

5. Show that, in the MKS system, the units of G can be expressed either as m^3 sec^{-2} kg^{-1} or newton m^2 kg^{-2}. What are they in the CGS system?

6. Does the numerical value of G depend on the size of the platinum-iridium cylinder which has been chosen as the unit of mass?

7. What is meant by the statement that equations 6-1 and 6-2 serve to define gravitational mass?

8. Does a spring balance measure gravitational or inertial mass? What about a chemical balance?

9. In figure 6-8, why do the forces shown not produce a forward acceleration of the crate and a backward acceleration of the man? Remember that the crate is too heavy for the man to move.

10. How are interatomic forces involved in (a) the pull of a muscle when we lift a weight, (b) the tug of a tractor dragging a log along the ground, (c) the upward force exerted by the top of a desk to support a book, (d) the buoyancy of the air raising a balloon.

11. When a baseball strikes your head, why are you aware of the total mass of the baseball, rather than just the mass of a few atoms in contact with your skin?

12. Taking into account the rotation of the earth, does a body released from rest really fall vertically? Describe the situation (a) from the point of view of an observer who is not rotating with the earth (b) from the point of view of the observer on earth who releases the falling body. In terms of the points of the compass, in which direction does the motion deviate from the vertical (a) at the equator, (b) at the north pole, and (c) at any latitude?

PROBLEMS

NEWTONIAN
MECHANICS

A

1. What is the magnitude of the force that produces an acceleration of 3.2 cm/sec^2 in a mass of 5 gm?

2. What is the magnitude in newtons of a force that produces an acceleration of 16 cm/sec^2 in a mass of 3 kg?

3. If a force of 5 newtons acts on a mass of 20 kg, what acceleration does it produce?

4. When a force of 25 dynes acts on a body it produces an acceleration of 1.5 m/sec². What is the mass of the body?

5. A mass of 2 gm moves in a circle of radius 5 cm with a constant speed of 7 cm/sec. What is the magnitude of the force acting on the mass?

6. A mass of 5 kg is 20 m away from a mass of 9 kg. What is the magnitude of the gravitational force exerted by one mass on the other?

7. A speck of dirt is 200 cm away from a stone with a mass of 0.1 kg. What is the gravitational acceleration of the speck due to the stone?

8. A mass of 10 gm is falling freely near the surface of the earth with a downward acceleration $g = 980$ cm sec^{-2}. What is the magnitude of the force exerted on it by the earth?

9. The earth exerts a force of 2×10^6 dyne on a body falling freely near its surface. What is the mass of the body?

10. If a mass m is falling freely with a downward acceleration **g**, write down an equation giving the force **F** pulling it toward the center of the earth.

B

11. What is the gravitational acceleration produced by a body of mass 6×10^{27} gm in another body a distance of 6.4×10^8 cm away from it?

12. What is the gravitational acceleration of the moon due to the attraction of the earth? Compare it with the acceleration due to gravity on the surface of the earth.

13. What is the gravitational acceleration of the earth due to the attraction of (a) the moon and (b) the sun? Compare them with one another and with the acceleration due to gravity on the surface of the earth.

14. What is the gravitational force exerted on the sun by the earth?

15. A star of mass 2×10^{32} kg is at a distance of 5×10^{10} m from another star of mass 5×10^{30} kg. What is the gravitational force between them?

16. If the electron in a hydrogen atom moves around the proton in a circle of radius 5.3×10^{-9} cm, what is the gravitational force exerted on the electron by the proton?

17. A specially designed billiard cue can measure the force exerted on the ball. It is found that, as long as the tip of the cue remains in contact with the ball, the force has a constant value of 0.7 newton. If the ball has a mass of 150 gm and is given a velocity of 0.6 m sec^{-1}, how long was the cue in contact with the ball?

18. A body of mass 5 kg is subject to horizontal forces of 3 newtons pointed due east and also 4 newtons pointed due north. What is the magnitude and direction of its acceleration?

19. A book with a mass of 800 gm is projected across the top of a bench with an initial velocity of 60 cm/sec. It comes to rest after traveling

15 cm. What is the force of friction exerted on the book by the top of the bench?

20. A mass of 3 kg is acted on by a force of 25 newtons. If it starts from rest, how far will it have traveled after 5 sec?

21. A mass of 2.5×10^{-3} gm is subject to a constant force. It starts with a velocity of 3×10^4 cm sec^{-1} in the same direction as the force. 1.5×10^{-6} sec later it has speeded up to a velocity of 1.2×10^5 cm sec^{-1}. What is the magnitude of the force?

22. A mass of 200 kg is subject to a constant force of 7.5 newtons in the positive direction. It starts with a velocity of 0.2 m sec^{-1} in the opposite direction to the force. Describe its motion during the next 15 sec, calculating all quantities of particular interest.

23. A planet of mass 3×10^{29} gm moves around a star with a constant speed of 2×10^6 cm sec^{-1} in a circle of radius 1.5×10^{14} cm. What is the gravitational force exerted on the planet by the star?

24. A mass of 12 gm on the end of a thread is being whirled around in a circle of radius 2 m. If the thread breaks when the force on it is 2000 dynes, what is the maximum speed the mass can have? Ignore gravity.

C

25. At what point in space does the gravitational force of the moon on a space ship exactly counteract the gravitational force of the earth? Do not assume that this point is on the line joining the earth to the moon, without first proving that it could not possibly lie to one side of this line. Could the point be on the far side of the moon from the earth, or on the far side of the earth from the moon? Why or why not?

26. If lengths were measured in meters and seconds, but a new unit of mass were chosen in order to make $G = 1$, how many kilograms would there be in this new unit of mass?

27. What would be the numerical value of G in the MKS system if (a) the kilogram were ten times larger than it is at present, with the units of length and time the same, (b) the meter were only one-tenth as long as it is at present, with the units of mass and time the same, (c) the second were three times longer than it is at present with the units of mass and length the same?

28. Referring to the diagram associated with this problem, what is the acceleration of the two masses? What is the force exerted on m_2 by the string joining it to m_1?

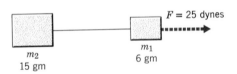

$F = 25$ dynes

m_2
15 gm

m_1
6 gm

Problem 6-28

29. Two bodies with masses M_1 and M_2 are initially at rest a distance R apart. They then move directly toward one another under the influence of their mutual gravitational attraction. The initial acceleration of M_1 is

$$a_1 = \frac{GM_2}{R^2}$$

Is it correct to say that the distance s_1 traveled by M_1 after t secs is

$$s_1 = \tfrac{1}{2}a_1 t^2$$
$$= \frac{GM_2 t^2}{2R^2} \text{?}$$

What is the ratio of the distance traveled by M_1 to the distance traveled by M_2?

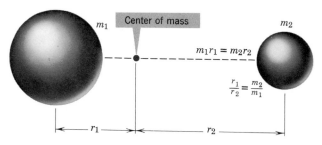

m_1 Center of mass m_2

$m_1 r_1 = m_2 r_2$

$\dfrac{r_1}{r_2} = \dfrac{m_2}{m_1}$

r_1 r_2

Problem 6-30

30. The **center of mass** of two particles is a point between the particles on the line joining them. By definition, if its distance from the particle of mass m_1 is r_1 and its distance from the particle of mass m_2 is r_2, then

$$\frac{r_1}{r_2} = \frac{m_2}{m_1}$$

(see diagram).

   (a) If the two particles are released from rest and then move toward one another under the influence of their gravitational attraction, show that the center of mass remains stationary.

   (b) Show that the result of part (a) is valid for any type of force obeying Newton's Third Law.

31. From a knowledge of the fact that the moon rotates around the earth once every 27.3 days in a circle of radius 3.80×10^{10} cm, calculate its acceleration. Compare it with the acceleration resulting from gravity at the surface of the earth and show that the inverse square law is not accurately obeyed. Why not?

32. Two astronauts have deserted their space ship in a region of space far from the gravitational attraction of any other body. Each has a mass of 100 kg and they are 100 m apart. They are initially at rest relative to one another. How long will it be before the gravitational attraction between them brings them 1 cm closer together? Express your answer in units which make it obvious how long a time this is (for example, is it about 1 minute? 1 day? 1 year?).

Some Illustrations of the Laws of Motion

7-1 The Earth's Gravitation; Weight and Mass

Figure 7-1 shows a small mass m falling near the surface of the earth. This situation is similar to figure 6-3 in which two bodies are shown exerting gravitational attractions on one another, except that in figure 6-3 we assumed that the diameters of the bodies were small compared with their distance apart. Clearly the radius of the earth is not small compared with the distance between the falling body and the center of the

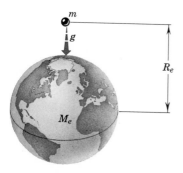

Figure 7-1 A body falling near the surface of the earth.

earth. However, it may be shown that a perfectly spherical body with uniform density behaves as though its total mass were concentrated at the center of the sphere. If we assume that the earth is a perfect sphere of uniform density, we may therefore apply equation 6-1 to the freely falling body m.

The acceleration of a freely falling body near the surface of the earth is called the **acceleration due to gravity** and is denoted by the symbol g. If M_e is the mass of the earth and R_e is the radius of the earth, which is practically the same thing as the distance of the falling body from the center of the earth, then equation 6-1 takes the form

$$g = \frac{GM_e}{R_e^2} \tag{7-1}$$

Equation 7-1 embodies a feature of gravitational effects which is well worth emphasizing. The acceleration g depends upon the universal constant G, the mass of the earth M_e, and the radius of the earth R_e, but it does not depend upon any property of the falling body. At a particular place on the earth's surface all bodies fall with the same acceleration (if the effect of the air on the motion is eliminated or allowed for). This important fact was first realized by Galileo and is immortalized in the famous story of how he dropped two very different bodies from the top of the Leaning Tower of Pisa and observed that they reached the ground at the same time. Since then the point has been tested by experiments of increasing complexity and sophistication, and we now know that g for various bodies of different size and chemical composition does not differ by more than 1 part in 10^{11}. More generally, as equations 6-1 and 6-2 imply, when a body is made to accelerate by the gravitational attraction of other bodies, the acceleration does not depend upon any property of the body that is being accelerated. This peculiar aspect of gravitational forces is not true of any other type of force. It is one of the fundamental principles underlying Einstein's general theory of relativity.

The earth is not a perfect sphere and its density is not uniform throughout, so equation 7-1 is only approximately true. In fact g varies from place to place on the earth's surface. At the top of a high mountain g is less than at sea level because one is farther from the center of the earth and the inverse square law comes into play. There is a tendency for g to increase as one goes from the equator to the poles. There are also local variations in g due to variations in the nature of the earth's crust. As we might expect, g changes when we approach a mountain, because of the gravitational attraction of the mountain, but the effect can be quite complicated because the mountain may have a "root" below sea level. Characteristic local variations in g have been used to help locate oil deposits. The approximate value of g is

$$g = 9.80 \text{ m sec}^{-2}$$
$$= 980 \text{ cm sec}^{-2} \tag{7-2}$$

and it varies by about 0.5% over the earth's surface. It can now be measured with such great accuracy that it is possible to observe the effect of the tides. The gravitational attraction of the ocean is slightly different at high and low tide, so that the value of g oscillates very slightly up and down twice a day.

So far we have concentrated on the *acceleration* of the falling body. The gravitational *force* of attraction exerted on the body by the earth is called the **weight** of the body, **W**. Applying Newton's second law to a freely falling body, the gravitational force on it is equal to the product of its mass and its acceleration.

Weight

In terms of the vector forces:

$$\mathbf{W} = m\mathbf{g} \tag{7-3}$$

In terms of the scalar magnitudes:

$$W = mg \tag{7-4}$$

Since the weight of a body is the gravitational force which the earth exerts on that body, it can be obtained from Newton's universal law of gravitation by inserting the appropriate symbols into equation 6-20 of the previous chapter:

$$W = \frac{GM_e m}{R_e{}^2} \tag{7-5}$$

This equation could also be obtained by inserting into equation 7-4 the value of g given by equation 7-1.

It is important to distinguish carefully between weight and mass. Mass is measured in kilograms or grams, whereas weight is a force measured in newtons or dynes. The mass of a body is an invariant property of that body and remains the same wherever the body is and whatever it is doing. (In the theory of relativity, mass varies with velocity, but we are ignoring this at present.) However, the weight of a body depends upon the local value of g and varies from place to place on the earth's surface. If the body is taken to the moon or one of the planets, its mass will remain unaltered, but its weight will suffer a drastic change, since the acceleration due to gravity will be very different. Strictly speaking, it is incorrect to talk about a weight of 10 grams. One should refer to a mass of 10 grams, which has a weight of 10g dynes, the exact value of this weight depending on the local value of g.

Example 7-1-1

At what distance from the center of the earth does the acceleration due to gravity have one half of the value that it has on the surface of the earth?

At the surface of the earth

$$g = \frac{GM_e}{R_e{}^2}$$

At a distance R from the center of the earth

$$g' = \frac{GM_e}{R^2}$$

If $g' = \frac{1}{2}g$

$$\frac{GM_e}{R^2} = \frac{1}{2}\frac{GM_e}{R_e{}^2}$$

$$R^2 = 2R_e{}^2$$

$$R = \sqrt{2}\,R_e$$

$$= 1.414 \times 6.38 \times 10^6 \text{ m}$$

$$= 9.02 \times 10^6 \text{ m}$$

The acceleration due to gravity is reduced to one half of its usual value at a distance of 9.02×10^6 m from the center of the earth. This is equivalent to a height of 2.64×10^6 m or 1640 miles above the surface of the earth.

Example 7-1-2

A newly discovered planet has twice the density of the earth, but the acceleration due to gravity on its surface is exactly the same as on the surface of the earth. What is its radius?

This problem must be approached carefully. We must express the acceleration due to gravity in terms of the *density* and the radius of the planet. If the radius is R and the mass of the planet M, then the acceleration due to gravity on its surface is

$$g_p = \frac{GM}{R^2}$$

Assuming the planet is spherical, its volume is the volume of a sphere of radius R:

$$V = \frac{4}{3}\pi R^3$$

Since Mass = Volume × Density,

$$M = \frac{4\pi R^3 \rho}{3}$$

where ρ (the Greek letter rho) is the density of the planet. Therefore

$$g_p = \frac{G\dfrac{4}{3}\pi R^3 \rho}{R^2}$$

$$= \frac{4\pi}{3} GR\rho$$

Similarly, the acceleration due to gravity on the surface of the earth is

$$g = \frac{4}{3}\pi GR_e\rho_e$$

where ρ_e is the density of the earth.

If $g_p = g$

then

$$\frac{4}{3}\pi GR\rho = \frac{4}{3}\pi GR_e\rho_e$$

Canceling $\frac{4}{3}\pi G$ on both sides,

$$R\rho = R_e\rho_e$$

If the density of the planet is twice that of the earth,

$$\rho = 2\rho_e$$

So $R2\rho_e = R_e\rho_e$

Whence $R = \frac{1}{2}R_e$

$$= \tfrac{1}{2} \times 6.38 \times 10^6 \text{ m}$$
$$= 3.19 \times 10^6 \text{ m}$$

The radius of the planet is one half of the radius of the earth, or 3.19×10^6 m.

7-2 A Falling Book and a Book Resting on a Desk

If I lift up a book of mass m and then let it fall to the floor, while it is falling it has a downward acceleration g. The only force acting on it is its weight $\mathbf{W} = m\mathbf{g}$, which is the force of gravitational attraction exerted on the book by the earth. In accordance with Newton's third law, the book must exert on the earth an upward force $\mathbf{W}'$, which is equal in magnitude but opposite in direction to $\mathbf{W}$ (figure 7-2a):

$$\mathbf{W}' = -\mathbf{W} \tag{7-6}$$

Also, while the book is falling downward toward the earth, the earth is falling upward toward the book with an acceleration

$$g_e = \frac{Gm}{R_e{}^2} \tag{7-7}$$

The ratio of this upward acceleration of the earth to the downward acceleration of the book is

$$\frac{g_e}{g} = \frac{m}{M_e} \tag{7-8}$$

This equation can be derived from equations 7-1 and 7-7. It may also be obtained by applying Newton's third law to equate the magnitudes

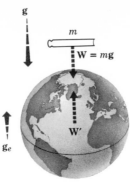
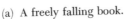

(a) A freely falling book.

(b) A book resting on a desk.

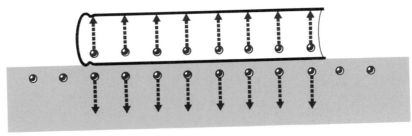

(c) The interatomic forces.

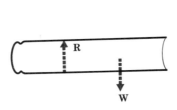

(d) A system consisting of the book only.

(e) A system consisting of the desk and the earth only.

Figure 7-2 Newton's laws applied to a book falling freely or resting on a desk. For the sake of clarity some of the forces have been displaced sideways.

of the forces on the earth and book and Newton's second law to express these forces as mass times acceleration:

$$M_e g_e = mg \tag{7-9}$$

Since the earth is much more massive than the book by a factor of about 10^{25}, the acceleration of the earth g_e is a factor 10^{25} smaller than g and is therefore negligible in practice, although it is present in principle.

If the book is placed on top of a desk it is still subject to the gravitational attraction of the earth, but it no longer accelerates downward because there is a compensating upward force exerted on it by the top of the desk (figure 7-2b). The atoms at the bottom of the book are in close proximity to the atoms on the top of the desk and they exert interatomic forces on one another (figure 7-2c). When the book is first placed on the desk its weight causes it to sink down, the atoms come closer and closer together and repel one another with stronger and stronger forces until the total upward repulsive force on the atoms at the bottom of the book is just sufficient to counteract the weight of the book. The interatomic forces obey Newton's third law and so, corresponding to the upward force $\mathbf{R}$ exerted on the book by the desk, there is an equal but opposite force $\mathbf{R}'$ exerted on the desk by the book:

$$\mathbf{R}' = -\mathbf{R} \tag{7-10}$$

Now consider the application of Newton's second law to this situation. To apply the second law correctly, we must first concentrate our attention on some well defined part of the universe. We then add together all the forces exerted on this part by the whole of the rest of the universe and equate the resultant force to the total mass of the separated part multiplied by its acceleration. Let us do this for the book (figure 7-2d). Since the book is at rest and has zero acceleration, the resultant force on it must be zero. This force is composed of two parts, a downward force $\mathbf{W}$ due to the gravitational attraction of the earth and an upward force $\mathbf{R}$ exerted by the atoms on top of the desk, so

$$\mathbf{W} + \mathbf{R} = 0 \tag{7-11}$$
$$\mathbf{W} = -\mathbf{R} \tag{7-12}$$

Similarly, if we isolate a system consisting of the earth and the desk, but not the book (figure 7-2e), this also has zero acceleration and the resultant force on it is zero. This resultant force has two parts, an upward force $\mathbf{W}'$ on the earth due to the gravitational attraction of the book and a downward force $\mathbf{R}'$ on the top of the desk due to the atoms at the bottom of the book, so

$$\mathbf{W}' + \mathbf{R}' = 0 \tag{7-13}$$
$$\mathbf{W}' = -\mathbf{R}' \tag{7-14}$$

This analysis may seem trivial and obvious, but it is important to distinguish between the roles played by the third law and the second law. The third law tells us that $\mathbf{W}$ is equal but opposite to $\mathbf{W}'$ and that $\mathbf{R}$ is equal but opposite to $\mathbf{R}'$. It is the second law that tells us that $\mathbf{W}$ is equal but opposite to $\mathbf{R}$ and that $\mathbf{W}'$ is equal but opposite to $\mathbf{R}'$, and then only because the accelerations are zero. We shall encounter an example in the very next section where it is wrong to assume that $\mathbf{W}$ and $\mathbf{R}$ counterbalance one another. It would be incorrect to adduce the third law to prove that $\mathbf{W}$ is equal and opposite to $\mathbf{R}$, because in the third law the forces that are equal and opposite act on different bodies, not on the same body. Notice also that $\mathbf{W}$ is a gravitational force and $\mathbf{R}$ is an electromagnetic force.

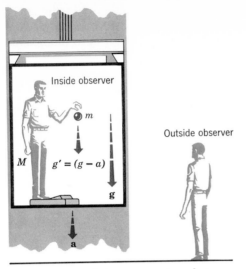

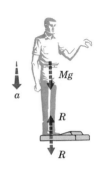

(a) Two points of view concerning what goes on inside an accelerating elevator.

(b) The forces on the man and the spring scales according to the *outside observer.*

Figure 7-3   An accelerating elevator.

7-3   An Accelerating Elevator; Apparent Weight

Consider the situation inside an elevator which is descending with a downward acceleration a, less than g (figure 7-3). Take the positive direction to be downward, so that both a and g are positive. The situation inside the elevator can be considered either from the point of view of an *Inside Observer* inside the elevator, or from the point of view of an *Outside Observer* with x-ray eyes who is standing on the ground floor watching the elevator fall and is able to see what is going on inside it.

Suppose that the man inside the elevator holds out a ball and then releases it. From the point of view of the *Outside Observer* the ball is like any other unsupported object and falls freely with a downward acceleration g. Relative to the elevator its downward acceleration is $(g - a)$. The *Inside Observer* therefore observes that the ball falls to the floor of the elevator with an acceleration

$$g' = g - a \qquad (7\text{-}15)$$

Suppose that the *Inside Observer* had been drugged, place inside the elevator and had awaked after the elevator had been given its downward acceleration. Finding himself inside a closed room, and being a physicist, he might perform some experiments to discover where he was. He would soon discover that bodies fell with an acceleration g' less than the usual value g. He might correctly assume that he was in an elevator falling with acceleration $a = (g - g')$, but he could equally well assume that he had been transported to a distant planet where the acceleration is g', and he would have no way of deciding which hypothesis was true. This is Einstein's **principle of equivalence**, which is at the foundation of the **general theory of relativity**. There is no way of distinguishing between gravitational effects and acceleration through space. It is a consequence of the peculiar feature of gravitational interactions that the acceleration does not depend upon any property of the body that is being accelerated.

Now suppose that the man inside the elevator steps onto spring scales. From the point of view of the *Outside Observer* two forces act on the man. First there is the gravitational attraction of the earth producing a downward force Mg, if M is the mass of the man. Secondly, the atoms on the top surface of the platform of the scales exert forces on the atoms on the bottoms of the soles of the man's shoes, producing a total upward force R on the man. The resultant downward force is $Mg - R$ and the man has a downward acceleration a, so Newton's second law tells us that

$$Mg - R = Ma \tag{7-16}$$
$$R = M(g - a) \tag{7-17}$$

Since interatomic forces obey Newton's third law, the soles of the man's feet exert a downward force R on the platform of the scales, and it is this force which the scales measure. The **apparent weight** of the man, W_a, is therefore

$$W_a = R \tag{7-18}$$
$$W_a = M(g - a) \tag{7-19}$$

which is less than his actual weight

$$W = Mg \tag{7-20}$$

which the scales would measure if the elevator were stationary.

From the point of view of the *Inside Observer*, an equally plausible description of the situation would be that he is located in a region of space where the acceleration due to gravity is g' and his weight is therefore

$$W_a = Mg'$$
$$= M(g - a) \tag{7-21}$$

An interesting situation arises if the cable breaks and the elevator falls freely with an acceleration $a = g$. Suppose that the *Inside Observer* holds out a ball and tries to drop it. Then, from the point of view of the *Outside Observer*, both the ball and the elevator are freely falling objects which fall in exactly the same way. The ball therefore stays in the same position relative to the elevator. The *Inside Observer* discovers that he is in a **gravity-free condition** and the ball does not fall to the floor of the elevator, but remains suspended at the same height above the floor. Under gravity-free conditions one might expect that objects would have no weight and equation 7-19 confirms that, when $g = a$, the apparent weight of the man, W_a, is zero.

This state of **weightlessness** is often discussed in connection with

Some
Illustrations
of the
Laws of
Motion

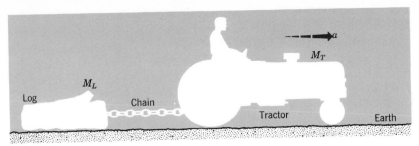

(a) The complete system.

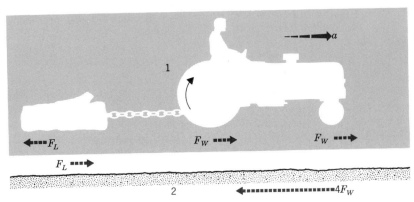

(b) Separating out (1) the tractor plus the log, (2) the earth.

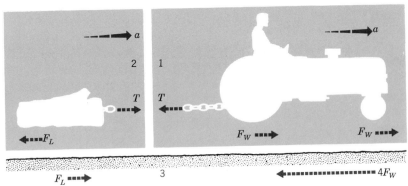

(c) Separating out (1) the tractor, (2) the log, (3) the earth.

Figure 7-4 A tractor dragging a log.

space travel. It can be achieved under three other types of circumstance. (a) An airplane can be put through a maneuver in which for several seconds during its flight its downward acceleration is exactly g. (b) A space rocket with its motor turned off or a satellite in orbit are effectively freely falling bodies, assuming that the surrounding atmosphere is too rarefied to exert any appreciable frictional force. From this point of view the definition of "freely falling" is that the only force on the body is the gravitational force, so that the acceleration is automatically the acceleration due to gravity. (c) In outer space, far from the gravitational effect of any massive body, the true weight and the apparent weight are both zero. In the solar system there are also special "null points" where the gravitational attractions of all the bodies in the solar system just happen to cancel one another exactly.

7-4   A Tractor Dragging a Log

This is illustrated in figure 7-4. The tractor has mass M_T, the log has mass M_L and the forward acceleration is a. Since a body cannot exert a resultant force on itself, how can the engine, which is part of the tractor, make the tractor and the log accelerate? The general procedure in a problem of this kind is to separate out some part of the universe and specify exactly what is included in this separated part. The next step is to think of *all* the *external* forces exerted on this separated part by the whole of the rest of the universe. The vector sum of all the external forces can then be equated to the product of the *total* mass of the separated part and its vector acceleration. (The internal forces between separate parts of the chosen system are irrelevant, because Newton's third law of motion ensures that these internal forces cancel in pairs.) If two vectors are equal, their components in any direction must be equal. It follows that the sum of the components of the external forces in any direction must be equal to the product of the total mass and the component of the acceleration in this direction.

To tackle the present problem let us separate out a system consisting of the tractor, the log, and the chain joining them (figure 7-4b). The rest of the universe then consists primarily of the earth. Since the tractor has no acceleration in a vertical direction, the vertical forces are of secondary interest. For completeness we can say that the combined weight of the tractor and the log, $(M_T + M_L)g$, is compensated by an upward thrust R_L of the ground on the log and upward thrusts R_W on each of the four wheels. (We are going to assume for simplicity that the forces are identical on each of the four wheels. This is a trivial point since we are only interested in the sum of the four forces.)

$$(M_T + M_L)g = R_L + 4R_W \qquad (7-22)$$

The horizontal forces are of primary interest and the only possible source of horizontal forces is the friction between the ground and the log or the ground and the wheels. A wheel is rotating in a clockwise direction in figure 7-4b and so the lowest point of the wheel in contact with the ground is trying to move to the left relative to the ground. The ground therefore exerts on it a frictional force F_W in the opposite direction, to the right. The log, however, is being dragged to the right and so the ground exerts on it a frictional force F_L to the left. These are the only horizontal forces (if the resistance of the air is neglected) and their resultant is $(4F_W - F_L)$ toward the right, in the direction of the acceleration a. The total mass is $(M_T + M_L)$ and Newton's second law therefore

Some
Illustrations
of the
Laws of
Motion

107

tells us that

$$4F_W - F_L = (M_T + M_L)a. \tag{7-23}$$

The forward acceleration of the tractor is a consequence of the fact that the earth is pushing it forward! The function of the engine is to rotate the wheels so that the part of the tire touching the ground tries to move *backward*, thus generating the *forward* frictional force F_W.

Notice that, as a consequence of Newton's third law, each wheel exerts a backward force F_W on the ground and the log exerts a forward force F_L on the ground. The forward acceleration of the tractor is therefore inevitably accompanied by a backward acceleration a_e of the whole earth given by

$$4F_W - F_L = M_e a_e \tag{7-24}$$

If the brakes were applied hard enough to prevent the wheels from rotating at all, the part of the tire in contact with the ground would then be moving forward relative to the ground and the frictional force would be backward. The engine would then be completely ineffective and the frictional forces would all be in a direction to retard the forward motion.

If we are asked to compute the force exerted on the log by the chain, then the universe must be divided into three parts: (1) the log, (2) the tractor and the chain, and (3) the rest of the universe, primarily the earth (figure 7-4c). At the point where the chain is joined to the log the atoms of the chain exert a resultant forward force T on the atoms of the log and, in accordance with Newton's third law, there is a backward force T on the rear end of the chain. The resultant forward force on the log is $T - F_L$ and Newton's second law tells us that

$$T - F_L = M_L a \tag{7-25}$$
$$T = M_L a + F_L \tag{7-26}$$

The resultant forward force on the tractor is $4F_W - T$ and Newton's second law applied to the tractor is

$$4F_W - T = M_T a \tag{7-27}$$

If we add equations 7-25 and 7-27 we obtain

$$4F_W - F_L = M_T a + M_L a \tag{7-28}$$

which is identical with equation 7-23 and everything is self-consistent.

7-5   A Satellite in Orbit

A man-made satellite will continue to orbit about the earth for many revolutions even though it is not powered by any fuel, whereas an airplane must use up fuel if it is to maintain its altitude. There are some important differences between the satellite and the airplane. The presence of the air is essential to the flight of an airplane, but this implies that its motion is inevitably opposed by the frictional resistance of the

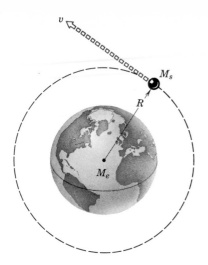

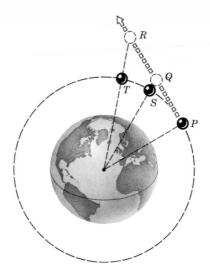

Figure 7-5 A satellite in orbit.

Figure 7-6 In a certain sense, the satellite *is* falling toward the earth.

air and it must use up fuel to overcome this resistance. A satellite, however, must be placed in orbit at a very high altitude where the atmosphere is so rarefied that it exerts very little frictional drag on the satellite. (Even so, it is this small frictional drag that eventually brings down the satellite after it has completed many revolutions around the earth.) Once it has reached this rarefied part of the atmosphere, the satellite moves in a stable orbit only if its velocity has exactly the right magnitude and direction for the orbit in question. The required magnitude of the velocity is greater than can be achieved by an airplane.

To avoid complicated mathematics and yet still illustrate the physical principles involved, consider the particularly simple case of a circular orbit of radius R (figure 7-5). Let M_e be the mass of the earth, M_s the mass of the satellite and v the velocity which the satellite must have if it is to remain in this orbit. The situation will be described from the point of view of an observer in space watching the satellite revolve around the earth (and also watching the earth rotate on its axis, although this is not relevant to the present problem). The gravitational attraction of the earth gives the satellite an acceleration in accordance with equation 6-1:

$$a_s = \frac{GM_e}{R^2} \tag{7-29}$$

By virtue of the fact that it is moving in a circle of radius R with velocity v, the satellite has an acceleration toward the center of the circle (section 5-4):

$$a_s = \frac{v^2}{R} \tag{7-30}$$

Some
Illustrations
of the
Laws of
Motion

109

The circle is a possible stable orbit only if the velocity has exactly the right value to make these two expressions for the acceleration equal:

$$\frac{GM_e}{R^2} = \frac{v^2}{R} \tag{7-31}$$

Whence

$$v = \sqrt{\frac{GM_e}{R}} \tag{7-32}$$

The radius of the orbit of Sputnik I was about 4,300 miles or 7×10^6 m. Its velocity must therefore have been

$$v = \sqrt{\frac{6.67 \times 10^{-11} \times 6.0 \times 10^{24}}{7 \times 10^6}}$$

$$= 7.5 \times 10^3 \text{ m/sec}$$

$$= 17,000 \text{ miles per hour}$$

Since the perimeter of its orbit was $2\pi R$, or about 4.4×10^7 m. the time taken for a complete revolution was

$$T = \frac{2\pi R}{v}$$

$$= \frac{4.4 \times 10^7}{7.5 \times 10^3} \text{ sec}$$

$$= 5.9 \times 10^3 \text{ sec}$$

$$= 1.64 \text{ hour}$$

The reader will easily be able to prove that the general expression for the time taken per revolution is

$$T = 2\pi \sqrt{\frac{R^3}{GM_e}} \tag{7-33}$$

Reverting to the layman's question, "Why does the satellite not fall down to the earth?", the only requirement that the earth's gravitational field imposes upon the satellite is that it shall always have an acceleration directed toward the center of the earth with the magnitude given by equation 7-29. A body released from rest at a height achieves this acceleration by falling directly downward. A satellite can achieve it equally well by moving continually in the same circle with just the right speed, since motion in a circle always involves an acceleration directed toward the center of the circle and given by equation 7-30. There is a sense in which the satellite *is* falling toward the earth. In the absence of the earth's gravitational attraction it would continue to move in a straight line tangential to the circle, in accordance with Newton's first law of motion. It would then occupy the successive positions, P, Q, R shown in figure 7-6. Actually, when it moves in a circular orbit it occupies the successive positions P, S, T, and this can be imagined to be the result of having fallen through the distances QS and RT.

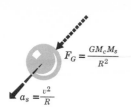
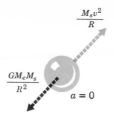

(a) The point of view of an observer who is not rotating.
(b) A possible point of view of an observer inside the satellite.

Figure 7-7 Two views of the forces on the satellite.

If we wish to discuss the motion of the satellite in terms of the gravitational force on it (figure 7-7a), Newton's universal law of gravitation (equation 6-20) tells us that this force is

$$F_G = \frac{GM_eM_s}{R^2} \tag{7-34}$$

By virtue of its motion in a circle the satellite has a radial acceleration

$$a_s = \frac{v^2}{R} \tag{7-35}$$

Newton's second law is

$$F_G = M_s a_s \tag{7-36}$$

or

$$\frac{GM_eM_s}{R^2} = M_s \frac{v^2}{R} \tag{7-37}$$

Cancelling M_s on both sides, this is identical with equation 7-31.

It is sometimes said that the gravitational force GM_eM_s/R^2 is "balanced" by a "centrifugal force" M_sv^2/R. This can be misleading, because the circular motion of the satellite inevitably implies an acceleration. If all the forces on the satellite were to balance out to zero, there could be no acceleration. From the point of view of an observer watching from outside, there is *only one true force*, the gravitational force F_G. However, an observer inside the satellite might take the egocentric point of view that he is at rest and the earth and all the fixed stars are revolving around him. In order to explain his lack of acceleration he would have to invent a centrifugal force M_sv^2/R to compensate the earth's gravitational pull (figure 7-7b). The meaning of v to him would be the velocity of the earth in its circular orbit around him. This point of view can be confusing and the student is advised not to adopt it. One observation is in favor of the egocentric observer inside the satellite. If he holds out a ball and releases it, it does not fall to the floor. Like the observer in the freely falling elevator, he seems to be in a gravity-free condition. From the point of view of the observer watching from out-

Some
Illustrations
of the
Laws of
Motion

111

side, the explanation is equally simple. The ball suspended in mid air inside the satellite has no motion relative to the satellite because both are moving in the same circular orbit with the same velocity and both have the required acceleration in the earth's gravitational field by virtue of this circular motion. Notice that equation 7-32 does not contain the mass of the satellite, M_s. So, if the velocity has the right value to keep the satellite in its circular orbit, it also has the right value to keep the ball in the same orbit.

Finally, let us realize that the arguments applied above to the motion of a man-made satellite around the earth are equally applicable to the motion of the moon around the earth, to the motion of any moon around any planet, or to the motion of any planet around the sun, as long as the appropriate radius and the appropriate masses are used.

QUESTIONS

1. A body of mass m is being whirled in a circle on the end of a string. The string suddenly snaps. Immediately afterward, what are the velocity and acceleration of the mass and what is the force on it, if this takes place (a) in outer space in a gravity-free region, (b) on the surface of the earth? In the latter case, does it matter whether the circle is horizontal or vertical?

2. Is the value of g at the surface of the earth affected by the gravitational attraction of the sun (or, for that matter, any body in the solar system)? Remember that g is measured by an observer on the surface of the earth.

3. If the planets in the solar system all had the same density, how would your weight on the surface of a planet vary with its radius?

4. A trapeze artist is hanging from a swinging trapeze at the end of a rope held between her clenched teeth. According to Newton's third law, which force is equal but opposite to her weight?

5. How does the force exerted by a passenger on the seat of an airplane depend upon the motion of the plane?

6. An astronaut is said to have a large *apparent* weight during take-off. Which of the following forces is a measure of his apparent weight? (a) The gravitational force exerted on him by the earth. (b) His mass multiplied by the acceleration of the rocket. (c) His mass multiplied by the acceleration due to gravity. (d) The force exerted by him on the chair in which he is sitting. (e) His mass multiplied by the square of his velocity and divided by the radius of the earth.

7. When its point of support is stationary near the surface of the earth, the time for one swing of a pendulum of length L is

$$T = 2\pi \sqrt{\frac{L}{g}}$$

What would be the time for one swing if the pendulum were (a) in a freely falling elevator, (b) in an elevator with a downward acceleration a less than g, (c) in an elevator with an upward acceleration a? What would happen if the elevator had a downward acceleration a greater than g?

8. Is it possible to deduce the mass of a planet by a careful study of its motion around the sun? Is it possible to deduce the mass of a planet by a careful study of the motion of the other planets?

9. A satellite is moving around the earth once every three hours in a circular orbit which passes over both the north and south poles. A hydrogen bomb hangs from a string below the satellite. When the satellite is over the north pole, the pilot cuts the string. The bomb is timed to explode 90 minutes later. Where will the explosion take place and who will suffer most from it?

10. Does the time taken for a satellite to go once round the earth in a stable circular orbit increase or decrease as the radius of the orbit is made larger?

11. The force of friction due to the very rarefied outer atmosphere gradually slows down a satellite. According to equation 7-31 a smaller velocity corresponds to a larger orbit. In fact, however, the satellite does not move outward but spirals inward toward the earth and eventually burns out in the more dense lower atmosphere. Resolve this dilemma.

PROBLEMS

A

1. Using the data on the inside of the cover of this book, find the weight of one standard kilogram. Would it have exactly this weight on top of Mount Everest?

2. The mass of Jupiter is 1.9×10^{27} kg and its diameter is 1.4×10^8 m. What is the acceleration due to gravity on the surface of Jupiter?

3. What would be the value of the acceleration due to gravity g if the earth had its present radius but twice as much mass?

4. What would be the value of the acceleration due to gravity g if the earth had its present mass but a radius equal to twice its present radius?

5. What is the acceleration due to gravity at a distance from the center of the earth equal to the diameter of the earth?

6. At what distance from the center of the earth is the acceleration due to gravity equal to 4.9 m/sec^2?

7. A mass of 2 kg has a weight of 16 newtons on a certain planet. What is the acceleration due to gravity on this planet?

8. What force is needed to give an *upward* acceleration of g to a mass of 3 kg near the surface of the earth?

9. A balloon is rising with an upward acceleration of 270 cm/sec^2. The upward force exerted on it by the buoyancy of the air is 2,500 dyne. What is the mass of the balloon?

10. In an elevator your apparent weight is one half of its normal value. What is the acceleration of the elevator and in which direction is it accelerating?

11. Use equation 7-33 to calculate how many days there would be in a year if the distance of the earth from the sun remained the same, but the sun had four times as much mass.

12. How many days would there be in a year if the mass of the sun remained the same, but the distance of the earth from the sun doubled?

13. A planet moves around a star in a circular orbit of radius 5×10^{13} cm with a speed of 3×10^7 cm/sec. What is the mass of the star?

B

14. At a certain place on the earth's surface g is exactly 980 cm/sec². What would its numerical value be if (a) the standard kilogram had been chosen to be twice as large as it is at present, or (b) the standard meter had been chosen to be one tenth as long as it is at present, or (c) the second had been chosen to be twice as long as it is at present?

15. What is the acceleration due to gravity at a height of 4×10^8 cm above the surface of the earth?

16. The acceleration due to gravity on the surface of a certain planet is 625 cm/sec². At a height of 3×10^8 cm above the surface, the acceleration due to gravity is 400 cm/sec². What is the radius of the planet?

17. If the mass of the earth were increased by a factor of 9, by what factor would its radius have to be changed to keep g the same?

18. If the radius of the earth were increased by a factor of 5, by what factor would its density have to be changed to keep g the same?

19. A satellite moves in an elliptical orbit around the earth. What is its acceleration when its distance from the center of the earth is equal to twice the radius of the earth?

20. What is the apparent weight of a 20 kg object in an elevator with an acceleration of (a) 4 m/sec² downward (b) 4 m/sec² upward (c) 9.8 m/sec² downward (d) 9.8 m/sec² upward (e) 15 m/sec² downward (f) 15 m/sec² upward?

21. A rocket is launched vertically with a constant acceleration. The astronaut's apparent weight is five times normal. What is the height of the rocket after 10 sec?

22. Referring to the diagram associated with this problem, what is the acceleration of the 250 gm block along the table? What is the upward force exerted on the block by the table?

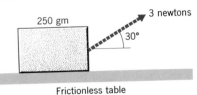

Frictionless table

Problem 7-22

23. A satellite of mass 1000 kg moves around a planet with a speed of 5×10^3 m/sec in a circular orbit of radius 1.6×10^7 m. If the radius of the planet is 7×10^6 m, what is the acceleration due to gravity on its surface?

24. Calculate the radius of the orbit of the satellite which always stays vertically above the same place on the earth's surface.

25. Using the data on the inside of the cover, calculate the velocity of the moon in its approximately circular orbit around the earth. What is the time taken to complete one revolution? Express your answer in days.

26. A tugboat is pulling a line of six identical barges (see diagram). The force F_1 exerted by the rope connecting the boat to the first barge is 1000 newtons. Calculate the forces in the other five ropes connecting the barges together. Justify your answer carefully. Is it necessary to assume that the acceleration is zero?

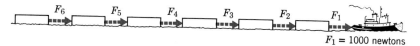

$F_1 = 1000$ newtons

Problem 7-26

C

27. The barges of the previous problem are now unequally loaded, as in the diagram. The frictional drag of the water on a barge is independent of its loading. The forward acceleration is 0.1 m/sec² and the force F_1 is 900 newtons. Calculate the forces in the other ropes and the frictional force exerted on each barge by the water.

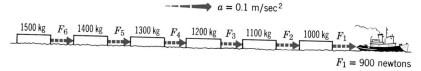

$a = 0.1$ m/sec²

$F_1 = 900$ newtons

Problem 7-27

28. Saturn has a total mass of 5.7×10^{26} kg and a mean density of 0.71 gm cm^{-3}. What is the acceleration due to gravity on its surface?

29. What is the percentage change in g between the surface of the earth and a point 100 miles above the surface?

30. Ignoring the effect of the atmosphere, what is the maximum possible velocity of an earth satellite in a stable circular orbit? Explain your answer.

31. A rocket of mass M_r is orbiting around a planet of mass M_p in a circle of radius R. It is burning fuel in order to maintain a steady speed twice as great as the speed that it would have if it were

Some Illustrations of the Laws of Motion

115

moving freely in the same orbit with its motor turned off. What force does the ejected fuel exert on the rocket and in which direction does it point?

32. A mass of 100 gm on the end of a rod is being whirled around with a constant speed of 300 cm/sec in a vertical circle of radius 50 cm near the surface of the earth. Calculate the force exerted by the rod on the mass in its highest position and its lowest position.

33. In the previous question, what are the magnitude and direction of the force exerted by the rod on the mass when the rod makes an angle θ with the vertical? (Define θ carefully.) Would it be possible to whirl a mass at constant speed in this way on the end of a *string* by carefully regulating the variation with time of the force on the other end of the string?

34. Two identical stars move in circular orbits under the influence of their mutual gravitational attraction. If the mass of each star is M and the distance between them is d, find an expression giving the time for one revolution. Justify every statement you make.

35. A galaxy contains about 10^{11} stars with an average mass close to the mass of the sun. The average distance between galaxies is about 10^{25} cm. Imagine two identical galaxies at this distance apart moving in circular orbits about their center of mass under their mutual gravitational attraction undisturbed by surrounding galaxies. Calculate the time for one revolution and compare it with current estimates of the age of the universe.

Momentum

8-1 Momentum and Force

As we have already pointed out, it is more painful to be hit on the head with a baseball than with a table tennis ball. The mass is not the only thing that matters, though. The velocity is also important. To be hit on the head by a baseball dropped from a height of one inch is much less painful than if the baseball were dropped from the roof of the house at a height of 20 feet and thereby given time to build up a greater velocity. The direction of the velocity is equally important. The real danger arises from a fast ball coming straight at you, whereas a glancing blow from the same ball can be comparatively harmless.

What matters is the product of the mass m and the vector velocity $\mathbf{v}$. This is called the **momentum** and is represented by the symbol $\mathbf{p}$.

Momentum,

$$\mathbf{p} = m\mathbf{v} \tag{8-1}$$

Since the velocity $\mathbf{v}$ is a vector, the momentum $\mathbf{p}$ is also a vector and is in the same direction as the velocity. The plural of momentum is momenta.

Newton's second law of motion can be expressed in a different form, using momentum. Suppose that a body of mass m has a velocity $\mathbf{v}_1$ at a time t_1 sec, and that it has a slightly different velocity $\mathbf{v}_2$ at a slightly later time t_2. The change in velocity is the final velocity minus the initial velocity, $(\mathbf{v}_2 - \mathbf{v}_1)$, and this change has taken place during a time interval $(t_2 - t_1)$ sec. The acceleration $\mathbf{a}$ is therefore, according to equation 5-4,

$$\text{Acceleration} = \frac{\text{Change in velocity}}{\text{Time taken}} \tag{8-2}$$

$$\mathbf{a} = \frac{(\mathbf{v}_2 - \mathbf{v}_1)}{(t_2 - t_1)} \quad \text{when } (t_2 - t_1) \text{ is very small} \tag{8-3}$$

Newton's second law of motion is

$$\text{Force} = \text{mass} \times \text{acceleration} \tag{8-4}$$

$$\mathbf{F} = m\mathbf{a} \tag{8-5}$$

$$= m\frac{(\mathbf{v}_2 - \mathbf{v}_1)}{(t_2 - t_1)} \tag{8-6}$$

$$= \frac{(m\mathbf{v}_2 - m\mathbf{v}_1)}{(t_2 - t_1)} \tag{8-7}$$

At the time t_1 the momentum $\mathbf{p}_1$ is $m\mathbf{v}_1$ and at time t_2 the momentum $\mathbf{p}_2$ is $m\mathbf{v}_2$. Therefore,

$$\mathbf{F} = \frac{(\mathbf{p}_2 - \mathbf{p}_1)}{(t_2 - t_1)} \qquad \text{when } (t_2 - t_1) \text{ is very small} \tag{8-8}$$

The new form of Newton's second law is

Force = Rate of change of momentum with time $\qquad$ (8-9)

In classical mechanics equations 8-4 and 8-9 are logically equivalent. However, in the theory of relativity the mass is no longer constant, but varies with the velocity. It is then found that the definition of force as mass multiplied by acceleration is unsatisfactory and that the definition as rate of change of momentum with time is much to be preferred. Moreover, as our exposition of physical theory gradually unfolds, it will be seen that the concept of momentum becomes more and more useful and develops more and more physical significance.

8-2 Conservation of Momentum

One reason why the concept of momentum is so useful is that, under the right circumstances, the total momentum of a system of interacting bodies remains constant as the motion proceeds.

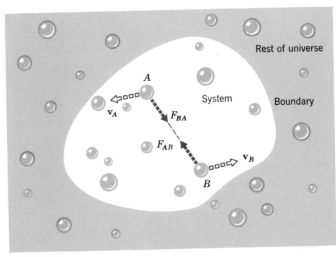

Figure 8-1 A system of bodies separated from the rest of the Universe.

Consider a system of bodies which is well defined in the sense that it is clearly understood which bodies belong to the system and which do not (figure 8-1). Let A and B be any two of these bodies. Then, according to Newton's third law of motion, the force exerted on A by B is equal and opposite to the force exerted on B by A:

$$\mathbf{F}_{AB} = -\mathbf{F}_{BA} \tag{8-10}$$

Since force is equal to the rate of change of momentum with time, the change in the momentum of A produced by B is always exactly counterbalanced by the change in the momentum of B produced by A. Therefore, the vectorial sum $(m_A\mathbf{v}_A + m_B\mathbf{v}_B)$ of the momenta of A and B is not changed by the mutual forces between A and B, any change in $m_A\mathbf{v}_A$ produced by B being exactly compensated by an opposite change in $m_B\mathbf{v}_B$ produced by A. This argument is clearly applicable to any pair of bodies in the system.

The **total momentum P** of the system is defined as the *vectorial* sum of the momenta of the individual bodies in the system:

Total momentum, $\mathbf{P} = m_A\mathbf{v}_A + m_B\mathbf{v}_B + \ldots$

$\ldots$ summed for all the bodies in the system. $\tag{8-11}$

Since all the *internal* forces between bodies inside the system occur in equal and opposite pairs, these internal forces cannot change the value of the total momentum, even though changes may occur in the individual terms of the sum. Any change in the total momentum must be caused by *external* forces exerted by bodies outside the system on bodies inside the system. In fact:

The rate of change of the total momentum **P** with time is equal to the resultant of all the *external* forces exerted on the system by the rest of the universe.

If the resultant of all the external forces on the system is zero, then the total momentum of the system cannot change with time. This is the law of conservation of momentum.

Law of conservation of momentum

If the resultant force exerted on a system of bodies by the rest of the universe is zero, then the vectorial sum of the momenta of all the bodies in the system is a constant vector which does not change with time.

It is difficult, if not impossible, to find a system of bodies on which the resultant force is *exactly* zero. (Think about this and discuss it with your colleagues and your instructor.) In practice, when we apply the law of conservation of momentum we have chosen a system on which the external force is so small that it is not able to produce a significant change of momentum during the time interval relevant to the problem.

If the whole universe can be considered to be a system of bodies subject to no external force, then presumably the total momentum of the universe is constant. A quantity like momentum which is conserved and cannot be created or destroyed is of particular interest in physics. Only a few such quantities are known and we shall take particular note of each one as it arises. In the next chapter we shall discover that energy is such a quantity (although, when the theory of relativity is taken into account, the law of conservation of energy is true only if mass is considered to be a form of energy).

The proof of the law of conservation of momentum is critically dependent upon the validity of Newton's third law of motion. In electromagnetism the third law is not always true and it is not obvious that momentum is conserved in this case. However, conservation of momentum is re-established by introducing the concept of the momentum of electromagnetic radiation in free space (see section 20-4) and it appears that conservation of momentum is more basic than the third law.

8-3   Simple Collisions

Figure 8-2 illustrates a lecture demonstration apparatus which has recently become a popular toy. It can even be found on the desks of tired executives, who play with it so that the rationality of the laws of physics may sooth their nerves which have been frayed by the irrationality of human affairs. Five steel balls, all of the same size and mass, hang side by side in a row. Imagine that the end ball on the right is pulled outward, raised through

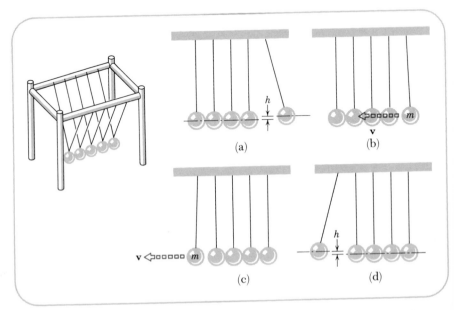

Figure 8-2   A popular toy demonstrates the law of conservation of momentum.

a height h and then released from rest (part a of figure 8-2). As it swings downward its velocity increases and just before it strikes the other balls (part b) it has acquired a velocity v which depends on h. All five balls then collide, but immediately after the collision it is observed that the four balls on the right are stationary and only the end ball on the left is moving (part c).

Moreover, it is moving with the same velocity v that the right-hand ball had before the collision because the left-hand ball rises through a height of exactly h before coming to rest (part d). It is easy to see why this follows by considering a single ball swinging like a pendulum as in figure 8-3. It falls through a height h and acquires a velocity v (a to b), but it is well known that this velocity is just right to enable it to rise through an equal height h on the other side (b to c). Returning to figure 8-2, the fact that the left-hand ball rises to a height h in part d therefore implies that it must have had a velocity v in part c.

Figure 8-2 therefore illustrates conservation of momentum in a simple, direct fashion. Just before the collision (part b) the momentum is entirely due to the right-hand ball which has a mass m and a velocity v, and consequently a momentum mv. Immediately after the complicated five ball collision, all the momentum is in the left-hand ball which also has a mass m, a velocity v, and a momentum mv (part c). If the use of algebraic symbols confuses you, interpret these remarks as meaning that m has a definite numerical value (say 20 gm), which is the same for all the balls, and v is a definite velocity (say 70 cm/sec toward the left). In both b and c the momentum is $20 \times 70 = 1400$ gm cm/sec toward the left. Momentum is therefore conserved during the collision.

The more complicated collisions of figure 8-4 can now be readily understood. In figure 8-4a two right-hand balls are released from rest at a height h and strike the other balls with a velocity v. The total momentum just before the collision is therefore $2mv$. After the collision the three right-hand balls are stationary and *two* left-hand balls fly away with velocity v and

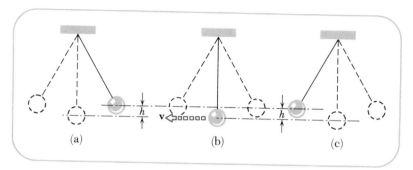

(a)   (b)   (c)

Figure 8-3   During the swing of a simple pendulum a fall through a height h from (a) to (b) produces a velocity v. This velocity causes the ball to rise through an equal height h on the other side during the motion from (b) to (c).

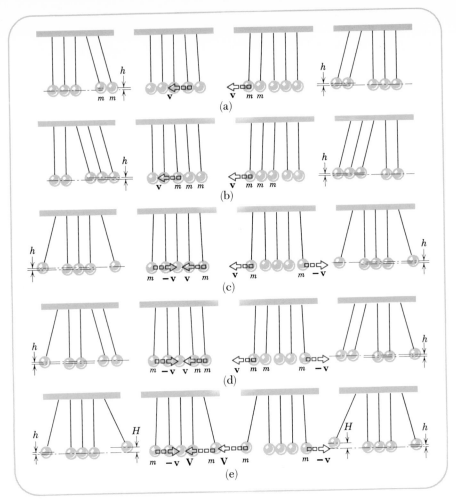

Figure 8-4 Various collisions that conserve momentum.

total momentum $2m\mathbf{v}$, exactly equal to the total momentum just before the collision.

In figure 8-4b three balls are dropped and three balls fly off on the other side. The interesting aspect of this is that the middle ball keeps going.

In figure 8-4c both end balls are simultaneously released from a height h. The right hand ball strikes with a velocity $\mathbf{v}$ and a momentum $m\mathbf{v}$. However, we must remember that velocity and momentum are vectors, and although the left-hand ball strikes with the same speed v it is moving in the opposite direction and its velocity is actually $-\mathbf{v}$ and its momentum is $-m\mathbf{v}$. The total momentum before the collision is therefore $m\mathbf{v} - m\mathbf{v} = 0$, because the two momentum vectors are in opposite directions and cancel one another. After the collision the three central balls remain stationary but the end balls are observed to rebound with their velocities exactly reversed. The right-hand ball has velocity $-\mathbf{v}$ and momentum $-m\mathbf{v}$, whereas the left-hand ball has velocity $+\mathbf{v}$ and momentum

$+mv$. The total momentum is then $-mv + mv = 0$, the same as before the collision.

In figure 8-4d two right-hand balls are released simultaneously with one left-hand ball at the same height. The two right-hand balls strike with a momentum $+mv$ and the left-hand ball strikes with a momentum $-mv$. The total initial momentum is therefore $+2mv - mv = +mv$. After the collision *one* right-hand ball flies off with a momentum $-mv$ and *two* left-hand balls fly off with momentum $+2mv$ and so the total momentum remains $-mv + 2mv = +mv$.

Finally, figure 8-4e introduces a new feature. Whereas the left-hand ball is still released from a height h and acquires a velocity $-\mathbf{v}$, the right-hand ball is released from a greater height H and acquires a velocity $\mathbf{V}$. The initial momentum is therefore $m\mathbf{V} - m\mathbf{v} = m(\mathbf{V} - \mathbf{v})$. The reader should by now be able to see that momentum is conserved if the left-hand ball flies off with a velocity $\mathbf{V}$ and rises to a height H, while the right-hand ball flies off with a velocity $-\mathbf{v}$ and rises to a height h.

A final comment should be made for the benefit of thoughtful students. In figure 8-2 the momentum in part b is equal to the momentum in part c. However, the total momentum in parts a and d is zero, because all the balls are then at rest. Thus, momentum is not conserved during the motion from a to b or from c to d. The reason for this is that during the motions from a to b and from c to d there are *external* forces, the gravitational forces responsible for the weights of the balls, and these forces have plenty of time to change the momenta of the swinging balls. However, during the transition from b to c, the multiball collision takes place very rapidly and the external forces are not given sufficient time to change the momentum by any significant amount.

8-4   More Complicated Collisions

Imagine two billiard balls colliding, as in figure 8-5. Suppose that their masses are m_1 and m_2, that their velocities before the collision are $\mathbf{v}_1$ and $\mathbf{v}_2$, and that their velocities after the collision are $\mathbf{u}_1$ and $\mathbf{u}_2$. These four velocities are now no longer all in the same direction. The system to be considered consists of the two balls and nothing else. Its total momentum

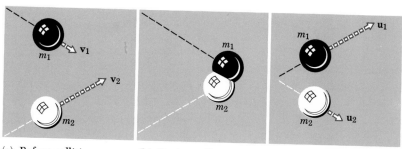

(a) Before collision.   (b) During collision.   (c) After collision.

Figure 8-5   The collision of two billiard balls.

Momentum

123

before the collision is $(m_1\mathbf{v}_1 + m_2\mathbf{v}_2)$ and its total momentum after the collision is $(m_1\mathbf{u}_1 + m_2\mathbf{u}_2)$. The law of conservation of momentum tells us that

$$m_1\mathbf{v}_1 + m_2\mathbf{v}_2 = m_1\mathbf{u}_1 + m_2\mathbf{u}_2 \tag{8-12}$$

This equation is not sufficient by itself to enable us to calculate the velocities after the collision. To do this we must know in more detail what occurs during the collision. When the balls are in close contact their shape is distorted and the atoms of one ball exert strong repulsive forces on the atoms of the other ball. The details are obviously very complicated, but the power of the law of conservation of momentum is that equation 8-12 can be written down without knowing any of these details, other than that the interatomic forces obey Newton's third law.

The law of conservation of momentum can be applied only when the resultant external force is zero or negligibly small. Let us convince ourselves that this is the case for the colliding billiard balls. On each ball there are two vertical external forces, the gravitational attraction of the earth mg and the upward thrust of the table R. Since there is no vertical acceleration these forces must exactly cancel one another (figure 8-6). Any external force which changes the momentum of the balls must be in a horizontal direction. Such a force might be provided by the gravitational attraction of a third billiard ball of mass m_3. Assuming that each ball has a mass of about 100 gm and the distance between m_2 and m_3 is about 50 cm, then the magnitude of the gravitational force exerted on m_2 by m_3 is

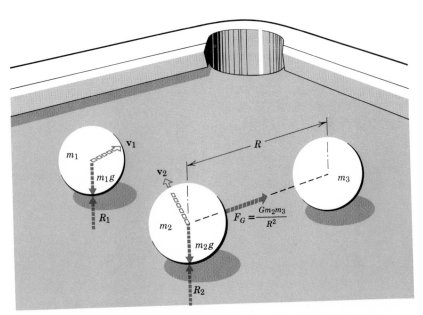

Figure 8-6 The billiard balls in relation to their environment.

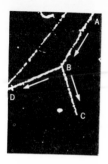

Figure 8-7 A bubble chamber photograph of the collision of two protons. Each white line is a trail of small bubbles left behind along the track of a fast moving charged particle. A proton entered at A and struck a stationary proton at B. Since it was stationary, this proton left no initial track in the chamber. BC and BD are the tracks of the two protons after the collision. Tracks of other charged particles not involved in this collision can also be seen on the photograph (Courtesy of the Lawrence Radiation Laboratory of the University of California.)

$$F_G = \frac{Gm_2m_3}{R^2}$$

$$= \frac{(6.67 \times 10^{-8}) \times (100) \times (100)}{50^2} \text{ dyne}$$

$$= 2.7 \times 10^{-7} \text{ dyne} \tag{8-13}$$

In one second this force could change the momentum of m_2 by about 2.7×10^{-7} gm cm sec^{-1} and therefore change its velocity by about 2.7×10^{-9} cm sec^{-1}. During a game of billiards the balls acquire velocities of several cm per sec and so the effect we have just calculated is clearly negligible. As an exercise the reader should also demonstrate that it is permissible to ignore the gravitational attraction of a player with a mass of 80 kg standing about 1 meter away from the balls.

Nuclear physics is very much concerned with collisions involving fundamental particles and nuclei. The law of conservation of momentum proves very valuable in the discussion of such collisions. In the discussion pertaining to figure 8-5, for example, m_1 and m_2 might be a proton and a neutron. Alternatively m_1 might be the nucleus of a helium atom, m_2 the nucleus of an aluminum atom, and so on. Figure 8-7 is a photograph of a collision between two protons.

Example 8-4-1

A lump of clay of mass 30 gm traveling with a velocity of 25 cm sec^{-1} collides head on with another lump of clay of mass 50 gm traveling with a velocity of 40 cm sec^{-1} in exactly the opposite direction. If the two lumps coalesce, what is the velocity of the combined lump after the collision, assuming that no external forces act on the system? (See figure 8-8.)

Although all the velocities are in the same straight line, it is important to remember that momentum is really a vector and to distinguish

Momentum

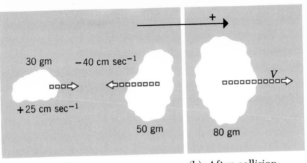

(a) Before collision.   (b) After collision.

Figure 8-8   Two lumps of clay colliding (example 8-4-1).

carefully between positive and negative directions. Choose the positive direction to be to the right, in the same direction as the initial velocity of the 30 gm lump of clay. Then,

$$\text{Initial momentum of 30 gm lump} = 30 \times (+25)$$
$$= +750 \text{ cgs units}$$
$$\text{Initial momentum of 50 gm lump} = 50 \times (-40)$$
$$= -2000 \text{ cgs units}$$
$$\text{Therefore, total initial momentum} = +750 - 2000$$
$$= -1250 \text{ cgs units}$$

If V is the velocity of the combined 80 gm lump after the collision,

Momentum after collision $= 80 \; V$

Equating the total momentum before to the total momentum after the collision,

$$-1250 = 80 \; V$$
$$V = -15.6 \text{ cm sec}^{-1}$$

The negative sign indicates that the combined lump is really traveling to the left in the diagram. Therefore, after the collision the combined lump has a velocity of 15.6 cm sec^{-1} in the same direction as the initial velocity of the 50 gm lump.

Conservation of momentum is sufficient to solve this problem completely. The reason for this is that considerable additional information about the nature of the collision is provided by the fact that the two bodies coalesce and move together after the collision.

Example 8-4-2

A star of mass 2×10^{30} kg moving with a velocity of 2×10^4 m sec^{-1} collides with a second star of mass 5×10^{30} kg moving with a velocity of 3×10^4 m sec^{-1} in a direction at right angles to the first star. If the two join together, what is their common velocity?

Figure 8-9 shows the situation before and after the collision. The

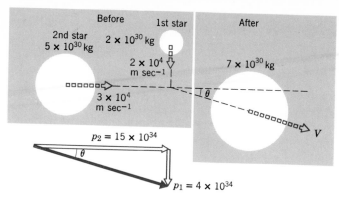

Figure 8-9 Two stars colliding (example 8-4-2).

momentum of the first star is

$$(2 \times 10^{30}) \times (2 \times 10^4) = 4 \times 10^{34} \text{ MKS units}$$

and is depicted in a vertical direction. The momentum of the second star is

$$(5 \times 10^{30}) \times (3 \times 10^4) = 15 \times 10^{34} \text{ MKS units}$$

and is depicted in a horizontal direction. The vector diagram shows how to add the initial momenta of the two stars. The total momentum of the system of two stars before the collision is represented by the resultant in the triangle of vectors. According to the law of conservation of momentum this also represents the momentum of the single star resulting from the coalescence.

Applying Pythagoras' theorem to the triangle of vectors, the magnitude of the total momentum is

$$(\sqrt{4^2 + 15^2}) \times 10^{34} = 15.52 \times 10^{34} \text{ MKS units}$$

Since the combined mass is 7×10^{30} kg, the velocity after coalescence is

$$\frac{15.52 \times 10^{34}}{7 \times 10^{30}} = 2.22 \times 10^4 \text{ m sec}^{-1}$$

To obtain the angle θ, notice that

$$\tan \theta = \frac{4 \times 10^{34}}{15 \times 10^{34}}$$

$$= 0.2667$$

Whence

$$\theta = 14.93°$$

Therefore, the single star resulting from the coalescence moves with a velocity of 2.22×10^4 m sec^{-1} in a direction making an angle of $14.93°$ with the direction in which the second star was initially moving.

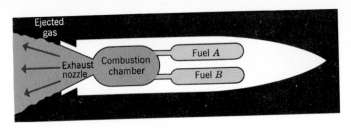

Figure 8-10 Rocket propulsion.

8-5 Rockets

Figure 8-10 illustrates the basic principles underlying the operation of a rocket. The two fuels are mixed in the combustion chamber, where they undergo a violent chemical reaction. The products of this reaction are hot, high pressure gases that escape through the exhaust nozzle with a high velocity and thereby acquire a large backward momentum. The total momentum of the system must remain constant and so the backward momentum acquired by the ejected gases must be compensated by an equal and opposite forward momentum given to the rocket and unspent fuel.

The above simple argument is valid if the rocket is in gravity-free space, but in the earth's gravitational field there is an external force on the system due to the gravitational attraction of the earth, and the law of conservation of momentum cannot then be applied. Consider a short interval of time of duration τ seconds, so short that the external force on the system does not have time to change appreciably. Define the system precisely as the rocket plus the fuel which remains unspent in the tanks at the beginning of the time interval. At the end of the time interval some of the fuel has been converted into hot gas, has been ejected from the rear of the rocket and has acquired a backward momentum. The external force on the system is its total weight $\mathbf{W}$. Since this force is equal to the change of momentum in one second, the change in the *total* momentum of the whole system during the time interval τ is

$$\text{Change in total momentum} = \mathbf{W}\tau \tag{8-14}$$

The ejected gases acquire an *extra* momentum $\mathbf{p}_g$ backward, while the rocket and the fuel still unspent at the end of the time interval acquire an *extra* forward momentum $\mathbf{p}_r$. From this point of view the change in total momentum is

$$\mathbf{p}_g + \mathbf{p}_r = \mathbf{W}\tau \tag{8-15}$$

This is shown as a vector diagram in figure 8-11. As long as the change in total momentum $\mathbf{W}\tau$ is small compared with the other vectors, it is still a reasonable description of the motion to say that the backward momentum given to the ejected gases is approximately compensated by an opposite and almost equal momentum acquired by the rocket.

Notice that $\mathbf{p}_g$ is the *extra* momentum acquired by the ejected gases.

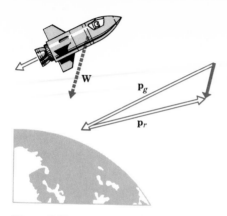

Figure 8-11 Momentum considerations when the rocket is subject to the gravitational attraction of the earth.

It is the difference between the momentum they have after ejection and the momentum of this part of the system at the beginning of the time interval when it was unspent fuel in the rocket tanks. This is not the same thing as the momentum of the ejected gases as measured by a stationary observer. The gases are ejected with a certain backward velocity *relative to the rocket*. If the rocket already has a high velocity it is conceivable that the ejected gases left behind might even move in the same direction as the rocket, though less rapidly.

It is not possible to give a simple, but complete, mathematical description of the firing of a rocket. We have just pointed out one difficulty; that the gases are ejected with a certain velocity relative to the rocket, but their velocity relative to the earth is continually changing. In addition, the fuel represents an appreciable part of the initial mass of the system. As the fuel is burnt up the mass of the rocket plus unspent fuel continually changes. Also, as the rocket leaves the earth the gravitational force on it weakens in accordance with the inverse square law. Moreover, in the lower atmosphere air friction cannot be ignored.

As a further illustration of the concept of momentum, let us consider the firing of a rocket from the point of view of a system which includes the rocket, the fuel, and the earth. The advantage of this system is that the gravitational forces between the rocket and the earth are *internal* to the system, whereas the external gravitational forces exerted by the other heavenly bodies are too weak to matter. It is therefore permissible to apply the law of conservation of momentum to this system. Assume, as is usually the case, that the fuel is all spent before the rocket leaves the earth's atmosphere. During the firing the rocket acquires an upward momentum and the ejected gases a downward momentum. However, the ejected gases mingle with the earth's atmosphere and eventually transfer their momentum to the earth. Therefore, when the firing is complete and the rocket has escaped from the earth's atmosphere, the earth has recoiled in the opposite direction to the motion of the rocket

Momentum

129

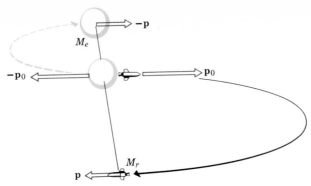

Figure 8-12 Conservation of momentum applied to the system that includes the rocket and the earth. The relative size of the earth's orbit has been exaggerated for the sake of clarity. It is actually very small indeed compared with the orbit of the rocket.

and has a momentum $-\mathbf{p}_0$ exactly equal and opposite to $\mathbf{p}_0$, the momentum of the rocket. The total momentum of the system earth plus rocket remains unchanged (figure 8-12). Subsequently the rocket moves in a curved trajectory as the earth's gravitational force acts upon it. However, according to Newton's third law, the force exerted by the earth on the rocket is exactly equal but opposite to the force exerted by the rocket on the earth. The total momentum of the combined system of earth plus rocket therefore remains constant. At any subsequent instant when the rocket has a momentum $\mathbf{p}$, the earth has an equal but opposite momentum $-\mathbf{p}$. The earth therefore describes an orbit which is a sort of inverted copy of the orbit of the rocket, but greatly reduced in size (figure 8-12). It is reduced in size because, although the momentum of the earth is always equal in magnitude to the momentum of the rocket, the mass of the earth is so much greater than the mass of the rocket that the velocity of the earth must be very much smaller than the velocity of the rocket:

$$M_e V_e = M_r V_r \tag{8-16}$$

so

$$\frac{V_e}{V_r} = \frac{M_r}{M_e} \tag{8-17}$$

QUESTIONS

1. Two identical lumps of clay traveling with the same speed in exactly opposite directions collide and stick together. What can you conclude from the law of conservation of momentum?

2. How does the momentum of a satellite in a circular orbit vary with time? How does its magnitude vary with the radius of the orbit?

3. A firework is projected vertically upward. At its highest position it explodes into three pieces with equal masses and equal speeds. What can be said about the directions of their velocities?

4. How does the law of conservation of momentum apply to a car accelerating along a straight horizontal highway?

5. How does the law of conservation of momentum apply to an athlete executing a high jump?

6. How does the law of conservation of momentum apply to the motion of the earth around the sun? (Hint: Is the sun really stationary at the center of the orbit? Suppose the sun and the earth had the same mass?)

7. Analyze the situation in the diagram from the point of view of conservation of horizontal and vertical momentum; (a) if the table is frictionless, (b) if it is not.

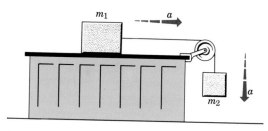

Question 8-7

8. A railroad tank car full of water has been disconnected from its train and shunted along a straight horizontal section of frictionless track (see diagram). A faucet has been left open and water streams out with a high velocity. Does the velocity of the car remain constant? Does the momentum of the car remain constant? Consider three cases. (a) The spout of the faucet points vertically downward. (b) The faucet is at the front end and its spout is horizontal. (c) The faucet is at the rear end and its spout is horizontal. What eventually happens to any horizontal momentum which the car loses? What about vertical momentum?

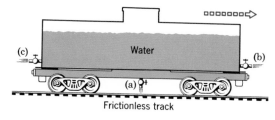

Momentum

Question 8-8

A

1. A mass of 7 gm has a speed of 25 cm sec^{-1}. What is the magnitude of its momentum.

2. What speed must be given to a mass of 5×10^4 gm in order that the magnitude of its momentum might be 6.5×10^7 gm cm sec^{-1}?

3. A body of mass 9 gm has an initial speed of 3 cm sec^{-1}. A constant force of 15 dynes is applied to it in the same direction as its velocity. What is the magnitude of its momentum after the force has been acting for 3 sec?

4. A constant force is applied to a mass of 150 gm in the same direction as its velocity. Its initial speed is 2×10^3 cm sec^{-1} but 5 sec later the speed has increased to 7×10^3 cm sec^{-1}. What is the magnitude of the force?

5. A mass m falls freely from rest near the earth's surface. Find an equation giving its momentum p after it has been falling for t sec.

6. A mass m falls freely from rest near the earth's surface. Find an equation giving its momentum p after it has fallen through a height h.

7. If a proton has a velocity of 3×10^6 cm/sec, what is its momentum (a) in CGS units, (b) in MKS units?

8. If an electron has a momentum of 2×10^{-27} kg m sec^{-1}, what is its velocity?

9. A spaceman is stationary in a gravity-free region of space. His total mass, including all his equipment, is 100 kg. He takes off his oxygen tank, which has a mass of 10 kg, and hurls it away with a velocity of 2 m/sec. With what velocity does he recoil?

10. Two spacemen are both floating with zero velocity in a gravity-free region of space. Spaceman A has a mass of 120 kg and spaceman B has a mass of 90 kg. A pushes B away from him toward the door of their space ship. If B's velocity is 0.5 m/sec, with what velocity does A recoil?

11. Two skaters are stationary in the center of a circular rink. They then push on one another so that they fly apart. One of the skaters has a mass of 90 kg and acquires an initial velocity of 0.8 m/sec. If the other skater has a mass of 75 kg, what is his initial velocity?

12. A lump of clay with a mass of 80 gm and a velocity of 200 cm/sec overtakes another lump of clay with a mass of 120 gm and a velocity of 150 cm/sec in exactly the same direction. What is their common velocity after they coalesce?

13. If a body of mass 300 gm is dropped from rest at a height of 2 m, what is its momentum just before it strikes the ground?

B

14. Two skaters stand in the center of a circular rink of frictionless ice and then push on one another until they fly apart. The skater of mass 75 kg reaches the edge of the rink after 12 sec. If the other

skater has a mass of 90 kg, how long does he take to reach the edge of the rink?

15. Two spacemen are floating in a gravity-free region of space with zero velocity *relative to one another*. They then push one another apart and acquire a relative velocity of 5 m/sec (the distance between them increases at the rate of 5 m/sec). If spaceman A has a mass of 120 kg and spaceman B has a mass of 90 kg, calculate the change in velocity of each one produced by the push exerted on him by the other.

16. Two spacemen are floating in a gravity-free region of space with zero velocity relative to one another. Spaceman A has a mass of 120 kg. Spaceman B has a mass of 90 kg, which includes a 5 kg ration box. B throws this box toward A with a velocity of 2 m/sec and A catches it. Calculate the change in velocity of each one produced by this procedure.

17. A lump of clay with a mass of 5 kg and a velocity of 1.5 m/sec collides head-on with another lump of clay with a mass of 2 kg and a velocity of 4 m/sec in exactly the opposite direction. What are the magnitude and direction of the velocity of the combined lump after they coalesce?

18. A missile has a mass of 300 kg and a horizontal velocity of 500 m/sec. A counter missile with a mass of 50 kg and a vertical velocity of 1000 m/sec strikes it and becomes firmly imbedded in it. If the impact takes place in a negligibly short time, determine the magnitude and direction of the velocity immediately afterwards. Why is it necessary to specify that the impact takes a very short time?

19. A rocket of mass 600 kg has a steady velocity of 800 m/sec and then breaks apart into two fragments. The fragment of mass 200 kg proceeds in the same direction with a speed of 600 m/sec. Calculate the magnitude and direction of the velocity of the other fragment.

20. A mass of 150 gm is thrown vertically downward with an initial velocity of 75 cm/sec from a height of 25 m. What is its momentum just before it strikes the ground?

21. A rubber ball of mass 75 gm is dropped from a height of 150 cm and bounces to a height of 120 cm. What is the ratio of the magnitudes of its momenta just before and just after it strikes the ground?

C

22. A lump of clay of mass 120 gm has a velocity of $+10$ cm/sec along the X axis. A second lump of mass 60 gm has a velocity of 15 cm/sec at an angle of 30° to the X axis (see diagram on following page). If the two lumps collide and coalesce, apply the law of conservation of momentum to the components of momentum (a) along the X axis (b) along the Y axis. Calculate the x and y components of the velocity of the combined lump. What is the magnitude and direction of the resultant velocity of the combined lump?

Momentum

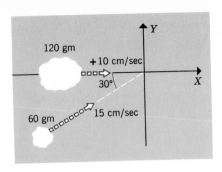

Problem 8-22

23. A ball of mass m is dropped from rest at a height h_1 and rebounds to a height h_2. From the point of view of an observer in outer space, relative to whom the earth was at rest before the ball was dropped, find expressions for the momentum of the *earth* (a) just before the ball strikes the ground (b) just after the ball strikes the ground (c) at the instant when the ball has rebounded to the height h_2. Take the downward direction, from the ball to the earth, to be positive.

24. An open-topped railroad car of mass M gm has an initial velocity v_0 along a straight horizontal track. Friction is completely negligible and v_0 remains constant until, at time $t = 0$, it suddenly starts to rain. The raindrops fall vertically with a velocity u and add a mass μ gm/sec of water to the car every second. Find an expression for the velocity of the car t secs after the rain starts, assuming it has not filled up with water. If the car holds up to V cm³ of water and the density of the rain is ρ gm/cm³, find the ultimate velocity of the car if the rain continues long enough.

25. Referring back to problem 8-16, perform a rough order of magnitude calculation to convince yourself that the result cannot be appreciably influenced by the gravitational attraction of the spacemen for one another, if they are any reasonable distance apart. Present each step in your argument carefully so that your final conclusion is logically unassailable.

26. Reconsider problem 6-30 in the light of the law of conservation of momentum.

27. At time t two particles, with masses m_1 and m_2, are at points (x_1, y_1, z_1) and (x_2, y_2, z_2). Their center of mass (x_c, y_c, z_c) is defined by the equations

$$x_c = \frac{m_1 x_1 + m_2 x_2}{m_1 + m_2}$$

$$y_c = \frac{m_1 y_1 + m_2 y_2}{m_1 + m_2}$$

$$z_c = \frac{m_1 z_1 + m_2 z_2}{m_1 + m_2}$$

Prove that the total momentum of the system is the same as if the total mass $(m_1 + m_2)$ were always concentrated at the center of mass. Extend the argument to a system containing an arbitrary number of particles. If Newton's second law of motion is applied to the system *as a whole*, give precise definitions of the quantities **F**, M and **a**.

Momentum

9

Energy

9-1   The Nature of Energy

Because the concept of energy is so familiar to us, its physical nature is correspondingly more difficult to understand. We know, for example, that gasoline contains energy and if we pour gasoline into a car this energy can be converted into energy of motion of the car. It is very tempting to visualize energy as though it were a sparkling fluid that is added to a body in order to make it move. Conservation of energy is then a simple consequence of the fact that the amount of fluid does not change. This attitude is too simple-minded and frequently very misleading. The best way to combat it is to emphasize the precise mathematical definition of the various forms of energy, although unfortunately it will not be possible to give rigorous mathematical proofs in a book at this level.

The point which must be made at the very beginning is that energy is another quantity like momentum which is conserved for a system of bodies subject to no external interference. The precise significance of "no external interference" will be defined later on. The energy of a system of bodies is the sum of several parts, such as kinetic energy, gravitational potential energy, electromagnetic potential energy, and so on, which are quite different from one another in character. The problem is to decide how precisely to define the various forms of energy so that the law of conservation of energy will apply. The energy of the gasoline is then not understood until it is realized that gasoline is composed of atoms and atoms are composed of negatively charged electrons and positively charged nuclei. The energy of the gasoline is ultimately discovered to be the kinetic energy and electromagnetic potential energy of these electrons and nuclei. However, since we have not yet broached the subject of electromagnetism in any detail, we shall initially concentrate on systems involving only gravitational forces and no electromagnetic forces, and we shall first define kinetic energy and gravitational potential energy. Various other forms of energy will be discussed in appropriate places later in the book.

9-2   Kinetic Energy

Kinetic energy is energy of motion. If a particle has mass m and velocity v, its kinetic energy is $\frac{1}{2}mv^2$.

| **Kinetic Energy** $= \frac{1}{2}mv^2$ | (9-1) |

Energy is a scalar quantity. There is no direction associated with it. Kinetic energy is always positive.

In the CGS system, when m is in gm and v in cm sec^{-1}, the unit of energy is the **erg.** Thus, a mass of 1 gm moving with a velocity of 1 cm sec^{-1} has a kinetic energy of $\frac{1}{2}$ erg. In the MKS system, when m is in kg and v in m sec^{-1}, the unit of energy is the **joule.** Thus, a mass of 1 kg moving with a velocity of 1 m sec^{-1} has a kinetic energy of $\frac{1}{2}$ joule.

Example 9-2-1

How many ergs are equivalent to 1 joule?
A mass of 2 kg moving with a velocity of 1 m sec^{-1} has a kinetic energy of

$$\frac{1}{2} \times 2 \times 1^2 = 1 \text{ joule}$$

But

$$2 \text{ kg} = 2000 \text{ gm}$$

$$1 \text{ m sec}^{-1} = 100 \text{ cm sec}^{-1}$$

A mass of 2000 gm moving with a velocity of 100 cm sec^{-1} has a kinetic energy of $\frac{1}{2} \times 2000 \times (100)^2 = 10^7$ erg.

Therefore,

$$1 \text{ joule} = 10^7 \text{ erg}$$

9-3   Gravitational Potential Energy

Kinetic energy is the energy that a single body possesses by virtue of its motion. Potential energy is a mutual property of a *pair* of bodies resulting from their proximity to one another and the fact that they interact with one another. Suppose, for example, that a body of mass M_1 is at a distance R from a body of mass M_2. Then their mutual gravitational potential energy is represented by Φ_G (the Greek letter capital phi with a subscript G to remind us that we are talking about gravitation). It is

Gravitational potential energy,

$$\Phi_G = -\frac{GM_1M_2}{R}$$

(9-2)

Notice that, in contradistinction to the gravitational force, the potential energy varies inversely as the distance apart, R, not as the square of R.

Notice also that gravitational potential energy is negative, whereas kinetic energy is positive. When the two bodies are infinitely far apart, their mutual potential energy is zero (the reciprocal of infinity is zero; see section A-8 of the Mathematical Appendix). As they approach closer

Energy

together, their mutual potential energy becomes more and more negative.

Gravitational potential energy is associated with gravitational interactions. There are corresponding types of potential energy associated with other interactions. For example, if two stationary electric charges, q_1 and q_2, are at a distance R apart, their mutual electrostatic potential energy is

Electrostatic potential energy,

$$\Phi_e = \frac{q_1 q_2}{R} \tag{9-3}$$

in the CGS system

If the charges are in motion, the situation is more complicated. We shall ignore electromagnetic energy throughout the rest of this chapter and deal only with bodies without electric or magnetic properties.

9-4 Conservation of Energy for a Pair of Particles

Consider two masses M_1 and M_2 that exert only gravitational forces on one another. Suppose that they are initially held firmly at rest at a distance R_0 apart (figure 9-1a) and are then released. Since they attract one another, they start to accelerate toward one another. As they approach closer and closer, they move faster and faster, and their kinetic energy, which is positive, becomes larger and larger. At the same time, as their distance apart R becomes smaller and smaller, their mutual gravitational potential energy, $(-GM_1M_2/R)$, becomes more and more negative. Starting with Newton's universal law of gravitation (equation 6-20), which gives the magnitudes of the accelerations, one can deduce how the velocities change as R changes. The mathematics is complicated, because the force is not constant but varies as R varies, and it is necessary to make use of the differential calculus. The final result is that the positive change in kinetic energy is always exactly compensated by the negative change in potential energy, so that the total energy re-

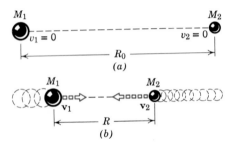

Figure 9-1 Two bodies moving directly toward one another under the influence of their mutual gravitational attraction.

mains constant. This is the law of conservation of energy applied to this case.

Consider the initial situation when the two masses are at rest (figure 9-1a).

Initial kinetic energy $= 0$ (9-4)

Initial potential energy $= -\dfrac{GM_1M_2}{R_0}$ (9-5)

Initial total energy $= -\dfrac{GM_1M_2}{R_0}$ (9-6)

Now consider the situation when the two masses are at a distance R apart. Suppose that their velocities are then v_1 and v_2 (figure 9-1b).

Kinetic energy $= \frac{1}{2}M_1v_1{}^2 + \frac{1}{2}M_2v_2{}^2$ (9-7)

Potential energy $= -\dfrac{GM_1M_2}{R}$ (9-8)

Total energy $= \frac{1}{2}M_1v_1{}^2 + \frac{1}{2}M_2v_2{}^2 - \dfrac{GM_1M_2}{R}$ (9-9)

Starting with the formula for the gravitational force, we can prove that this last quantity (the sum of the kinetic energies and the potential energy) has the same numerical value at all instants of time throughout the motion. It therefore always has the value that it had initially,

$$\frac{1}{2}M_1v_1{}^2 + \frac{1}{2}M_2v_2{}^2 - \frac{GM_1M_2}{R} = -\frac{GM_1M_2}{R_0} \tag{9-10}$$

The discussion has so far been confined to motion along the line joining the two masses, but the result can be made more general. Figure 9-2 shows the two masses moving along curved paths under the influence of one another's gravitational attraction. At a particular instant of time,

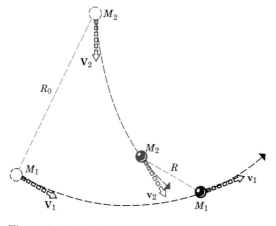

Energy

Figure 9-2 Two bodies moving along curved paths and exerting gravitational forces on one another.

139

suppose that their distance apart is R_0 and their velocities are V_1 and V_2. Their total energy is then

$$\tfrac{1}{2}M_1V_1^2 + \tfrac{1}{2}M_2V_2^2 - \frac{GM_1M_2}{R_0}$$

At any other arbitrary instant, suppose that their distance apart is R and their velocities are v_1 and v_2. Their total energy is then

$$\tfrac{1}{2}M_1v_1^2 + \tfrac{1}{2}M_2v_2^2 - \frac{GM_1M_2}{R}$$

It can be proved that this last quantity has the same numerical value at all instants of time throughout the motion:

$$\tfrac{1}{2}M_1v_1^2 + \tfrac{1}{2}M_2v_2^2 - \frac{GM_1M_2}{R}$$

$$= \text{a constant, independent of time}$$

$$= \tfrac{1}{2}M_1V_1^2 + \tfrac{1}{2}M_2V_2^2 - \frac{GM_1M_2}{R_0} \qquad (9\text{-}11)$$

Example 9-4-1

Two masses of 2.0×10^7 kg and 3.0×10^7 kg are initially at rest a distance of 400 m apart. They move toward one another under the influence of their mutual gravitational attraction. Find their velocities when the distance between them has fallen to 200 m.

When they are initially at rest, they have no kinetic energy and their mutual gravitational potential energy in the MKS system is

$$\frac{-GM_1M_2}{R_0} = -\frac{(6.67 \times 10^{-11}) \times (2 \times 10^7) \times (3 \times 10^7)}{400}$$

$$= -100 \text{ joules}$$

Since this is the total initial energy, the total energy of the system must have this value at all subsequent instants of time during the motion. When the distance between the two masses has fallen to 200 m (one half of the initial distance), the mutual gravitational potential energy is easily seen to be -200 joules. Since the total energy must still be -100 joules, the kinetic energy must have increased to $+100$ joules:

$$\tfrac{1}{2}M_1v_1^2 + \tfrac{1}{2}M_2v_2^2 = +100 \qquad (9\text{-}12)$$

To obtain v_1 and v_2 separately, we can make use of the Law of Conservation of Momentum. Initially, when the two masses are at rest, the total momentum is zero and it must remain zero at all subsequent instants. The two masses therefore move toward one another with equal but opposite momenta:

$$M_1v_1 = M_2v_2 \qquad (9\text{-}13)$$

$$(2 \times 10^7)v_1 = (3 \times 10^7)v_2$$

$$v_1 = 1.5v_2 \qquad (9\text{-}14)$$

Equation 9-12 may therefore be rewritten

$$\tfrac{1}{2}(2 \times 10^7)(1.5v_2)^2 + \tfrac{1}{2}(3 \times 10^7)v_2{}^2 = 100$$
$$(2.25 \times 10^7)v_2{}^2 + (1.5 \times 10^7)v_2{}^2 = 100$$
$$(3.75 \times 10^7)v_2{}^2 = 100$$

$$v_2 = \frac{1}{\sqrt{37.5 \times 10^4}}$$
$$= \frac{1}{6.12 \times 10^2}$$
$$= 1.63 \times 10^{-3} \text{ m sec}^{-1}$$

From equation 9-14

$$v_1 = 1.5 \times (1.63 \times 10^{-3})$$
$$= 2.44 \times 10^{-3} \text{ m sec}^{-1}$$

Therefore, when the two masses are 200 m apart, the 2×10^7 kg mass has a velocity of 2.44×10^{-3} m sec^{-1}, and the 3×10^7 kg mass has a velocity of 1.63×10^{-3} m sec^{-1}.

9-5 The Escape Velocity of a Rocket

With what velocity must a rocket be fired in order to escape completely from the earth's gravitational attraction? This is called the "escape velocity."

The earth has a mass M_e and the rocket a mass M_r. When the firing is complete and all the fuel is spent, the distance of the rocket from the center of the earth is R_e, which is not very different from the radius of the earth. Situation 1 (figure 9-3a) is the instant just before the firing starts. The total *momentum* is then zero because the earth and the rocket are at rest. Situation 2 (figure 9-3b) is the instant just after the firing is complete and all the fuel is spent. The rocket then has a velocity V_r and the earth is recoiling with a velocity V_e. As we explained in the last chapter, momentum is conserved during the firing and

$$M_r V_r = M_e V_e \tag{9-15}$$

Let us compare the kinetic energy of recoil of the earth with the kinetic energy given to the rocket.

$$\frac{\text{Kinetic energy of earth}}{\text{Kinetic energy of rocket}} = \frac{\tfrac{1}{2}M_e V_e{}^2}{\tfrac{1}{2}M_r V_r{}^2}$$
$$= \frac{M_e{}^2 V_e{}^2}{M_r{}^2 V_r{}^2} \frac{M_r}{M_e}$$
$$= \frac{M_r}{M_e} \text{ since } (M_e V_e)^2 = (M_r V_r)^2 \tag{9-16}$$ **Energy**

The mass of the rocket is very small compared with the mass of the earth. So the kinetic energy of the earth is very small compared with

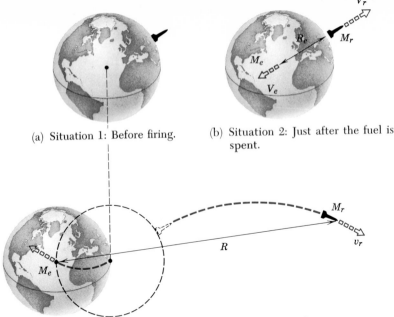

(a) Situation 1: Before firing.

(b) Situation 2: Just after the fuel is spent.

(c) Situation 3: The rocket is at a distance R from the earth.

Figure 9-3 A rocket escaping from the earth's gravitational attraction.

the kinetic energy of the rocket. In fact, a simple extension of the argument shows that the kinetic energy of the earth may always be neglected in comparison with the kinetic energy of the rocket at any time during the subsequent flight.

Therefore, we shall not introduce any serious error if we take the total kinetic energy in situation 2 to be equal to the kinetic energy of the rocket alone.

$$\text{Kinetic energy in situation } 2 = \tfrac{1}{2} M_r V_r^2 \tag{9-17}$$

The burning of the fuel takes place quickly and the rocket has not traveled very far, so the distance of the rocket from the earth is still very nearly R_e and the mutual gravitational potential energy of the rocket and the earth in situation 2 is

$$\text{Potential energy in situation } 2 = -\frac{GM_e M_r}{R_e} \tag{9-18}$$

The total energy is the sum of the kinetic and potential energies.

Total energy in situation 2

$$= \frac{1}{2} M_r V_r^2 - \frac{GM_e M_r}{R_e} \tag{9-19}$$

As the rocket climbs away from the earth, the earth's gravitational attraction tries to pull it back and reduces its velocity. If it has been launched with too small an initial velocity V_r, the pull of the earth eventually turns it around and it starts to fall back again. In situation 3 (figure 9-3c) the rocket has climbed to a distance R from the center of the earth and its velocity has been reduced to v_r. Under these circumstances,

Kinetic energy in situation $3 = \frac{1}{2}M_r v_r^2$

$$\text{(9-20)}$$

Potential energy in situation $3 = -\dfrac{GM_e M_r}{R}$

$$\text{(9-21)}$$

Total energy in situation 3

$$= \frac{1}{2}M_r v_r^2 - \frac{GM_e M_r}{R} \tag{9-22}$$

During the motion that converts situation 2 into situation 3 the earth and the rocket constitute a system subject to no external interference and the total energy of this system must be conserved. The total energy in situation 2 must equal the total energy in situation 3.

$$\frac{1}{2}M_r V_r^2 - \frac{GM_e M_r}{R_e} = \frac{1}{2}M_r v_r^2 - \frac{GM_e M_r}{R} \tag{9-23}$$

To understand what is going on, let us put in some numbers. Suppose that the initial kinetic energy $\frac{1}{2}M_r V_r^2$ is $+700$ ergs and the initial potential energy $(-GM_e M_r/R_e)$ is -1000 ergs. (These numbers are chosen for convenience of exposition. Realistic numbers for actual rockets would be very much larger.) Then the initial total energy is $+700 - 1000 = -300$ ergs. When the rocket has doubled its distance from the earth, $R = 2R_e$, and the negative potential energy is halved and becomes -500 ergs. Since the total energy must remain -300 ergs, the kinetic energy $\frac{1}{2}M_r v_r^2$ must have been reduced to $+200$ ergs, because $+200 - 500 = -300$. In general, as the rocket moves away v_r becomes smaller, the positive kinetic energy $\frac{1}{2}M_r v_r^2$ becomes smaller and the positive energy thus made available is needed to make the potential energy less negative.

If the rocket is to escape completely, it must be possible for R to become infinitely large. The potential energy $(-GM_e M_r/R)$ is then 0, whereas in situation 2 just after launching it was $(-GM_e M_r/R_e)$. The potential energy therefore increases by $(+GM_e M_r/R_e)$ and this increase is achieved at the expense of the kinetic energy. For this to be possible the initial kinetic energy must be at least $(+GM_e M_r/R_e)$. The minimum speed of launching before the rocket can escape is therefore given by the equation

$$\frac{1}{2}M_v V_r^2 = \frac{GM_e M_r}{R_e} \tag{9-24}$$

Whence

$$V_r^2 = \frac{2GM_e}{R_e} \tag{9-25}$$

$$V_r = \sqrt{\frac{2GM_e}{R_e}} \tag{9-26}$$

The values of the gravitational constant G, the mass of the earth M_e, and the radius of the earth R_e may be found inside the cover of this book. Inserting them into equation 9-26 we find that

$$\text{Escape velocity} = \sqrt{\frac{2 \times 6.67 \times 10^{-8} \times 6.0 \times 10^{27}}{6.4 \times 10^8}} \tag{9-27}$$

$$= 1.11 \times 10^6 \text{ cm/sec} \tag{9-28}$$

$$= 24{,}800 \text{ mph} \tag{9-29}$$

Finally, notice that we cannot relate situation 1 to situation 2 by applying the law of conservation of energy in the form of an equation similar to equation 9-23. This is because, during the burning of the fuel, the chemical energy of the fuel is being converted into kinetic energy of the rocket. We then have a situation involving forms of energy other than kinetic energy and gravitational potential energy.

9-6 Conservation of Energy for a System of Many Bodies

Consider a collection of several bodies (figure 9-4) that exert only gravitational forces on one another. Suppose that the external forces exerted on this system by the rest of the universe can be neglected. Let the various masses be M_1, M_2, M_3, M_4, etc., and their velocities v_1, v_2, v_3, v_4, etc. Then the total kinetic energy of the system is the sum of terms like $\frac{1}{2}M_1v_1^2$ for all the bodies of the system.

$$\text{Total kinetic energy} = \frac{1}{2}M_1v_1^2 + \frac{1}{2}M_2v_2^2 + \frac{1}{2}M_3v_3^2 + \cdots \text{ with one term for each body in the system.} \tag{9-30}$$

The total potential energy contains a contribution from each pair of bodies in the system. For example, the contribution from M_1 and M_2 is

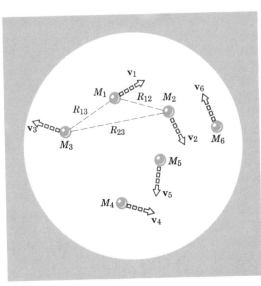

Figure 9-4 A system of bodies exerting gravitational forces on one another.

$(-GM_1M_2/R_{12})$, where R_{12} is the distance between M_1 and M_2. There is also a contribution $(-GM_1M_3/R_{13})$ from the pair M_1 and M_3, and a contribution $(-GM_2M_3/R_{23})$ from the pair M_2 and M_3.

Total mutual gravitational potential energy

$$= -\frac{GM_1M_2}{R_{12}} - \frac{GM_1M_3}{R_{13}} - \frac{GM_2M_3}{R_{23}} - \cdots$$

with one term for every possible pair of bodies in the system.   (9-31)

Notice that the number of pairs is much larger than the number of bodies. For a system of 10 bodies, for example, the expression for the kinetic energy contains 10 terms, but the expression for the potential energy contains 45 terms. Each body in the system must be paired off with every other body in the system.

The total energy of the system is the sum of the kinetic energy and the potential energy.

Law of conservation of energy

Total Energy

$$= \tfrac{1}{2}M_1v_1{}^2 + \tfrac{1}{2}M_2v_2{}^2 + \tfrac{1}{2}M_3v_3{}^2$$

+ one similar term for each body

$$-\frac{GM_1M_2}{R_{12}} - \frac{GM_1M_3}{R_{13}} - \frac{GM_2M_3}{R_{23}}$$

— one similar term for each pair of bodies

= a constant, independent of time.   (9-32)

The law of conservation of energy says that the expression given in equation 9-32 has a numerical value that remains the same at all instants of time, however complicated the motion and however much the individual terms may vary, always assuming that the system is sub-ject to no external interference.

Let us look at this from a different point of view. Imagine that Newton's laws of motion are known, but that the concept of energy has not been introduced into physics. Start by writing down an equation for each body, in which the product of its mass and acceleration is equated to the sum of all the gravitational forces exerted on it by all the other bodies. These forces are given by the universal law of gravitation (equation 6-20). By suitable mathematical manipulation of the resulting equations, it is possible to prove with mathematical rigor that the expression in equation 9-32 is a constant, independent of time. It would then be reasonable to attach special importance to this particular mathematical quantity and to give it a name—energy. One would also notice that equation 9-32 contains terms of two different types, terms of the form $\tfrac{1}{2}Mv^2$, which could be called kinetic energy, and terms of the form $(-GM_1M_2/R_{12})$, which could be called potential energy.

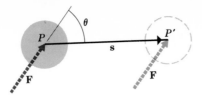

Figure 9-5 A force doing work on a body.

Energy always has this significance. In a particular physical situation, which might be much more complicated than the one we have considered, the equations of motion can always be manipulated to yield a mathematical expression that retains a constant numerical value throughout the motion. The various different parts of this expression can then be identified with various types of energy.

9-7 Work

When the system of bodies interacts strongly with the rest of the universe, the total energy of the system is no longer constant but changes with time. The change in total energy can be shown to be equal to a quantity called "the work done by the external forces."

Consider a body which is acted on by a force **F** as in figure 9-5. Suppose that, during a short interval of time, the body is displaced through a vectorial distance s from P to P'. If the time interval is short enough, the motion of the body is essentially a straight line and the force **F** does not have time to change appreciably. **F** and s are not necessarily in the same direction and the angle between them is θ, as in the figure. The work done by the force during this small displacement is defined as

$$\textbf{Work} = Fs \cos \theta \tag{9-33}$$

Work, like energy, is a scalar quantity and has the same units as energy. Thus, in the MKS system, if a force of 1 newton displaces a body through a distance of 1 meter parallel to the force, the work done is 1 **joule**. In the CGS system, if a force of 1 dyne displaces a body through a distance of 1 cm parallel to itself, the work done is 1 **erg**.

If **F** and s are in the same direction, $\theta = 0$, $\cos \theta = 1$, and the work done is just Fs (Work = Force × Distance). However, if **F** and s are at right angles, $\theta = 90°$, $\cos \theta = 0$, and the work done is zero. A force

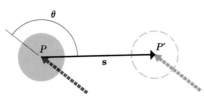

Figure 9-6 In this case, the work is negative.

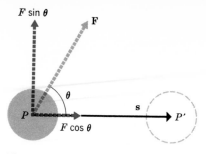

Figure 9-7   The work is done by the component parallel to s.

does no work if the body on which it acts moves in a direction perpendicular to the force.

The work done on a body can be negative. If, as in figure 9-6, θ lies between 90° and 270°, $\cos \theta$ is negative and the work done is negative (see section C-4 of the Mathematical Appendix).

An alternative approach is to replace the force **F** by its rectangular components parallel and perpendicular to the displacement **s** (figure 9-7). The magnitude of the component parallel to **s** is $F \cos \theta$, and the work done is the magnitude of this component multiplied by the distance s through which the body is displaced:

Work = Component of force parallel to displacement × Displacement

$$(9\text{-}34)$$

(a) The work done is positive.

(b) The work done is negative.

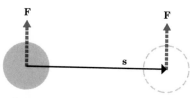

(c) The work done is zero.

Energy

Figure 9-8   The work done by a force depends upon the angle between it and the displacement.

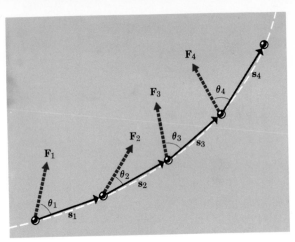

Figure 9-9 The work done by a variable force on a body moving along a curved path.

If the component points in the same direction as the displacement **s**, the work done is positive (figure 9-8*a*). If the component points in the opposite direction to the displacement **s**, the work done is negative (figure 9-8*b*). If the resultant force **F** is perpendicular to the displacement **s**, $\theta = 90°$, $\cos \theta = 0$, there is no component of **F** in the direction of **s** and the work done is zero (figure 9-8*c*).

The above definition of work assumes that the displacement **s** is along a straight line and that the force **F** remains constant throughout the displacement. If the body moves along a curved path and the force **F** varies as it does so, we may proceed in the following way. The curved path can be considered to be equivalent to a large number of very short, straight segments s_1, s_2, s_3 and so on (figure 9-9). If the segment s_1 is short enough, the force $\mathbf{F}_1$ varies very little while the segment is being traversed. The work done during this segment is therefore $F_1 s_1 \cos \theta_1$. The total work done during the complete path is the sum of all the contributions from all the small segments.

$$\text{Total work done} = F_1 s_1 \cos \theta_1 + F_2 s_2 \cos \theta_2$$
$$+ F_3 s_3 \cos \theta_3 + \text{similar contributions from all the}$$
$$\text{other segments.} \quad (9\text{-}35)$$

Reverting again to the total energy of a *system* of bodies, in general the force on any one body such as M_1 can be divided into an internal force $\mathbf{F}_1^{int}$ due to the other bodies in the system and an external force $\mathbf{F}_1^{ext}$ due to the rest of the universe. To calculate the work done on the system of bodies, the internal force is ignored and the work done by the external force alone is calculated. The work done by the external forces is then added for all the bodies of the system. It can be proved that:

Work and energy

During any interval of time, not necessarily short, the change in the total energy of a system of bodies is equal to the total work done by the *external* forces acting on the bodies.

Conservation of energy applies to this situation in the following sense. If the total energy of the system of bodies increases by a certain amount, the total energy of the rest of the universe decreases by an exactly equal amount. The rest of the universe does an amount of work on the system which is equal to the change in total energy of the system. At the same time the system does an amount of work on the rest of the universe which is equal in magnitude but has the opposite sign. This produces in the rest of the universe a change of total energy which is equal in magnitude but opposite in sign to the change in total energy of the system. The amount of energy in the whole universe therefore does not change.

It is now possible to state precisely under what circumstances the law of conservation of energy applies to a system. The work done by the external forces must be zero or too small to matter. This is obviously true if the external forces are zero or very small. The work done can also vanish under special circumstances, such as when the motion of a body is perpendicular to the external force acting on it.

Example 9-7-1

A single body of mass M in free space is acted on by a constant force F in the same direction in which it is moving. Show that the work done by the force is equal to the increase in kinetic energy of the body.

This is a case of motion in a straight line with constant acceleration. Referring to figure 9-10, suppose that at the zero of time the body is at the origin and is moving with a velocity v_0. Suppose that at time t it has moved through a distance x and its velocity has changed to v. If a is the constant acceleration, then equation 4-23 of chapter 4 is applicable:

$$v^2 = v_0^2 + 2ax \qquad (9\text{-}36)$$

The work done by the force F in moving the body from the origin through a distance x is

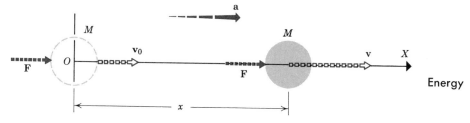

Figure 9-10   Illustrating example 9-7-1.

$$\text{Work done} = Fx \tag{9-37}$$

From equation 9-36

$$x = \frac{(v^2 - v_0{}^2)}{2a} \tag{9-38}$$

So

$$\text{Work done} = \frac{F(v^2 - v_0{}^2)}{2a} \tag{9-39}$$

But, from Newton's second law

$$F = Ma$$

$$\frac{F}{a} = M \tag{9-40}$$

So

$$\text{Work done} = \tfrac{1}{2}M(v^2 - v_0{}^2) \tag{9-41}$$

$$= \tfrac{1}{2}Mv^2 - \tfrac{1}{2}Mv_0{}^2 \tag{9-42}$$

$$\text{Work done} = \text{Final kinetic energy} - \text{Initial kinetic energy} \tag{9-43}$$

9-8 A Body Near the Surface of the Earth

Figure 9-11a shows a body of mass m resting on the surface of the earth. Its distance from the center of the earth is R_e, the radius of the earth. As before, the mass of the earth is denoted by M_e. The mutual gravitational potential energy of the earth and the body m is

$$\Phi_G = - \frac{GM_e m}{R_e} \tag{9-44}$$

Now suppose that the body is raised through a height h (figure 9-11b),

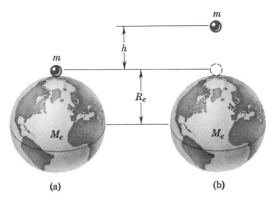

(a)   (b)

Figure 9-11 Increasing the potential energy of a body by raising it above the surface of the earth.

so that its distance from the center of the earth becomes $(R_e + h)$. Then the new potential energy is

$$\Phi'_G = -\frac{GM_em}{(R_e + h)} \qquad (9\text{-}45)$$

The increase in potential energy is

$$\begin{aligned}
\Phi'_G - \Phi_G &= -\frac{GM_em}{(R_e + h)} - \left(-\frac{GM_em}{R_e}\right) \\
&= GM_em\left(\frac{1}{R_e} - \frac{1}{R_e + h}\right) \\
&= GM_em\frac{[(R_e + h) - R_e]}{R_e(R_e + h)} \\
&= \frac{GM_emh}{R_e(R_e + h)} \qquad (9\text{-}46)
\end{aligned}$$

In many problems involving bodies moving in the vicinity of the surface of the earth the height h is very small compared with the radius of the earth, R_e. Then very little error is introduced if the expression $(R_e + h)$ in the denominator is replaced by R_e (see section A-7 of the Mathematical Appendix). The increase in potential energy then becomes

$$\Phi'_G - \Phi_G = \frac{GM_e}{R_e{}^2}\,mh \qquad (9\text{-}47)$$

In section 7-1 of chapter 7 it was proved that the acceleration due to gravity near the surface of the earth is

$$g = \frac{GM_e}{R_e{}^2} \qquad (9\text{-}48)$$

Therefore, $\Phi'_G - \Phi_G = gmh$ $\qquad (9\text{-}49)$

Changes in gravitational potential energy near the surface of the earth

When a body of mass m near the surface of the earth is *raised* through a height h small compared with the radius of the earth, the *increase* in gravitational potential energy is mgh.

This may also be proved in a slightly different way. Suppose that the mass m is resting on the surface of the earth and a giant from outer space approaches, grasps the earth firmly in his left hand and holds it still, while raising the mass m with his right hand (figure 9-12a). To hold the earth still he must exert a downward force mg on it to counteract the gravitational attraction of the body m. To raise the body m he must exert on it an upward force just a little bit bigger than mg. If this force is larger than mg by an infinitesimally small amount, the resultant force on m is very small. Its acceleration is therefore very small and, while it

Energy

151

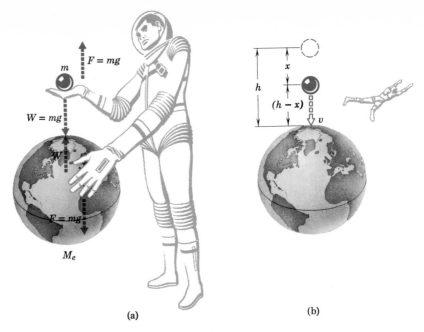

Figure 9-12 A giant from outer space intervenes.

is being raised, it never acquires an appreciable velocity, so that its kinetic energy can always be neglected. The work done by the giant on the system containing the earth and the mass m is the force, mg, multiplied by the distance through which the force raises m. So, when m has been raised to a height h,

Work done by giant $= mgh$ $\qquad$ (9-50)

As we pointed out in the previous section, the external work done on the system is equal to the increase in total energy of the system. Since the giant performs the operation in such a way that no kinetic energy is produced, this increase in energy is entirely potential energy. The increase in potential energy is therefore equal to the work done, which is equal to mgh.

Now suppose that the giant releases the earth and the body m and goes away (figure 9-12b). The mass m falls down to the earth and the earth falls a very small distance upward toward the mass m. The system is now subject to no external interference and the law of conservation of momentum is applicable. The momentum of the earth falling upward is equal but opposite to the momentum of m falling downward. An argument very similar to that used in section 9-5 shows that the kinetic energy of the earth can always be neglected in comparison with the kinetic energy of m. At the instant when the giant releases m it is at rest and has no kinetic energy, so the total energy of the system is the same as its potential energy:

Initial total energy $= -\dfrac{GM_e m}{R_e} + mgh$ $\qquad$ (9-51)

The subsequent motion is motion in a straight line with constant acceleration g. Equation 4-26 of chapter 4 tells us that, after the body has fallen a distance x, its velocity is given by

$$v^2 = 2gx \tag{9-52}$$

Its kinetic energy is therefore

$$\tfrac{1}{2}mv^2 = mgx \tag{9-53}$$

Its height above the ground is $(h - x)$ and its potential energy is

$$-\frac{GM_em}{R_e} + mg(h - x) \tag{9-54}$$

The total energy is therefore

$$-\frac{GM_em}{R_e} + mg(h - x) + mgx \tag{9-55}$$

$$= -\frac{GM_em}{R_e} + mgh \tag{9-56}$$

This is the same as the initial total energy (equation 9-51) and is a constant throughout the fall. We have therefore proved the law of conservation of energy in this simple case.

When the body strikes the ground and embeds itself into the earth, the kinetic energy is again zero, but the gravitational energy is only $(-GM_em/R_e)$. An amount of energy mgh seems to have been lost. The explanation is that, while the body is embedding itself into the earth, an amount of energy mgh is converted into heat, a form of energy that we have not yet included in the conservation law.

Example 9-8-1

A skier takes off from a ski jump with a velocity of 50 m sec^{-1} at an unknown angle to the horizontal and lands at a point whose vertical distance below the point of take-off is 100 m (figure 9-13). What is his speed just before landing, if the friction of the air can be ignored?

In the absence of friction we can assume that the sum of the kinetic energy and the potential energy is constant, and as before the kinetic energy of the earth can be ignored. At take-off the kinetic energy of the skier is

$$\tfrac{1}{2}mv_0^2 = \tfrac{1}{2}m \times 50^2$$

$$= 1250m \text{ joule}$$

where m is the mass of the skier in kg.

During the jump the skier falls through a height of 100 m and *loses* an amount of gravitational potential energy given by

$$mgh = m \times 9.8 \times 100 \text{ joule}$$

$$= 980m \text{ joule}$$

This energy must reappear as an increase in his kinetic energy, which then has the value

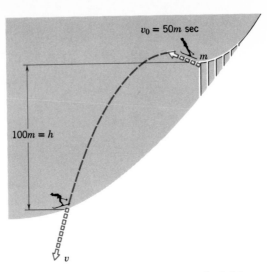

Figure 9-13   The ski jumper of example 9-8-1.

$$\tfrac{1}{2}mv^2 = 1250m + 980m$$
$$= 2230m$$
$$v^2 = 2 \times 2230$$
$$= 4460$$
$$v = 66.8 \text{ m sec}^{-1}$$

Just before landing the skier has a speed of 66.8 m sec^{-1}.

The power of the law of conservation of energy lies in the fact that we can derive this result without knowing the angle of take-off, the horizontal distance traveled, or the exact nature of the curved trajectory.

QUESTIONS

1. Can you imagine circumstances under which the total momentum of a system of bodies is conserved, but its total energy is not?
2. Can you imagine circumstances under which the total energy of a system of bodies is conserved, but its total momentum is not?
3. If the total energy of a system of bodies remains constant, is the resultant external force necessarily zero?
4. How does the kinetic energy of a satellite in a stable circular orbit vary with the radius of the orbit?
5. A boy sleds down a slope and ends up in a snow bank. Where did his kinetic energy come from and where does it go to?
6. A boy is pushing a bicycle up a hill. Where is the energy produced by the work that he does?
7. An automobile is traveling at a steady speed along a straight horizontal road. What happens to the energy released from the gasoline which is used up?
8. If you lift a book from the floor and place it on a desk, its gravita-

tional potential energy increases. Where did the energy come from? Trace back the source of the energy as far as you can, in the hope of finding the ultimate source of all energy on the surface of the earth.

9. In section 9-8 it was required of the giant that he should lift the mass m with so small an acceleration that it never acquires an appreciable velocity. However, if he does so, it takes a very long time to lift the mass. It is not clear that even a very small acceleration will not produce a large velocity if given long enough. Convince yourself that a very small constant acceleration does indeed imply a very small velocity at height h.

PROBLEMS

A

1. If a mass of 3 gm has a velocity of 4 cm/sec, what is its kinetic energy?
2. If a mass of 0.7 kg has a kinetic energy of 140 joule, what is its velocity?
3. If an electron has a kinetic energy of 10^{-10} erg, what is its velocity?
4. If a proton has a momentum of 10^{-18} gm cm sec^{-1}, what is its kinetic energy?
5. By what factor is the momentum of a body multiplied when its velocity is doubled?
6. By what factor is the kinetic energy of a body multiplied when its velocity is doubled?
7. What is the mutual gravitational potential energy of two stars of masses 5×10^{29} kg and 3.5×10^{30} kg when the distance between them is 2×10^{12} m.
8. When two bodies are 150 cm apart their mutual gravitational potential energy is -700 ergs. What is their mutual gravitational potential energy when they are 200 cm apart?
9. A force of 25 dyne pushes a mass of 6 gm through a distance of 16 cm in the direction of the force. How much work does it do?
10. A body of mass 0.5 kg is subject to a force of 7.5 newton in a direction exactly opposite to the direction of its velocity. What is the work done by the force after the body has traveled 6 m?
11. A horizontal force of 50 dyne pointing toward the northeast acts on a body which moves 3 cm toward the east. How much work is done by the force?
12. A 200 gm book is carried from the street to a second floor room 4.5 m higher. What is the change in its gravitational potential energy?
13. Niagara Falls is approximately 50 m high and about one million kilograms of water flows over it per second. How much potential energy is lost per second? An electric light bulb requires 100 joules of energy per second. How many such bulbs could be lit if all the energy of Niagara Falls could be utilized? At an average of one bulb per person, what sort of community would be supplied with electric power for lighting?

14. A mass of 400 gm is dropped from rest at a height of 15 m. What is its kinetic energy just before striking the ground? (It is not necessary to calculate the velocity of the body.)

B

15. If the momentum of a body is doubled, what is the ratio of its final to its initial kinetic energy? If the kinetic energy of a body is doubled, what is the ratio of its final to its initial momentum?

16. What is the mutual gravitational potential energy of the earth and the moon?

17. What is the mutual gravitational potential energy of yourself and the earth? What velocity in mph would you have to be given to have a kinetic energy of equal magnitude (but opposite sign, of course)?

18. What is the escape velocity from the moon?

19. When a satellite is in a stable circular orbit, what is the ratio of its kinetic energy to its potential energy?

20. A tractor is dragging a 400 kg log. The chain from the tractor to the log is at 30° to the horizontal and exerts a force of 120 newtons. How much work is done by the tractor in dragging the log through a distance of 25 m?

21. How much work is done in pushing a mass of 50 kg a distance of 5 m up a frictionless ramp at 30° to the horizontal?

22. A 250 gm book is slid across a rough table with an initial velocity of 50 cm/sec. How much work is done by the frictional force in bringing it to rest? If the magnitude of this frictional force is 0.15 newton, how far does the book slide before coming to rest?

23. Assuming that the earth's orbit around the sun is exactly circular, how much work is done on the earth in one year by the gravitational attraction of the sun?

24. Two equal masses of 150 gm approach one another with equal speeds of 25 cm/sec in exactly opposite directions. If they coalesce upon collision, what is the difference between the total kinetic energy before and after the collision? What happens to the "lost" kinetic energy?

25. A lump of clay of mass 20 gm traveling with a velocity of 15 cm sec^{-1} collides head-on with another lump of clay of mass 35 gm traveling with a velocity of 50 cm sec^{-1} in exactly the opposite direction. If the two lumps coalesce, how much kinetic energy is "lost"?

26. A rubber ball is dropped from rest at a height of 10 m. If it loses 30% of its kinetic energy when it bounces, to what height does it rebound?

27. A star of mass 2×10^{32} gm and another star of mass 4×10^{31} gm are initially at rest a long way away from one another. They then move directly toward one another under the influence of their mutual gravitational attraction. Calculate the velocity of each star when the distance between their centers is 10^{15} cm.

28. Imagine that the mass of the earth remains constant but that its radius gradually shrinks. For what value of the radius would the

escape velocity become equal to 3×10^{10} cm sec^{-1}, which is the velocity of light? What would the density of the earth then be? What would be the average distance between the protons and neutrons in the earth? (You may assume that the arguments given in section 9-5 remain valid up to the velocity of light. In fact, the theory of relativity requires that these arguments should be drastically modified.)

C

29. What is the total energy of a satellite moving in a stable circular orbit? Express your answer in terms of G, the mass of the earth M_e, the mass of the satellite M_s, the radius of the orbit R and *nothing else*. If the frictional drag of the outer atmosphere reduces the total energy, does the radius of the orbit increase or decrease?

30. If a rocket is fired vertically upward with a velocity $\frac{1}{2}V_e$, prove that its farthest distance from the center of the earth is $(4/3)R_e$. R_e is the radius of the earth and V_e is the escape velocity.

31. If a rocket is fired in any direction with a velocity V_r less than V_e, prove that its distance from the center of the earth can never exceed R_m, where

$$\frac{1}{R_m} = \frac{1}{R_e} - \frac{V_r^2}{2GM_e}$$

32. A rocket is fired at the moon. It leaves the surface of the earth with a velocity of 1.25×10^6 cm sec^{-1} and from then on never fires its motors again. With what velocity does it strike the moon?

33. A rocket is fired vertically upward with a velocity of 8×10^5 cm sec^{-1}. To what height does it rise?

34. With what velocity must a rocket be fired from the earth in order to escape from the whole solar system? Compare your result with the velocity to escape from the earth as calculated in section 9-5.

35. From the point of view of the work done per second, shoveling snow can be compared with climbing up stairs. Roughly what is the equivalent number of flights of stairs climbed per minute. You must decide or discover for yourself the magnitudes of all relevant quantities, but express your answer in such a way that it gives you a proper feeling for the rigors of shoveling snow.

36. Extend the discussion of section 5-6 to calculate the speed of the projectile when it is at a height y above the point of projection. Hence show that energy is conserved in this case.

37. The **center of mass** of two particles was defined in problem 8-27. Show that the total kinetic energy of the two particles may be considered to be the sum of two parts. The first part is the kinetic energy of the total mass $(m_1 + m_2)$ concentrated at the center of mass and moving with the velocity of the center of mass. The second part is the "internal kinetic energy," which is the total kinetic energy of the two particles from the point of view of an observer stationed at and moving with the center of mass.

Energy

157

10

Rotation

10-1 The Moment of a Couple

Grasp the bottom right hand corner of this book in your right hand and grasp the bottom left hand corner in your left hand. With the right hand push the book directly away from you and at the same time pull it directly towards you with the left hand, using a force equal in magnitude but exactly opposite in direction to the force of the right hand (figure 10-1). Since the two forces exactly compensate one another the resultant force on the book is zero, its acceleration is zero, and so this procedure cannot start the book moving bodily away from you. What happens, of course, is that the book rotates in a counterclockwise direction. The reason is that, although the forces are equal but opposite, they are displaced sideways from one another and therefore produce a turning effect.

A pair of equal but opposite forces displaced sideways from one another is called a **couple.** If the magnitude of each force is F and their perpendicular distance apart is d, the **moment** of the couple is defined in the following way.

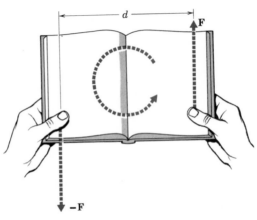

Figure 10-1   A couple rotating a book.

Moment, or torque, of a couple,

$$\tau = Fd \tag{10-1}$$

It will be represented by τ, which is the Greek letter tau. An alternative name for the moment of a couple is the **torque** exerted by the couple. The moment, or torque, plays the same role for rotational motion that force plays for motion along a straight line. During the rest of this chapter we shall develop this analogy between rotational motion and motion along a straight line.

Consider any system of bodies. It can be proved that the sum of all the external forces on the system is equivalent to a resultant force plus a couple. The resultant force produces an acceleration of the system as a whole, in accordance with Newton's first law. The couple accelerates the rotation of the system.

10-2 Rotation of a Rigid Body about a Fixed Axis

A rigid body is a system composed of atoms that always remain in the same positions relative to one another. If the body suffers a simple displacement, all the atoms undergo the same displacement. If the body is rotated through a certain angle about an axis, every single atom is rotated through the same angle about this axis. The reader should visualize the earth rotating about a line through the north and south poles, or an automobile wheel rotating about an axle. In any complicated motion the rigid body can be shown to be instantaneously rotating about a straight line axis, such as the north-south line for the earth or the axle of the wheel. In many cases the position of this axis changes with time, as in the case of a precessing top tilted at an angle to the vertical (figure 10-2a). In order to avoid some difficult mathematics, we shall confine our attention to rotation about an axis which remains fixed in

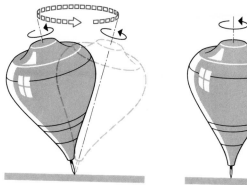

(a) A precessing top.

(b) A top spinning about a fixed vertical axis.

Figure 10-2 A spinning top is a rotating rigid body.

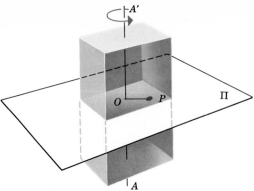

(a) A rotating rigid body of arbitrary shape and its cross-section in a plane Π.

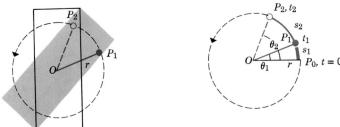

(b) The rotation as it appears in the plane Π.

(c) The circular motion of the small portion P.

Figure 10-3 A rigid body rotating about a fixed axis AA'.

space. This situation would be realized for a top with its axis vertical (figure 10-2b).

Figure 10-3a shows a rigid body of arbitrary shape rotating about a fixed axis AA'. P represents any small portion of this body with negligible size and a small mass m. As the body rotates P describes a circle contained in the plane Π, which is perpendicular to the axis AA'. The center of the circle O is a point on the axis AA' where it intersects the plane Π and the line OP is always perpendicular to AA'. Figure 10-3b shows the cross-section of the body in the plane Π as seen by an observer looking from A' in the direction $\overrightarrow{A'A}$. The two rectangles are consecutive positions of the outline of this cross-section as it rotates in a counterclockwise direction. P_1 and P_2 are corresponding positions of the small portion of the body we are directing our attention toward. The broken line is the circle described by P with its center O on the axis AA'.

This circle is reproduced again in figure 10-3c. P_0 is the position of P at time $t = 0$. P_1 is its position at a later time t_1 and P_2 its position at a still later time t_2. Between times 0 and t_1 the line OP has rotated through an angle θ_1, and the point P has traveled a distance s_1 measured along the perimeter of the circle. As we have already explained in section 5-3, the definition of the angle θ_1 measured in *radians* is

$$\theta_1 = \frac{s_1}{r} \qquad\qquad (10\text{-}2)$$

where r is the radius of the circle, which is the same thing as the perpendicular distance from P to the axis AA'. Similarly, the angle turned through between times 0 and t_2, when P has reached P_2, is

$$\theta_2 = \frac{s_2}{r} \qquad\qquad (10\text{-}3)$$

Since the body is rigid the angle turned through in any interval of time is the same for all parts of the body. The angles θ_1 and θ_2 might be regarded in a more general way as the angles turned through by the whole body after times t_1 and t_2. The **angular velocity** of the body at time t_1 is represented by ω (the Greek letter omega) and is defined as

Angular velocity,

$$\omega = \frac{(\theta_2 - \theta_1)}{(t_2 - t_1)} \qquad\qquad (10\text{-}4)$$

in the limit when $(t_2 - t_1)$ is sufficiently small.

This equation should be compared with equation 4-2, which defines velocity along a straight line. The angle θ in rotational motion is the analogue of the coordinate x in linear motion. The angular velocity ω, which is the rate of change of θ with time, is the analogue of the linear velocity v, which is the rate of change of x with time. Angular velocity has already been discussed in section 5-3 in the special case of uniform circular motion, for which ω is a constant independent of time.

ω is measured in radians per second or rad sec^{-1}. Rates of rotation are sometimes given in revolutions per minute or rpm. Since one complete rotation is 2π radians, rpm can clearly be converted into rad sec^{-1} by multiplying by the factor $2\pi/60$.

Suppose that the angular velocity is ω_1 at time t_1, but that it has changed to ω_2 at a slightly later time t_2. The angular acceleration is represented by α (the Greek letter alpha) and is defined as

Angular acceleration,

$$\alpha = \frac{(\omega_2 - \omega_1)}{(t_2 - t_1)} \qquad\qquad (10\text{-}5)$$

in the limit when $(t_2 - t_1)$ is sufficiently small.

Rotation

Angular acceleration is the rate of change of angular velocity with time and is measured in rad sec^{-2}. It is the analogue of the linear acceleration a, as defined by equation 4-7.

10-3 Moment of Inertia

Moment of inertia plays the same role in angular motion that is played by mass in linear motion. The small portion P of the body makes a contribution m to the total mass of the body. Its contribution to the moment of inertia about the axis AA' is mr^2. The total moment of inertia about this axis is obtained by adding together all terms such as mr^2 contributed by each small portion of the body. It is represented by the symbol I. The recipe for calculating I is

Moment of inertia

Divide the body into a large number of very small portions. Suppose that a typical portion has a mass m and is at a perpendicular distance r from the axis AA'. Then its contribution to the moment of inertia I is mr^2 and

I = Sum of all contributions such as mr^2 from all portions of the body   (10-6)

Notice that r depends upon the position of the axis of rotation AA'. The moment of inertia is not a fixed constant for a body; it is different for different axes of rotation. The units of moment of inertia are gm cm^2 or kg m^2.

Suppose that a couple τ acts on the rigid body and that the forces constituting this couple both lie in a plane perpendicular to the axis AA' and therefore accelerate the rotation about this axis (figure 10-4). With a little mathematical effort it can be shown that

$$\tau = I\alpha \qquad\qquad (10\text{-}7)$$

Torque = Moment of inertia $\times$ Angular acceleration

This is the rotational analogue of Newton's second law of motion for linear motion

$$F = Ma \qquad\qquad (10\text{-}8)$$

The torque τ is the analogue of the force F; the moment of inertia I is the analogue of the mass M; and the angular acceleration α is the analogue of the linear acceleration a. If no couple is applied to the body it continues to rotate with a constant angular velocity. This is the analogue of Newton's first law of motion.

10-4 Kinetic Energy of a Rotating Body

Since I is the analogue of M and ω is the analogue of v, one might expect that the kinetic energy of a rotating body would be $\frac{1}{2}I\omega^2$, by analogy with $\frac{1}{2}Mv^2$ for linear motion. This particular result is quite easily proved.

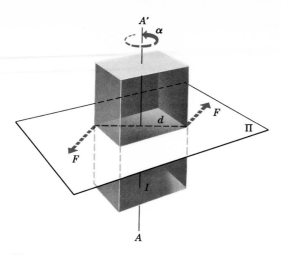

Figure 10-4 A couple $\tau = Fd$ produces an angular acceleration α.

Consider the small portion P of the body moving in a circle of radius r as shown in figure 10-3c. Its angular velocity is ω and, as was shown in section 5-3, its instantaneous linear velocity along the tangent to the circle is

$$v = \omega r \tag{10-9}$$

The kinetic energy of P is

$$\tfrac{1}{2}mv^2 = \tfrac{1}{2}m\omega^2 r^2$$
$$= \tfrac{1}{2}\omega^2(mr^2) \tag{10-10}$$

The kinetic energy of the whole body is therefore the sum of $\tfrac{1}{2}\omega^2(mr^2)$ for all portions of the body. But, since the angular velocity ω is the same for all parts of the body, this could also be written

$\tfrac{1}{2}\omega^2$(sum of mr^2 for all portions of the body)

However, the quantity inside the parentheses is the moment of inertia I of the body, and so

Kinetic energy of a rotating body
$$= \tfrac{1}{2}I\omega^2 \tag{10-11}$$

In linear motion, when a force F moves a body through a distance x, the work done is Fx. When a couple τ rotates a body through an angle θ,

Work done by a couple $= \tau\theta \tag{10-12}$

Rotation

As might be expected, this work is equal to the increase in the kinetic energy $\tfrac{1}{2}I\omega^2$, assuming that no other couples act on the body.

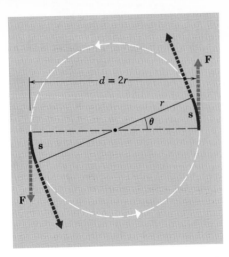

Figure 10-5   The work done by a couple.

Equation 10-12 can readily be proved with the help of figure 10-5. If the couple rotates the body through a *very small* angle θ, the displacement of each force is

$$s = r\theta \qquad (10\text{-}13)$$

Since the angle is very small, this displacement is along a straight line in almost the same direction as **F**, and also the direction of **F** changes very little during the displacement. The work done by a single force is therefore

$$Fs = Fr\theta \qquad (10\text{-}14)$$

(Do not confuse the angle of rotation θ with the angle also called θ in equation 9-33. The latter is approximately 0 in the present case.) The total work done by both forces is

$$\text{Work done} = 2Fr\theta \qquad (10\text{-}15)$$

However, $2r = d$, the distance between the two forces. Also the moment, or torque, of the couple is defined as

$$\tau = Fd$$
$$\tau = 2Fr \qquad (10\text{-}16)$$

Therefore

$$\text{Work done} = \tau\theta \qquad (10\text{-}17)$$

Although this has been proved for the case of a very small θ, it remains valid for a large θ if the couple τ remains constant throughout the rotation.

By definition, the angular momentum A of a particle of mass m moving with speed v in a circle of radius r has a magnitude mvr.

Angular momentum of a particle moving in a circle

$$A = mvr \tag{10-18}$$

Since

$$v = \omega r \tag{10-19}$$
$$A = mr^2\omega \tag{10-20}$$

A rotating rigid body may again be considered as a collection of small particles and its total angular momentum is

$$A = (\text{sum of } mr^2 \text{ for all portions of the body})\omega \tag{10-21}$$
$$= I\omega$$

Angular momentum of a rotating body

$$A = I\omega \tag{10-22}$$

Linear momentum p is Mv. I is the analogue of M and ω is the analogue of v. The angular momentum $I\omega$ is therefore the obvious rotational analogue of the linear momentum Mv.

Since

$$\tau = I\alpha$$
$$= I \times (\text{rate of change of } \omega \text{ with time})$$
$$= \text{Rate of change of } I\omega \text{ with time}$$

Torque = Rate of change of angular
$$\text{momentum with time} \tag{10-23}$$

This is clearly analogous to the fact that *force* is equal to the rate of change of *linear* momentum with time (equation 8-8).

If there is no external torque, the total angular momentum remains constant. This is the **law of conservation of angular momentum,** which can be demonstrated in the following way. A man stands on a slowly rotating turntable mounted on ball bearings that introduce very little friction (figure 10-6a). Initially his arms are outstretched and each hand holds a mass m at a distance R from the axis of rotation. The two masses make a contribution $2mR^2$ to the moment of inertia of the rotating system. He then bends his arms and pulls in the two masses until their distance from the axis has the much smaller value r (figure

Rotation

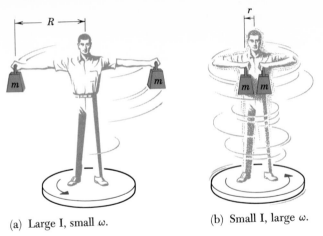

(a) Large I, small ω. (b) Small I, large ω.

Figure 10-6 A demonstration to illustrate the law of conservation of angular momentum.

10-6b). Their contribution to the moment of inertia is thereby reduced to $2mr^2$. In addition the mass of the man's arms is closer to the axis of rotation and its contribution to the moment of inertia is also reduced. Since the moment of the couple exerted on the system by the frictional forces due to the ball bearings is negligibly small, the angular momentum of the system cannot change. The product $I\omega$ must therefore remain constant. Pulling the two masses inward reduces the value of I. The value of ω must therefore increase. The result of the maneuver is that the man suddenly rotates much faster. If he extends his arms again, the rotation immediately slows down. This principle is used by a figure skater executing a spin (figure 10-7). If she starts her spin slowly with outstretched arms, she can increase her rate of rotation suddenly by folding her arms across her chest. She can slow down by extending her arms again.

Linear momentum is a vector with direction as well as magnitude.

Figure 10-7 A figure skater increasing her rate of rotation by reducing her moment of inertia.

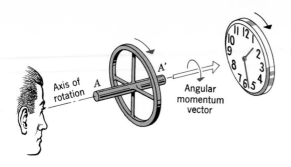

Figure 10-8 Representation of angular momentum by a vector.

Angular momentum can also be associated with a direction, the direction of the axis about which the body is rotating. The vector representing angular momentum therefore has a length $I\omega$ and a direction parallel to the axis of rotation. There are two such directions, opposite to one another, and it is necessary to establish a convention for choosing between them. The convention is that the body shall be seen to be rotating clockwise if we look along the axis of rotation in the direction of the arrowhead on the angular momentum vector. This is illustrated in figure 10-8.

The vectorial nature of angular momentum can be demonstrated in the following way. The man on the turntable holds a bicycle wheel which is rotating about a horizontal axis (figure 10-9a). Initially the turntable is not rotating and the total angular momentum of the system is represented by a *horizontal* vector. The vertical component of the angular momentum is zero. Since the ball bearings cannot exert an appreciable couple about a vertical axis, this vertical component must subsequently remain zero. If the man turns the wheel so that it has an angular momentum about a vertical axis (figure 10-9b), then the man and the turntable start to rotate in the opposite direction to the wheel.

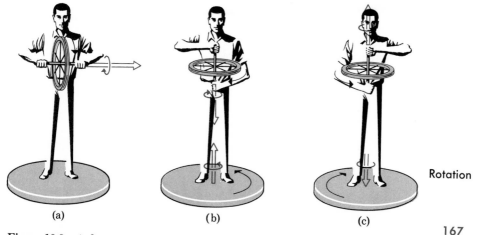

Rotation

(a)   (b)   (c)

Figure 10-9 A demonstration of the vectorial nature of angular momentum.

The vertical vector representing the angular momentum of the man and the turntable must be exactly equal and opposite to the vector representing the angular momentum of the wheel. If the man then turns the wheel through 180°, he reverses his own direction of rotation so that his angular momentum still cancels that of the wheel (figure 10-9c).

What happens to the component of angular momentum along a horizontal axis? In figure 10-9a there is such a horizontal component, whereas in parts (b) and (c) it seems to have disappeared. The explanation is that the ball bearings of the turntable can exert a couple about a horizontal axis. We cannot therefore apply the law of conservation of angular momentum to rotation of the system about a horizontal axis. However, if we consider a different system including the wheel, the man, the turntable, *and the earth*, we can apply the conservation law to this system. When the man turns the wheel from position (a) to position (b) it loses angular momentum in a horizontal direction. During the maneuver, the earth exerts on the turntable a couple about a horizontal axis which destroys the original horizontal angular momentum of the wheel. Meanwhile, in accordance with Newton's third law of motion, the turntable exerts on the earth an equal but opposite couple and the whole earth is made to rotate with an angular momentum equal to that lost by the wheel!

If a system contains several bodies rotating about different axes, the angular momentum vectors of the various bodies point in various directions. The total angular momentum of the system is the vector sum of the angular momentum vectors of the separate bodies. The law of conservation of angular momentum refers to this resultant vector.

Law of conservation of angular momentum

If no external torque acts on a system of bodies, the vector sum of the angular momentum vectors of the bodies must remain constant in direction and magnitude.

Example 10-5-1

A satellite of mass M_s is placed in a stable circular orbit of radius R around the earth. What is its angular momentum about an axis through the earth perpendicular to the plane of its orbit (figure 10-10)?

The motion of a satellite was discussed in section 7-5 and its velocity was shown to be

$$v = \sqrt{\frac{GM_e}{R}} \tag{10-24}$$

where M_e is the mass of the earth. In accordance with the definition of angular momentum given in equation 10-18, its angular momentum is

$$A = M_s v R \tag{10-25}$$

$$= M_s R \sqrt{\frac{GM_e}{R}}$$

$$= M_s \sqrt{GM_e R} \tag{10-26}$$

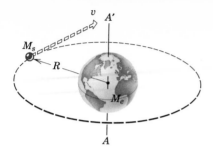

Figure 10-10 Illustrating example 10-5-1.

The previous analysis has tacitly assumed that the satellite rotates about the axis through the earth as though it were a solid body. If the reader will imagine a solid rod firmly fixed to the satellite and extending from it to the axis, he will realize that this implies that the satellite must always present the same face to the earth, as the moon is well known to do in its motion around the earth. In fact, however, the satellite may be spinning quite rapidly about an axis passing through its own center. We shall now show that this cannot possibly make any appreciable difference to our result, thus illustrating a type of argument which is very useful in physics. Sometimes, when we are completely ignorant of some aspect of a situation, such as the rotation of the satellite about an axis through its own center, we can nevertheless show that this aspect could never make any appreciable difference to some other quantity we are calculating, such as the angular momentum about an axis through the earth.

The moment of inertia of the satellite about an axis *through its own center* can be put equal to $M_s k^2$. This is merely a definition of the length k which is called the *radius of gyration*. Its exact value depends upon the shape of the satellite and the way in which its mass is distributed throughout it, but its magnitude is not very different from the observable "size" of the satellite. We shall take k to be about 1 m. If the satellite is spinning around its own axis with an angular velocity ω', this produces an angular momentum of

$$A' = M_s k^2 \omega' \tag{10-27}$$

We wish to show that this is necessarily small compared with the value previously calculated for the angular momentum about an axis through the earth (equation 10-26).
We wish to show that

$$\frac{A'}{A} = \frac{M_s k^2 \omega'}{M_s \sqrt{GM_e R}} = \frac{k^2 \omega'}{\sqrt{GM_e R}} \tag{10-28}$$

is very much less than one.

Rotation

The worst possible case is realized by making A as small as possible. This involves making R as small as possible, but R cannot possibly be smaller than the radius of the earth (6.37×10^6 m). The worst possible case is therefore given by

169

$$\frac{A'}{A} = \frac{1^2\omega'}{\sqrt{6.67 \times 10^{-11} \times 6.0 \times 10^{24} \times 6.37 \times 10^6}}$$

$$= 2 \times 10^{-11}\omega' \qquad (10\text{-}29)$$

For A' to be comparable with A, ω' would have to be about 10^{10} rad sec^{-1}, which corresponds to 10^{11} revolutions per minute. This is an enormous rate of rotation and long before it could be realized the material of the satellite would give way and the satellite would disintegrate. A value of 10^4 for ω' is just conceivable, in which case A' would represent a correction to A of only 2 parts in 10^7.

Example 10-5-2

A turntable of mass M and radius R is rotating with angular velocity ω_a on frictionless bearings. A spider of mass m falls vertically on to the rim of the turntable. What is the new angular velocity ω_b? The spider then slowly walks in toward the center of the turntable. What is the angular velocity ω_c when the spider is at a distance r from the center? Assume that, apart from a negligibly small inward velocity along the radius, the spider has no velocity relative to the turntable.

Consider the system which includes the turntable and the spider. Since the bearing is frictionless and the resistance of the air is to be ignored, no external couple acts on this system and its angular momentum must always remain the same. Just before the spider lands on the rim (figure 10-11a) the spider has no angular motion about the axis AA' and the angular momentum is contained entirely in the turntable. The moment of inertia of the turntable is known to be

$$I_t = \tfrac{1}{2}MR^2 \qquad (10\text{-}30)$$

The angular momentum is therefore

$$A = I_t\omega_a$$

$$= \tfrac{1}{2}MR^2\omega_a \qquad (10\text{-}31)$$

When the spider is standing on the rim (figure 10-11b) he takes up the motion of the turntable and both have an angular velocity ω_b. The moment of inertia of the spider is

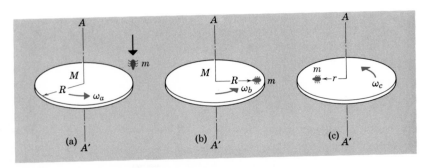

(a)   (b)   (c)

Figure 10-11   Illustrating example 10-5-2.

$$I_{sb} = mR^2 \tag{10-32}$$

The total moment of inertia of the system is

$$I_b = I_t + I_{sb}$$
$$= \tfrac{1}{2}MR^2 + mR^2$$
$$= \tfrac{1}{2}(M + 2m)R^2 \tag{10-33}$$

The angular momentum is

$$A = I_b\omega_b$$
$$= \tfrac{1}{2}(M + 2m)R^2\omega_b \tag{10-34}$$

Applying the law of conservation of angular momentum and equating the angular momenta before and after the spider lands,

$$\tfrac{1}{2}(M + 2m)R^2\omega_b = \tfrac{1}{2}MR^2\omega_a \tag{10-35}$$

$$\omega_b = \frac{M}{M + 2m}\omega_a \tag{10-36}$$

When the spider is at a distance r from the center (figure 10-11c), the angular velocity of both the spider and the turntable is ω_c. The moment of inertia of the spider is then

$$I_{sc} = mr^2 \tag{10-37}$$

The total moment of inertia is

$$I_c = I_t + I_{sc}$$
$$= \tfrac{1}{2}MR^2 + mr^2 \tag{10-38}$$

The angular momentum is

$$A = I_c\omega_c$$
$$= (\tfrac{1}{2}MR^2 + mr^2)\omega_c \tag{10-39}$$

Applying the law of conservation of angular momentum

$$(\tfrac{1}{2}MR^2 + mr^2)\omega_c = \tfrac{1}{2}MR^2\omega_a \tag{10-40}$$

$$\omega_c = \frac{\tfrac{1}{2}MR^2}{(\tfrac{1}{2}MR^2 + mr^2)}\omega_a$$

$$= \frac{\omega_a}{\left(1 + \dfrac{2mr^2}{MR^2}\right)} \tag{10-41}$$

Check that this agrees with equation 10-36 when $r = R$. Convince yourself that, as the spider walks inward and r decreases, the angular velocity increases. When the spider reaches the center and $r = 0$

$$\omega_c = \omega_a \text{ when } r = 0 \tag{10-42}$$

It is interesting to consider what happens to the kinetic energy in this problem. Suppose that the spider falls from so small a height that the kinetic energy acquired during the fall is negligible. Then the kinetic energy just before the spider lands is entirely due to the turntable and is

$$E_a = \tfrac{1}{2}I_t\omega_a^2 = \tfrac{1}{4}MR^2\omega_a^2 \tag{10-43}$$

After the spider lands the kinetic energy is

$$E_b = \tfrac{1}{2}(I_t + I_{sb})\omega_b^2 = \tfrac{1}{4}(M + 2m)R^2\omega_b^2 \tag{10-44}$$

Using the value of ω_b given by equation 10-36

$$E_b = \tfrac{1}{4}(M + 2m)R^2\left(\frac{M}{M+2m}\right)^2\omega_a^2$$

$$= \tfrac{1}{4}MR^2\omega_a^2\left(\frac{M}{M+2m}\right)$$

$$= E_a\left(\frac{M}{M+2m}\right) \tag{10-45}$$

This is clearly smaller than E_a and kinetic energy is therefore lost when the spider strikes the table. The reason is that initially the table moves faster than the spider's legs and the frictional force between the table and his legs eventually drags him up to the same speed as the table. The lost kinetic energy is turned into heat generated while the table is rubbing against the spider's legs.

When the spider is at a distance r from the center, the kinetic energy is

$$E_c = \tfrac{1}{2}(I_t + I_{sc})\omega_c^2$$

$$= \tfrac{1}{2}(\tfrac{1}{2}MR^2 + mr^2)\omega_c^2$$

$$= \tfrac{1}{4}MR^2\left(1 + \frac{2mr^2}{MR^2}\right)\omega_c^2 \tag{10-46}$$

From equations 10-36 and 10-41 it is easy to show that

$$\omega_c = \frac{(M+2m)}{M}\frac{\omega_b}{\left(1 + \dfrac{2mr^2}{MR^2}\right)} \tag{10-47}$$

So

$$E_c = \tfrac{1}{4}MR^2\left(1 + \frac{2mr^2}{MR^2}\right)\left(\frac{M+2m}{M}\right)^2\frac{\omega_b^2}{\left(1 + \dfrac{2mr^2}{MR^2}\right)^2}$$

$$= \tfrac{1}{4}(M + 2m)R^2\omega_b^2\frac{(M+2m)}{M\left(1 + \dfrac{2mr^2}{MR^2}\right)} \tag{10-48}$$

Substituting the value of E_b from equation 10-44

$$E_c = E_b\frac{(M+2m)}{\left(M + 2m\dfrac{r^2}{R^2}\right)} \tag{10-49}$$

Since r is less than R, it is clear that the expression in parentheses in the numerator is greater than the expression in parentheses in the denominator. E_c is therefore always greater than E_b.

As the spider walks in toward the center, the kinetic energy increases. Where does this extra kinetic energy come from? In order to walk on the turntable the spider must apply forces to it with his legs and these forces accelerate the turntable, increasing its kinetic energy. The energy used to do this was stored in the leg muscles of the spider and was originally derived from the chemical energy of the food eaten by the spider. If the spider were to walk outward toward the rim again, the kinetic energy of the turntable and spider would decrease. The reason for this is that as the spider walks outward he has to tense his leg muscles in order to hang on to the turntable, and the lost kinetic energy is put back into storage in his leg muscles.

10-6 Rotation in the Universe

Rotation plays a pervasive role in the universe on both the small and the large scales. On the small scale many of the fundamental particles, in particular electrons, protons, and neutrons are found to be spinning like tops. Moreover, the angular momentum of this spin is not arbitrary, but has a fixed value related to the fundamental physical constant of quantum mechanics, Planck's constant h (to be discussed in chapter 26).

Basic Unit of Angular Momentum

$$= \frac{h}{2\pi}$$

$$= 1.0546 \times 10^{-27} \text{ gm cm}^2 \text{ sec}^{-1} \tag{10-50}$$

An electron, a proton, or a neutron always spins with precisely one-*half* a unit of angular momentum. Other particles, such as the particle of light, the photon, spin with exactly one whole unit of angular momentum. A very important theoretical distinction can be made between the particles called **bosons**, which spin with an integral number of units of angular momentum, and **fermions**, which spin with a half-integral number of units of angular momentum.

When an electron is inside an atom, it revolves around the nucleus and, in addition to its intrinsic spin of one half a unit, its orbital motion has an angular momentum which is always an exact integral number of basic units. When atoms come together to form a molecule, the molecule as a whole rotates with an angular momentum which is again an exact integral number of basic units.

We shall return to these matters and discuss them in greater detail later in this book.

Rotation

On an astronomical scale the earth is well known to be rotating about its north-south axis once a day, producing a velocity at the equator of about

Figure 10-12 The "Whirlpool" galaxy.

900 mph. The earth revolves around the sun with an orbital velocity of about 70,000 mph. The sun itself is spinning at a rate which varies with latitude on the sun but corresponds to about one revolution per month. There is evidence that most stars are rotating in a similar way. Many stars join together in pairs with the two members of the pair rotating around one another.

Our sun is a member of a disk-shaped group of some one hundred billion stars, called the Galaxy with a capital G, and this is also rotating. Although it takes two hundred million years to go around once, this nevertheless imparts to the sun, which is about two thirds of the way out from the center of the disk, an enormous velocity of about one half a million miles per hour. The universe contains many such groups of stars, called galaxies with a little g, and there is good evidence that most of them are rotating. Figure 10-12 makes this easily believable.

The rotation of stars shows up dramatically in the recently discovered **pulsars.** A pulsar is an unusual star from which we receive on earth short bursts (pulses) of radio waves repeated at regular intervals. For a particular pulsar the time interval between pulses is remarkably constant and lies in the range from $\frac{1}{30}$ second up to several seconds. The currently accepted explanation is that the star has developed a novel mechanism which is not yet fully understood, but the end result of which is the emission of a highly directional beam of radio waves, rather like the light beam from a search-light. This beam rotates with the star, and the radio signal is received by us each time the beam sweeps over the earth (figure 10-13).

For the most rapidly blinking pulsar the radio pulses arrive 30 times a second and so the star must be making 30 complete revolutions every second. This is very fast indeed compared with one revolution per month

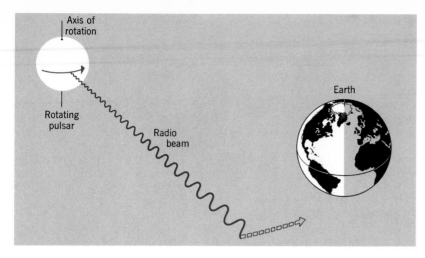

Figure 10-13 The rotating radio "searchlight" beam from a pulsar. The earth receives a pulse of radio waves once during each revolution when the beam sweeps across it.

for a star like the sun and the intriguing question is how the pulsar comes to be rotating so rapidly. Although the details are still controversial the answer is understood in principle and it provides an interesting example of the law of conservation of angular momentum.

A star tends to collapse under its own weight because its inner regions exert strong inward gravitational forces on its outer regions. This is normally prevented because the heat generated at the center of the star "blows up" the star for somewhat the same reason that a rubber balloon expands if the air inside it warms up and therefore exerts a greater pressure. (This will be explained more fully in chapter 11.) Eventually, however, the star exhausts its fuel and **gravitational collapse** sets in. If the star is massive enough, nothing can stop this collapse until the star has shrunk to a diameter of only a few miles and has been compressed to an enormously high density. While it is shrinking it is not subject to any external couple, and so its angular momentum remains constant. It speeds up for exactly the reasons applied to the man on the turntable of figure 10-6 or the skater of figure 10-7.

Referring to figure 10-14a, we see that if the star is a uniform sphere with a mass M and a radius R, its moment of inertia can be shown to be

$$I = \tfrac{2}{5}MR^2 \tag{10-51}$$

If its angular velocity is ω, its angular momentum is

$$A = I\omega$$
$$= \tfrac{2}{5}MR^2\omega \tag{10-52}$$

Rotation

After it has shrunk to a radius r (figure 10-14b) and speeded up to an angular velocity ω', its angular momentum is

175

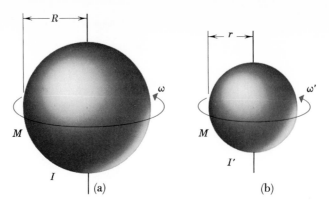

Figure 10-14 A collapsing star.

$$A' = \tfrac{2}{5}Mr^2\omega' \tag{10-53}$$

Since the numerical value of the angular momentum does not change,

$$\tfrac{2}{5}Mr^2\omega' = \tfrac{2}{5}MR^2\omega \tag{10-54}$$

Whence

$$r^2\omega' = R^2\omega \tag{10-55}$$

or

$$\frac{\omega'}{\omega} = \frac{R^2}{r^2} \tag{10-56}$$

The angular velocity is inversely proportional to the square of the radius of the star. The radius of a star like the sun is about 10^{11} cm, whereas the radius of the pulsar is about 10^6 cm. The radius has therefore shrunk by a factor of about 10^{-5}, and the angular velocity has consequently increased by a factor of about $(10^5)^2$, or 10^{10}. Since the sun rotates once in about 3×10^6 seconds, a pulsar formed by gravitational collapse of a star similar to the sun might be expected to rotate once in about 3×10^{-4} sec, or about 3000 times every second!

The point has been made, but we have obtained too high a rate of rotation because the arguments are crude and there are many complications. A star which will eventually collapse right down to a pulsar must probably be more massive than the sun and, of course, its initial angular velocity may not be exactly that of the sun. During the collapse the star is likely to explode, throwing off its outer layers and temporarily becoming one of the brightest stars in the sky, a **supernova.** Many pulsars are in fact found in the locations of past supernovae and are observed to be surrounded by the debris of the explosion. Also, pulsars are observed to be gradually slowing down and so their angular velocities were appreciably greater when they were first formed. Nevertheless, although the detailed theory will undoubtedly be very complicated, a simple argument has enabled us to

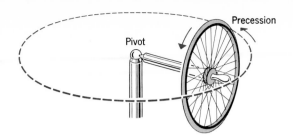

Figure 10-15 Precession of a gyroscope.

understand why very high rates of rotation might be a natural occurrence in dying stars.

10-7 Precession of a Gyroscope

Figure 10-15 illustrates a common method of demonstrating the precession of a gyroscope. The bicycle wheel is spinning rapidly about a horizontal axis. It is supported by a pivot which lies on the axis of rotation but at some distance from the center of the wheel. The axis of rotation remains horizontal but slowly changes its direction so that, in addition to the rapid rotation about the horizontal axis, there is a slow rotation about a vertical axis. Sometimes there is also a pronounced up and down wobble of the horizontal axis of rotation, but a skilled demonstrator can start the motion in such a way that this wobble is inconspicuous. The rotating bicycle wheel is the **gyroscope.** The slow change in direction of the main horizontal axis of rotation is called **precession.** The wobble is called **nutation.**

This demonstration is quite impressive. It is unlikely that the precession would be predicted by an observer who had never seen it before and, even after it has been observed, it has no obvious explanation. However, it can be explained plausibly in terms of the concept of angular momentum as a vector. A completely satisfactory explanation requires advanced mathematics.

Suppose that the wheel is *not rotating,* but is released from rest with its axle horizontal (figure 10-16). Under these circumstances, the wheel merely falls down, the center of the wheel describing a circle in a vertical plane about the pivot as center. This is because the weight of the wheel **W** acts downward through its center. Although the pivot is able to provide an upward supporting force **R**, this is displaced sideways from the weight and the two forces constitute a couple. With Cartesian axes in the directions shown in the figure, this couple tries to turn the wheel about a horizontal axis parallel to the y direction.

The moment of the couple τ is equal to the rate of change of angular momentum with time (equation 10-23). In a very short time interval $(t_2 - t_1)$, the couple therefore produces a change in angular momentum of magnitude

$$A_{extra} = \tau(t_2 - t_1) \qquad (10\text{-}57)$$

Rotation

177

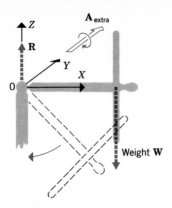

Figure 10-16   Vertical fall of a non-rotating wheel.

It is important to realize that, since the couple produces a turning effect about the y axis, this small change in angular momentum is represented by a vector in the y direction. It corresponds to the circular falling motion of the center of the wheel in a plane perpendicular to the y axis. Since the motion is clockwise when we look in the positive y direction, A_{extra} is positive.

Now return to the case when the wheel is spinning rapidly about a horizontal axis which, at the initial time t_1, is parallel to the x axis. If the angular velocity is ω and the relevant moment of inertia is I, the initial angular momentum at time t_1 is represented by a vector $\mathbf{A}_1$ in the positive x direction of magnitude:

$$A_1 = I\omega \tag{10-58}$$

The weight of the wheel still produces a torque trying to turn the wheel about the y axis and, in the time interval $(t_2 - t_1)$, this still produces an angular momentum A_{extra} in the y direction given by equation 10-57. Therefore, at the end of the time interval (at time t_2), the new angular momentum $\mathbf{A}_2$ is the vector sum of the original angular momentum $\mathbf{A}_1$ along the x axis and the added angular momentum $\mathbf{A}_{extra}$ along the y axis:

$$\mathbf{A}_2 = \mathbf{A}_1 + \mathbf{A}_{extra} \tag{10-59}$$

This vector addition is shown in figure 10-17, and it is clear that the new angular momentum $\mathbf{A}_2$ is still in a horizontal direction but has rotated through a small angle θ. This means that the wheel is still rotating about a horizontal axis. The center of the wheel has not fallen vertically, as it did when the wheel was not rotating, but has precessed in a horizontal direction.

If $(t_2 - t_1)$ is very short and θ consequently very small, the value of θ in radians is very nearly equal to tan θ, so

$$\theta = \frac{A_{extra}}{A_1} \tag{10-60}$$

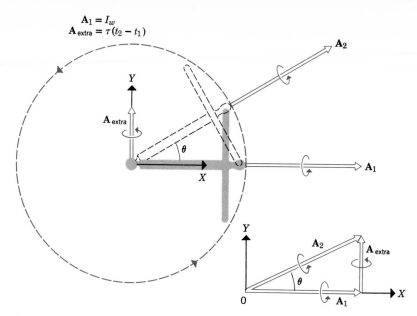

Figure 10-17 The vector diagram (this is a "bird's-eye view" of the gyroscope looking down the z axis from above).

Making use of equations 10-57 and 10-58,

$$\theta = \frac{\tau(t_2 - t_1)}{I\omega} \tag{10-61}$$

The angular velocity of the *precessional* motion about the vertical z axis is

$$\Omega = \frac{\theta}{(t_2 - t_1)} \tag{10-62}$$

$$\Omega = \frac{\tau}{I\omega} \tag{10-63}$$

The explanation of the precession of a gyroscope is therefore somewhat similar to the explanation given in section 7-5 of why a satellite remains in orbit. It was pointed out that the satellite must have an acceleration

$$a_s = \frac{GM_e}{R^2} \tag{10-64}$$

in the earth's gravitational field. If it starts from rest, it can achieve this by falling directly toward the center of the earth. Alternatively, it can achieve this acceleration by continually moving in a circular orbit with exactly the correct speed so that

$$\frac{v^2}{R} = \frac{GM_e}{R^2} \tag{10-65}$$

Rotation

179

Similarly, there is a torque τ acting on the gyroscope which requires that in a time $(t_2 - t_1)$ it must acquire an extra angular momentum

$$A_{extra} = \tau(t_2 - t_1) \tag{10-66}$$

about a horizontal axis (the y axis) perpendicular to the line joining the pivot to the center of the wheel. If the wheel is not rotating, it can achieve this by falling in a vertical plane. Alternatively, if the wheel already has a horizontal angular momentum A_1 in the x direction, it can achieve the same result if this angular momentum vector precesses through a small angle θ and thereby develops a small component A_{extra} in the y direction. However, it must precess at exactly the correct rate, as given by equation 10-63.

10-8 The Analogy between Rotation and Linear Motion

Table 10-1 summarizes the analogy between various quantities that are relevant to linear motion and corresponding quantities that play a similar role in rotational motion. This table will also serve to remind the reader of many of the important laws and concepts that have been introduced during our discussion of Newtonian mechanics.

Table 10-1

Linear Motion	Rotational Motion
Distance x	Angle θ
Velocity v	Angular velocity ω
Acceleration a	Angular acceleration α
Mass M	Moment of inertia I $I = $ Sum of mr^2
Force F	Moment, or torque, of a couple τ
$F = Ma$	$\tau = I\alpha$
Kinetic energy $= \frac{1}{2}Mv^2$	Kinetic energy $= \frac{1}{2}I\omega^2$
Work done $= Fx$	Work done $= \tau\theta$
Momentum, $p = Mv$	Angular momentum, $A = I\omega$
Force = Rate of change of momentum with time	Torque = Rate of change of angular momentum with time
Law of Conservation of Momentum, if there is no resultant external force	Law of Conservation of Angular Momentum, if there is no resultant external couple.

QUESTIONS

1. If a body is rotating, must there necessarily be a couple with a non-zero moment acting on it?

2. Pursuing the analogy between linear motion and angular motion show that

$$\omega = \omega_0 + \alpha t$$
$$\theta = \omega_0 t + \tfrac{1}{2}\alpha t^2$$
$$\omega^2 = \omega_0{}^2 + 2\alpha\theta$$

3. Which of the following numbers is nearest to the CGS value in gm cm^2 of the moment of inertia of a dime about an axis through its center perpendicular to its faces? (a) 10^5, (b) 10^2, (c) 10^{-1}, (d) 10^{-3}, (e) 10^{-7}.

4. When a figure skater is executing a spin, is her moment of inertia greater or less than 100 kg m^2?

5. If the angular momentum of the satellite in figure 10-10 is represented by a vector, in which direction does this vector point?

6. If the angular momentum of the earth due to its daily rotation is represented by a vector, does this vector point from the south toward the north or from the north toward the south?

7. Can you think of a simple argument to obtain equation 10-42 without having to derive equation 10-41 first?

8. Why is it difficult to design a helicopter with only one propeller?

9. How can the law of conservation of angular momentum be applied to an automobile accelerating along a straight horizontal highway?

10. When you switch on an electric fan, where does its angular momentum come from?

11. Consider the demonstration of figure 10-6 from the point of view of conservation of *energy*.

12. Two equal masses are connected by a massless string which passes over a pulley, as in the diagram. Discuss the motion from the point of view of conservation of energy, linear momentum and angular momentum. Consider two cases (a) the bearing of the pulley is frictionless and the frictional drag of the air can be ignored, (b) friction cannot be ignored. How would your conclusions be modified if one mass were larger than the other?

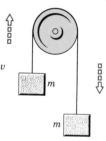

Rotation

Question 10-12

13. Two identical gyroscopes are connected back to back, as in the diagram, and are pivoted at a point midway between them. Will the system precess if the two gyroscopes are given exactly equal angular velocities (*a*) in the same direction, (*b*) in opposite directions, as in the diagram?

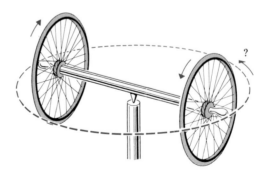

Question 10-13

PROBLEMS

The moment of inertia of a sphere of mass M and radius R about an axis through its center is $\frac{2}{5}MR^2$. The moment of inertia of a circular disk of mass M and radius R about an axis through its center perpendicular to its flat faces is $\frac{1}{2}MR^2$.

A

1. Dick pulls vertically upward on one end of a teeter-totter with a force of 200 newtons, and Tom pushes vertically downward on the other end also with a force of 200 newtons. If the board is 4 meters long and is at an angle of 30° to the horizontal, what is the moment of the couple exerted by the two boys?

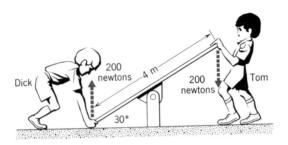

Problem 10-1

2. A mass of 15 gm moves in a circle of radius 7 cm with a speed of 35 cm/sec. What is (*a*) the angular velocity, (*b*) the moment of inertia, (*c*) the angular momentum, (*d*) the kinetic energy? (*c*) and (*d*) may be calculated by using either the speed v or the angular velocity ω. Check that you obtain the same answer in both cases.

3. A mass of 20 kg is falling vertically at the north pole with a speed of 200 cm/sec. What is its angular momentum about a point on the equator?

4. Referring to the diagram for this problem, the two small 3 gm balls are connected by a stout wire of negligibly small mass. A large 2 kg ball exerts a gravitational force of attraction on a nearby small ball. The gravitational force which it exerts on the small ball furthest away from it can be neglected. Calculate the couple on the system consisting of the two small balls and the wire joining them.

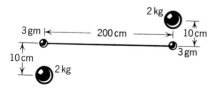

Problem 10-4

5. Four balls, each of mass 20 gm, are located at the four corners of a square with a side of length 40 cm. Calculate the total moment of inertia about an axis through the center of the square perpendicular to the plane of the square. The diameter of a ball can be considered to be very small compared with 40 cm.

6. A flywheel has a moment of inertia of 2×10^4 gm cm² about its axis of rotation. When its angular velocity is 2.5 rad sec⁻¹, what is its kinetic energy?

7. What is the angular momentum of the flywheel in the previous question?

8. Calculate the moment of the couple which must be applied to a wheel with a moment of inertia of 2.5 kg m² to accelerate it from rest to 1000 rpm in 5 sec.

9. A phonograph record is rotating at $33\frac{1}{3}$ rpm. A fly of mass 0.2 gm is stuck to the record at a distance of 8 cm from the center. What is the angular momentum of the fly?

10. A racing car of mass 800 kg is traveling around a circular track of radius 200 m with a steady speed of 60 mph. Find its angular momentum in MKS and CGS units.

B

11. An electron moves in a circle of radius 4 cm. What must its speed be if its angular momentum about the center of the circle is $h/2\pi = 1.05 \times 10^{-27}$ CGS units?

12. What is the angular momentum about an axis through the north and south poles of a 1 kg mass resting on the ground at the equator?

13. What is the kinetic energy of the earth due to its daily rotation? Take the earth to be a perfect sphere of uniform density.

14. What is the angular momentum of the earth that results from its daily rotation? Compare the answer with the angular momentum of

the moon that is caused by its orbital motion around the earth. The moon takes 27.3 days to orbit once around the earth.

15. Assuming that time can be measured with an accuracy of 1 part in 10^{11}, what mass would have to be transferred from the north pole to the equator to produce a measurable difference in the length of the day? Assuming a mean density of 3 gm/cm^3 for the crust of the earth, convert this mass into a volume. Calculate the length in miles of the side of a cube with the same volume. Is this a feasible experiment?

16. The hydrogen molecule, H_2, consists of two hydrogen atoms, each having a mass of 1.67×10^{-24} gm, separated by a distance of 7.4×10^{-9} cm. What is its moment of inertia about an axis midway between the two nuclei and perpendicular to the line joining them? Assume that all the mass is concentrated in the nucleus, which has a negligibly small diameter. If the angular momentum about this axis is $h/2\pi = 1.05 \times 10^{-27}$ gm cm^2 sec^{-1}, what is the kinetic energy of rotation?

C

17. As the tides sweep across the surface of the earth, they exert a frictional drag which gradually slows down the rotation of the earth. How large a couple do they exert if the length of the day increases by 0.0016 sec every century? In order to conserve angular momentum under these circumstances, should the moon slowly increase or decrease the radius of its orbit about the earth? (The moon revolves about the earth in the same direction that the earth rotates on its axis.)

18. Suppose that the eventual effect of the tides is to make the time for one rotation of the earth exactly equal to the time for one revolution of the moon about the earth. While this is happening, the earth-moon system is to be considered free from any external influence. Calculate the length of the day under these circumstances, assuming the second to be defined in terms of an atomic clock (section 2-4) rather than in terms of the rotation of the earth. What would be the distance of the moon from the earth? (Hint: First convince yourself that, when the length of the day is equal to the length of the month, the angular momentum of rotation of the earth is negligibly small compared with the orbital angular momentum of the moon).

19. The diagram shows a flywheel (mass M, radius R, and moment of inertia $\frac{1}{2}MR^2$) which can rotate about a horizontal axis on a frictionless bearing. One end of a rope is attached to a point on its rim. The rope is wrapped around the wheel and its other end supports a hanging mass m. If the angular velocity of the wheel is ω, convince yourself that the downward velocity of m is $v = \omega R$. Using the law of conservation of energy, show that the velocity of m, after it has

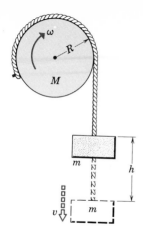

Problem 10-19

fallen through a height h from rest, is

$$v = \sqrt{\frac{4mgh}{2m + M}}$$

Neglect the mass and thickness of the rope.

20. A hoop of mass m and radius r rolls down a bank. Using the law of conservation of energy, find its velocity when it is a vertical distance h below the point from which it started at rest. Work done against frictional forces is to be considered negligible. The condition for rolling is that the point of the hoop touching the ground is instantaneously at rest.

21. A carousel with a moment of inertia of 10,000 kg m² is rotating freely with an angular velocity of 0.500 rad sec⁻¹. A chimpanzee of mass 30 kg is standing on the edge of the platform at a distance of 2 m from the axis of rotation. If the chimpanzee jumps vertically upward into the arms of a spectator standing on the ground by the side of the platform, what is the new angular velocity of the carousel? (Assume that the velocity of the chimpanzee relative to the platform has a zero horizontal component at the instant his feet leave the platform.)

22. A turntable is rotating with angular velocity ω_0 on frictionless bearings. A man standing at a distance R from the axis drops a lump of clay from rest, relative to himself, at a height h above the turntable. Upon striking the turntable, the clay sticks to it at the point of impact. The mass of the lump of clay is m. The combined moment of inertia of the man and the turntable, excluding the clay, is I. Find the angular velocity of the turntable (*a*) immediately after the lump of clay is released by the man (*b*) after the clay has stuck to the turntable.

23. Repeat problem 22 with the modification that the man throws the lump of clay with a horizontal velocity v_r which, in his frame of reference, is radially outward.

Rotation

185

C

ATOMS
AND
HEAT

11

The
Ideal
Monatomic
Gas

11-1 Atoms and Chemistry

The laws of mechanics were discovered in the seventeenth century. The eighteenth century was devoted mainly to developing the mathematical techniques that enabled these laws to be applied to various problems. This was the age of the great mathematical physicists d'Alembert, Bernoulli, Euler, Lagrange, Laplace, and many others. They were interested in such problems as the detailed behavior of the solar system, the motion of a spinning top, and the flow of liquids. The mathematics was very difficult and the success of their solutions produced overwhelming confidence in the validity of Newtonian mechanics. However, all these successes were concerned with the behavior of macroscopic bodies, and many important questions were left unanswered. The laws of mechanics predicted that a lead ball would fall to the earth at the same rate as an iron ball, but the difference between lead and iron was not understood and it was not known why one cubic centimeter of lead has a mass of 11.37 grams, whereas one cubic centimeter of iron has a mass of 7.87 grams. The first major step towards providing answers to questions of this kind was the emergence of the atomic theory of matter at the beginning of the nineteenth century.

The idea that matter is composed of atoms was proposed by the Greek philosophers Empedocles and Democritus between 500 and 400 B.C., but the establishment of a scientific theory involves more than a bright idea. It is necessary to show that this idea is useful in explaining known phenomena and in predicting new phenomena. In the case of atoms the initial evidence came mainly from chemistry. The evidence was first presented by Dalton at the beginning of the nineteenth century and the situation was finally clarified by Cannizzaro in the middle of the century. (This interval of fifty years between the presentation of an important idea and its proper understanding is a striking illustration of the slow progress of science in the past compared with its present explosive development.) Once the chemical evidence had convinced scientists of the existence of atoms, the atomic theory was found to provide a very successful explanation of the properties of gases and later of solids and liquids also. In this chapter we shall give a very brief review of the chemical evidence and then proceed to discuss the properties of gases.

Hydrogen and oxygen combine to form water. Hydrogen is a **chemical element** and is therefore composed of atoms that are all of the same kind and are called hydrogen atoms. Actually the hydrogen atoms join together in pairs to form hydrogen **molecules.** Since a hydrogen molecule contains two atoms it is said to be *diatomic.* Oxygen is also a chemical element and is also diatomic, but its molecule is composed of two oxygen atoms that are different in many respects from hydrogen atoms. In particular an oxygen atom is approximately 16 times more massive than a hydrogen atom. When hydrogen and oxygen react chemically, two hydrogen molecules come together with one oxygen molecule and the atoms are reshuffled to form two water molecules, each of which contains one oxygen atom and two hydrogen atoms (figure 11-1). Since the molecule of water contains atoms of more than one kind, water is a **chemical compound.** Four atoms of hydrogen combine with two atoms of oxygen and the mass of the oxygen atom is 16 times that of the hydrogen atom. This provides a ready explanation of the experimental observation that a given mass of hydrogen combines with 8 times its mass of oxygen to give 9 times its mass of water.

At the time of this writing, 104 different chemical elements are known, corresponding to 104 different kinds of atoms. A molecule is obtained by selecting some of these atoms and joining them together in some spatial arrangement. The number of possible selections is large and

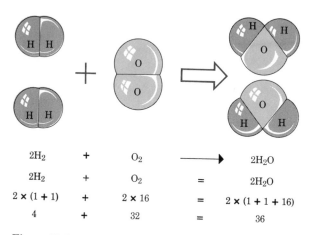

$2H_2$	+	O_2	$\longrightarrow$	$2H_2O$
$2H_2$	+	O_2	=	$2H_2O$
$2 \times (1 + 1)$	+	2×16	=	$2 \times (1 + 1 + 16)$
4	+	32	=	36

The Ideal
Monatomic
Gas

Figure 11-1 Two molecules of hydrogen combine with one molecule of oxygen to form two molecules of water.

Figure 11-2 A model of a sugar molecule. This particular sugar molecule is composed of six carbon atoms, twelve hydrogen atoms, and six oxygen atoms. Its formula is therefore $C_6H_{12}O_6$. The carbon atoms are represented by the black balls, the hydrogen atoms by the white balls, and the oxygen atoms by the gray balls. (Courtesy of the Physical Science Study Committee and D. C. Heath and Co.)

the number of possible spatial arrangements is large, thus explaining the large variety of chemical compounds in nature. Figure 11-2 shows a model of a molecule of sugar.

A chemical reaction occurs when several molecules come together and reshuffle their atoms to produce different molecules. As in the case of the combination of hydrogen and oxygen to form water, if it is known which atoms are involved in the composition of each molecule and if the relative masses of the atoms are known, it is easy to calculate the proportions by mass in which the reactants must be mixed in order to react completely and to calculate the mass of each product formed. It was this success of the atomic theory in predicting the masses involved in chemical reactions which first convinced chemists and physicists of the existence of atoms.

11-2 Attempts To See Atoms

Atoms are very small. The diameter of an atom is only about 10^{-8} cm, which means that if atoms were placed side by side along a line touching one another, it would take about one hundred million of them to form a line only one centimeter long.

An atom is too small to be seen directly, even with a microscope. A fundamental rule of microscopy is that two objects cannot be distinguished from one another if their distance apart is much less than the wavelength, λ, of the light illuminating them. (The wavelength was defined in section 1-1. See also figure 1-1.) An object smaller than λ is seen as a blur of light with a diameter of about λ. This is a consequence of the diffraction phenomena which will be discussed in Chapter 21. Since the smallest wavelength of visible light is 4×10^{-5} cm and the diameter of an atom is about 10^{-8} cm, we obviously cannot hope to look at an

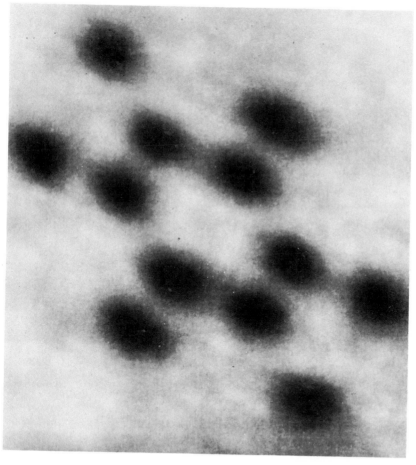

Figure 11-3 A "photograph" of a hexamethylbenzene molecule. The chemical formula is $C_6(CH_3)_6$ or $C_{12}H_{18}$. The dark black blobs are the twelve carbon atoms; the hydrogen atoms do not show up clearly. This is not a photograph of an individual molecule. It was obtained by gathering x-rays from a large number of molecules and electronically converting the x-ray signal into a visible signal. (From Jay Orear, *Fundamental Physics*, John Wiley & Sons, Inc. The photograph was obtained by Dr. M. L. Huggins of the Kodak Research Laboratory.)

The Ideal Monatomic Gas

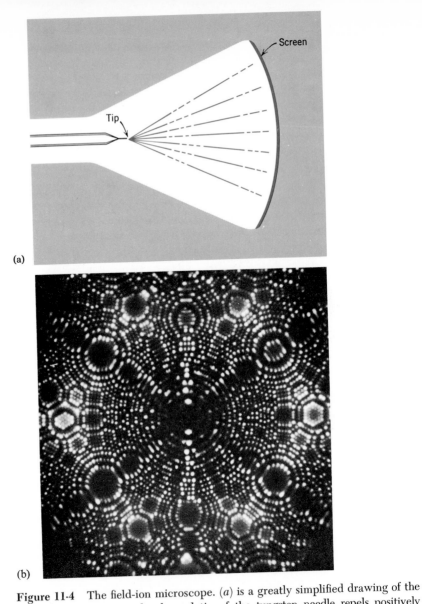

(a)

(b)

Figure 11-4 The field-ion microscope. (*a*) is a greatly simplified drawing of the microscope. The positively charged tip of the tungsten needle repels positively charged helium atoms (helium ions) in its immediate vicinity. More ions leave from a point near an atom on the surface than from a point between two atoms. The ions travel radially outward in straight lines to the fluorescent screen, where they produce a pattern of illumination which reproduces the arrangement of atoms on the tip. (*b*) is a photograph of the fluorescent screen and each bright dot is caused by an atom on the tip (or possibly, in some cases, by a group of two or three atoms). (*c*) is a model of the tip of the needle made by stacking cork balls, each of which represents a tungsten atom. The simulated needle points directly toward the camera, and its tip is in the center of the photograph. The cork balls that represent those atoms that are particularly effective in repelling ions were coated with luminescent paint and are seen in part (*d*), which was obtained by photographing the model (*c*) in the dark. The similarity between (*d*) and (*b*) is impressive. To illustrate how sharp the tip of the needle is, a magnified photograph of it is shown in the lower part of (*e*) and is compared with the tip of an ordinary pin. (Photographs courtesy of Dr. Erwin Muller. (*c*) Photograph by Paul Weller.)

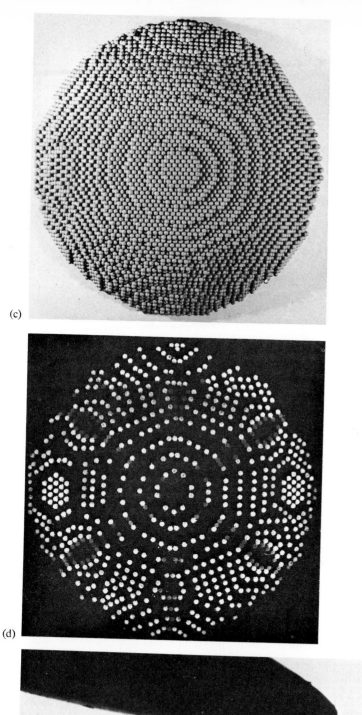

(c)

(d)

(e)

atom and examine its structure. (The student who is still not familiar with powers of ten should pause and consider that 4×10^{-5} is 4,000 times as big as 10^{-8}.)

What about the lesser prospect of seeing an individual atom as a blur of light with a diameter of about 4×10^{-5} cm? This is not possible for an atom in a solid or a liquid, since the atoms in solids and liquids are separated from one another by distances of about 3×10^{-8} cm. The individual blurs would overlap and obscure one another. We must look at a collection of atoms whose distances apart are in excess of 4×10^{-5} cm. This condition is satisfied by a gas at a very low pressure. Unfortunately, the atoms of a gas are moving in straight lines with very high velocities. At room temperature an atom stays inside a region of size 4×10^{-5} cm for only about 2×10^{-9} sec! The atoms can be slowed down by lowering the temperature, but at the very lowest temperature technically attainable a suitable atom can be made to stay inside a region of size 4×10^{-5} cm for only 2×10^{-8} sec. Therefore, if we attempted to look at an individual atom of a gas using the best possible design of optical microscope, the blur would move around too rapidly to be followed with the naked eye.

X-rays are similar to light, but they have much shorter wavelengths that can be much smaller than the diameter of an atom. It is therefore possible to design an x-ray "microscope" which will reveal the structure of the atom. It is still not possible to see an individual atom, but if the x-rays are scattered by an orderly array of atoms in a solid, it is possible to deduce from the detailed nature of the scattering the spatial arrangement of the atoms in the solid. Figure 11-3 is a photograph of a complicated organic molecule reconstructed from such an x-ray experiment.

One of the most impressive ways of "seeing" atoms is the *field ion microscope* invented by E. W. Muller (figure 11-4). The explanation of its action may not be completely understood by the reader until he has read some of the later chapters in this book, but the picture of atoms on the surface of the metal tip is easily appreciated (figure 11-4b).

Nuclear physicists have invented several techniques for detecting the presence of an atomic *nucleus*, even though the nucleus cannot be seen directly. In the cloud chamber photograph of figure 11-5, each white track is a line of water droplets resulting from the passage of an electrically charged nucleus.

To illustrate the vast number of atoms in even a small particle of matter, let us suppose that a microscope had been invented that could distinguish the individual atoms. Suppose that you attempted to use this microscope to count the atoms in a rain drop only 0.1 cm in diameter. The psychological limit on the rate of counting could hardly be in excess of 5 per second. At this rate, it would take 3×10^{11} years to count all the atoms! Many astronomers believe that the universe was created about 3×10^{10} years ago. Therefore, if you had started to count these atoms the very instant the universe came into being, you would still be far from having completed the job.

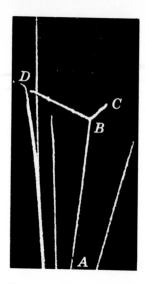

Figure 11-5 A cloud chamber photograph of the tracks of atomic nuclei. Each white line is a trail of small water droplets left behind along the track of a nucleus. In the case of the Y-shaped track, a helium nucleus entered at A and struck a stationary oxygen nucleus at B. The oxygen nucleus was projected along BC and the helium nucleus recoiled along BD.

11-3   Atomic Masses and Avogadro's Number

Atoms contain protons, neutrons, and electrons. The mass of the neutron is very slightly larger than the mass of the proton, and the mass of the electron is very small compared with either. The lightest atom is hydrogen and its nucleus is just a single proton. An atom whose nucleus contains p protons and n neutrons is therefore approximately $(p + n)$ times more massive than the hydrogen atom. However, although the ratio of the mass of the atom to the mass of the hydrogen atom is very nearly an integer, it is not quite so. The small departure from an exactly integral ratio is easily measurable and in Chapter 30 we shall see that it has an important physical significance.

The choice of the unit of atomic mass is complicated, involving experimental convenience, the relative accuracy of different atomic mass determinations, and the differing requirements of chemists and physicists. Until quite recently the atomic mass unit was defined in such a way that the commonest isotope of oxygen had an atomic mass of exactly 16 units. This isotope of oxygen has 8 protons and 8 neutrons in its nucleus and its atom is therefore approximately, but not exactly, 16 times more massive than the hydrogen atom. It is denoted by the symbol $_8O^{16}$. The subscript 8 is the number of protons in the nucleus, and the superscript 16 is the total number of protons and neutrons in the nucleus.

However, at an international meeting held in 1960 it was decided that the definition should be changed so that the commonest isotope of carbon, $_6C^{12}$, which has six protons and six neutrons in its nucleus, should have an atomic mass of exactly 12. The atomic mass of oxygen is then 15.995 and the atomic mass of hydrogen is 1.0078. This means that the ratio of the mass of the hydrogen atom, $_1H^1$, to the mass of the carbon atom, $_6C^{12}$, is 1.0078/12.0000. The ratio of the mass of the oxygen atom, $_8O^{16}$, to the mass of the carbon atom is 15.995/12.0000. In general, the *atomic mass* M_a of an atom is the ratio of the mass of the atom to $\frac{1}{12}$ the mass of the $_6C^{12}$ atom. The *atomic mass unit* is $\frac{1}{12}$ of the mass of a $_6C^{12}$ atom and is represented by the symbol u.

$$1 \text{ atomic mass unit} = 1 \text{ u} = 1.66 \times 10^{-24} \text{ gm} \tag{11-1}$$

The atomic mass is a ratio and is not the same as the mass of a single atom. Thus, the mass of a single $_6C^{12}$ atom is 12 u or $12 \times 1.66 \times 10^{-24}$ gm $= 19.92 \times 10^{-24}$ gm. The mass of a single $_8O^{16}$ atom is 15.995 u or $15.995 \times 1.66 \times 10^{-24}$ gm $= 26.55 \times 10^{-24}$ gm. The mass of an atom with atomic mass M_a is M_a u or $M_a \times 1.66 \times 10^{-24}$ gm.

The *molecular mass* M_m of a chemical compound is the ratio of the mass of its molecule to $\frac{1}{12}$ the mass of a $_6C^{12}$ atom. The mass of a single molecule is therefore M_m u or $M_m \times 1.66 \times 10^{-24}$ gm. The molecular mass is equal to the sum of the atomic masses of all the atoms contained in the molecule. Even in the case of a pure chemical element, the molecule frequently contains several atoms. Molecules of oxygen, for example, usually contain two atoms and are therefore said to be *diatomic*. A diatomic molecule of oxygen is represented by the formula O_2. Its molecular mass is obviously twice the atomic mass of oxygen and is therefore $2 \times 15.995 = 31.99$. The mass of a single diatomic oxygen molecule is 31.99 u or $31.99 \times 1.66 \times 10^{-24}$ gm. Oxygen can also exist in the form of ozone, O_3, which has three atoms in its molecule and is said to be *triatomic*. If the molecule contains only one atom, as is the case for helium, He, the element is said to be *monatomic*. The molecular mass and the atomic mass are then the same thing.

One *mole* of a substance is a quantity of the substance which has a mass in grams numerically equal to the molecular mass or, if the substance is monatomic, a mass in grams numerically equal to the atomic mass. One mole of $_6C^{12}$ is 12.0000 gm. One mole of diatomic oxygen is 31.99 gm.

Since

$$\text{Mass of 1 mole} = M_m \text{ gm} \tag{11-2}$$

$$\text{Mass of 1 molecule} = M_m \times 1.66 \times 10^{-24} \text{ gm} \tag{11-3}$$

Number of molecules in 1 mole:

$$N_A = \frac{M_m}{M_m \times 1.66 \times 10^{-24}} \tag{11-4}$$

$$N_A = \frac{1}{1.66 \times 10^{-24}} \tag{11-5}$$

$N_A = 6.02 \times 10^{23}$ (11-6)

This result is obviously the same for all substances. The number N_A that we have just calculated is an important constant called **Avogadro's number**. It is the number of molecules in 1 mole, that is, M_m gm, of any substance. As a special case, it is the number of atoms in 1 mole, that is, M_a gm, of any monatomic substance. For example, there are N_A atoms in 12 gm of $_6C^{12}$ or in 1.0078 gm of $_1H^1$. Notice that the numerical value of N_A is the reciprocal of the atomic mass unit in grams.

1 **atomic mass unit** $= \frac{1}{12}$ of the mass of a $_6C^{12}$ atom (11-7)

$$= 1 \text{ u} = 1.66 \times 10^{-24} \text{ gm} \quad \text{(11-8)}$$

Atomic mass, $M_a = \dfrac{\text{Mass of the atom}}{\frac{1}{12} \text{ of the mass of a } _6C^{12} \text{ atom}}$ (11-9)

Mass of a single atom $= M_a \times 1.66 \times 10^{-24}$ gm (11-10)

Molecular mass, $M_m = \dfrac{\text{Mass of the molecule}}{\frac{1}{12} \text{ of the mass of a } _6C^{12} \text{ atom}}$ (11-11)

Mass of a single molecule $= M_m \times 1.66 \times 10^{-24}$ gm (11-12)

One mole has a mass of M_m gm (11-13)

One mole of a monatomic element has a mass of M_a gm (11-14)

Avogadro's number, N_A, is the number of molecules in one mole or M_m gm. For a monatomic element it is the number of atoms in M_a gm.

$$N_A = 6.02 \times 10^{23} \quad \text{(11-15)}$$

Example 11-3-1

At a counting rate of 5 atoms per second, how long would it take to count the atoms in a spherical droplet of water 0.1 cm in diameter?

The volume of a sphere of *radius* $r = \frac{4}{3}\pi r^3$

Volume of droplet $= \frac{4}{3}\pi(0.05)^3$

$$= 5.24 \times 10^{-4} \text{ cm}^3$$

The density of water is 1 gm per cubic centimeter, and so the mass of the droplet is 5.24×10^{-4} gm.

A molecule of water contains two atoms of $_1H^1$ and one atom of $_8O^{16}$ and its molecular mass is therefore

$(2 \times 1.0078) + 15.995 = 18.0$ to sufficient accuracy

Therefore, 18 gm of water contains $N_A = 6.02 \times 10^{23}$ molecules.

5.24×10^{-4} gm of water contains

$\dfrac{5.24 \times 10^{-4}}{18} \times 6.02 \times 10^{23}$ molecules

The Ideal Monatomic Gas

Since each molecule contains 3 atoms, the number of atoms in the water droplet is

$$3 \times \frac{5.24 \times 10^{-4}}{18} \times 6.02 \times 10^{23} = 5.22 \times 10^{19} \text{ atoms}$$

At a rate of 5 per second the time taken to count these atoms would be

$$1.04 \times 10^{19} \text{ sec}$$

There are approximately 3.16×10^7 sec in a year, so the time taken would be

$$\frac{1.04 \times 10^{19}}{3.16 \times 10^7} = 3.3 \times 10^{11} \text{ years}$$

11-4   The Ideal Monatomic Gas

The ideal monatomic gas is the simplest state of matter and can be discussed with relatively simple mathematics. As we have already explained, **monatomic** means that the molecules are merely single atoms. This includes helium (He), neon (Ne), argon (Ar), mercury vapor (Hg), sodium vapor (Na), potassium vapor (K), but it excludes diatomic gases such as oxygen (O_2), nitrogen (N_2), hydrogen chloride (HCl), and also excludes triatomic gases such as ozone (O_3), steam (H_2O), and carbon dioxide (CO_2).

A gas is said to be **ideal** if there are no forces between its molecules. Actually all molecules exert forces on one another, but these forces decrease very rapidly as the distance between the molecules is increased. If the gas is expanded into a very large volume so that its density becomes very small and the molecules are very far apart, the forces between the molecules become negligibly small. An ideal gas is therefore the limiting case of an actual gas when its volume becomes very large and its density becomes very small.

An ideal monatomic gas can be visualized as a collection of small hard spheres (the atoms) that are separated by distances much larger than their diameters and that exert no forces on one another. These atoms are in rapid motion and move in straight lines until they rebound from the wall of the container, or very occasionally collide with one another (figure 11-6). The distribution of the atoms in space is perfectly random and the velocities are uniformly distributed over all possible directions. The atoms do not all have the same speed. They have all possible speeds from zero upward. However, there is an average speed, v_a, which we shall soon define more precisely. Most of the atoms have speeds that are not very different from v_a and there are only a few atoms with speeds very much smaller than v_a or very much larger than v_a. This average speed v_a is related to the temperature of the gas. The higher the temperature, the greater is the average speed.

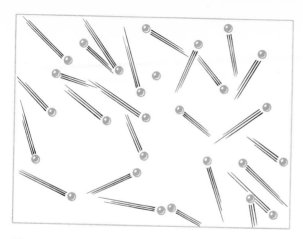

Figure 11-6   The kinetic theory picture of an ideal monatomic gas. The tails are intended to create an impression of the motion of the atoms and do not imply that the atom leaves any sort of wake behind it.

11-5   The Internal Energy of an Ideal Monatomic Gas

The atoms of an ideal gas exert no forces on one another and so there is no mutual potential energy of these atoms. The only source of energy is the kinetic energy of motion of the atoms. An atom of mass m moving with a velocity v has a kinetic energy $\frac{1}{2}mv^2$ and so the total kinetic energy, U, of all the atoms is

Total kinetic energy, U = Sum of $\frac{1}{2}mv^2$ for all the atoms

of the ideal monatomic gas   (11-16)

U is called the **internal energy** of the gas. It is a form of energy present within the gas that is not immediately obvious until we realize that the gas consists of atoms in motion. If we take an object such as a baseball with mass M and throw it with a velocity V, then in mechanics its kinetic energy is said to be $\frac{1}{2}MV^2$. However, if the baseball is visualized as a collection of atoms in rapid motion in all directions, the total kinetic energy of all these atoms is greatly in excess of $\frac{1}{2}MV^2$. In fact, the velocity V is unlikely to be much greater than 5×10^3 cm/sec, whereas the internal velocities could easily be as large as 5×10^5 cm/sec. The internal kinetic energy is therefore at least 10,000 times as great as the "obvious" term $\frac{1}{2}MV^2$.

Fortunately, the total kinetic energy can be divided into two parts, a part $\frac{1}{2}MV^2$ which is due to the motion of the baseball as a whole, and an "internal kinetic energy" that is the same as the total kinetic energy of the random motion of the atoms when the baseball as a whole is not moving. As long as the temperature remains constant the "internal kinetic energy" remains constant and all our attention can be directed toward changes in the term $\frac{1}{2}MV^2$, as in the previous chapters. But if energy is added in the form of heat the internal kinetic energy increases and the temperature rises. If, for example, the baseball strikes the

The Ideal
Monatomic
Gas

ground and embeds itself in a patch of loose earth, the kinetic energy $\frac{1}{2}MV^2$ is destroyed and at least part of it is converted into extra internal kinetic energy, so that the atoms move faster and the baseball warms up.

There are some complications involved in the internal energy of a solid such as a baseball, so let us return to the ideal monatomic gas. The average speed v_a can be defined in such a way that the internal energy U would be the same if all the atoms had this speed. If there are N atoms in the gas, v_a is *defined* by

$$U = N\tfrac{1}{2}mv_a{}^2 \tag{11-17}$$

$$N\tfrac{1}{2}mv_a{}^2 = \text{Sum of } \tfrac{1}{2}mv^2 \text{ for all the atoms} \tag{11-18}$$

Cancelling $\frac{1}{2}m$ on both sides,

$$Nv_a{}^2 = \text{Sum of } v^2 \text{ for all the atoms} \tag{11-19}$$

Notice that v_a is a rather special sort of average. It is really the kinetic energy that is being averaged.

For an ideal monatomic gas there is a direct relationship between the internal energy and the concept of **absolute temperature, T.**

Internal energy and absolute temperature

$$U = N\tfrac{1}{2}mv_a{}^2$$
$$= N\tfrac{3}{2}kT \tag{11-20}$$

The constant k is a fundamental constant called

Boltzmann's constant

$$k = 1.380 \times 10^{-16} \text{ CGS units} \tag{11-21}$$

Equation 11-20 is one way of defining the temperature T, which is called the absolute temperature and is written $T\,^\circ\text{K}$. This is referred to verbally as "degrees absolute" or "degrees Kelvin" (in honor of the physicist Lord Kelvin who was responsible for clarifying the concept of temperature during the middle years of the nineteenth century). The absolute scale of temperature is much more commonly used in physics than the Centigrade or Fahrenheit scales, because the zero of the absolute scale is really the lowest possible temperature that can ever be attained. Equation 11-20 implies that when $T = 0$, $U = 0$ and $v_a = 0$. The **absolute zero of temperature, $0\,^\circ\text{K}$,** is therefore the temperature at which the ideal monatomic gas has lost all its internal energy and its atoms have come to rest. Clearly, we cannot do any better than this.

In the next chapter we shall return to the concept of absolute temperature and explain the significance of Boltzmann's constant k. We shall also see that there is a relationship between the temperature $T\,^\circ\text{K}$ measured on the absolute scale and the temperature $t\,^\circ\text{C}$ measured on the more familiar Centigrade scale.

$$T\,^{\circ}\text{K} = t\,^{\circ}\text{C} + 273.15 \qquad\qquad (11\text{-}22)$$

It follows that the absolute zero of temperature is

$$0\,^{\circ}\text{K} = -273.15\,^{\circ}\text{C} \qquad\qquad (11\text{-}23)$$

The temperature at which ice melts under a pressure of one standard atmosphere (the ice point) is

$$0\,^{\circ}\text{C} = 273.15\,^{\circ}\text{K} \qquad\qquad (11\text{-}24)$$

The temperature at which water boils under a pressure of one standard atmosphere (the steam point) is

$$100\,^{\circ}\text{C} = 373.15\,^{\circ}\text{K} \qquad\qquad (11\text{-}25)$$

The most familiar scale of temperature is the Fahrenheit scale, because of its use in meteorology and medicine. On the Fahrenheit scale, the ice point is $32\,^{\circ}\text{F}$, the steam point is $212\,^{\circ}\text{F}$ and the absolute zero is $-459.67\,^{\circ}\text{F}$. However, the Fahrenheit scale is very rarely used in physics and the absolute scale is much to be preferred to it or the Centigrade scale.

Finally, it is important to emphasize that the simple expression of equation 11-20 for the internal energy is true only for an ideal monatomic gas. For all other substances the relationship between internal energy U and temperature T is much more complicated.

11-6 The Pressure of an Ideal Monatomic Gas

The atoms of a gas are in rapid motion and are continually striking the walls of the container. This steady bombardment of atoms is equivalent to a force pushing the wall outward, and is the origin of the pressure of the gas. More precisely, the pressure is the force exerted on unit area of a plane wall. However complicated the shape of the container (figure 11-7), select a small area of the wall which is small enough to be essentially plane and which has an area A. The bombardment of atoms exerts on it an outward force F which is perpendicular to the small portion of the wall. The pressure p of the gas is then

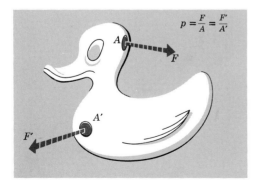

$$p = \frac{F}{A} = \frac{F'}{A'}$$

The Ideal

Monatomic

Gas

Figure 11-7   The pressure inside a balloon with a complicated shape.

Pressure = force per unit area

$$p = F/A \qquad\qquad (11\text{-}26)$$

Assuming that the temperature is constant throughout the whole of the gas, and ignoring some very small effects due to the earth's gravitational field, the pressure has the same value on all parts of the wall. Notice that pressure is measured in dynes per square centimeter or newtons per square meter. For the benefit of readers who have encountered pressures measured in cm of mercury or standard atmospheres, let us emphasize that such units are secondary and nonfundamental. The equations given in this book are true only when the pressure is expressed in dyne cm^{-2} or newton m^{-2}. The conversion factors are

$$1 \text{ cm of mercury} \equiv 1.33 \times 10^4 \text{ dyne } cm^{-2} \qquad\qquad (11\text{-}27)$$

$$1 \text{ standard atmosphere} \equiv 76 \text{ cm of mercury} = 1.01 \times 10^6 \text{ dyne } cm^{-2} \qquad\qquad (11\text{-}28)$$

The reader should prove that

$$1 \text{ newton } m^{-2} = 10 \text{ dyne } cm^{-2} \qquad\qquad (11\text{-}29)$$

When an atom of the gas comes very close to an atom of the wall they exert strong repulsive forces on one another. The atom of the gas is given an acceleration which deflects it back into the body of the container. The atom of the wall stays where it is because the neighboring atoms of the wall exert still stronger forces on it which hold it in position. In the case of a rubber balloon the forces between the atoms of the rubber are initially incapable of doing this and the bombardment of gas atoms pushes the rubber atoms outward. The balloon expands and the rubber sheet stretches, thereby increasing the strength of the forces between the rubber atoms until, at a certain equilibrium size, these forces are strong enough to compensate for the bombardment of gas atoms. Actually, a similar thing happens even in the case of a steel container, but only an extremely small expansion of the container is needed to increase the interatomic forces sufficiently. If the pressure of the gas is increased, a point is eventually reached at which the forces between the atoms of the wall have been pushed to their limit. The balloon or steel container then explodes.

The forces between a gas atom and an atom of the wall are electromagnetic in character and are related to the electromagnetic interactions between the electrons and protons in the atoms. At first sight it might seem that it would be impossible to calculate the pressure of a gas without going into the complicated details of this interaction. However, as sometimes happens in physics, the value of the pressure of an ideal monatomic gas turns out to be independent of the exact details of the mechanism giving rise to it. A complete, rigorous proof of the expression for the pressure is a formidable mathematical task. We shall

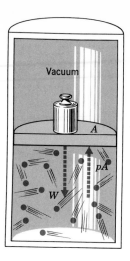

Figure 11-8   A gas supporting a frictionless piston inside a cylinder. (*W* and *pA* really act along the same straight line, but *pA* has been displaced sideways for the sake of clarity.)

therefore indicate the nature of the calculation, and try to make the result of it appear plausible.

Imagine the gas is contained in a cylinder with a freely moving piston (figure 11-8). The piston must fit tightly enough to prevent any gas from escaping, but it must be so well lubricated that the wall of the cylinder exerts no frictional force on the piston. If the cross-sectional area of the face of the piston is A and the pressure of the gas is p, the gas exerts an upward force pA on the lower face of the piston. The space above the piston is assumed to be a vacuum and there is then no force on the upper face of the piston. Weights are placed on top of the piston until

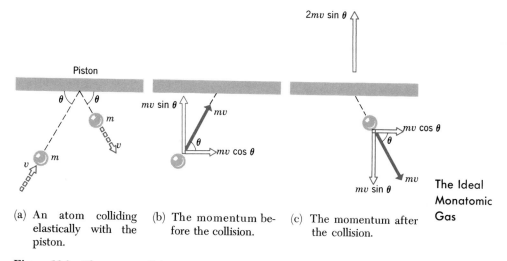

(a) An atom colliding elastically with the piston.

(b) The momentum before the collision.

(c) The momentum after the collision.

The Ideal Monatomic Gas

Figure 11-9   The atoms colliding with the piston transfer momentum to it.

the combined weight W of the piston and added weights is just sufficient to balance the upward thrust of the gas

$$W = pA \qquad (11\text{-}30)$$

Now consider an elastic collision between a gas atom and the piston (figure 11-9). Before the collision the momentum of the atom can be resolved into two components, a horizontal component $mv \cos \theta$ to the right and a vertical component $mv \sin \theta$ *upward*. After the collision the components are a horizontal component $mv \cos \theta$ to the right and a vertical component $mv \sin \theta$ *downward*. The horizontal component has not changed, but the vertical component has reversed its direction. Since momentum is a vector and its direction is as important as its magnitude, we must realize that the atom has acquired an extra *downward* momentum of $2\,mv \sin \theta$. Applying the law of conservation of momentum, it follows that the piston has acquired an equal but opposite *upward* momentum of $2\,mv \sin \theta$.

Every time an atom collides with the piston it transfers momentum to it, and there are many such collisions resulting in a steady transfer of upward momentum from the gas atoms to the piston. In Chapter 8 we saw that the force on a body is the same thing as the change in its momentum in 1 second (equation 8-9). It follows that the upward force pA exerted on the piston by the gas is equal to the momentum transferred from the gas atoms to the piston by all the collisions taking place in 1 second. Actually, of course, the piston does not acquire an upward momentum because it is also subject to a downward gravitational force W which gives it a downward momentum W every second. Since $pA = W$ the bombardment of the gas atoms transfers an *upward* momentum pA to the piston every second, which exactly compensates the *downward* momentum W which the piston would acquire every second if it were free to fall unhindered in the earth's gravitational field.

To calculate the pressure p, therefore, we need only calculate the momentum transferred in *one second* to *unit area* of the piston by the gas atoms. This is a laborious calculation when done properly, but in making it we do not need to know anything about the forces between the gas atoms and the atoms of the piston except that they obey Newton's third law of motion and that the law of conservation of momentum is therefore applicable. The result of the calculation is

Pressure of an ideal monatomic gas

$$p = \frac{1}{3}\frac{N}{V}mv_a{}^2 \qquad (11\text{-}31)$$

p is the pressure of the gas in dyne cm^{-2}. N is the number of atoms of the gas. V is the volume occupied by the gas in cm^3. m is the mass of an atom in gm. v_a is the average speed of the atoms in cm sec^{-1} as defined by equation 11-19.

This result may be understood in the following way. In a single collision the momentum transferred to the wall is proportional to mv. The colliding atoms have all velocities from zero upward but the average momentum is approximately mv_a. The number of collisions per second is proportional to the number of atoms per unit volume in the vicinity of the wall, that is to N/V. Also, the faster the atoms are moving, the more often they collide with the wall and the number of collisions per second is proportional to v_a. Putting these three factors together, the momentum transferred to the wall per second is proportional to

$$(mv_a) \times \frac{N}{V} \times (v_a) = \frac{N}{V} mv_a^2$$

This differs from equation 11-31 only by a factor of $\frac{1}{3}$, which comes from making the correct averaging procedure.

11-7 The Ideal Gas Equation

The absolute temperature of an ideal monatomic gas was introduced by equation 11-20. From this equation it follows very simply that

$$\tfrac{1}{3}mv_a^2 = kT \tag{11-32}$$

Substituting this into equation 11-31 we obtain

$$p = \frac{N}{V} kT \tag{11-33}$$

The ideal gas equation in terms of the total number of atoms.

$$pV = NkT \tag{11-34}$$

This equation may be put into a slightly different form. If the total mass of the gas is $M(=Nm)$ and the atomic mass is M_a, the number of moles is M/M_a. One mole contains N_A atoms, where N_A is Avogadro's number, and so the total number of atoms is

$$N = \frac{M}{M_a} N_A \tag{11-35}$$

The ideal gas equation then becomes

$$pV = \frac{M}{M_a} N_A kT \tag{11-36}$$

The Ideal
Monatomic
Gas

Defining a new constant R, called the *gas constant*, the ideal gas equation can be expressed in terms of quantities which can be directly measured, rather than in terms of the total number of atoms.

Gas constant

$$R = N_A k$$
$$= 8.32 \times 10^7 \text{ erg deg}^{-1} \text{ mole}^{-1} \tag{11-37}$$

Ideal gas equation

$$pV = \frac{M}{M_a} RT \tag{11-38}$$

Remember that the pressure p is in dyne cm^{-2}, the volume V is in cm^3, M is the mass of the gas in gm, M_a is the atomic mass, and T is the absolute temperature in °K.

Now let us pause and consider what we have achieved. We started with a model of an ideal monatomic gas as a collection of atoms in random motion. The pressure of the gas was then calculated by considering the bombardment of gas atoms on the walls of the container. The pressure was found to depend upon the number of atoms per unit volume and the average kinetic energy of the atoms (equation 11-31). The concept of temperature was introduced as a measure of the average kinetic energy (equation 11-20). Finally the ideal gas equation was expressed in a form that involves only macroscopically measurable variables and which contains no reference to the behavior of the atoms (equation 11-38).

Actually, we have reversed the historical order of events. Originally, crude temperature scales were based on the fact that most substances expand as the temperature is raised, as in the familiar mercury-in-glass thermometer. On the basis of such crude temperature scales, the macroscopic form of the ideal gas equation was discovered experimentally. It was then realized that a more fundamental temperature scale could be based on the properties of gases at very low densities. Finally, the atomic theory provided a microscopic explanation of the ideal gas equation and also revealed the relationship between absolute temperature and internal energy for an ideal monatomic gas.

If a fixed mass of gas is kept at a fixed temperature, everything on the right hand side of equation 11-38 is constant and so

$$pV = K = \text{a constant} \tag{11-39}$$
$$p = K/V \tag{11-40}$$
$$V = \frac{K}{p} \tag{11-41}$$

This is **Boyle's Law** and was discovered experimentally by the English physicist Robert Boyle as early as the seventeenth century, a few years before Newton discovered the laws of motion. It tells us that if we keep the temperature of a sample of gas constant and decrease its volume, the pressure increases in inverse proportion to the volume. To obtain a vivid picture of what is happening, the reader may imagine the atoms to be

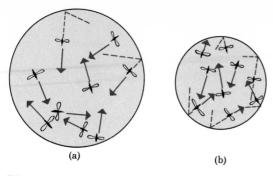

(a)   (b)

Figure 11-10   An analogy between Boyle's Law and the behavior of a swarm of mosquitoes inside a rubber balloon.

analogous to a swarm of angry mosquitoes, trapped inside a rubber balloon and continually flying from side to side and battering themselves against the rubber in an attempt to escape (figure 11-10). When the balloon has a large volume (figure 11-10a) the mosquitoes spend most of their time flying across the balloon and only occasionally hit the rubber; but when the volume is small (figure 11-10b) they hit the rubber much more frequently and exert a greater outward pressure on it.

If the volume of the gas is kept constant and its temperature is varied, then

$$p = \frac{M}{M_a} \frac{R}{V} T \tag{11-42}$$

$$p = K'T \tag{11-43}$$

where K' is another constant. The pressure of the gas is directly proportional to the absolute temperature. This is essentially **Charles' Law,** although the French physicist Jacques Charles did not express it in quite this way, since he was not familiar with the concept of absolute temperature. He proposed this law at the end of the eighteenth century, more than a century after the discovery of Boyle's law and a few years before Dalton's atomic theory. Its pictorial interpretation is that raising the temperature is equivalent to making the mosquitoes angrier, so that they fly faster, hit the rubber more often and with a greater impact and therefore exert a greater pressure on it.

Equation 11-34 may be rewritten

$$N = \frac{pV}{kT} \tag{11-44}$$

Consider two different ideal monatomic gases corresponding to two different chemical elements. If they have the same pressure p, the same temperature T and occupy equal volumes V, then everything on the right hand side of equation 11-44 is the same for the two gases. They must therefore contain the same number of atoms N. This is **Avogadro's Law.** Equal volumes of two ideal monatomic gases at the same temperature and pressure contain the same number of atoms.

The Ideal
Monatomic
Gas

All gases are ideal when expanded to a very large volume V, because the molecules are then very far apart and the forces between them are negligibly small. So let us next consider gases that are ideal because they occupy a large volume V, but which are polyatomic, that is, are composed of molecules containing more than one atom. The calculation of the pressure by considering the momentum transferred to the wall per second proceeds exactly as before and equation 11-41 again results.

$$p = \frac{1}{3}\frac{N}{V}mv_a^2 \text{ is true for an ideal polyatomic gas} \tag{11-45}$$

It is still permissible to relate the absolute temperature to the average kinetic energy as in equation 11-32.

$$\tfrac{1}{3}mv_a^2 = kT \quad \text{is true for an ideal polyatomic gas} \tag{11-46}$$

Putting these last two equations together, the ideal gas equation results.

$$pV = NkT \quad \text{is still true for an ideal polyatomic gas} \tag{11-47}$$

but N is now the number of *molecules* contained in the sample of gas. The ideal gas equation may be written in the form

$$pV = \frac{M}{M_m}RT \tag{11-48}$$

but M_m is now the *molecular* mass.

Equation 11-47 leads to a more general form of Avogadro's law. If two different gases have the same temperature and pressure and occupy equal volumes, then they contain equal numbers of *molecules*. This law has an immediate application to chemical reactions between gases. Consider the combination of hydrogen and oxygen to form water

$$2H_2 + O_2 = 2H_2O \tag{11-49}$$

It is obvious that N molecules of oxygen combined with 2N molecules of hydrogen to give 2N molecules of water. If the N molecules of oxygen occupy a volume V, then, at the same temperature and pressure, the 2N molecules of hydrogen occupy a volume 2V. Thus, if the hydrogen and oxygen are at the same temperature and pressure, a given volume of oxygen combines with twice its volume of hydrogen. Simple relationships of this kind were easily discovered experimentally and their ready explanation in terms of atomic theory led to the acceptance of this theory.

So far we have discovered no difference between an ideal polyatomic gas and an ideal monatomic gas, inasmuch as both obey the ideal gas equation. Why, then, was so much initial emphasis placed on the gas being *monatomic*? The answer is that the total internal energy of a polyatomic gas is a complicated thing and is not given by equation 11-20. It is still true that the *kinetic energy* associated with the motion of the molecules is related to the absolute temperature as in equation 11-20.

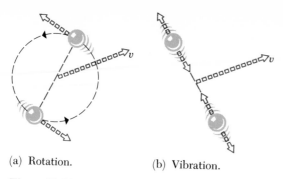

(a) Rotation.   (b) Vibration.

Figure 11-11 Internal modes of motion of a diatomic molecule.

Sum of $\frac{1}{2}mv^2$ for all the molecules

$$= N\frac{1}{2}m{v_a}^2$$

$$= N\frac{3}{2}kT \tag{11-50}$$

However, this is not the only source of internal energy, because the individual atoms are able to move inside the molecule. In figure 11-11a the molecule as a whole is moving with a velocity v, but in addition the two atoms are rotating about one another. In figure 11-11b the two atoms are vibrating along the line joining their centers. These modes of motion of the atoms inside the molecule represent extra kinetic energy and an extra contribution to the internal energy. So the total internal energy of a polyatomic gas is

$$U = N\frac{3}{2}kT$$

$$+ \text{rotational energy}$$

$$+ \text{vibrational energy} \tag{11-51}$$

The rotational and vibrational energies increase as the temperature is raised, but they are by no means directly proportional to T. In fact the situation is quite complicated and these additional forms of energy can be dealt with satisfactorily only by using quantum mechanics.

When the volume is decreased and the molecules (or atoms) come close enough together that we cannot ignore the interatomic forces, the gas is non-ideal. The ideal gas equation 11-48 is then no longer applicable. Neither is the average velocity of the molecules given by equation 11-50. Moreover, there is yet another type of contribution to the internal energy arising from the mutual potential energy of pairs of atoms corresponding to the interatomic forces. This is analogous to the gravitational potential energy of equation 9-31, except that the interatomic forces are electromagnetic and more complicated.

The air around us is a little too dense to be a good ideal gas. If one makes accurate measurements on it, it is easy to detect departures from the equations applicable to an ideal gas. However, the equations for the ideal gas are still approximately obeyed and are adequate for many purposes, if one is not asking for high accuracy.

The Ideal
Monatomic
Gas

Example 11-8-1

About how many molecules are there in 1 cm³ of air and what is their average distance apart?

Use equation 11-44.

$$N = \frac{pV}{kT}$$

The pressure of the air is approximately $p = 10^6$ dyne cm⁻² (equation 11-28). The temperature of the air is approximately $300°K$. If $V = 1$ cm³,

$$N = \frac{10^6 \times 1}{(1.38 \times 10^{-16}) \times 300}$$

$$= 2.5 \times 10^{19}$$

In 1 cm³ of air there are approximately 2.5×10^{19} molecules. Imagine the 1 cm³ to be divided up into little cubes of side a, each of which contains a molecule. Then the volume of each cube is a^3 and there are $1/a^3$ cubes.

$$\frac{1}{a^3} = 2.5 \times 10^{19}$$

$$a^3 = \frac{1}{2.5 \times 10^{19}}$$

$$= 4 \times 10^{-20}$$

$$a = 3.4 \times 10^{-7} \text{ cm}$$

This is the average distance apart of the molecules and is about 20 times the size of an oxygen or nitrogen molecule.

Example 11-8-2

The best vacuum that can be produced corresponds to a pressure of about 10^{-10} dyne cm⁻². How many molecules remain in 1 cm³?

$$N = \frac{pV}{kT}$$

$$= \frac{10^{-10} \times 1}{(1.38 \times 10^{-16}) \times 300}$$

$$= 2,500$$

There is still a large number of molecules left. Their average distance apart is a little less than a millimeter.

Example 11-8-3

What is the average velocity of the molecules of the air?

From equation 11-50,

$$mv_a{}^2 = 3kT$$

Multiply both sides by Avogadro's number N_A

$$N_A m v_a{}^2 = 3 N_A k T$$

But $N_A m = M_m$ and $N_A k = R$, so

$$v_a{}^2 = \frac{3RT}{M_m}$$

Air consists mainly of nitrogen, which is diatomic, and its effective molecular mass is approximately twice the atomic mass of nitrogen. So

$$M_m \approx 2 \times 14 = 28$$

$$v_a{}^2 = \frac{3 \times 8.32 \times 10^7 \times 300}{28}$$

$$= 26.8 \times 10^8$$

Taking the square root

$$v_a = 5.2 \times 10^4 \text{ cm sec}^{-1}$$

This is equivalent to 1,160 miles per hour!

Example 11-8-4

A volume of 50 liters is filled with helium at 15°C to a pressure of 100 standard atmospheres. Assuming that the ideal gas equation is still approximately true at this high pressure, calculate approximately the mass of helium required.

Rearranging equation 11-38,

$$M = M_a \frac{pV}{RT}$$

Careful attention must be paid to the units. The atomic weight of helium is 4.0. One standard atmosphere is 1.01×10^6 dyne cm^{-2} and so p is 1.01×10^8 dyne cm^{-2}. One liter is 10^3 cm^3 and so V is 5×10^4 cm^3. R is 8.32×10^7 in CGS units. T is the absolute temperature, not the Centigrade temperature, and is therefore $15 + 273.15 = 288.15°$K. So

$$M = \frac{4.0 \times (1.01 \times 10^8) \times (5 \times 10^4)}{(8.32 \times 10^7) \times 288.15}$$

$$= 840 \text{ gm}$$

The helium has a mass of 840 gm.

11-9 Temperature in the Universe

The temperature of the human body is 310°K (37°C, 98°F) and it is very efficiently regulated to stay within a degree or so of this value. The temperature of the human environment varies from about 185°K (−88°C, −127°F) in the Antarctic to about 330°K (57°C, 135°F) in places like Death Valley or the Sahara Desert. There are many reasons why we must live at this kind of temperature rather than, say, 1°K or 1,000,000°K, but the easiest to appreciate is that the human body is composed largely of

The Ideal
Monatomic
Gas

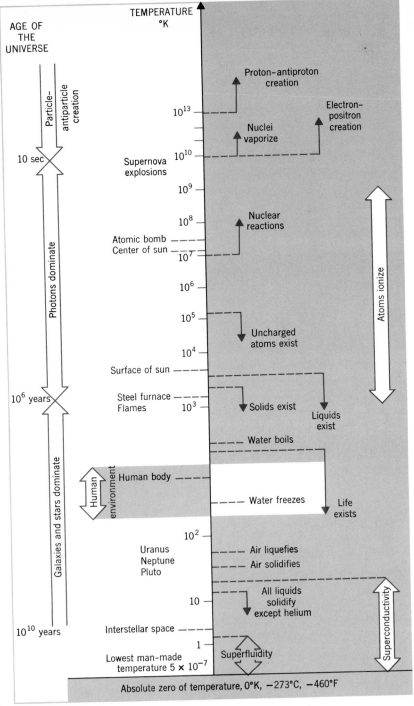

Figure 11-12 Temperature in the universe. The temperature scale has been distorted in order to fit in the information conveniently.

water which freezes at 273°K (0°C, 32°F) and boils at 373°K (100°C, 212°F). On the other hand, because we live at these temperatures we have certain prejudices about the nature and behavior of matter. If we could experience an environment at 1°K or 1,000,000°K we would find things very different. Therefore, with the help of figure 11-12, let us consider the temperatures at various places and times in the universe and the corresponding states of matter.

Starting at the temperature of the human body, 310°K, and going to lower temperatures, we see that the first major change occurs when water freezes at 273°K. Without some means of maintaining their temperature above that of their environment, living organisms cannot remain active below this temperature, although organisms which have been very carefully dehydrated or very carefully frozen can remain dormant down to the very lowest temperatures and will revive when carefully warmed up again. At about 80°K (−193°C, −315°F) air liquefies and at about 70°K (−203°C, −333°F) it solidifies. Temperatures in this vicinity probably exist on the outer planets, Uranus, Neptune, and Pluto. Below 14°K (−259°C, −434°F) all liquids solidify with the exception of liquid helium which can remain liquid right down to 0°K unless it is encouraged to solidify by the application of a little pressure. The temperature of outer space, far from any stars, is 3°K (−270°C, −454°F). The lowest man-made temperature is 5×10^{-7}°K but that, of course, is achieved under very special circumstances inside a complicated physical apparatus.

At very low temperatures strange new phenomena occur. Quantum mechanical effects, which are usually important only on the scale of atoms or individual fundamental particles, can then manifest themselves on a macroscopic scale for collections of large numbers of atoms or particles. Below 20°K many metals become **superconducting** and present no resistance to the flow of electricity. The large number (10^{20}, say) of superconducting electrons in a ring of superconducting metal can flow cooperatively around the ring to produce an electric current which apparently persists unaltered forever (figure 11-13a). In many ways this is analogous to the "perpetual motion" of an electron orbiting around a nucleus in an atom.

Below 2°K liquid helium becomes **superfluid** and behaves as though it had no viscosity. It can flow with speeds of several centimeters per second through cracks as narrow as 10^{-5} cm (figure 11-13b). In fact, there is a tendency for the speed to increase as the crack becomes narrower. When placed in an open container, liquid helium forms a film about 100 atoms thick on the wall and climbs out through this thin film (figure 11-13c). There is evidence that this flow can still take place when the film is only one or two atoms thick. If liquid helium is set into rotation by the walls of a rotating container and the container is then brought to rest, the liquid goes on rotating forever (figure 11-13d). Moreover, the angular momentum per superfluid helium atom is exactly one basic quantum mechanical unit of angular momentum, $h/2\pi$ (section 10-6 and chapter 26).

Going upward from the temperature of the human body, water boils at 373°K and active life ceases to exist before this temperature is reached. The temperature inside a flame is about 1000°K, inside a steel furnace about

The Ideal
Monatomic
Gas

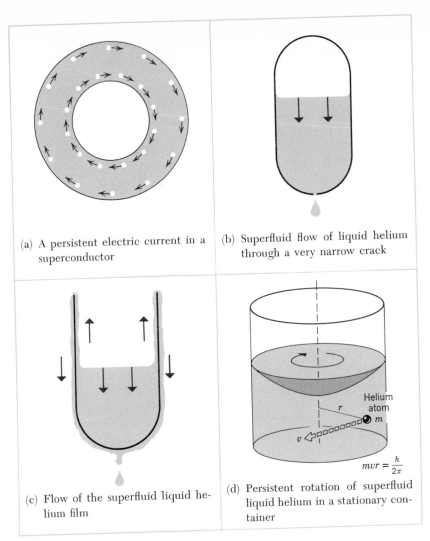

(a) A persistent electric current in a superconductor

(b) Superfluid flow of liquid helium through a very narrow crack

(c) Flow of the superfluid liquid helium film

(d) Persistent rotation of superfluid liquid helium in a stationary container

$$mvr = \frac{h}{2\pi}$$

Figure 11-13 Superconductivity and superfluidity.

2000°K, and on the surface of the sun about 6000°K. (At these higher temperatures the numerical value of the temperature on the Centigrade scale is not significantly different and the Fahrenheit scale is rarely used.) Solids cease to exist under ordinary pressure at about 4000°K and liquids cease to exist at about 5000°K. The reason for this is not unlike the argument used to derive the escape velocity of a rocket in section 9-5. At a high enough temperature the positive kinetic energy of an atom in a solid or liquid exceeds the negative potential energy which it has by virtue of the attractive forces between it and neighboring atoms. The atom can then escape from the solid or liquid, converting part of its kinetic energy into potential energy as it climbs out, but still having some kinetic energy left over to keep going when it is well outside.

The surface of the sun at $6000°K$ is therefore essentially a hot gas, but at this temperature yet another change has already begun. When the fast-moving atoms collide, they sometimes strike one another hard enough to chip off one or two electrons, leaving behind a positively charged **ion** with too few electrons to cancel completely the positive charge on its nucleus. The material at the surface of the sun is therefore a mixture of uncharged atoms, charged ions, and free electrons. As the temperature is raised the ionization progresses and at temperatures above $100,000°K$ ($10^5°K$) there are almost no uncharged atoms left. As the temperature is raised still further, the electrons are stripped off the atoms one by one, but it requires a temperature as high as one billion degrees absolute ($10^9°K$) to remove the last electron from the heaviest element.

When ionization is well advanced, matter is in the form of a **plasma** of negatively charged free electrons and positively charged ions or, ultimately, electrons and bare nuclei. Above about $10^7°K$ **nuclear reactions** take place. The nuclei collide with one another and reshuffle their neutrons and protons to form new nuclei. In section 31-3 we shall discuss this in further detail and explain why it cannot happen at lower temperatures. The center of the sun is at about 20 million degrees absolute ($2 \times 10^7°K$) and is a dense plasma of free electrons, free protons (hydrogen nuclei) and bare helium nuclei. The source of the sun's energy is a series of nuclear reactions which convert the protons into helium nuclei (section 31-3). Temperatures as high as $4 \times 10^7°K$ probably occur during the explosion of an atomic bomb.

At temperatures above 10 billion degrees absolute ($10^{10}°K$) nuclei begin to vaporize. The neutrons and protons are able to escape from the nucleus in the same way that an atom escapes from a solid or liquid at lower temperatures. Temperatures of this magnitude may well occur for short periods of time during the explosion of a star to produce a supernova (sections 10-6 and 28-5).

Above $10^{10}°K$ an entirely new type of process becomes possible. The large amounts of energy available can create new matter in the form of electrons and their antiparticles, or opposites, positrons (section 32-3). Above $10^{13}°K$ protons and antiprotons, or neutrons and antineutrons can be created (section 32-4). There is no evidence that such very high temperatures exist in the present universe, but a very highly regarded theory of the evolution of the universe maintains that it started out at an extremely high temperature and cooled off. For the first one-tenth of one-thousandth of a second the temperature exceeded $10^{13}°K$ and space was full of a hot dense "soup" of all kinds of particles and antiparticles jostling up against one another. When the temperature fell below this value, the protons and neutrons and their antiparticles disappeared, but the electrons and positrons stayed around for another 10 seconds until the temperature fell below $10^{10}°K$. For the next one million years the particles of light, photons, dominated the universe until the temperature fell to about $4000°K$. Photons left over from this era still fill the whole of space and can be detected by sensitive radar receivers. In fact, when the TV channel closes down at night and your TV screen turns black with white dots running over it, these

The Ideal
Monatomic
Gas

dots may well be caused in part by the photons left over from the early universe. For the major part of its history, lasting some 10 billion years, the universe has contained galaxies and stars and has had very different temperatures in different places. The primeval photons, however, have slowly cooled off to 3°K, which is what is meant here by the present temperature of interstellar space.

QUESTIONS

1. Why, do you think, was the factor $\frac{3}{2}$ introduced into equation 11-20? (A reasonable alternative might be to define a different Boltzmann constant $k' = \frac{3}{2}k$.)
2. The density of diamond is 3.5 gm/cm³. A diamond composed of 1 mole of $_6C^{12}$ would be about as big as (a) a house, (b) a horse, (c) a football, (d) an apple, (e) a cherry, (f) a pea, (g) a poppy seed and (h) a speck of dust?
3. What would you guess is meant by one mole of electrons? What would its mass be?
4. If the same temperature is represented by t_F°F on the Fahrenheit scale, by t°C on the Centigrade scale and by T°K on the absolute scale, prove that

$$t_F = \tfrac{9}{5}t + 32$$
$$t_F = \tfrac{9}{5}T - 459.67$$

5. Imagine a container of gas isolated in gravity-free outer space. If the molecules of the gas transfer momentum to the walls of the container, does the container thereby acquire a steady acceleration?
6. An ideal monatomic gas is compressed in such a way that its temperature does not change. How does its internal energy vary with with its volume?
7. Imagine that the temperature of a fixed mass of an ideal monatomic gas is raised from T_1 to T_2 at a constant pressure p and its volume increases from V_1 to V_2. If the change in temperature $(T_2 - T_1)$ is very small and the increase in volume $(V_2 - V_1)$ is therefore also very small, the *coefficient of expansion at constant pressure* may be defined as

$$\alpha_p = \frac{(V_2 - V_1)}{V_1(T_2 - T_1)}$$

Is α_p independent of temperature? Express α_p in terms of p, V_1 and T_1.

8. Criticize the following argument. Atmospheric pressure exerts a force of 15 pounds on every square inch of surface. The surface area of a man is about 4,000 square inches. Therefore, the force of the atmosphere on a man is 15 × 4,000 pounds or 30 tons.

A

1. What is the molecular mass of (a) diatomic hydrogen (H_2), (b) ozone (O_3), (c) carbon dioxide (CO_2), (d) hydrogen peroxide (H_2O_2)?
2. What is the mass of a molecule of (a) ozone (O_3), (b) steam (H_2O), (c) helium (He), (d) nitrogen (N_2)?
3. How many moles are there in (a) 1 gm of $_6C^{12}$, (b) 15.995 gm of diatomic oxygen, (c) 2 gm of helium, (d) 3 kg of carbon dioxide?
4. How many atoms are there in (a) 1 gm of $_6C^{12}$, (b) 1 mole of diatomic oxygen, (c) 3 kg of carbon dioxide?
5. How many molecules are there in (a) 4 gm of helium, (b) 8 gm of diatomic oxygen, (c) 1 gm of steam?
6. A gas at a pressure of 5×10^4 dyne/cm² is contained inside a cubical box. If the length of a side of the box is 25 cm, what is the force exerted by the gas on one face of the box?
7. If the cylinder of figure 11-8 has an inside radius of 5 cm and the total mass of the piston is 150 gm, what is the pressure of the gas?
8. Neon gas is contained inside a box of volume 500 cm³ at a pressure of 5×10^4 dyne/cm² and a temperature of 150°K. What is the average kinetic energy per atom? (Neon is monatomic).
9. If one mole of a gas is contained in a tank of volume 10^6 cm³ at a temperature of 300°K, what is its pressure?
10. If 10^{-3} gm of helium is introduced into a box with a fixed volume of 25 cm³, to what temperature must it be heated to raise its pressure to 10^6 dyne/cm².
11. An ideal monatomic gas at 400°K exerts a pressure of 10^5 dyne/cm². If it is compressed to half its original volume and, as a result, its temperature changes to 480°K, what is its new pressure?
12. An ideal monatomic gas is at a temperature T_1, exerts a pressure p_1 and occupies a volume V_1. If its volume is changed to V_2 and its pressure to p_2, find an algebraic expression for its new temperature T_2.
13. If an ideal monatomic gas occupies a volume of 5×10^4 cm³ and exerts a pressure of 3×10^3 dyne/cm² at a temperature of 250°K, how many atoms are there?

B

14. Obtain an approximate value of the density of the air surrounding you at this moment.
15. Most gases liquefy long before they can be cooled to a temperature very near 0°K. The most favorable case is helium, which can probably be studied in its gaseous form down to about 0.2°K. What is the average speed of the helium atoms at this temperature?
16. What mass of diatomic hydrogen is needed to fill a 10 liter container to a pressure of 1 atmosphere at 18°C?
17. Assuming that interstellar space contains one atom per cubic centimeter and is at a temperature of 3°K, what is the pressure there?

The Ideal
Monatomic
Gas

217

Compare it with the best vacuum that can be produced in the laboratory (example 11-8-2).

18. In figure 11-8, the cylinder has an inside radius of 1.5 cm and contains 10^{-3} gm of helium. It is immersed in boiling water to ensure that the gas is always at a temperature of 100°C. (a) If the total mass of the piston is 300 gm, what is the height of the column of gas? (b) If the apparatus is then immersed in ice, what is the new height? (c) How much mass must be removed from the piston to restore the original height?

19. What are the extreme temperatures produced by your local climate? Suppose that you inflate your automobile tires on the coldest day in winter to the correct pressure of 26 pounds per square inch as recommended by the manufacturer. Assuming that no air leaks out and that the volume remains constant, what will the pressure be on the hottest day in summer? Ignore heating of the tire by frictional contact with the road. (When the gauge reads 0 pounds per square inch, the pressure inside the tire is one atmosphere, which is 14.7 pounds per square inch.)

C

20. What mass of helium gas is needed to fill a spherical meteorological balloon with a radius of 1.5 m to a pressure of 8 cm of mercury *above* atmospheric pressure when the temperature is 55°F? If the sun's rays warm the balloon to 100°F and the pressure inside it rises to 12 cm of mercury above atmospheric pressure, by how much does the radius increase?

21. Some helium gas is contained in a box of negligible weight at the ambient temperature of the room, which may be taken to be 300°K. If all the internal energy could be abstracted from the gas and used to lift the box, through what height could it be raised? Express your answer in miles.

22. At what temperature would the average speed of the molecules of the air be equal to the escape velocity from (a) the earth, (b) the moon? Discuss carefully the question of why the earth has retained its atmosphere, but the moon has not.

23. Assuming that the force exerted by the atmosphere on a small area of the surface of the earth is equal to the total weight of the atmosphere directly above that area, calculate the total mass of the atmosphere and compare it with the mass of the earth and the mass of the oceans (problem 2-23). If the whole of the atmosphere were at 300°K and had the same density at all heights, what would be its height in miles?

24. Make a rough estimate of the mass of air needed to inflate an automobile tire.

25. An ideal gas of monatomic neon is contained in a cubical box of height 1 meter at room temperature (300°K). A molecule with the average speed, v_a, travels from the floor to the ceiling of the box without

encountering another molecule. What is the fractional change in velocity produced by gravity?

26. An astronomical object is believed to be a rarefied cloud of hydrogen. It contains both monatomic and diatomic molecules. What is the ratio of their average speeds?

27. One mole of an ideal monatomic gas A at $200°$K is mixed with two moles of a different ideal monatomic gas B at $400°$K. They come into thermal equilibrium with one another without losing any internal energy. What is their final common temperature?

28. At $0.01°$K helium vapor in contact with liquid helium has a pressure of about 4×10^{-312} dyne cm^{-2}. Calculate the average distance between helium atoms in light years.

29. If you were able to solve problem 9-37, attempt to generalize the proof to a system containing any number of particles. Is there any difficulty in defining the temperature of an ideal monatomic gas when the container has a velocity V comparable with the average velocity v_a of the atoms?

The Ideal
Monatomic
Gas

12

The Nature of Heat

12-1   Average Properties of a Large Number of Atoms

In Chapter 2 it was suggested that a complete description of the universe would involve a description of the motion of each atom, electron, proton, or neutron in the universe. Since the whole lifetime of the universe does not give us enough time to count the atoms in a rain drop, a full description of the motion of these atoms through all time is clearly a psychologically impossible task. We must settle for less. Fortunately, the fact that even a small sample of matter contains an enormous number of atoms introduces a simplification, inasmuch as many of the interesting properties of the sample of matter depend only upon the *average* behavior of all the atoms. When we considered the ideal monatomic gas, we discovered that its pressure depended only upon the number of atoms per unit volume and upon the *average* kinetic energy of these atoms. It was certainly not necessary to follow the path of each atom in detail. Neither was it necessary to know in detail the nature of the forces between the gas atoms and the atoms of the wall, even though these forces are the basic cause of the pressure.

However, we did make the assumption that the temperature of the gas was the same everywhere. This assumption will now be put in a rather different light by emphasizing that the atoms must be in **statistical equilibrium** with one another. Suppose that we start with a fan inside the gas stirring it up into turbulent motion and that we apply heat rapidly to the top right hand corner of the box, making the temperature there higher than elsewhere (figure 12-1a). This is not a case of statistical equilibrium. If the fan is taken away and the hot flame is removed, the gas then settles down into a state of statistical equilibrium. The flow of gas produced by the fan dies down as the atoms collide with

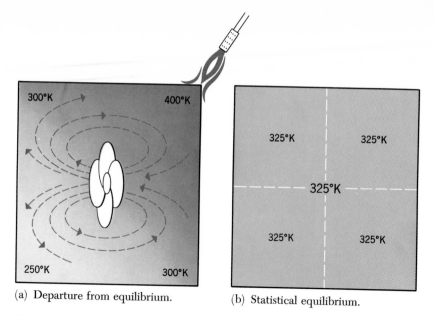

(a) Departure from equilibrium.　　　　(b) Statistical equilibrium.

Figure 12-1 The concept of statistical equilibrium.

the walls and lose their forward momentum. The hot gas atoms in the top right hand corner initially have a higher average velocity, but they move into the bulk of the gas, collide with the colder atoms and eventually distribute their excess kinetic energy among all the atoms. When statistical equilibrium has been reached (figure 12-1*b*), a small sample of gas anywhere in the box has no net velocity of flow in any particular direction. Also, the average velocity of its atoms is the same as for a sample taken from any other part of the box, which is equivalent to saying that the temperature is the same everywhere. Before the gas comes into equilibrium, there is a flow of energy (heat) from the top right hand corner into the rest of the gas. In the state of statistical equilibrium there is no tendency for energy to flow from one part of the system into another part.

The present chapter is concerned with large collections of atoms in statistical equilibrium and with such properties as pressure, temperature, and internal energy, which depend only on the average behavior of the atoms. This subject is called **thermodynamics.**

12-2　The Zeroth Law of Thermodynamics and the Concept of Temperature

For a substance other than an ideal monatomic gas there is no simple relationship between the temperature and the average kinetic energy of the atoms. We must therefore look for a more general definition of temperature.

Figure 12-2*a* shows a bath of hot oil with a glass beaker of cold water partially immersed in it. The fast-moving molecules of the hot oil strike

The Nature
of Heat

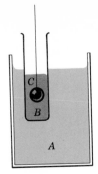

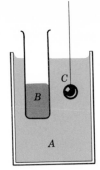

(a) The water *B* comes into statistical equilibrium with the hot oil *A*.

(b) The copper ball *C* comes into statistical equilibrium with the water *B*.

(c) The copper ball *C* is found to be in statistical equilibrium with the hot oil *A*.

Figure 12-2 The zeroth law of thermodynamics.

the slow-moving molecules of the glass beaker and transfer kinetic energy to them. The molecules of the glass beaker are thereby made to move faster and, when they strike the slow-moving molecules of water, energy is transferred to the water. In everyday language, we would say that the glass conducts heat from the hot oil into the water. The cooling of the oil slows down its molecules and the heating of the water speeds up its molecules. Eventually, the process stops and no more energy is transferred from the oil to the water. The water has then come into statistical equilibrium with the oil.

Now immerse a cold copper ball in the water. Energy is transferred from the water molecules to the copper atoms until the copper ball is in statistical equilibrium with the water. If this cools the water a little, energy flows from the oil into the water until the oil and water are again in equilibrium. When everything settles down and there is no more transfer of energy, the copper ball is in statistical equilibrium with the water and the water is in statistical equilibrium with the oil (figure 12-2*b*).

Now rapidly transfer the copper ball from the water to the oil (figure 12-2*c*). It will be found that the copper ball is already in statistical equilibrium with the oil and there is no further transfer of energy from the copper to the oil or vice versa.

The zeroth law of thermodynamics

If body *A* is in statistical equilibrium with body *B* and body *B* is in statistical equilibrium with body *C*, then *C* is in statistical equilibrium with *A* and if *C* is placed in contact with *A* there is no transfer of energy between them.

In our illustration *A* is the oil, *B* is the water, and *C* is the copper ball. In everyday language we would say that the oil and water come to the same temperature. The copper ball then assumes the same temperature as the water, but this automatically implies that the ball has the same temperature as the oil, and so when it is immersed in the oil there is no flow of heat. However, the concept of temperature is useful and valid only because the zeroth law is true. Suppose it were not true and the copper, water, and oil were allowed to come to the same temperature as in figure 12-2b. If we then immersed the ball into the oil and a flow of heat took place, we would have to assume that the ball and oil were at different temperatures, which would be a very contradictory situation.

The basic idea which emerges from all this is as follows. The essential aspect of the concept of temperature is that two bodies are at the same temperature if they can be placed in intimate contact with one another without any heat flowing from one to the other. Independently of the precise way in which the temperature scale is defined, the same number can be assigned to the temperature of two bodies if, and only if, they can be placed in contact without any flow of heat.

12-3   Fluctuations

The zeroth law has the appearance of being trivially obvious. The fact that bodies behave in this way is so much a part of our everyday experience that we take it for granted. However, before underestimating this law the student should consider how he would *prove* it by considering the interactions between the copper, water, and oil molecules which lead to statistical equilibrium. The proof is in fact a major problem in pure mathematics. Moreover, it is not difficult to think of a situation in which the law is not true. Imagine an ideal monatomic gas in statistical equilibrium. If we wish to determine the temperature of a small sample of this gas we need only measure the total kinetic energy of the atoms in the sample and use equation 11-16.

$$\text{Total kinetic energy} = N\tfrac{3}{2}kT \tag{12-1}$$

Suppose that we apply this equation to the whole of the gas and find that $T = 100°K$. If we then examine a very small sample of gas, containing say 10,000 atoms, we might discover to our surprise that its temperature is $101°K$. If we examine another small sample of gas containing 10,000 atoms just to one side of the first sample, we might discover that the temperature of this second sample is $99°K$. We might also observe that there is a flow of energy from the first sample to the second sample, even though the gas as a whole is supposed to be in statistical equilibrium (figure 12-3a). The reason is that the concepts of temperature and statistical equilibrium are appropriate only for very large numbers of atoms, and 10,000 is not a large enough number.

To see why there is no difficulty when we are dealing with a large number of atoms, suppose that the gas as a whole consists of 0.13 gm of helium, which contains about 2×10^{22} atoms. Make an imaginary division into two equal volumes on opposite sides of the container, each side containing 10^{22} atoms, and then measure the temperature of each

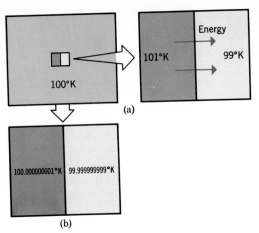

(a)

(b)

Figure 12-3 Temperature fluctuations of small and large samples of an ideal monatomic gas.

side separately. If we could determine the total kinetic energy with very great accuracy we might detect a temperature difference of one billionth of a degree (10^{-9}°K) between the two sides (figure 12-3b). This is much smaller than the sensitivity of practical thermometers and need never worry us. The larger the number of atoms we take, the smaller are the fluctuations in temperature.

We use the expression "fluctuations in temperature" because the sample of 10,000 atoms does not stay at 101°K. It is losing energy to the neighboring sample which is at 99°K and after a fraction of a second its temperature may have fallen to 100°K or 99°K or even occasionally

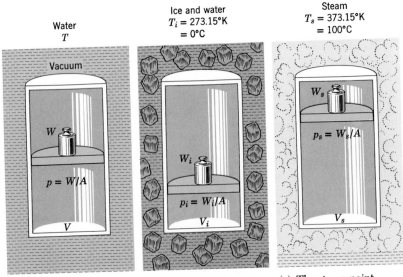

(a) Water at a tempera-
ture of T°K.

(b) The ice point.

(c) The steam point.

Figure 12-4 Establishing the absolute scale of temperature.

98°K. In fact its temperature is rapidly fluctuating backward and forward by about one degree on either side of 100°K. This is because the atoms of the sample are making random collisions with neighboring atoms and sometimes by chance they acquire more kinetic energy than their share and sometimes less kinetic energy than their share. However, if we were to average the temperature (i.e., kinetic energy) over a long period of time, the average temperature would be very close to 100°K and would get closer and closer the longer the time over which the average was taken. The temperature of 100°K can be obtained either by measuring the average kinetic energy of a very large number of atoms at a single instant of time, or by observing the fluctuations of the kinetic energy of a small number of atoms over a very long period of time and taking a suitable average.

12-4   Temperature

Now let us consider how to measure temperature. The ideal monatomic gas provides us with a very good thermometer. The device shown in figure 11-8, although not used in practice, will serve to illustrate the principles involved. Immerse this device in a bath of water whose temperature T is to be measured (figure 12-4a). Water is a complicated substance, and there is no simple way of determining its temperature by measuring its macroscopic properties, such as pressure and density. However, if the water and the ideal monatomic gas are in intimate thermal contact and have come into statistical equilibrium with one another, we can be certain that the gas has the same temperature as the water. The temperature of the ideal monatomic gas can be deduced from its observable properties.

A rearrangement of the ideal gas equation (equation 11-38) gives

$$T = \frac{M_a}{M} \frac{pV}{R} \tag{12-2}$$

The atomic mass of the gas M_a is known and its mass M and volume V can be easily measured. As already explained on page 192, the pressure p can be found by measuring the combined weight W of the piston and its load and by using equation 11-30 ($W = pA$, where A is the area of the piston). The only other quantity in equation 12-2 is R. We could, if we wished, have an international agreement on the exact value to be assigned to R. It is more convenient in practice to fix the value of some standard temperature, and the one chosen is the temperature of a mixture of pure ice and water at a pressure of one standard atmosphere. This temperature is called the **ice point** and, by international agreement:

The absolute temperature of the ice point is defined as

$$T_i = 273.15°K \tag{12-3}$$

Let us therefore place our ideal gas thermometer in a mixture of pure ice and water at one standard atmosphere and again measure the pressure, p_i, and the volume, V_i (figure 12-4b). Then

$$T_i = \frac{M_a}{M} \frac{p_i V_i}{R} \qquad (12\text{-}4)$$

Dividing equation 12-2 by equation 12-4

$$\frac{T}{T_i} = \frac{pV}{p_i V_i} \qquad (12\text{-}5)$$

or

$$T = 273.15 \frac{pV}{p_i V_i} \qquad (12\text{-}6)$$

To determine the unknown temperature T, we need only measure p and V at the temperature T and at the ice point.

Why was the ice point assigned the peculiar value 273.15°K? Place the ideal gas thermometer in the steam immediately above the surface of water boiling under a pressure of one standard atmosphere, and measure this temperature, which is called the **steam point** (figure 12-4c).

$$T_s = 273.15 \frac{p_s V_s}{p_i V_i} \qquad (12\text{-}7)$$

The result of the measurement will be

$$T_s = 373.15°\text{K} \qquad (12\text{-}8)$$

which is 100 degrees above the ice point. On the Centigrade scale (now frequently called the Celsius scale) the ice point is 0°C and the steam point is 100 degrees higher at 100°C. If, and only if, the ice point is defined as 273.15°K, the steam point is 100 degrees higher on the absolute scale also. The size of the degree is then the same on the absolute and Centigrade scales and the only difference between them is a shift of 273.15 degrees in the zero:

$$T°\text{K} = t°\text{C} + 273.15 \qquad (12\text{-}9)$$

The Centigrade scale had been used for many years before the absolute scale was invented. It is still extensively used in engineering, the biological sciences, and (in Europe) meteorology. It was therefore thought desirable to define the absolute scale in such a way that it had a simple relationship to the Centigrade scale, as in equation 12-9.

Once the ice point has been defined as 273.15°K, the gas constant R may be determined by measurements on a gas held at the ice point. Equation 12-4 tells us that

$$R = \frac{M_a}{M} \frac{p_i V_i}{273.15} \qquad (12\text{-}10)$$

The value of Boltzmann's constant k is then obtained from $R = N_A k$. The final point to emphasize is that one of the constants, T_i, R or k is at our disposal to define arbitrarily. The decision actually made is far from

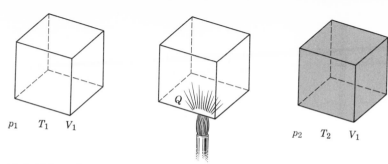

(a) Initial conditions.   (b) Heat is added.   (c) Final conditions.

Figure 12-5 Addition of heat to an ideal monatomic gas at constant volume.

fundamental and depends upon the historical tradition that there shall be 100 degrees between the ice point and the steam point.

12-5   Heat, Energy, and Work

Heat is a form of energy. The heat added to a system is the same thing as the energy added to the system, but a little care must be taken not to neglect some part of this energy.

Consider first the simplest case of heat added to an ideal monatomic gas in a container of fixed volume V_1 (figure 12-5). Suppose that the initial temperature and pressure are T_1 and p_1 and that the addition of an amount of heat Q raises the temperature and pressure to T_2 and p_2. Initially the total kinetic energy of the gas atoms is

$$U_1 = \tfrac{3}{2}NkT_1 \tag{12-11}$$

The addition of heat raises the total kinetic energy to

$$U_2 = \tfrac{3}{2}NkT_2 \tag{12-12}$$

The heat added is equal to the increase in the energy of the system:

$$Q = U_2 - U_1 \tag{12-13}$$

or

$$Q = \tfrac{3}{2}Nk(T_2 - T_1) \tag{12-14}$$

Now suppose that heat is added to an ideal monatomic gas that does not have a fixed volume but is kept at a fixed pressure. In the arrangement of figure 12-6 the constancy of the pressure is automatically assured if the total weight of the piston and its load is left unaltered. If the initial pressure, volume, and temperature are p_1, V_1, and T_1, the addition of an amount of heat Q' raises the temperature to T_2 and expands the volume to V_2, the pressure meanwhile remaining at p_1. The internal energy depends only on the temperature and the increase in internal energy is again

$$U_2 - U_1 = \tfrac{3}{2}Nk(T_2 - T_1) \tag{12-15}$$

The Nature of Heat

227

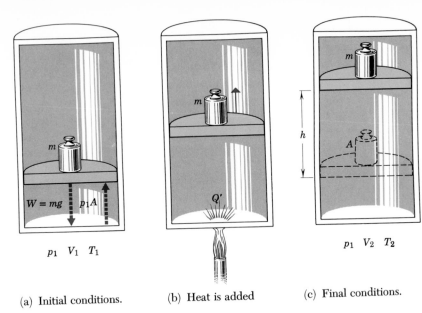

(a) Initial conditions.   (b) Heat is added   (c) Final conditions.

Figure 12-6 Addition of heat to an ideal monatomic gas at constant pressure.

However, an additional amount of energy must be supplied to the gas to enable it to do the work necessary to raise the piston as the gas expands. If the piston has a cross-sectional area A and is raised through a height h, the extra volume made available to the gas is a cylinder of base area A and height h. The volume of this cylinder is

$$(V_2 - V_1) = Ah \tag{12-16}$$

The external force exerted on the piston by the gas is p_1A. In the process of moving the piston through a distance h parallel to this force, the work done against the piston is:

$$\text{Work done} = \text{Force} \times \text{distance}$$
$$= (p_1A) \times h \tag{12-17}$$
$$= p_1Ah \tag{12-18}$$
$$= p_1 \times (Ah) \tag{12-19}$$
$$= p_1(V_2 - V_1) \tag{12-20}$$

The heat added to the system does two things. It increases the internal energy of the gas and it also raises the piston. Therefore,

$$\text{Heat added} = \text{Increase in internal energy of gas}$$
$$+ \text{Work done by gas on piston} \tag{12-21}$$

$$\text{or } Q' = \tfrac{3}{2}Nk(T_2 - T_1) + p_1(V_2 - V_1) \tag{12-22}$$

In the above analysis we were considering the addition of energy to a system consisting of the gas alone. We therefore had to consider a simultaneous loss of energy as the gas did work on its surroundings by raising the piston. It may prove helpful to reconsider the matter from the point of view of the system shown in figure 12-7. This includes the gas, its container, the piston and its load, and the earth, but excludes the source of heat (which might, for example, be a burner and a tank of fuel). The heat supplied to this system by the burner achieves two things. Firstly, it increases the kinetic energy of the gas atoms in accordance with equation 12-15. Secondly, it expands the gas, raises the piston and increases the mutual gravitational potential energy of the piston and the earth. If the mass of the piston is m, the increase in gravitational potential energy is mgh, but

$$mg = W$$
$$= p_1 A \tag{12-23}$$

So, the increase in potential energy is

$$mgh = Wh$$
$$= p_1 A h$$
$$= p_1(V_2 - V_1) \tag{12-24}$$

This is identical with the work done by the gas in raising the piston. However, from the point of view of the new system, which includes the earth, this quantity is no longer the work done by the system against its surroundings. Instead the mutual potential energy of the earth and the piston can now be regarded as part of the internal energy of the new system. The quantity $p_1(V_2 - V_1) = mgh$ can be considered to be an

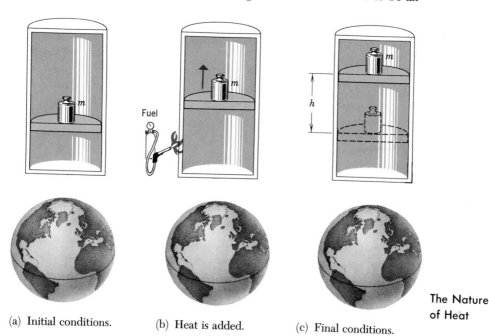

(a) Initial conditions.   (b) Heat is added.   (c) Final conditions.

Figure 12-7   When the earth is included in the system, the work done by the expanding gas is seen to be equivalent to an increase in gravitational potential energy.

increase in the *internal energy* of the new system. When we are discussing the addition of heat to a system, we must clearly understand what is included in the system, what can rightfully be regarded as part of the internal energy of the system and what must be regarded as work done by the system on its surroundings.

Example 12-5-1

If a small amount of heat q is added to M gm of a substance and raises its temperature from T_1 to T_2, the *specific heat per gm* of the substance is defined as

$$C = \frac{q}{M(T_2 - T_1)} \tag{12-25}$$

It is the heat added per gram per degree rise in temperature. The conditions which prevail while the heat is being added must be specified carefully. Find expressions for the specific heat of an ideal monatomic gas if: (*a*) its *volume* is kept constant while the heat is added; (*b*) its *pressure* is kept constant while the heat is added.

(*a*) Equation 12-14 gives an expression for the heat added to an ideal monatomic gas when its temperature rises from T_1 to T_2 at constant volume

$$q = \tfrac{3}{2}Nk(T_2 - T_1) \tag{12-26}$$

The *specific heat at constant volume* is therefore

$$C_v = \frac{\tfrac{3}{2}Nk(T_2 - T_1)}{M(T_2 - T_1)}$$

$$C_v = \frac{3Nk}{2M} \tag{12-27}$$

Since there are N_a atoms in M_a gm of the gas, the numbers of atoms in 1 gm is

$$\frac{N}{M} = \frac{N_a}{M_a} \tag{12-28}$$

So

$$C_v = \frac{3N_a k}{2M_a} \tag{12-29}$$

Since

$$N_a k = R \tag{12-30}$$

$$C_v = \frac{3R}{2M_a} \tag{12-31}$$

(*b*) Equation 12-22 gives an expression for the heat added to an ideal monatomic gas when its temperature rises from T_1 to T_2 at constant pressure

$$q' = \tfrac{3}{2}Nk(T_2 - T_1) + p_1(V_2 - V_1) \tag{12-32}$$

For an ideal gas

$$pV = NkT \tag{12-33}$$

Therefore

$$p_1(V_2 - V_1) = Nk(T_2 - T_1) \tag{12-34}$$

and so

$$q' = \tfrac{5}{2}Nk(T_2 - T_1) \tag{12-35}$$

The *specific heat at constant pressure* is therefore

$$C_p = \frac{\tfrac{5}{2}Nk(T_2 - T_1)}{M(T_2 - T_1)} \tag{12-36}$$

This is the same as equation 12-27 except for the factor $\tfrac{5}{2}$. The same procedure that was used in part (a) of this problem will therefore give us

$$C_p = \frac{5R}{2M_a} \tag{12-37}$$

12-6 The First Law of Thermodynamics

Many of the equations that appeared in the previous section are valid only for an ideal monatomic gas. However, equation 12-21 is fundamental and is true for the most general system even if it contains non-ideal gases, liquids or solids.

The first law of thermodynamics

The heat added to a system is equal to the increase in the internal energy of the system plus the work done by the system against its surroundings.

The work done against the surroundings may take various forms. The change that takes place in the system may necessitate the performance of work against external forces of a gravitational, electrical or magnetic character. In all cases, however, by enlarging the system to include other systems on which the work is done, it is possible to eliminate the work and talk only about internal energy. Under these circumstances, the first law is clearly seen to be nothing more than another statement of the law of conservation of energy. The internal energy of the source of heat decreases, but an exactly equal amount of internal energy appears in the system to which heat is being added. This statement is meaningful only after we have realized that kinetic and potential energy can reside in the atoms of which matter is composed.

To illustrate the complexity of the internal energy, let us consider a

The Nature
of Heat

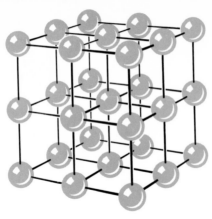

Figure 12-8 A solid with a simple cubic lattice. The maroon spheres represent the atoms and the straight lines joining them are imaginary. For the sake of clarity, the size of the atom has been reduced in comparison with the distance between atoms. In actual fact, the atoms are almost touching. Also, they oscillate rapidly about the positions shown.

simple solid. Its atoms are arranged in an orderly geometric pattern to form a crystal lattice. Figure 12-8 shows one of the simplest possible lattices, a simple cubic lattice in which the atoms are placed at the corners of imaginary cubes stacked side by side like building blocks. In actual fact the atoms are not stationary but oscillate about their mean positions in all directions. However, each atom remains within the vicinity of its own particular lattice site and very rarely wanders through the body of the lattice. As the temperature is raised the atoms vibrate more rapidly and swing further away from their lattice sites. In classical physics the absolute zero would be the temperature at which the oscillations ceased. Rather strangely, as we shall see later, quantum mechanics will not permit this and in fact the atoms are still vibrating quite rapidly at $0°K$. As the temperature T is raised the kinetic energy associated with these vibrations steadily increases, but it varies in a complicated way with T and is certainly not equal to $\frac{3}{2}NkT$.

In addition, the mutual potential energy of the atoms must be included in the internal energy. The atoms exert electromagnetic forces on one another and these forces hold the atoms in position on the solid lattice. The forces vary in a complicated way with the distance apart of the atoms, but they nevertheless give rise to a negative potential energy between each pair of atoms. This is similar in character, if not in mathematical form, to the mutual gravitational potential energy discussed in section 9-3. The total potential energy is negative, and could be defined in the following way. Ignore the vibrations and place each atom at rest on its lattice site. To pluck out one atom and pull it out of the solid against the attraction of the other atoms would require the performance of work, i.e., the addition of energy to the system. The total potential energy might be defined as the energy needed to pluck out all the atoms

Suppose that we started with the situation of figure 12-9b and the container then received a small jolt which rotated it through a small angle (figure 12-10). Then the atoms would no longer bounce straight back from the wall, but would bounce off at an angle. There is good reason to believe that the simplicity of the motion would soon be destroyed and it would rapidly revert to complete randomness as in figure 12-9c. Can the same thing happen in reverse? Can a disordered motion develop into an ordered motion? It obviously can, because of a fundamental principle of Newtonian mechanics known as **time reversal,** which says that for any possible motion, there is another possible motion which would be obtained by running the first motion backward in time. Suppose I drop a body from rest at a height h and it hits the ground with a velocity v. The time reversed motion corresponds to throwing the body upward from the ground with velocity v so that it rises to a height h before it comes to rest. If I were to film the falling motion, then by running the film *backward* through the projector I could reproduce in all details the time reversed motion in which the ball was thrown upward.

Now return to the above case of the ordered motion of figure 12-9b degenerating into the disordered motion of figure 12-9c. Halt the process at some instant of time when disorder has been reached, and then reverse the velocity of each atom without changing its magnitude. The subsequent motion will be an exact time reversal of the preceding motion and will continue until the ordered motion of figure 12-9b has been established.

In practice this never happens. The reason is that a disordered motion can be realized by assigning positions and velocities to the atoms in an enormously large number of ways. Of this very large number of disordered motions only a very few are of the very special kind that would develop into an ordered motion such as that in figure 12-9b within a reasonably short time. It is conceivable that any random motion, if given sufficient time, would eventually develop into an ordered motion. However, in the case of an overwhelming majority of random motions the time taken would be enormously large, greatly in excess of the "age of the universe." Moreover, even when the ordered motion was realized it would persist for only a fraction of a second. The situation is similar to taking a new pack of cards that is carefully ordered into suits, and shuffling it. The initial order is very soon destroyed, but it would be a very long time indeed before the shuffling reproduced the initial order again. There are very many ways of arranging a pack in a random fashion, but there is only one way of arranging it in the order in which it comes from the manufacturer.

What we are saying then is that the ordered arrangement of figure 12-9b is quite consistent with the laws of mechanics, but it is very unlikely to occur in practice. On the other hand, the disordered arrangement of figure 12-9c, *or some similar random arrangement,* is very likely to be found in practice. Moreover, a random arrangement is *overwhelmingly* more probable. A student confronted with these ideas for the first time is likely to argue that although the ordered arrangement is improbable it is possible and he may ask why it does not sometimes occur. Such a student should sit down with a pack of cards and try shuffling them until they are arranged into suits. His patience with his own argument will soon become exhausted.

12-8   Entropy; the Second and Third Laws of Thermodynamics

It is possible to put these ideas on a firm mathematical basis and to define a quantity S, called the **entropy**, which is a measure of the probability that a particular type of motion will occur. Since a disordered situation is also a more probable situation, S is also a measure of disorder. The situation in figure 12-9*b* is an ordered situation and an improbable situation, so it corresponds to a small value of the entropy S. The situation in figure 12-9*c* is a disordered situation, but a more probable situation, so it corresponds to a larger value of the entropy S. Subject to the slightest perturbation, the ordered situation would rapidly develop into the disordered situation with larger entropy. The chance of the disordered situation ever developing into the ordered situation is so very small that it can be neglected. This is an example of a very general law.

The second law of thermodynamics

When a system containing a large number of particles is left to itself, it assumes a state with maximum entropy, that is, it becomes as disordered as possible.

Since entropy is a measure of probability, all that this says is that "what will occur is what is most likely to occur." The probability of what is likely to occur is so overwhelmingly large compared with any other possibility that this other possibility can be completely discounted.

The stipulation that the system is to be "left to itself" means that no heat must be added or subtracted and no work must be done on the system, so that its internal energy cannot change. Under these circumstances the system comes into a state of statistical equilibrium. What we are now saying is that the state of statistical equilibrium is the state which has the maximum possible entropy that can be achieved for this particular value of the internal energy. It is the most probable state and also the most disordered.

Let us now concentrate on the value of the entropy when statistical equilibrium has been reached and ignore non-equilibrium states. Start with a system at temperature T with an entropy S and add a small quantity of heat q so that the temperature increases to $T + t$ and the entropy of the new equilibrium state is $S + s$. The amount of heat added must be very small, so that t is very small compared with T, and it must be added slowly and carefully so that the system passes through a sequence of equilibrium states. It can be shown that:

The practical definition of entropy

$$s = \frac{q}{T} \qquad\qquad (12\text{-}39)$$

Increase in entropy $= \dfrac{\text{Heat added}}{\text{Temperature}}$

Thus, although entropy is a subtle mathematical concept related to the laws of probability, as long as we confine ourselves to equilibrium states, this equation provides us with a direct macroscopic method of measuring changes in entropy. It also shows that, as the temperature is increased, the entropy increases. The system is more disordered at higher temperatures. Conversely, as the temperature is lowered the entropy decreases and the system becomes more ordered. When the system is perfectly ordered, it is not possible to go any further and the absolute zero of temperature has been reached.

The third law of thermodynamics

A system in equilibrium at the absolute zero of temperature is in a state of perfect order and has zero entropy.

Let us add, though, that the third law, when stated in this particular form, must be applied with caution. It is very easy to obtain a system at a temperature near $0°K$ which is not really in statistical equilibrium, but which appears to be so because the rate at which it is changing in the direction of equilibrium is extremely slow. Such a system might be very disordered and have a high entropy. Given enough time, it would eventually come into equilibrium with its low temperature surroundings. It would become perfectly ordered and would have zero entropy. However, this might take a very long time indeed.

Perfection is difficult to achieve and the absolute zero of temperature is impossible to achieve. One inescapable consequence of these laws of thermodynamics is that it is impossible to make an apparatus that will reach the absolute zero of temperature in a finite number of operations.

In the simple case of an ideal monatomic gas at $0°K$ the atoms are all at rest and this is obviously a very orderly arrangement as far as the distribution of velocities amongst the atoms is concerned. However, once the atoms have been set in motion at a finite temperature, it is not immediately obvious why raising the temperature and making the atoms move faster should also be equivalent to producing more disorder. This point is more clearly demonstrated by the behavior of the electrons which are responsible for the magnetic properties of an iron bar magnet. As we shall explain in more detail in Chapters 17 and 29, an electron is itself a small bar magnet. Most materials are not magnetic because the electrons cancel one another in pairs, and for every electron pointing

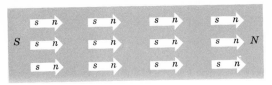

(a) T = 0°K. Perfect order. S = 0.

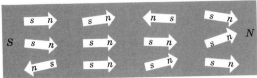

(b) T greater than 0°K but less than the Curie temperature. Partial order. S greater than 0 but small.

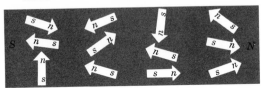

(c) T greater than the Curie temperature. Disorder. S large.

Figure 12-11   Disordering of the electrons in an iron bar magnet as the temperature is increased.

its bar magnet in a particular direction there is another electron pointing in exactly the opposite direction. However, in the case of iron a certain fraction of the electrons all line up in the same direction and the specimen of iron has net magnetic properties.

At 0°K these electrons are all completely lined up as in figure 12-11a. As far as the electron alignment is concerned, this is a state of perfect order with zero entropy (S = 0). It also produces a strong bar magnet. At a temperature slightly above 0°K the thermal agitation misaligns some of the electrons (figure 12-11b). There is then only partial order, the entropy is greater than zero and the strength of the bar magnet is diminished. As the temperature is steadily increased the misalignment becomes worse, the electrons become more disordered, the entropy increases, and the strength of the bar magnet decreases. At a certain temperature, called the Curie temperature, the misalignment becomes complete, and as many electrons point one way as the opposite way. Above this temperature the electrons are completely disordered (figure 12-11c) and the entropy associated with the alignment of the electrons approaches its maximum value. Moreover, the iron is no longer a permanent bar magnet.

Example 12-8-1

Helium gas is contained in a cylinder with a frictionless piston. Initially the temperature of the gas is 50°K, its pressure is 30 dyne cm^{-2}, and it occupies a volume of 150 cm^3. Calculate the increase in entropy if the temperature is raised by one millionth of a degree at constant pressure.

Since the fractional change in temperature is small, we can use equation 12-39:

$$s = \frac{q}{T} \tag{12-40}$$

The heat q added to change the temperature at constant pressure is given by equation 12-22:

$$q = \tfrac{3}{2}Nk(T_2 - T_1) + p_1(V_2 - V_1) \tag{12-41}$$

At the initial temperature T_1, the ideal gas equation takes the form

$$p_1 V_1 = NkT_1 \tag{12-42}$$

At the final temperature T_2, the pressure p_1 is the same and so

$$p_1 V_2 = NkT_2 \tag{12-43}$$

Therefore

$$p_1(V_2 - V_1) = Nk(T_2 - T_1) \tag{12-44}$$

and

$$q = \tfrac{5}{2}Nk(T_2 - T_1) \tag{12-45}$$

We are given $(T_2 - T_1) = 10^{-6}\,°\mathrm{K}$, but we are not told the number of atoms N. However, from equation 12-42 we may readily deduce that

$$Nk = \frac{p_1 V_1}{T_1} \tag{12-46}$$

and the quantities p_1, V_1 and T_1 are all known. Therefore,

$$
\begin{aligned}
q &= \frac{5}{2}\frac{p_1 V_1}{T_1}(T_2 - T_1) \tag{12-47}\\
&= \frac{5}{2}\frac{30 \times 150 \times 10^{-6}}{50}\\
&= 2.25 \times 10^{-4} \text{ ergs}
\end{aligned}
$$

The increase in entropy of the gas is therefore

$$
\begin{aligned}
s &= \frac{q}{T}\\
&= \frac{2.25 \times 10^{-4}}{50}\\
&= 4.5 \times 10^{-6} \text{ ergs deg}^{-1}
\end{aligned}
$$

12-9   The Arrow of Time and the Fate of the Universe

The second law of thermodynamics can be illustrated by a familiar example. Suppose that a body at a temperature T_1 is connected to a

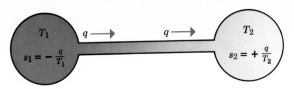

Figure 12-12 When heat flows from a hot body to a cold body, entropy increases.

body at a lower temperature T_2 by a rod which can conduct heat (figure 12-12). This situation is a departure from statistical equilibrium and it is well known that heat flows from the hot body to the cold body until their two temperatures are equal. Suppose that a small amount of heat q leaves the hot body and enters the cold body. Then the entropy of the hot body decreases by q/T_1 and the entropy of the cold body increases by q/T_2. The change in the total entropy of the system is therefore:

$$\text{Change in entropy} = -\frac{q}{T_1} + \frac{q}{T_2}$$

$$= q\left(\frac{1}{T_2} - \frac{1}{T_1}\right) \tag{12-48}$$

Since q is positive and T_2 is less than T_1, it is easy to convince yourself that this change is positive. Heat flows from a body at a high temperature to a body at a lower temperature because this increases the total entropy of the system, as required by the second law of thermodynamics. If the heat q were to flow from the cold body to the hot body, the change in entropy would be $q\left(\frac{1}{T_1} - \frac{1}{T_2}\right)$ which is negative. The combined entropy of the two bodies would therefore decrease and the second law would be violated.

In principle, of course, we can imagine a film running backward through a projector and showing a flow of heat from the cold body to the hot body. Such an "uphill" flow of heat is dynamically possible, but the exact set of initial circumstances that would make it happen is so very unlikely to occur in practice that it can be completely discounted.

Everyday experience contains many examples of processes that are very unlikely to occur in reverse. Milk poured into coffee soon distributes itself uniformly by diffusion, but it is inconceivable that it should later separate out again into a stream of pure milk leaping up into the jug. A log readily burns to form ash and various gases which mix with the air, but it is inconceivable that these gases should return from the atmosphere to the flame and recombine with the ash to form a new log. Can you imagine any process that would reconvert an omelet into an egg capable of hatching into a chicken? Wind, rain, and frost can reduce an ancient city to rubble, but are unlikely ever to rebuild it again.

All these examples involve the interaction of a large number of atoms, and they are all instances of systems that are initially not in statistical equilibrium but that move toward equilibrium. The process that actually occurs starts from a state of order with low entropy and develops into a state of greater disorder with higher entropy, in accordance with the second law of thermodynamics. The time reversed process, which never occurs in nature, produces an increase in order from a state of high entropy to a state of lower entropy.

A film running backward will therefore not deceive us for long. We shall soon notice the custard pie reforming on the face of the comedian and returning to the hand of the thrower. The correct forward direction of time is unambiguously determined as the direction in which order diminishes and entropy increases. For this reason, entropy has been described as "time's arrow," pointing from the past toward the future.

More important than the direction of time is the implication that the universe is heading toward a death of disordered chaos. Life as we know it requires the existence of states of high order and departures from statistical equilibrium such as temperature differences. A living organism is itself a highly ordered entity with low entropy which cannot come into statistical equilibrium with its surroundings except by dying. The functioning of any machine relies upon the availability of sources of energy not in equilibrium with their surroundings. A fuel is a chemical substance of low entropy, but the burning of the fuel produces chemical substances with higher entropy which are of no further use as fuels.

When we have used up all the fuels available to us on earth, we might make use of the energy of the tides. The tides are caused by the gravitational forces exerted by the moon on the oceans. The frictional drag of the tides is gradually slowing down the rotation of the earth. Eventually the length of the day will be equal to the length of the month, the moon will always stay in the same position above a fixed point on the earth's surface, and there will be no more lunar tides.

We might then use the energy of sunlight or harness the winds, which derive their energy from sunlight. Ultimately, however, the sun will burn itself out and will probably end up as some cold inactive object at the same temperature as the space around it. This, in fact, may well be the ultimate fate of the whole universe, which may well become a uniform mixture of lifeless objects all in equilibrium with one another and with the radiation filling the space around them. Nothing that we would consider to have any interest or any value could then occur except as a highly improbable, short-lived, small fluctuation away from equilibrium.

There is a possible reprieve. We do not yet understand the behavior of the universe as a whole. There may be some aspect of this behavior which overrides the second law of thermodynamics. We still have too poor an understanding of these matters to predict the distant future with any confidence.

QUESTIONS

1. If two bodies are in statistical equilibrium with one another, do they have (*a*) the same T, (*b*) the same U, (*c*) the same S?

2. Can two bodies at different pressures be in statistical equilibrium with one another? If you think so, quote an example.

3. Are the various bodies in the solar system in statistical equilibrium with one another? Explain your answer fully.

4. Is a saucepan of water steadily boiling on a stove a case of statistical equilibrium?

5. Would a universe in complete statistical equilibrium be an interesting place to live in? List all the ways in which it would differ from the actual universe.

6. Comment on the following definitions of the ice-point. (a) The temperature of ice. (b) The temperature of a mixture of ice and water. (c) A temperature exactly 100 centigrade degrees below the boiling point of water. (d) The temperature at which one mole of an ideal monatomic gas at a pressure of one atmosphere contains exactly N_A atoms. (e) Exactly $273.150000°$K.

7. Comment on the following definitions of the absolute zero of temperature. (a) The temperature at which all atoms come to rest. (b) The temperature at which the internal energy of all substances is zero. (c) The temperature at which all substances have zero entropy. (d) The temperature at which a body in statistical equilibrium has its maximum possible order.

8. Comment on the following definitions of heat. (a) Another name for energy. (b) The change in the total kinetic energy of random motion of the atoms of a body. (c) The change in the total internal energy of a body. (d) Energy in motion from a high temperature to a low temperature.

9. Which of the following are universally applicable under all possible circumstances? For the cases that are not universally applicable, state the restricted circumstances under which they can be applied.

(a) The first law of thermodynamics.
(b) The second law of thermodynamics.
(c) The third law of thermodynamics.
(d) $S = 0$ at $0°$K.
(e) $v_a = 0$ at $0°$K.
(f) $U = 0$ at $0°$K.
(g) $U = \frac{3}{2}NkT$.
(h) $pV = \dfrac{M}{M_a} RT$.
(i) Heat always flows from a high temperature to a low temperature. (Is this true for two neighboring *very small* samples of a gas?)

10. What is the value of the specific heat of water at $15°$C?

11. Give several examples familiar in everyday life of (a) work being converted into heat, (b) heat being converted into work.

12. Give several examples familiar in everyday life of entropy increasing toward its maximum possible value.

13. In a device such as the one shown in figure 12-6a, the piston is slowly pushed downward with a steady velocity. As the gas is com-

pressed it warms up. Explain this by considering the collisions of the gas atoms with the moving piston.

14. Describe qualitatively how the first law of thermodynamics can be applied to (a) warming a poker in a fire, (b) boiling away a saucepan of water on a stove, (c) the boy scout method of lighting a fire by rubbing together two sticks.

15. What are the units of entropy?

16. Can mankind invalidate the second law of thermodynamics by intervening and attempting to create order out of chaos?

17. A balloon containing helium gas is punctured and the helium diffuses throughout the whole room. How does this illustrate the second law of thermodynamics?

18. A container is divided into two parts by a thin partition. One side contains hydrogen gas and the other side contains helium gas. If the partition is broken, does the entropy of the system increase? If both parts contain helium, does the entropy increase?

PROBLEMS

A

1. Show that equation 12-14 may also be written

$$Q = \tfrac{3}{2}V_1(p_2 - p_1)$$

2. Show that equation 12-22 may also be written

$$Q' = \tfrac{5}{2}p_1(V_2 - V_1)$$

3. A gas expands from a volume of 5×10^3 cm^3 to a volume of 7×10^3 cm^3 at a constant pressure of 5×10^4 dyne/cm^2. How much work does it do?

4. What volume is occupied by one mole of an ideal gas at the ice-point when its pressure is one atmosphere?

5. One mole of an ideal monatomic gas is warmed from 300°K to 350°K without changing the volume of its container. How much heat is added?

6. If 1 joule of heat is added to 1 gm of helium gas at an initial temperature of 200°K without changing its volume, what is its final temperature?

B

7. One mole of an ideal monatomic gas is warmed from 300°K to 350°K *at a constant pressure*. How much heat is added?

8. If one joule of heat is added to 1 gm of helium gas at an initial temperature of 200°K without changing its *pressure*, what is its final temperature?

9. A balloon is being inflated in such a way that the pressure inside it is always 1.5×10^6 dyne cm^{-2}. How much work is done by the balloon against its surrounding while its radius increases from 1 m

to 2 m? Is your calculation valid even though the mass of gas inside the balloon does not remain constant?

10. A container of negligible mass contains helium gas. It falls from a height of 100 m and, when it strikes the ground, all of its kinetic energy is used to warm up the gas. Through how many degrees does its temperature rise?

11. One mole of an ideal monatomic gas is heated from $100°K$ to $100.01°K$ without changing its volume. What is the increase in its entropy?

C

12. Imagine that an international commission has defined a new temperature scale on which the absolute zero is $0°N$ and the ice point is $100°N$. What is the new value of Boltzmann's constant, k, in ergs per degree N? What is the steam point on this new scale?

13. Prove that the specific heats at constant volume and constant pressure of an ideal polyatomic gas are related by the equation

$$C_p - C_v = \frac{R}{M_m}$$

even though the values of C_p and C_v separately cannot be derived by any simple theory.

14. In the arrangement of figure 12-6a, suppose that the mass on the piston is always kept too small by an infinitesimal amount and the gas therefore slowly expands. No heat is allowed to enter or leave the system during this expansion. As the gas expands, its temperature falls. Why? If there are two moles of an ideal monatomic gas and the temperature falls from $300°K$ to $250°K$, how much work has been done in raising the piston?

15. An ideal monatomic gas consists of N_1 atoms of mass m_1 and N_2 different atoms of mass m_2. Remembering that the atoms of an ideal monatomic gas rarely encounter one another, find equations for the mixture which are analogous to equations 11-20, 11-34, 12-14, and 12-22.

16. Equation 12-39 is valid even when there is no change in temperature. Find an expression for the change in entropy of an ideal monatomic gas which undergoes a small, slow expansion from p_1, V_1 to p_2, V_2 at constant temperature T. Compared with V_1, $(V_2 - V_1)$ is to be considered very small. Evaluate the expression when $T = 200°K$, $V_1 = 300$ cm³, and $(p_1 - p_2) = 2 \times 10^{-2}$ dyne cm^{-2}.

D

ELECTRICITY AND MAGNETISM: PARTICLES AND FIELDS

13

Stationary Electric Charges

13-1   Electricity

In the previous chapters most of the emphasis has been placed on gravitational forces and very little has been said about electromagnetic forces. Whenever interatomic forces had to be introduced, we were reminded of the existence of electromagnetic forces, but the details of the electromagnetic character of interatomic forces cannot be explained properly without quantum mechanics. As we shall soon see, electromagnetic forces are more complicated than gravitational forces and our discussion of them will lead us to important new concepts such as field theory, the special theory of relativity, and eventually quantum mechanics. Although interatomic forces play an important part in everyday experience, there are no common everyday experiences that make us directly aware of the basic character of electromagnetic forces in the same way that the behavior of falling bodies makes us directly aware of the character of gravitational forces. To discover the basic facts about electromagnetic forces we must do some deliberate experimentation such as the following.

Rub a glass rod with a silk cloth. Suspend the rod at the end of a length of string, as in figure 13-1a, and place the silk cloth near one end of it. The rod will be attracted to the silk. Rubbing the glass and silk together has conditioned them in such a way that they exert additional forces on one another. Moreover, an estimate of the magnitude of these forces shows that they are comparable with the weight of the glass rod. This is an important point, because the weight of the rod results from the gravitational attraction of an extremely large body, the earth, whereas the new "electrical force" is produced by a comparatively small body, the silk cloth. Electrical forces, once they have been produced, seem to be very strong.

Now rub a second glass rod with a second silk cloth and bring the second rod up to the first suspended rod. This time there will be repul-

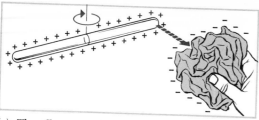

(a) The silk attracts the glass rod.

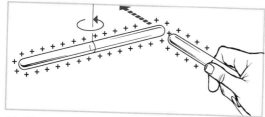

(b) The two glass rods repel one another.

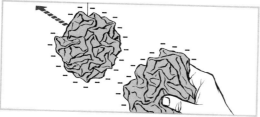

(c) The two silk cloths repel one another.

Figure 13-1 The basic experimental facts of electricity.

sion (figure 13-1*b*). Similarly, if one of the silk cloths is wadded into a ball and suspended, and the second cloth is brought up to it, again there is repulsion (figure 13-1*c*). Gravitational forces are always attractive, but this new force seems to be sometimes attractive and sometimes repulsive.

These facts were known to the ancient Greeks, but the first attempt at a correct explanation of them was made by Benjamin Franklin and his contemporaries in the eighteenth century. They postulated that electricity is some sort of fluid which exists in two forms, positive and negative. Most bodies are *electrically neutral* because they contain equal amounts of positive and negative electricity that exactly cancel one another. When the silk cloth is rubbed against the glass rod, electricity flows from one to the other in such a way that the glass rod is left with an excess of positive electricity and the silk cloth therefore has an equal excess of negative electricity. It is assumed that each type of fluid can be neither created nor destroyed. Since the glass and the silk each contained equal amounts of positive and negative electricity before they were rubbed together, it is easy to see that, if a certain amount of excess positive electricity appears on the glass, an equal amount of negative electricity must appear on the silk, assuming that suitable experimental precautions have been taken to prevent any of the electricity from leaking away.

In figure 13-1 these ideas have been represented by placing plus signs near the glass rod and minus signs near the silk cloth. The experimental facts illustrated in this figure prove that positive electricity repels positive electricity, negative electricity repels negative electricity, but positive and negative electricity attract one another.

Modern physics explains all this in the following way. Atoms are composed of neutrons, protons, and electrons. Neutrons are electrically neutral, so we may ignore them at the moment. A proton is a unit of positive electricity and an electron is a unit of negative electricity. These two units have equal but opposite electrical effects, so that a negative electron placed exactly on top of a positive proton would cancel it electrically. An atom contains a certain number of protons in a central nucleus and an equal number of electrons orbiting around this nucleus. Except in the immediate vicinity of the atom, this turns out to be equivalent to putting the electrons on top of the nucleus and the atom as a whole is therefore electrically neutral. When the silk is rubbed against the glass, some of the electrons near the outside of the glass atoms (silicon and oxygen) are rubbed off and migrate into the silk. Electrons prefer to be in silk rather than glass for reasons that are now understood but are too complicated to describe here. The silk cloth therefore acquires an excess of electrons and a net negative charge. The glass rod has a deficiency of electrons, or, if you prefer, an excess of protons, and this gives it a net positive charge.

This model gives a facile explanation of the nature of electricity, but let us extract from it the following basic facts, which are still not properly understood. As far as electrical effects are concerned, fundamental particles come in three different kinds, neutral, positive, and negative. Neutral particles show no electrical effects. Positive particles repel positive particles. Negative particles repel negative particles. A positive particle and a negative particle attract one another. The effect of a negative particle is exactly equal but opposite to the effect of a positive particle.

The idea that electricity cannot be created or destroyed may seem to imply that electrons and protons cannot be created or destroyed. We now know that there are processes in nature during which particles are created or destroyed. However, particles are always created in pairs, one negative and one positive. Conversely, in a destruction process a positive particle and a negative particle come together and destroy one another. The sum of positive and negative electricity therefore remains constant. This is known as **conservation of charge**. We might even go a little further than what we are quite certain of, and guess that the number of positive particles in the universe is always exactly equal to the number of negative particles.

Here we shall repeat in more detail some of the things that have already been said in sections 6-4 and 6-6. Throughout the discussion of electromagnetic phenomena, and for that matter throughout most of the rest of the book, the main emphasis will be placed on CGS units rather than MKS units. The main reason for this is that the basic formulas with which we are concerned are simpler in the CGS system and it is easier to emphasize the fundamental physical principles involved. The MKS system introduces certain complications which tend to obscure the basic issues. Actually, the introduction of these complications at an early stage results in a simplification of the equations at a more advanced stage of the theory. The MKS system is therefore preferred by electrical engineers and others whose approach to the subject is a practical one. We shall not be primarily concerned with this advanced aspect of the theory and we will confine ourselves to a brief summary of the MKS approach after the main ideas have been presented, mainly in order to introduce practical units such as the ampere and the volt.

Suppose that a body contains n_+ fundamental particles with positive electric charge and n_- fundamental particles with negative electric charge. Then the total electric charge, q, on the body is

$$q = (n_+ - n_-)e \tag{13-1}$$

In the CGS system the unit of charge is either called the **statcoulomb,** or sometimes the **electrostatic unit of charge,** abbreviated to esu. The quantity e is a fundamental constant of nature. It is a measure of the electric charge on any single charged fundamental particle, but for historical reasons it is called **the charge on the electron.** Its value is

$$e = 4.802 \times 10^{-10} \text{ statcoulomb} \tag{13-2}$$

For definiteness, we shall always take e to be a positive number. Then, in the case of a positively charged body n_+ is greater than n_- and the total charge q is a positive quantity. In the case of a negatively charged body n_+ is less than n_- and q is a negative quantity.

Suppose that a body with a charge q_1 is at a distance R from a body with a charge q_2 and that they are both stationary (figure 13-2). As in the case of gravitation, it is necessary to assume that the size of each body is small compared with their distance apart. In the CGS system each body experiences an electrostatic force given by the equation

Figure 13-2   Illustrating Coulomb's Law.

249

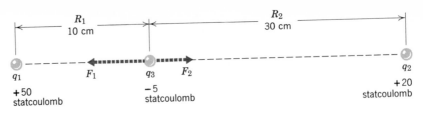

Figure 13-3   Illustrating example 13-2-1.

Coulomb's law

$$F_e = \frac{q_1 q_2}{R^2}$$

(13-3)

R is measured in cm, q_1 and q_2 in statcoulombs, and F_e is then in dynes. The forces on the two bodies act along the line joining their centers. The two forces are equal in magnitude, but opposite in direction, obeying Newton's third law. If both charges are positive, or if both are negative, the forces are repulsive. If one of the charges is positive and the other is negative, the forces are attractive.

The stipulation that the charges must be stationary is very important. If the charges are in motion there are additional forces which produce the phenomena of magnetism. The study of the electrical effects of *stationary* charges is called **electrostatics.**

Example 13-2-1

Calculate the resultant force on the charge q_3 in figure 13-3.
The force exerted by q_1 on q_3 is

$$F_1 = \frac{q_1 q_3}{R_1{}^2}$$

$$= \frac{50 \times 5}{10^2}$$

$$= 2.5 \text{ dyne}$$

(13-4)

Since q_1 is positive and q_3 is negative, this force is attractive and is directed to the left toward q_1.
The force exerted on q_3 by q_2 is

$$F_2 = \frac{q_2 q_3}{R_2{}^2}$$

$$= \frac{20 \times 5}{30^2}$$

$$= \frac{100}{900}$$

$$= 0.111 \text{ dyne}$$

(13-5)

Since q_2 is positive and q_3 is negative, this force is attractive and is directed to the right toward q_2.

The resultant force on q_3 is clearly

$$F_R = 2.5 - 0.111$$
$$= 2.389 \text{ dyne} \tag{13-6}$$

and is directed to the left.

Example 13-2-2

Calculate the magnitude of the electrostatic force exerted by the proton on the electron in a hydrogen atom and compare it with the weight of the electron.

In a hydrogen atom the nucleus is a single proton and a single electron moves around it at an average distance away of 0.53×10^{-8} cm. The charge on each particle is e and the electrostatic force is

$$F_e = \frac{e^2}{R^2}$$
$$= \frac{(4.8 \times 10^{-10})^2}{(0.53 \times 10^{-8})^2}$$
$$= 8.2 \times 10^{-3} \text{ dyne} \tag{13-7}$$

The mass of an electron is

$$m_e = 9.11 \times 10^{-28} \text{ gram} \tag{13-8}$$

Its weight is therefore

$$m_e g = 9.11 \times 10^{-28} \times 980$$
$$= 8.9 \times 10^{-25} \text{ dyne} \tag{13-9}$$

The ratio of the electrostatic force to the weight is

$$\frac{F_e}{m_e g} = \frac{8.2 \times 10^{-3}}{8.9 \times 10^{-25}}$$
$$= 9.2 \times 10^{21} \tag{13-10}$$

The electrostatic force is therefore overwhelmingly larger than the weight. We never have to worry about gravitational forces when we are considering the theory of the hydrogen atom.

Example 13-2-3

Calculate the magnitude of the electrostatic force of repulsion between two protons in a nucleus and the mass of material that this force would be able to support in the gravitational field at the surface of the earth.

The distance between two protons in a nucleus is about 5×10^{-13} cm. The electrostatic force between them is therefore

$$F_e = \frac{e^2}{R^2}$$

Stationary.
Electric
Charges

251

$$= \frac{(4.8 \times 10^{-10})^2}{(5 \times 10^{-13})^2}$$

$$= 9.2 \times 10^5 \text{ dyne} \tag{13-11}$$

This could support the weight of a mass m given by

$$mg = F_e \tag{13-12}$$

$$m = \frac{F_e}{g}$$

$$= \frac{9.2 \times 10^5}{980}$$

$$= 940 \text{ gram} \tag{13-13}$$

The force is so large that it could support a weight of about 1 kilogram! One kilogram of matter contains about 6×10^{26} protons and neutrons!

This large electrostatic force is repulsive and tends to blow the nucleus apart. The *nuclear* forces which hold the nucleus together must be even stronger.

13-3 The Relative Magnitudes of Gravitational and Electrostatic Forces

The last two examples illustrate how electrostatic forces can become enormously large compared with gravitational forces. Let us consider this matter directly by comparing the electrostatic force with the gravitational force for a proton at a distance R from an electron. The electrostatic force is

$$F_e = \frac{e^2}{R^2} \tag{13-14}$$

The gravitational force is

$$F_G = G \frac{m_e m_p}{R^2}$$

where m_e and m_p are the masses of the electron and proton respectively. The ratio is

$$\frac{F_e}{F_G} = \frac{e^2}{G m_e m_p} \tag{13-15}$$

$$= \frac{(4.8 \times 10^{-10})^2}{(6.67 \times 10^{-8}) \times (9.11 \times 10^{-28}) \times (1.67 \times 10^{-24})}$$

$$= 2.3 \times 10^{39} \tag{13-16}$$

This is an enormously large number! The electrostatic interaction is very much stronger than the gravitational interaction.

Since gravitational forces are so puny compared with electrostatic forces, it is interesting to consider why gravitation is so important in our everyday experience. The reason, of course, is that electricity comes in positive and negative forms, which can cancel one another out. Most matter is electrically neutral and produces no electrostatic forces, so that the gravitational forces are given a chance to reveal themselves. In a world composed entirely of negative electrons the gravitational forces would be completely overpowered by the electrostatic forces.

It seems strange that there should be two different kinds of interaction, that they should be similar in the sense that they both vary as $1/R^2$, and yet one is so very much stronger than the other. The ratio F_e/F_G depends only on the fundamental constants e, G, m_e and m_p. Moreover, these constants are combined in such a way that the ratio is independent of the units of mass, length, and time. The electrostatic force is 2.3×10^{39} times stronger than the gravitational force, and this fact would not be influenced in any way if the standard meter were defined in terms of an entirely different platinum-iridium bar, the standard kilogram in terms of a different platinum-iridium cylinder, or if the rate of rotation of the earth changed overnight, thus altering the definition of the second of time. Many scientists believe that there must be a fundamental significance in a number such as F_e/F_G which is related to two of the fundamental interactions of nature, but which is independent of the arbitrariness of human decision.

The British astronomer and physicist, Sir Arthur Stanley Eddington, has suggested that F_e/F_G is directly related to certain properties of the universe as a whole. Modern theories of the universe strongly suggest that there is a limit to the observable universe. There is reason to believe that, however good our telescopes, we shall not be able to see beyond a distance R_u, which we shall call the **radius of the observable universe** and which is approximately 10^{28} cm. The radius of an electron, r_e, is approximately 3×10^{-13} cm. Therefore

$$\frac{\text{Radius of universe}}{\text{Radius of electron}} = \frac{R_u}{r_e} = \frac{10^{28}}{3 \times 10^{-13}}$$

$$= 30 \times 10^{39} \text{ approximately} \qquad (13\text{-}17)$$

This number is also independent of the unit of length, is related to very fundamental properties of the universe and is very large, yet its magnitude is not very different from the ratio F_e/F_G. Here is a hint that the factors that determine the ratio of the electrostatic to the gravitational interaction also determine the ratio of the radius of the universe to the radius of the electron.

An interesting explanation of this numerical coincidence has been provided by the American physicist and astronomer, R. H. Dicke. He approaches the question from the point of view of the present *age* of the universe, T_u. The present *size* of the *observable* universe, R_u, is determined by the distance which light travels in a time T_u, because light originating from a greater distance than R_u could not yet have reached us even if it had started out at the beginning of time and, thus, we could not yet have seen that distant region of the universe. If c is the speed of light,

$$R_u = cT_u \qquad (13\text{-}18)$$

(This equation is not quite accurate because of the continual expansion of the universe, but that will not make any real difference to the present rough arguments.)

The time t_e for light to travel a distance equal to the radius of an electron is given by

$$r_e = ct_e \qquad (13\text{-}19)$$

This time interval t_e can be used as a basic unit of time relevant to the whole of atomic physics and is free from the arbitrariness of the conventional unit of time, the second, which depends upon the rate of rotation of the particular planet on which we happen to live. In terms of this basic "atomic unit of time," the age of the universe is T_u/t_e and equations 13-18 and 13-19 tell us that

$$\frac{T_u}{t_e} = \frac{R_u}{r_e} \qquad (13\text{-}20)$$

This is the very quantity that interests us because it has roughly the same magnitude as F_e/F_G.

The universe must be at least old enough to have given life a chance to originate; otherwise, we would not be here to discuss the matter. Living organisms in general and our bodies in particular contain elements such as carbon, oxygen, nitrogen and phosphorus. The early universe is believed to have contained mainly hydrogen, with perhaps a little helium, and the more complicated elements are believed to have been manufactured by nuclear reactions in the interiors of certain short-lived stars during the later phases of their evolution. A limit to how young the universe can be is therefore obtained by developing a theory of the evolution of stars and estimating the shortest possible time in which a suitable star can manufacture the elements needed by life.

A limit on how old the universe can be is provided by the need to have an active star (the sun) pouring out energy to warm a planet (the earth) and to provide a suitable environment for the existence of life (us). The theory of stars should therefore also be used to calculate the life span of a comparatively long-lived star like the sun.

The stability of a star and the rate at which it uses up its fuel are determined by a delicate balance between two opposing effects. The gravitational attraction between its inner regions and its outer regions would make the star collapse under its own weight if it were not counteracted by the "blowing up" effect of the thermal motion of the atoms. Gravity therefore plays a decisive role and the order of magnitude of the time taken to exhaust the nuclear fuel and produce heavy elements can be shown to be roughly

$$T_s = \frac{e^2}{Gm_e m_p} t_e \qquad (13\text{-}21)$$

which is about one billion years. The shortest-lived stars are the most massive stars, and they produce heavy elements in a time somewhat less than T_s. If these elements are then incorporated into a planetary system with a central star like the sun, this star will keep the planets warm for a time somewhat greater than T_s. (Our solar system is known to be about five billion years old). If we are not too particular about factors of 10 or 100, we can therefore say that the age of our universe, T_u, is about as big as T_s. Equation 13-21 then tells us that

$$\frac{T_u}{t_e} = \frac{e^2}{Gm_e m_p} \quad \text{roughly} \tag{13-22}$$

From equations 13-15 and 13-20 this is seen to be equivalent to

$$\frac{R_u}{r_e} = \frac{F_e}{F_G} \quad \text{roughly} \tag{13-23}$$

There is another numerical coincidence which may have an even more profound significance. From the astronomical observations on that part of the universe which is nearest to us, we can make a rough estimate of the average density of matter in the universe and hence deduce the total number of fundamental particles in the universe. We find that, very approximately, the universe contains about 10^{79} protons, neutrons, and electrons. Note that

$$\left(\frac{F_e}{F_G}\right)^2 = 0.5 \times 10^{79} \tag{13-24}$$

The number of fundamental particles in the present observable universe is approximately equal to the *square* of the large number, F_e/F_G, that we have just been discussing.

This second numerical coincidence is partly dependent upon what we have previously said about the present age of the universe, but it cannot be completely explained in that way. It can be explained by applying the General Theory of Relativity to the universe as a whole and making certain assumptions about the large scale geometrical properties of space, but the physical significance of this approach is not yet entirely clear. There is a strong possibility that the ultimate explanation will be related to Mach's Principle, which will be discussed in chapter 25. This principle says that the inertial mass of a particle like an electron is determined by a novel type of interaction, probably gravitational in character, between the electron and all the particles in the universe. This obviously might lead to a connection between the number of particles in the universe and the constants which are relevant to local physics.

Example 13-3-1

A shower of protons from outer space deposits equal charges $+q$ on the earth and the moon, and the electrostatic repulsion then exactly counterbalances the gravitational attraction. How large is q and what is the fractional change in the number of protons in the moon?

If R is the distance between the earth and the moon, the electrostatic force is

$$F_e = \frac{q^2}{R^2} \tag{13-25}$$

Stationary
Electric
Charges

If M_e and M_l are the masses of the earth and the moon respectively, the gravitational force is

$$F_G = \frac{GM_eM_l}{R^2} \qquad (13\text{-}26)$$

Since the two forces are equal,

$$\frac{q^2}{R^2} = \frac{GM_eM_l}{R^2} \qquad (13\text{-}27)$$

$$q = \sqrt{GM_eM_l} \qquad (13\text{-}28)$$

Inserting $G = 6.67 \times 10^{-8}$ CGS units, $M_e = 6.0 \times 10^{27}$ grams, and $M_l = 7.3 \times 10^{25}$ grams,

$$q = 1.7 \times 10^{23} \text{ statcoulombs} \qquad (13\text{-}29)$$

The number of protons that must be deposited on the moon to produce this charge is

$$\frac{q}{e} = \frac{1.7 \times 10^{23}}{4.8 \times 10^{-10}}$$

$$= 3.5 \times 10^{32} \qquad (13\text{-}30)$$

The moon is composed of electrons, protons, and neutrons. The mass of the electrons is small compared with the mass of the protons and neutrons. The mass of a neutron is almost the same as the mass of a proton and in the nuclei of the atoms of the moon the number of protons is approximately equal to the number of neutrons. The number of protons in the moon is therefore approximately $M_l/2m_p$, where M_l is the mass of the moon and m_p is the mass of a proton (1.67×10^{-24} grams). So,

$$\text{Number of protons in moon} = \frac{7.3 \times 10^{25}}{2 \times 1.67 \times 10^{-24}}$$

$$= 2.2 \times 10^{49} \text{ approximately} \qquad (13\text{-}31)$$

Hence,

$$\frac{\text{Number of protons added}}{\text{Number of protons already present}} = \frac{3.5 \times 10^{32}}{2.2 \times 10^{49}}$$

$$= 1.6 \times 10^{-17} \qquad (13\text{-}32)$$

This emphasizes how delicate is the balance between protons and electrons in the solar system. If only a very small proportion of the protons were not neutralized by electrons, this would make a very big difference to the motion of the planets and their moons.

13-4 Electrostatic Potential Energy

When a mass M_1 is at a distance R from a mass M_2, the *gravitational* force between them is

$$F_G = \frac{GM_1M_2}{R^2} \tag{13-33}$$

Their mutual *gravitational* potential energy is

$$\Phi_G = -\frac{GM_1M_2}{R} \tag{13-34}$$

When a charge q_1 is at a distance R from a charge q_2, the *electrostatic* force between them is

$$F_e = \frac{q_1q_2}{R^2} \tag{13-35}$$

Since this has a similar mathematical form to the gravitational force, the reader should not find it difficult to believe that:

The mutual electrostatic potential energy of two charges

$$\Phi_e = \frac{q_1q_2}{R} \tag{13-36}$$

If q_1 and q_2 are measured in statcoulombs and R in cm, the potential energy is measured in ergs. Notice that, as in the case of gravitational forces, the zero of potential energy corresponds to the situation when the two charges are infinitely far apart. When $R = \infty$, $\Phi_e = 0$.

The sign of the potential energy requires careful consideration. If either q_1 or q_2 is negative and the other is positive, the electrostatic force is attractive. The analogy with gravitation is complete and the potential energy should be negative. This is obviously so, because in equation 13-36 R is intrinsically positive and if either q_1 or q_2 is negative and the other is positive, the whole expression for Φ_e is negative.

If q_1 and q_2 are both positive, or both negative, the mutual potential energy is a positive quantity. This is so because the force between the charges is repulsive. Imagine two positive charges q_1 and q_2 initially at rest at a distance R_0 apart as in figure 13-4a. (Similar arguments apply

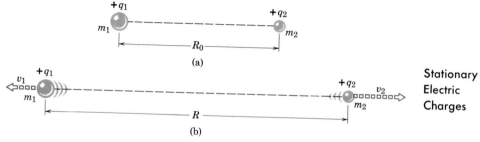

Figure 13-4 Conservation of energy for two charges repelling one another.

Stationary Electric Charges

to two negative charges). Then

$$\text{Initial kinetic energy} = 0 \tag{13-37}$$

$$\text{Initial potential energy} = +\frac{q_1 q_2}{R_0} \tag{13-38}$$

The two charges repel one another and move directly away from one another. When they are at a distance R apart, suppose that they have acquired velocities v_1 and v_2, as in figure 13-4b. Under these circumstances

$$\text{Kinetic energy} = \tfrac{1}{2}m_1 v_1^2 + \tfrac{1}{2}m_2 v_2^2 \tag{13-39}$$

$$\text{Potential energy} = +\frac{q_1 q_2}{R} \tag{13-40}$$

Applying the law of conservation of energy and equating the initial and final values of the total energy,

$$\tfrac{1}{2}m_1 v_1^2 + \tfrac{1}{2}m_2 v_2^2 + \frac{q_1 q_2}{R} = \frac{q_1 q_2}{R_0} \tag{13-41}$$

Since R is greater than R_0 and the potential energy is positive, the final potential energy $q_1 q_2/R$ is less than the initial potential energy $q_1 q_2/R_0$. As the motion proceeds, the potential energy decreases and the energy thus made available appears as kinetic energy, which is intrinsically positive.

The discussion of this example has been oversimplified. Actually, as the charges accelerate away from one another they lose energy by radiating heat and light. For slow accelerations the effect is unimportant and can be ignored, but for rapid accelerations it completely invalidates equation 13-41. We are beginning to see how electromagnetic interactions are much more complicated than gravitational interactions.

Example 13-4-1

Calculate the total mutual potential energy of all the three charges in figure 13-5.

We must include the mutual potential energy of each pair of charges. The mutual potential energy of q_1 and q_2 is

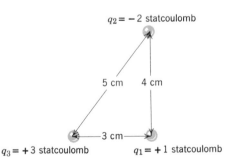

$q_2 = -2$ statcoulomb

5 cm   4 cm

$q_3 = +3$ statcoulomb   3 cm $q_1 = +1$ statcoulomb

Figure 13-5   Illustrating example 13-4-1.

$$\Phi_{12} = \frac{(+1) \times (-2)}{4} \tag{13-42}$$

$$= -0.5 \text{ erg}$$

The mutual potential energy of q_2 and q_3 is

$$\Phi_{23} = \frac{(-2) \times (+3)}{5} \tag{13-43}$$

$$= -1.2 \text{ erg}$$

The mutual potential energy of q_3 and q_1 is

$$\Phi_{31} = \frac{(+3) \times (+1)}{3}$$

$$= +1.0 \text{ erg} \tag{13-44}$$

The total mutual potential energy is

$$\Phi_{123} = \Phi_{12} + \Phi_{23} + \Phi_{31} \tag{13-45}$$

$$= -0.5 - 1.2 + 1.0$$

$$= -0.7 \text{ erg} \tag{13-46}$$

QUESTIONS

1. If you were given a charged body, how would you decide whether its charge was positive or negative?
2. Does the mass of a body change when it is given a charge? If so, does it increase or decrease?
3. A charged body attracts small *uncharged* objects such as dust particles. Can you imagine why?
4. Given N protons and N electrons, can you arrange them in some geometrical pattern such that the resultant electrostatic force on every single particle is zero? Think about this for N equal to 1, 2 and 3 and N very large. What conclusions are suggested about the structure of matter?
5. Consider the following attempt to explain gravity in terms of electrical forces. "The earth has a net positive charge. Falling bodies contain equal numbers of protons and electrons but the negative charge on the electron is slightly larger than the positive charge on the proton." In particular consider the following points. (a) Why does this theory not explain the fact that all bodies fall with the same acceleration? (b) Can you think of an assumption about the nature of the neutron which would make all bodies fall with the same acceleration? (c) Can the motion of the moon around the earth be explained by the theory? (d) Are there any difficulties in explaining the motions of all the bodies in the solar system on this theory? You may choose the charge on each body in the solar system to have any value that is needed to fit the facts. (e) What experiments might be performed to test the theory?

Stationary.
Electric
Charges

PROBLEMS

A

1. What is the charge on 1 gm of electrons?
2. What is the charge on 1 mole of a material if one electron is removed from each molecule? (This quantity of electricity is called 1 Faraday.)
3. A charge of $+5$ statcoulombs is 2 cm away from a charge of $+7$ statcoulombs. What is the electrostatic force on each charge? What is their mutual electrostatic potential energy?
4. A charge of -3 statcoulombs is 1.5 m away from a charge of $+9$ statcoulombs. What is the electrostatic force on each charge? What is their mutual electrostatic potential energy?
5. What must be the distance between two charges of 15 statcoulombs and 40 statcoulombs if the electrostatic force on either is to be 6 dynes?
6. An electron and a proton are stationary at a distance apart of 2×10^{-4} cm. What is the electrostatic force on (a) the proton (b) the electron? What is their mutual electrostatic potential energy?
7. A mass of 5 gm carrying a charge of 200 statcoulombs is at a distance of 15 cm from a mass of 9 gm carrying a charge of 40 statcoulombs. Calculate the acceleration of each mass.
8. An electron and a proton are instantaneously at rest at a distance apart of 5×10^{-9} cm. What is the acceleration of (a) the proton (b) the electron?
9. Make a rough estimate of the electrostatic potential energy of a hydrogen atom.

B

10. Referring to figure 13-3, calculate the direction and magnitude of the resultant forces on q_1 and q_2. What is the mutual potential energy of all three charges?
11. Referring to figure 13-5, calculate the resultant electrostatic force on q_1.
12. What equal charges must be placed on the sun and the earth to counterbalance the gravitational force between them?
13. Three equal charges of $+4$ statcoulomb are placed at the vertices of an equilateral triangle, each side of which is 2 cm long. Calculate the electrostatic force on any one of these charges. What is their total mutual potential energy?
14. Make a rough estimate of the electrostatic potential energy of a lithium nucleus which contains three protons.
15. Using x and y coordinates in two dimensions and measuring all lengths in cm, a charge of $+3$ statcoulomb is placed at the point $(-3, 0)$. A charge of $+5$ statcoulomb is placed at the point $(0, -5)$. Calculate the magnitude and direction of the electrostatic force on a charge of $+2$ statcoulomb placed at the origin. What is the total mutual potential energy of all three of these charges?

16. Again we are using x and y coordinates and measuring lengths in cm. A charge of $+2.5$ statcoulomb is placed at the point $(+2, 0)$. A charge of -9 statcoulomb is placed at the point $(+7, 0)$. Where must a charge of $+1.5$ statcoulomb be placed so that the resultant electrostatic force on the -9 statcoulomb is zero?

17. A charge of $+5$ statcoulomb is placed at the origin and a charge of $+45$ statcoulomb is placed at the point $(+10, 0)$. Where must a third charge be placed so that the resultant electrostatic force on it is zero?

18. A charge of $+5$ statcoulomb is placed at the origin and a charge of -45 statcoulomb is placed at the point $(+10, 0)$. Where must a third charge be placed so that the resultant electrostatic force on it is zero?

C

19. One ton (9.07×10^5 gm) of protons is taken from the earth, loaded on to a rocket and landed on the moon. At the same time the velocity of the moon is adjusted so that it remains in the same stable circular orbit. How many seconds difference does this make to the length of the month? Is it an increase or a decrease?

20. A proton and an electron are initially at rest at a very great distance from one another. They then move toward one another under the influence of their mutual electrostatic attraction. When they are 1 cm apart calculate (*a*) the sum of their kinetic energies, (*b*) the ratio of their velocities, (*c*) their individual velocities. Assume that there is no appreciable loss of energy by radiation.

21. If the earth had its present mass, but were composed entirely of protons, what would be the escape velocity of an electron? (Use the formulas of classical physics. The velocity calculated in this way greatly exceeds the velocity of light and is not possible according to the theory of relativity.)

22. A single proton approaches a nucleus containing Z protons and N neutrons. If the nucleus can be regarded as firmly held in position, what is the minimum velocity which the proton must have at a great distance from the nucleus in order to approach within 10^{-12} cm of it? About how much difference would it make if the nucleus were completely free to move? Assume that the only forces acting on the proton are electrostatic.

23. If an electron moves around a proton in a stable circular orbit of radius r, find an expression for its velocity in terms of r, e and the mass of the electron, m_e. What is the ratio of its kinetic energy to its potential energy?

24. In the figure the dipole consists of two bodies carrying charges $+q$ and $-q$ connected by a rigid rod of length l. The charge $+Q$ is at a distance R from the center of the rod and all three charges lie on the same straight line. Prove that, if R is very much greater than l, the force on Q due to the dipole is approximately equal to $2Qql/R^3$.

Stationary
Electric
Charges

261

Problem 13-24

25. Four protons are placed at the corners of a regular tetrahedron with a side of length 10^{-8} cm. Calculate exactly the electrostatic potential energy of the arrangement.

26. Eight protons are placed at the corners of a cube with a side of length 10^{-8} cm. Calculate exactly the electrostatic potential energy of the arrangement.

27. Calculate the total number of particles, N_s, in the sun, assuming that it is composed entirely of hydrogen. If

$$N_s = \left(\frac{F_e}{F_G}\right)^n$$

(see section 13-3) calculate the value of the index n and comment.

The Concept of a Field

14-1   A New Look at the Universe

So far we have been developing the idea that the universe is a collection of discrete particles moving through empty, featureless space. We have assumed that, in the ultimate analysis, the behavior of the universe must be described in terms of the motion of these particles, and that this motion can be explained and predicted if we know certain laws that tell us the acceleration given to a particle by its interactions with the other particles. We have introduced the first two of these laws, Newton's law of universal gravitation and Coulomb's law for electrostatic forces.

The next step is to consider a third type of force which exists between two *moving* charges and which is the origin of magnetism. Unfortunately the laws governing magnetic forces cannot be expressed as simply as the laws of gravitation and electrostatics. The next step in our program is therefore a difficult one. In the early nineteenth century, when the laws of magnetism were being discovered, the discovery was considerably aided by a new attitude to the nature of the universe embodied in the concept of electric and magnetic fields. Towards the end of the century, Clerk Maxwell was able to explain all the known phenomena of electromagnetism by a set of simple, elegant equations describing the behavior of the electric and magnetic fields. Simplicity and success are two powerful advocates for any scientific theory, so we must now turn our attention towards this new idea of a **field.**

Our previous attitude implies that two bodies at a great distance from one another, with only empty, featureless space between them, can nevertheless influence one another's motion. This is called **action-at-a-**

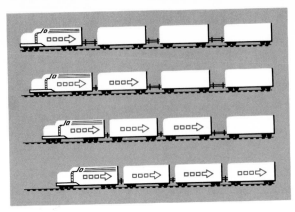

Figure 14-1   Transmission of an interaction along a line of railroad cars.

distance and it worried many scientists and philosophers, who felt that two bodies could interact only if they were in "direct contact" with one another. To appreciate this point of view, imagine a railroad engine backing up to a line of railroad cars. The shock of impact is transmitted along the line as the first car hits the second, the second hits the third, and so on, until the disturbance eventually reaches the caboose (figure 14-1). However, in a situation such as the gravitational attraction between the earth and the moon, there are no visible, tangible bodies in between to provide the connecting links (figure 14-2). To overcome this difficulty, it was suggested that the whole of space is filled with an invisible material called the **ether,** having properties similar to a jelly. If you press your finger into a jelly, the strain is transmitted to all parts of the jelly and a pressure is exerted on a body embedded in the jelly some distance away (figure 14-3). In the same way, the earth was imagined to produce a strain in the ether which extended to the moon and transmitted a force to the moon (figure 14-4).

   This idea of a material ether pervading the whole of space no longer appears to be tenable. Some of the most forceful arguments against it will be presented later in our discussion of the theory of relativity. Nevertheless, this line of thought is fruitful in the sense that it has taken our attention away from the material bodies and discrete particles and has raised the important question of whether space itself does not possess

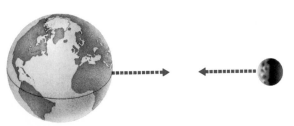

Figure 14-2   The moon and the earth interact with one another even though the space between them is apparently empty.

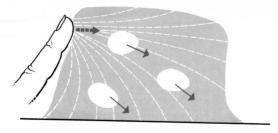

Figure 14-3 Strain is transmitted through a jelly to objects embedded in the jelly.

properties of interest to the physicist. Our original prescription for describing the instantaneous state of the universe was to define the position of each elementary particle by giving its Cartesian coordinates. Now let us consider a possible alternative description in which physical properties are assigned to each point in space even if there is no particle located at that point. The procedure is to designate each point by giving the values of its x, y, and z coordinates and then to associate with this point certain other numbers that are a measure of such physical properties of the point as the gravitational field, the electric field, and the magnetic field. Exactly how this is done and what it means physically will be discussed throughout the rest of this chapter. The important point to make at this stage is that this procedure is not only possible, but extremely useful. In particular, it clarifies and simplifies the complicated subject of electromagnetism. Later on it will help us to understand the nature of light and other electromagnetic radiations. It is an essential aspect of quantum mechanics and the general theory of relativity.

14-2   Analogy with a Flowing Stream

Since the above discussion has been somewhat abstract, a concrete analogy might prove helpful. Imagine a rapidly flowing trout stream containing several whirlpools that dimple its surface (figure 14-5). A complete description of the flow could be obtained by giving the velocity of the water at each point in the stream. Mathematically this would involve first setting up three Cartesian axes and then determining

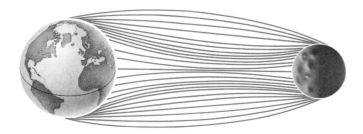

The
Concept
of a Field

Figure 14-4   Is there a jelly-like ether between the earth and the moon? Is strain transmitted through this jelly to produce an attraction between the earth and the moon?

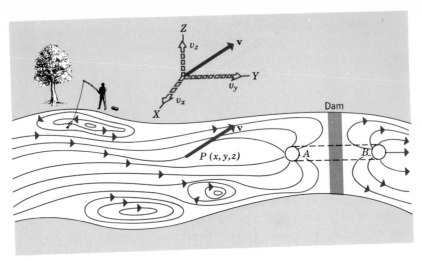

Figure 14-5 The velocity field of a flowing stream.

the Cartesian coordinates x, y, and z of the point P relative to these axes. The velocity **v** of a small sample of water located at P could then be given as a set of three numbers v_x, v_y, and v_z, the X, Y, and Z components of this velocity (see section 3-6). The resulting description of the flow might be referred to as a **velocity field**. Associated with each point (x, y, z) in the water would be a set of three numbers (v_x, v_y, v_z), the physical significance of which would be that they were the Cartesian components of the velocity of the water at that point.

A pictorial representation of this velocity field is obtained from the stream lines of figure 14-5. Each stream line represents the path followed by a small sample of water as it flows down the stream. (This statement is true only if the pattern of stream lines does not change with time. We are idealizing the flow of an actual trout stream.) At any point P the velocity **v** of the water is tangential to a stream line.

In the vicinity of a whirlpool the water is moving around in circles. Each whirlpool produces a dimple on the surface of the water. Are these dimples the analogs of elementary particles? A dimple is a highly localized entity which can move over the surface of the stream in the same way that an elementary particle moves through space. However, there is no material body located at the dimple. It is merely a point at which the stream lines behave in a peculiar manner. This suggests to us an extreme swing of the pendulum in our attitude toward describing the universe. Now the emphasis is placed on assigning properties to all the points in space so that space is filled with various fields. An elementary particle is then a point at which a field behaves in a peculiar way. In the case of the whirlpool the stream lines go round in circles. Another possible situation, which is perhaps more akin to an elementary particle, occurs when the stream lines all converge on a point. Suppose that a dam is built across the stream and the flow is taken under the dam by drilling a very narrow tunnel below the bed of the stream, starting at a point A some distance upstream of the dam and emerging at a point B

ELECTRICITY
AND
MAGNETISM:
PARTICLES
AND FIELDS

266

some distance downstream of the dam. The stream lines all converge upon A and then diverge from B (see figure 14-5). As we shall see later in this chapter, the behavior of the streamlines in the vicinity of A or B is similar to the behavior of a gravitational or electric field in the vicinity of an elementary particle.

14-3 The Gravitational Field

To describe completely the gravitational field of the earth at any point in the vicinity of the earth, it is sufficient to know the gravitational acceleration that a body would have if placed at that point. Since this acceleration is the same for all sufficiently small bodies, independently of their size, shape, or chemical composition (section 7-1), it is a unique property of the point, which can be calculated and associated with the point even if there is in fact no accelerating body there. The gravitational acceleration is a vector which will be represented by the symbol **g**. The gravitational field therefore associates a *vector* with each point in space. The gravitational field at some representative points is shown in figure 14-6. At a point P_1 near the surface of the earth the magnitude of **g** is approximately 980 cm sec^{-2}, but the magnitude falls off inversely as the square of the distance from the center of the earth and is much smaller at a point such as P_2. The direction of **g** is approximately toward the center of the earth, but, at a point such as P_3 near a mountain M_1, the gravitational attraction of the mountain tilts the direction of **g** toward the mountain. Sometimes, as at P_4, the mountain M_2 has a "root" with a density less than the average density of the earth, in which case **g** may tilt away from the mountain. The gravitational field is therefore quite complicated.

If a small mass m is placed at any point, the gravitational force on it is

$$\mathbf{F}_G = m\mathbf{g} \tag{14-1}$$

or,

$$\mathbf{g} = \frac{\mathbf{F}_G}{m} \tag{14-2}$$

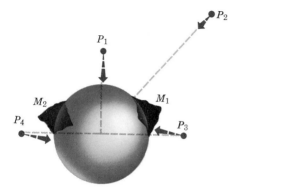

The Concept of a Field

Figure 14-6   The gravitational field **g** at some representative points in the vicinity of the earth.

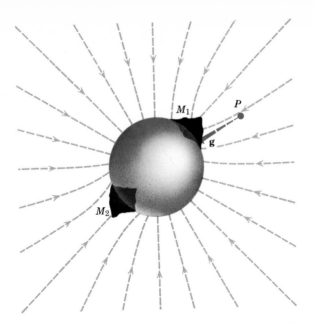

Figure 14-7 Field lines of the earth's gravitational field.

The gravitational field at a point could therefore be defined as the gravitational force *per unit mass* on a body placed at the point. It can therefore be called a "field of force," as long as it is clearly understood that the force acts on unit mass.

> **The gravitational field** at any point in space is the gravitational acceleration **g** which a small body would have if placed at that point. It might also be defined as the gravitational force per unit mass on a small body placed at the point

In the same way that the flow of the stream was represented by stream lines, the gravitational field can be represented by **field lines,** as shown in figure 14-7. The significance of a field line is that, at any point such as *P*, the tangent to the field line is in the same direction as the gravitational acceleration. The field line does *not* represent the path taken by a freely falling body; this is a much more complicated matter. Notice how the field lines converge upon the mountain M_1, but veer away from the mountain M_2 with a low density root.

If the earth were a uniform sphere of mass M_e and radius R_e, the gravitational field at its surface would be (section 7-1)

$$g_s = \frac{GM_e}{R_e{}^2} \tag{14-3}$$

(To avoid confusion, we are now adopting the symbol g_s for the acceleration due to gravity at the *surface* of the earth, and *g* is used for the

ELECTRICITY
AND
MAGNETISM:
PARTICLES
AND FIELDS

268

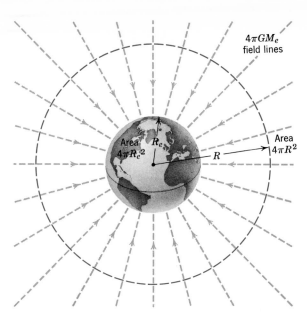

Area $4\pi R_e{}^2$

R_e

R

Area $4\pi R^2$

$4\pi GM_e$ field lines

Figure 14-8 The field lines for a perfectly uniform spherical earth. The number of lines crossing unit area is a measure of the strength of the field.

acceleration due to gravity at any arbitrary distance R from the center of the earth.) For a uniform spherical earth the field lines would be radial straight lines, as shown in figure 14-8. Suppose that these lines are uniformly distributed over the surface of the earth and that the number of lines is chosen to be $4\pi GM_e$. Then the lines are distributed over an area $4\pi R_e{}^2$ and the number of lines entering unit area of the surface of the earth is

$$\frac{4\pi GM_e}{4\pi R_e{}^2} = \frac{GM_e}{R_e{}^2}$$

$$= g_s \tag{14-4}$$

At a distance R from the center of the earth, the $4\pi GM_e$ lines cross a spherical surface of area $4\pi R^2$. The number of field lines crossing unit area of this sphere is

$$\frac{4\pi GM_e}{4\pi R^2} = \frac{GM_e}{R^2}$$

$$= g \tag{14-5}$$

The magnitude of the acceleration is always equal to the number of field lines crossing unit area of the spherical surface.

This last result can be shown to be true for any gravitational field, however complicated, due to any arbitrary distribution of masses throughout space. The field lines can always be drawn in such a way that the number of lines entering a body of mass M is $4\pi GM$. The magnitude of the gravitational field g at any point P can then be found in the

The Concept of a Field

269

following way (refer to figure 14-9). Consider a small surface of area A which contains P and is at right angles to the field lines near P. If N field lines pass through this surface, then the magnitude of the field is equal to the number of field lines per unit area, or

$$g = \frac{N}{A} \tag{14-6}$$

The direction of the field $\mathbf{g}$ is, of course, tangential to the field lines. If the area A is small enough, the direction of this tangent is the same for all the lines passing through A.

The field lines therefore provide a very vivid pictorial representation of the field. When the lines are close together the field is strong, but when the lines are far apart the field is weak. This is very obvious in the simple case of figure 14-8. Near the surface of the earth, where g is large, the lines crowd together, but as we move away from the earth and g becomes smaller, the field lines spread out.

14-4   The Gravitational Potential at a Point

In section 9-3 the mutual gravitational potential energy of two masses M_1 and m a distance R_1 apart was stated to be

$$\Phi_G = -\frac{GM_1 m}{R_1} \tag{14-7}$$

The gravitational potential at a point P a distance R_1 from a mass M_1 is defined as the mutual potential energy if unit mass ($m = 1$) were placed at P. The potential *at a point* will be designated by ϕ, a small Greek letter phi, as compared with Φ, a capital phi, for the *mutual* potential energy. Clearly the potential at P is

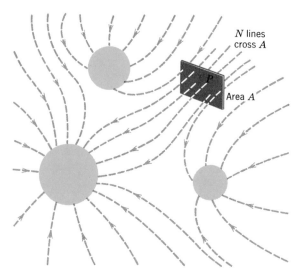

Figure 14-9   In the most general case, the number of field lines crossing unit area of a surface perpendicular to the lines is a measure of the strength of the field.

(a) The mutual gravitational potential energy of M_1 and m at a distance R_1 apart.

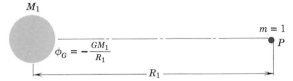

(b) The gravitational potential at a point P a distance R_1 from M_1.

Figure 14-10 The distinction between (a) the mutual gravitational potential energy of two bodies and (b) the potential at a point.

Gravitational potential at a point distant R_1 from a mass M_1

$$\phi_G = -\frac{GM_1}{R_1} \qquad (14\text{-}8)$$

The units of gravitational potential are ergs per gram or joules per kilogram. The distinction between Φ_G and ϕ_G is illustrated in figure 14-10.

Now consider the more general case when P is at a distance R_1 from a mass M_1, a distance R_2 from a mass M_2, a distance R_3 from a mass M_3 and so on for any number of masses we care to include (figure 14-11). If a mass m were placed at P, its *mutual* gravitational potential energy with respect to all the masses would be

$$\Phi_G = -\frac{GM_1m}{R_1} - \frac{GM_2m}{R_2} - \frac{GM_3m}{R_3}$$

— and so on until all the masses are included. (14-9)

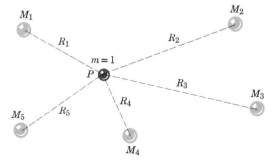

The Concept of a Field

Figure 14-11 The gravitational potential at a point P surrounded by several masses.

The gravitational potential at the point P is obtained by putting $m = 1$

Gravitational potential due to several masses

$$\phi_G = -\frac{GM_1}{R_1} - \frac{GM_2}{R_2} - \frac{GM_3}{R_3}$$

— and so on until all the masses are included. (14-10)

The gravitational potential is a property of the point P that can be associated with this point even if there is no mass m there. If a mass m were placed at P its potential energy would be

$$\Phi_G = m\phi_G \qquad (14\text{-}11)$$

The **gravitational potential field** is completely defined if each point in space has associated with it a number ϕ_G, the physical significance of which is that, if a mass m were placed at that point, its mutual potential energy with respect to all the other masses would be $m\phi_G$. Since the potential ϕ_G is a scalar, the gravitational potential field associates a scalar quantity with each point in space, whereas the field **g** associates a vector with each point in space.

If we define the gravitational field in terms of the acceleration **g**, then each point has three numbers associated with it, the X, Y, and Z components of **g**. The potential field associates only one number, ϕ_G, with each point in space, and is therefore much simpler. It is relevant to ask whether one gains this simplicity at the expense of not having complete information about the field. In fact there is no lack of completeness, since we shall now prove that, if ϕ_G is known at all points in space, the value of **g** at all points can be deduced from it.

Imagine a body of mass m located at the point A, as in figure 14-12. The field exerts a force $m\mathbf{g}$ on this body. To hold the body in position at A it is necessary to apply an external force **F** of equal magnitude in the opposite direction. If the magnitudes F and g of **F** and **g** are taken to be intrinsically positive,

$$F = mg \qquad (14\text{-}12)$$

The reader can imagine that he is holding m with his hand and applying this force **F** to it. Now allow the body to move slowly to a neighboring point B very close to A, so close that the force $m\mathbf{g}$ changes very little between A and B. Then, in accordance with the definition of work given in section 9-6;

The work done by the externally applied force **F**

$= $ (Component of **F** along $\overrightarrow{AB}$) $\times$ (Length of AB) (14-13)

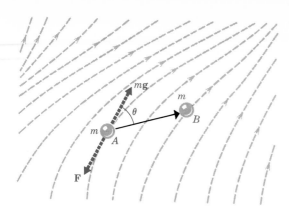

Figure 14-12 The method of finding the relationship between **g** and ϕ_G.

But,

$$\text{Component of } \mathbf{F} \text{ along } \overrightarrow{AB} = -F \cos \theta \qquad (14\text{-}14)$$
$$= -mg \cos \theta \qquad (14\text{-}15)$$

The negative sign is necessary because the component of **F** is in the opposite direction to the vector $\overrightarrow{AB}$, which goes from A to B. So,

$$\text{The work done by the externally applied force } \mathbf{F} = -mg \cos \theta \times AB$$
$$(14\text{-}16)$$

If ϕ_G^A and ϕ_G^B are the gravitational potentials at A and B, then the potential energy of the system has changed from $m\phi_G^A$ to $m\phi_G^B$. Since the motion was performed very slowly and m was not allowed to acquire any kinetic energy,

$$\text{Increase in energy of the system} = m\phi_G^B - m\phi_G^A \qquad (14\text{-}17)$$

But, as was explained in section 9-7, the increase in the energy of the system is equal to the work done by the external force. So,

$$-mg \cos \theta \times AB = m\phi_G^B - m\phi_G^A \qquad (14\text{-}18)$$

Cancelling m throughout and changing the sign of both sides,

$$g \cos \theta \times AB = (\phi_G^A - \phi_G^B) \qquad (14\text{-}19)$$

or

$$g \cos \theta = \frac{(\phi_G^A - \phi_G^B)}{AB} \qquad (14\text{-}20)$$

The Concept of a Field

Now notice that $g \cos \theta$ is the component of **g** in the direction of $\overrightarrow{AB}$.

Since the sign of this equation can cause confusion, notice that the direction of the component is from A to B, the potential at A comes first inside the bracket and AB is always considered positive. If ϕ_G^A is less than ϕ_G^B (which includes the case where both are negative but ϕ_G^A is more negative than Φ_G^B) then the right hand side of the equation is negative and the component of **g** is really in the direction from B to A. One way to remember all this is that the field points in the direction from the higher potential to the lower potential.

Now let us return to the question of finding **g** if we are given the values of ϕ_G at all points in space. To find **g** at any point A, deduce the x-component of g from equation 14-21 by taking a neighboring point B such that $\overrightarrow{AB}$ is in the same direction as the X axis. Similarly, find the y-component by taking $\overrightarrow{AB}$ parallel to the Y axis, and find the z-component by taking $\overrightarrow{AB}$ parallel to the Z axis. The vector **g** is then completely determined when its three rectangular components are known (see section 3-6).

14-5   The Electric Field

If a mass m is placed at a point P where the gravitational field is **g**, the gravitational force on it is $m\mathbf{g}$.

$$\mathbf{F}_G = m\mathbf{g} \qquad (14\text{-}22)$$

The gravitational field is therefore the gravitational force *per unit mass*.

Similarly, if a charge q is placed at a point P where the electric field is **E**, the electric force on it is $q\mathbf{E}$.

The electric field is therefore the electric force *per unit charge*. If q is a positive charge, $\mathbf{F}_e$ and **E** are in the same direction, but if q is negative they point in opposite directions. This is taken into account in equation

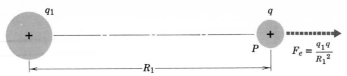

(a) The electrostatic force between charges q_1 and q a distance R_1 apart.

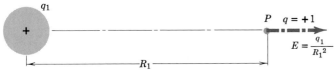

(b) The electric field at a distance R_1 from a charge q_1.

Figure 14-13 The electric field due to a single point charge.

14-23 if the scalar quantity q is given the appropriate sign. The electric field **E**, like the gravitational field **g**, associates a vector with each point in space. The units of E are **dynes per statcoulomb**.

A gravitational field and an electric field may exist simultaneously at the point P. If a body of mass m carrying a charge q is placed at P, the resultant force on it is then the vector sum of the gravitational force $m\mathbf{g}$ and the electric force $q\mathbf{E}$.

Consider, as in figure 14-13, a point P at a distance R_1 from a charge q_1. Place a small positive charge q at P. The electrostatic force on q is

$$F_e = \frac{q_1 q}{R_1^2} \tag{14-24}$$

The electric field E at P is such that

$$F_e = qE \tag{14-25}$$

Electric field E at a distance R_1 from a point charge q_1

$$E = \frac{q_1}{R_1^2} \tag{14-26}$$

Note carefully that if q_1 is positive **E** points away from q_1, but if q_1 is negative **E** points toward q_1.

Now suppose P is at a distance R_1 from a charge q_1, at a distance R_2 from a charge q_2, at a distance R_3 from a charge q_3, and so on. The electric field at P is the *vectorial* sum of a number of contributions similar to equation 14-26. This is illustrated in figure 14-14.

The electric field may be represented pictorially by **electric field lines.** These are such that the direction of **E** at any point P is tangential to the field line passing through P. Figure 14-15 shows the field lines for a single charge. They are radial straight lines which diverge from a positive charge and converge upon a negative charge. If the number of lines

The
Concept
of a Field

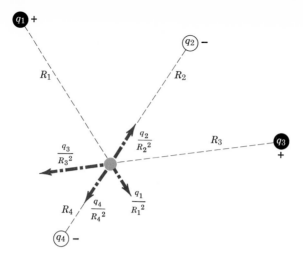

Figure 14-14 The electrostatic field due to a distribution of stationary point charges. The resultant electric field at P is the vector sum of the various contributions represented by arrows.

leaving or entering the point charge q is $4\pi q$, at a distance R from the charge these lines are distributed over a spherical surface of area $4\pi R^2$ and the number of lines per unit area is

$$\frac{4\pi q}{4\pi R^2} = \frac{q}{R^2}$$

$$= E, \text{ the value of the electric field at a distance } R \text{ from } q \quad (14\text{-}27)$$

Consider the electric field due to any arbitrary distribution of positive and negative charges. The field lines can be drawn so that the number of lines *leaving* any positive charge q is $4\pi q$ and the number of lines

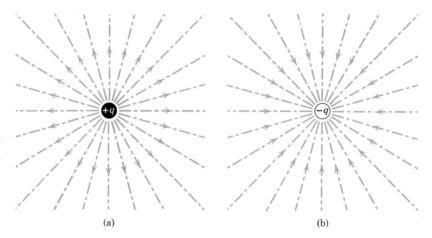

(a)   (b)

Figure 14-15 The field lines for (a) a single point positive charge (b) a single point negative charge.

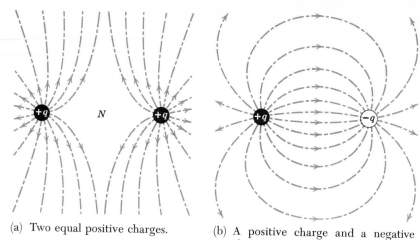

(a) Two equal positive charges.

(b) A positive charge and a negative charge of equal magnitude.

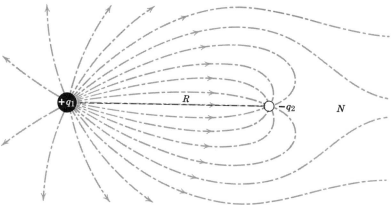

(c) A large positive charge and a small negative charge.

Figure 14-16 Some typical field patterns.

entering any negative charge q' is $4\pi q'$. The strength of the field at any point P is then equal to the number of lines crossing unit area of a small surface at right angles to the field lines. When the field lines are close together the field is strong, and when the lines are far apart the field is weak.

Figure 14-16 shows how the field lines behave in some simple cases. These pictures are very graphic. One can imagine the two positive charges of 14-16*a* pushing one another apart and the positive and negative charges of 14-16*b* pulling one another together. The points marked N in 14-16*a* and *c* are points where the electric field is zero. Notice how the field lines veer away from these points. In (*a*) the point N is midway between the two charges, since it is obvious that at this point the two charges make exactly equal but opposite contributions to the field.

Look at figure 14-16 and ask yourself the following question. Suppose

The
Concept
of a Field

277

the charges were fundamental particles such as protons or electrons. Is there really "something" at the point were the particle is located, or is it just a point from which all the field lines diverge or on which all the field lines converge? Should the emphasis be placed on the particles located at a few points in space or should the emphasis be on the field at all points in space? Present day physics talks about both particles and fields. Exactly how the future will resolve this issue is an open question.

Example 14-5-1

The square $ABCD$ has a side of 10 cm length. Equal positive charges of $+50$ statcoulombs are placed at A and B and equal negative charges of -100 statcoulombs are placed at C and D. Calculate the electric field at P, the center of the square. (See figure 14-17.)

P is the point where the two diagonals AC and BD cross. Let $AP = BP = CP = DP = R$. Then, applying Pythagoras' theorem to the right angle triangle APB,

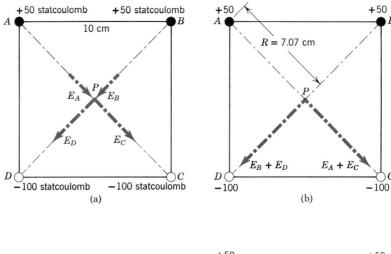

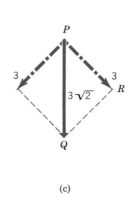

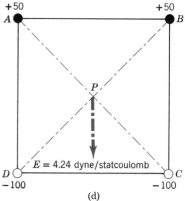

ELECTRICITY
AND
MAGNETISM:
PARTICLES
AND FIELDS

278   **Figure 14-17**   Illustrating example 14-5-1.

$$AP^2 + BP^2 = AB^2$$
$$2R^2 = 10^2$$
$$R^2 = 50$$
$$R = 7.07 \text{ cm} \tag{14-28}$$

The charge q_A at A produces a contribution E_A to the field at P given by

$$E_A = \frac{q_A}{R^2}$$
$$= \frac{50}{50}$$
$$= 1 \text{ dyne per statcoulomb} \tag{14-29}$$

Since the charge at A is positive, $\mathbf{E}_A$ points directly *away* from A (see figure 14-17a). Similarly, the charge at B produces a contribution

$$E_B = 1 \text{ dyne per statcoulomb} \tag{14-30}$$

which points directly away from B. The negative charge q_C at C produces a contribution

$$E_C = \frac{q_C}{R^2}$$
$$= \frac{100}{50}$$
$$= 2 \text{ dyne per statcoulomb} \tag{14-31}$$

Since q_C is negative this contribution points directly *toward* C. Similarly, the charge at D produces a contribution

$$E_D = 2 \text{ dyne per statcoulomb} \tag{14-32}$$

which points directly toward D.

Adding together E_A and E_C, which point in the same direction, we get a vector of magnitude 3 dyne per statcoulomb pointing toward C. Adding E_B to E_D we get a vector of magnitude 3 dyne per statcoulomb pointing toward D. This is shown in figure 14-17b. Figure 14-17c gives the triangle of vectors PQR used to add the two vectors shown in diagram (b).

The resultant field $\mathbf{E} = \overrightarrow{PQ}$ is obviously vertically downward and its magnitude is obtained from Pythagoras' theorem.

$$PQ^2 = PR^2 + RQ^2$$
$$E^2 = 3^2 + 3^2$$
$$= 18$$
$$E = 4.24 \text{ dyne per statcoulomb.} \tag{14-33}$$

The
Concept
of a Field

The resultant field at P is therefore 4.24 dyne per statcoulomb pointing in the direction $\overrightarrow{AD}$.

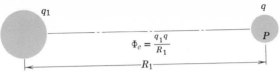

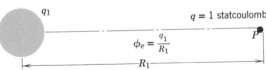

(a) The mutual electrostatic potential energy of charges q_1 and q at a distance R_1 apart.

(b) The electric potential at a point P a distance R_1 from a charge q_1.

Figure 14-18 The mutual potential energy of two charges and the electric potential at a point.

14-6   The Electric Potential at a Point

If a small *positive* charge q is located at a point P, a distance R_1 from a charge q_1, the mutual electrostatic potential energy of the two charges is

$$\Phi_e = \frac{q_1 q}{R_1} \tag{14-34}$$

This is positive if q_1 is positive and negative if q_1 is negative. The electric potential at the point P due to the charge q_1 is obtained by putting $q = 1$ statcoulomb (see figure 14-18).

Electric potential at a distance R_1 from a single point charge q_1

$$\phi_e = \frac{q_1}{R_1} \tag{14-35}$$

If the charge q_1 is positive, ϕ_e is positive. If q_1 is negative, ϕ_e is negative. The units of ϕ_e are clearly ergs per statcoulomb. This unit is called a **statvolt**. Its relationship to the more familiar unit of potential, the volt, will be explained later.

When the point P is at a distance R_1 from a charge q_1, a distance R_2 from a charge q_2, a distance R_3 from a charge q_3, and so on, the total electric potential is:

Electric potential due to any distribution of point charges.

$$\phi_e = \frac{q_1}{R_1} + \frac{q_2}{R_2} + \frac{q_3}{R_3} + \text{ and so on} \tag{14-36}$$

The contributions from the positive charges are positive and the contributions from the negative charges are negative. Note that the potential is a scalar and we are concerned with ordinary algebraic addition, rather than vector addition, which makes the calculation of the potential somewhat easier than the calculation of the electric field.

If the point P is at an infinitely large distance from all the charges, the denominator in each term is infinite and each term is zero. The electric potential "at infinity" is therefore zero.

If a point charge q is placed at a point P where the electric potential is ϕ_e, its total mutual potential energy with respect to all the other charges is

$$\Phi_e = q\phi_e \qquad \text{(14-37)}$$

In this equation we must give careful attention to the sign of both q and ϕ_e to obtain the sign of Φ_e correctly. q is in statcoulombs, ϕ_e is in statvolts and Φ_e is then in ergs.

The electric field is completely determined if ϕ_e is known at all points in space. The field vector $\mathbf{E}$ can be derived from ϕ_e in the same way that the gravitational acceleration $\mathbf{g}$ can be derived from the gravitational potential ϕ_G. The student may find it instructive to turn back to the proof given in section 14-4, modify it for the electrical case, and then check against the discussion following this paragraph.

Imagine a positive charge q located at the point A of figure 14-19. The force exerted on this charge by the electric field is $q\mathbf{E}$. To hold the body in position at A it is necessary to apply an equal and opposite external force $\mathbf{F}$

$$\mathbf{F} = q\mathbf{E} \qquad \text{(14-38)}$$

Allow the charge to move slowly to a point B, which is so close to A that the electric field is almost the same at B as at A. During this motion,

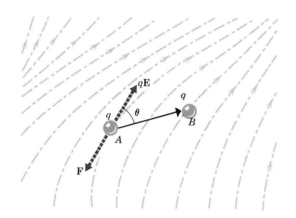

Figure 14-19   The method of finding the relationship between $\mathbf{E}$ and ϕ_e.

Work done by the externally applied force **F**

$$= (\text{Component of } \mathbf{F} \text{ along } \overrightarrow{AB}) \times AB \quad (14\text{-}39)$$

But,

$$\text{Component of } \mathbf{F} \text{ along } \overrightarrow{AB} = -F \cos \theta \quad (14\text{-}40)$$

$$= -qE \cos \theta \quad (14\text{-}41)$$

The negative sign is necessary because the component of **F** points from B to A, whereas in the expression for the work done we require the component to point from A to B. The work done may be rewritten:—

Work done by the externally applied force $\mathbf{F} = -qE \cos \theta \times AB$

$$(14\text{-}42)$$

The potential energy of the system has changed from $q\phi_e^A$ to $q\phi_e^B$, with no development of kinetic energy. So,

$$\text{Increase in energy of system} = q\phi_e^B - q\phi_e^A \quad (14\text{-}43)$$

This must be equal to the work done by the external force:

$$-qE \cos \theta \times AB = q\phi_e^B - q\phi_e^A \quad (14\text{-}44)$$

or

$$E \cos \theta = \frac{(\phi_e^A - \phi_e^B)}{AB} \quad (14\text{-}45)$$

To find the electric field, E, given the electric potential, ϕ_e

If A and B are two points very close to one another in an electric field **E**, then

$$\text{Component of } \mathbf{E} \text{ at } A \text{ in the direction } \overrightarrow{AB} = \frac{\phi_e^A - \phi_e^B}{AB}$$

$$(14\text{-}46)$$

The direction of this component is from the point at the higher potential toward the point at the lower potential.

Choosing $\overrightarrow{AB}$ successively in the X, Y, and Z directions, one can find the three rectangular components of **E**, which is then completely determined.

In equation 14-46 the units of E are dynes per statcoulomb, the units of ϕ_e are statvolts, and the length AB is in cm. This equation shows that the electrostatic units of electric field might also be expressed as **statvolts per cm.**

Example 14-6-1

Four equal charges of $+100$ statcoulombs are placed at the four corners of a square with a side of length 10 cm. What is the electric

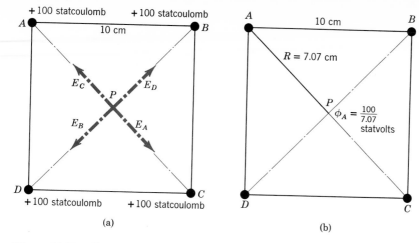

Figure 14-20 Illustrating example 14-6-1.

field at the center of the square? What is the electric potential at the center?

The electric field is easy. Referring to figure 14-20a, the contribution E_A from the charge at A is exactly compensated by the contribution E_C from the charge at C. Also, the contribution E_B from the charge at B is exactly compensated by the contribution E_D from the charge at D. The resultant field is therefore zero.

Turning now to figure 14-20b, the center of the square P is at a distance $R = 7.07$ cm from A. The contribution of the charge q_A to the potential at P is therefore

$$\text{Contribution to } \phi_e \text{ from } q_A = \frac{q_A}{R}$$

$$= \frac{+100}{7.07}$$

$$= +14.14 \text{ statvolt} \tag{14-47}$$

The contribution is the same from each charge and each contribution is positive. Since the potential is a scalar, the total potential is four times each contribution, so

$$\text{Potential at } P = 4 \times 14.14$$

$$= 56.56 \text{ statvolt} \tag{14-48}$$

Although the electric field is zero, the electric potential is not zero.

Example 14-6-2

In the simple case of the field due to a single point charge q, check the method for obtaining $\mathbf{E}$ from ϕ_e (figure 14-21).

Choose the points A and B to be on the same radius, in which case the component of $\mathbf{E}$ in the direction $\overrightarrow{AB}$ is the total field. Let A be at a

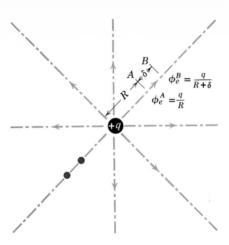

Figure 14-21 Illustrating example 14-6-2.

distance R from q and B at a distance $(R + \delta)$, where δ is very small. The potential at A is

$$\phi_e^A = \frac{q}{R} \tag{14-49}$$

The potential at B is

$$\phi_e^B = \frac{q}{R + \delta} \tag{14-50}$$

Using equation 14-46 to calculate the electric field at A

$$E = \frac{\phi_e^A - \phi_e^B}{AB} \tag{14-51}$$

$$= \frac{\dfrac{q}{R} - \dfrac{q}{R + \delta}}{\delta} \tag{14-52}$$

$$= \frac{1}{\delta} \left[\frac{q(R + \delta)}{R(R + \delta)} - \frac{qR}{(R + \delta)R} \right]$$

$$= \frac{1}{\delta} \frac{(qR + q\delta - qR)}{R(R + \delta)}$$

$$= \frac{q}{R(R + \delta)}$$

$$= \frac{q}{R^2 \left(1 + \dfrac{\delta}{R}\right)} \tag{14-53}$$

Since δ must be made very small, δ/R can be made very small indeed compared with 1 and can be neglected. So,

$$E = \frac{q}{R^2} \tag{14-54}$$

which agrees with equation 14-26.

Example 14-6-3

This example is based on Millikan's famous experiment to measure the charge on the electron, e. A small oil drop carrying a single excess electron was suspended motionless in mid-air by a downward vertical electric field which exerted an upward electrical force on it exactly equal but opposite to its weight (see figure 14-22). If m was the mass of the droplet and E was the strength of the electric field, the equality of electrical and gravitational forces tells us that

$$eE = mg \tag{14-55}$$

$$e = \frac{mg}{E} \tag{14-56}$$

The mass of the droplet was obtained from the density ρ of the oil and the radius r of the droplet. The volume of the spherical droplet was

$$\text{Volume} = \frac{4\pi r^3}{3} \tag{14-57}$$

Its mass was therefore

$$m = \rho \times \text{Volume}$$

$$= \rho \frac{4\pi r^3}{3} \tag{14-58}$$

The radius of the droplet could not be measured directly and was therefore derived in the following subtle way. The electric field was switched off and the droplet was observed to fall with a constant velocity. This is not a case of a body falling freely in a vacuum because a small body falling through the air quickly accelerates up to a constant limiting velocity, which then remains unchanged for the rest of the fall. At this limiting velocity the gravitational attraction of the earth on the droplet is exactly counterbalanced by the frictional force exerted by the air through which the droplet is moving. The net force is then zero, and the acceleration is zero. The theory of this limiting velocity is too complicated to explain here, but it is well understood. It turns out that the limiting velocity is very sensitive to the radius of the droplet. The radius can therefore be deduced from the steady velocity of fall.

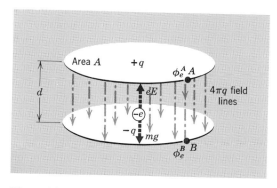

The
Concept
of a Field

Figure 14-22 Millikan's oil drop experiment to measure the charge on the electron.

The other quantity which must be measured to obtain e from equation 14-56 is the electric field E. The electric field was produced by two parallel, horizontal, circular metal plates carrying equal but unlike charges (see figure 14-22). To produce an upward force on a negative charge, a downward electric field is needed. The upper plate was therefore positively charged and the lower plate was negatively charged. (If you prefer, the negative electron was attracted to the positive charge on the upper plate and repelled by the negative charge on the lower plate.) The quantity directly measured was not the charge, but the potential difference between the two plates. Potential differences are measured by a very commonly used instrument called a voltmeter. If A is a point on the upper plate and B a point on the lower plate directly below A, the voltmeter measured the quantity

$$V = \phi_e^A - \phi_e^B \tag{14-59}$$

But, according to equation 14-46, the electric field is given by

$$E = \frac{(\phi_e^A - \phi_e^B)}{AB} \tag{14-60}$$

$$= \frac{V}{d} \tag{14-61}$$

where d is the vertical separation of the plates. Putting equation 14-56 in the form

$$e = \frac{mgd}{V} \tag{14-62}$$

everything in this equation could be measured and the value of e could then be deduced.

What charge q must be placed on the plates to produce the potential difference V and the necessary field E? When two large parallel plates are very close together and carry equal but opposite charges, the field lines go straight from one plate to the other in a direction perpendicular to the plates. If $4\pi q$ field lines leave the top plate and enter the bottom plate, we can obtain the electric field from the number of lines crossing unit area, as explained in section 14-5. If A is the area of a plate, the number of field lines per unit area is

$$\frac{4\pi q}{A} = E \tag{14-63}$$

or

$$q = \frac{EA}{4\pi} \tag{14-64}$$

Using equation 14-61 to put this in terms of the potential difference V,

$$q = \frac{VA}{4\pi d} \tag{14-65}$$

A pair of oppositely charged parallel plates like this is called a *capacitor*. Its capacity C is defined as

$$C = \frac{q}{V} \tag{14-66}$$

$$= \frac{A}{4\pi d} \tag{14-67}$$

Similar devices are extensively used in electronic circuits.

QUESTIONS

1. Are you more attracted to the idea of a universe consisting of discrete particles separated by empty space or to the idea of a continuous universe in which every point of space has physical properties? Try to justify your preference (a) on the basis of the physics you already know (b) by reference to your personal impression of the nature of the universe derived from direct observation of your surroundings.

2. Can you think of any other examples of fields not quoted in the text? In each case define the field carefully.

3. Which of the following is the best definition of a field? (a) The force on unit mass at any point. (b) The force on unit charge at any point. (c) A quantity which has a unique value at every point in space. (d) The tangent to a field line. (e) The work done in bringing unit mass or unit charge from infinity to a point.

4. Can two field lines intersect at an angle?

5. Explain carefully why a body moving freely in the earth's gravitational field does not necessarily follow a field line. Give examples. Would it do so if it were released from rest?

6. Make a rough sketch of the pattern of gravitational field lines due to the earth and the moon. (Hints: Consider the character of the field very near the earth, very near the moon, and at a very great distance from both. Is there a point where the field is zero? Of all the lines coming from infinity, what fraction enters the earth and what fraction enters the moon?)

7. If the number of electric field lines entering a closed surface is exactly equal to the number leaving, what can be said about the electric charges inside the surface?

8. If the number of electric field lines leaving a closed surface exceeds the number entering by N, what can be said about the electric charges inside the surface?

9. Referring to figure 14-16c, find an expression giving the position of the point N in terms of q_1, q_2 and their distance apart, R. What happens if the magnitude of q_2 is greater than the magnitude of q_1? Where is N if the two charges have equal magnitudes but opposite sign (as in figure 14-16b)?

10. Explain the connection between equation 14-21 and equation 9-49 of chapter 9.
11. Equation 14-46 establishes a procedure for finding **E** given ϕ_e. Suppose that **E** is known at all points in space but ϕ_e is not. Establish a procedure for finding the difference in ϕ_e between two points, (*a*) if the two points are very close together, (*b*) if the two points are far apart. How could the absolute value of ϕ_e at a point be determined?

PROBLEMS

A

1. What is the gravitational potential at a point 2 cm from a mass of 100 gm?
2. What is the gravitational field at a distance of 2 m from a mass of 100 kg?
3. Find the gravitational field and the gravitational potential at a point which is 10^{10} cm from the center of the earth. (Ignore the moon, the sun, and all the other bodies in the solar system.)
4. A mass of 10 gm carrying a charge of $+2$ statcoulomb is located at a point where the electric field is 50 statvolt/cm. What force acts on it?
5. A mass of 2 kg carrying a charge of $+0.01$ statcoulomb is located at a point where the electric field is 3 statvolt/cm. What is its acceleration?
6. What is the acceleration of a proton in an electric field of 10^3 statvolt/cm?
7. What is the magnitude of the electric field at a distance of 5 cm from a point charge of 600 statcoulomb?
8. What is the electric potential at a distance of 10^{-3} cm from a point charge of -3×10^{-5} statcoulomb?
9. Find the electric field and the electric potential at a point 10^{-8} cm from an electron.
10. A charge of $+5$ statcoulomb is 1 m from a charge of -9 statcoulomb. Find the electric field and the electric potential at the midpoint of the line joining the two charges.
11. What is the electric potential at the center of the square in figure 14-17?
12. What is the potential energy of an electron located at a point where $\phi_e = +1000$ statvolt?
13. Two points A and B are on the same electric field line. The distance between them is 5×10^{-4} cm. The potential at A is 5.123 statvolt and the potential at B is 5.158 statvolt. Find the magnitude and direction of the electric field at A.
14. A parallel plate capacitor, such as the one shown in figure 14-22, has a gap of 0.1 cm between its plates and the electric field between them is 1.5 statvolt/cm. What is the potential difference between the two plates?

ELECTRICITY
AND
MAGNETISM:
PARTICLES
AND FIELDS

288

15. What is the difference in gravitational potential between the floor and the ceiling of a room 3 m high?

16. Find the contribution to the gravitational potential at the surface of the earth due to (*a*) the earth itself, (*b*) the moon, (*c*) the sun. Compare the magnitudes of these three contributions.

17. Calculate the value of the vertical electric field which can support the weight of an electron.

18. A mass of 3 gm carrying a charge of $+5$ statcoulomb is in a uniform electric field, which has the same magnitude and direction everywhere. It starts from rest and 2 sec later has traveled a distance of 10 cm. Calculate the magnitude of the field.

19. An electron is initially at rest in a uniform electric field of 2.5 statvolt/cm. What is its velocity 10^{-9} sec later?

20. A proton has an initial velocity of 10^8 cm/sec in exactly the opposite direction to a uniform electric field of 0.2 statvolt/cm. How far does it travel before coming to rest?

21. Two 1 gm masses are at rest relative to one another. An equal number n of electrons is removed from each so that their gravitational attraction is exactly counterbalanced by their electrostatic repulsion? What is n?

22. Three charges, each of value $+7$ statcoulomb, are placed at the three corners of a square with its sides 3 cm long. Calculate the electric field and the electric potential at the unoccupied corner.

23. Charges of $+3$ statcoulomb and -4 statcoulomb are placed at opposite ends of a diagonal of a square with its sides 1.5 cm long. What is the potential difference between (*a*) the two unoccupied corners (*b*) one unoccupied corner and the center.

24. How much work is done in slowly taking a charge of -15 statcoulomb from a point where the potential is $+1.5$ statvolt to a point where the potential is -2.5 statvolt?

25. How much work is done in slowly bringing an electron from an infinite distance up to a point where the potential is $+500$ statvolt?

C

26. Calculate the magnitude of a uniform electric field from the data in the table, which gives the potential in statvolts at points with given Cartesian coordinates.

Point (lengths in cm)	(0, 0, 0)	(0.1, 0, 0)	(0, 0.5, 0)	(0, 0, 0.2)
Potential in statvolt	2.31	2.42	2.03	2.40

27. A dipole consists of two equal but opposite charges $+q$ and $-q$ a distance l apart. Find the electric field produced by this dipole at a point on the line joining the charges a distance R from the point midway between them. Simplify the expression in the case when R is very much larger than l. Find a similar expression for the potential at the point.

The Concept of a Field

289

28. A capacitor consists of two parallel circular disks, each of radius 2 cm, a distance 0.05 cm apart. What is its capacity? If the potential difference between its plates is 0.3 statvolt, what is the charge on either plate?

29. Use the concept of field lines to determine the electric field at all points inside and outside a thin uniform spherical shell of charge.

30. From the results of problem 29, deduce the electric *potential* at all points inside and outside a thin uniform spherical shell of charge.

31. From the results of problem 29, find the electric *field* at all points inside and outside a uniform spherical distribution of charge.

32. A tunnel is drilled through the center of the earth to emerge at the antipodes. A ball is dropped into the tunnel. Find an algebraic expression for the gravitational force on the ball when it is at a distance r from the center of the earth. First express the force in terms of G, the mass of the ball, m, the mass of the earth, M_e, the radius of the earth, R_e, and r. Then eliminate G and M_e and express the force in terms of m, r, R_e, and g_s, the acceleration due to gravity at the surface of the earth.

ELECTRICITY
AND
MAGNETISM:
PARTICLES
AND FIELDS

290

15

Electric Currents

15-1   Electric Currents and Batteries

The practical applications of electricity are numerous and well known, but they hardly ever make use of the electrostatic forces between charged bodies. They rely instead upon the effects produced by an electric current flowing through a metal wire that has no net electric charge. When an electrical appliance is plugged into a wall socket, electrons flow in through one pin of the plug, circulate through the wires inside the appliance and leave via the other pin of the plug. As many electrons leave as enter and there is never any net accumulation of charge inside the appliance. Actually, since the wall socket usually supplies 60 cycle alternating current, the electrons surge backward and forward sixty times a second. This is a non-essential complication and it will be simpler to discuss the direct current produced by a storage battery of the kind used to operate the ignition system of an automobile.

Figure 15-1 shows a storage battery with a copper wire connected between its terminals. The battery consists essentially of two metal plates immersed in dilute sulfuric acid. One of the plates is coated with sponge lead and the other is coated with lead dioxide. In section 13-1 we stated that when a glass rod is rubbed with a silk cloth the electrons have a preference for being in the silk rather than the glass and therefore some electrons migrate from the glass to the silk. Similarly, electrons prefer to be in lead rather than lead dioxide, and by a series of elaborate processes electrons are transferred from the lead dioxide plate through the sulfuric acid to the lead plate. The lead dioxide plate therefore has a deficiency of electrons, an excess of protons, and a positive charge. It is the positive terminal of the battery. The lead plate acquires an excess of electrons, is negatively charged, and is the negative terminal of the battery.

The electric potential ϕ_+ at the positive terminal is higher than the potential ϕ_- at the negative terminal. The battery operates in such a

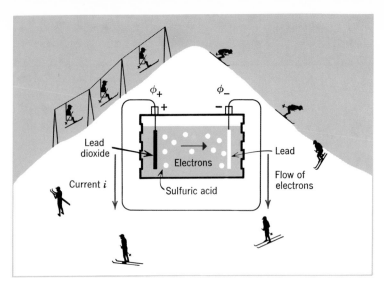

Figure 15-1   A storage battery with a copper wire connected between its terminals compared with a ski lift and a ski run.

way that the potential difference V between its terminals is almost constant.

$$V = \phi_+ - \phi_- = 0.67 \times 10^{-2} \text{ statvolt, approximately} \tag{15-1}$$

(Later we shall see that this is equal to 2 volts in the practical units which may be more familar to the reader.) V decreases a little if a large current is drawn from the battery, and V also steadily decreases with time as the battery "runs down" because of chemical processes that contaminate its plates. However, these changes are never more than a few per cent and we will simplify the discussion by assuming that the current drawn from the battery is small and that the potential difference V between its terminals remains constant.

If a copper wire is connected between the terminals, electrons flow continuously from the negative terminal through the wire to the positive terminal and then through the battery back to the negative terminal. In copper, as in other metals, some of the electrons on the outside of the copper atom are very loosely held and are easily shaken free. These electrons do not remain attached to a particular copper atom, but wander freely throughout the whole of the copper wire. They are repelled by the negative terminal and attracted by the positive terminal and are therefore continually accelerating toward the positive terminal. This does not mean that the electrons move faster and faster the nearer they get to the positive terminal. They frequently collide with the copper atoms and, although an electron acquires an extra velocity toward the positive terminal in between collisions, this extra velocity is destroyed at each collision. However, in between collisions the electron does get a chance to move toward the positive terminal, and the net result is a slow drift of the electrons toward this terminal.

In the absence of the storage battery, with no current flowing, free electrons move through the copper with very high velocities of the order of magnitude of 10^8 cm/sec. (This is about 2 million miles per hour!) Since these velocities are randomly distributed in all directions, there is no net flow of electrons one way or another. When the storage battery produces a potential difference between the ends of the wire, the net result is equivalent to adding a small extra "drift velocity" v_d to each electron. This drift velocity points along the wire toward the positive terminal and is in the same direction and has the same magnitude for all the free electrons. The random chaotic motion therefore acquires a slight bias in this direction and more electrons move toward the positive terminal than away from it. However, v_d is always very small compared with the random velocity. A typical value of v_d might be 0.1 cm/sec, which is a billion times smaller than the random velocity of 10^8 cm/sec. One should therefore imagine an electron to be moving with a very high velocity, frequently colliding with the copper atoms and changing the direction of this velocity, but making slow progress in the direction of the positive terminal (see figure 15-2).

If a piece of wood were connected between the terminals, no current would flow. All the electrons in the wood are firmly held to their atoms and are not free to move from one terminal to the other. Wood is an **insulator** because it will not carry an electric current. Copper is a **conductor** of electricity. All metals are conductors.

When the electrons flowing through the copper wire arrive at the positive terminal they must be returned to the negative terminal, otherwise the charges on the two terminals would soon be neutralized and the current would cease to flow. The transfer of electrons through the sulfuric acid to the negative terminal against the repulsion of the negative charge on this terminal is a consequence of certain complicated chemical processes that occur inside the battery. In figure 15-1 a storage battery is compared with a ski lift and a ski run. The chemical processes inside the battery that lift the electrons from the positive terminal to the negative terminal are analogous to the ski lift that lifts the skiers from the bottom to the top of the mountain. The electron being pulled through the copper wire by the electrical attraction of the positive terminal is analogous to the skier gliding down the ski run by taking advantage of the gravitational attraction of the earth. The electron maintains a steady drift velocity v_d because of its numerous collisions with the copper atoms. Similarly, the skier can maintain a steady velocity down the run by making skillful use of the frictional forces exerted on him by the snow.

15-2 The Quantitative Definition of Electric Current

The electric current i is defined as the net rate of flow of charge across any cross-section of the wire. Suppose that, in t seconds, n_r electrons

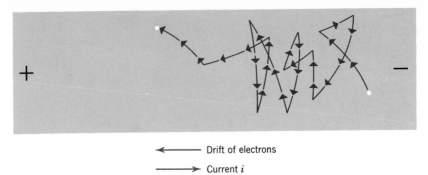

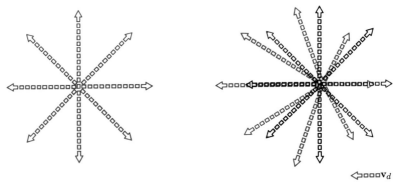

———————————— Drift of electrons

————————————→ Current i

(a) On top of the chaotic motion of the electron in all directions there is superimposed a slow drift toward the positive terminal.

(b) In the absence of a current, these eight vectors, representing the velocities of eight electrons, are equally distributed in all directions.

(c) In the presence of a current, the superposition of a drift velocity v_d on the same eight vectors produces a bias toward the left.

Figure 15-2   The motion of the free electrons in a wire carrying a current.

cross over from right to left and a smaller number n_l cross over from left to right (figure 15-3). There is a net flow of $(n_r - n_l)$ electrons from right to left, corresponding to the flow of a negative charge

$$-q = -(n_r - n_l)e \qquad (15\text{-}2)$$

from right to left. The current is the charge flowing in one second.

$$i = \frac{q}{t} \qquad (15\text{-}3)$$

$$= \frac{(n_r - n_l)e}{t} \qquad (15\text{-}4)$$

If e and q are in statcoulombs and t in seconds, the current i is measured in **statamperes**. The current is due to the flow of negatively charged electrons from right to left. This is equivalent to a flow of positive charge

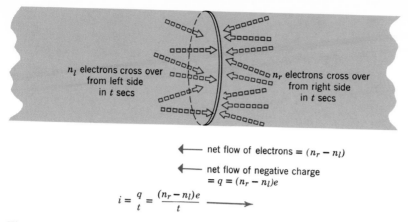

net flow of electrons = $(n_r - n_l)$

net flow of negative charge
= $q = (n_r - n_l)e$

$$i = \frac{q}{t} = \frac{(n_r - n_l)e}{t}$$

Figure 15-3 Illustrating the definition of electric current.

from left to right and it is traditional to assign the direction of the current on the assumption that positive charge is flowing. The direction of the current is therefore always opposite to the direction of net flow of the electrons.

Let us ignore the random motion and assume that all the electrons are moving with a steady drift velocity v_d from right to left. A thorough mathematical analysis justifies this procedure. In figure 15-4, imagine that the wire is stationary, but that the electrons are fixed to a strip of paper which is moving to the left with a velocity v_d. In t seconds the strip moves a distance $v_d t$ and all the electrons inside the cylinder with a broken outline move through the cross section we are considering. If the cross-sectional area is A, the volume of this cylinder is $Av_d t$. If there are n_0 free electrons per unit volume, the number of electrons crossing over is $n_0 A v_d t$. The charge flowing across in t seconds is therefore

$$q = e n_0 A v_d t \tag{15-5}$$

The current is therefore

$$i = \frac{q}{t} \tag{15-6}$$

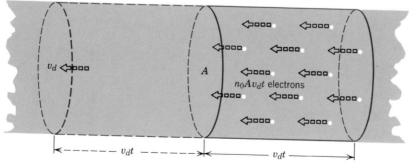

Figure 15-4 Expressing the current in terms of the drift velocity.

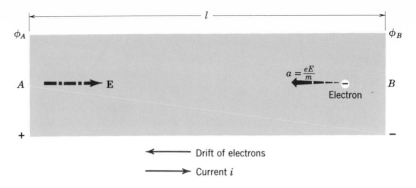

Figure 15-5 Illustrating the discussion of Ohm's Law.

$$i = en_0 A v_d \tag{15-7}$$

15-3 Ohm's Law

Now imagine a straight wire AB of length l with a potential difference $\phi_A - \phi_B = V$ between its ends (figure 15-5). It might, for example, be connected directly across the terminals of a storage battery, in which case V would be $(\phi_+ - \phi_-)$. Since there is a potential difference between the ends of the wire, there is an electric field E inside the wire given by equation 14-46

$$E = \frac{\phi_A - \phi_B}{AB} \tag{15-8}$$

$$E = \frac{V}{l} \tag{15-9}$$

If A is positive with respect to B, this field points from A to B.

According to equation 14-23, this field exerts a force on an electron given by

$$F = eE \tag{15-10}$$

$$F = \frac{eV}{l} \tag{15-11}$$

Since the electron is negatively charged, this force is in the opposite direction to E and points from B to A. If m is the mass of the electron, it has an acceleration a in the direction B to A given by

$$a = \frac{F}{m} \tag{15-12}$$

$$a = \frac{eV}{ml} \tag{15-13}$$

Immediately after a collision with an atom, the electron has no preferred velocity in the direction of flow. If a time τ sec elapses before the next collision, during this time the electric field gives the electron an extra velocity $a\tau$ in the direction B to A. However, this extra velocity is

ELECTRICITY
AND
MAGNETISM:
PARTICLES
AND FIELDS

296

completely destroyed by the subsequent collision and the electron starts all over again to build up its drift velocity from 0 to $a\tau$. The average drift velocity is $\frac{1}{2}a\tau$.

$$v_d = \tfrac{1}{2}a\tau \tag{15-14}$$

Substituting the value of a from equation 15-13

$$v_d = \frac{eV\tau}{2ml} \tag{15-15}$$

Inserting this value of v_d into equation 15-7,

$$i = en_0A\left(\frac{eV\tau}{2ml}\right) \tag{15-16}$$

or

$$i = \frac{e^2n_0\tau A}{2ml}V \tag{15-17}$$

Introducing a new quantity R, defined by

$$R = \frac{2ml}{e^2n_0\tau A} \tag{15-18}$$

then

Ohm's law

$$i = \frac{V}{R} \tag{15-19}$$

R depends only on the dimensions of the wire and the properties of the free electrons. For a given wire at a fixed temperature, it is a constant known as the **electrical resistance** of the wire. Ohm's law states that the current is directly proportional to the potential difference between the ends of the wire and inversely proportional to the resistance. Note also, from equation 15-17, that the current is proportional to the cross-sectional area A of the wire and inversely proportional to its length, l. The resistance R is inversely proportional to A and proportional to l. In the CGS system the units of electrical resistance are **statohms**.

Ohm's law is true for metals at a fixed temperature and many other materials, but it is not universally true. Our proof is dependent on the assumptions we have made concerning the behavior of the flowing electrons. In some materials and in some special devices, the behavior of the electrons is much more complicated and the current is not linearly proportional to the potential difference. Certain special devices known as rectifiers are designed in such a way that the resistance is small when the current flows in one direction, but is very large when it flows in the opposite direction.

Electric
Currents

297

15-4 Joule Heat

When an electron is at the end B of the wire, its potential energy is $-e\phi_B$. When it reaches A, its potential energy is $-e\phi_A$, which is more negative. The potential energy of the electron has decreased by

$$e(\phi_A - \phi_B) = eV \tag{15-20}$$

If the electron were free to move unhindered from B to A, this loss in potential energy would reappear as an increase in kinetic energy of the electron. When the electron is continually colliding with the atoms of the wire, its average kinetic energy is not allowed to increase. In between collisions the electric field accelerates the electron and increases its kinetic energy, but the ensuing collision transfers this extra kinetic energy to the copper atoms which are thereby made to vibrate more rapidly. The energy therefore finally appears as heat in the copper wire. It is well known that a wire warms up when an electric current flows through it. The heat generated in this way is known as **joule heat**.

In t sec a charge

$$-q = -it \tag{15-21}$$

leaves B and an equal charge arrives at A. The charge leaving B initially had potential energy $-q\phi_B$, but the charge arriving at A has a potential energy of $-q\phi_A$. The loss in potential energy is therefore

$$q(\phi_A - \phi_B) = qV \tag{15-22}$$
$$= itV$$

The heat appearing in the wire is therefore

$$H = itV \tag{15-23}$$

The joule heat developed per second is

Joule heat

$$\frac{H}{t} = iV \tag{15-24}$$

Heat developed per sec = Current $\times$ potential difference

If i is in statamperes and V in statvolts, the heat is in ergs/sec.

Using Ohm's law (equation 15-19), the joule heat might also be expressed as

$$\frac{H}{t} = \frac{V^2}{R} \tag{15-25}$$

or

$$\frac{H}{t} = i^2R \tag{15-26}$$

It is interesting to consider the energy changes taking place when a wire is connected between the terminals of a storage battery (figure 15-1).

The original source of the energy is the **chemical energy** of the materials constituting the battery. When current is drawn from the battery, the chemical reaction taking place is essentially

$$\text{Pb} \quad + \quad \text{PbO}_2 \quad + \quad 2\text{H}_2\text{SO}_4$$

$$\left[\left(\begin{smallmatrix}\text{One atom} \\ \text{of lead}\end{smallmatrix}\right) + \left(\begin{smallmatrix}1 \text{ molecule of} \\ \text{lead dioxide}\end{smallmatrix}\right) + \left(\begin{smallmatrix}2 \text{ molecules of} \\ \text{sulfuric acid}\end{smallmatrix}\right)\right]$$

$$\longrightarrow \quad 2\text{PbSO}_4 \quad + \quad 2\text{H}_2\text{O}$$

(15-27)

change into $\left[\left(\begin{smallmatrix}2 \text{ molecules of} \\ \text{lead sulfate}\end{smallmatrix}\right) + \left(\begin{smallmatrix}2 \text{ molecules} \\ \text{of water}\end{smallmatrix}\right)\right]$

Any molecule appearing in this equation is a structure made from electrons, protons, and neutrons. The energy of the molecule is obtained by adding the kinetic energies of these particles to the mutual potential energies of the particles taken in pairs. The most important contribution to the potential energy is the electrostatic potential energy of the charged particles. The combined energy of the molecules on the left hand side of equation 15-27 is greater than the combined energy of the molecules on the right hand side. Therefore, when the chemical reaction takes place, energy is made available. Inside the battery this energy is used to lift electrons from the positive terminal to the negative terminal, where their electric potential energy is greater.

As the electrons are accelerated toward the positive terminal through the copper wire, their potential energy is first of all converted into extra kinetic energy. Collisions with the copper atoms immediately transfer this energy to the copper atoms, which vibrate more rapidly and the wire warms up. The joule heat appearing in the wire was originally present as kinetic and potential energy of the fundamental particles in lead, lead dioxide and sulfuric acid molecules.

A final transformation of energy occurs if the wire becomes so hot that it glows like the filament of an electric light bulb, emitting light and heat radiation. There is then a partial conversion of heat energy into the energy of electromagnetic waves traveling through empty space. The reader will understand this better after he has read chapter 20.

15-5 Practical Units

For our present purposes, it will be sufficient to define the practical units in terms of the CGS units we have used so far.

Charge: 1 **coulomb** $= 3 \times 10^9$ statcoulombs   (15-28)

Current: 1 **ampere** $= 1$ coulomb per second   (15-29)

                $= 3 \times 10^9$ statcoulombs per second

                $= 3 \times 10^9$ statamperes   (15-30)

Potential difference: 1 **volt** $= \dfrac{1}{300}$ statvolt   (15-31)

Electrical resistance: 1 **ohm** $= \dfrac{1}{9 \times 10^{11}}$ statohm.   (15-32)

In electrostatics the practical units require rather cumbersome formulas that we shall mention only to illustrate the difficulty. Coulomb's law becomes

$$F = \frac{1}{4\pi\varepsilon_0} \frac{q_1 q_2}{R^2} \tag{15-33}$$

The force is in newtons, q_1 and q_2 are in coulombs, and R is in meters, because one always uses the MKS system in connection with practical units. ε_0 is a constant of nature

$$\varepsilon_0 = 8.85415 \times 10^{-12} \text{ coulomb}^2 \text{ newton}^{-1} \text{ m}^{-2} \tag{15-34}$$

The mutual electrostatic potential energy of two charges is

$$\Phi_e = \frac{1}{4\pi\varepsilon_0} \frac{q_1 q_2}{R} \tag{15-35}$$

With q_1 and q_2 in coulombs and R in meters, Φ_e is in joules, the unit of energy in the MKS system.

If a charge q coulombs is placed at a point where the electric field is E, and the force on it is F newtons, then E is defined as

$$E = \frac{F}{q} \text{ newtons/coulomb} \tag{15-36}$$

The field due to a distribution of charges is

$$E = \text{Vector sum of contributions of magnitude} \frac{q_1}{4\pi\varepsilon_0 R_1^2} \tag{15-37}$$

with q_1 in coulombs and R_1 in meters.

The electric potential at a point due to a distribution of charges is

$$\phi_e = \text{Sum of terms such as} \frac{1}{4\pi\varepsilon_0} \frac{q_1}{R_1} \tag{15-38}$$

With q_1 in coulombs and R_1 in meters, ϕ_e is in volts.

We shall make very little use of equations (15-33) to (15-38). However, the following equations will be more useful.

If a charge q coulombs is placed at a point where the electric potential is ϕ_e volts, its potential energy in joules is

$$\Phi_e = q\phi_e \tag{15-39}$$

(joules) = (coulombs) × (volts)

If an electric current is due to the flow of q coulombs of charge across any cross section in t seconds, the measure of the current in amperes is

$$i = \frac{q}{t} \tag{15-40}$$

(amperes) = (coulombs)/(secs)

In Ohm's law,

$$i = \frac{V}{R}$$

(15-41)

if i is in amperes and V in volts, then R is in ohms.

If a current of i amperes is flowing in a wire with a potential difference of V volts between its ends, then the heat developed must be measured in joules/sec or watts.

1 watt = 1 joule per sec

(15-42)

Then

$$\frac{H}{t} = i \times V$$

(15-43)

(watts) = (amperes) $\times$ (volts)

Example 15-5-1

The type of wall socket commonly found in the house is capable of delivering a current of 5 amperes. If this current flows through a copper wire with a diameter of 0.1 cm, what is the drift velocity v_d of the electrons?

We are going to use equation 15-7 and so we shall need to know n_0, the number of free electrons per cm³. The copper atom is known to have one loosely bound electron, so we shall assume that one free electron is provided by each copper atom. n_0 is therefore equal to the number of copper atoms in 1 cm³. The atomic mass of copper is 64 and so 64 gm of copper contains $N_A = 6.0 \times 10^{23}$ atoms (This is Avogadro's number. See section 11-3.) 1 cm³ of copper has a mass of 9.0 gm and the number of atoms in 1 cm³ is therefore

$$n_0 = 6.0 \times 10^{23} \times \tfrac{9}{64}$$
$$= 8.4 \times 10^{22}$$

(15-44)

The current is

$$5 \text{ amps} = 5 \times 3 \times 10^9 \text{ statamps}$$
$$i = 1.5 \times 10^{10} \text{ statamps}$$

(15-45)

The cross-sectional area of the wire is

$$A = \pi r^2$$
$$= \pi \times \left(\frac{0.1}{2}\right)^2$$
$$= 7.854 \times 10^{-3} \text{ cm}^2$$

(15-46)

The charge on the electron is 4.80×10^{-10} statcoulomb. From equation 15-7

$$i = en_0 A v_d$$

(15-47)

the drift velocity is

$$v_d = \frac{i}{en_0A} \tag{15-48}$$

$$= \frac{1.5 \times 10^{10}}{(4.80 \times 10^{-10}) \times (8.4 \times 10^{22}) \times (7.854 \times 10^{-3})}$$

Drift velocity, $v_d = 4.74 \times 10^{-2}$ cm/sec $\tag{15-49}$

If the wire is 1 meter long, the time taken for an electron to drift from one end to the other is $100/(4.74 \times 10^{-2})$ sec $= 2.11 \times 10^3$ sec $= 35$ minutes!

Notice that we chose to work in CGS units throughout. In this particular example we could have obtained the correct answer if we had left i in amps and expressed e in coulombs. To be consistent, we would then have had to express n_0 in atoms per m^3, A in m^2 and v_d in m/sec. We would, however, have obtained the correct answer if we had left n_0, A and v_d in CGS units, but this cannot be relied upon without careful consideration of the problem. The student is advised never to mix systems of units and then he can be certain of not running into trouble.

Example 15-5-2

An automobile battery produces a potential difference (or "voltage") of 12 volts between its terminals. (It really consists of six 2 volt batteries following one after the other.) A headlight bulb is to be connected directly across the terminals of the battery and dissipate 40 watts of joule heat. What current will it draw and what must its resistance be? Using equation 15-43,

$$\frac{H}{t} = iV \tag{15-50}$$

$$(40 \text{ watts}) = (i \text{ amps}) \times (12 \text{ volts}) \tag{15-51}$$

$$i = \frac{40}{12}$$

$$= 3.33 \text{ amps} \tag{15-52}$$

The bulb draws 3.33 amps from the battery.

From Ohm's law (equation 15-41)

$$i = \frac{V}{R} \tag{15-53}$$

$$R = \frac{V}{i} \tag{15-54}$$

$$R \text{ ohms} = \frac{12 \text{ volts}}{3.33 \text{ amps}} \tag{15-55}$$

$$R = 3.6 \text{ ohms} \tag{15-56}$$

The resistance of the bulb must be 3.6 ohms.

Alternatively, using equation 15-25,

$$\frac{H}{t} = \frac{V^2}{R} \tag{15-57}$$

$$40 \text{ watts} = \frac{(12 \text{ volts})^2}{R \text{ ohms}} \tag{15-58}$$

$$R = \frac{(12)^2}{40} \text{ ohms}$$

$$= \frac{144}{40}$$

$$= 3.6 \text{ ohms} \tag{15-59}$$

15-6 Electrons in a Vacuum: The electron volt

A discharge tube consists of two metal plates, or *electrodes*, inside an evacuated glass tube (figure 15-6). A potential difference of V volts is established between the two electrodes by some device such as a large number of storage batteries connected one after the other. If ϕ_+ volts is the potential at the positive electrode, and ϕ_- volts the potential at the negative electrode, then

$$V = (\phi_+ - \phi_-) \tag{15-60}$$

Consider an electron in the immediate vicinity of the negative electrode. It is accelerated toward the positive electrode. Unlike the electrons in a copper wire which frequently collide with the copper atoms, this electron is in a vacuum and experiences no collision, so its velocity continually increases until it reaches the positive electrode.

We can apply the law of conservation of energy to the electron and equate the increase in its kinetic energy to the decrease in its potential energy. At the negative electrode,

Initial potential energy $= -e\phi_-$ $\tag{15-61}$

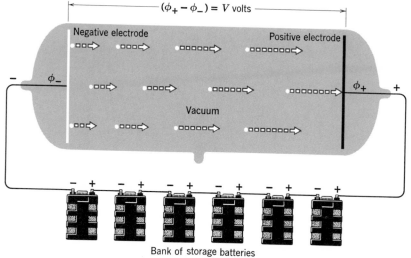

Electric Currents

Figure 15-6 A discharge tube accelerating electrons through a potential difference of V volts.

At the positive electrode,

Final potential energy $= -e\phi_+$ (15-62)

Therefore, the decrease in potential energy is

Initial potential energy − final potential energy

$$= -e\phi_- - (-e\phi_+) = e(\phi_+ - \phi_-) \quad (15\text{-}63)$$
$$= eV \text{ from equation } 15\text{-}60. \quad (15\text{-}64)$$

The potential energy lost reappears as a gain in kinetic energy, so

Increase in kinetic energy $= eV$ (15-65)

If v_1 and v_2 are the initial and final velocities of the electron, the increase in kinetic energy may sometimes be expressed in the form $(\frac{1}{2} mv_2^2 - \frac{1}{2} mv_1^2)$, where m is the mass of the electron. As we shall see in chapter 24, this is true only when v_1 and v_2 are small compared with the velocity of light. Otherwise, the theory of relativity requires a more complicated expression for the kinetic energy.

Whatever the exact expression for the kinetic energy, equation 15-65 tells us that the increase in kinetic energy is simply eV and it is proportional to the potential difference or *voltage* through which the electron has been accelerated. Given the voltage, the increase in energy can be calculated without any additional information. In the early days of research in atomic and nuclear physics, many experiments were performed in which electrons or other charged particles were accelerated to higher energies in a discharge tube or some similar device. It was found very convenient to specify the energies of these particles just by quoting the voltage through which they had been accelerated *from rest*, and this became part of the language of physics. A particle with a charge of $+e$ or $-e$, which has been accelerated from rest through a potential difference of V volts, is said to have a kinetic energy of V **electron volts**. The electron volt is a unit of energy, like the erg or the joule.

Since the potential difference is given in volts, the charge on the electron must be expressed in coulombs

$$e = 1.60 \times 10^{-19} \text{ coulomb} \quad (15\text{-}66)$$

The energy in equation 15-65 is then in joules. The change in potential energy of an electron that has been accelerated through V volts is therefore:

Change in potential energy $= eV$ joule (15-67)
$$= 1.60 \times 10^{-19} V \text{ joule} \quad (15\text{-}68)$$

or, since 1 joule $= 10^7$ erg (15-69)

Change in potential energy $= 1.60 \times 10^{-12} V$ erg (15-70)

Putting V equal to 1 volt, it follows that

$$1 \text{ electron volt} = 1.60 \times 10^{-19} \text{ joule} \qquad (15\text{-}71)$$
$$= 1.60 \times 10^{-12} \text{ erg} \qquad (15\text{-}72)$$

Equations 15-71 and 15-72 provide the best definition of the electron volt. Strictly speaking, it is not correct to say that, when an electron moves through a potential difference of V volts, its gain in kinetic energy is eV. An accelerating electron loses energy by radiating light and other forms of electromagnetic energy, although this does not become of practical importance until V is very large.

One electron volt is abbreviated to 1 eV, 10^3 eV is abbreviated to 1 KeV, 10^6 eV to 1 MeV and 10^9 eV to 1 BeV in the United States, or to 1 GeV in Europe.

So far we have been considering particles carrying a single electronic charge, e. Sometimes experiments are performed with particles having a charge ne, where n is an integer. Such particles might be nuclei with n protons or atoms that have lost n electrons and therefore have n excess protons. (An atom that has lost one or more electrons is called an *ion*.) In this case, if the particle is accelerated through V volts, it acquires an extra kinetic energy of nV electron volts.

The electron volt is a unit of energy, related to the erg and the joule by the conversion factors given in equations 15-71 and 15-72. It can be used for energies other than the kinetic energy of an accelerated charged particle. For example, the kinetic energy of an atom of a gas could be expressed in electron volts, as could the mutual potential energy of two charged particles or the energy needed to remove an electron from inside an atom to a point well outside the atom (the ionization energy).

Example 15-6-1

In a hydrogen atom, the electron is at a distance of 0.53×10^{-8} cm from the proton. Express their mutual electrostatic potential energy in electron volts.

This problem will be solved by two different methods. Both methods will employ CGS units. The mutual potential energy of two charges is

$$\Phi_e = \frac{q_1 q_2}{R} \qquad (15\text{-}73)$$

Putting $q_1 = -q_2 = e = 4.80 \times 10^{-10}$ statcoulomb, and $R = 0.53 \times 10^{-8}$ cm,

$$\Phi_e = -\frac{(4.80 \times 10^{-10})^2}{0.53 \times 10^{-8}} \text{ erg} \qquad (15\text{-}74)$$
$$= -4.35 \times 10^{-11} \text{ erg} \qquad (15\text{-}75)$$

Electric
Currents

Using the conversion factor of equation 15-72 to express this as V electron volts.

305

$$V = -\frac{4.35 \times 10^{-11}}{1.60 \times 10^{-12}} \text{ electron volts} \tag{15-76}$$

$$V = -27.1 \text{ eV} \tag{15-77}$$

In the second method, we imagine the electron to be brought from infinity to its position in the hydrogen atom. At infinity the electric potential due to the proton is zero, but at the final position of the electron it is

$$\phi = +\frac{e}{R} \tag{15-78}$$

$$= \frac{4.8 \times 10^{-10}}{0.53 \times 10^{-8}} \text{ statvolt} \tag{15-79}$$

$$= 9.06 \times 10^{-2} \text{ statvolt} \tag{15-80}$$

But, according to equation 15-31, 1 statvolt is 300 volts, and so

$$\phi = 9.06 \times 10^{-2} \times 300 \text{ volt} \tag{15-81}$$

$$= +27.1 \text{ volt} \tag{15-82}$$

When the electron is brought from infinity to its position in the atom, it moves from a point where the potential is zero to a point where the potential is $\phi = +27.1$ volt, and so the change in potential energy of the system is

$$V = -27.1 \text{ eV} \tag{15-83}$$

QUESTIONS

1. Develop an analogy between the flow of electricity through a circuit and the flow of water through a pipe. For every quantity relevant to the flow of the electricity find a corresponding quantity for the flow of the water. What are the analogs of Ohm's law and Joule heat?
2. Although the drift velocity of the electrons is very slow, an electric light bulb turns on almost immediately when the switch is closed. Explain this. (Hint: The analogy with the flow of water through a pipe may prove helpful.)
3. Is the resistance of a 150 watt light bulb greater or less than that of a 60 watt bulb? Remember that the voltage applied across either is approximately 110 volt.
4. Is equation 15-24 valid for a conductor which does not obey Ohm's law? What about equations 15-25 and 15-26?
5. If a metal were composed of positively charged electrons (positrons) and negatively charged protons (antiprotons), would there be any way of discovering this by studying the phenomena described in this chapter?
6. Is the distance traveled by an electron in between collisions equal to $v_d \tau$?

PROBLEMS

A

1. If a current of 5 statampere flows through a wire, what charge passes a point in the wire in one minute?
2. If a charge of 50 coulomb flows past a point in a wire in one second, what is the value of the electric current in statamp?
3. If 10^{15} electrons flow past a point in a wire in one second, what is the value of the electric current in (*a*) statamperes (*b*) in amperes?
4. If a current of 10^{-15} amp flows through a wire, what is the net number of electrons passing any point in the wire per second?
5. The arrangement shown in the figure consists of eight small charged bodies uniformly spaced around a circle of radius 1 m. Each body has a charge of $+200$ statcoulomb. If the circle is rotated at 800 rpm, what is the equivalent current (*a*) in statamp (*b*) in amp?

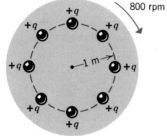

Problem 15-5

6. If a resistance of 120 ohm is connected across the terminals of a 6 volt battery, what current flows in it? How much heat is developed in it per sec.?
7. If a voltage of 12 volt causes a current of 2.5 amp to flow through a metal wire, what is its resistance? How many ergs of joule heat are developed in it in one hour?
8. What is the resistance of a 110 volt 60 watt electric light bulb? (It is true that the current through the bulb alternates backward and forward, but the ratings quoted are appropriate averages which satisfy the formulas of this chapter.)
9. If a small body carrying a charge of 10^{-6} coulomb moves freely through a potential difference of 500 volt, starting from rest, calculate its final kinetic energy in electron volts.

B

10. A 110 volt 60 watt electric light bulb is connected between the terminals of a bank of storage batteries producing 110 volt. How many electrons enter the bulb per second?

11. A 12 volt storage battery promises to deliver 1 amp for 100 hours. If it does so, how many coulombs of charge will have passed between its terminals? How many electrons is this? How much energy has been provided by the battery? Assuming complete utilization of this energy, how many 50 kg crates could be lifted through a height of 1 m?

12. Given the conversion factors between amps and statamps and between volts and statvolts, deduce the conversion factor between ohms and statohms.

13. A 100 ohm heater carrying 0.03 amp is immersed in helium gas at a low pressure. After 5 minutes the temperature has risen from 15°C to 95°C. Assuming that all the joule heat is used to increase the internal energy of the gas, calculate the mass of gas present.

14. If the average kinetic energy of a molecule of an ideal gas is expressed in the form αT electron volts, what is the numerical value of α? What temperature is "equivalent" to 30 BeV?

15. Calculate the velocity of a 110 electron volt electron. Convert your answer into mph. Compare it with the drift velocity calculated in example 15-5-1 and explain the difference.

16. Without questioning the formulas of classical physics that have been presented so far, calculate the potential difference in volts through which an electron would have to fall from rest to acquire a velocity equal to the velocity of light (3×10^{10} cm/sec).

17. Through what potential difference would an α-particle (2 protons plus 2 neutrons) have to fall from rest to acquire a kinetic energy of 1 MeV? What would its velocity then be?

18. Two protons in a nucleus are 5×10^{-13} cm apart. Calculate their mutual electrostatic potential energy in electron volts.

C

19. A long cylindrical celestial object is believed to be a cloud of free electrons. Its diameter is 4.5×10^7 cm and it is observed to be drifting in the direction of its length with a velocity of 1.5×10^7 cm sec^{-1}. From the magnetic effects in its vicinity it is seen to be equivalent to a current of 2×10^{14} statamp. Calculate the average number of electrons per cm^3 in the cloud.

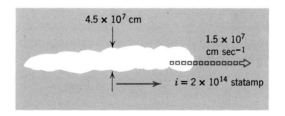

Problem 15-19

20. Two wires with electrical resistances R_1 and R_2 are connected *in*

series as in the diagram. Find the effective resistance between the points A and B.

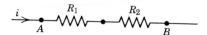

Problem 15-20

21. As shown in the diagram associated with this problem, the two resistances are now connected *in parallel.* Find the effective resistance between the points C and D.

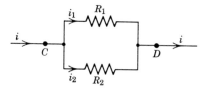

Problem 15-21

22. A copper wire has a resistance of 10 ohm. It is melted down and made into a new wire with only one half the original length, but with a correspondingly larger diameter. What is its new resistance?

23. When a wire 20 cm long is connected between the terminals of a 6 volt battery, the drift velocity of the electrons is 0.6 cm sec^{-1}. What is the time between collisions, τ?

24. A copper wire 1 meter long with a diameter of 0.1 cm has a resistance of 0.02 ohm. Calculate τ, the average time between collisions of an electron with the copper atoms. About how far does an electron travel between collisions? Compare this distance with the average distance between adjacent copper atoms and comment on the comparison.

25. A capacitor consists of two circular metal plates of radius 4 cm separated by a gap 0.05 cm wide. One plate holds a charge of $+120$ statcoulombs and the other plate holds an equal negative charge. A copper wire with a resistance of 10 ohms is connected between the two plates. What time will elapse before 0.01 statcoulomb has leaked from one plate to the other?

Electric
Currents

Moving Charges

16-1   Forces Between Moving Charges

Figure 16-1 illustrates a simple experiment to demonstrate the forces between electric currents. The vertical wires W_1 and W_2 are freely hinged at the top and their lower ends dip into mercury pools. This leaves the wires free to move sideways, but also, since mercury is a metal and conducts electricity, enables electric currents to be fed into the wires from the two storage batteries shown. When the storage batteries are connected so that the two currents flow in the same direction, the wires move toward one another. When the currents flow in opposite directions the wires move apart. Parallel currents moving in the same direction attract one another. Parallel currents moving in opposite directions repel one another. One way to memorize this is to notice that it is the other way round from electrostatic forces. Like charges repel one another and unlike charges attract one another.

In this experiment the wires always contain as many protons as electrons and do not accumulate any net charge. The forces between the

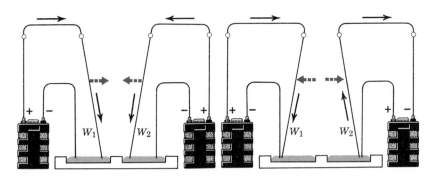

Figure 16-1   An experiment to demonstrate the forces between two parallel wires carrying electric currents.

wires are not the electrostatic forces discussed in Chapter 13, they are a consequence of the fact that electrons are moving inside the wires. When two charged particles are both in motion, they exert on one another a new kind of force that depends upon their velocities and is zero if either velocity is zero. This new force is responsible for the magnetic properties of materials such as iron bar magnets and also for the magnetic effects of electric currents.

If we were to proceed as we have done previously, we would consider two charges q_1 and q_2 a distance R apart moving with velocities v_1 and v_2, as in figure 16-2a. We would then write down a law, similar to Newton's law of gravitation, or Coulomb's law of electrostatic forces, expressing the force in terms of the charges, their velocities and their distance apart. Unfortunately, it is not possible to formulate such a law in any simple form. Even a reasonable approximation to the law is very cumbersome. At this point, the concept of the **electromagnetic field** makes the discussion much easier.

To find the force exerted by q_1 on q_2 we proceed as follows. The moving charge q_1 produces at each point in space not only an electric field $\mathbf{E}$, but also a **magnetic field B** (figure 16-2b). These fields can be calculated in terms of q_1 and v_1, and the position of the point relative to q_1. The values of $\mathbf{E}$ and $\mathbf{B}$ at the point P_2 where the second charge q_2 is located enable the force on q_2 to be calculated (figure 16-2c). The electric field $\mathbf{E}$ exerts a force $q_2\mathbf{E}$ which is independent of the motion of q_2 (see equation 14-23). In a manner that we shall explain below, the magnetic field $\mathbf{B}$ enables us to calculate that part of the force which depends on the velocity of q_2.

To find the force exerted by the second charge q_2 on the first charge q_1, we proceed in a similar way. The charge q_2 produces electric and magnetic fields at the point P_1 where q_1 is located. These fields can be expressed in terms of q_2 and v_2 and the relative positions of q_1 and q_2. The force on q_1 can then be calculated from the values of $\mathbf{E}$ and $\mathbf{B}$ at P_1.

Looked at from this point of view, the magnetic field $\mathbf{B}$ is a mathematical trick which helps us to calculate the forces between moving charges. The observational fact is that moving charges exert forces on one another which depend upon their velocities. With sufficient mathematical ingenuity we might explain this fact without ever introducing the concept of a magnetic field. As a matter of history, however, the concept arose quite naturally at the beginning of the nineteenth century, when Faraday, Ampère, Oersted, Gauss, Biot, Savart, and others were discovering the basic experimental facts about the magnetic effects of electric currents and magnetic materials. One thing is certain: the concept certainly helps to simplify the discussion.

Before embarking upon a more detailed treatment, the following preliminary remarks are necessary. First, although the laws governing the magnetic effects of electric currents in metal wires were discovered in the early nineteenth century, the proper understanding of the forces between two moving charged particles required Einstein's theory of rela-

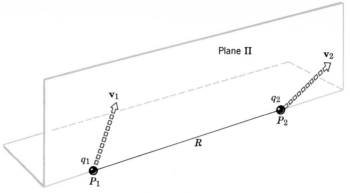

(a) Two charges moving in arbitrary directions.

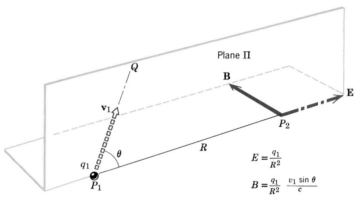

$$E = \frac{q_1}{R^2}$$

$$B = \frac{q_1}{R^2} \frac{v_1 \sin \theta}{c}$$

(b) The electric and magnetic fields at a point P_2 due to a moving charge q_1 at P_1.

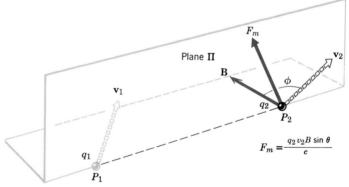

$$F_m = \frac{q_2 v_2 B \sin \theta}{c}$$

(c) The magnetic force $\mathbf{F}_m$ on a moving charge q_2 at P_2 due to the magnetic field $\mathbf{B}$ produced by q_1.

Figure 16-2 The forces between two moving charges.

tivity. In the ensuing discussion we shall take advantage of our present more complete understanding of the situation. Second, we shall assume that the charges are moving with *constant* velocities, because the existence of accelerations makes the situation even more complicated. Third, we shall assume that the velocities of the charges are small. The exact significance of "small" will be made clear in the next section. The drift velocities of electrons responsible for electric currents in metal wires are certainly quite small enough.

16-2   The Magnetic Field due to a Moving Charge

Imagine a *positive* charge q_1 at P_1 moving with a velocity v_1. Let us consider the magnetic field **B** produced by this motion at any point P_2 a distance R from P_1 (figure 16-2b). The angle between $\overrightarrow{P_1P_2}$ and v_1 is θ. It is not necessary that another charge should be located at P_2, although for convenience we shall later place q_2 at P_2.

The magnitude of **B** is

$$B = \frac{q_1}{R^2}\frac{v_1 \sin\theta}{c} \tag{16-1}$$

We are again confining ourselves to CGS units. q_1 is in statcoulombs, R is in cm and v_1 is in cm/sec. The units of B are called **gauss**. The quantity c is a constant of nature. Its units are cm/sec, the units of a velocity, and its value is

$$c = 2.9979 \times 10^{10} \text{ cm/sec} \tag{16-2}$$

This is precisely the speed of light! It is one of the most important constants of physics. As we are now discovering, it enters in a fundamental way into the laws of electromagnetism. As we shall explain in a later chapter, it is the speed of light because light is an electromagnetic wave governed by the laws of electromagnetism. It plays a fundamental role in the theory of relativity. It is the maximum speed with which information can be transferred from one part of the universe to another.

The magnitude of B depends upon $\sin\theta$. When $\theta = 0°$ or $180°$, $\sin\theta = 0$ and the magnetic field is zero. This means that, if the charge q_1 is moving directly toward or directly away from the point P_2, it produces no magnetic field at that point. For a fixed R the field is greatest when $\theta = 90°$ and $\sin\theta = 1$, that is, at points whose direction with respect to q_1 is at right angles to the velocity v_1 of q_1. Notice also that $B = 0$ if $v_1 = 0$. A charge produces no magnetic field unless it is in motion.

The *direction* of **B** is one of the complications which requires careful attention. **B** is perpendicular to both v_1 and $\overrightarrow{P_1P_2}$. If v_1 and $\overrightarrow{P_1P_2}$ are both in the plane Π of figure 16-2, then **B** is perpendicular to this plane. It is still necessary to formulate a rule to tell us whether **B** points outward from the plane (toward the reader) or inward (away from the reader). In figure 16-2b the direction of the magnetic field is inward, away from the reader. Figure 16-3 illustrates the following rule for obtaining the direction of **B**.

Right hand rule number 1

Hold the right hand flat with the four fingers side by side in the same plane as the palm and with the thumb sticking out, but still in this plane. Point the thumb in the direction in which the charge q_1 is moving, that is, along $\mathbf{v}_1$. Point the fingers in the direction $\overrightarrow{P_1P_2}$, from the moving charge toward the point at which the field is being calculated. Then the magnetic field $\mathbf{B}$ at P_2 is perpendicular to the palm in the direction from the back of the hand toward the front, that is, the direction in which the hand would push.

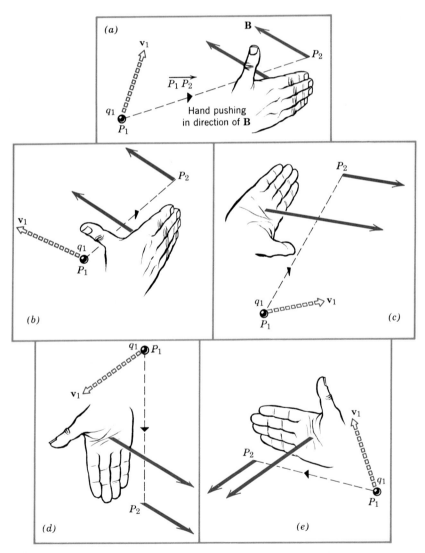

Figure 16-3   Right hand rule number 1. Part *a* corresponds to figure 16-2.

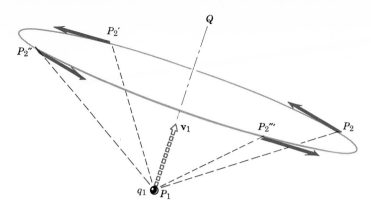

Figure 16-4 Rotating figure 16-3a around an axis in the direction of $\mathbf{v}_1$, one obtains a circular magnetic field line.

A careful study of the diagrams in figure 16-3 will help the reader to understand this rule.

The above rule has been formulated on the assumption that the moving charge q_1 is positive. If q_1 is negative the magnetic field **B** is in exactly the opposite direction. A consistent way to deal with negative charges in electromagnetism is as follows:

Negative charges in motion

If a moving charge is negative, reverse the direction of its velocity and then pretend that it is positive.

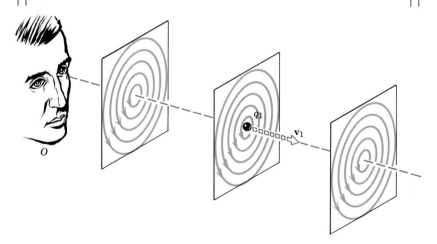

Figure 16-5 A charge moving with a constant velocity produces circular magnetic field lines. These magnetic field lines fill the whole of space, but the diagram shows only some typical lines in three particular planes. With $\mathbf{v}_1$ in the direction shown here, the lines go around counterclockwise from the point of view of the reader looking directly at the page, but they go around clockwise from the point of view of the observer O who is looking in the direction of $\mathbf{v}_1$. The charge q_1 is assumed to be positive.

Moving
Charges

315

The reader should verify for himself, in situations such as those shown in figure 16-3, that this will always give a magnetic field in the opposite direction to that obtained for a positive charge.

The diagram of figure 16-3a can be rotated about a line P_1Q through q_1 in the direction of v_1 (figure 16-4). In any of its orientations, this diagram still provides us with a valid method of finding the direction of **B**. The point P_2 describes a circle with its center on P_1Q, and **B** is everywhere tangential to this circle. A line which has the property that its tangent at any point is in the direction of the magnetic field at that point is called a **magnetic field line.** The magnetic field lines for a moving charge are clearly circles centered on a line through the charge in the direction of its motion (figure 16-5). An observer looking in the direction of v_1 sees the field lines go round in a clockwise direction. This provides us with an alternative rule for finding the direction of **B,** which some students may prefer (see figure 16-6).

The clock rule

Look in the direction in which the charge is moving. Then the magnetic field lines are circles which go round in the same way that the hands of a clock rotate. (If the charge is negative, its velocity should first be reversed, which means that one must look in the opposite direction to that in which the negative charge is really moving).

The *electric* field at P_2 due to the charge q_1 is almost the same as if q_1 were stationary. It points in the direction $\overrightarrow{P_1P_2}$ and its magnitude is given approximately by equation 14-26.

$$E = \frac{q_1}{R^2} \text{ approximately} \qquad\qquad (16\text{-}3)$$

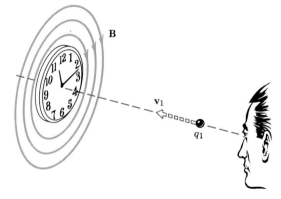

ELECTRICITY
AND
MAGNETISM:
PARTICLES
AND FIELDS

Figure 16-6   The clock rule. An observer looking in the direction in which the positive charge is moving sees the magnetic field lines going around clockwise.

However, the approximation is good only when the velocity v_1 of q_1 is small. We are now in a position to explain that "small" means "small compared with the speed of light c." If v_1 is not small compared with c, the equations for B and E are more complicated.

$$B = \frac{q_1}{R^2} \frac{v_1 \sin \theta}{c} \frac{\left(1 - \dfrac{v_1^2}{c^2}\right)}{\left(1 - \dfrac{v_1^2 \sin^2 \theta}{c^2}\right)^{3/2}} \tag{16-4}$$

$$E = \frac{q_1}{R^2} \frac{\left(1 - \dfrac{v_1^2}{c^2}\right)}{\left(1 - \dfrac{v_1^2 \sin^2 \theta}{c^2}\right)^{3/2}} \tag{16-5}$$

The student is not expected to remember these equations or to make use of them. They are quoted here to illustrate two points. Firstly, electromagnetism is clearly a complicated subject. Secondly, it is interesting that the electric field $\mathbf{E}$ depends slightly on the velocity $\mathbf{v}_1$. The magnetic field $\mathbf{B}$ is, of course, very dependent on v_1, since $\mathbf{B} = 0$ if $v_1 = 0$. The student who feels that his mathematical ability is up to the task should show that equations 16-4 and 16-5 are practically the same as equations 16-1 and 16-3 if v_1/c is very much less than 1.

16-3   The Force on a Moving Charge in a Magnetic Field

We have just discussed the magnetic field $\mathbf{B}$ at P_2 due to the charge q_1 at P_1 moving with a velocity $\mathbf{v}_1$. Now suppose that there is a *positive* charge q_2 at P_2 moving with a velocity $\mathbf{v}_2$, and let us consider the force $\mathbf{F}_m$ exerted on it by the magnetic field $\mathbf{B}$. Let ϕ be the angle between $\mathbf{B}$ and $\mathbf{v}_2$ (figure 16-7). Then the magnitude of the force is

$$F_m = \frac{q_2 B v_2 \sin \phi}{c} \tag{16-6}$$

If q_2 is in statcoulombs, B in gauss, v_2 and c in cm/sec, then the force F_m is in dynes.

If $v_2 = 0$, $F_m = 0$. The magnetic field exerts no force on a stationary charge. If $\phi = 0°$ or $180°$, then $\sin \phi = 0$ and $F_m = 0$. If the charge is

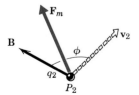

Figure 16-7   The magnetic force $\mathbf{F}_m$ on a charge q_2 moving with a velocity $\mathbf{v}_2$ in a magnetic field $\mathbf{B}$.

moving along the magnetic field lines, in the direction of **B** or in exactly the opposite direction to **B**, the magnetic field exerts no force on it. The force is greatest when $\phi = 90°$ and $\sin \phi = 1$, that is, when the motion is perpendicular to **B**.

The *direction* of the force $\mathbf{F}_m$ is perpendicular to both **B** and v_2 and is given by another right hand rule (figure 16-8).

Right hand rule number 2

Hold the right hand flat as before. Point the thumb in the direction in which the charge q_2 is moving, that is, along v_2. Point the fingers in the direction of the magnetic field **B**. Then the magnetic force $\mathbf{F}_m$ is in the direction in which the hand would push.

Lest the existence of two right hand rules should confuse you, notice carefully that in both cases the thumb points in the direction of the velocity of the charge and the push is in the initially unknown direction which is being determined. If q_2 is a negative charge, its velocity should first be reversed and the rule then applied. Verify that the force is then in the opposite direction from a positive charge.

The value of **B** given by equation 16-1 can be inserted in equation 16-6 to yield

$$F_m = q_2\left(\frac{q_1}{R^2}\frac{v_1}{c}\sin\theta\right)\frac{v_2}{c}\sin\phi \tag{16-7}$$

or

$$F_m = \left(\frac{q_1 q_2}{R^2}\right)\left(\frac{v_1}{c}\frac{v_2}{c}\right)(\sin\theta\sin\phi) \tag{16-8}$$

Let us compare this with the electrostatic force

$$F_e = \frac{q_1 q_2}{R^2} \tag{16-9}$$

The factor $(\sin\theta\sin\phi)$ lies somewhere between 0 and 1 and is not relevant to our subsequent remarks. The important difference between

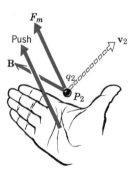

Figure 16-8   Illustrating right hand rule number 2.

F_m and F_e is the factor $(v_1 v_2 / c^2)$. In most practical cases v_1 and v_2 are very much less than c and so the magnetic force is much smaller than the electric force. If, for example, v_1 and v_2 are drift velocities of electrons in wires and have magnitudes of the order of 0.1 cm/sec, since c is 3×10^{10} cm/sec, the factor $(v_1 v_2 / c^2)$ is approximately 1.1×10^{-23}!

If magnetic forces are usually so small compared with electric forces, why can we not ignore them and concentrate on the much larger electric forces? The reason is the same as the reason why we cannot ignore gravitational forces even though they are much weaker than electric forces (see section 13-3). Since charges can be both positive and negative, most matter is electrically neutral and electric forces are frequently cancelled out. In the experiment of figure 16-1, the wires contain equal numbers of protons and electrons and there are no net *electric* fields present. However, the electrons are drifting along the wire, whereas the protons have no net drift velocity. The *magnetic* effects of the moving electrons are therefore not cancelled by the protons.

It is interesting to speculate whether there are *gravitational* forces which depend on the velocities of the gravitating masses. If such forces exist they are probably weaker than ordinary gravitational forces by a factor such as $(v_1 v_2 / c^2)$, where v_1 and v_2 are the velocities of the gravitational masses. But gravitational mass is always positive, and it is not possible to cancel out the main part of the gravitational force to enable the weak velocity-dependent part to reveal itself. The general theory of relativity makes it very plausible that there are velocity-dependent gravitational forces, but it is extremely difficult to devise an experiment to detect them.

The magnetic field **B** at P_2 allows us to calculate that part, $\mathbf{F}_m$, of the force on q_2 which depends on its velocity $\mathbf{v}_2$. The electric field **E** enables us to calculate the part, $\mathbf{F}_e$, of the force which is independent of the velocity and is still present if $\mathbf{v}_2 = 0$ and q_2 is at rest. In accordance with equation 14-23, the electric force is

$$\mathbf{F}_e = q_2 \mathbf{E}$$

(16-10)

This is quite independent of v_2, but it does depend slightly on v_1, as shown by equation 16-5. If we are calculating the additional force between the charges due to their motion, the change in the electric force due to the slight change in **E** can be just as important, or even more important, than the magnetic force due to **B**. This complication might have made the discovery of the laws of electromagnetism very difficult indeed, except for the fortunate circumstance that experiments with magnetic materials and electric currents employ only neutral bodies and the electric field is thereby eliminated.

To calculate the force on q_2 we need know only **E** and **B** at the point P_2. So far we have assumed that these fields are due solely to q_1 at P_1, but they may be due to the combined effect of several charges in the vicinity of P_2. We then calculate the contribution of each charge to **E** or **B**, using the method of the previous section, and add the various contributions, using, of course, *vector* addition.

Let us reiterate one aspect of our discussion to clarify the significance

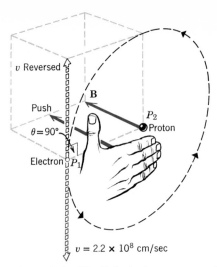

Figure 16-9 The Bohr hydrogen atom of example 16-3-1.

of the electric field **E** and the magnetic field **B** at any point P in space. If a charge is placed at P, the electric field enables us to calculate that part of the force on it which is independent of its velocity, using equation 16-10. The magnetic field **B** enables us to calculate that part of the force which depends on the velocity of the charge, using equation 16-6 and right hand rule number 2. This aspect of the calculation does not require that the velocity of the charge at P be small compared with the speed of light c. The fields **E** and **B** at P are due to the surrounding charges and can be calculated from equations such as 16-1 and 16-3. In general, however, these calculations can be cumbersome. If the velocities of the surrounding charges are not small compared with c, we must use equations 16-4 and 16-5. If we relax the condition that the charges have zero acceleration, the calculation becomes very difficult indeed.

Example 16-3-1

In the Bohr theory of the hydrogen atom, the electron moves around the proton in a circle of radius 0.53×10^{-8} cm with a velocity of 2.2×10^8 cm/sec. What is the magnetic field it produces at the proton? This problem is illustrated in figure 16-9.

In equation 16-1,

$$B = \frac{q_1}{R^2} \frac{v_1}{c} \sin \theta \tag{16-11}$$

q_1 is the charge on an electron, or 4.802×10^{-10} statcoulomb. P_1P_2 is the radius of the electronic orbit, so $R = 0.53 \times 10^{-8}$ cm. v_1 is 2.2×10^8 cm/sec and $\theta = 90°$, so that $\sin \theta = 1$. Therefore, the field at the proton is

$$B = \frac{4.802 \times 10^{-10}}{(0.53 \times 10^{-8})^2} \times \frac{2.2 \times 10^8}{3 \times 10^{10}} \times 1 \tag{16-12}$$

$$B = 1.25 \times 10^5 \text{ gauss} \tag{16-13}$$

Figure 16-9 shows the application of right hand rule number 1 to determine the direction of **B**. Since the charge on the electron is negative, its velocity has first been reversed. **B** is perpendicular to both v_1 and $\overrightarrow{P_1P_2}$ and is therefore perpendicular to the plane of the circular orbit. When the electron is rotating in a counterclockwise direction as shown, **B** goes *into* the plane of the paper.

Example 16-3-2

Proton number 1 has a velocity of 10^9 cm/sec in a horizontal plane in a southeasterly direction. Proton number 2 is at a distance of 10^{-4} cm in an easterly direction and has a velocity of 2×10^9 cm/sec in a northwesterly direction (figure 16-10a). Calculate the magnitude and direction of the *magnetic* force on proton number 2.

First calculate the magnetic field produced by proton number 1 at the site of proton number 2. Using equation 16-1,

$$B = \frac{q_1}{R^2} \frac{v_1}{c} \sin \theta \tag{16-14}$$

q_1 is $e = 4.802 \times 10^{-10}$ statcoulomb, R is 10^{-4} cm, v_1 is 10^9 cm/sec and $\theta = 45°$, so $\sin \theta = 0.707$. Therefore

$$B = \frac{4.802 \times 10^{-10}}{(10^{-4})^2} \times \frac{10^9}{3 \times 10^{10}} \times 0.707 \tag{16-15}$$

or

$$B = 1.13 \times 10^{-3} \text{ gauss} \tag{16-16}$$

Figure 16-10b shows the application of right hand rule number 1 to find the direction of **B**. The plane containing v_1 and $\overrightarrow{P_1P_2}$ is a horizontal plane and, since **B** is perpendicular to both of them, it must be in a vertical direction. The diagram shows the push of the hand to be in a vertical *upward* direction away from the center of the earth.

Now calculate the force exerted by this magnetic field on proton number 2 (figure 16-10c). Using equation 16-6,

$$F_m = q_2B \frac{v_2}{c} \sin \phi \tag{16-17}$$

q_2 is $e = 4.802 \times 10^{-10}$ statcoulomb, B has been shown to be 1.13×10^{-3} gauss, v_2 is 2×10^9 cm/sec. Since v_2 is in a horizontal plane and **B** is vertical, the two are at right angles, so $\phi = 90°$ and $\sin \phi = 1$. Therefore,

$$F_m = (4.802 \times 10^{-10}) \times (1.13 \times 10^{-3}) \times \left(\frac{2 \times 10^9}{3 \times 10^{10}}\right) \times 1 \tag{16-18}$$

$$F_m = 3.62 \times 10^{-14} \text{ dyne} \tag{16-19}$$

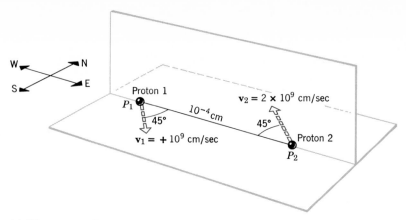

(a) The two moving protons.

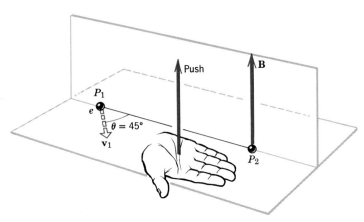

(b) The direction of **B** at proton 2.

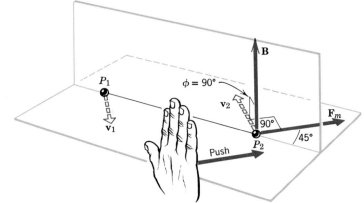

(c) The direction of $\mathbf{F}_m$.

Figure 16-10 Illustrating example 16-3-2.

Figure 16-10c shows the use of right hand rule number 2 to find the direction of $\mathbf{F}_m$. Since $\mathbf{F}_m$ is perpendicular to $\mathbf{B}$, which is vertical, $\mathbf{F}_m$ must be in a horizontal plane. Since $\mathbf{F}_m$ is perpendicular to $\mathbf{v}_2$, which is in a northwesterly direction, $\mathbf{F}_m$ must be in either a southwesterly or a northeasterly direction. The direction of the push of the hand shows it to be northeasterly.

Example 16-3-3

A charged fundamental particle has a velocity $\mathbf{v}$ in a direction perpendicular to a uniform magnetic field $\mathbf{B}$ (uniform means that $\mathbf{B}$ has the same magnitude and direction at all points throughout the region of space inside which the particle moves). The electric field $\mathbf{E}$ is zero everywhere. Show that the subsequent path of the particle is a circle and calculate its radius. If the particle is a proton with an energy of 1 MeV, what size field would be needed to bend its path into a circle of radius 10 cm?

As we have already explained, the magnetic force is always perpendicular to the velocity. When a particle moves in a direction perpendicular to the force acting on it, the force does no work (section 9-7). The energy of the particle therefore remains constant. Since the energy in this case is entirely kinetic energy, the magnitude of the velocity must remain constant. Only the direction of the velocity can change. The force, and therefore the acceleration, are always perpendicular to the velocity. These are exactly the conditions for uniform circular motion (section 5-4).

Figure 16-11 shows the particle moving in a circle of radius r in

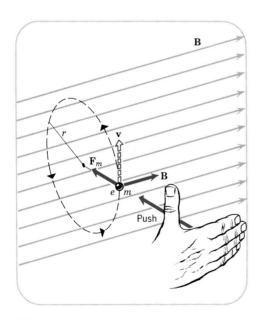

Figure 16-11 The circular path of a charged particle in a uniform magnetic field. (example 16-3-3).

a plane perpendicular to **B.** Application of right hand rule number 2 in various positions of the particle demonstrates that the magnetic force **F**$_m$ is always directed toward the center of the circle. Convince yourself that, if you look in the direction of **B,** a positive charge goes round in a counterclockwise direction, and a negative charge in a clockwise direction. The direction in which a charge goes round its circle can always be found from the requirement that the magnetic force must point toward the center of the circle.

From equation 16-6 the force is

$$F_m = \frac{evB}{c} \tag{16-20}$$

When a particle moves uniformly in a circle of radius r with velocity v, its acceleration is v^2/r (equation 5-27). If its mass is m, Newton's second law tells us that the force is the mass times the acceleration, or

$$F_m = \frac{mv^2}{r} \tag{16-21}$$

Combining equations 16-20 and 16-21

$$\frac{evB}{c} = \frac{mv^2}{r} \tag{16-22}$$

which can be rearranged to give the radius

$$r = \frac{mvc}{eB} \tag{16-23}$$

The faster the particle moves, the larger is the radius. The larger the applied field **B,** the smaller is the radius. Heavy particles are more difficult to bend into circular paths than light particles.

Consider now a proton with a kinetic energy of 1 MeV $= 10^6$ eV. Using equation 15-72, this kinetic energy is

$$\tfrac{1}{2}mv^2 = 10^6 \times 1.6 \times 10^{-12} \text{ erg} \tag{16-24}$$
$$= 1.6 \times 10^{-6} \text{ erg}$$

Since the mass of the proton is

$$m = 1.67 \times 10^{-24} \text{ gm} \tag{16-25}$$

$$v = \sqrt{\frac{2 \times 1.6 \times 10^{-6}}{1.67 \times 10^{-24}}} \tag{16-26}$$
$$= 1.38 \times 10^9 \text{ cm/sec} \tag{16-27}$$

Rearranging equation 16-23,

$$B = \frac{mvc}{er} \tag{16-28}$$

Inserting the known values of the quantities on the right hand side, the

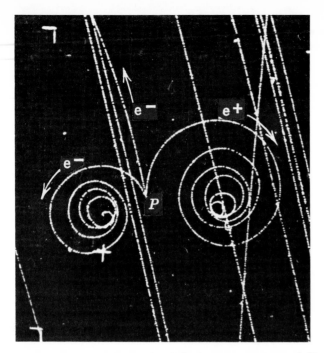

Figure 16-12 Curved tracks of electrons in a magnetic field that is perpendicular to the page and points toward the reader. At P a certain process suddenly produces two negatively charged electrons and a positron (which is identical with an electron except that it is positively charged). A white track is a trail of bubbles left behind after the passage of a charged particle. The electron e⁻ which moves off to the left follows a circular path in the magnetic field, but slows down as it loses energy in collisions with the atoms of the liquid through which it is passing. As its velocity decreases, the radius of its circular path decreases (equation 16-23) and so it spirals inward. The positron e⁺ which moves off to the right has the opposite charge and its path is curved in the opposite direction. The electron e⁻ in the center has a very high initial velocity and so the radius of its circular path is very large. (Courtesy of the Lawrence Radiation Laboratory of the University of California.)

field needed to make $r = 10$ cm is

$$B = \frac{(1.67 \times 10^{-24}) \times (1.38 \times 10^{9}) \times (3 \times 10^{10})}{(4.8 \times 10^{-10}) \times 10} \tag{16-29}$$

$$B = 1.44 \times 10^4 \text{ gauss} \tag{16-30}$$

Figure 16-12 shows the curved tracks of electrons moving in a magnetic field.

Example **16-3-4**

A copper wire of length l is perpendicular to a magnetic field **B** and has a velocity **v** in a direction perpendicular to both the wire and **B** (figure 16-13). Show that a potential difference exists between the two ends of the wire and calculate its magnitude.

The copper atoms are forced to move with a constant velocity **v** by some driving mechanism. (For example, the wire might be mounted on a toy engine moving with steady speed along a straight track.) However, there are free electrons inside the copper which can move freely through the wire. These electrons have a velocity v perpendicular to a magnetic field B and therefore experience a magnetic force

$$F_m = \frac{evB}{c} \tag{16-31}$$

Reversing the velocity of the negative electron and then applying right hand rule number 2, this force is seen to be downward in figure 16-13. The free electrons therefore move downward along the wire. The top of the wire, A, loses electrons and becomes positively charged, while the bottom of the wire, B, gains electrons and becomes negatively charged.

The top of the wire therefore acquires a potential ϕ_A which is more positive than the potential ϕ_B at the bottom of the wire. In accordance with equation 14-46, this difference in potential produces an electric field E in the wire, given by

$$E = \frac{\phi_A - \phi_B}{l} \tag{16-32}$$

Since the field goes from the more positive toward the more negative potential, it is *downward*. It exerts a force

$$F_e = eE \tag{16-33}$$

on the electron, which is *upward* because the electron is negatively

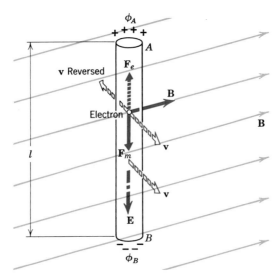

Figure 16-13 Illustrating example 16-3-4. A copper wire moving perpendicularly to a magnetic field.

charged. (The positively charged top of the wire attracts the electron and the negatively charged bottom of the wire repels it.)

The electric force $\mathbf{F}_e$ is seen to be in the opposite direction to the magnetic force $\mathbf{F}_m$. As more and more electrons move to the bottom of the wire, $\mathbf{F}_e$ increases until it is large enough to counterbalance $\mathbf{F}_m$

$$F_e = F_m \tag{16-34}$$

The resultant force on the electrons is then zero and there is no further tendency for the electrons to migrate. Equilibrium is achieved, but only by virtue of the existence of a steady potential difference, $\phi_A - \phi_B$, between the ends of the wire.

Substituting into equation 16-34 the value of F_e from equation 16-33 and the value of F_m from equation 16-31

$$eE = \frac{evB}{c} \tag{16-35}$$

or

$$E = \frac{vB}{c} \tag{16-36}$$

Now insert the value of E from equation 16-32

$$\frac{(\phi_A - \phi_B)}{l} = \frac{vB}{c} \tag{16-37}$$

or

$$(\phi_A - \phi_B) = \frac{vBl}{c} \tag{16-38}$$

which is the required expression for the potential difference between the ends of the wire. If v and c are in cm/sec, B in gauss and l in cm, $(\phi_A - \phi_B)$ is in statvolts.

QUESTIONS

1. A dense beam of electrons all moving in the same direction has a tendency to spread out as it advances. Two overlapping beams of protons and electrons of equal density moving in opposite directions have a tendency to shrink. Explain this.

2. In which of the following cases is there no magnetic force between two electrons? (a) When they move directly toward one another. (b) When they move directly away from one another. (c) When they have the same velocity in a direction perpendicular to the line joining them. (d) When they have equal and opposite velocities in a direction perpendicular to the line joining them?

3. If an electron moves horizontally toward the north in a horizontal field pointing toward the west, in which direction is the magnetic force on it?

Moving
Charges

327

4. I am looking down on a proton which is moving horizontally in a uniform vertical magnetic field pointing up at me. Do I see it going round its circular orbit in a clockwise or counterclockwise direction?

5. When I look toward an electron, I see it moving in a counterclockwise circular orbit in a vertical plane. Specify the direction of the magnetic field responsible for this. (Hint: In this problem it is the direction of the magnetic field which is unknown initially. However, right hand rule number 2 is based on the assumption that you are trying to find the direction of the force, given the directions of the velocity and the field.)

6. Imagine that you have invented an apparatus that can project an electron with any velocity in any direction and can then measure the force on it with very great accuracy. Explain precisely how this apparatus could be used to determine **E** and **B** at any point in space.

7. Try to discover a case in which the magnetic forces between two moving charges do not obey Newton's third law of motion.

8. A student trying to find the directions of the forces between two moving charges always uses his left hand by mistake. Does he get the right answer?

9. A uniform horizontal magnetic field points toward the north. A horizontal copper rod is perpendicular to this field. If the rod is allowed to fall freely from rest, which end of it becomes positively charged.

10. A horizontal copper rod is perpendicular to a uniform horizontal magnetic field. The rod is moved up and down in a vertical plane in an oscillatory fashion. Ignoring the friction of the air, does this operation require a continuous expenditure of work? If so, where does the work go to?

B

Question 16-10

11. When an electric current flows in a long straight wire perpendicular to a magnetic field, a potential difference is produced across the wire (the Hall effect). Why?

PROBLEMS

A

1. What is the magnitude of the magnetic field produced by a charge of 7 statcoulomb at a distance of 5 cm in a direction at right angles to its velocity of 3×10^4 cm/sec?

2. What is the magnitude of the magnetic field produced by a charge of 3×10^{-4} statcoulomb at a distance of 2×10^{-3} cm in a direction at $45°$ to its velocity of 5×10^7 cm/sec?

3. What is the magnitude of the magnetic force exerted by a magnetic field of 15 gauss on a charge of 3 statcoulomb with a velocity of 3×10^9 cm/sec in a direction (a) parallel to the field, (b) perpendicular to the field (c) at an angle of $45°$ to the field.

4. A charge of 6×10^{-9} statcoulomb moving with a velocity of 9×10^8 cm/sec in a direction perpendicular to a magnetic field experiences a magnetic force of 9×10^{-6} dyne. What is the magnitude of the magnetic field?

5. A charged body moving with a velocity of 8×10^5 cm/sec at an angle of $30°$ to a magnetic field of 0.1 gauss experiences a magnetic force of 0.5 dyne. What is the magnitude of the charge on the body?

6. Eight equal charges, each of value $+0.2$ statcoulomb, are spaced uniformly around a circle of radius 6 cm. If the circle rotates with an angular velocity of 5×10^4 radians/sec, what is the magnetic field at the center of the circle?

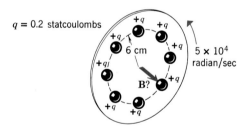

Problem 16-6

7. A charge of $+5$ statcoulomb has a horizontal velocity of 4×10^7 cm/sec toward the north. What are the magnitude and direction of the magnetic field which it produces at a point 2 cm due west of itself?

8. An electron has a velocity of 1.5×10^8 cm/sec vertically upward. What are the magnitude and direction of the magnetic field which it produces at a point 10^{-6} cm due north of itself.

9. A charge of -0.1 statcoulomb has a velocity of 2×10^4 m/sec toward the east in a horizontal magnetic field of 300 gauss pointing due south. What are the magnitude and direction of the magnetic force on the charge?

Moving
Charges

10. A proton has a velocity of 1.5×10^9 cm/sec due north in a magnetic field of 5×10^5 gauss pointing vertically upward. What are the magnitude and direction of the magnetic force on it?

11. Two electrons are 10^{-7} cm apart. They have identical velocities of 2×10^8 cm/sec in a direction perpendicular to the line joining them. Find the magnitude of the magnetic force between them. What is the ratio of the magnetic force to the electric force?

B

12. A charge of $+3.5$ statcoulomb has a horizontal velocity of 2.5×10^7 cm/sec toward the northeast. What are the magnitude and direction of the magnetic field at a point 3 cm due west of the charge?

13. A 2 MeV proton is moving vertically upward. What are the magnitude and direction of the magnetic field it produces at a point 3×10^{-7} cm southwest of itself?

14. An electron has a horizontal velocity of 3×10^8 cm/sec toward the southeast in a horizontal magnetic field of 10,000 gauss pointing due north. What are the magnitude and direction of the magnetic force on it?

15. An electron is 2.5×10^{-7} cm due east of a proton. The proton has a horizontal velocity of 1.5×10^8 cm/sec due north. The electron has a horizontal velocity of 2×10^9 cm/sec toward the northwest. Find the magnitude and direction of the magnetic force on (a) the electron, (b) the proton.

16. What are the magnitude and direction of a magnetic field which could support the weight of an electron with a horizontal velocity of 1 cm/sec due east?

17. A proton has a velocity of 1.5×10^8 cm/sec perpendicular to a magnetic field of 15,000 gauss. What is the radius of its circular orbit?

18. An electron moves in a circular orbit of radius 1.5 m in a magnetic field of 0.2 gauss. Find (a) its velocity (b) its momentum (c) its energy in electron volts.

19. In a certain region of interstellar space, there are free electrons with kinetic energies of 10^{-3} eV moving in circular orbits of radius 2.5×10^4 m. What is the magnitude of the magnetic field which causes this motion?

20. A fundamental particle of mass m moves in a circular orbit perpendicular to a magnetic field B. Find an expression for the time taken to complete one revolution. How does it depend upon (a) the velocity v (b) the radius r (c) the mass m (d) the field B?

C

21. An α-particle is the nucleus of a helium atom. It contains two protons and two neutrons and has a mass of 6.64×10^{-24} gm. An α-particle is accelerated by a potential drop of 2×10^3 volts and then moves into a region where its velocity is perpendicular to a

magnetic field of 200 gauss. Calculate the radius of its circular orbit.

22. An electron moves with a velocity v in a circle of radius r. What are the instantaneous magnitude and direction of the magnetic field which it produces at a point P on the axis of rotation a distance l from the center of the circle? What are the magnitude and direction of the average magnetic field, averaged over a time during which the electron has made many revolutions?

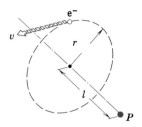

Problem 16-22

23. In example 16-3-4, if B is the horizontal component of the earth's magnetic field and has a magnitude of 0.2 gauss, what velocity v is needed to produce a potential difference of 1 volt between the ends of a wire 10 cm long? Express the answer in cm sec^{-1} and in mph.

24. If a charged body has a velocity $\mathbf{v}$, how can an electric field and a magnetic field be applied to it in such a way that the electric force is exactly counterbalanced by the magnetic force? Can $\mathbf{E}$ have any direction relative to $\mathbf{v}$, if the direction of $\mathbf{B}$ is then chosen to suit? Can $\mathbf{B}$ have any direction relative to $\mathbf{v}$, if the direction of $\mathbf{E}$ is then chosen to suit it? What other restrictions are there on $\mathbf{E}$ and $\mathbf{B}$? What may be chosen arbitrarily?

25. What would be the percentage change in the magnitude of the magnetic force calculated in example 16-3-2 if we were to use equation 16-4 for the magnetic field? (Hint: Since the difference is small, you will have to make your calculations very carefully.)

26. In example 16-3-2 calculate the electric force between the two charges, using equation 16-5. What is the difference between this and the value the force would have if the charges were stationary? Compare this difference with the magnetic force.

27. A thin plane sheet of metal foil of area A and thickness d has a velocity $\mathbf{v}$ which is parallel to its own plane. There is a magnetic field $\mathbf{B}$ which is also parallel to the plane of the foil, but is perpendicular to $\mathbf{v}$. If the net charge on the foil is zero, find an expression for the magnitude of the charge on one face of the foil.

28. A long thin metal rod has a velocity $\mathbf{v}$ at an angle θ to a magnetic field $\mathbf{B}$. The orientation of the rod is quite arbitrary, and you may specify it in any way you find convenient. Show how to find the potential difference between the ends of the rod.

29. A vertical copper disk is rotating about a horizontal axis in a uniform magnetic field parallel to this axis. Describe qualitatively how

Moving
Charges

331

the free electrons distribute themselves over the disk? Does it matter which way the disk rotates? (Hint: Calculate the electric field at any distance, r, from the axis. Then consider the electric field lines and the fact that they must start and end on charges.)

Problem 16-29

ELECTRICITY
AND
MAGNETISM:
PARTICLES
AND FIELDS

332

17

Magnetism

17-1 Magnetic Fields due to Electric Currents

In the previous chapter we indicated how magnetic effects are a consequence of the behavior of moving charges. In an ultimate analysis all magnetic phenomena can be related to the interactions between charged fundamental particles. We now turn to the case of the magnetic effects of electric currents in metal wires, which was historically important to the development of the theory of electromagnetism and which is still very important because of its engineering applications to electrical machinery and electrical devices. These magnetic effects are a consequence of the presence of free electrons drifting steadily along the wire. A great simplification is introduced by the fact that the drift velocity is always very much smaller than the speed of light. Also, the number of protons in the wire is equal to the number of electrons and they cancel out each other's *electric* field. However, although a proton is moving chaotically in all directions, it stays in the vicinity of one point and has no net drift velocity to produce a magnetic field which would cancel the magnetic field of the drifting electrons. We can therefore concentrate entirely on the magnetic field $\mathbf{B}$ and assume that the electric field $\mathbf{E}$ is zero.

With the help of figure 17-1, we shall consider the problem of calculating the magnetic field $\mathbf{B}$ at a point P due to a current i_1 flowing in a wire bent into any arbitrary shape. Concentrate upon a short segment CD of the wire, which is so short that it can be considered to be straight. If we can calculate the contribution of this segment to the value of $\mathbf{B}$ at P, then the wire can be completely divided up into such short lengths and the contributions from all of them added vectorially to give the total field at P. Suppose that l_1 is the length of CD, A is the cross-sectional area of the wire, and R is the distance from CD to P.

The current i_1 is due to free electrons drifting with a velocity v_d in the opposite direction to i_1. In accordance with our convention, let us reverse this drift velocity and pretend that the drifting charges are positive and move in the same direction as the current. Let θ be the angle between the direction of the reversed velocity and the line joining the electron to the point P. Since CD is so very short, θ is the same for all

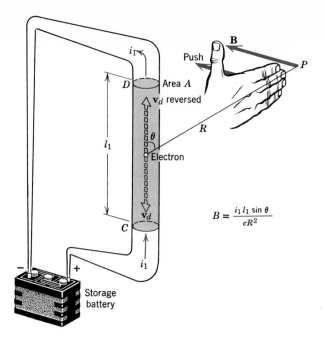

Figure 17-1   The Biot-Savart law.

the electrons and is the same as the angle between CD and the line joining the center of CD to P.

The magnetic field produced at P by a single electron is obtained from equation 16-1, putting $q_1 = e$ and $v_1 = v_d$,

$$\text{Contribution to B from 1 electron} = \frac{e}{R^2}\frac{v_d}{c}\sin\theta \tag{17-1}$$

The volume of the segment of wire is l_1A and, if there are n_0 free electrons in 1 cm³, it contains $n_0 l_1 A$ drifting electrons. The combined effect of all these electrons is

$$\text{Contribution to B from } CD = n_0 l_1 A\frac{e}{R^2}\frac{v_d}{c}\sin\theta \tag{17-2}$$

The electric current i_1 is introduced by means of equation 15-7

$$i_1 = e n_0 A v_d \tag{17-3}$$

Since the combination $e n_0 A v_d$ appears in equation 17-2, this equation can be rewritten in the following way.

Biot–Savart law

Contribution to B from the short length of wire CD

$$= \frac{i_1 l_1}{cR^2}\ \sin\theta \tag{17-4}$$

i_1 should be in statamperes, l_1 and R in cm, c in cm/sec and **B** is then in gauss. This law was proposed by Biot and Savart in the early nineteenth century to provide a consistent explanation of numerous experiments on the magnetic effects of currents, long before the discovery of the electron by J. J. Thomson in 1897.

The direction of the contribution to **B** from CD is clearly given by a slight modification of right hand rule number 1. Instead of the direction of the velocity of a positive charge or the reversed direction of the velocity of a negative charge, we use the direction of flow of the current i_1, which is opposite to the direction of the drift velocity of the electrons. The rule is therefore: hold the right hand flat. Point the thumb in the direction of the current through CD. Point the fingers in the direction from CD to the point P. Then the contribution of CD to **B** is in the direction in which the hand would push.

Even when the Biot-Savart law has been used to calculate the contribution from each segment of the wire, there remains the problem of adding together these contributions. This is usually a difficult mathematical problem beyond the scope of this book. We shall merely quote the result in the particular case of a very long straight wire. With the help of figure 17-2a, it is easy to see that the contribution to **B** at a point P from *any* segment CD of the wire is perpendicular to a plane containing P and the wire. The total magnetic field **B** must therefore be perpendicular to this plane. If the length of a perpendicular from P to the wire is r, the value of B can be shown to be

$$B = \frac{2i_1}{cr} \tag{17-5}$$

Figure 17-2a can be rotated about the wire as an axis to obtain cir-

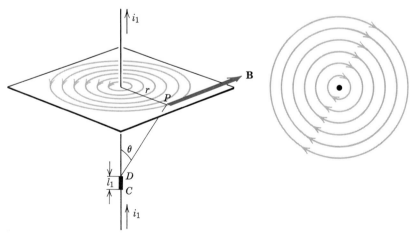

(a) Circular magnetic field lines due to a current in a straight wire.

(b) Looking in the direction in which the positive current is flowing.

Magnetism

Figure 17-2   Magnetic field due to the current in a straight wire.

335

cular magnetic field lines as in the case of a moving charge (figure 16-5). Figure 17-2b shows the appearance of these field lines in a plane through P perpendicular to the wire. Exactly the same pattern of lines would be obtained in any other plane perpendicular to the wire, intersecting the wire anywhere along its length. The clock rule of the previous chapter can be reformulated to apply to the present situation. If we look along the wire in the direction in which the current is flowing (opposite to the direction in which the electrons are drifting), the magnetic field lines are circles going round in a clockwise direction.

Figure 17-3 was obtained by sprinkling iron filings on a sheet of paper perpendicular to a straight wire carrying a current. Iron filings are long and thin and align themselves in the direction of the magnetic field, in the same way that a compass needle points along the earth's magnetic field. The circles centered on the wire are easily seen in this photograph. This technique was known to Faraday and helped him to form his very vivid picture of a magnetic field as a pattern of lines in space.

The magnetic field lines produced by a current flowing in a circular loop are shown in figure 17-4. Although the lines are no longer perfect circles, it is still true that, if we stand at any point on the wire and look in the direction of the current, the lines go round the wire in a clockwise direction.

The reader may already have noticed one important difference be-

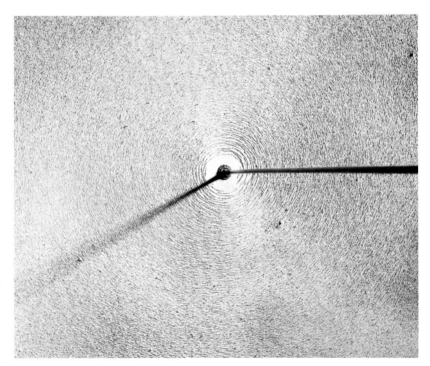

ELECTRICITY
AND
MAGNETISM:
PARTICLES
AND FIELDS

336

Figure 17-3 The circular magnetic field lines surrounding a straight wire carrying a current, as revealed by the iron filings technique. (Photograph by Berenice Abbott.)

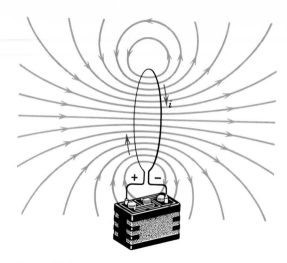

Figure 17-4 Magnetic field lines due to a current in a circular loop of wire.

tween electric fields and magnetic fields. Electric field lines sometimes start on a positive charge and end on a negative charge. Magnetic field lines never start or end, but always form closed loops.

Example **17-1-1**

Calculate **B** at the center of a circular loop of wire (figure 17-5).

Apply the Biot–Savart law to calculate the contribution from a short segment of the wire of length l_1. R is now the radius of the circle. θ is $90°$ and $\sin \theta = 1$. So,

$$\text{Contribution to } \mathbf{B} \text{ from } l_1 = \frac{i_1 l_1}{cR^2} \tag{17-6}$$

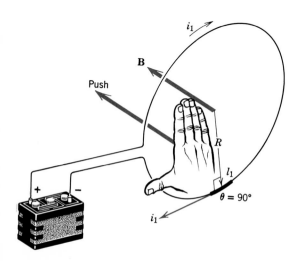

Magnetism

Figure 17-5 Illustrating example 17-1-1.

Right hand rule number 1, as illustrated in the figure, tells us that the contribution is perpendicular to the plane of the circle. If the current goes round clockwise as shown, it points into the plane of the paper.

The contribution from any other segment of the wire is in the same direction and so the contributions may be added directly, without worrying about vector addition. If the wire is divided into segments of length l_1, l_2, l_3 and so on, the total field is

Sum of contributions to **B** from all segments of the wire

$$= \frac{i_1 l_1}{cR^2} + \frac{i_1 l_2}{cR^2} + \frac{i_1 l_3}{cR^2} + \text{ and so on} \tag{17-7}$$

$$= \frac{i_1}{cR^2} (l_1 + l_2 + l_3 + \text{ and so on}) \tag{17-8}$$

But the sum of the lengths of all the segments is the total length of the circumference of the wire,

$$(l_1 + l_2 + l_3 + \text{ and so on}) = 2\pi R \tag{17-9}$$

Whence

$$B = \frac{i_1}{cR^2} 2\pi R \tag{17-10}$$

or

$$B = \frac{2\pi i_1}{cR} \tag{17-11}$$

Its direction is perpendicular to the plane containing the circular loop of wire. Looking in the direction of **B**, the current i_1 goes round clockwise.

17-2   The Force on a Current in a Magnetic Field

Now consider the problem of calculating the force exerted by a magnetic field on a loop of wire of arbitrary shape carrying a current i_2. Again the procedure is to divide the wire into segments that are so short that they can be considered to be straight. The force on each segment is calculated by the method given below and the forces are then added vectorially.

In figure 17-6, EF is a typical segment of length l_2 and cross-sectional area A. The magnetic field in the vicinity of EF is **B** at an angle ϕ to EF. The current i_2 in EF is due to electrons drifting with velocity v_d in the opposite direction to i_2. Reverse the velocity of an electron and then use equation 16-6 to calculate the force exerted on it by **B**, putting $q_2 = e$ and $v_2 = v_d$.

Force on a single drifting electron in $EF = \dfrac{eBv_d \sin \phi}{c}$ $\tag{17-12}$

The volume of EF is $l_2 A$ and it contains $n_0 l_2 A$ drifting electrons, so

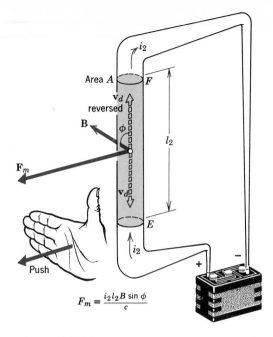

Area A

i_2

F

v_d reversed

$\mathbf{B}$

ϕ

l_2

$\mathbf{F}_m$

v_d

E

i_2

Push

$$F_m = \frac{i_2 l_2 B \sin \phi}{c}$$

Figure 17-6 Force on a current in a magnetic field.

Force on all the drifting electrons in $EF = n_0 l_2 A \dfrac{eBv_d \sin \phi}{c}$ (17-13)

According to equation 15-7, the current through EF is

$i_2 = e n_0 A v_d$ (17-14)

Equation 17-13 can therefore be simplified to give:

> **Force exerted by a magnetic field on a short segment of wire carrying a current i_2**
>
> $$F_m = \frac{i_2 l_2 B \sin \phi}{c}$$ (17-15)

If i_2 is in statamperes, l in cm, B in gauss, and c in cm/sec, then F_m is in dynes.

The direction of the force $\mathbf{F}_m$ is given by the following slight modification of right hand rule number 2. Hold the right hand flat. Point the thumb in the direction of the current through EF. Point the fingers in the direction of the magnetic field. Then the force is in the direction in which the hand would push.

Magnetism

Example 17-2-1

Two very long straight parallel wires carry currents i_1 and i_2 and are a distance r apart. Calculate the force on a length l_2 of the wire with current i_2. Verify that the force is attractive when the currents are in the same direction, but repulsive when they are in opposite directions (see section 16-1).

Figure 17-7a is the case of the two currents flowing in the same direction. The magnetic field lines due to i_1 are circles that go around clockwise for an observer looking in the direction of i_1. According to equation 17-5 the magnetic field produced by i_1 at a point on the other wire is

$$B = \frac{2i_1}{cr} \tag{17-16}$$

This field is perpendicular to i_1 and to the radius r drawn perpendicularly to both wires. Since $\mathbf{B}$ is also perpendicular to i_2, $\phi = 90°$, $\sin \phi = 1$. Equation 17-15 for the force exerted by the field on a length l_2 of the wire with current i_2 takes the form

$$F_m = \frac{i_2 l_2 B}{c} \tag{17-17}$$

Combining equations 17-16 and 17-17, the required force is

$$F_m = \frac{2i_1 i_2 l_2}{c^2 r} \tag{17-18}$$

(a) Currents in same direction.

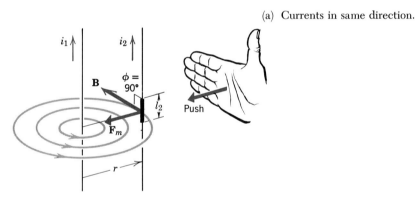

(b) Currents in opposite directions.

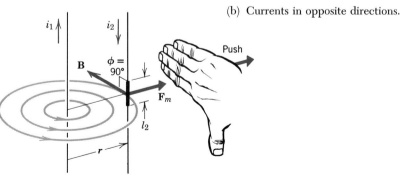

Figure 17-7   Illustrating example 17-2-1.

In figure 17-7a the magnetic field is into the plane of the paper. Applying right hand rule number 2 with the thumb pointing upward in the direction of i_2 and the fingers pointing into the paper in the direction of **B**, the right hand would push to the left, toward i_1. This confirms our previous statement in section 16-1 that parallel currents flowing in the same direction attract one another.

Figure 17-7b is the case of currents flowing in opposite directions. i_1 is still upward and **B** is still into the plane of the paper. The right hand must now be placed with the thumb pointing *downward* in the direction of i_2 and the fingers pointing into the plane of the paper in the direction of **B**. The hand would then push to the right, so the force is repulsive.

17-3 Current Loops

In preparation for our discussion of magnetic materials, let us now consider the forces between two parallel circular loops carrying currents that go round in the same direction. In figure 17-8a the magnetic field lines due to the current in loop I are shown. The field on a segment at the top of loop II tilts slightly upward. The force on this segment is perpendicular to both the field and the current. Application of right hand rule number 2 shows that the force tilts slightly backward from an upward vertical direction. This force therefore has a horizontal component directed toward loop I. The situation is similar for all segments of loop II and the force on any segment has a component directed *toward* loop I. Loop II is therefore attracted toward loop I. The argument can be turned round to show that loop I is also attracted toward loop II. When the currents flow around the loops in the same direction, the loops attract one another.

The situation when the currents flow in opposite directions is shown in figure 17-8b. Now the force on a segment at the top of loop II is tilted slightly forward from the downward direction and has a horizontal component directed *away from* loop I. The two loops now repel one another.

Two parallel circular loops attract one another if their currents go round in the same direction, but repel one another if their currents go round in opposite directions.

The other situation we need to understand is a current in a loop at an angle to a uniform magnetic field. This magnetic field does not include the field produced by the loop itself but is an "externally applied" magnetic field due to the presence of electric currents elsewhere. For simplicity we shall consider a rectangular loop $ABCD$ (figure 17-9a) carrying a current i_2. Applying right hand rule number 2, the force on AB acts through the center of AB and is vertically upward. The force on CD acts through the center of CD and is vertically downward. These two forces are equal and opposite and act along the same line, so they cancel one another. The forces on BC and DA are also equal and

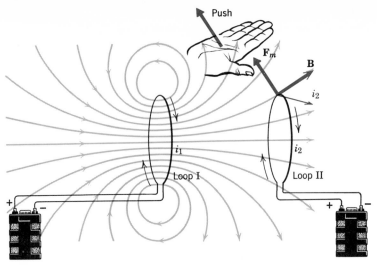

(a) Currents in same direction.

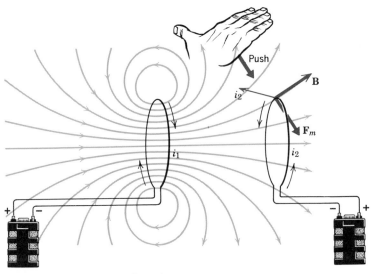

(b) Currents in opposite directions.

Figure 17-8 The forces between currents in parallel circular loops.

opposite, but, if we look at the loop from the top in the direction $\overrightarrow{BC}$ (figure 17-9b), these two forces are displaced sideways from one another and constitute a couple. This couple rotates the loop until its plane is perpendicular to the applied magnetic field.

Figure 17-10 shows various possible orientations of the coil as seen from above. The current in DA is always upward out of the paper, so the force on DA is always in the same direction, whatever the orientation of the coil. Similarly, the force on BC is always in the same direction. A careful study of these diagrams reveals that the couple always tends

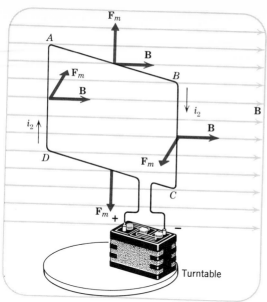

(a) A rectangular loop free to rotate in a magnetic field, showing the directions of the forces on the four straight sides.

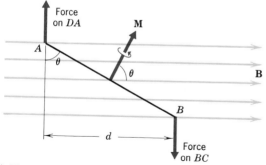

(b) Top view, looking down on AB.

Figure 17-9 Couple on a rectangular loop carrying a current in a magnetic field.

to rotate the coil until its plane is perpendicular to the field. In addition, however, when the coil is lined up the current must go round the loop clockwise when we look along the direction of the magnetic field (e and j). If the coil is aligned in the "wrong" direction (a and f), with the current going round counterclockwise when we look along the field, the couple on it is zero, but it is unstable. A very small rotation to one side (b or g) produces a couple that increases the rotation and starts the coil rotating through 180° until it reaches the "right" direction (e or j). However, a slight rotation to one side from the "right" direction (d or i) produces a couple which returns the coil to the right direction.

The result we have obtained is independent of the exact shape of the coil. As an exercise, the student may wish to convince himself that the

Magnetism

343

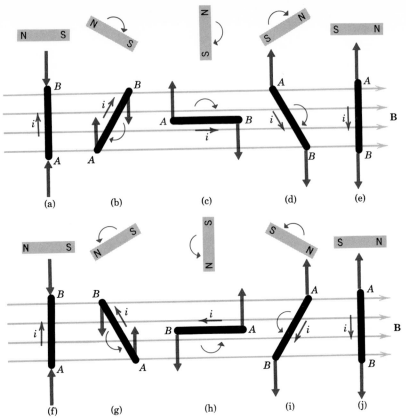

Figure 17-10 A loop carrying a current in a magnetic field rotates until the current is going around clockwise for an observer looking along the magnetic field lines. A bar magnet behaves in a similar way. (In this diagram, the rectangular loop of figure 17-9 is being viewed from above.)

resultant couple on a circular coil is also always in a direction that produces the stated effects.

> If a coil carrying a current is free to rotate in a magnetic field, it turns until its plane is perpendicular to the field and the current is in a clockwise direction from the point of view of an observer looking in the direction of the field.

The couple on the rectangular loop is easily calculated. Using equation 17-15, the force on either DA or BC is, ($\phi = 90°$, $\sin \phi = 1$)

$$F = \frac{i_2(\text{length of } BC)B}{c} \qquad (17\text{-}19)$$

The perpendicular distance between the two forces is seen from figure 17-9b to be

$$d = (\text{length of } AB) \sin \theta \qquad (17\text{-}20)$$

where θ is the angle between the field and a direction perpendicular to the plane of the coil. θ is also the angle between the force and AB. Remembering the definition of the moment of a couple given in equation 10-1,

$$\text{Moment of couple, } \tau = Fd \qquad (17\text{-}21)$$

$$= \frac{i_2(\text{length of } BC)B}{c}(\text{length of } AB) \sin \theta \qquad (17\text{-}22)$$

But $(\text{length of } BC) \times (\text{length of } AB) = \text{Area of rectangle} \qquad (17\text{-}23)$

$$= \mathcal{A}$$

So, moment of couple $\tau = \dfrac{i_2 \mathcal{A} B \sin \theta}{c} \qquad (17\text{-}24)$

This equation can be shown to be true for a coil of any shape with area $\mathcal{A}$.

The quantity

$$M = \frac{i_2 \mathcal{A}}{c} \qquad (17\text{-}25)$$

is called the **magnetic moment** of the coil. It can be treated as a vector **M,** with the magnitude quoted and a direction perpendicular to the plane of the coil such that, if one looks in this direction, the current is going round in a clockwise direction (see figure 17-9b). The moment of the couple on the coil may be rewritten

$$\tau = MB \sin \theta \qquad (17\text{-}26)$$

Its sense is always such as to rotate **M** into the same direction as **B.**

17-4 Magnetic Materials

To most people magnetism is not a property of electric currents, but of bar magnets. The following facts about the behavior of bar magnets are well known. If the magnet is freely suspended, it turns until it is pointing in an approximately north-south direction. This is the principle of the magnetic compass. The end of the bar pointing in a northerly direction is called the **north pole** of the magnet and the other end is called the **south pole.** If the magnet is suspended in any magnetic field, its north pole points in the direction of a field line. The earth produces a magnetic field with field lines running approximately, but not exactly, from south to north, and it is this magnetic field which lines up a magnetic compass. If the north pole of a magnet is brought near the south pole of another magnet, they attract one another. But if two north poles or two south poles are brought together, they repel one another. We shall now show how all these phenomena can be explained in terms of microscopic electric currents inside the material of the bar magnet.

The magnetic properties of a material are due to the behavior of some of its electrons. An electron can produce a magnetic field in two

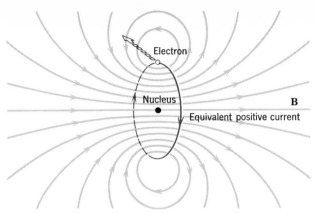

(a) An electron moving in a circular orbit is equivalent to a circular loop of wire with the current flowing in the opposite direction to the electron.

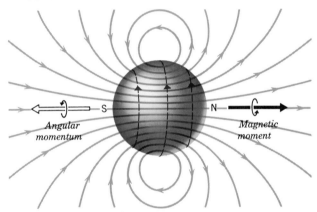

(b) A spinning electron. Any small piece of the electron moves in a circular orbit and produces a magnetic field like a current flowing in a circular loop.

Figure 17-11   Two ways in which an electron can produce a magnetic field.

separate ways. An electron moving around a nucleus in a circular orbit is equivalent to a circular loop of current (figure 17-4) and produces a similar magnetic field (figure 17-11*a*), although we must remember that the electron is negatively charged and moves in the opposite direction to the current. In addition, even a stationary electron produces a magnetic field, because it is spinning like a top. The electron may be visualized as a small spherical cloud of negative charge, which is spinning about an axis such as *SN* in figure 17-11*b*. Any small portion of the electron describes a circular path about *SN* and is equivalent to a current in a circular loop of wire. The combined effect of the circular motions of all portions of the electron is a magnetic field which is again similar to that produced by a current in a circular loop. In a bar magnet, most of the magnetism is due to electron spin and the bodily motion of electrons in circular orbits plays a very minor role which we shall henceforth ignore. Let us add, though, that circular orbits are important in many weakly magnetic materials unsuitable for bar magnets.

Most materials show no net magnetic effect because, for every electron moving in a circular orbit of a certain kind, there is another electron moving in an orbit identical with the first, except that the electron moves the other way round it and the two orbits exactly cancel one another. Also, for every electron spinning about an axis in a certain direction, there is another electron spinning about a parallel axis but rotating the opposite way round, so that the two electron spins cancel one another. The important exceptions are the **ferromagnetic** materials, which are metals such as iron, cobalt, and nickel, and the **ferrimagnetic** materials, which are non-metallic compounds such as magnetite (the material used by the Chinese when they invented the magnetic compass). In these materials there is an excess of electrons all spinning about the same axis in the same direction and adding together to produce a large combined magnetic effect. This occurs even when there is no externally applied magnetic field. A bar magnet is usually made of iron alloyed with certain other elements that confer upon it the property that the number of electrons with their spins lined up changes only very slightly however badly the material is treated, and so it can justly be called a "permanent" magnet.

Each spinning electron produces a magnetic field of the kind shown in figure 17-11*b*. The combined effect of all the electrons is to produce the total field shown in figure 17-12, which is qualitatively similar to the field due to a single electron or the field due to a circular loop of wire carrying a current (figure 17-4). Figure 17-13 is a photograph of the field due to an actual bar magnet, as revealed by the iron filings technique. The line joining the south pole to the north pole is the axis of rotation of the electrons. If we look in the direction from the south pole toward the north pole, the electrons are spinning counterclockwise and are equivalent to loops of positive current going around clockwise.

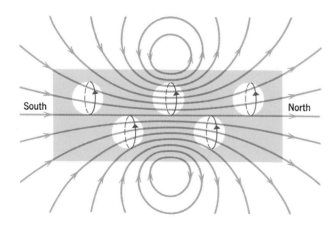

Figure 17-12 The magnetic field of a bar magnet. Five spinning electrons are shown. In a bar magnet of normal size, there would be about 10^{24} spinning electrons.

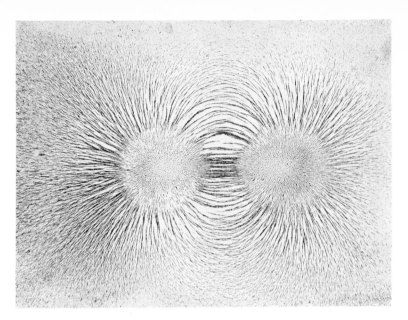

Figure 17-13 The magnetic field of a bar magnet as revealed by the iron filings technique. (Photograph by Berenice Abbott.)

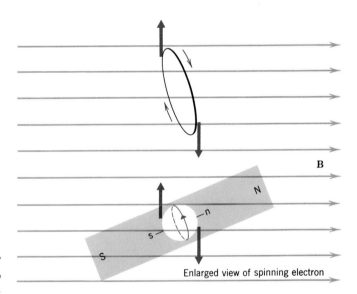

Enlarged view of spinning electron

Figure 17-14 The alignment of a bar magnet along an external magnetic field is similar to the alignment of a current loop. Looking along the field, the current in the loop must go round clockwise. Looking along the field, the electrons must be spinning counterclockwise and therefore be equivalent to positive current going around clockwise. This means that, when the bar magnet is lined up, looking along the external field is the same as looking from the south to the north pole.

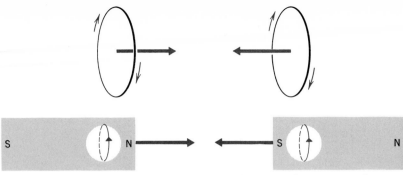

(a) A north pole attracting a south pole is analogous to the attraction between parallel current loops in the same direction.

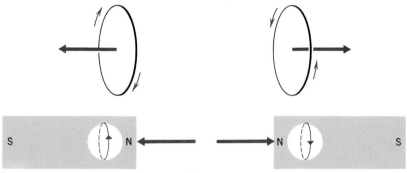

(b) A north pole repelling a north pole is analogous to the repulsion between parallel current loops in opposite directions.

Figure 17-15 Forces between bar magnets interpreted in terms of the forces between spinning electrons.

In accordance with our previous discussion in section 17-1, this means that *the magnetic field lines enter the south pole and leave the north pole.*

We shall now treat a bar magnet as a collection of circular loops of current which go around clockwise if we look from the south pole toward the north pole. We shall show that the properties of a bar magnet are explicable by analogy with the behavior of circular loops of current. For example, we have already shown that, when such a loop is placed in a magnetic field, there is a couple acting on it which rotates it until the plane of the loop is perpendicular to the field and the current is clockwise when we look in the direction of the field. Applying these considerations to a bar magnet placed at an angle to an external magnetic field (figure 17-14), a couple acts on each spinning electron and the magnet is rotated until a vector pointing from the south pole toward the north pole is in the same direction as the magnetic field. In figure 17-10, a small bar magnet has been included in each diagram to illustrate the analogy.

When the north pole of one magnet is brought up to the south pole of another magnet (figure 17-15a), the electrons in the two magnets are

Magnetism

349

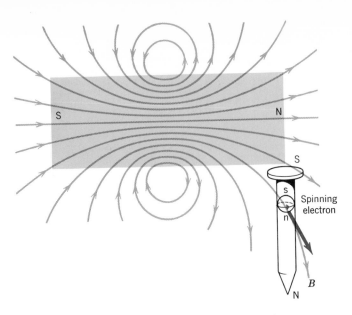

Figure 17-16 A bar magnet picking up an iron nail.

spinning in the same direction. The attraction of the north pole for the south pole is therefore a consequence of the fact that parallel loops with currents in the same direction attract one another (figure 17-8a). Conversely, when the north pole of one magnet is brought up to the north pole of another magnet (figure 17-15b), the electrons in the two magnets are spinning in opposite directions. Parallel loops with currents in opposite directions repel one another (figure 17-8b). A similar argument applies to adjacent south poles.

A well-known property of a bar magnet is that it will pick up an iron nail. The electrons responsible for the magnetic properties of an iron nail are very easily turned over without the nail as a whole moving. If the north pole of a bar magnet is brought up to the nail (figure 17-16), these electrons turn round until they are spinning counterclockwise from the point of view of an observer looking along the magnetic field produced by the bar magnet at the nail. We have already said that the electrons are spinning counterclockwise when we look from the south pole toward the north pole. It follows, with the help of figure 17-16, that the end of the nail nearest to the north pole of the bar magnet is the south pole of the nail. The north pole of the bar magnet attracts the south pole of the nail and the nail moves toward the bar magnet and usually sticks to it.

QUESTIONS

1. If the earth can be considered to be a huge bar magnet, is the "north pole" of this magnet in the vicinity of the geographic north pole or the geographic south pole?

2. If the earth's magnetism is assumed to be due to a large circular loop of current in the interior of the earth, what is the plane of this loop and what is the direction of the current around it?
3. How could a careful study of the magnetic field over the whole surface of the earth decide between the two possibilities that the earth's magnetic field is caused primarily by a circular loop of current (*a*) in the interior of the earth, or (*b*) in the upper atmosphere?
4. In the arrangement shown in the diagram, what is the direction of the force on the current-carrying wire?

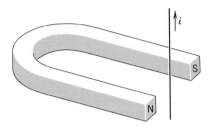

Question 17-4

5. Make rough sketches of the pattern of magnetic field lines produced by two parallel infinitely long straight wires carrying equal currents (*a*) in the same direction, (*b*) in opposite directions.
6. Does an infinitely long vertical straight wire carrying a current exert a net magnetic force on an infinitely long horizontal straight wire carrying a current? The wires do not necessarily intersect. Prove your answer rigorously.
7. A circular loop of wire carrying a current hangs with its plane vertical and is free to rotate about a vertical axis (see diagram). What will it do when the north pole of a magnet approaches it from each of the positions shown?

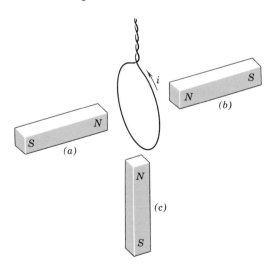

Magnetism

Question 17-7

8. You are locked in a room without windows and told to determine approximately the direction of north. You are provided with the following and nothing else: (a) an ample supply of copper wire of all possible diameters, (b) a pair of wire cutters made of nonmagnetic steel, (c) a storage battery with its terminals marked + and −. Describe precisely what you would do.

9. Oersted performed one of the first experiments to reveal a connection between magnetism and electric currents. He placed a straight horizontal wire carrying a current directly above a freely pivoted magnetic compass needle. The wire was originally parallel to the needle (see diagram). What did the compass needle do? Explain its behavior in terms of the forces acting on a microscopic current loop (a spinning electron) inside the needle.

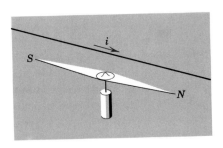

Question 17-9

10. By analogy with equation 17-26, how might one define the magnetic moment of a bar magnet?

11. Suppose that any one of the figures in this chapter were held up to a mirror. Apart from the reversal of the lettering, what would be wrong with the mirror image?

PROBLEMS

A

1. A straight segment of wire 0.1 cm long carries a current of 10^{10} statamp. What magnetic field does it produce at a distance of 150 cm in a direction perpendicular to itself?

2. A straight segment of wire 2 cm long with a resistance of 0.15 ohm has 6 volt applied between its ends. What magnetic field does it produce 25 m away in a direction perpendicular to itself?

3. A hoop of insulating material with a radius of 15 cm carries a charge of +2 statcoulomb distributed uniformly around it. If it rotates with an angular velocity of 10,000 radian/sec about an axis through its center perpendicular to its plane, what magnetic field is produced at its center?

ELECTRICITY
AND
MAGNETISM:
PARTICLES
AND FIELDS

352

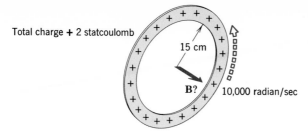

Total charge **+** 2 statcoulomb

15 cm

B?

10,000 radian/sec

Problem 17-3

4. A straight vertical wire 1 m long carrying a current of 5×10^9 statamp flowing vertically upward is in a uniform horizontal magnetic field of 0.5 gauss pointing due north. What are the magnitude and direction of the magnetic force on it?

5. A straight horizontal wire 35 cm long carrying a current of 5 amp is at an angle of 60° to a horizontal magnetic field of 800 gauss. What are the magnitude and direction of the magnetic force on it?

6. A horizontal wire 10 cm long carries a current of 10 amp. Find the magnitude and direction of the smallest magnetic field which can support the weight of the wire, assuming that its mass is 2.5 gm.

7. A circular ring of wire of radius 10 cm carries a current of 2.5 amp. What is the magnitude of the magnetic force exerted on the ring by a uniform magnetic field of 500 gauss in a direction perpendicular to the plane of the ring? In which direction does this force act?

B

8. A long straight wire carries a current of 5×10^{10} statamp. If the radius of the wire is 0.1 cm, what is the largest value of the magnetic field at any point outside the wire?

9. An astronomical object consists of a long thin cylindrical cloud of electrons. Its length is 3×10^{12} cm and its radius is 5×10^8 cm. Its density is 20 electrons per cm^3 and these electrons are all drifting in the direction of the length of the cloud with a velocity of 10^6 cm/sec. What is the magnetic field produced by this cloud at a distance of 10^{10} cm from its axis?

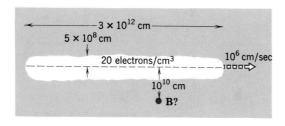

3×10^{12} cm

5×10^8 cm

20 electrons/cm^3

10^6 cm/sec

10^{10} cm

B?

Magnetism

Problem 17-9

10. Two long straight wires are parallel and 20 cm apart. They carry currents of 5×10^9 statamp and 2×10^{10} statamp respectively. What is the magnetic force on 1 cm of either wire?

11. For the situation described in the previous question, find the magnetic field at a point mid-way between the two wires if their currents are (a) in the same direction (b) in opposite directions.

12. An electron has a velocity of 6.82×10^8 cm/sec parallel to a long straight wire carrying a current of 45.6 amp. If the perpendicular distance of the electron from the wire is 12.3 cm, calculate the force on the electron.

13. A long horizontal straight wire carries a current of 3 amp flowing from the south toward the north. At a perpendicular distance of 1.2 m due east of the wire is a proton with a horizontal velocity of 5×10^7 cm sec^{-1} toward the northeast. Find the magnitude and direction of the magnetic force on the proton.

14. An electron has a velocity of 3×10^6 cm sec^{-1} vertically upward. At a perpendicular distance of 3 m from the electron is a long straight vertical wire carrying a current of 10^{10} statamperes flowing upward. Find the magnitude and direction of the magnetic force on 0.1 cm of the wire (a) at the same horizontal level as the electron, (b) at a point on the wire 4 m higher than the electron.

C

15. A long vertical straight wire and a long horizontal straight wire both carry a current i statamp. The perpendicular distance between them is $2r$ cm. Find the magnitude and direction of the magnetic field at the midpoint of their common perpendicular (see diagram).

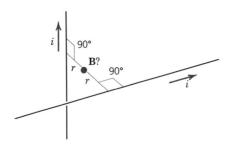

Problem 17-15

16. Two long parallel straight wires carry equal currents of 2×10^{10} statamp. The distance between them is 8 cm. Find the magnitude and direction of the magnetic field at a point which is at a perpendicular distance of 8 cm from either wire if the currents are (a) in the same direction (b) in opposite directions.

17. For the situation shown in the diagram, find the magnitude and direction of the resultant magnetic force on the rectangular loop. Express this force in terms of the magnetic moment of the loop.

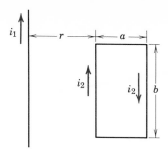

Problem 17-17

18. What is the magnetic moment of a current of 2×10^8 statamp flowing round a circular loop of radius 3 cm?

19. What is the magnetic moment of an electron moving around a circle of radius r with a velocity v. What is the ratio of the magnetic moment to the angular momentum? (Hint: The current is the charge flowing past any point in one second. How many times does the electron go past any point in one second?)

Magnetism

18

Fields that Change with Time

18-1 Changing Magnetic Fields

A changing magnetic field produces an electric field. This was discovered independently by Faraday and Henry in a series of experiments similar to the ones we shall now describe. Loop I of figure 18-1 is connected to a storage battery and carries a current that can be switched on or off by the switch S. Loop II is parallel to loop I and is connected to a galvanometer, which is a device for detecting the presence of a current flowing

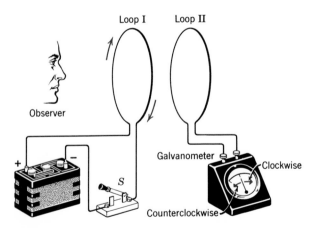

Figure 18-1 Electromagnetic induction. When the current in loop I is changing, a current is induced in loop II. When the current in loop I settles down to a steady value, there is no current in loop II.

through loop II. A pointer, which is normally at the center of the dial of this galvanometer, moves to the right when a current flows clockwise around II, and moves to the left when a current flows counterclockwise around II (clockwise and counterclockwise will always refer to the point of view of an observer looking in the direction from I toward II).

There is no battery connected to II to produce a current through it. When a steady current flows through I it produces a steady magnetic field in the vicinity of II, but this field cannot affect the electrons in II unless they are already in motion. Under these circumstances, therefore, there is no reason for a current to flow through II and, in fact, the galvanometer is observed to show no deflection. However, when the current in loop I is switched on or off, the galvanometer shows a momentary deflection which quickly returns to zero. A current is induced in loop II while the current in loop I is changing. This phenomenon is known as **electromagnetic induction.**

When the current in loop I is changing, the drift velocity of the electrons in loop I is changing and these electrons have an acceleration. So far we have avoided accelerating electrons. Now we are forced to conclude that, when an electron in I is accelerating, it exerts a force on a stationary electron in II and sets it in motion. Since this force acts on a *stationary* electron, it is presumably due to an *electric* field produced at II. When the electrons in I have a steady velocity, their electric field is exactly cancelled by the protons in I. But when the electrons in I have an acceleration they produce an extra electric field at II, which is not compensated by the nonaccelerating protons.

This explanation of the experiment implies action-at-a-distance. An alternative, and eventually more fruitful, approach is as follows. The

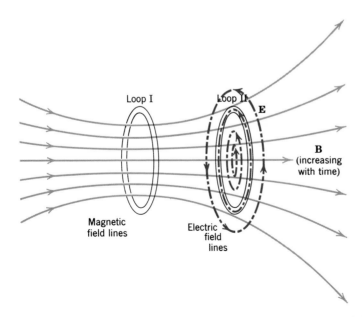

Figure 18-2 A changing magnetic field produces circular electric field lines.

moving electrons in loop I produce a magnetic field at loop II. When the current in loop I is changing, the magnetic field at loop II is changing. A changing magnetic field is surrounded by circular electric field lines (figure 18-2). Some of these lines lie inside the metal wire of loop II and exert forces on the electrons in this wire, driving them around the loop.

18-2 Faraday's Law of Electromagnetic Induction

Faraday discovered a simple law applicable to all cases of electro-magnetic induction. First consider the simplest case of a circular loop of radius r, area $A = \pi r^2$ and circumference $C = 2\pi r$ (figure 18-3). Suppose that the magnetic field is everywhere perpendicular to the plane of the loop and has the same value at all points inside the loop. Initially, at time t_1 sec, this field has a value B_1, but at a slightly later time t_2 sec has changed to B_2. Then this changing field is surrounded by circular electric field lines. One of these lines lies inside the loop. Suppose that the electric field has the value E for this particular line. The simplified form of Faraday's law in this case is

Simplified form of Faraday's law

$$CE = - \frac{A(B_2 - B_1)}{c(t_2 - t_1)} \qquad (18\text{-}1)$$

or

$$E = - \frac{r(B_2 - B_1)}{2c(t_2 - t_1)} \qquad (18\text{-}2)$$

for a circular loop perpendicular to a uniform magnetic field B.

Equation 18-2 is obtained from 18-1 by inserting $C = 2\pi r$ and $A = \pi r^2$. E is proportional to $(B_2 - B_1)/(t_2 - t_1)$, which is the rate of change of B. It is assumed that $(t_2 - t_1)$ is small enough so that this rate

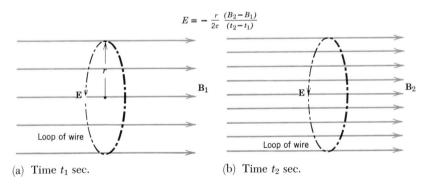

$$E = - \frac{r}{2c} \frac{(B_2 - B_1)}{(t_2 - t_1)}$$

ELECTRICITY
AND
MAGNETISM:
PARTICLES
AND FIELDS

(a) Time t_1 sec.   (b) Time t_2 sec.

Figure 18-3   Illustrating the simple form of Faraday's law. Area of loop, $A = \pi r^2$. Circumference of loop, $C = 2\pi r$. The circular electric field line shown coincides with the loop of wire.

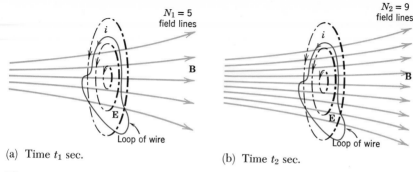

$N_1 = 5$
field lines

$N_2 = 9$
field lines

B

B

i

i

E

E

Loop of wire

Loop of wire

(a) Time t_1 sec.

(b) Time t_2 sec.

Figure 18-4 Illustrating the more general form of Faraday's law.

of change can be considered constant during the time interval (compare the definition of velocity in section 4-1). Notice that the constant c (the speed of light) again enters into the equation. If r is in cm, c in cm/sec, B in gauss, and t in secs, then E is in dynes per statcoulomb (or, if you prefer, statvolts per cm). The minus sign on the right hand side of each equation is connected with Lenz's law, which will be explained in the next section, but the full significance of this minus sign cannot be made properly clear without entering into some advanced mathematics.

The magnetic field lines may be drawn so that the number of lines crossing unit area perpendicularly is equal to the strength of the field, B. In that case the quantity

$$N = AB \tag{18-3}$$

is the total number of lines passing through the full area A of the loop and is called the **magnetic flux** through the loop. If the magnetic flux at time t_1 is $N_1 = AB_1$ and at time t_2 is $N_2 = AB_2$, then Faraday's law, as given by equation 18-1, may be rewritten

$$CE = -\frac{1}{c}\frac{(N_2 - N_1)}{(t_2 - t_1)} \tag{18-4}$$

Even if the magnetic field is not perpendicular to the loop and varies in magnitude from place to place (figure 18-4), the number N of field lines passing through the loop is still a well defined quantity. Equation 18-4 is then the more general form of Faraday's law applicable to this case.

The force on an electron inside the circular loop is eE in a direction tangential to the loop. While this force drives the electron around the loop, it always remains parallel to the direction in which the electron is moving. The work done by the force is therefore the product of the force and the total length of its path ($\theta = 0$ and $\cos\theta = 1$ in equation 9-33). When the electron has made one complete circuit, the length of the path is $C = 2\pi r$, so:

Work done by the induced electric field as it drives an electron around one complete circuit

$$= eEC \tag{18-5}$$

Fields
that Change
with Time

359

The quantity which appears on the left hand side of equation 18-1 or 18-4 is therefore

CE = Work done by the induced field as it drives
unit charge round one complete circuit   (18-6)

When the loop is not circular and does not coincide with an electric field line, this work is still a physically meaningful quantity and the most general expression of Faraday's law is found to be:

General form of Faraday's law

Work done by the induced field as it drives unit charge once around the loop

$$= \frac{1}{c} \frac{(N_2 - N_1)}{(t_2 - t_1)}$$ (18-7)

$$= \frac{1}{c} \text{ multiplied by the rate of change of the magnetic flux}$$

through the loop.

This work is converted into joule heat as the electrons collide with the atoms of the wire. In one second the total charge driven round the loop is equal to the current i. The work done in one second is therefore

$$\text{Work done in 1 sec} = \frac{i}{c} \frac{(N_2 - N_1)}{(t_2 - t_1)}$$ (18-8)

According to equation 15-26,

$$\text{Joule heat developed per second} = i^2 R$$ (18-9)

where R is the resistance of the loop. Equating the work done to the joule heat it produces,

$$i^2 R = \frac{i}{c} \frac{(N_2 - N_1)}{(t_2 - t_1)}$$ (18-10)

The current induced in the loop is therefore

$$i = \frac{1}{cR} \frac{(N_2 - N_1)}{(t_2 - t_1)}$$ (18-11)

Since we are using CGS units, i is in statamperes, c in cm/sec, N in gauss cm^2, t in secs and R in *statohms*.

18-3   Lenz's law

We have already stated that the induced electric field lines are circles around the changing magnetic field, but we have not yet given a law to help decide whether they go clockwise or counterclockwise. This law is:

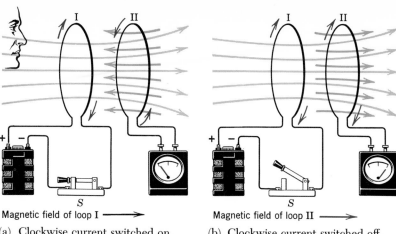

(a) Clockwise current switched on.   (b) Clockwise current switched off.

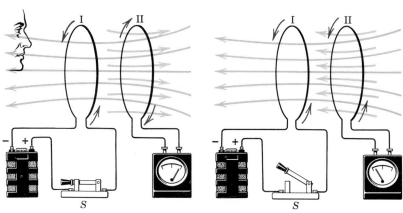

(c) Counterclockwise current switched on.   (d) Counterclockwise current switched off.

Figure 18-5   Illustrating Lenz's law.

Lenz's law

The current induced in the loop by the changing magnetic flux itself produces a magnetic flux in such a direction as to oppose the change in the original flux.

Figure 18-5 illustrates the application of this law to the apparatus of figure 18-1. If the current in loop I is flowing clockwise (from the point of view of an observer looking from I toward II), it produces magnetic field lines going through the loops from left to right (see figure 17-4). When this current is switched on, as in figure 18-5a, a magnetic field from left to right grows inside loop II. To counteract this, the induced current in loop II flows counterclockwise and produces a magnetic field from right to left, which *partially* cancels the field produced by loop I.

When the clockwise current in I is switched off (figure 18-5*b*), a magnetic field from left to right disappears in II. The induced current in II then flows clockwise, producing a magnetic field from left to right in an attempt to replace the field that is disappearing. Similarly, if a counterclockwise current in I is switched on (figure 18-5*c*), a magnetic field from right to left appears in II. The induced current in II is then clockwise in an attempt to cancel this field. If the counterclockwise current in I is switched off (figure 18-5*d*) the magnetic field from right to left disappears and a counterclockwise current is induced in II in an attempt to replace it.

The flux through loop II may also be made to change by moving loop I when it carries a steady current. In figure 18-6*a*, the steady clockwise

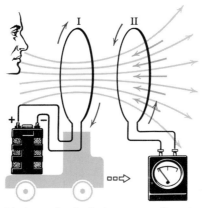 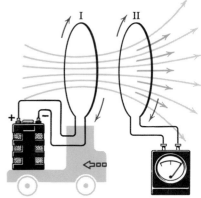

(a) A steady clockwise current flows. in loop I, but loop I moves toward loop II.

(b) Loop I moves away from loop II.

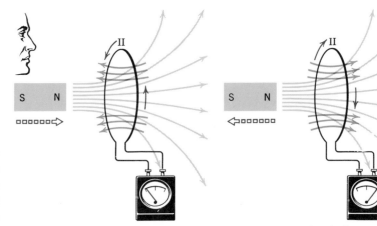

(c) The north pole of a bar magnet is brought up to loop II.

(d) The north pole of a bar magnet recedes from loop II.

Figure 18-6 Currents induced by moving the source of the magnetic field.

current in loop I produces a magnetic field from left to right. When loop I moves nearer to loop II the strength of this field inside loop II increases and, as is clear from the diagram, the number of field lines through II increases. A counterclockwise current therefore flows round loop II and produces a field from right to left in an attempt to prevent the flux inside II from increasing. Notice that the induced current in II is in the opposite direction from the current in I. Loop II therefore repels loop I (see figure 17-8b) and tries to prevent it from approaching any nearer. This is another respect in which the induced current tries to prevent the change in magnetic field that is inducing it.

In figure 18-6b the clockwise current in loop I is receding from loop II. The flux through loop II is therefore decreasing and a clockwise current flows in II, producing a magnetic field from left to right in an attempt to prevent the decrease. The currents now flow in the same direction round the two loops and II attracts I (figure 17-8a) in an attempt to prevent it from receding.

A bar magnet produces a similar magnetic field to a loop of current. A current can therefore be induced in loop II by moving a bar magnet in its vicinity (figure 18-6c and d). The student will now be able to verify the direction of the induced current shown in these diagrams and also verify that this induced current exerts a force on the magnet which resists its motion.

Example 18-3-1

The current i in a long straight wire is increasing at a steady rate

$$i = (1 + 2t) \times 10^8 \text{ statamps} \tag{18-12}$$

A small circular loop of wire of radius $a = 0.1$ cm is in a plane through the wire and its center is a distance $r = 100$ cm from the wire (figure 18-7). If the resistance of the loop is $R = 10^{-15}$ statohms, what is the induced current i_l flowing round it and in which direction does it flow?

The current i in the straight wire produces circular magnetic field lines. In the vicinity of the loop these lines are perpendicular to the plane of the loop. According to equation 17-5 the magnitude of the magnetic field at the center of the loop is

$$B = \frac{2i}{cr} \tag{18-13}$$

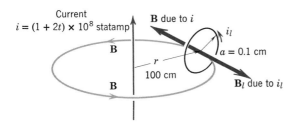

Figure 18-7 Illustrating example 18-3-1.

Fields
that Change
with Time

363

Since the radius a of the loop is very small compared with its distance r from the wire, it is reasonable to assume that the magnetic field has this value at all points inside the loop. The flux through the loop is therefore

$$N = BA \tag{18-14}$$

$$= \frac{2i\pi a^2}{cr} \tag{18-15}$$

At time t_1 secs $\quad i = (1 + 2t_1) \times 10^8$ and this flux is

$$N_1 = \frac{2\pi a^2}{cr}(1 + 2t_1) \times 10^8 \tag{18-16}$$

At time t_2 secs the flux has changed to

$$N_2 = \frac{2\pi a^2}{cr}(1 + 2t_2) \times 10^8 \tag{18-17}$$

Subtracting equation 18-16 from equation 18-17, the change in flux is

$$N_2 - N_1 = \frac{2\pi a^2}{cr}2(t_2 - t_1) \times 10^8 \tag{18-18}$$

According to equation 18-11, the current in the loop is

$$i_l = \frac{1}{cR}\frac{(N_2 - N_1)}{(t_2 - t_1)} \tag{18-19}$$

Inserting the value of $(N_2 - N_1)$ from equation 18-18

$$i_l = \frac{1}{cR}\frac{4\pi a^2}{cr} \times 10^8 \tag{18-20}$$

or

$$i_l = \frac{4\pi a^2}{c^2 Rr} \times 10^8 \tag{18-21}$$

$$= \frac{4\pi \times (0.1)^2 \times 10^8}{(3 \times 10^{10})^2 \times 10^{-15} \times 100}$$

$$i_l = 0.14 \text{ statampere} \tag{18-22}$$

The magnetic field produced by the straight wire goes through the loop *into* the page in figure 18-7 and it is increasing with time. The current i_l in the loop tries to counteract this by producing a magnetic field coming *out* of the page. Reference to figure 17-4 reveals that the induced current must therefore flow counterclockwise round the loop.

18-4   Circuits Moving in a Steady Magnetic Field

Suppose that loop I carries a steady current and is stationary, but loop II moves toward it with a velocity **v** as in figure 18-8a. From the point of view of Observer I, who is stationary relative to loop I, the magnetic flux through loop II increases as loop II moves into regions of stronger and stronger magnetic field. Thus, according to Faraday's law, there should

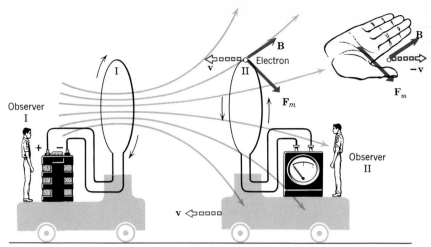

(a) The point of view of Observer I. Loop II moves in the steady magnetic field of loop I.

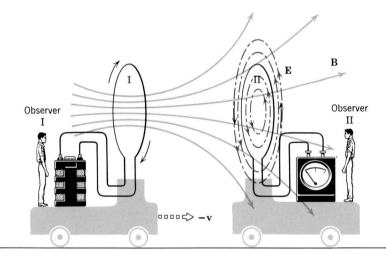

(b) The point of view of Observer II. As loop I moves, the magnetic field at loop II changes.

Figure 18-8 Two kinds of electromagnetic induction.

be an induced current in loop II. However, since nowhere in space is the magnetic field due to loop I changing with time, there is no induced electric field to produce this induced current. Should we conclude that there is in fact no induced current and that Faraday's law is not applicable to this situation?

Reconsider the situation from the point of view of Observer II, who is stationary relative to loop II (figure 18-8b). He sees loop I and its magnetic field moving toward him with a velocity −**v**. This is exactly the situation of figure 18-6a, and in that case we concluded that there *was* an induced current.

Fields
that Change
with Time

365

One way to resolve this paradox would be to argue that we should accept the point of view of the observer who is *really* at rest because it depends upon which loop is *really* moving. Such an approach would be in flat contradiction to the Theory of Relativity, which maintains that it is not permissible to assume that there is a specially preferred state of absolute rest. Only relative motion can have any physical significance and, since the relative velocity of the two loops is the same in parts *a* and *b* of figure 18-8, exactly the same physical effects should occur in both cases. We shall discuss these matters more fully in chapter 22, but we are now obtaining an interesting preview of the nature of the problem.

An appeal to experiment, which must always be the final arbiter, reveals that there is an induced current in loop II which has exactly the same value whether we move loop I or loop II. Faraday's law of electromagnetic induction is valid in both cases, and it is in fact only the relative velocity of the two loops that matters. Both observers see exactly the same current induced in loop II, but they explain its occurrence in different ways.

As we have already pointed out, Observer I (figure 18-8a) sees a magnetic field which remains constant at any point fixed in his space and therefore does not produce circular electric field lines to drive the electrons around loop II. With sufficient ingenuity, however, Observer I is able to give another explanation of the generation of the current. Consider a free electron at the top of loop II. It is moving to the left in the magnetic field **B** due to loop I. This field exerts a magnetic force on it which, after reversing its velocity and applying right hand rule number 2, is seen to be perpendicular to the page, toward the reader, and therefore tangential to the loop. A similar force acts at all points on the loop and so the free electrons are driven around the loop in a clockwise direction, which is equivalent to a counterclockwise current. A detailed analysis reveals that Faraday's law is obeyed.

Observer II argues differently (figure 18-8b). From his point of view, the electrons in loop II are not moving to the left in the magnetic field and therefore experience no magnetic force. Instead, as loop I moves toward him, its magnetic field at any point fixed in his space does change with time and does produce circular electric field lines driving the electrons around loop II.

Observer I therefore argues that the current is driven by a magnetic force, whereas Observer II argues that it is driven by an electric force. Moreover, Observer II sees circular electric field lines, whereas Observer I sees no electric field. This emphasizes that an electric or magnetic field is a mathematical convenience rather than a real concrete "thing" located in space. The hard physical fact, about which there can be no disagreement whatever the detailed theoretical explanation of a particular observer, is that the electrons do drift around the loop.

An analogous case is a rectangular loop rotating in a steady uniform magnetic field (figure 18-9a). Again there are no circular electric field lines, but the number of magnetic field lines passing through the loop is clearly changing. Let the magnetic field be horizontal and the axis of rotation vertical. When the plane of the loop is parallel to the field **B**, the vertical sides *BC* and *DA* of the loop are moving with a horizontal

ELECTRICITY
AND
MAGNETISM:
PARTICLES
AND FIELDS

366

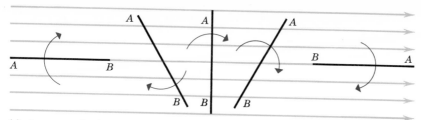

(a) A rectangular loop rotating in a magnetic field, showing how the magnetic flux through the loop depends upon its orientation. In this part of the diagram, the loop is viewed edgewise from the top.

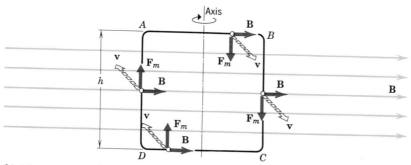

(b) The magnetic forces on an electron in various parts of the loop when the plane of the loop is parallel to the magnetic field.

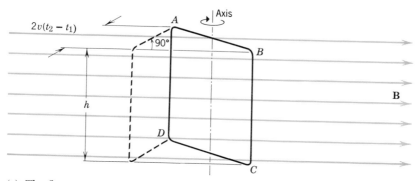

(c) The flux passing through the loop after it has rotated through a small angle.

Figure 18-9 The current induced in a rotating loop.

velocity **v** perpendicular to **B** (figure 18-9b). The free electrons in these sides therefore experience magnetic forces of magnitude

$$F_m = \frac{evB}{c} \tag{18-23}$$

Fields that Change with Time

These forces are parallel to the wires, in the directions shown. The work done by the magnetic force in moving an electron from D to A is

$$F_m h = \frac{evBh}{c} \tag{18-24}$$

where $h = DA = BC$ is the height of the loop. An equal amount of work is done in moving the electron from B to C. However, in the horizontal sides AB and CD the force on an electron is vertical and perpendicular to the wire, so it does no work when the electron is moved from A to B or C to D. The total work done in moving an electron once around the loop is therefore:

Work done by the magnetic forces in driving an electron once around the loop

$$= \frac{2evBh}{c} \tag{18-25}$$

Consequently the work done in driving unit charge once around the loop

$$= \frac{2vBh}{c} \tag{18-26}$$

At time t_1, when the plane of the loop is parallel to the magnetic field, no magnetic flux passes through the loop.

$$N_1 = 0 \tag{18-27}$$

At a slightly later time t_2, DA and BC have each moved a distance $v(t_2 - t_1)$ in a direction perpendicular to the field (figure 18-9c). The flux passing through the loop is therefore due to the magnetic field lines passing perpendicularly through a rectangle of height h, width $2v(t_2 - t_1)$, and area $2v(t_2 - t_1)h$. So,

$$N_2 = 2v(t_2 - t_1)hB \tag{18-28}$$

Therefore,

$$\frac{1}{c} \frac{(N_2 - N_1)}{(t_2 - t_1)} = \frac{2vhB}{c} \tag{18-29}$$

Comparing equations 18-26 and 18-29, it is clear that Faraday's law is applicable in the form

Work done by the magnetic forces in driving unit charge once around the loop

$$= \frac{1}{c} \frac{(N_2 - N_1)}{(t_2 - t_1)} \tag{18-30}$$

We have just described the principle of the **electric generator,** which is used to generate the electricity supplied to our homes and factories. If a coil is made to rotate in a steady magnetic field a current is induced in the coil and flows through any external circuit connected to the coil. This is illustrated in a very simple form in figure 18-10.

Example 18-4-1

In figure 18-11, the copper rod AB slides without friction on the metal rails DA and CB. The length of the rod is l and its velocity is v. A uniform magnetic field B is perpendicular to the plane containing the rails and the rod. Prove Faraday's law in this case.

ELECTRICITY
AND
MAGNETISM:
PARTICLES
AND FIELDS

368

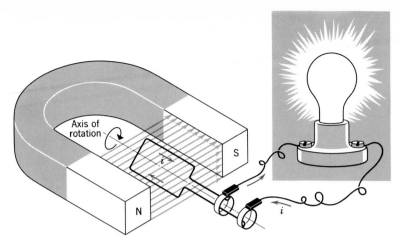

Figure 18-10   The principle of the electric generator.

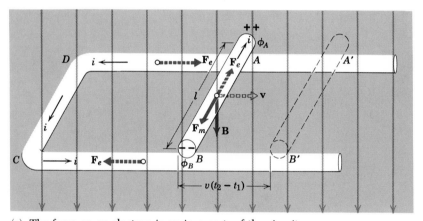

(a) The force on an electron in various parts of the circuit.

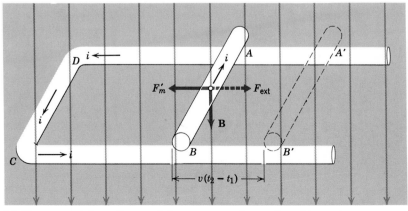

(b) The external force F_{ext} needed to maintain the steady velocity of AB.

Fields
that Change
with Time

369

Figure 18-11   Illustrating example 18-4-1.

This situation is similar to example 16-3-4, with the following important difference. There is a magnetic force

$$F_m = \frac{evB}{c} \tag{18-31}$$

driving the free electrons in the copper rod from A to B. However, when they arrive at B they are able to flow along the rail from B to C to D and back again to A. There is therefore a steady current i flowing around the circuit in a direction opposite to the flow of the electrons.

Positive charge accumulates at A and negative charge accumulates at B and there is a steady difference of electric potential $(\phi_A - \phi_B)$ between A and B. The condition for equilibrium is not that the electric field in AB should cancel the magnetic force on the electrons, as it did in example 16-3-4. In fact the steady electric field in AB is too small to do this, and a current does flow. However, when a steady state has been reached, it is important that the current flowing into A from B should exactly equal the current flowing out from A toward D, otherwise charge would continue to accumulate at A. The potential difference $(\phi_A - \phi_B)$ therefore adjusts itself until it can force a certain current around the branch $ADCB$ and allow the same current to flow from B to A. In the branch $ADCB$ it is the electric field due to $(\phi_A - \phi_B)$ that drives the current. In the branch BA the magnetic force F_m drives the current, experiencing some opposition from the electric field. The value of $(\phi_A - \phi_B)$ needed to achieve a steady state depends on the electrical resistances of the two branches.

The electric field makes no contribution to the work done in taking unit charge once round the circuit. If a positive unit charge is moved from B to A along the rod, an amount of work $(\phi_A - \phi_B)$ must be done on it to increase its potential energy. During the return journey along the rails, from A to D to C to B, the potential energy falls again and an equal amount of work is made available. The only work done during the circuit is done by the magnetic force that exists in the rod AB but not elsewhere. This work is:

Work done in taking one electron round the circuit

$$= F_m l \tag{18-32}$$

$$= \frac{evBl}{c} \tag{18-33}$$

So:

Work done in taking unit charge round the circuit

$$= \frac{vBl}{c} \tag{18-34}$$

Now consider the rate of change of flux through the loop $ABCD$. In a time interval $(t_2 - t_1)$ the rod moves a distance $v(t_2 - t_1)$ to the position $A'B'$. The area of the loop increases by the area of the rectangle $AA'B'B$, which is $lv(t_2 - t_1)$. The increase in flux is therefore

$$N_2 - N_1 = lv(t_2 - t_1)B \tag{18-35}$$

Therefore

$$\frac{1}{c} \frac{(N_2 - N_1)}{(t_2 - t_1)} = \frac{lvB}{c} \tag{18-36}$$

According to equation 18-34 the right hand side is the work done in taking unit charge around the loop. Equation 18-36 is therefore Faraday's law.

The work done in taking charge around the circuit is eventually transformed into joule heat. What is the origin of this energy? In one second a charge i flows around the circuit. The joule heat developed is therefore obtained by multiplying equation 18-34 by i.

$$\text{Joule heat developed per sec} = \frac{ivBl}{c} \tag{18-37}$$

If a current i flows in AB the magnetic field exerts on it a force given by equation 17-15

$$F_m' = \frac{ilB}{c} \tag{18-38}$$

Application of right hand rule number 2 reveals that this force is to the left, as shown in figure 18-11b. F_m' is the magnetic force on the electrons due to their drift motion *parallel to AB*. It should be distinguished from F_m, which is the magnetic force on the electrons due to their velocity v *perpendicular to AB*.

If the rod AB is to maintain a steady velocity v without accelerating, Newton's second law (equation 6-12) tells us that there must be no resultant force on the wire. It is therefore necessary to apply an external force $\mathbf{F}_{ext}$, equal but opposite to $\mathbf{F}_m'$

$$F_{ext} = F_m' \tag{18-39}$$

$$= \frac{ilB}{c} \tag{18-40}$$

In one second the rod moves a distance v parallel to $\mathbf{F}_{ext}$. The work done by F_{ext} in one second is therefore:

Work done by F_{ext} in 1 sec

$$= F_{ext}v \tag{18-41}$$

$$= \frac{ilBv}{c} \tag{18-42}$$

This is identical with the expression in equation 18-37 for the joule heat developed in one second. It is clear that this joule heat comes from the work done by the externally applied force which maintains the steady velocity of the rod AB.

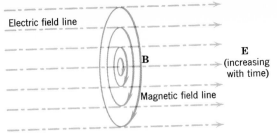

Electric field line

B

E
(increasing
with time)

Magnetic field line

Figure 18-12 A changing electric field produces circular magnetic field lines (compare figure 18-2).

18-5 Changing Electric Fields

The last few sections have been concerned with the basic fact that a changing *magnetic* field is encircled by *electric* field lines. It is also true that a changing *electric* field is encircled by *magnetic* field lines (figure 18-12). Moreover, the equations describing the two phenomena are very similar. Consider a circular magnetic field line of radius r and suppose that the electric field is the same at all points inside it. If this electric field has a value E_1 at time t_1, and a value E_2 at time t_2, then

Magnetic field produced by a changing electric field

$$B = +\frac{r}{2c}\frac{(E_2 - E_1)}{(t_2 - t_1)} \qquad (18\text{-}43)$$

This is similar to equation 18-2 except that there is a plus sign instead of a minus sign. This fact is related to the direction of **B**. If we look in the direction of $(\mathbf{E}_2 - \mathbf{E}_1)$, which is the change in field during the time interval, the magnetic field lines go round *clockwise*. Lenz's law is equivalent to exactly the opposite rule for the electric field produced by a changing magnetic field. If we look in the direction of the change in field $(\mathbf{B}_2 - \mathbf{B}_1)$, the electric field lines go round *counterclockwise* (go back to section 18-3 and convince yourself of this).

If the field is not constant inside the circular magnetic field line, then we must define the **electric flux** N' as the number of electric field lines passing through the circular magnetic field line. If the electric flux is N_1' at time t_1, and N_2' at time t_2, by analogy with equation 18-4

$$2\pi rB = +\frac{1}{c}\frac{(N_2' - N_1')}{(t_2 - t_1)} \qquad (18\text{-}44)$$

We cannot proceed any further to obtain the analogue of equation 18-7, because the force exerted by a *magnetic* field on an electron is perpendicular to the velocity of the electron and does no work on the electron.

As we explained in section 17-1 and figure 17-2, a current also produces circular magnetic field lines. There is a strong analogy between a current and a changing electric field. Figure 18-13 will help to make

ELECTRICITY
AND
MAGNETISM:
PARTICLES
AND FIELDS

372

this clear. The capacitor shown consists of two parallel plates of area A, on one of which there is a positive charge $+q$ and on the other a negative charge $-q$. A wire is connected between the two plates and the charge leaks from one plate to the other producing a current i in the wire. Suppose that the charge on the plates is q_1 at time t_1 and q_2 at a slightly later time t_2. During the time interval $(t_2 - t_1)$, a charge $(q_1 - q_2)$ leaks off the positively charged plate and, therefore, from the definition of current (equation 15-3),

$$i = \frac{(q_1 - q_2)}{(t_2 - t_1)} \tag{18-45}$$

The capacitor was discussed in example 14-6-3 and it was pointed out that the number of electric field lines leaving the positive plate and entering the negative plate is 4π multiplied by the charge on a plate. Also, the number of field lines per unit area is equal to the electric field E between the plates and the total number of lines (the electric flux N') is EA. Therefore, at time t_1 the electric flux is

$$N_1' = E_1A = 4\pi q_1 \tag{18-46}$$

At time t_2 the electric flux is

$$N_2' = E_2A = 4\pi q_2 \tag{18-47}$$

So,

$$(q_1 - q_2) = \frac{(N_1' - N_2')}{4\pi} \tag{18-48}$$

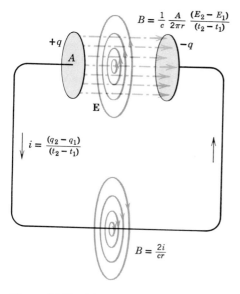

$$B = \frac{1}{c} \frac{A}{2\pi r} \frac{(E_2 - E_1)}{(t_2 - t_1)}$$

$$i = \frac{(q_2 - q_1)}{(t_2 - t_1)}$$

$$B = \frac{2i}{cr}$$

Figure 18-13   The magnetic field produced by a changing electric field is similar to the magnetic field produced by a current.

and

$$i = \frac{1}{4\pi} \frac{(N_1' - N_2')}{(t_2 - t_1)} \qquad (18\text{-}49)$$

or

$$i = \frac{1}{4\pi} \frac{(E_1 - E_2)}{(t_2 - t_1)} A \qquad (18\text{-}50)$$

According to equation 17-5 the current i in the wire is surrounded by circular magnetic field lines and the field at a distance r from the wire is

$$B = \frac{2i}{cr} \qquad (18\text{-}51)$$

According to equation 18-44, the changing electric field between the plates of the capacitor is surrounded by circular magnetic field lines and the magnetic field is given by

$$B = \frac{1}{2\pi rc} \frac{(N_2' - N_1')}{(t_2 - t_1)} \qquad (18\text{-}52)$$

$$= \frac{2}{cr} \left[\frac{1}{4\pi} \frac{(N_2' - N_1')}{(t_2 - t_1)} \right] \qquad (18\text{-}53)$$

This is the field that would be produced by a current

$$i' = \frac{1}{4\pi} \frac{(N_2' - N_1')}{(t_2 - t_1)} \qquad (18\text{-}54)$$

This "equivalent current" has exactly the same magnitude as the current i which actually flows in the wire connecting the two plates (equation 18-49), although it is opposite in sign. This question of sign is merely related to the fact that N_2' is less than N_1' because the charge on a plate is decreasing, and so the expression in equation 18-54 is negative. Instead of worrying about the sign, let us concentrate on the more important question of the direction of the magnetic field lines, as shown in figure 18-13. The magnetic field lines encircling the wire must go round clockwise as we look in the direction of the current (compare figure 17-2). Between the plates, the electric field is decreasing. The change in field $(E_2 - E_1)$ during the time interval $(t_2 - t_1)$ is therefore a vector pointing from the negative plate toward the positive plate. The rule illustrated in figure 18-12 therefore implies that, as we look from the negative plate toward the positive plate, the magnetic field lines go round clockwise.

These conclusions can be summarized in the following statement. The magnetic field produced by the changing electric field is exactly the same as if the capacitor were removed and the ends of the wire joined together, allowing the current i to flow around a closed loop.

18-6 Maxwell's Equations—The Field Triumphant

It is not possible to do full justice to Maxwell's equations without expressing them in their elegant mathematical form, but this requires advanced mathematics. We shall have to be satisfied with a simplified statement of each equation and a qualitative discussion of their physical significance, which is profound.

The essential feature of Maxwell's equations is that they are concerned with the behavior of electric and magnetic fields at points in empty space. Moreover, the behavior at a point in space is not related to charged bodies at a distance from this point, but to the behavior of the electric and magnetic fields at other points in the *immediate vicinity* of the first point. However, if a charge happens to be located at the point in question, then this is taken into account by extra terms in the equations. This is a complete swing of the pendulum away from our earlier action-at-a-distance approach to the universe, in which we assumed interactions between discrete particles at a distance from one another and did not worry about the space in between them. The present attitude might be crudely described as the "handing on" of interactions from a point in space to a neighboring point.

The four Maxwell equations are collected together in figure 18-14 and each is illustrated by a simple diagram. These diagrams do not necessarily represent the most general situation to which an equation applies, but are included in the hope that they will help the reader to understand what each equation is about.

Equation I is concerned with the fact that electric field lines sometimes start on positive charges and end on negative charges. If P is the point in space which interests us, surround P by a *very small* spherical surface. If there is no charge at P, then all the electric field lines entering this sphere pass through it and the number of lines leaving is equal to the number of lines entering. However, if there is a positive charge $+q$ inside the sphere, then electric field lines diverge from this charge and the number of lines leaving the sphere exceeds the number of lines entering by $4\pi q$ (see section 14-5). If there is a negative charge $-q$ inside the sphere, field lines converge on this charge and the number *entering* the sphere exceeds the number leaving by $4\pi q$.

Although it is not immediately obvious, equation I is a mathematical consequence of Coulomb's law (see section 14-5 and the discussion associated with figure 14-15).

We now know that a charge located at a point would have to be a fundamental particle with a charge of 4.8×10^{-10} statcoulomb. It would give rise to $4\pi \times 4.8 \times 10^{-10} = 5.47 \times 10^{-9}$ field lines, which is silly! This difficulty is easily overcome. Postulate that the number of field lines leaving a charge $+q$ or entering a charge $-q$ is $4\pi Lq$, where L is a number much larger than 10^9. Then a large number of lines leave a charge e. If this number is not integral, the fractional part may be ignored without serious error if the number is made large enough. The number of lines crossing perpendicularly over a surface of unit area at any point in space is then L times E.

It is wrong to think that electric field lines *must* always start on a positive charge and end on a negative charge. In electromagnetic induction a changing magnetic field produces circular electric field lines that do not start or end, but form closed loops.

Fields
that Change
with Time

375

Figure 18-14 Maxwell's equations.

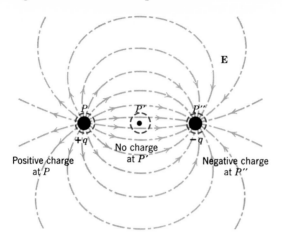

Positive charge at P

No charge at P'

Negative charge at P''

Equation I

If electric field lines diverge from a point or converge on a point, there is a charge located at that point. If the lines are drawn so that the number crossing unit area represents the magnitude of **E**, then $4\pi q$ lines originate at a charge $+q$ and terminate at a charge $-q$.

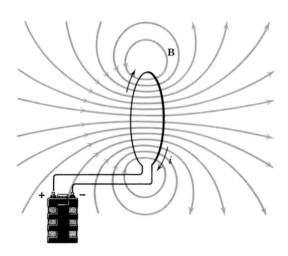

Equation II

Magnetic field lines cannot start at a point or end at a point. They always form closed loops.

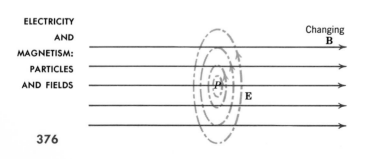

Changing **B**

Equation III

A changing magnetic field produces an electric field. The field changing inside a *small* circle of radius r produces an electric field E going round the circle given by

$$E = -\frac{r}{2c}\frac{(B_2 - B_1)}{(t_2 - t_1)}$$

Figure 18-14 (*continued*).

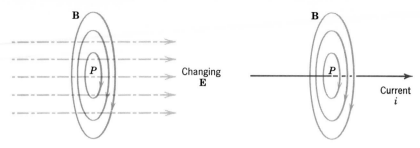

Changing
E

Current
i

Equation IV

A changing electric field produces circular magnetic field lines

$$B = + \frac{r}{2c} \frac{(E_2 - E_1)}{(t_2 - t_1)}$$

A current i also produces circular magnetic field lines

$$B = \frac{2i}{cr}$$

When both a changing electric field and a current are present in the vicinity of P

$$B = \frac{2i}{cr} + \frac{r}{2c} \frac{(E_2 - E_1)}{(t_2 - t_1)}$$

Significance of E and B

The force exerted by **E** on a charge q at P is

$$\mathbf{F}_e = q\mathbf{E}$$

and is independent of the motion of q.

The force exerted by **B** on a charge q at P moving with a velocity v at an angle ϕ to **B** is

$$F_m = \frac{qvB \sin \phi}{c}$$

The direction of F_m is given by right hand rule number 2.

Equation II states that *magnetic* field lines *always* form closed loops. There is no such thing as a "magnetic charge" and so magnetic field lines never start at a point or end at a point.

Equation III is related to Faraday's law of electromagnetic induction and the fact that a changing magnetic field produces an electric field. Suppose that the magnetic field in the immediate vicinity of P changes from $\mathbf{B}_1$ at time t_1 to $\mathbf{B}_2$ at a slightly later time t_2. This produces circular

Fields
that Change
with Time

377

electric field lines centered on P in a plane perpendicular to $(\mathbf{B}_2 - \mathbf{B}_1)$. Consider one such circle of radius r, which is so small that it is reasonable to assume that the value of $\mathbf{B}$ at all points inside this circle differs by a negligible amount from the value at P. Then the electric field E going around this circle and produced by the changing magnetic field inside the circle, is

$$E = -\frac{r}{2c}\frac{(B_2 - B_1)}{(t_2 - t_1)} \tag{18-55}$$

Circular electric field lines surround each point where the magnetic field is changing. These small circles add together to give a resultant field that may bear no obvious resemblance to the original small circles. In the electromagnetic induction experiment of figure 18-1, the magnetic field is changing at all points inside loop II. The small circles add together to give large circular electric field lines centered on the axis of the loop.

Equation IV expresses the fact that a magnetic field may be caused either by a changing electric field or by a current of electricity. The effect of a changing electric field is analogous to the effect of a changing magnetic field. Suppose that the electric field in the immediate vicinity of $\overset{\cdot}{P}$ changes from $\mathbf{E}_1$ at time t_1 to $\mathbf{E}_2$ at a slightly later time t_2. This produces circular magnetic field lines centered on P in a plane perpendicular to $(\mathbf{E}_2 - \mathbf{E}_1)$. Consider one such circle of radius r, which is so small that it is reasonable to assume that the value of $\mathbf{E}$ at all points inside this circle differs by a negligible amount from the value at P. Then the magnetic field B going around this circle, and produced by the changing electric field inside the circle, is

$$B = +\frac{r}{2c}\frac{(E_2 - E_1)}{(t_2 - t_1)} \tag{18-56}$$

Suppose there is an electric current i in the vicinity of P, so that a charge $i(t_2 - t_1)$ passes perpendicularly through the small circle of radius r in the time interval $(t_2 - t_1)$. According to equation 17-5 this also produces circular magnetic field lines and

$$B = \frac{2i}{cr} \tag{18-57}$$

This is related to the Biot-Savart law (equation 17-4).

If both a changing electric field and an electric current exist at P, then

$$B = \frac{2i}{cr} + \frac{r}{2c}\frac{(E_2 - E_1)}{(t_2 - t_1)} \tag{18-58}$$

Algebraic equations have significance in physics only when the physical meaning of the symbols is properly understood. In the case of Maxwell's equations, it is important to have two further equations which tell us the physical significance of $\mathbf{E}$ and $\mathbf{B}$. If a charge q is placed at P,

E determines that part of the force on q which is independent of its motion,

$$\mathbf{F}_e = q\mathbf{E} \tag{18-59}$$

B determines that part of the force on q which does depend upon its motion. If q has a velocity **v** at an angle ϕ to **B**, the magnetic force is

$$F_m = \frac{qvB \sin \phi}{c} \tag{18-60}$$

The direction of F_m is given by right hand rule number 2.

The four Maxwell equations, supplemented by equations 18-59 and 18-60, enable us to solve any problem in classical electromagnetism.

18-7   A New Attitude and a New Concept

In order to appreciate the new attitude to physical phenomena underlying Maxwell's equations, let us consider the magnetic field produced by a moving charge (figure 18-15). In equation 16-1 we simply wrote down the expression for the field produced by the moving charge at a distance R in a direction determined by θ, implicitly assuming a sort of action-at-a-distance. We shall now point out that the magnetic field at P may be looked upon as a consequence of a changing electric field at P. As the charge moves past P, its distance from P and its direction relative to P are continually changing. The electric field it produces at P is therefore continually changing in magnitude and direction. According to Maxwell's fourth equation, this produces a magnetic field in the vicinity of P, which is nothing more than the magnetic field of equation

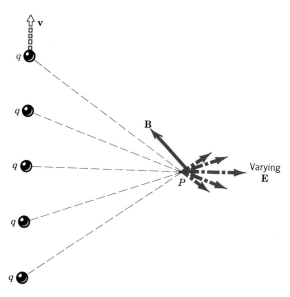

Fields
that Change
with Time

Figure 18-15   As the charge q moves past P, the electric field it produces at P varies with time and this gives rise to a magnetic field at P.

16-1. It is interesting to realize that, since the rate of change of **E** varies as the charge moves past, **B** also changes with time. This is obvious from equation 16-1, because R and θ change with time. A changing **B** produces an electric field, so there is an extra electric field over and above the one given by Coulomb's law. This is the origin of the small correction that must be applied to the electric field of a charge when it is in motion (equation 16-5).

We have still not entered into the true spirit of Maxwell's equations, because we have talked about the *electric* field at P as though it were produced by a charge *at a distance*. We ought really to describe the behavior of **E** and **B** at P only in terms of the behavior of **E** and **B** at points in the immediate vicinity of P. We should really proceed somewhat as follows. In the immediate vicinity of q there are diverging electric field lines originating on q (Maxwell's first equation). There are also circular magnetic field lines, firstly because a charge in motion is equivalent to an electric current and secondly because the electric field changes as the charge moves (Maxwell's fourth equation). Since the charge is moving, the electric and magnetic fields it produces in its immediate vicinity change with time. This produces electric and magnetic fields at points a little further removed from the charge, which in their turn produce electric and magnetic fields still further removed from the charge (Maxwell's third and fourth equations). In this way the disturbance travels out from the charge until it reaches P. In fact, it travels out as an **electromagnetic wave** with the speed of light, c, and takes a finite time to reach P.

This new concept that we have just encountered has revolutionary importance to our thinking about the nature of the universe. Electromagnetic interactions are not instantaneous, but travel from one particle to another with a finite speed, the speed of light. This was not noticeable in the classical experiments on electricity and magnetism that we have described so far. In these experiments on magnetic fields produced by currents and currents induced by changing magnetic fields, one part of the apparatus was rarely more than 100 cm from any other part. An electromagnetic disturbance traveling at 3×10^{10} cm/sec covers 100 cm in only 3.33×10^{-9} sec! Such a short delay in time could never have been detected in the experiments. However, there was one phenomenon well known to nineteenth century physicists in which these considerations could not be ignored. Light is itself an electromagnetic wave which obeys Maxwell's equations and travels with a speed of 3×10^{10} cm/sec. We shall now turn our attention to the nature of waves and the character of electromagnetic radiation.

ELECTRICITY
AND
MAGNETISM:
PARTICLES
AND FIELDS

380

QUESTIONS

1. In figure 18-5, would the direction of the current in loop II change if loop I were placed to the right of loop II, but everything else were left unaltered? In figure 18-6, would the direction of the current in loop II change if loop I or the magnet were to the right of loop II, but still moved in the same direction?

2. In figure 18-6c, what happens if the magnet moves right through loop II and emerges on the other side? Is there an induced electric field in the loop when the center of the magnet is exactly at the center of the coil? If so, what is its direction? (Hint: The important quantity is the rate of change of the *flux* through the coil. Two magnetic field lines passing through the coil in opposite directions cancel one another.)

3. Suppose that loop II is cut and the two ends are pulled slightly apart to prevent a current from flowing across the gap. What happens when the magnetic flux through the loop is changing? Write down a formula for the electric potential difference produced across the gap.

4. In figure 18-5, when the loops are stationary and the current in loop I is switched on or off, how does the force between the loops attempt to counteract the change inducing the current?

5. If two loops are side by side in the same plane, determine the direction of the current in loop II when (*a*) the current in loop I is switched on (*b*) the current in loop I is switched off, (*c*) there is a steady current in loop I and it is brought nearer to loop II, (*d*) there is a steady current in loop I and it is taken further away from loop II.

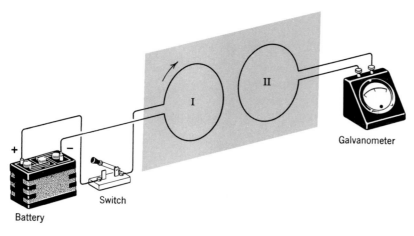

Question 18-5

6. You are locked in a room without windows and told to determine the approximate direction of north. You are provided with the following and nothing else: (*a*) an ample supply of copper wire of all possible diameters, (*b*) a pair of wire cutters made of nonmagnetic steel, (*c*) a galvanometer which deflects both ways with clear instructions telling you which way the current flows between its terminals when the needle deflects in one direction. Describe precisely what you would do.

7. The conditions of the previous question are repeated except that there is no way of deducing the direction of flow of the current through the galvanometer from the deflection of the needle. How-

Fields
that Change
with Time

381

ever, you are also given a bar magnet with its ends marked N and S. How could you then proceed *without suspending the bar magnet and using it as a compass?* Under these circumstances, if you were given a storage battery with its terminals unmarked, how could you determine which was the positive terminal?

8. Invent a method of moving rod AB of example 18-4-1 with a constant velocity. In your method, what is the ultimate source of the energy which eventually appears as joule heat in the circuit $ABCD$?

9. In example 18-4-1, would exactly the same result be obtained if AB were fixed and the rails moved with a velocity v to the left? Repeat all the details of the discussion of this example for this new situation.

10. According to the analysis of a rotating loop in a steady magnetic field (section 18-4 and figure 18-9), there are magnetic forces to produce currents in the vertical wires DA and BC, but not in the horizontal wires AB and CD. Obviously, a steady current cannot flow in the vertical wires unless it also flows in the horizontal wires. A current cannot flow in any wire unless there is a force to overcome the resistance of the wire. What happens? Does this invalidate the mathematical analysis leading to equation 18-26?

11. In an electromagnetic induction experiment, the induced electric field lines are circles, but the loop consists of wire of constant cross-section bent into a complicated shape. Use the equations of section 15-3 to show that the induced electric field tries to produce different currents in different parts of the wire. What happens?

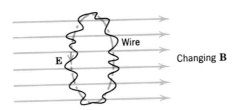

Question 18-11

12. In part (*a*) of the diagram a rotating bar magnet is situated on the axis of a circular loop. Is there a current induced in the loop? In part (*b*) of the diagram the magnet is stationary and the loop

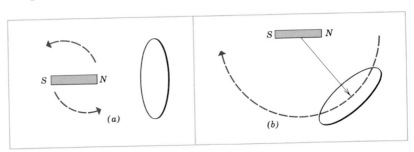

Question 18-12

rotates around it. Is there now a current induced in the loop? Why? Discuss these two situations as fully as you can.

13. In part (a) of the diagram a rotating magnet is near an infinitely long straight wire perpendicular to the plane in which the magnet rotates. In part (b) the magnet is stationary, but the wire rotates around it. (The horizontal parts AB and CD of the wire may be considered to be infinitely far away from the magnet.) In both cases discuss fully the question of whether a current is induced in the wire. Consider several typical orientations of the magnet relative to the wire.

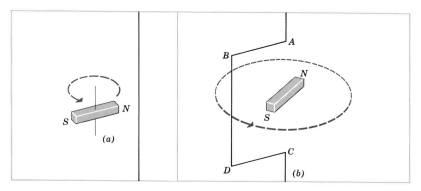

Question 18-13

14. In the discussions associated with figures 18-9 and 18-11 simple proofs were given of the applicability of Faraday's law of electromagnetic induction, but these proofs contain a logical flaw. The magnetic force on a moving charge is always perpendicular to the velocity of the charge and so can never do work on the charge. Reconsider more carefully exactly what is happening to an electron in the loop of figure 8-9 or the copper rod of figure 8-11 and try to remedy this deficiency in the proofs.

PROBLEMS

A

1. A uniform magnetic field of 2.5×10^5 gauss is perpendicular to the plane of a circular loop of radius 1.2 cm. What is the magnetic flux through the loop?

2. If the current is measured in amperes, resistance in ohms and flux in gauss cm², show that equation 18-11 takes the form

$$i = \frac{10^{-8}}{R} \frac{(N_2 - N_1)}{(t_2 - t_1)}$$

3. A magnetic field of 1000 gauss is perpendicular to the plane of a circular loop of radius 5 cm. One millionth of a second later it has decreased to 998 gauss. Find the electric field in the loop.

4. A circular loop of wire of radius 3.5 cm with a resistance of 2.5×10^{-13} statohm is perpendicular to a magnetic field of 10^3 gauss. The field decreases steadily to zero in 2×10^{-3} sec. What current flows round the loop while this is happening?

B

5. A capacitor has plates of area 450 cm^2 with a separation of 0.02 cm. If a charge of 3×10^{-6} coulombs is placed on the capacitor, what is the total electric flux between its plates?

6. Several times in this chapter it was said that a current i statamp flowing round a loop is equivalent to taking a charge of i statcoulomb completely round the loop every second. Suppose, for example, that the drift velocity is 0.1 cm/sec and the circumference of the loop is 10 cm. Then no electron succeeds in going completely round the loop in one second. Demonstrate to your own complete satisfaction that the charge taken round the loop per second may nevertheless be equated to i, even though the change in magnetic field takes place in much less than one second.

7. A circular loop of wire with a resistance of 0.15 ohm is stretched round the equator of a spherical balloon with a radius of 20 cm. A magnetic field of 1000 gauss is perpendicular to the plane of the loop. The balloon bursts and the radius of the strained wire decreases by 0.1 cm in 10^{-3} sec. What is the average current which flows in the wire while this is happening?

8. Repeat example 18-3-1 with $i = 0.1 + 0.05t$ amp, $a = 1$ cm, $r = 10$ m and $R = 20$ ohm. Express i_l in amp.

C

9. In the situation of example 18-4-1, the rod AB, the rails DA and CB, and the cross-connection CD are all made of thin copper wire with an electrical resistance of 6×10^{-3} ohm per cm of its length. $AB = CD = l = 4$ cm. $DA = CB = 10$ cm. The velocity of the rod $v = 100$ cm sec^{-1}. The magnetic field $B = 1000$ gauss. (a) Calculate the current i flowing around $ABCD$. (b) Deduce the potential difference $(\phi_A - \phi_B)$ between A and B. (c) Find the magnitudes of the electric force F_e and the magnetic force F_m on an electron in AB. Show that F_e is exactly $\frac{1}{7}$ of F_m and discuss the significance of this. (d) Find the value of the external force F_{ext} needed to maintain the steady velocity v. (e) Make a rough estimate of the magnitude of the magnetic field B_s at the center of the loop $ABCD$ due to the current i. Compare B_s with B and comment on a possible complication if B_s were comparable with, or larger than, B.

10. A small loop of wire of height h cm, breadth b cm and resistance R statohms is at a distance of r cm from a straight wire carrying a steady current i statamps (see the diagram). The breadth b is very small compared with r. The loop is rotating with angular velocity ω

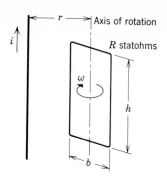

Axis of rotation

R statohms

Problem 18-10

about an axis through its center parallel to the wire. Use Faraday's law to show that the current induced in the loop at the instant when its plane is parallel to the magnetic field is

$$i_l = \frac{2ihb\omega}{c^2 rR}$$

11. Repeat the previous problem for the case when b is not small compared with r.

12. A long straight wire carries a current i. A rectangular loop is in the same plane as the wire. The side of length a is parallel to the wire and the side of length b is perpendicular to the wire. The resistance of the loop is R statohms. If the loop has a velocity v perpendicular to the wire, find the current induced in it when the nearest side is at a distance r from the wire.

13. The two circular loops shown in the diagram are concentric and lie in the same plane. The radius r_1 of the larger loop is very much greater than the radius r_2 of the smaller loop. The two loops are both made of wire with a resistance of ρ statohms per cm length of wire. A steadily increasing potential difference, $V = V_0(1 + \alpha t)$, is applied to a small gap in the larger loop. Find the current in the smaller loop.

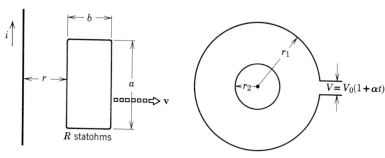

Problem 18-12　　　　**Problem 18-13**

Fields
that Change
with Time

WAVES

19

The Nature of Waves

19-1　Simple Harmonic Motion

In a wave something oscillates, moving first in one direction, then in the opposite direction, and then back again to the original direction. The motion is repeated periodically at exactly equal intervals of time. The simplest form of such a periodic motion is called simple harmonic motion. Before discussing the nature of waves, we must first consider some aspects of simple harmonic motion.

Figure 19-1 shows three typical examples of simple harmonic motion: (*a*) is the backward and forward swing of a pendulum, (*b*) is the up and

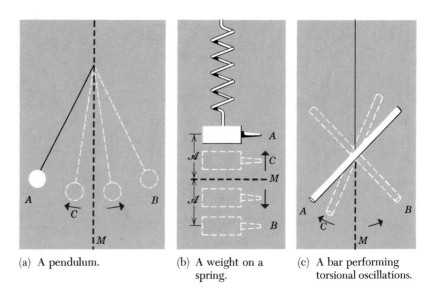

(a)　A pendulum.

(b)　A weight on a spring.

(c)　A bar performing torsional oscillations.

　Figure 19-1　Three examples of simple harmonic motion.

down motion of a weight hanging on a spring, and (c) is the twisting and untwisting motion of a horizontal bar suspended by a wire. These motions are called periodic motions because they obviously repeat themselves periodically. The **period** T is defined as the time taken to complete one oscillation, that is the time taken to go from A to B and back again to A. The starting point is irrelevant. The same time T is taken to start from C, go to B, then to A, and back to C.

The **frequency** is defined as the number of oscillations in one second and is denoted by ν (the Greek letter nu). Obviously, the frequency is the reciprocal of the period.

$$\nu = \frac{1}{T} \tag{19-1}$$

If one oscillation takes $\frac{1}{10}$ second, then 10 complete oscillations are performed in 1 second. Frequency is measured in *cycles per second,* written c/sec or, if you wish, sec^{-1}. 1 kilocycle/sec is 10^3 c/sec and 1 megacycle/sec is 10^6 c/sec.

The **amplitude** of the oscillations, A, is the distance between the central position and an extreme position. For example, in figure 19-1b the amplitude is the distance from the mid-position of the motion, M, to the highest position or the lowest position. It is *not* the distance from the lowest position to the highest position, which is $2A$. Imagine that the weight is originally hanging at rest, is pulled downward through a distance A and then let go. The central position of the motion, M, would then be the original rest position and the weight would oscillate between a point a distance A below this rest position and a point a distance A above the rest position.

In simple harmonic motion there is always an equilibrium position at which the body could remain at rest. If the body is displaced from this position the system exerts a *restoring force* on it, which is in such a direction as to return it to its equilibrium position. Moreover, the magnitude of the force is directly proportional to the distance the body has been displaced from its equilibrium position. This direct proportionality between restoring force and displacement is the distinguishing feature of simple harmonic motion as compared with other, more complicated periodic motions.

Consider a mass m sliding on a perfectly smooth, frictionless table and attached to one end of a spring, the other end of which is held fixed by a peg P. In figure 19-2a the mass is shown at rest. The spring then has its natural, unextended length L and exerts no force on m. In 19-2b the mass has been given a **displacement** x to the right, increasing the length of the spring from L to $L + x$. It is a well known characteristic of a spring that, when it is extended, it exerts a force in an attempt to regain its original length. The greater the extension the greater is the restoring force. In fact the restoring force, F, is directly proportional to the extension, x.

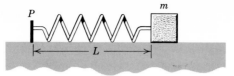

(a) Spring at rest with its natural length L.

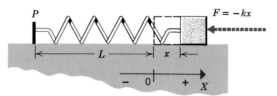

(b) Extended spring.

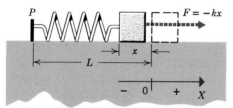

(c) Compressed spring.

Figure 19-2   Illustrating Hooke's law.

Hooke's law

$$F = -kx \qquad (19\text{-}2)$$

This was discovered by Robert Hooke in the seventeenth century, and is known as Hooke's law. The constant k is called the **force constant** of the spring.

The minus sign in equation 19-2 ensures that the force is always a restoring force. If the origin of the X axis is taken to be the equilibrium position of m, and x is a positive displacement to the *right*, then F is negative, which indicates that it points to the *left*. However, if the spring is compressed (figure 19-2c), then x is negative and F is positive, which indicates that it points to the right.

The acceleration a of m is obtained by applying Newton's second law of motion

$$a = F/m \qquad (19\text{-}3)$$

or

$$a = -\frac{k}{m}x \qquad (19\text{-}4)$$

From this equation, with the aid of the integral calculus, it is possible to deduce how the displacement x varies with time t.

Equation of simple harmonic motion

$$x = A \cos \frac{2\pi}{T} t \tag{19-5}$$

$$= A \cos 2\pi \nu t$$

A is the amplitude and T is the period, which can be shown to be

$$T = 2\pi \sqrt{\frac{m}{k}} \tag{19-6}$$

Notice that the period is independent of the amplitude. The body takes as long to swing backward and forward over a short distance as it does over a long distance.

Equation 19-5 is analogous to equation 4-22 of Chapter 4, inasmuch as it shows how the x coordinate of the moving body m depends upon the time t, but the form of equation 19-5 is very different. Figure 19-3 is a graph of x against t for simple harmonic motion. The reader will easily be able to verify the points at $t = 0$, $T/4$, $T/2$, $3T/4$ and T if he remembers that

$$\cos 0 = +1 \tag{19-7}$$

$$\cos \frac{\pi}{2} = \cos 90° = 0 \tag{19-8}$$

$$\cos \pi = \cos 180° = -1 \tag{19-9}$$

$$\cos \frac{3\pi}{2} = \cos 270° = 0 \tag{19-10}$$

$$\cos 2\pi = \cos 360° = \cos 0° = +1 \tag{19-11}$$

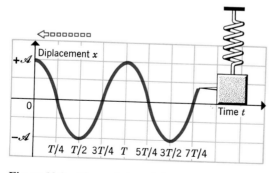

Figure 19-3   The variation of displacement with time for simple harmonic motion. The weight W is oscillating on the end of a spring. A pen attached to W traces a line on a piece of graph paper which is pulled past it from right to left at a steady rate.

In this way it is easy to see that the motion repeats itself every T seconds and that x oscillates between $+A$ and $-A$.

According to equation 19-5, x has its maximum positive value $+A$ when $t = 0$. There is no reason why we should not take the zero of time at any instant during the motion, when x has some value other than $+A$. The equation of simple harmonic motion must then be generalized to

Generalized equation of simple harmonic motion

$$x = A \cos\left(\frac{2\pi}{T}t + \phi\right) \cdots \tag{19-12}$$

$$= A \cos(2\pi\nu t + \phi)$$

ϕ is called the **phase constant** and

$$x = A \cos\phi \text{ when } t = 0 \tag{19-13}$$

The position that the body has reached in its motion is determined by the **phase angle**, θ

$$\theta = \frac{2\pi}{T}t + \phi \tag{19-14}$$

These considerations are important when we are dealing with two simple harmonic motions which have the same period but different phase constants, and are therefore *out of phase*.

$$x_1 = A_1 \cos\left(\frac{2\pi t}{T} + \phi_1\right) \tag{19-15}$$

$$x_2 = A_2 \cos\left(\frac{2\pi t}{T} + \phi_2\right) \tag{19-16}$$

The difference in phase angles is always the same as the difference in phase constants

$$\theta_1 - \theta_2 = \phi_1 - \phi_2 \tag{19-17}$$

One motion always lags behind the other by the same amount.

This concept of phase is illustrated in figure 19-4 by two weights hanging side by side from springs. In figure 19-4a the two oscillations are *in phase* and $\phi_1 = \phi_2$. In 19-4b the two oscillations are 180° *out of phase* and $\phi_1 - \phi_2 = \pi = 180°$.

Example 19-1-1

A vertical spring has an unstretched length L. When a mass m hangs at rest from its lower end, its length increases to $L + l$. Find the period of small vertical oscillations of m (figure 19-5).

When the mass is hanging at rest (figure 19-5b), the extension of the spring is l and the force exerted by it on m is, according to Hooke's law,

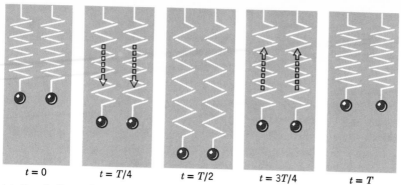

(a) Two balls oscillating in phase.

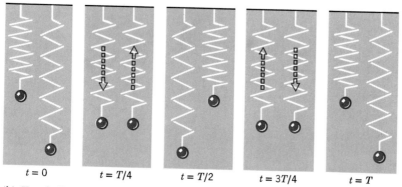

(b) Two balls oscillating 180° out of phase.

Figure 19-4 Illustrating the concept of phase.

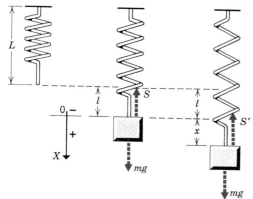

(a) No weight.   (b) Weight at rest.   (c) Weight oscillating.

Figure 19-5 Illustrating example 19-1-1.

$$S = -kl \tag{19-18}$$

The positive direction is downward, but S is an upward force and therefore negative. Since the mass has no acceleration, this upward force exerted by the spring must be exactly counterbalanced by the weight, mg.

$$S = -mg \tag{19-19}$$
$$\text{So} \quad mg = kl \tag{19-20}$$

Suppose that, during the oscillation, the mass is a distance x below its equilibrium position, so that the extension of the spring is $l + x$ (figure 19-5c). The force exerted on m by the spring is then

$$S' = -k(l + x) \tag{19-21}$$
$$= -kl - kx \tag{19-22}$$

The total force F on m is the sum of S' and the weight mg.

$$F = S' + mg \tag{19-23}$$
$$F = -kl - kx + mg \tag{19-24}$$

According to equation 19-20, $-kl$ cancels $+mg$, so

$$F = -kx \tag{19-25}$$

This is exactly the same as equation 19-2 and therefore represents a simple harmonic motion with a period given by equation 19-6

$$T = 2\pi \sqrt{\frac{m}{k}} \tag{19-26}$$

The problem was formulated in such a way that k was not given, but l was. From equation 19-20, however,

$$k = mg/l \tag{19-27}$$

Substituting this value of k into equation 19-26 and cancelling the m which appears in the numerator and denominator

$$T = 2\pi \sqrt{\frac{l}{g}} \tag{19-28}$$

19-2 Waves as Means of Communication

When a stone is thrown into a quiet lake, circular ripples spread out from the point of impact and travel over the surface of the lake. The disturbance eventually reaches some distant part of the lake and produces physical effects there, even though there is no actual transfer of water from the point of impact of the stone to the distant point. If the wave passes over a floating piece of wood, the wood bobs up and down but stays in the vicinity of its original position and does not move forward with the advancing wave front. Similarly, if the motion of a small droplet of water near the surface could be studied, it would be found to

WAVES

394

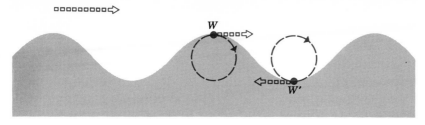

Figure 19-6 The water wave travels from left to right, but a small droplet of water, such as W or W', moves in a vertical circle with a stationary center.

be going round and round in a vertical circle but the center of the circle would remain fixed (figure 19-6).

An important aspect of waves is that they provide a means whereby one part of the universe can influence another part without the bodily transfer of matter between the two parts. Consider, for example, how you might communicate with a friend on the opposite bank of the lake (figure 19-7). One way would be to shoot over an arrow with a message attached to it. This would involve the bodily transfer of matter. A second way would be to throw a stone into the lake. When the ripples reached the float of his fishing line and made it bob up and down, this would be a prearranged signal to your friend that it was time to go to lunch. Yet another possibility would be to shout across the lake. This would send a sound wave traveling through the air, as we shall explain in a later section of this chapter. In neither of the last two methods would there be any transfer of matter from you to your friend. Still another method

(a) Bodily transfer of matter.

(b) Waves traveling over the surface of the lake.

(c) Sound waves traveling through the air.

(d) Electromagnetic waves (light).

Figure 19-7 Some methods of communication.

would be to switch on a flashlight, with the understanding that, as soon as your friend saw the light, he would pack up and go to lunch. This method involves the propagation of electromagnetic waves (light waves) through space. It is the most interesting of all the methods because, as we shall see, not only does it not require the transfer of matter across the lake, but it does not even require the presence of any matter in the intervening space. Light can travel through a vacuum.

19-3   Transverse Waves

Water waves are very unsuitable for a detailed discussion because they are more complicated than most waves. We have already mentioned the unnecessary complication that the particles of water near the surface move in circles. We shall discuss instead the case of a transverse wave

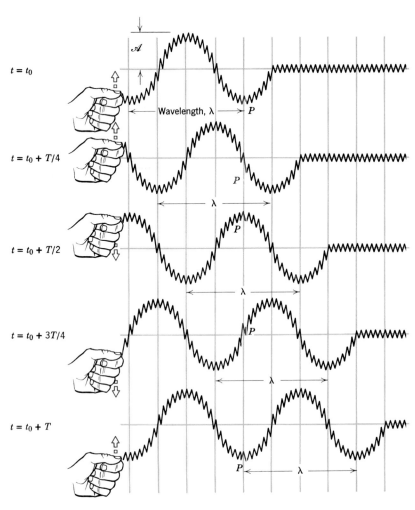

Figure 19-8   A transverse wave on a long spring.

on a long spring, which is more closely analogous to the electromagnetic waves that will eventually be our main interest.

Imagine a long horizontal spring with one end attached to the wall and the other end held in the hand and pulled taut. By moving the hand up and down rapidly two or three times, a wave can be made to travel along the spring as shown in figure 19-8. One of the coils of the spring P is painted maroon so that its motion can be easily observed. Although the wave travels along the spring in a horizontal direction, P oscillates up and down on a vertical straight line and does not move in a horizontal direction. The motion of P is, in fact, simple harmonic. A wave is called a **transverse wave** if the oscillatory motion of any part of the system, such as P, is at right angles to the direction in which the wave is traveling.

The shape of the spring at a fixed instant of time is a *sine wave function* similar to figure 19-3. However, the difference between figures 19-3 and 19-8 should be clearly understood. If the vertical displacement of a particular coil, such as P, were plotted against time, a curve such as figure 19-3 would be obtained. Figure 19-8, on the other hand, shows the displacements of all the coils at a fixed instant of time, and is therefore a plot of displacement against distance along the spring at constant t. In figure 19-3 the distance between adjacent peaks or adjacent valleys of the curve in the direction of the time axis represents the period T of the oscillation. In figure 19-8 the horizontal distance between adjacent peaks or adjacent valleys is a real distance in space. It is called the **wavelength** of the wave and is usually represented by the Greek letter lambda, λ.

Now let us consider the motion of the coil P as the wave travels along the spring. Careful study of figure 19-8 reveals that, while one wavelength is passing over P, the coil goes through one complete oscillation, which takes a time T. The wave therefore travels forward a distance λ in a time T. The **wave velocity** V must be

$$V = \frac{\lambda}{T} \tag{19-29}$$

The frequency of the wave is the number of oscillations which P makes in one second and is related to the period by equation 19-1. So

$$\nu = \frac{1}{T} \tag{19-30}$$

$$V = \nu\lambda \tag{19-31}$$

Wave velocity = Frequency × Wavelength $\qquad$ (19-32)

This important equation is true for all types of waves.

19-4 Longitudinal Waves: Sound

Let us repeat the experiment with a long horizontal spring but now oscillate its free end backward and forward in a horizontal direction

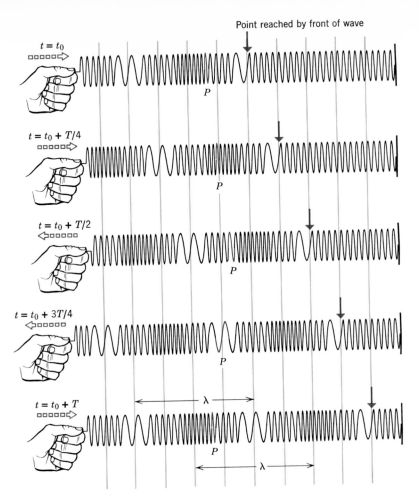

Figure 19-9 A longitudinal wave on a long spring.

(figure 19-9). The spring remains straight, but a wave of alternating compressions and extensions travels along its length. The painted coil performs simple harmonic motion in a horizontal direction. The oscillatory motion of a part of the system is in the same direction that the wave is traveling and so this is a **longitudinal wave.** The wavelength is the distance from a point on the spring where the coils are closest together to the next nearest point where the coils are closest together. The reader should convince himself that equation 19-32 also applies to this case.

Sound is usually transmitted as a longitudinal wave in the air. Consider what happens when you shout to your friend across a lake, as in figure 19-7c. Air is expelled from your lungs through your windpipe, past two elastic membranes, known as *vocal cords*, inside the windpipe (figure 19-10). The flow of air past the vocal cords causes them to vibrate, and the size of the opening between them varies periodically.

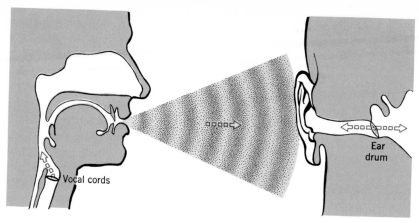

Figure 19-10 Speech, sound, and hearing. The anatomy is simplified in order to illustrate the basic principles.

The air is consequently expelled in "puffs" that follow one another in rapid succession. The number of puffs per second determines the frequency (or *pitch*) of the sound emerging from your mouth.

These puffs of air do not travel all the way across the lake to your friend's ear. As a puff emerges it increases the pressure of the air in the immediate vicinity of the mouth. In between puffs the molecules of air near the mouth move further apart to relieve the excess pressure, but they collide with neighboring molecules and force them closer together so that a region of high pressure and density is created a little farther from the mouth. In fact the process continues and the region of high pressure moves outward from the mouth with the *speed of sound*. Each puff creates such a region of high pressure moving outward. The sound wave therefore consists of a train of high pressure regions following one another in rapid succession with the frequency of the sound and separated by regions of lower pressure (figure 19-10). The individual molecules do not travel with the high pressure regions, they merely oscillate backward and forward parallel to the direction in which the sound is traveling. The behavior of the molecules is similar to the behavior of the coils of the spring in figure 19-9. A region of high pressure and high density is analogous to a part of the spring where the coils are close together. A region of low pressure is analogous to a part of the spring where the coils are far apart.

When the high pressure regions impinge upon a membrane known as the *ear drum* in your friend's ear, they cause it to oscillate backward and forward with the same frequency as your vocal cords. The mechanism of your friend's ear converts this oscillation into a train of pulses of electrical current traveling along one of his nerves to his brain. The frequency of arrival of the electrical pulses is exactly the same as the frequency of vibration of your vocal cords. He is thereby made aware of what you are saying.

The Nature
of Waves

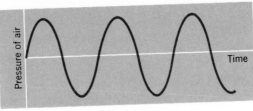

(a) A "pure" note of fixed frequency.

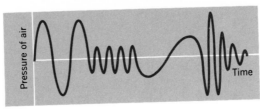

(b) Speech.

Figure 19-11   Variation with time of the pressure at a point in the air for (*a*) a "pure" note of constant pitch, and (*b*) speech of variable frequency.

The frequency of vibration of the vocal cords is determined by the force with which air is expelled from the lungs and by certain muscles in the throat that vary the tautness of the vocal cords and the size and shape of the neighboring cavities. If these factors remain constant, the sound produced is very nearly a "pure note" of definite frequency. The variation with time of the pressure of the air at a fixed point in space is then almost a sine wave (figure 19-11*a*). However, during intelligible speech the muscles of the throat are continually varying their behavior and the sound produced is continually changing its frequency and amplitude, as in figure 19-11*b*. If you shout "lunch time" across the lake, there is a particular pattern of behavior of the throat muscles corresponding to this phrase and a particular sequence of frequencies in the sound produced. This sequence of frequencies is accurately reproduced by the vibration of your friend's ear drum and the pattern of electrical pulses traveling to his brain. This is how he is able to interpret what you are saying.

Sound can also travel through solids and liquids. If you completely submerge your ears in the bath, you can still hear the sound of a running faucet directly propagated through the water. As in the case of the air, the sound consists of a train of regions of high pressure and high density traveling through the water, while the water molecules oscillate backward and forward in the same direction. Such a longitudinal wave can propagate through a solid. A solid can also support a *transverse* sound wave, in which the atoms are oscillating in a direction at right angles to the direction in which the sound is traveling.

WAVES

The lowest frequency audible to the human ear is about 16 sec^{-1} and the highest frequency is about 2×10^4 sec^{-1}, though this varies somewhat with the individual. However, the word *sound* is used loosely to describe a wave of any frequency propagating through a material body and involving oscillations of the atoms of that body. **Ultrasonics** is the study of sound waves with frequencies ranging from the upper limit of audibility (about 2×10^4 sec^{-1}) to about 5×10^8 sec^{-1}. **Hypersonics** is concerned with the very highest sound frequencies, from about 5×10^8 sec^{-1} upward. In example 19-4-2 we shall show that there is a practical upper limit of about 10^{14} sec^{-1} to the frequency of a sound wave. At the time of this writing the highest frequency which has been excited experimentally is 2.5×10^{10} sec^{-1}.

Example 19-4-1

The frequency of middle C is 256 sec^{-1}. What is its wavelength in air?
The velocity of sound in air varies with temperature. At $18°C$, which is a common room temperature, it is 3.4×10^4 cm/sec. Using equation 19-32,

$$\text{Wave velocity} = \text{Frequency} \times \text{Wavelength} \qquad (19\text{-}33)$$

we see that

$$\text{Wavelength} = \frac{\text{Wave velocity}}{\text{Frequency}} \qquad (19\text{-}34)$$

$$= \frac{3.4 \times 10^4}{256} \qquad (19\text{-}35)$$

$$= 133 \text{ cm} \qquad (19\text{-}36)$$

This is approximately 4 feet. When the pianist plays middle C, the high pressure regions traveling toward a listener through the air are about 4 feet apart, but they are moving with the very high speed of 3.4×10^4 cm/sec (760 mph) and so 256 of them arrive at the listener's ear every second.

Example 19-4-2

Estimate the upper limit to the frequency of "sound" waves in ordinary matter.
Equation 19-34 makes it clear that the higher the frequency the shorter is the wavelength. We are therefore asking what is the shortest wavelength sound can have. The answer is that, when the wavelength becomes shorter than the distance apart of the atoms, the concept of a wave breaks down because there is nothing in between the atoms to oscillate (figure 19-12). The atoms are closest together in solids and highly compressed gases. However, when two atoms are about 10^{-8} cm apart they repel one another very strongly and it is difficult to force them any nearer. Consequently, in no material are the atoms or molecules found to be closer together than about 10^{-8} cm. This is therefore the order of magnitude of the shortest wavelength of sound.
The highest value found for the velocity of sound is about 10^6 cm/sec. Using equation 19-32

The Nature
of Waves

(a) The behavior of the atoms for a transverse wave with a wavelength longer than the distance between atoms.

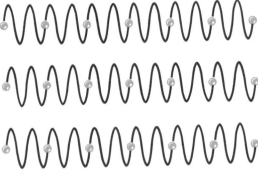

(b) There is no meaning in a wavelength shorter than the distance between atoms because there is nothing in between the atoms to oscillate.

Figure 19-12 Illustrating example 19-4-2.

$$\text{Highest frequency} = \frac{\text{Largest wave velocity}}{\text{Shortest wavelength}} \tag{19-37}$$

$$= \frac{10^6}{10^{-8}} \text{ approximately} \tag{19-38}$$

$$= 10^{14} \text{ sec}^{-1} \text{ approximately} \tag{19-39}$$

This is much larger than the highest frequency yet produced experimentally, which is 2.5×10^{10} sec^{-1}.

The reader should note the character of these arguments, which are not precise calculations, but rather intelligent guesses. We have made it very plausible that the highest frequency is not very much bigger than 10^{14} sec^{-1}. It is conceivable that, by applying a very high pressure to the right gas or solid, we could force its atoms a little closer than 10^{-8} cm and achieve a velocity of sound a little larger than 10^6 cm/sec. In this way we might perhaps realize a frequency of, say, 2×10^{14} sec^{-1}. We should be very surprised, though, if we ever pushed the frequency up to 10^{15} sec^{-1}. In fact we know that there is a limit to the procedure, because there is reason to believe that, when the pressure on a substance becomes extremely large, its atoms break up into separate electrons and nuclei.

QUESTIONS

1. In the simple harmonic motions of figure 19-1, in which of the positions A, C or M is the velocity (a) greatest, (b) zero? In which of these positions is the acceleration (a) greatest, (b) zero?
2. What are the phase angles of simple harmonic motion (a) in the two extreme positions (b) in the mid-position? Assume that the simple harmonic motion is described by equation 19-12.
3. Draw diagrams similar to figure 19-4 for two oscillations which are out of phase by (a) 90°, (b) −90°, (c) 45°, (d) 270°, (e) 360°.
4. Referring to figure 19-5, the mass m is attached to the spring in its unstretched position and is then released with zero initial velocity. What is the amplitude of the subsequent oscillation?
5. In its mid-position the pendulum of figure 19-1a has kinetic energy. In its extreme position the pendulum has come to rest and has no kinetic energy. Where has the kinetic energy gone?
6. In its mid-position the oscillating weight of figure 19-1b has kinetic energy. In its highest position the weight is at rest and has no kinetic energy. Where has the kinetic energy gone? (Before giving too hasty an answer, ask the same question for the situation in figure 19-2.)
7. The velocity of sound in a material usually has the same order of magnitude as the average velocity of its molecules. Why does this seem reasonable to you?
8. When a jet airplane "breaks the sound barrier" the resulting sonic boom is due to a ridge of very high pressure traveling through the air with a velocity in excess of the average velocity of the molecules of the air. How is this possible?
9. Can you think why it is not possible to have a *transverse* sound wave in a liquid or a gas? Why is it possible in a solid?

PROBLEMS

Any spring mentioned in any problem can be assumed to have a negligible mass.

A

1. What is the period of a sound wave with a frequency of 2.5×10^3 sec^{-1}?
2. What is the frequency of a light wave with a period of 1.6×10^{-15} sec?
3. What is the wavelength of the light wave in the previous question?
4. A sound wave in a liquid has a frequency of 4.5×10^6 sec^{-1} and a wavelength of 1.2×10^{-2} cm. What is the wave velocity?

5. A sound wave in a solid has a wave velocity of 1.5×10^5 cm sec^{-1} and a wavelength of 3×10^{-6} cm. What is its frequency?

6. What is the period of a light wave with a wavelength of 10^{-13} cm in a vacuum?

7. A force of 2 newtons extends a spring 0.3 cm. What is the force constant of the spring?

8. A mass of 15 gm hanging on the end of a vertical spring extends it by 0.25 cm. What is the force constant of the spring?

9. Write down the equation expressing the displacement as a function of time for a simple harmonic motion with a period of 3.5 sec if the distance between the two extreme positions is 7.4 cm.

10. Assuming that the lowest audible note has a frequency of 16 sec^{-1}, and the highest audible note has a frequency of 2×10^4 sec^{-1}, find the corresponding wavelengths in air. In each case think of a familiar object which has a size comparable with the wavelength in question. (Take the velocity of sound in air to be 3.4×10^4 cm sec^{-1}.)

11. The pendulum of a grandfather clock makes the floor vibrate with a period of $\frac{1}{2}$ sec. What would be the wavelength in air of the sound generated by this vibration. (Velocity of sound in air $= 3.4 \times 10^4$ cm sec^{-1}.)

B

12. When a mass m hangs on the end of a spring it extends it l cm. The spring is then cut into two identical halves. If the same mass is hung from the end of one half, what will the extension be? Write out a proof of your answer with meticulous logic so that each statement is an obvious deduction from the previous statement.

13. A force of 10^4 dyne extends a spring by 4 cm. If a mass of 100 gm hangs from the spring, calculate the period of small vertical oscillations.

14. A mass of 20 gm hangs from a vertical spring and extends it by 15 cm. Calculate the frequency of small vertical oscillations.

15. A proton in a complicated molecule can oscillate with simple harmonic motion in a certain direction. If the proton is displaced 10^{-9} cm from its equilibrium position in this direction, the restoring force is 4.5×10^{-4} dyne. What is the frequency of the proton's oscillation.

16. A mass M gm hangs from a vertical spring and extends it by l cm. If a mass of m gm is gently added to the M gm, find the amplitude and period of the ensuing oscillation.

C

17. If a body is performing simple harmonic motion with a period of 10 sec and an amplitude of 1.5 cm, find the maximum value of its acceleration. (Hint: You do not need to know k and m separately. The ratio k/m is sufficient.)

18. Find the period for the arrangement shown in the diagram. The two springs have force constants k_1 and k_2.

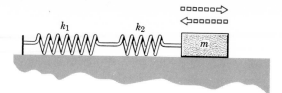

Problem 19-18

19. Find the period for the arrangement shown in the diagram. Discuss qualitatively what would happen if the mass m were displaced in a horizontal direction perpendicular to the line joining the two pegs and then released.

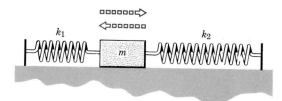

Problem 19-19

20. A body is simultaneously given a simple harmonic motion parallel to the X-axis and another simple harmonic motion with the same period and amplitude parallel to the Y-axis. This means that its x and y coordinates at any time t are given by

$$x = A \cos (2\pi\nu t + \phi_1)$$
$$y = A \cos (2\pi\nu t + \phi_2)$$

Describe as precisely as you can the nature of the two-dimensional motion when (a) the phase angles ϕ_1 and ϕ_2 are equal

(b) $\phi_1 = \phi_2 + \dfrac{\pi}{2}$ (c) $\phi_1 = \phi_2 - \dfrac{\pi}{2}$

(Hint: Suppose that the body is at the point P at time t and O is the origin. Find the length OP and the angle between OP and the X-axis.)

21. A point P moves around a circle of radius r with a constant angular velocity ω (see diagram). Q is the foot of a perpendicular from P to any diameter. Prove that:

(a) The motion of Q is simple harmonic

$$OQ = x = A \cos \omega t$$

(b) The amplitude of the simple harmonic motion, $A = r$.

(c) The period of the simple harmonic motion is the time for one complete revolution of P, or $T = \dfrac{2\pi}{\omega}$.

(d) The velocity of Q at any time is $v = -\omega A \sin \omega t$.

(e) The maximum velocity of Q occurs when it is at O and is
$$\omega A = \frac{2\pi A}{T}.$$
(f) The acceleration at any time is $a = -\omega^2 x$.
(g) The maximum acceleration of Q occurs in the extreme positions and is $\omega^2 A$.
(h) $k = m\omega^2$

(Hint: The velocity and acceleration of Q are the x-components of the velocity and acceleration of P.)

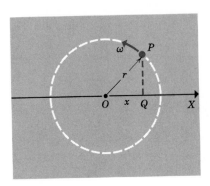

Problem 19-21

22. This is a continuation of problem 14-32, which in its turn is based on problems 14-29 and 14-30. A straight tunnel is drilled through the center of the earth to emerge at the antipodes. A ball is dropped into the tunnel. Ignoring friction, prove that its motion is simple harmonic and evaluate the period.

23. A straight tunnel is drilled from a point on the earth's surface to another point on the earth's surface not necessarily at the antipodes. A ball is dropped into the tunnel. Ignoring friction, prove that its motion is simple harmonic and evaluate the period.

WAVES

Electromagnetic Waves

20-1   The Nature of an Electromagnetic Wave

When an electric charge accelerates, it radiates an electromagnetic wave. In order of increasing frequency and decreasing wavelength, the important types of electromagnetic radiation are radio waves, infrared radiation (sometimes called heat radiation, but see section 20-3), visible light, ultraviolet light, x-rays, and γ-rays. These will be described in more detail in section 20-3. Their production can always be ultimately related to the acceleration of electrons, protons, or other charged fundamental particles.

To illustrate the production and character of electromagnetic waves, let us consider the transmission and reception of a radio wave (figure 20-1). The transmitting aerial is a long metal rod or wire carrying an electric current oscillating backward and forward with a frequency which, in a typical case, might be a million times per second. In section 15-1 an electric current was shown to be due to the motion of the free electrons in a metal. In the present instance each free electron performs a simple harmonic motion with a frequency of one million cycles per second (10^6 sec^{-1}). During its simple harmonic motion, the electron is accelerating and it therefore radiates an electromagnetic wave. The oscillating electron is continually changing its position and velocity and the electric and magnetic fields it produces in its immediate vicinity are continually changing. In accordance with Maxwell's equations and the considerations of chapter 18, these changing fields produce other fields in their vicinity and the disturbance travels outward with the speed of light. (See the discussion on page 380.)

When Maxwell's equations are applied to this situation and solved, it is found that, except in the immediate vicinity of the oscillating electron, the outgoing electromagnetic wave has the form shown in figure 20-1*a*. In this diagram we are considering the electric and magnetic fields at a single instant of time at *points along a line* which, for convenience, is taken to be in a direction perpendicular to the motion of the electron.

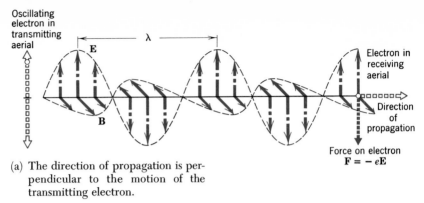

(a) The direction of propagation is per-
    pendicular to the motion of the
    transmitting electron.

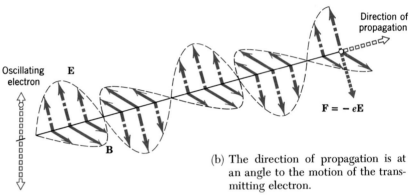

(b) The direction of propagation is at
    an angle to the motion of the trans-
    mitting electron.

Figure 20-1 The electromagnetic wave produced by an oscillating electron. In
the interests of clarity, the diagram ignores the fact that, as the wave travels away
from the electron, its energy is spread over a greater area and the magnitudes of **E**
and **B** decrease.

At each point the electric and magnetic fields are represented by two
vectors. The electric field **E** is seen to be parallel to the motion of the
electron, but its direction and magnitude vary along the line in such a
way that the tips of the vectors trace out a sine wave function. The magnetic field **B** is perpendicular to the electric field and to the direction of
travel and it traces out a similar sine wave function. In fact, in the CGS
system which we are now using almost exclusively, the magnitude of the
magnetic field at any point is exactly equal to the magnitude of the electric field.

$$E = B \tag{20-1}$$

The diagram in figure 20-1 represents the electric and magnetic fields
at points along the line *at a fixed instant of time*. The whole pattern
should now be imagined to move along the line away from the electron
with the speed of light. As various parts of the pattern pass over a fixed
point, the vector representing the electric field oscillates between a maximum value in an upward direction and an equal value in a downward
direction. Similarly, the magnetic field oscillates in and out of the plane

of the page. This is similar to the behavior of the spring in figure 19-8. It is important to realize, however, that we are describing the electric and magnetic fields in empty space at *points* along a straight line and that nothing is moving in a direction perpendicular to this line. The significance of the picture is that, if an electron *were* present at any point in space, the oscillating electric field at that point would exert an oscillating force on it and the electron would be made to perform a simple harmonic motion similar to the motion of the electrons in the transmitting aerial. This is what actually happens to the free electrons in the receiving aerial. It is the mechanism whereby the behavior of the electrons in the transmitting aerial is reproduced by the electrons in the receiving aerial, so that communication becomes possible through the intervening empty space.

Figure 20-1b shows the form of the wave along a line which is not perpendicular to the motion of the transmitting electron. *An electromagnetic wave is always a transverse wave*, because the electric and magnetic vectors are always perpendicular to the direction in which the wave is traveling. The motion of the transmitting electron, the electric vector at any point, and the line joining that point to the electron (which is the direction of propagation) all lie in the same plane. The magnetic vector is perpendicular to this plane. The electric and magnetic vectors have the same magnitude and are in phase. They both assume a maximum value at the same time and are zero at the same time. Looking in the direction of propagation, when the electric vector **E** points upward, the magnetic vector **B** points to the right.

20-2   The Velocity of Light

The solution of Maxwell's equations shows that the velocity of an electromagnetic wave is the constant c which was introduced in chapter 16 in connection with the magnetic forces between moving charges. The value of c can be deduced from measurements of the forces between charges. Suppose that we take two charges q_1 and q_2 and measure the electrostatic force between them when they are stationary. From the equation giving the electrostatic force

$$F_e = \frac{q_1 q_2}{R^2} \tag{20-2}$$

the product $q_1 q_2$ can be deduced. Now set the charges in motion and measure the magnetic force. If the velocities are v_1 and v_2 in directions defined by the angles θ and ϕ (as explained in Chapter 16) the magnetic force is

$$F_m = \frac{q_1 q_2}{R^2} \frac{v_1 v_2}{c^2} \sin \theta \sin \phi \tag{20-3}$$

The product $q_1 q_2$ was determined in the electrostatic experiment and everything else in this equation can be measured except c^2, which can therefore be deduced from the equations. In fact

$$c^2 = \frac{F_e}{F_m} v_1 v_2 \sin \theta \sin \phi \qquad (20\text{-}4)$$

The actual experiments are, of course, much more complicated and cannot be properly understood without a more detailed knowledge of physics, but they are based on the principle just described. These experiments have no apparent connection with electromagnetic waves, but the value they yield for c should, according to Maxwell's theory, be the velocity of propagation of electromagnetic waves. When Hertz produced radio waves in 1885, twelve years after Maxwell had published the final form of his equations, he measured their wavelength λ and frequency ν and deduced their velocity from the equation

$$c = \nu\lambda \qquad (20\text{-}5)$$

The result was in excellent agreement with the value deduced from the electrostatic and magnetic measurements.

The really striking aspect of the situation, though, was that the velocity of an electromagnetic wave c was exactly the same as the velocity of *light*. The velocity of light had been deduced from astronomical observations by Roemer in 1675 and by Bradley in 1729 and had been measured in a terrestrial experiment by Fizeau in 1849. At the beginning of the nineteenth century the experiments of Young and Fresnel on interference and diffraction of light (to be described in Chapter 21) had convinced scientists that light is a wave. One of Maxwell's outstanding achievements was the realization that light is an *electromagnetic* wave. Figure 20-1 also serves to illustrate the nature of light. The only difference between light and radio waves is that the frequency of light is much greater and its wavelength is much shorter.

The principle of the method used by Fizeau to measure the velocity of light is illustrated in figure 20-2. A beam of light is periodically interrupted by the teeth of a rotating toothed wheel. A burst of light passing

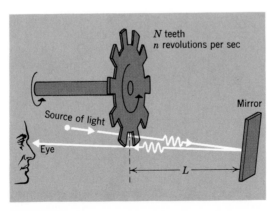

Figure 20-2   Fizeau's method of measuring the velocity of light.

through the gap between two teeth travels a distance L to a mirror and is reflected back again. It returns to the wheel after a time $2L/c$ seconds has elapsed in flight. If during this time the wheel has rotated far enough so that the position originally occupied by a gap is occupied by a tooth, the light is blocked out and is not seen by an observer on the other side of the wheel. If, however, the wheel has rotated so that the next gap occupies the position originally occupied by the first gap, the light passes through and is seen. If there are N gaps and the wheel performs n revolutions per second, the time needed for one gap to replace its neighboring gap is $1/Nn$ secs. Therefore, if

$$\frac{2L}{c} = \frac{1}{Nn} \tag{20-6}$$

the light will be seen. The procedure is to increase the rate of rotation from zero until the light disappears and then appears again. If n is the number of revolutions per second when the light reappears at its brightest,

$$c = 2LNn \tag{20-7}$$

The velocity of light has now been measured by many different methods, some using light and others using radio waves. These methods all agree within the limits of their experimental accuracy. The accepted value at the time of writing is

$$c = 2.99792 \times 10^{10} \text{ cm sec}^{-1} \tag{20-8}$$

with an uncertainty of

$$0.00001 \times 10^{10} \text{ cm sec}^{-1} \tag{20-9}$$

The difference between this and 3×10^{10} cm sec^{-1} is so small that we shall use the approximately correct value of 3×10^{10} cm sec^{-1} throughout the rest of this book.

20-3 Different Kinds of Electromagnetic Waves

Table 20-1 lists the various kinds of electromagnetic radiation with the range of frequency and wavelength covered by each. The limits are only approximate and the ranges frequently overlap, so that radiation of a particular frequency might be described as either of two overlapping types depending upon the method of production. In figure 20-3 the wavelengths are compared with the sizes of various well-known objects.

The electric power supplied by the utility company takes the form of an alternating current with a frequency of 60 cycles per second. If this is looked upon as a source of electromagnetic radiation, the wavelength is:

$$\text{Wavelength} = \frac{\text{Velocity}}{\text{Frequency}} \tag{20-10}$$

$$= \frac{3 \times 10^{10}}{60} \tag{20-11}$$

$$= 5 \times 10^8 \text{ cm} \tag{20-12}$$

Table 20-1 Types of Electromagnetic Radiation

Type	Frequency cycles per sec	Wavelength cm	Method of Production
Electric Power Utility	60	5×10^8	Electrical machinery
AM Radio	0.5×10^6 to 2×10^7	6×10^4 to 1.5×10^3	Electronic
TV and FM Radio	4×10^7 to 2×10^8	7.5×10^2 to 1.5×10^2	Electronic
Microwaves	10^9 to 3×10^{11}	30 to 0.1	Special vacuum tubes
Infrared	3×10^{11} to 4.3×10^{14}	0.1 to 7×10^{-5}	Hot bodies
Visible Light	4.3×10^{14} (Red) to 7.5×10^{14} (Violet)	7×10^{-5} to 4×10^{-5}	Lamps, sun
Ultraviolet	7.5×10^{14} to 10^{16}	4×10^{-5} to 3×10^{-6}	Special lamps, very hot bodies
x-rays	10^{16} to 3×10^{20}	3×10^{-6} to 10^{-10}	Collision of electrons with targets
γ-rays and Bremsstrahlung	10^{18} to ? (at least 5×10^{24})	3×10^{-8} to 6×10^{-15} (or lower)	Nuclear reactions, accelerators

This is comparable with the size of the earth and so it is not surprising that we do not normally concern ourselves with the wave properties at this low frequency.

The frequencies used in AM broadcasting range from about 0.5×10^6 to 2×10^7 sec^{-1}. The wavelengths lie in the range from about one half a mile down to the size of a house. Somewhat higher frequencies are used for FM radio and television, extending up to 2×10^8 sec^{-1}. The shortest wavelength is about the size of a man. All these **radio waves** are produced by oscillating electrons in metal aerials. The procedure for inducing the electrons to oscillate involves electronic vacuum tubes and electronic circuits.

Radio waves with frequencies between 10^9 and 3×10^{11} sec^{-1} are called **microwaves**. Their wavelengths range from the size of a cat to the

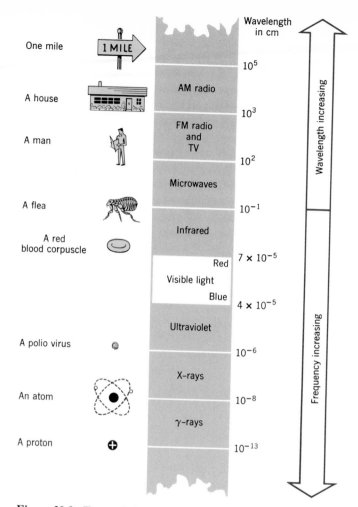

Figure 20-3 Types of electromagnetic radiation. The wavelength scale has been distorted in order to fit in the information conveniently.

size of a flea. They are produced by special vacuum tubes called magnetrons and klystrons. Microwaves are used in radar, the principle of which is illustrated in figure 20-4. The transmitter is switched on for only a short time and therefore sends out a "pulse" of radiation or a "wave-packet." This wave-packet is reflected from the object being viewed and returns to the radar set. The time spent in flight enables the distance of the object to be deduced. This technique is now being used to measure the distance between the earth and the planets and to map out the solar system more accurately.

Electromagnetic radiation visible to the human eye covers only a small range of wavelengths in the vicinity of 5×10^{-5} cm. This is the size of the smallest object that can be seen in an optical microscope. This is no coincidence, because there is a fundamental theory of the behavior of microscopes that reveals that the smallest object that can be

Electro-
magnetic
Waves

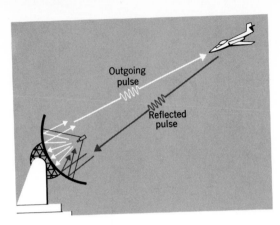

Figure 20-4 Radar.

seen is comparable in size with the wavelength of the electromagnetic radiation being used. The longest wavelength of **visible light** is red. As the wavelength is decreased, the color changes through orange, yellow, green, blue to violet, which has the shortest wavelength. An interesting question is why these particular wavelengths of electromagnetic radiation should have been chosen for human vision. Part of the answer is that these particular frequencies are strongly emitted by the sun but are not appreciably absorbed by the earth's atmosphere. There are radio waves that also satisfy this condition, but their very long wavelength makes them unsuitable for human vision because of the general principle that two objects cannot be distinguished from one another if their distance apart is much smaller than the wavelength of the radiation being used to view them.

Infrared radiation has wavelengths intermediate between microwaves and red light. It is sometimes called "heat radiation" because it is strongly emitted by a fire or a central heating radiator and is readily absorbed by the human body to produce a sensation of warmth. A radiator emits no visible light and it cannot be seen in an unlit room, but the heat radiation from it can easily be felt. However, the term "heat radiation" is not a good one because all frequencies of electromagnetic radiation are emitted by a hot body and all can be absorbed by matter and warm it up. Imagine a steel ball hanging in a dark room. If the ball is at the same temperature as the room (about 300° K) it is not visible. It is actually emitting electromagnetic radiation of all frequencies from zero to infinity, but very little is emitted in the visible region. Most of its radiation has a frequency in the vicinity of a certain "frequency of maximum radiation," ν_m, which in this case is about 1.8×10^{13} sec^{-1}, corresponding to a wavelength of 1.7×10^{-3} cm. This is in the "far" infrared, well below the frequency of red light (4.3×10^{14} sec^{-1}). If the temperature of the ball is increased, the radiation it emits at *any* frequency increases, but ν_m moves toward higher frequencies. The majority of its radiation is emitted in the vicinity of this frequency, ν_m, which is linearly proportional to the absolute temperature

$$\nu_m = 5.9 \times 10^{10} T \qquad\qquad (20\text{-}13)$$

Therefore, as the temperature is raised above room temperature, the ball soon begins to emit light near the red end of the spectrum and becomes "red hot." At still higher temperatures ν_m is located in the middle of the visible region. The ball then emits all colors strongly and becomes "white hot." At very high temperatures, blue light is emitted preferentially and the ball is "blue hot."

Radiation with frequencies greater than 7.5×10^{14} sec^{-1}, beyond the violet end of the visible spectrum, is called **ultraviolet light.** It is produced by special lamps or very hot bodies. The sun emits a large amount of ultraviolet light, but most of it is absorbed by the earth's atmosphere.

X-rays are produced when a beam of fast moving electrons hits a metal target. Figure 20-5 is a diagram of an x-ray tube. An evacuated glass tube has sealed into it two metal electrodes, between which is applied a large electric potential difference, usually several thousand volts, but sometimes as much as a million volts. Electrons leaving the negative electrode are accelerated toward the positive electrode and arrive there with very high velocities. Upon colliding with the atoms of the positive electrode, sometimes called the "target," the electrons are brought abruptly to rest. The consequent large negative acceleration results in the radiation of very high frequency electromagnetic waves. The upper limit of about 3×10^{20} sec^{-1} for the frequency is determined by the highest voltage that it is practicable to apply across such a tube. X-ray wavelengths range from about 3×10^{-6} cm to 10^{-10} cm. The size of an atom is about 10^{-8} cm, which lies in the middle of this range of wavelengths.

The well-known property of x-rays to "see inside" solid bodies, such as the human body, depends upon their ability to pass through matter which is opaque to visible light. The intensity of x-rays is attenuated

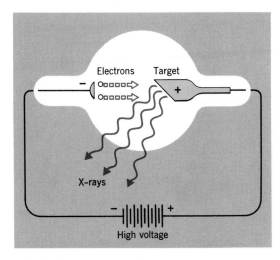

Electro-
magnetic
Waves

Figure 20-5 An x-ray tube (simplified).

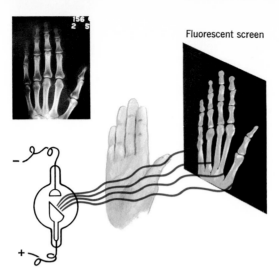

Figure 20-6 X-ray radiography.

slightly upon passage through matter and the absorption is greater for substances that are more dense and contain elements of greater atomic mass. If a beam of x-rays passes through the hand, there is very little absorption by the flesh but much more by the bones. If the beam then falls upon a screen coated with a suitable material, the x-rays cause this material to fluoresce and emit visible light, but a shadow is cast by the bones (figure 20-6).

γ-rays are emitted by radioactive nuclei. Their frequencies range from about 10^{18} sec^{-1} upward and therefore overlap with x-rays. The large machines that accelerate electrons or protons to enormously high energies have produced γ-rays with frequencies up to about 5×10^{24} sec^{-1}. This corresponds to a wavelength of 6×10^{-15} cm, which is about $\frac{1}{20}$ the diameter of a proton.

The division of the electromagnetic spectrum into regions in this manner is largely a matter of convenience depending upon the experimental technique used. Radiation with a wavelength of 0.1 cm produced by a Klystron (a type of electronic vacuum tube) would be described as microwave radiation, but identical radiation separated out from the radiant heat of a hot body would be described as far infrared. Radiation with a wavelength of 3×10^{-6} cm could be produced by a voltage of 100 volts across an x-ray tube and would then be called a "soft" x-ray, but it could equally well be produced by an ultraviolet lamp. A voltage of 100,000 volts across an x-ray tube would produce x-rays with a wavelength of 3×10^{-9} cm, but identical radiation might be emitted by a radioactive substance and it would then be called γ-rays. When very high frequency radiation is produced by the deceleration of electrons with energies of 10^9 electron volts, it is a moot point whether it should be called x-rays or γ-rays. Actually the term x-rays is reserved for lower frequency radiation produced by a tube similar to that shown in figure 20-5. Very high frequency radiation is usually described as γ-rays. The γ-rays produced by deceleration of high energy charged particles are often called "bremsstrahlung" (German for "braking radiation").

20-4 Energy and Momentum of Electromagnetic Radiation

Since the radiation from the sun warms us up, electromagnetic waves clearly carry energy with them. They also carry momentum. Let us consider how this comes about.

Reverting to the problem of the magnetic forces between two moving charges (Chapter 16), suppose that the charge q_1 has a velocity v_1 perpendicular to the line joining the two charges, whereas q_2 has a velocity v_2 along this line (figure 20-7). First calculate the magnetic force exerted by q_1 on q_2. According to section 16-2, the magnetic *field* produced at q_2 by q_1 is

$$B_{12} = \frac{q_1}{R^2} \frac{v_1 \sin \theta}{c} \tag{20-14}$$

θ is the angle between $\mathbf{v}_1$ and the line joining q_1 to q_2, and is therefore 90°. So $\sin \theta = 1$ and

$$B_{12} = \frac{q_1 v_1}{R^2 c} \tag{20-15}$$

Right hand rule number 1 (page 314) tells us that $\mathbf{B}_{12}$ goes perpendicularly into the plane of the paper. According to section 16-3, the force exerted by this field on q_2 is

$$F_{12} = q_2 B_{12} \frac{v_2}{c} \sin \phi \tag{20-16}$$

ϕ is the angle between $\mathbf{v}_2$ and $\mathbf{B}_{12}$, and is 90°. So $\sin \phi = 1$ and

$$F_{12} = q_2 B_{12} \frac{v_2}{c} \tag{20-17}$$

Substituting the value of B_{12} from equation 20-15

$$F_{12} = \frac{q_1 q_2}{R^2} \frac{v_1 v_2}{c^2} \tag{20-18}$$

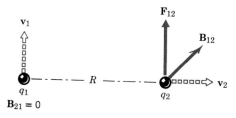

$\mathbf{F}_{12}$

$\mathbf{v}_1$

$\mathbf{B}_{12}$

q_1

$B_{21} = 0$

— R — q_2 $\mathbf{v}_2$

Electro-
magnetic
Waves

Figure 20-7 Illustrating the failure of Newton's third law of motion for magnetic forces.

417

Right hand rule number 2 (page 318) tells us that this force acts upward toward the top of the page.

Now consider the magnetic force exerted by q_2 on q_1. The magnetic field $\mathbf{B}_{21}$ produced at q_1 by the motion of q_2 is given by an equation analogous to equation 20-14. However, θ is now the angle between v_2 and the line joining q_2 to q_1, and this angle is $180°$. Since $\sin 180° = 0$,

$$B_{21} = 0 \tag{20-19}$$

and q_2 exerts no magnetic force on q_1!

The magnetic force exerted by q_1 on q_2 is given by equation 20-17 and is not zero and is therefore not equal and opposite to the force exerted by q_2 on q_1. Magnetic forces do not obey Newton's third law (section 6-6). Force is equal to rate of change of momentum (section 8-1). The magnetic force exerted by q_1 on q_2 therefore gives q_2 an extra upward momentum, but there is no compensating change in the momentum of q_1. The law of conservation of momentum is apparently violated. For similar reasons there is also an apparent violation of the law of conservation of energy.

The way out of this dilemma is as follows. Suppose that the two charges are free to move subject to no external influence other than the forces that they exert on one another. Because they do exert forces on one another, they both accelerate, and an accelerating charge radiates electromagnetic waves. The electromagnetic radiation carries away the energy and momentum that was apparently lost in the previous analysis. We can reinstate the laws of conservation of energy and momentum if we make the following assumptions about the energy and momentum carried away by the radiation. Remembering the picture of an electromagnetic wave given in figure 20-1, suppose that the instantaneous electric and magnetic fields produced by the wave near a point P in space are $\mathbf{E}$ and $\mathbf{B}$. The energy and momentum are being transported in the direction in which the wave is traveling, which is perpendicular to both $\mathbf{E}$ and $\mathbf{B}$. Let S be the energy and P the momentum moving in *one second* across a surface of *unit area* near P perpendicular to the direction of propagation. Then:

Transfer of energy and momentum by an electromagnetic wave

Energy crossing unit area per second

$$S = \frac{cEB}{4\pi} \tag{20-20}$$

Momentum crossing unit area per second

$$P = \frac{S}{c} \tag{20-21}$$

$$= \frac{EB}{4\pi} \tag{20-22}$$

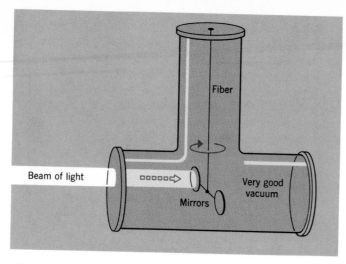

Figure 20-8 A demonstration of the pressure of light.

For the simple type of wave illustrated in figure 20-1, $E = B$ and we may write

$$S = \frac{cE^2}{4\pi} = \frac{cB^2}{4\pi} \tag{20-23}$$

$$P = \frac{E^2}{4\pi} = \frac{B^2}{4\pi} \tag{20-24}$$

When electromagnetic radiation falls on a material body and is absorbed by it, energy is transferred to the body and warms it up. At the same time momentum is also transferred to the body and a force is exerted on it. This is often called the **pressure of light.** It can be verified experimentally by direct measurement of the force produced when light falls on a surface. In the apparatus illustrated in figure 20-8, the two mirrors are suspended on a thin fiber inside a very good vacuum. If a beam of light falls on one of the mirrors, the pressure it exerts is seen to twist the suspended system.

The following simple example will give a good idea of why electromagnetic waves carry momentum. Figure 20-9 shows an electromagnetic wave incident on a proton and producing fields **E** and **B** in its vicinity. The field **E** exerts an upward force on the proton and gives it an upward velocity **v**. Once it has acquired this velocity, the magnetic field **B** exerts a magnetic force on the proton. Application of right hand rule number 2 shows that this force is in the direction in which the wave is traveling. It is in fact the pressure of light. The proton acquires momentum in the direction of propagation and this must have come from the electromagnetic wave. The reader should verify that the force is always in the forward direction of propagation even when the electric field is downward and the magnetic field goes into the plane of the paper. If the charge in figure 20-9 were negative (an electron, say), it would be given

Electro-
magnetic
Waves

419

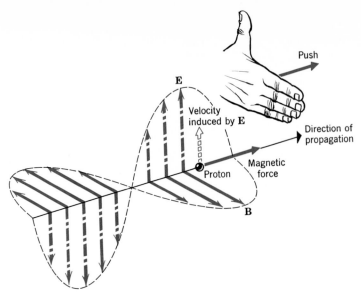

Push

E

Velocity
induced by E

Direction of
propagation

Magnetic
force

Proton

B

Figure 20-9 The origin of the pressure of light.

a downward velocity by **E**, but **B** would still exert a forward force on this
downward moving negative charge.

Example 20-4-1

Calculate the pressure exerted by a 100 watt light bulb on black wallpaper 100 cm away from it. Assume that all the electric power is converted into electromagnetic radiation which is emitted equally in all
directions.

The energy emitted in 1 second is 100 joules or 10^9 ergs. After traveling a distance R, this is spread uniformly over a spherical surface of
area $4\pi R^2$. Putting $R = 100$ cm, the energy falling on unit area of the
wallpaper per second is

$$S = \frac{10^9}{4\pi R^2} \tag{20-25}$$

$$= \frac{10^9}{4\pi \times 10^4} \tag{20-26}$$

$$= 7.96 \times 10^3 \text{ erg sec}^{-1} \text{ cm}^{-2} \tag{20-27}$$

The momentum falling on unit area per second is

$$P = \frac{S}{c} \tag{20-28}$$

$$= \frac{7.96 \times 10^3}{3 \times 10^{10}} \tag{20-29}$$

$$= 2.65 \times 10^{-7} \text{ CGS units} \tag{20-30}$$

WAVES

420   Black wallpaper was chosen because it absorbs all the radiation incident

on it. P is therefore the momentum given up to unit area of the wall per second, which is the same thing as the force per unit area, or the pressure (page 201).

Pressure of radiation on wall

$$= 2.65 \times 10^{-7} \text{ dyne cm}^{-2} \tag{20-31}$$

The pressure exerted by the air on the wall (atmospheric pressure) is about 10^6 dyne cm^{-2}. The pressure of the lamp is negligible compared with this (but see problem 20-10).

QUESTIONS

1. An electromagnetic wave has a horizontal velocity directly toward you. At a certain instant the electric field at a point is vertically downward. What is the direction of the magnetic field?
2. In an experiment similar to Fizeau's for measuring the velocity of light (figure 20-2), the rate of rotation of the wheel is slowly increased from zero and the light disappears and then reappears again. What will happen if the rate of rotation is steadily increased beyond this point?
3. Which of the following has the lowest frequency? (a) γ-rays, (b) blue light, (c) infrared light, (d) ultraviolet light.
4. Which of the following has the shortest wavelength? (a) radio waves, (b) x-rays, (c) red light, (d) ultraviolet light?
5. If the wavelength were comparable to the size of an apple, what type of electromagnetic radiation would it be?
6. The unit of length is now defined in terms of the wavelength of a certain spectral line and the unit of time in terms of the period of a different spectral line. Is this equivalent to *defining* the value of the velocity of light, c? What is the significance of an experimental determination of the value of c? Would it make any difference if the units of length and time were defined in terms of the same spectral line?
7. Is Newton's third law of motion applicable to the magnetic forces between two electrons moving in the same direction with different speeds v_1 and v_2 (a) if the line joining them is perpendicular to their velocities, (b) if the line joining them is not perpendicular to their velocities?
8. As the sun radiates away energy does it also lose momentum?
9. Two billiard balls collide inelastically. A small portion of each ball near the point of contact is warmed by the collision. Suppose that the red ball rapidly radiates away the heat before it has had time to spread to the rest of the ball. The white ball is less efficient at radiating heat and the heat spreads uniformly through the ball before it is radiated away. Can you think of a reason why the collision might appear to violate the law of conservation of momentum? Can you think of any situation in classical mechanics in which momentum

is rigorously conserved and no objection of this kind can be advanced even if one is very ingenious?

10. Continuing the discussion of the situation in figure 20-7, the magnetic force on a charge is always perpendicular to its velocity and therefore cannot do work. The failure of the magnetic forces to obey Newton's third law of motion is therefore irrelevant to the question of whether energy is conserved. Show that there is still reason to doubt conservation of energy if one uses the exact expression of equation 16-5 to calculate the electric forces. (Strictly speaking the analysis would still be inadequate because equations 16-4 and 16-5 apply only to charges with constant velocities. Even more complicated expressions must be used for accelerating charges. However, the final result is that energy and momentum are not necessarily conserved unless the electromagnetic radiation is taken into account.)

PROBLEMS

A

1. For an electromagnetic wave similar to the one shown in figure 20-1 the electric field at a point can be represented by

$$E = 0.03 \cos\left(2\pi\nu t + \frac{\pi}{6}\right) \text{ statvolt/cm}$$

Find a similar expression for the magnetic field at this point in gauss.

2. Approximately how long does it take an electromagnetic wave to travel (a) from this book to your eyes, (b) one city block, (c) to your radio receiver from a transmitting station ten miles away, (d) from New York to Los Angeles, (e) once round the equator, and (f) from the sun to the earth?

3. In a Fizeau type apparatus for measuring the velocity of light (figure 20-2), the wheel has 256 teeth. The light first reappears at maximum brightness when the angular velocity is 1.5×10^4 rad sec^{-1}. What must be the distance from the mirror to the wheel?

4. What must be the temperature of the surface of a hot body if the radiation which it emits consists mainly of (a) radio waves with wavelengths near 1 kilometer, (b) microwaves with wavelengths near 10 cm, (c) infrared radiation with wavelengths near 10^{-3} cm, (d) blue light with wavelengths near 5×10^{-5} cm, (e) x-rays with wavelengths near 10^{-8} cm, (f) γ-rays with wavelengths near 10^{-13} cm?

5. The electric vector of a simple electromagnetic wave (figure 20-1) has a magnitude of 1.5×10^{-2} statvolt/cm. Calculate (a) the energy and (b) the momentum crossing unit area per second.

6. As you lie on the beach sunbathing, a typical energy flow for the sunlight might be 10^6 erg cm^{-2} sec^{-1}. Calculate the corresponding electric field.

7. If the energy flow of the sunlight is 10^6 erg cm^{-2} sec^{-1}, calculate the pressure exerted on a surface which is perpendicular to the

sun's rays and which absorbs all the light falling on it. Compare it
with atmospheric pressure.

B

8. What would the magnitude of the electric field of a light wave have
to be if it exerted a pressure equal to atmospheric pressure on a
totally absorbing surface perpendicular to the direction of propagation.

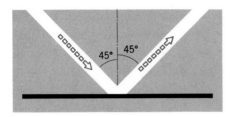

Problem 20-9

9. A beam of light is incident on a surface at an angle of 45° and is
totally reflected. The angle of reflection is equal to the angle of incidence (see diagram). If the electric field is 0.1 statvolt/cm, what is
the pressure on the surface?
10. Using a device known as a laser it is possible to produce very intense radiation with electric fields of the order of 10^8 volt/cm. If
this radiation falls perpendicularly on to a totally absorbing surface,
calculate the pressure in atmospheres.

C

11. In the vicinity of the earth the energy flow of electromagnetic
radiation from the sun is 1.4×10^6 erg cm^{-2} sec^{-1}. Calculate the
pressure in dyne cm^{-2} and in atmospheres on a totally absorbing
surface perpendicular to the light near the surface of the *sun*.
12. An astronaut of mass 100 kg has fallen out of his space ship and is
stranded in gravity-free space. He possesses only a flashlight with a
6 watt bulb and an inexhaustible supply of batteries. If he keeps the
flashlight turned on and always points it in the same direction, how
long will it take him to increase his velocity by 60 mph. You may
assume that all the energy going into the bulb is transformed into
electromagnetic radiation and that the reflector behind the bulb
produces a parallel beam with negligible spread. Express your
answer in whatever units seem appropriate to assess the feasibility
of the maneuver.

Electro-
magnetic
Waves

21

Behavior
Characteristic
of
Waves

21-1   Passage of a Wave Through a Very Narrow Slit

Newton believed light was a stream of particles. His point of view was the more popular one until the beginning of the nineteenth century, when the opposing point of view, that light is a wave, was supported by a series of experiments of the kind we shall describe in this chapter. This controversy concerning the nature of light is related to the two different views we have presented of the nature of the universe. Newton's point of view is consistent with the idea that the universe is nothing more than a collection of particles located at points in space. A wave, however, occupies a whole region of space. Since a light wave can travel through a vacuum, it must either be described by assigning properties such as the electric and magnetic fields $\mathbf{E}$ and $\mathbf{B}$ to all the points in the region of space, or else even a vacuum is filled with some mysterious substance such as the jelly-like ether mentioned in section 14-1.

The properties of waves are very conveniently demonstrated with a ripple tank (figure 21-1). The ripples are formed on shallow water in a tank with a glass bottom. The tank is illuminated from above and the ripples cast a shadow on a screen of white paper below the glass bottom. Those parts of the screen in line with the crests of the wave are bright, whereas parts in line with the troughs are dark. A convenient way to excite the ripples is to dip a thin rod into the water and oscillate it up and down with the required frequency. The wave is then seen on the screen as a series of alternating black and white circular bands centered on the rod (figure 21-2). These bands move outward from the rod with the velocity of the wave.

If the lamp above the tank is switched on and off with the same period as the wave, then in between flashes of light one crest has just

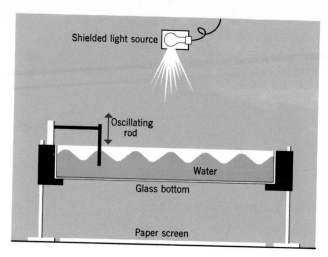

Figure 21-1 A ripple tank.

enough time to move into the position originally occupied by the crest ahead of it. The bands of light then appear to be stationary.

A single wave crest is a circle joining all adjacent points in the surface where the upward displacement of the water is greatest. A trough is a circle joining all adjacent points in the surface where the *downward* dis-

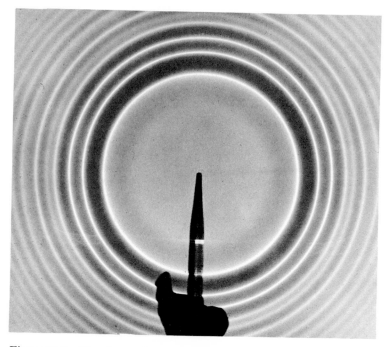

Figure 21-2 Circular ripples on a ripple tank. Actually, this photograph was not obtained by dipping an oscillating rod into the water, but by allowing a single drop of water to fall into the tank. (Photograph by Berenice Abbott.)

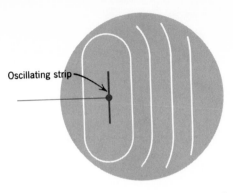

Figure 21-3 A wavefront that is not circular.

placement of the water is greatest. A circle of this kind is called a **wave-front**. It may be defined more generally as a line joining all adjacent points in the surface which have the same phase (see section 19-1). The wavefronts travel outward from the source of the wave with the wave velocity. For a three dimensional wave the wavefronts are surfaces, not lines. If a bomb is exploded in mid-air, the sound of the explosion travels outward in all directions. Sound consists of oscillations in the pressure of the air. Adjacent points in the air at which the pressure has reached its maximum value all lie on a spherical surface that expands outward from the point of the explosion. Wavefronts do not have to be circular or spherical. If a vertical strip of metal dips into the ripple tank and oscillates up and down, it sends out wavefronts that are almost straight in the center but bend near the edges of the strip (figure 21-3).

Now suppose that two barriers are placed in the ripple tank in line with one another, but with a very narrow opening between them (figure 21-4). Those parts of a wavefront which strike the barriers are reflected backward, but the very small region of the wavefront incident on the gap passes through. However, it does not pass through undisturbed.

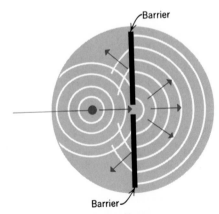

Figure 21-4 Passage of a circular ripple through a very narrow opening.

Whatever the shape of the incident wavefronts, the wavefronts on the other side are circles. Moreover these circles are centered on the gap, not on the original source. This property of a wave which makes it spread out after passing through a narrow opening is called **diffraction** and is one of the most distinctive characteristics of the behavior of waves.

A good way to visualize what is happening is as follows. Imagine each *very small* portion of a wavefront to be a disturbance that acts as the source of an outgoing circular **wavelet**. The barriers of figure 21-4 block out everything except the wavelet from the small portion of the wavefront at the gap, and only this wavelet is seen on the other side of the gap. In the absence of the barriers, however, each small portion of the wavefront sends out its own circular wavelet and the wavelets from all parts of the wavefront add together to produce a new wavefront centered on the original source (figure 21-5). This is, in fact, the way in which the original wavefront is maintained on its outward motion.

The wavelets that have just been described were invented by Christiaan Huygens and are frequently called **Huygens' wavelets**. The wave theory of light was advocated by Huygens and Robert Hooke in the seventeenth century, at the same time that Newton was putting forward his particle theory of light. The phenomenon of diffraction had already been discovered by Francesco Maria Grimaldi in the early part of the century, but this fact was largely ignored in the controversy. To Huygens, the major problem was to explain the fact that light appears to travel in straight lines. Objects cast sharp shadows. It is not possible to see round a corner, although it is possible to hear a source of sound out of sight

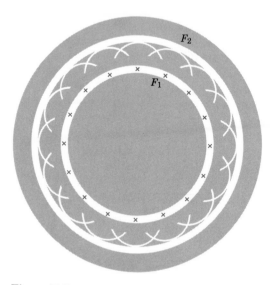

Figure 21-5 Huygens' wavelets adding together to form a new wavefront. F_1 is the original wavefront. Points such as those marked X on F_1 act as sources of Huygens' wavelets. These wavelets add together to form the new wavefront F_2, which has propagated ahead of F_1.

Figure 21-6 The pedestrian cannot see the newspaper vendor, but he can hear him. The sound waves bend around the corner. The light apparently does not.

round the corner (figure 21-6). Huygens used his wavelets in an attempt to show how waves might appear to be traveling in straight lines. As we shall properly appreciate later in this chapter, the key to this difficulty is the very short wavelength of visible light. The idea that light travels in straight lines is called the **rectilinear propagation of light**. In this chapter we shall be concerned with phenomena in which the rectilinear propagation of light breaks down and the behavior of the light is characteristic of a wave. At the same time we shall be concerned to show

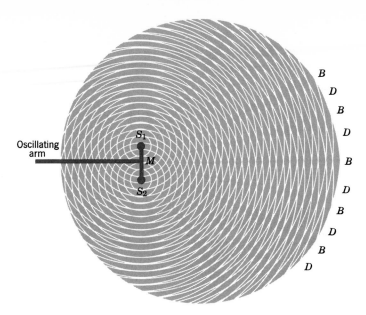

Figure 21-7 Interference of ripples from two neighboring rods S_1 and S_2 attached to the same oscillating arm.

how, under most circumstances, the rectilinear propagation of light is approximately true.

21-2   Passage of a Wave Through Two Parallel Slits: Interference

The wave nature of light was established by a very dramatic experiment performed by Thomas Young in 1801. To understand this experiment, let us again consider the ripple tank with two probes oscillating in unison side by side (figure 21-7). Each probe is the center of a set of circular ripples, but the two sets of ripples overlap and interfere with one another. When two crests coincide they augment one another and produce a crest twice as high as either. When two troughs coincide they produce a trough twice as deep as either. However, if a crest from one source coincides with a trough from the other source, they cancel one another and the surface of the water is neither raised nor lowered.

The net result is shown in figure 21-7, which is based on a diagram given by Young in his original account. If this diagram is viewed by placing the eye near the right edge of the page and looking across the diagram at a grazing angle, it gives the appearance of a bunch of alternately bright and dark beams radiating outward from a point M midway between the two sources. A line starting at M and ending at a point marked B passes through the middle of a bright beam. Along such a line crests coincide with crests and troughs coincide with troughs. The bright beams therefore represent regions of the surface that are oscillating up and down with enhanced amplitude. A line starting at M and ending at a point marked D passes through the middle of a dark beam.

Behavior
Characteristic
of Waves

429

Along such a line crests always coincide with troughs. The dark beams therefore represent regions of the surface where the wave amplitude is reduced. The result of interference between the two sources is therefore to produce beams moving radially outward from a point midway between the two sources.

Figure 21-8 is a photograph of a ripple tank with two such sources. This photograph shows very clearly the diverging beams of ripples separated by regions in which the surface is almost undisturbed.

Young's achievement was to perform this experiment with light. To avoid non-essential complications, we shall not describe Young's original experiment, but a simplified version of it (figure 21-9). Light from a lamp is passed through a filter that transmits only one color, giving an electromagnetic wave of definite frequency. The light then passes through a long, very narrow slit S_0 and, in a similar way to the ripples of figure 21-4, emerges with cylindrical wavefronts. It then passes on to a screen in which there are two *narrow*, parallel slits S_1 and S_2 very close together. Each slit acts as a source of cylindrical wavefronts, but the waves from the two sources interfere with one another in the same way as the ripples from the two sources in figure 21-7.

Figure 21-8 Photograph of the interference between the ripples from two neighboring sources in a ripple tank. The striped beams, in which the white crests and black troughs are clearly seen, correspond to the bright beams of figure 21-7. The uniformly grey beams, where no crests or troughs can be seen, correspond to the dark beams of figure 21-7. (Reproduced with permission from *Physics*, by the Physical Science Study Committee, D. C. Heath and Co.)

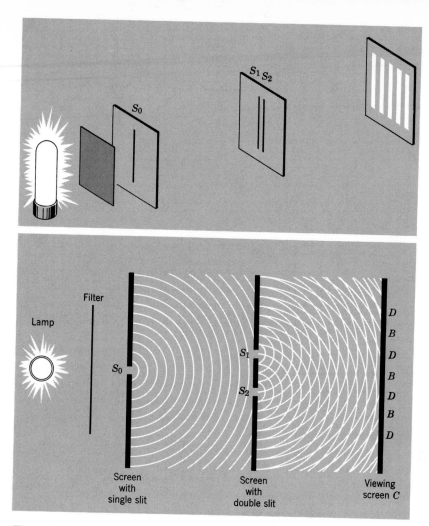

Figure 21-9 Young's double slit experiment.

Figure 21-9 may also be viewed at grazing incidence in the same way as figure 21-7. The diverging bright and dark beams will again be clearly seen. A bright beam represents a region where the oscillating electric fields originating from S_1 and S_2 are in the same direction and reinforce one another to produce an intense electromagnetic wave. When a bright beam hits the screen C at a point marked B, it illuminates it and produces a bright strip on the screen. However, the dark beams represent regions where the electric fields from S_1 and S_2 are in opposite directions and cancel one another. A point D on the screen struck by a dark beam is therefore dark. The interference pattern seen on the screen consists of a set of parallel bright strips corresponding to the points B, separated by dark strips corresponding to the points D. These are called **interference fringes.** A photograph of a set of interference fringes is shown in figure 21-10. The formation of such fringes is a

Behavior
Characteristic
of Waves

431

Figure 21-10   Interference fringes from a double slit. (Photograph by Dr. Brian Thompson.)

characteristic property of waves, but it is very difficult to explain it in terms of the particle theory of light.

21-3   Theory of the Double Slit Experiment

Referring to figure 21-11, suppose that the two slits, or sources, S_1 and S_2, are a distance d apart. The distance from the slits to the screen C is L. M is the point midway between S_1 and S_2 and N is the point on the screen C directly opposite M. T_1 and T_2 are points on the screen directly opposite S_1 and S_2.

Consider what is happening at a point P on the screen a distance y from the midpoint N. The wavelet arriving at P from S_1 has traveled a distance S_1P, but the wavelet from S_2 has traveled a greater distance S_2P. The wavelet from S_2 therefore arrives later than the wavelet from S_1 and the way in which the two combine depends upon this time lag, as shown in the inserts to the right of the screen in figure 21-11.

The wavelets arriving at the midpoint N have traveled the same distance and they arrive at the same time. They are therefore in phase and augment one another to produce a bright fringe at N. As P moves upward from N, the *path difference* $(S_2P - S_1P)$ increases until, when P is at D_1, it is equal to $\lambda/2$. The wavelet from S_2 then has to travel an extra half wavelength and arrives half a period behind the wavelet from S_1. The two wavelets are then exactly $180°$ *out of phase* and cancel one another, as shown in the insert to the right of D_1. D_1 is the center of a dark fringe. As P moves further away from N, it reaches a point B_1 where $(S_2P - S_1P)$ is equal to one whole wavelength, λ. As the insert shows, the two wavelets again augment one another and B_1 is the center of a bright fringe. Continuing this type of argument, at D_2 the wavelet from S_2 travels an extra distance $3\lambda/2$ and here there is a dark fringe. At B_2 the path difference is 2λ and there is a bright fringe, and so on. Similar arguments apply to the formation of fringes below N, except that in this region of the screen the wavelet from S_1 arrives later than the wavelet from S_2.

In order to determine the positions of the bright and dark fringes, it is clear that we must calculate the path difference $(S_2P - S_1P)$. Applying Pythagoras' theorem to the right angled triangle S_2PT_2

WAVES

$$S_2P^2 = T_2P^2 + S_2T_2^2 \tag{21-1}$$

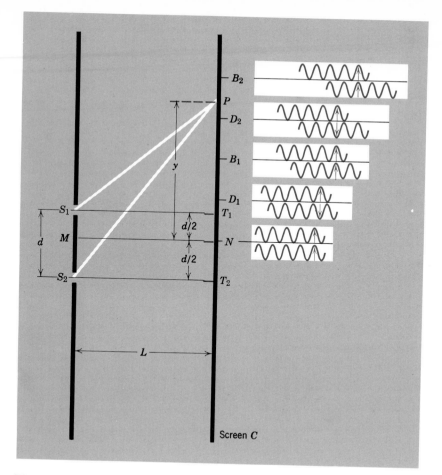

Figure 21-11 Illustrating the theory of the double slit experiment.

Similarly, for the right angled triangle S_1PT_1

$$S_1P^2 = T_1P^2 + S_1T_1^2 \qquad (21\text{-}2)$$

Therefore,

$$S_2P^2 - S_1P^2 = (T_2P^2 + S_2T_2^2) - (T_1P^2 + S_1T_1^2) \qquad (21\text{-}3)$$

But $\quad T_2P = NP + T_2N \qquad (21\text{-}4)$

$$= y + \frac{d}{2} \qquad (21\text{-}5)$$

And $\quad T_1P = NP - NT_1 \qquad (21\text{-}6)$

$$= y - \frac{d}{2} \qquad (21\text{-}7)$$

Also $\quad S_2T_2 = S_1T_1 = L \qquad (21\text{-}8)$

So equation 21-3 may be rewritten

$$S_2P^2 - S_1P^2 = \left(y + \frac{d}{2}\right)^2 + L^2 - \left(y - \frac{d}{2}\right)^2 - L^2 \qquad (21\text{-}9)$$

This simplifies to

$$S_2P^2 - S_1P^2 = 2\,dy \qquad (21\text{-}10)$$

But

$$S_2P^2 - S_1P^2 = (S_2P - S_1P)(S_2P + S_1P) \qquad (21\text{-}11)$$

So

$$S_2P - S_1P = \frac{2\,dy}{(S_2P + S_1P)} \qquad (21\text{-}12)$$

In a double slit interference experiment with visible light, the slits must be close together and d is usually about 0.1 cm or less. The fringe pattern is small and the fringes are close together, so y is never much in excess of 0.1 cm. However, the distance L from the slits to the screen has to be very large and is usually at least 100 cm. Under these circumstances, S_2P and S_1P are each very nearly equal to L (figure 21-12) and very little error is introduced if we insert into equation 21-12

$$S_2P + S_1P = 2L \qquad (21\text{-}13)$$

Finally, then, the path difference we are trying to calculate is:

$$\text{Path difference, } (S_2P - S_1P) = \frac{dy}{L} \qquad (21\text{-}14)$$

There is a central bright fringe at N. The first bright fringe above N is at B_1, a distance y_1 from N, where $(S_2P - S_1P)$ is equal to λ. At this point

$$(S_2P - S_1P) = \frac{dy_1}{L} \qquad (21\text{-}15)$$

$$= \lambda \qquad (21\text{-}16)$$

$$\text{Whence } y_1 = \frac{\lambda L}{d} \qquad (21\text{-}17)$$

The second bright fringe above N is at B_2, a distance y_2 from N, where the path difference is equal to 2λ. At B_2

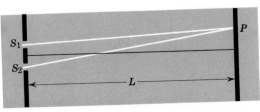

Figure 21-12 $(S_2P + S_1P)$ is not very different from $2L$.

$$(S_2P - S_1P) = \frac{dy_2}{L} \tag{21-18}$$

$$= 2\lambda \tag{21-19}$$

Whence $y_2 = 2\frac{\lambda L}{d}$ $\hspace{2cm}$ (21-20)

The third bright fringe corresponds to a path difference of 3λ:

$$(S_2P - S_1P) = \frac{dy_3}{L} \tag{21-21}$$

$$= 3\lambda \tag{21-22}$$

$$y_3 = 3\frac{\lambda L}{d} \tag{21-23}$$

In general, the nth bright fringe corresponds to a path difference of $n\lambda$

$$(S_2P - S_1P) = \frac{dy_n}{L} \tag{21-24}$$

$$= n\lambda \tag{21-25}$$

$$y_n = n\frac{\lambda L}{d} \tag{21-26}$$

The fringes are uniformly spaced and the distance between any two adjacent bright fringes, δ, is

Distance between adjacent bright fringes

$$\delta = \frac{\lambda L}{d} \tag{21-27}$$

Notice that the distance between the fringes is proportional to L and increases as the screen is moved further away. This is obvious if we look at figure 21-9 at grazing incidence, since the "beams" move further apart as the screen is moved further away from the slits.

The distance apart of the fringes increases as the wavelength λ increases. This is illustrated in figure 21-13, which shows two ripple tank photographs for two different wavelengths of the ripples. When the wavelength is very small the fringes are very close together and may be difficult to observe. The wavelength of visible light is about 5×10^{-5} cm, which explains why interference effects with light are not an obvious part of everyday experience and emphasizes the fact that Young's experiment was not easy and had to be thoughtfully designed and carefully executed.

Another interesting aspect of equation 21-27 is that the distance between the fringes is inversely proportional to d. *As the slits are moved closer together, the fringes move further apart.*

Behavior
Characteristic
of Waves

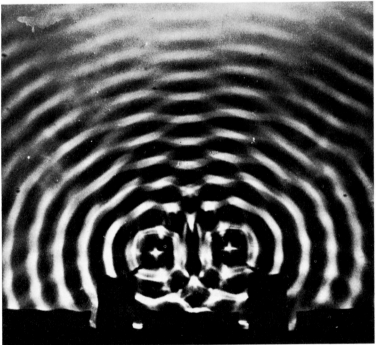

Figure 21-13 Two ripple tank photographs of interference with two different wavelengths. Notice that, when the wavelength is longer, the beams spread further apart. (Reproduced with permission from *Physics*, by the Physical Science Study Committee, D. C. Heath and Co.)

A simple rearrangement of equation 21-27 gives

$$\lambda = \frac{d\delta}{L} \qquad (21\text{-}28)$$

By measuring δ, the distance between the fringes; d, the separation of the slits; and L, the distance from the slits to the screen, it is possible to deduce the wavelength of the particular color of light being used.

Example 21-3-1

In a double slit interference experiment the distance between the slits is 0.05 cm and the screen is 2 meters from the slits. The light is yellow light from a sodium lamp and it has a wavelength of 5.89×10^{-5} cm. What is the distance between the fringes?

In the formula of equation 21-27

Distance between fringes,

$$\delta = \frac{\lambda L}{d} \qquad (21\text{-}29)$$

$$\lambda = 5.89 \times 10^{-5} \text{ cm} \qquad (21\text{-}30)$$

$$L = 200 \text{ cm} \qquad (21\text{-}31)$$

$$d = 0.05 \text{ cm} \qquad (21\text{-}32)$$

$$\text{So } \delta = \frac{5.89 \times 10^{-5} \times 200}{0.05} \qquad (21\text{-}33)$$

$$= 0.233 \text{ cm.}$$

21-4 Passage of a Wave Through a Single Slit of Any Width: Diffraction

At the beginning of this chapter we explained that a wave on a ripple tank incident on a very narrow opening emerges as a circular wave spreading in all directions. However, sunlight incident on a large hole in a blind passes through as a well defined beam and casts on the floor a patch of light that has sharp edges and reproduces the shape of the hole (figure 21-14). Is this evidence against the wave theory of light? The sunbeam is very easy to explain if the sunlight consists of a stream of particles traveling directly in straight lines from the sun through the hole in the blind to points on the floor. It is easy to understand why Newton favored a particle theory of light and why Huygens was concerned to prove that a wave theory is not inconsistent with the rectilinear propagation of light.

It is important to realize that, although the gap through which the ripples pass may be the same size as the hole in the blind, the wavelength of the ripples is very much longer than the wavelength of light. As we shall soon see, the wave will spread out in all directions, as in figure 21-14a, when the width of the slit is smaller than the wavelength of the wave. So let us consider what happens to a wave passing through a slit if the width of the slit is varied in comparison with the wavelength.

Behavior Characteristic of Waves

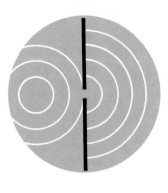
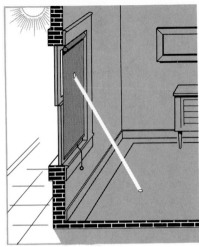

(a) Spreading of a ripple as it passes through a narrow gap.

(b) Sunlight streaming through a hole in a blind appears to travel in straight lines. (The hole is assumed to be about one inch wide.)

Figure 21-14   The contrast between diffraction and rectilinear propagation.

Diffraction of light by a slit was studied by Augustin Jean Fresnel during a series of experiments which followed soon after Young's research in the early nineteenth century, and which put the wave theory of light on a firm foundation. The principle of Fresnel's experiment is shown in figure 21-15. Light of a single color is incident on the narrow slit S_0 and emerges with cylindrical wavefronts, which are diffracted by the slit S_1. The resulting diffraction pattern is viewed on the screen C. If light consisted of particles moving in straight lines, the slit would cast a very sharp image on the screen C. The broken line on the screen is the outline of this image. In fact, however, the light spreads outside the bound-

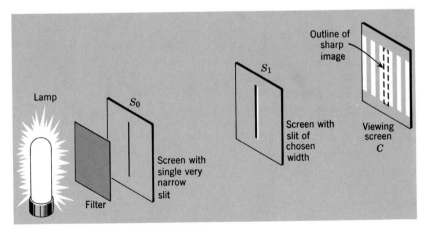

WAVES

438   **Figure 21-15**   Fresnel's experiment on diffraction by a single slit.

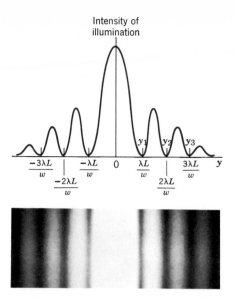

Figure 21-16 The diffraction pattern produced by a single slit. For the sake of clarity, the graph has been distorted. The ratio of the intensity of a side fringe to the intensity of the central fringe is actually even smaller than it appears to be on the graph. (Photograph by Brian Thompson.)

ary of this image and produces a set of **diffraction fringes.** The central fringe is bright and is wider than the sharp image. It is accompanied on either side by a set of fringes somewhat similar to interference fringes, except that the side fringes decrease noticeably in intensity as their distance from the central fringe increases. Moreover, the central fringe is twice as wide as the side fringes. Figure 21-16 is a photograph of such a set of fringes. The graph above this photograph shows how the intensity of illumination of the screen varies with distance from the center of the pattern.

A wide slit must not be treated as the source of a single Huygens' wavelet. Instead we must divide up the wavefront incident upon the slit into a very large number of very narrow strips, each of which can then be considered to give rise to a wavelet. Figure 21-17 shows all these wavelets traveling to the midpoint N of the diffraction pattern. For clarity the slit has been divided into only five strips, but it should really be divided into many more than this. Since the distance L to the screen is very large compared with the width of the slit w all the wavelets travel approximately the same distance and all arrive at N with the same phase. They therefore augment one another and the center of the pattern is very bright.

Figure 21-18a shows what happens at the center of the first dark fringe, D_1. It is convenient to divide the slit into two halves, each of width $w/2$, and, of course, each half must be divided into a very large number of very narrow strips. The strip at the top of the upper half is s_1 and the strip at the top of the lower half is t_1. The wavelet from t_1

Behavior
Characteristic
of Waves

439

arriving at D_1 has traveled further than the wavelet from s_1. By analogy with the calculation performed in section 21-3, leading to equation 21-14 the path difference is

$$t_1D_1 - s_1D_1 = \frac{\left(\frac{w}{2}\right)y_1}{L} \qquad (21\text{-}34)$$

$$= \frac{wy_1}{2L} \qquad (21\text{-}35)$$

The distance d between the slits in equation 21-14 has been replaced by the distance from s_1 to t_1, which is $w/2$, half the width of the single slit. The quantity y_1 is, of course, the distance of the point D_1 from the center of the diffraction pattern.

D_1 is the center of the first dark fringe if the wavelet from t_1 is exactly half a wavelength behind the wavelet from s_1. The two wavelets then arrive exactly out of phase and cancel one another. An identical argument shows that the wavelet from the strip s_2 just below s_1 is canceled by the wavelet from the strip t_2 just below t_1. In fact each strip in the upper half of the slit is canceled by a corresponding strip in the lower half of the slit and the net effect at D_1 is zero.

The condition that D_1 should be the center of the first dark fringe is therefore

$$t_1D_1 - s_1D_1 = \frac{\lambda}{2} \qquad (21\text{-}36)$$

$$= \frac{wy_1}{2L} \qquad (21\text{-}37)$$

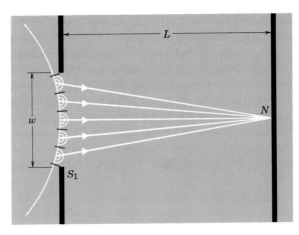

Figure 21-17   The wave-front incident on the slit S_1 is divided into narrow strips, each of which is the source of a Huygens' wavelet. For simplicity, only five strips are shown, but there should be many more than this. In an experiment with light, L is very much greater than w and the wavelets arriving at the midpoint N of the screen have all traveled approximately the same distance.

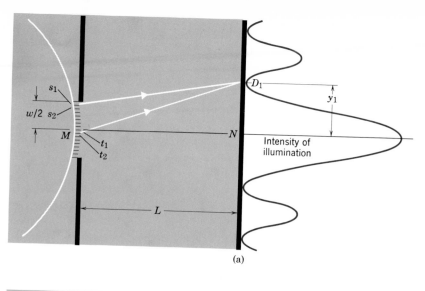

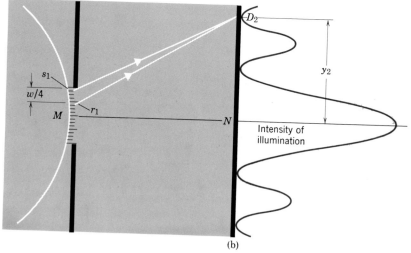

Figure 21-18 The procedure for considering how the Huygens' wavelets add together at (a) the center of the first dark fringe, and (b) the center of the second dark fringe.

or
$$y_1 = \frac{\lambda L}{w} \tag{21-38}$$

Now consider D_2, the center of the second dark fringe. The slit should then be divided into four equal parts and each part divided into a large number of very narrow strips (figure 21-18b). s_1 is again the strip at the top of the upper quarter and r_1 is the strip at the top of the next highest quarter. The path difference for the wavelets from s_1 and r_1 is

$$s_1 D_2 - r_1 D_2 = \left(\frac{w/4}{L}\right) y_2 \tag{21-39}$$

If this path difference is $\lambda/2$, s_1 cancels r_1 and every strip in the top quarter cancels a strip in the next highest quarter. Similarly, every strip in the third quarter is canceled by a strip in the lowest quarter. The condition for the second dark fringe is therefore

$$\frac{wy_2}{4L} = \frac{\lambda}{2} \qquad (21\text{-}40)$$

$$\text{or} \quad y_2 = \frac{2\lambda L}{w} \qquad (21\text{-}41)$$

To obtain the position of the center of the third dark fringe, the slit should be divided into six parts. A similar argument shows that the distance y_3 of the center of the third dark fringe from N, the center of the pattern, is

$$y_3 = \frac{3\lambda L}{w} \qquad (21\text{-}42)$$

The procedure can be continued to obtain the positions of all the fringes and, of course, the same procedure can be applied to the lower half of the diffraction pattern.

The values of y corresponding to the centers of the dark fringes have been marked on the graph in figure 21-16. The two following statements are readily understood after examining this graph carefully.

Diffraction by a slit

Width of central bright region

$$= 2y_1 = \frac{2\lambda L}{w} \qquad (21\text{-}43)$$

Width of a side fringe

$$= y_2 - y_1 = y_3 - y_2 \text{ etc.} \qquad (21\text{-}44)$$

$$= \lambda L/w \qquad (21\text{-}45)$$

The central bright fringe is twice as wide as the outer bright fringes. As in the case of interference, the spread of the pattern is proportional to the distance from the slit to the screen. The following two features of the pattern deserve special emphasis.

The spread of the diffraction pattern is proportional to the wavelength.
The spread of the diffraction pattern is inversely proportional to the width of the slit.

As the slit is made narrower, the image to be expected on the basis of the particle theory also becomes narrower, but the diffraction pattern actually observed becomes broader!

21-5 Diffraction and Rectilinear Propagation

As the slit is made narrower and narrower, the diffraction pattern becomes broader and broader until the central bright fringe occupies the whole of the screen. Consider the special case when $w = \lambda$. If we put $w = \lambda$ in equation 21-38, the distance of the first dark fringe from the center of the pattern would be

$$y_1 = L \qquad\qquad\qquad (21\text{-}46)$$

However, this calculation is not permissible, because the theory was worked out on the assumption that y_1 is very small compared with L. Figure 21-19 shows what really happens. Divide the slit into two equal parts of width $\lambda/2$, following the procedure of figure 21-18a. The first dark fringe occurs when the wavelet from the top strip of the upper half is $\lambda/2$ behind the wavelet from the top strip of the lower half. The diagram makes it clear that this can happen only when the wavelets travel in a direction at right angles to the direction of travel of the original wavefront. The first dark fringe has moved out to infinity and the whole of the screen is illuminated! The diffracted wave emerging from the slit spreads out in all directions, in the same way as the ripples of figure 21-14a. The condition for this is now seen to be that the width of the slit should be equal to, or less than, the wavelength. Since the

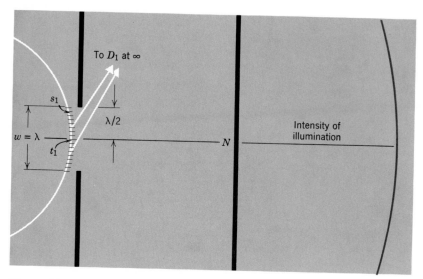

Figure 21-19 When the width of the slit $w = \lambda$, the center of the first dark fringe is at an infinite distance from the center of the screen. The wavelets from s_1 and t_1 can cancel only in a direction perpendicular to the direction in which the original wavelet is traveling.

Behavior Characteristic of Waves

443

(a)   (b)   (c)

Figure 21-20 Diffraction of light by an aperture with the shape of a long thin isosceles triangle. Part *b* is the image that would be formed if the light traveled strictly in straight lines. Part *a*, taken with a short exposure, shows the formation of crossed diffraction fringes inside the triangular image. Part *c*, taken with a longer exposure, shows the faint outer fringes. Considering the aperture to be a tapered slit, notice how the diffraction fringes move further apart as the slit becomes narrower at the top. (Photographs by Brian Thompson.)

wavelength of light is about 5×10^{-5} cm, it is very difficult to make a slit narrow enough to realize this condition for light waves.

Now let us consider the opposite extreme of a very wide slit. As the slit is made wider the diffraction pattern becomes narrower. We must remember, though, that even if the light traveled strictly in straight lines, the slit would cast an image of definite width on the screen. Under these circumstances, what is seen on the screen is an image of the slit with very closely spaced fringes near its edge. As the slit is made wider, the image becomes wider, but the fringes move closer together and eventually become very difficult to distinguish. This is the situation in figure 21-14*b*, where the width of the hole in the blind is very much larger than the wavelength of the light. Strictly speaking the edge of the spot of light on the floor is not perfectly sharp. However, the wavelength of light is so very small compared with the width of the hole, that the diffraction fringes at the edge of the spot are very close together and extend over so narrow a region that they cannot be seen except in a carefully designed experiment.

As long as the width of the hole in the blind is large compared with the wavelength of light, the shape of the spot of light on the floor bears a close resemblance to the shape of the hole. However, if the width of the hole were comparable with the wavelength, then the light on the floor would be a broad-spreading diffraction pattern bearing little resemblance to the shape of the hole. This always happens with any device, such as an optical microscope, in which light is used to view an object. When the size of the object is not very different from the wavelength of the light, diffraction effects occur and the shape of the object is blurred. Similarly, two objects closer together than the wavelength of light cannot be distinguished from one another. In general, when electromagnetic radiation of wavelength λ is being used to examine an object, it is not possible to see fine details much smaller in size than λ.

Diffraction patterns can have considerable aesthetic appeal. Figure 21-20 is the diffraction pattern of a tapered slit, which is narrow at the top and wide at the bottom. This illustrates admirably the fact that, as the slit becomes wider, the diffraction fringes move closer together. Figure 21-21 is the diffraction pattern of a rectangular slit, for which the length and width are not very different. It is similar to the pattern for a very long, very narrow slit, but shows fringes in both the horizontal and vertical directions. The shape of the rectangular slit is shown in the bottom right hand corner. In accordance with the general principle that the diffraction pattern is broader the narrower the slit, the long side of a rectangle in the diffraction pattern is parallel to the short side of the slit.

21-6   Reflection and Refraction

In a discussion of reflection and refraction and the principles underlying optical equipment such as microscopes and telescopes, it is usually permissible to assume that the light travels in straight lines, or rays, from one part of the apparatus to another. A ray is best visualized as a narrow beam of light whose width is very small compared with the character-

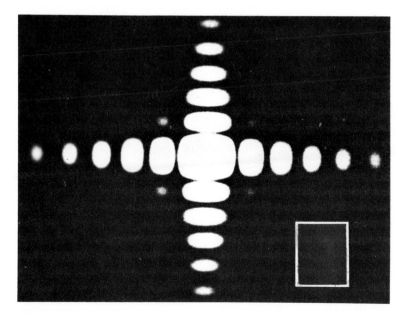

Figure 21-21   The diffraction pattern of a rectangular aperture 0.8 cm × 0.7 cm. A magnified drawing of the aperture is shown in the bottom right hand corner of the photograph. Notice that the diffraction pattern is more spread out in a direction parallel to the narrower side of the aperture. (Photograph by Brian Thompson, from Wolf and Born, *Principles of Optics*, Pergamon Press, Ltd. 1959).

Behavior
Characteristic
of Waves

445

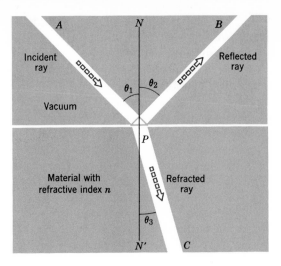

Figure 21-22 Reflection and refraction.

istic dimensions of the apparatus, but very large compared with the wavelength of light, so that diffraction effects are unimportant.

Consider such a beam incident on the plane surface of a transparent material, such as glass or water (figure 21-22). The *incident ray AP* divides up into a *reflected ray PB* and a *refracted ray PC*. The line *NPN'* is perpendicular to the surface. The angle θ_1 between *PN* and the incident ray *AP* is called the *angle of incidence*. The angle θ_2 between *PN* and the reflected ray *PB* is called the *angle of reflection*. These two angles are always equal.

Law of reflection

Angle of reflection = Angle of incidence

$$\theta_2 = \theta_1 \tag{21-47}$$

The angle θ_3 between *N'P* and the refracted ray *PC* is called the *angle of refraction*. The relationship between θ_3 and θ_1 depends upon a quantity n, called the *refractive index* of the material.

Law of refraction (Snell's law)

$$\frac{\text{Sine of angle of incidence}}{\text{Sine of angle of refraction}} = \text{Refractive index} \tag{21-48}$$

$$\frac{\sin \theta_1}{\sin \theta_3} = n \tag{21-49}$$

WAVES

The refractive index n is characteristic of the material and varies from one material to another. It also varies with the frequency of the light, as we shall explain later.

Let us try to understand these laws by considering what happens when an electromagnetic wave passes through a material body. The body is composed of negatively charged electrons and positively charged nuclei. The oscillating electric field of the wave exerts an oscillating force on the charged particles and makes them perform simple harmonic motion. The oscillating charges then radiate new electromagnetic waves. However, the incident wave arrives at the various atoms at different times and the atoms reradiate with different phases. The way in which all the electromagnetic waves add together is a very complicated mathematical problem and we can only quote the result here. It turns out that the wave travels through the body with a velocity u, which is less than the velocity c of light in a vacuum. A time t after arriving at the surface of the body, the incident ray, if left to itself, would have traveled a distance ct. However, because of the interference from the radiating atoms, it actually travels only a distance ut. At distances between ut and ct, the waves reradiated by the atoms cancel out the incident wave!

The frequency of the wave inside the body is the same as the frequency, ν, of the incident wave in a vacuum. It is the frequency with which the incident wave makes the electrons and nuclei oscillate. However, since the velocity of the wave is less inside the body, the wavelength is smaller than it was in a vacuum. In the vacuum the wavelength is obtained from equation 19-32.

$$\lambda = \frac{c}{\nu} \tag{21-50}$$

Inside the body the wavelength is

$$\lambda_{in} = \frac{u}{\nu} \tag{21-51}$$

With the help of figure 21-23, consider a wavefront PQ incident on the surface of a transparent material. The point P reaches the surface first and immediately sends out a wave into the material with velocity u. The point Q meanwhile continues to travel through the vacuum with velocity c and eventually strikes the surface at Q' after a time

$$t = \frac{QQ'}{c} \tag{21-52}$$

During this time the wave from P has traveled to P' with velocity u and

$$PP' = ut \tag{21-53}$$

$$= \frac{uQQ'}{c} \tag{21-54}$$

Rearranging this equation,

$$\frac{QQ'}{PP'} = \frac{c}{u} \tag{21-55}$$

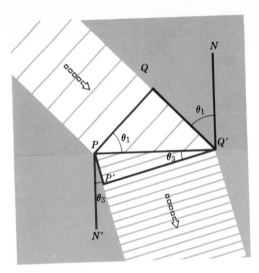

Figure 21-23   Refraction of a wavefront.

Consider the right angled triangle PQQ' and notice that

$$\text{Angle } QPQ' = 90° - \text{angle } QQ'P \tag{21-56}$$
$$= \theta_1 \tag{21-57}$$

So
$$\sin \theta_1 = \frac{QQ'}{PQ'} \tag{21-58}$$

Similarly, for right angled triangle $PP'Q'$,

$$\text{Angle } PQ'P' = 90° - \text{angle } P'PQ' \tag{21-59}$$
$$= \theta_3 \tag{21-60}$$

So
$$\sin \theta_3 = \frac{PP'}{PQ'} \tag{21-61}$$

Dividing equation 21-58 by equation 21-61,

$$\frac{\sin \theta_1}{\sin \theta_3} = \frac{QQ'}{PQ'} \frac{PQ'}{PP'} \tag{21-62}$$

$$\frac{\sin \theta_1}{\sin \theta_3} = \frac{QQ'}{PP'} \tag{21-63}$$

Replacing QQ'/PP' by its value as given by equation 21-55,

$$\frac{\sin \theta_1}{\sin \theta_3} = \frac{c}{u} \tag{21-64}$$

WAVES   This is identical with Snell's law if

$$\frac{c}{u} = n \tag{21-65}$$

$$\textbf{Refractive index} = \frac{\text{Velocity of light in a vacuum}}{\text{Velocity of light in the material}} \qquad (21\text{-}66)$$

This equation has been verified by measuring the velocity u, using a method similar to the one shown in figure 20-2 and interposing the material in the light beam.

The velocity of an electromagnetic wave inside a material varies with frequency. This is called **dispersion**. For example, the velocity of blue light in glass is less than the velocity of red light. It follows from equation 21-66 that the refractive index is greater for blue light than for red light. It then follows from Snell's law that, when white light is incident on glass, the blue component is bent more than the red component (figure 21-24a). Referring to equation 21-49, for a fixed θ_1 a larger value of n means a smaller value of $\sin \theta_3$, which is equivalent to a smaller value of θ_3. Figure 21-24b shows how this leads to the formation of a spectrum by a glass prism.

As far as the reflected wave is concerned, it may be shown that the radiating atoms inside the material cancel one another and the only radiating atoms that matter are the ones near the surface of the material. With the help of figure 21-25, consider a wavefront PQ reflected from a

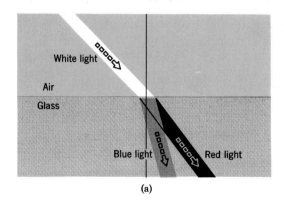

(a)

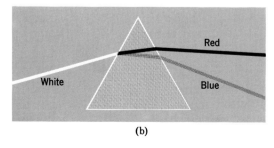

(b)

Figure 21-24 Resolution of white light into its component colors by a glass prism.

Behavior
Characteristic
of Waves

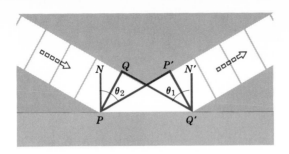

Figure 21-25 Reflection of a wavefront.

surface. P reaches the surface first and the oscillating electric vector at P sets the electrons and nuclei near the surface in motion and they reradiate back into the vacuum. By the time Q has reached the surface at Q', the wave radiated from P has reached P'. However, both waves are traveling in a vacuum with velocity c, so

$$PP' = QQ' \tag{21-67}$$

The two right angled triangles PQQ' and $Q'P'P$ have a common side PQ'. Also the side PP' of one is equal to the side QQ' of the other. These triangles are therefore congruent and all their angles and sides are equal in pairs. In particular

$$\text{Angle } QQ'P = \text{Angle } P'PQ' \tag{21-68}$$

But

$$\text{Angle } QQ'P = 90° - \theta_1 \tag{21-69}$$
$$\text{Angle } P'PQ' = 90° - \theta_2 \tag{21-70}$$

Whence

$$\theta_1 = \theta_2 \tag{21-71}$$

which is the law of reflection.

QUESTIONS

1. In Young's double slit experiment, why must d be small and L large?
2. If Young's double slit experiment were performed with white light, containing all possible colors, what would the interference fringes look like?
3. The arrangement shown in the diagram is known as a Lloyd's mirror experiment. The very narrow slit is very near the mirror and parallel to the plane of the mirror. Explain why such an arrangement produces interference fringes?

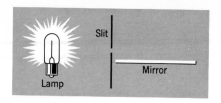

Question 21-3

4. In a Young's double slit experiment a thin film of a material with a refractive index greater than 1 is placed directly in front of one of the slits. What effect does this have on the interference fringes?

5. If the apparatus for a Young's double slit experiment were completely immersed in water, what difference would this make to the fringes?

6. In connection with diffraction by a slit, it was stated that the wave-front must be divided up into a large number of very narrow strips. How narrow must the strips be for this method to work?

7. Describe the nature of the diffraction pattern for the following slit widths: (a) $\lambda/2$, (b) λ, (c) 2λ, (d) 10λ.

8. In the analysis leading to equation 21-43 the width of the image of the slit on the screen was ignored. When is this permissible?

9. When photographing a small object through a microscope, is it better to use red or blue light? Assume that the film is equally sensitive to both colors.

10. If electromagnetic radiation could be used to study the structure of a proton, what wavelengths would be suitable? What type of radiation would this be?

11. A radio telescope operates on a frequency of 8×10^7 cycles per second. Assuming any advance in techniques permitted by the laws of physics, could it ever be used to take a photograph of two astronauts shaking hands on the moon?

12. What is the extension of Snell's law to the case of a ray of light crossing a plane boundary from a material of refractive index n_1 to a material of refractive index n_2?

13. Suppose that the refracted ray in figure 21-22 is reversed and the light passes from the medium out into the vacuum. Show that there is no ray emerging into the vacuum if θ_3 exceeds a certain value and find this value. (Under these circumstances all the light is reflected back into the medium.)

PROBLEMS

A

1. In a Young's double slit experiment the light has a frequency of 6×10^{14} sec^{-1}, the distance between the slits is 0.08 cm and the distance from the slits to the screen is 1.2 m. What is the distance between centers of adjacent fringes?

Behavior
Characteristic
of Waves

451

2. In a Young's double slit experiment using blue light with a wavelength of 4.5×10^{-5} cm, the size of the room does not permit the distance between the slits and the screen to exceed 2.5 m. What is the maximum permissible distance between the slits if the fringes must be separated by at least 0.1 cm?

3. In a Young's double slit experiment, the distance between centers of adjacent fringes is 0.01 cm. If the distance from the screen to the slits is doubled, the distance between the slits is halved and the wavelength of the light is changed from 6.5×10^{-5} cm to 4×10^{-5} cm, what is the new distance between fringes?

4. In a Young's double slit experiment, the distance between the slits is 0.062 cm and the distance from the slits to the screen is 472 cm. If the distance between centers of adjacent fringes is found to be 0.39 cm, what is the wavelength of the light?

5. What is the velocity of light in a material with a refractive index of 1.65?

6. If infrared radiation with a frequency of 2.6×10^{14} sec^{-1} is passing through a material with a refractive index of 1.37, what is its wavelength?

B

7. Light with a wavelength of 5.5×10^{-5} cm passes through a slit 0.15 cm wide and forms a diffraction pattern on a screen 2.5 m away. What is the width of the central fringe? Take the outer boundary of the central fringe to be the point D_1 in figure 21-18 where the intensity of illumination is zero.

8. If the apparatus described in the previous question were immersed in water, which has a refractive index of 1.33, what would then be the width of the central fringe?

9. A beam of light is incident at an angle of 45° on the plane surface of a pool of water. What is the angle of refraction? (The refractive index of water is 1.33.)

10. In a Fizeau type experiment to measure the velocity of light (figure 20-2), the light first reappears at maximum brightness when the rate of rotation of the toothed wheel is 2,210 revolutions per minute. A tank of liquid is then inserted into the light beam and occupies 35% of the distance between the wheel and the mirror. The rate of rotation of the wheel must be adjusted to 1,730 revolutions per minute to restore the fringes to maximum brightness. What is the refractive index of the liquid?

C

11. In the double slit experiment of problem 1 a thin film is placed in front of one of the slits. Its thickness is 1.2×10^{-4} cm and it is made of a material with a refractive index of 1.22. Through what distance do the fringes move sideways?

12. In a Young's double slit experiment, the filter passes red light and blue light, each with a sharply defined frequency. It is found that

the fifth blue fringe coincides exactly with the fourth red fringe. (The central fringe is taken as the first fringe.) What conclusion can you draw? If you were given the distance between the slits and the distance from the slits to the screen, could you draw any further conclusions?

13. When a light ray passes from a vacuum into a medium of refractive index n, the angle of incidence is found to be twice the angle of refraction ($\theta_1 = 2\theta_3$). Find the angle of incidence in terms of n. For what range of values of n is this a possible situation?

14. A ray of light is incident upon a mirror at some arbitrary angle of incidence θ_1. The mirror is rotated through an angle ϕ about an axis in the plane of the mirror perpendicular to the incident ray. Through what angle does the reflected ray rotate?

15. A ray of light passes from a vacuum into a slab of glass with a refractive index of 1.57. The angle of incidence is 55°. The slab is then rotated through an angle of 20° about an axis in the plane of its surface perpendicular to the incident ray in the direction which reduces the angle of incidence. Through what angle does the refracted ray rotate?

Behavior
Characteristic
of Waves

RELATIVITY

22

The
Problem
and
Its
Solution

22-1   Communication Between Different Parts of the Universe

The reader should now refer back to section 18-7, where the discussion of the electromagnetic field culminated in a new point of view. Action-at-a-distance was abandoned and replaced by the concept that interactions are "handed on" from one point in an electromagnetic field to a neighboring point. This "handing on" process is, in fact, an electromagnetic wave. We have digressed in the three previous chapters to consider its properties in some detail. We shall now return to a consideration of a very basic aspect of the situation, which is that interactions propagate through empty space with a well-defined velocity c.

With the help of figure 22-1a, consider two charges q_1 and q_2 moving along complicated paths under the influence of the electromagnetic forces they exert on one another, but subject to no other forces. The approach of Newtonian mechanics is that, if we are given the initial positions and velocities of the two charges and if we know the exact nature of the electromagnetic forces, the two paths ABC and DEF can be calculated. However, if the electromagnetic forces are correctly described in terms of Maxwell's equations, a new and very important aspect of the situation arises to complicate matters. *The forces between the two charges are not exerted instantaneously but are transmitted from one charge to the other with the speed of light.*

To understand this, suppose that the charge q_1 collides with a neutral body at the point B of its path (figure 22-1b). It is easy to arrange matters so that this neutral body exerts a negligible force on q_1 until the distance between them is very small. The previous calculation of the

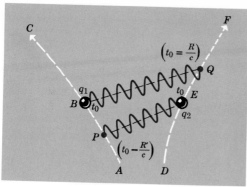

(a) Two charges q_1 and q_2 describing paths through space subject only to the forces they exert on one another.

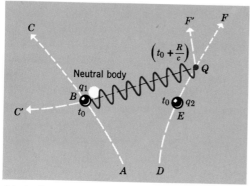

(b) At B q_1 collides with a neutral body.

Figure 22-1 The electromagnetic forces between two charges are not instantaneous but experience a time lag determined by the velocity of light.

first part of the path AB is still valid, but the path BC' after the collision is very different from the anticipated path BC. During the collision q_1 undergoes a rapid change in velocity and has a very large acceleration. It therefore radiates an electromagnetic wave, part of which sets off in the direction of q_2. Meanwhile, however, q_2 continues along its previously calculated path *as though the collision had not occurred.* It is not until the electromagnetic wave reaches q_2 that it becomes aware of the collision and starts to deviate from the calculated path.

Suppose that the collision occurs at time t_0 secs, when q_1 is at B. At this same time, t_0 secs, q_2 is at E. The electromagnetic wave from the collision travels a distance R before it reaches q_2, which has meanwhile traveled on to the point Q. q_2 therefore becomes aware of the collision at a time $\left(t_0 + \dfrac{R}{c} \right)$ secs, and its subsequent path QF' is then related to BC' rather than BC.

The collision was introduced to dramatize the situation, but a similar time lag is present even in the undisturbed paths of figure 22-1a. When

The Problem and Its Solution

q_2 is at E at time t_0 secs, the force exerted on it by q_1 is not dependent upon the simultaneous position and motion of q_1 at B. It depends instead upon the position and motion of q_1 when it was at P at an earlier time $\left(t_0 - \dfrac{R'}{c}\right)$, where R' is the distance an electromagnetic disturbance travels from q_1 at P to q_2 at E. The position and motion of q_1 at B at time t_0 initiate an electromagnetic disturbance that reaches q_2 when it is at the point Q at time $\left(t_0 + \dfrac{R}{c}\right)$. The force on q_2 over the path EQ between times t_0 and $\left(t_0 + \dfrac{R}{c}\right)$ therefore depends on the motion of q_1 over the path PB between times $\left(t_0 - \dfrac{R'}{c}\right)$ and t_0. This is why q_2 moves on to Q before it becomes aware of the collision experienced by q_1 at B.

The emphasis has been placed on the force exerted on q_2 by q_1, but exactly similar considerations apply to the force exerted on q_1 by q_2. This force does not depend on the simultaneous position and motion of q_2 but upon its position and motion at an earlier instant, allowing time for an electromagnetic wave to travel from q_2 to q_1.

This time lag for the transfer of forces is an important feature of the behavior of the universe. In Chapter 6, when we were discussing Newtonian mechanics, we wrote down Newton's law of universal gravitation (equation 6-20) and Coulomb's law for electrostatic forces (equation 6-26) as though the forces were transmitted *instantaneously* from one body to the other. We now see that this is not true for electromagnetic forces. It is probably not true for gravitational forces either, although it must be admitted that our present knowledge of gravitation is very incomplete and the propagation of gravitational disturbances through space is still an unsolved problem.

22-2   Time at Distant Places

Now let us consider some of the far-reaching consequences of this fact that two parts of the universe are not in immediate contact with one another, but can only influence one another via disturbances which travel between them with the speed of light. In the constellation Andromeda there is a celestial object called the Andromeda galaxy which is visible to the unaided human eye. It is a group of stars which lies far away from our own local group of stars. It is so far away that light takes about 2 million years to travel from it to us. We see it, not as it is at the moment, but as it was 2 million years ago, at a time when man, if he existed at all, was very ape-like. We cannot completely exclude the possibility that one million years ago the Andromeda nebula exploded and is now dispersed throughout space. We should have to wait another million years to discover this.

Suppose that I step out into my garden on a clear moonlit night and first notice a moth hovering round the lamp near my door. I then glance at the moon, transfer my attention to the brightest star in the sky, Sirius, and finally look at the Andromeda galaxy. From my point of view as a local observer, observing what happens at the particular spot on the planet Earth where I find myself, the time sequence of these events is unambiguous. It is: moth, moon, Sirius, Andromeda galaxy. If, however, I ask what time the events occurred at the distant places where they originated, the sequence is reversed. Taking into account the time taken for the light to travel from the event to me, I must conclude that I saw light that was emitted by the Andromeda galaxy 2 million years ago, light that left Sirius 8.7 years ago, sunlight that was reflected from the moon 1.28 seconds ago, and lamplight that was reflected from the moth one billionth of a second ago.

Even so, the situation has been oversimplified. The earth and the Andromeda galaxy are moving toward one another with a relative velocity of about 3×10^7 cm sec^{-1}, which is about one thousandth of the velocity of light. Under these circumstances, should we assume that the light emitted by the galaxy travels away from it with a velocity of 3×10^{10} cm sec^{-1} *relative to the galaxy*, and that the earth moves forward to meet this light with a velocity of 3×10^7 cm sec^{-1}, so that the velocity of the light relative to the earth is $3 \times 10^{10} + 3 \times 10^7 = 3.003 \times 10^{10}$ cm sec^{-1}? Or should we assume that space is filled with an ether (see section 14-1) and that light travels with a velocity of 3×10^{10} cm sec^{-1} *relative to the ether*? In the latter case we should have to know the velocity of the earth relative to the ether before we could calculate the time taken for the light to travel from the galaxy to the earth. The two assumptions about the behavior of light could easily lead to a difference of a thousand years or more in the two estimates of the time.

Now suppose that the skeleton of an early man has been discovered and, by means of refined dating techniques, the year of his death has been accurately determined. I wish to dramatize the situation by the statement, "If you look at a certain star in the Andromeda galaxy tonight, the light that enters your eye left this star exactly one year after this prehistoric man died." However, this statement might be based on the assumption that the velocity of light is measured relative to the galaxy. On the alternative assumption that the velocity of light should be measured relative to an ether, the statement might have to be modified to, "If you look at a certain star in the Andromeda galaxy tonight, the light that enters your eye left the star nine hundred and ninety-nine years before this prehistoric man died." The point at issue is whether, at the instant the light left the star, the man had already died, or whether he had not yet even been born!

The concepts of past, present, and future are taken very much for granted. The point that has been made in this section is that, when we are trying to decide the relative times of two events occurring in different parts of the universe, we cannot do this without making the correct assumptions about the behavior of light. Einstein's theory of relativity tells us what are the correct assumptions and how we should make our

The Problem and Its Solution

459

calculations. The test of "correctness" is that all the predictions of the theory agree with experimental observations. In order to understand this correct procedure, it is necessary to abandon many of our cherished prejudices and to be prepared for some astonishing consequences.

22-3   The Michelson-Morley Experiment

The difficulties of the last section depended upon the answer to the question: "Relative to what should the velocity of light be measured? The source? The observer? The ether?" In the case of *sound* the answer to a similar question is well understood. Sound travels with a certain velocity *relative to the material through which it is traveling.* Suppose that the velocity of sound through the air at the ambient temperature is 750 mph. If I speak to you with a 50 mph gale blowing from behind me directly toward you (figure 22-2), the sound emerging from my mouth travels at 750 mph relative to the air and so it travels from me to you with a speed of $750 + 50 = 800$ mph. On the other hand, if you speak to me against the direction of the wind, the sound travels from you to me with a speed of $750 - 50 = 700$ mph.

Perhaps the situation is the same for light. Perhaps, as was suggested in connection with the concept of a field, space is filled with a mysterious, invisible, imponderable, intangible, jellylike substance called the **ether.** Perhaps light travels through the ether in the same way that sound travels through the air and the velocity of light c is 2.9979×10^{10} cm/sec *relative to the ether.* If that were so, absolute rest could mean "at rest relative to the ether." There would be a special frame of reference stationary relative to the ether and we could recognize this frame because it would be the only one in which the velocity of light would be c. In frames moving relative to the ether, the velocity of light would be different. For example, we could chase after the light with a velocity $\frac{1}{4}c$ relative to the ether and the light would then appear to be moving away from us with a velocity $\frac{3}{4}c$ (figure 22-3). As we shall soon see, this is *not* what happens.

Figure 22-2 The velocity of sound is 750 mph relative to the air through which the sound is propagated.

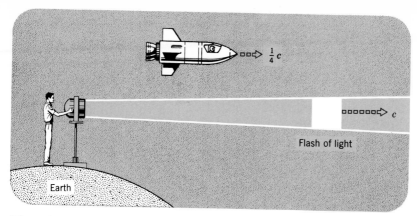

Figure 22-3 According to prerelativistic ideas, if there were an ether, and if light always had a velocity c relative to this ether, a spaceship chasing the flash of light with a velocity $\frac{1}{4}c$ relative to the ether would see the light moving away with a velocity $\frac{3}{4}c$.

In 1887 Michelson and Morley performed a very famous experiment to investigate whether or not light has a velocity c relative to the ether. It is illustrated in figure 22-4. A beam of light falls upon a lightly silvered glass plate P and is divided into a transmitted beam B_1 and a reflected beam B_2 of approximately equal intensity. The two beams travel at right angles to one another through the same distance l and are then reflected back by the mirrors M_1 and M_2. Upon returning to P they join together again and produce interference fringes similar to those discussed in section 21-2, but somewhat more complicated in detail. If for any reason one of the beams takes a longer time to return to P, the fringes are displaced. Suppose, for example, one beam experiences an extra time delay of half a period and arrives half a wavelength behind the other beam. Then, at those points where crests previously coincided with crests and there was

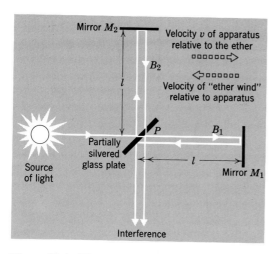

Figure 22-4 The Michelson-Morley experiment.

a bright fringe, a crest of the delayed beam coincides with a trough of the other beam and there is a dark fringe. The fringes have moved over so that dark fringes replace bright fringes. This is a very sensitive way of measuring a difference in the times taken by the two beams to travel to their mirrors and back.

If the two beams travel exactly the same distance $2l$, and if their velocity is c *relative to the apparatus*, they should arrive back at exactly the same time. Suppose, however, that the velocity of light ought to be measured *relative to the ether*, and that the whole apparatus is moving through the ether with a velocity v parallel to the beam B_1. Then, on its outward journey the beam B_1 is traveling against the "ether wind" and its velocity *relative to the apparatus* is $(c - v)$. The time taken on its outward journey is therefore $l/(c - v)$. On its return journey it is moving with the "ether wind" and has a velocity $(c + v)$ relative to the apparatus. The return journey therefore takes a time $l/(c + v)$. The total time t_1 for the outward and return journeys is therefore

$$t_1 = \frac{l}{(c - v)} + \frac{l}{(c + v)} \cdots \tag{22-1}$$

$$= \frac{2lc}{(c^2 - v^2)} \tag{22-2}$$

To estimate the time t_2 taken by beam B_2, let us suppose that we are stationary relative to the ether and are watching the apparatus move past us with velocity v. Then the path of the light beam appears as shown in figure 22-5. When the light reaches the mirror M_2 at time $\frac{1}{2}t_2$, the mirror has moved a distance $\frac{1}{2}vt_2$ relative to the ether. When the light returns to P, P has moved a total distance vt_2 relative to the ether. The total distance traveled by the light is $AB + BC$. Since the situation is being viewed by an observer stationary relative to the ether, this path is traversed with a velocity c. It follows that

$$AB = BC = \tfrac{1}{2}ct_2 \tag{22-3}$$

Applying Pythagoras' theorem to triangle ABD

$$AB^2 = BD^2 + AD^2 \tag{22-4}$$

Substituting the values of AB, BD, and AD

$$(\tfrac{1}{2}ct_2)^2 = l^2 + (\tfrac{1}{2}vt_2)^2 \tag{22-5}$$

$$\text{or} \quad \tfrac{1}{4}c^2t_2{}^2 = l^2 + \tfrac{1}{4}v^2t_2{}^2 \tag{22-6}$$

$$\tfrac{1}{4}t_2{}^2(c^2 - v^2) = l^2 \tag{22-7}$$

$$t_2{}^2 = \frac{4l^2}{(c^2 - v^2)} \tag{22-8}$$

Finally, the time taken by beam B_2 is seen to be

$$t_2 = \frac{2l}{\sqrt{c^2 - v^2}} \tag{22-9}$$

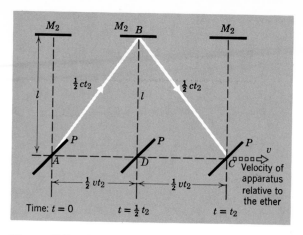

Figure 22-5 The Michelson-Morley experiment. The path of beam B_2 as seen by an observer stationary relative to the ether.

This is not the same as the time t_1 taken by beam B_1. In fact t_1 is slightly longer than t_2. With a little algebraic manipulation and some permissible approximations if v is very much less than c, we can show that the time difference is

$$t_1 - t_2 = \frac{lv^2}{c^3} \tag{22-10}$$

In the first instance the magnitude and direction of the velocity of the earth relative to the ether was not known. The apparatus was therefore floated on a vat of mercury and slowly rotated. According to the "ether wind" hypothesis, in one of its orientations the light beam B_1 would have been parallel to $\mathbf{v}$. The above analysis would then have applied, and t_1 would have been greater than t_2 by the amount given in equation 22-10. However, after rotating the apparatus through 90° from this position, the other beam B_2 would have been parallel to $\mathbf{v}$. The role of the two beams would then have been interchanged and t_2 would have been greater than t_1 by the amount given in equation 22-10. The total variation in $(t_1 - t_2)$ upon rotating through 90° would have been twice this amount and there would have been a corresponding shift of the interference fringes.

When the experiment was performed no significant shift of the fringes was observed, even though the apparatus was capable of detecting a shift as small as one two hundredth of the distance between fringes corresponding to a time difference of about 10^{-17} sec!

Either the ether theory of the experiment was wrong, or the velocity of the earth relative to the ether was less than 10^6 cm sec^{-1} and the effect was therefore too small to be detected. To investigate the latter possibility, the experiment was performed again six months later. The velocity of the earth in its orbit around the sun is about 3×10^6 cm sec^{-1}.

The Problem and Its Solution

463

Over a period of six months this velocity reverses its direction. If the velocity of the earth relative to the ether had been less than 10^6 cm sec^{-1} when the experiment was first performed, six months later it might have been expected to be about 6×10^6 cm sec^{-1}. This would have been easily detectable, but there was still no observable shift of the fringes. The experiment has been performed many times since by different observers and its accuracy is continually being improved, but no effect has ever been observed.

22-4   Further Evidence

One suggestion put forward to explain the Michelson-Morley experiment was that the earth drags the ether along with it and the velocity of the earth relative to the ether in its immediate vicinity is therefore inevitably zero. This suggestion is inconsistent with a well-known astronomical phenomenon known as the aberration of starlight. As the earth moves in its orbit around the sun, all the stars, even the most distant ones, appear to move in small flattened circles (ellipses) in the sky. The basic reason for this is illustrated in figure 22-6. Part (a) shows the passage of light from the star through a stationary telescope. Part (b) shows that, if the telescope is moving, but still points in the same direction, the light may miss the bottom of the telescope and the star will not be seen. Part (c) shows that this can be avoided by tilting the telescope at a suitable angle such that, as the telescope moves, the light remains in the center of the telescope at all times. The light from the star appears to be coming from a direction at a small angle α to its actual direction. This angle depends upon the velocity of the telescope and hence the velocity of the earth, and changes throughout the year as the earth moves around the sun.

The detailed theory of this effect obviously depends upon the very issues that concern us at the moment. With respect to what should the velocity of light be measured? Let us assume that there is an ether and that the ether in the immediate vicinity of the earth is dragged along with it and therefore moves relative to the stars. This point of view can be reconciled with the observational facts on aberration only by making very special, and highly implausible, assumptions about the way in which the velocity of the ether varies near the surface of the earth. These special assumptions are excluded by two observations. The Michelson-Morley experiment still gave a null result when performed on top of a high mountain and in a balloon! Also, Michelson showed that the velocity of light relative to the earth is independent of height above sea level.

Another suggestion to explain the Michelson-Morley experiment was that the velocity of light should be measured *relative to the source of the light*. This is obviously adequate, because the source moves with the apparatus, and the assumption automatically ensures that the velocity of light is c for an observer moving with the apparatus. However, this suggestion was excluded when Miller performed the Michelson-Morley

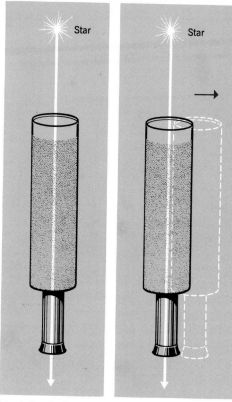
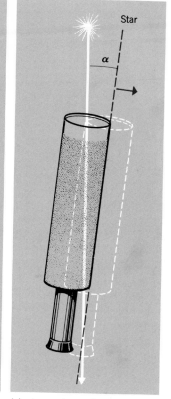

(a) A fixed telescope.

(b) A moving telescope pointing in the same direction.

(c) A moving telescope tilted through an appropriate angle α.

Figure 22-6 The aberration of light.

experiment using light *from the sun* and still obtained a null result. Moreover, de Sitter pointed out that the suggestion was inconsistent with observations on binary stars. Figure 22-7 shows a small star B rotating around a larger star A. In the position P the star B has a velocity V away from the earth and, according to the assumption under discussion, the light from it should travel toward the earth with a velocity $(c - V)$ relative to the earth. Half a revolution later B is at Q and has a velocity V toward the earth. Light from B should then travel toward the earth with a velocity $(c + V)$ relative to the earth. There is even a possibility that, in some cases, the light emitted at Q might catch up with the light emitted at P and we should see the star at Q before seeing it at P! In any case the apparent motion of B around A would be seriously affected by the assumption. De Sitter was able to show that effects of this kind are not present in the observational data on binary stars.

A most ingenious suggestion to explain the Michelson-Morley experiment was made by Fitzgerald and elaborated by Lorentz. They suggested that, when a body moves through the ether, its length is con-

The Problem and Its Solution

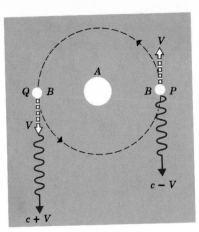

Figure 22-7 Illustrating de Sitter's discussion of a binary star.

tracted in the direction of motion. The atoms of a body are held together by electromagnetic forces. If the ether is the medium through which these forces are transmitted, motion through the ether might modify the forces in such a way as to make the atoms move closer together in the direction of motion.

Let us assume that a body of length l when at rest relative to the ether has a contracted length $l\sqrt{1 - \dfrac{v^2}{c^2}}$ when moving with a velocity v relative to the ether in the direction of this length. In the theory of the Michelson-Morley experiment (figure 22-4), the Fitzgerald-Lorentz contraction means that, when the apparatus is in motion, the distance traveled by beam B_1 is not l but $l\sqrt{1 - \dfrac{v^2}{c^2}}$. Substituting this new value of the distance into equation 22-2, the time taken by beam B_1 is

$$t_1' = \frac{2cl\sqrt{1 - \dfrac{v^2}{c^2}}}{(c^2 - v^2)} \tag{22-11}$$

$$= \frac{2l}{\sqrt{c^2 - v^2}} \tag{22-12}$$

This is exactly the same as the time taken by beam B_2, as given in equation 22-9. The Fitzgerald-Lorentz contraction therefore ensures that the two beams arrive back at exactly the same time, whatever the velocity of the apparatus relative to the ether.

This ingenious possibility was excluded by an experiment performed by Kennedy and Thorndike. Their apparatus was similar to that of Michelson and Morley, except that the beam B_1 traveled a distance l_1 to the mirror M_1, whereas the beam B_2 traveled a very different distance l_2 to the mirror M_2. Assuming the Fitzgerald-Lorentz contraction, the time taken by beam B_1 was

$$t_1' = \frac{2l_1}{\sqrt{c^2 - v^2}} \tag{22-13}$$

The time taken by beam B_2 was

$$t_2 = \frac{2l_2}{\sqrt{c^2 - v^2}} \qquad (22\text{-}14)$$

The difference

$$t_1 - t_2 = \frac{2(l_1 - l_2)}{\sqrt{c^2 - v^2}} \qquad (22\text{-}15)$$

is not zero and depends upon the velocity v relative to the ether. Again, however, no effect was observed.

A possible misunderstanding of the significance of the Kennedy-Thorndike experiment should be avoided. As we shall see later, the theory of relativity predicts a contraction of a moving body very similar to the Fitzgerald-Lorentz contraction, whereas we have said that the experiment excludes this. The point to be made is that the theory of relativity predicts the Fitzgerald-Lorentz contraction *plus other effects*, whereas the Kennedy-Thorndike experiment proves that the Fitzgerald-Lorentz contraction *alone* is not adequate.

22-5 Einstein's Solution of the Problem

The last two sections give some idea of the interplay of experiment, hypothesis, counter-hypothesis and counter-experiment which took place in the closing years of the nineteenth century and the early years of the twentieth century. The evidence was against the idea that light travels through an ether and has a velocity c relative to it. Neither did it seem permissible to assume that the velocity of light is c relative to the source that emits it. In the year 1905 Albert Einstein published his first paper on the theory of relativity, and out of a state of confusion there emerged a new insight into the nature of the universe and the nature of space and time. The basic ideas underlying this theory will be described in the following pages. On the subject of immediate concern, the velocity of light, Einstein made a clear and unequivocal statement. The velocity of light in a vacuum always has the same value c *relative to the observer*.

Whenever an experiment is performed to measure the velocity of light in a vacuum, the resulting numerical value (2.9979×10^{10} cm sec^{-1}) is always the same, wherever the experiment is performed and whatever the velocity of the apparatus and observer relative to other bodies. When Michelson measured the velocity of light at different heights above sea level, he always obtained the same answer. In the Michelson-Morley experiment (figure 22-4) the velocity of the two beams is always the same relative to the apparatus (and hence the observer) and so the beams inevitably arrive back at the same time. The changing velocity of the earth as it moves around the sun is irrelevant. The velocity of the light emitted by either member of a pair of binary stars (figure 22-7) has the same value relative to the observing telescope, independently of the motion of the stars around one another.

The hypothesis that the velocity of light is always the same relative

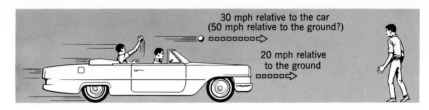

Figure 22-8 A problem in the addition of velocities.

to the observer is so simple, and provides so ready an explanation of the Michelson-Morley experiment and related experiments, that it is important to realize why it leads to such a revolution in our ideas. Let us first consider the "common sense" attitude to the following situation. A baseball player can throw a ball horizontally with a speed of 30 mph. Suppose that he stands up in a car which is traveling at 20 mph and throws the ball forward (figure 22-8). The ball is given a speed of 30 mph relative to the car, but the car itself is moving in the same direction at 20 mph. An observer standing still and watching this event "obviously" sees the ball fly past him at 30 + 20 = 50 mph.

Now consider an analogous situation involving light (figure 22-9). Space ship B is traveling past space ship A at the very high speed of 2×10^{10} cm/sec relative to A. B sends out flashes of light in the forward direction. The pilot of B measures the velocity of this light by erecting a mirror at a known distance in front of the lamp and measuring the time taken for the reflected light to return to the lamp. He obtains the usual value of 3×10^{10} cm/sec since, according to Einstein's hypothesis, this is the velocity of light relative to any observer, independently of his position or velocity. The pilot of A can also measure the velocity of the light flashes by, for example, setting up two detecting devices in the light beam at a known distance apart and measuring the time interval

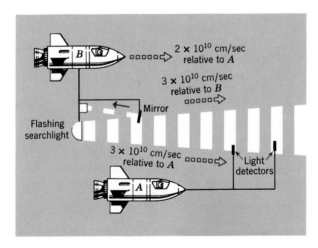

Figure 22-9 Both A and B measure a velocity of 3×10^{10} cm/sec for the light beam, even though B has a velocity of 2×10^{10} cm/sec relative to A.

between the arrival of a flash at the two devices. The arguments which were applied to the baseball and the car suggest that the pilot A should obtain a value of $3 \times 10^{10} + 2 \times 10^{10} = 5 \times 10^{10}$ cm/sec. However, this is contrary to Einstein's hypothesis, since *all* observers, measuring the velocity of *any* light beam, must obtain a value of 3×10^{10} cm/sec.

The natural reaction to this discussion is that it proves the absurdity of Einstein's hypothesis. Nevertheless, physicists now believe, with a high degree of confidence, that pilot A would obtain a value of 3×10^{10} cm/sec. It is common sense that is wrong. The outstanding achievement of Einstein's genius was the realization that the arguments applied to the car and the ball are based upon certain subtle fallacies which will be revealed in the next chapter. We now believe that the velocity of the ball relative to the stationary observer would not be 50 mph but 49.999999999999933 mph (see section 24-1). The correction introduced by the theory of relativity is so very small that it is clear why we are not aware of it in our everyday experience of cars and baseballs. The reason for this, as we shall see later, is that the velocities of the car and the ball are very small compared with the velocity of light. In the example of the two space ships and the beam of light, the velocities are not small compared with c.

Reconsider the situation of figure 22-3 in terms of these new ideas. We must first abandon the idea that the light is traveling through an ether. The light flashes have a velocity c relative to the observer operating the searchlight, not because he is stationary relative to the ether, or even because he is stationary relative to the source of the light, but because the light flashes must appear to *any* observer to have a velocity c. In particular, the observer in the spaceship also sees the light flashes moving away from him with a velocity c, not $\frac{3}{4}c$. In fact, however fast the spaceship chases after the light, the light flashes will always appear to the observer in the spaceship to be gaining on him with a velocity c! Chasing light is the ultimate frustration.

What Einstein clearly realized was that we cannot start to understand a concept such as velocity without devising some procedure for determining time at distant places. We must know where the body is at what time. If we look at it, we are relying on the light traveling from the body to our eyes and the journey takes a definite time. We do not see the body where it is now, but where it was at an earlier time when the light left it. To correct for this, it is necessary to know how fast the light traveled, and this immediately raises all the questions about the velocity of light with which we have already been preoccupied in this chapter.

22-6 Frames of Reference

Einstein's theory shifts the emphasis away from the hypothetical, and perhaps non-existent ether and places it instead on the observer. Let us therefore consider in what respects different observers, moving relative to one another, obtain different views of the behavior of the universe. Suppose that our baseball player, standing in a car traveling at 20 mph,

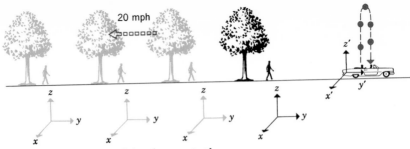

(a) The point of view of the observer in the car.

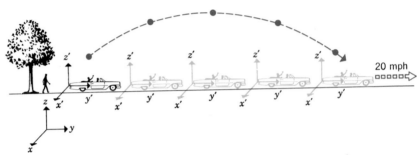

(b) The point of view of the observer standing at the side of the road.

Figure 22-10 The motion of a ball seen from two different frames of reference.

throws his ball vertically upward so that it has no horizontal velocity relative to the car. If he concentrates his attention on the ball, he sees it rising above him and falling again, but always remaining directly above his head (figure 22-10a). Meanwhile, out of the corner of his eye he sees the road, and all objects fixed on it, moving toward his rear at 20 mph. On the other hand, an observer standing still at the side of the road says that the ball initially had a horizontal velocity of 20 mph which it retained throughout its motion, and therefore it followed the curved path (a parabola) shown in figure 22-10b. If either observer wished to provide a mathematical description of the motion, he could set up as his **frame of reference** a set of Cartesian axes, as described in section 2-3. However, one set of axes would be rigidly attached to terra firma and the other set would be rigidly attached to the car and would move with a velocity of 20 mph relative to the first set.

The two observers have very different views of the motion of the ball and other moving bodies, but they agree in one important respect. They both concede the validity of Newton's laws of motion (Chapter 6) and they both agree that the acceleration of the ball is directed toward the center of the earth and has the value

$$g = \frac{GM_e}{R_e{}^2} \tag{22-16}$$

as given by Newton's universal law of gravitation (sections 6-3 and 7-1).

The realization that the fundamental laws of mechanics are independent of the velocity of the observer was one of the major achievements of Galileo and Newton and it may be called the classical principle of relativity.

The classical principle of relativity

The fundamental laws of mechanics are independent of the velocity of the observer.

It is not obvious, however, that this principle also applies to the laws of electromagnetism. As we have already explained in Chapter 16, electromagnetic forces do depend upon velocity. Suppose that the observer at the side of the road sets up two equal electric charges q which are firmly attached to the road and are therefore stationary relative to himself. Let the distance between the charges be R and suppose that the line joining them is perpendicular to the direction of motion of the car (figure 22-11a). The observer standing at the side of the road says that

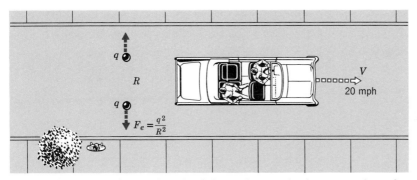

(a) The observer standing at the side of the road sees only electrostatic forces between the two charges.

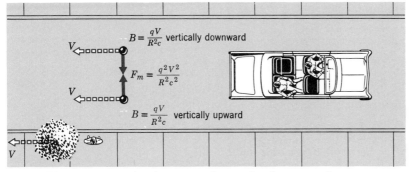

(b) In addition to seeing the electrostatic forces, the observer in the car sees magnetic forces between the two charges.

Figure 22-11 Two observers moving relative to one another have very different views on the character of the forces between two electric charges.

The Problem and Its Solution

471

the only electromagnetic force between the stationary charges is the electrostatic force, given by Coulomb's law (section 13-2).

$$F_e = \frac{q^2}{R^2} \tag{22-17}$$

However, the observer in the car sees that the two charges are moving away from him with a velocity V of 20 mph (figure 22-11b). He therefore concludes that each moving charge produces a magnetic field

$$B = \frac{q}{R^2}\frac{V}{c} \tag{22-18}$$

in the vicinity of the other charge (section 16-2). This magnetic field exerts on the moving charge a magnetic force (section 16-3)

$$F_m = \frac{qBV}{c} \tag{22-19}$$

$$= \frac{q^2}{R^2}\frac{V^2}{c^2} \tag{22-20}$$

The observer in the car sees a magnetic force between the charges, whereas the observer standing at the side of the road does not! Moreover, the observer in the car sees a magnetic field and an electric field, whereas the other observer sees only an electric field!

Can both observers be right, or is the point of view of one of them to be preferred over the other? Do the laws of electromagnetism, if properly formulated, apply in all frames of reference or only in a special frame of reference which is at rest (not at rest relative to an observer but *absolutely* at rest for some fundamental reason)? The frame of reference fixed with respect to the road seems the more reasonable of the two, for it would appear rather conceited for the observer in the car to imagine that he is always at rest and all the rest of the world is moving relative to him. However, as we well know, the road is not really at rest, but is in rapid motion as the earth rotates on its axis, as the earth orbits around the sun and as the sun moves relative to the other stars.

We might therefore ask the astronomers to investigate these motions and correct for them. However, an astronomer can only investigate the motion of his telescope relative to other objects and he usually chooses these objects to be the "fixed stars." We believe that the earth is rotating because the stars appear to perform a daily rotation in the sky. We believe that the earth revolves about the sun because the sun changes its apparent position relative to the stars throughout the year. Moreover, we believe that the sun is moving in a certain direction relative to the stars because, if we look in this direction, we notice that the stars, although they are moving in random directions relative to one another, appear to be moving as a body toward us. In the opposite direction the stars appear to be moving as a body away from us.

The astronomers might, in this way, decide upon a frame of reference at rest relative to the stars. However, most of the stars in the sky belong to a group of stars known as the Galaxy, with a capital G. Far outside this local group of stars are other groups of stars known as galaxies, with a little g. Relative to these objects the Galaxy is rotating at a rate that gives the sun an extra velocity of about 2.5×10^7 cm sec^{-1}. This is much greater than its velocity relative to nearby stars. The "fixed stars" are therefore not really fixed, and so the method must be extended to measure velocities relative to extragalactic objects. Clearly, in order to eliminate completely a possible motion of nearby matter, the observations should be extended to include all the matter in the observable universe. Unfortunately, our telescopes have still revealed to us only a fraction of what we believe is the whole observable universe, and the further we look the more difficult it is to obtain reliable observations. Moreover, modern theories of the nature of the universe, independently of whether they are right or wrong, strongly suggest that observations spanning these enormous distances must not be lightly interpreted in terms of laws which are valid over short distances in our immediate vicinity.

It is conceivable that astronomy might eventually provide us with a rather special frame of reference which, in some sense or other, might be described as being "at absolute rest." It does not follow, though, that this frame will have any special significance as far as the laws of physics are concerned. To resolve this problem, we must revert to the question of whether it is possible to perform experiments in mechanics or electromagnetism that will single out for us a special frame of reference at absolute rest.

It is easy to decide whether or not a frame of reference is *accelerating*, if we assume the validity of Newton's first law of motion (section 6-1). According to this law an isolated body, free from the influence of other bodies, moves in a straight line with constant velocity and therefore has zero acceleration. If, from the point of view of a particular frame of reference, such a body appears to have an acceleration, then it must be the frame of reference itself which has an acceleration of equal magnitude in the opposite direction. Correcting for this, one obtains a frame of reference which is not accelerating and which is called an *inertial frame*, or sometimes a *Galilean frame*. From the point of view of such a frame all isolated bodies move in straight lines with constant velocities.

An inertial frame found in this way is not unique. Any other frame moving with a constant velocity relative to the first frame is also an inertial frame. From the point of view of the second frame all isolated bodies move with constant velocities, although these velocities do not have the same values they have from the point of view of the first frame. Moreover, the classical principle of relativity tells us that the laws of mechanics are the same in all inertial frames. Therefore, no experiment in mechanics enables us to single out one of these frames as being different from the others and at absolute rest. Which brings us back again to the question of whether an experiment in electromagnetism can tell us that a particular frame is at absolute rest. If the ether exists, absolute rest might mean "at rest relative to the ether."

Light is an electromagnetic phenomenon but, as we have explained earlier in this chapter, all attempts to relate the velocity of light to an absolute frame of reference have failed. In section 18-4 it was shown

The Problem
and Its
Solution

473

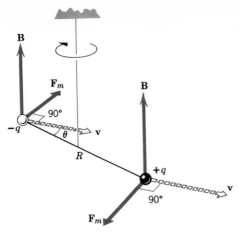

Figure 22-12 The Trouton-Noble experiment.

that the phenomena of electromagnetic induction are independent of which part of the apparatus is considered to be in motion. (At this point the reader would be well advised to read this section again.) Many electromagnetic experiments have been performed with the hope of detecting an absolute frame of reference, but they have all failed. A particularly good example is an experiment performed by Trouton and Noble in 1903. In principle their apparatus consisted of two equal and opposite electric charges, $-q$ and $+q$, held at a fixed distance R apart in a horizontal plane, but suspended on a fiber so that they were free to rotate (figure 22-12). Suppose that such an arrangement is moving relative to the ether with a velocity v as shown. Then, in accordance with the ideas presented in Chapter 16, each moving charge produces a magnetic field in the vicinity of the other charge. Applying right hand rule number 1, this field acts upward on either charge. The magnetic fields exert forces on the moving charges which, according to right hand rule number 2, are in the directions shown. The two magnetic forces constitute a couple (section 10-1), which twists the fiber and rotates the apparatus through a small angle. No such effect was observed. The initial assumption that the charges have a velocity relative to the ether appears not to be meaningful.

The observational evidence therefore clearly indicates that the principle of relativity should be extended to the laws of electromagnetism, and perhaps to all the laws of physics.

The modern principle of relativity

The fundamental laws of physics have the same form in all inertial frames of reference. It is impossible to perform an experiment in physics that will single out one particular frame as being at absolute rest.

This principle was first suggested by Poincaré, and Lorentz solved the mathematical problem of how the laws of electromagnetism could be made the same in all inertial frames, but it was Einstein who first fully understood the physical significance of the principle and who worked out its startling consequences.

The fundamental laws of electromagnetism are embodied in Maxwell's equations (section 18-6). These equations involve the fundamental constant c, which can be shown to be the velocity of an electromagnetic wave and consequently the velocity of light. This constant must be the same in all inertial frames, and so the velocity of light is the same in all inertial frames.

Constancy of the velocity of light

The velocity of an electromagnetic wave in a vacuum has the same value in all inertial frames of reference.

In formulating the modern principle of relativity and the principle of the constancy of the velocity of light, we have used the concept of an inertial frame of reference, thereby excluding accelerating frames. The **special theory of relativity**, which is our immediate concern, confines itself to inertial frames and is primarily concerned with electromagnetism and some aspects of mechanics. The introduction of accelerating frames of reference leads to the **general theory of relativity**, which is primarily a theory of gravitation.

QUESTIONS

1. Does the rotation of the earth on its axis give the apparatus a large enough velocity to affect the Michelson-Morley experiment?
2. Show that the absence of the anticipated effect in the Michelson-Morley experiment could be explained if the velocity of the ether were in a vertical direction, perpendicular to both light beams? How could this possibility be eliminated?
3. Sketch a rough graph to show qualitatively how the expected time difference in the Michelson-Morley experiment varies with the angle as the apparatus is rotated through 360°. In how many positions is the time difference zero and where are they? Assume that the velocity of the ether is horizontal and that the two light beams travel exactly equal distances in the laboratory frame of reference ($l_1 = l_2$).
4. Explain carefully how a repetition of the Michelson-Morley experiment using light from the sun excludes the possibility that the velocity of light has a fixed value c relative to its source.
5. If the telescope of figure 22-6 were filled with water, would you expect the angle of aberration, α, to become larger or smaller or remain unaltered?

The Problem and Its Solution

6. As the baseball player in a car traveling at 20 mph throws his ball vertically upward relative to the car, a monkey hanging from the branch of a tree lets go and falls to the ground. Describe the motion of the ball from the point of view of a frame of reference attached to the falling monkey.

7. Explain carefully why a laboratory on the surface of the earth is not really in an inertial frame of reference. Can you suggest why it is frequently permissible to assume that it is in an inertial frame of reference?

8. Can you give a *completely conclusive* reason for believing that it is the earth that rotates rather than the stars? Try hard to criticize and demolish every argument that might be advanced.

9. In the Trouton-Noble experiment, why is it not sufficient to make one single observation at a particular time on a particular day? Suggest a more decisive procedure.

10. "An observer moving relative to an electric field also sees a magnetic field." Discuss this in connection with the situation of figure 22-11. How is the direction of the magnetic field related to the directions of the electric field and the relative velocity?

11. "An observer moving relative to a magnetic field also sees an electric field." Discuss the case of a copper wire moving in a magnetic field (figure 16-13). Consider the point of view of an observer moving with the wire. How is the direction of the electric field related to the directions of the magnetic field and the relative velocity?

PROBLEMS

A

1. A river is flowing smoothly with a uniform velocity of 50 cm/sec at all points in the water. A buzzer is sounded under the water and the sound is reflected from two rocks each at a distance of 75 m from the buzzer, one directly downstream and the other in a direction at right angles to the flow. What is the time interval between the arrival of the two echoes back at the buzzer? The velocity of sound in stationary water is 1.4×10^5 cm/sec.

2. A gale is blowing with a velocity of 4×10^3 cm/sec. How long will it take for the sound of a siren to travel to a point 500 meters away in the direction (a) from which the wind is blowing, (b) toward which the wind is blowing, (c) perpendicular to the velocity of the wind? The velocity of sound in air may be taken to be 3.3×10^4 cm/sec.

3. In an experiment similar to the one performed by Michelson and Morley, the distance l from the plate P to either mirror is 2×10^3 cm. The wavelength of the light used is 5×10^{-5} cm. What time difference, $t_2 - t_1$, might be expected to result from a velocity of 3×10^6 cm/sec

relative to the ether. Compare this with the period of the light used.

B

4. In an experiment similar to the one performed by Michelson and Morley, the distance l from the plate P to either mirror is 2590 cm. The wavelength of the light used is 5.89×10^{-5} cm and the apparatus is capable of detecting a time difference equal to one-hundredth of the period of this light. If no effect is observed when the apparatus is rotated through $90°$, what is the maximum possible value of the velocity of the apparatus relative to the ether?

C

5. If the earth has a velocity v in a direction perpendicular to the line joining it to a star, show that the angle of aberration, α, is given by

$$\tan \alpha = \frac{v}{c}$$

6. Derive equation 22-10 from equations 22-2 and 22-9 using the approximations discussed in section A7 of the Mathematical Appendix.

7. The sun is rotating about an axis which passes through its center and is very nearly perpendicular to the line joining the sun and the earth. If light traveled with a velocity c relative to its source, the time for which a sunspot was visible on earth would not be exactly equal to the time for which it was hidden behind the sun. Calculate the difference for a sunspot on the equator of the sun, given that the radius of the sun is 7×10^{10} cm and one complete rotation takes 24.7 days. The distance from the sun to the earth is 1.49×10^{13} cm. How far away would an observer have to be to see the sunspot simultaneously on opposite sides of the sun (assuming the original hypothesis about the behavior of light)?

8. Derive an expression for the moment of the couple expected in the Trouton-Noble experiment, assuming that the velocity of the apparatus relative to the ether is horizontal and is at an angle θ to the line joining the two charges (figure 22-12). For what values of θ is the moment of the couple (a) zero (b) largest?

The Problem
and Its
Solution

Time
and
Space

23-1   Prejudices and Observations—A Philosophic Entr'acte

The essential aspects of the last chapter may be briefly summarized as follows. Different parts of the universe do not interact instantaneously, but with a time lag determined by the speed of light. Consequently, the time of a distant event cannot be decided without making the correct assumption concerning the velocity of light relative to the observer. Experiment indicates that the correct assumption is that the velocity of light is the same for all unaccelerated observers, independently of their velocities. This is a consequence of a more general principle that the laws of physics have the same form for all unaccelerated observers.

In spite of the overwhelming experimental evidence in its favor, the assumption of the constancy of the velocity of light comes into direct conflict with our preconceived ideas about addition of velocities (section 22-5 and the discussion associated with figure 22-9). Einstein's outstanding contribution was his realization that this difficulty can be resolved by a reformulation of our concepts of time and space. We must abandon our prejudice that time and space are absolute and the same for all observers. The only consistent way an observer can determine the times of distant events is by assuming the constancy of the velocity of light. It is then found that different observers moving relative to one another may disagree about these times. One observer may decide that two events occurred at exactly the same time, whereas another observer may decide that one of the events preceded the other by a definite time interval. It is even possible for the first observer to decide that an event 1 preceded an event 2, while the other observer decides that 2 preceded 1. As a consequence of their disagreement about time, the two observers may also disagree about the distance between two objects at a fixed instant of time, or about the length of an object.

We are not normally aware of this disagreement between different observers, because the frames of reference we normally deal with have relative velocities much less than the velocity of light. The discrepancies between moving observers are then immeasurably small and the theory of relativity gives almost exactly the same results as classical physics. However, as soon as we deal with velocities comparable with the velocity of light, only the theory of relativity is consistent with the observations. The theory of relativity is "unreasonable" only in the sense that it violates some of our unjustifiable prejudices. In the important respect of correctly predicting the results of experimental observations, the theory of relativity has been tested in many ways and has not yet failed.

Consider how even our conventional ideas of space are really dependent upon an elaborate comparison of theory with observations. Suppose that I enter the classroom to give a lecture on relativity and, by a supreme mental effort, I have abandoned all my prejudices concerning the nature of space. I am determined to rely only upon direct observations and to accept only those theories that are consistent with all the observations. I am startled to discover that the students sitting in the front row are approximately ten times taller than the students sitting in the back row. I then formulate two hypotheses, between which I must decide by suitable experiments. Either students come in varying sizes and the university has conveniently provided smaller chairs in the rear for the smaller students; or else the space in the room has the very peculiar property that it shrinks a student as he moves to the rear. I therefore ask a student in the back row to move to the front row, and discover that, as he does so, he increases in size until he is comparable with the students already in the front row. My second hypothesis is confirmed!

However, as a scientist I have been trained not to accept a theory without subjecting it to every possible test. I therefore move to the rear of the room and discover that the students in the back row are then ten times taller than the students in the front row. My second hypothesis is shattered. In spite of my post-Renaissance upbringing, which has taught me not to look upon myself as the center of the universe, I am beginning to hypothesize that I possess some extraordinary power whereby students increase in stature the nearer they approach to me.

While I am still standing at the rear of the room, my colleague Professor X enters at the front and I solicit his help. He tells me that the students in the front row are obviously ten times taller than the students in the back row. I am reluctant to concede to my colleague the same powers that I possess myself, but even if I did we should presumable cancel one another out and all students would become the same size.

Professor X and I consult on these matters and come up with a very unorthodox theory. We conclude that our observations have been deceiving us. All students are really more or less the same size, but, due to some imperfection in our method of vision, they *appear* to grow smaller as they move away from us. We invent a device called a meter rule, with which we can measure the height of a student. We discover that, as a student holding a meter rule moves away from us, both student and rule shrink in the same way and maintain the same ratio to one another. We are even able to invent a system of geometry which provides a mathematical description of the situation. Upon later referring to the library, we are disappointed to find that we must concede prior publication to a certain Dr. Euclid.

Most of us perform this important piece of scientific research at a very early age, and for the rest of our lives accept its conclusions intuitively. Unfortunately, the upbringing of our children is such that they are confined to frames of reference whose relative velocities are small compared with the velocity of light. If we could take our babies and propel them through space with large variable velocities, we should presumably raise a generation to whom the results of the theory of relativity would be intuitively obvious.

23-2 Procedure

We shall therefore proceed as follows. We shall try to avoid making any assumptions about time and space. Instead we shall always consider an experimental procedure whereby an observer can measure a time interval or a distance and compare his measurements with another observer moving relative to him. However, in the interpretation of these measurements we shall assume without question the validity of the modern principle of relativity. We shall assume that the fundamental laws of physics are the same for all unaccelerated observers, and that it is impossible to make any measurement which singles out one frame of reference as being at absolute rest. In particular, the velocity of light will be assumed to be the same for all observers. This procedure is not intended to prove that time and space *ought to* have certain properties, but merely to examine the consequences of the assumptions. The final test, then, will be whether these consequences are all consistent with experimental observations, or whether it is possible to deduce consequences so obviously violating our experience of time and space as to be absurd.

Imagine two physicists A and B, each in his own space ship equipped with a laboratory, traveling through empty space with different constant velocities. Each physicist must set up apparatus to measure time and length *in his own laboratory* and to investigate the laws of physics *in his own laboratory*. According to our working assumption, they are going to agree in all fundamental respects in this phase of the procedure. As we shall see, the interesting disagreements occur when an observer looks outside his space ship at the other observer's apparatus and compares it with his own. Further disagreements occur when the observers look at distant events and try to decide how far away they are and when they happened.

To measure lengths either observer could proceed in the conventional way already described in section 2-4. He could decide that a convenient unit of length is the wavelength of the orange light emitted by krypton

86 atoms. He could then construct a standard meter rule consisting of a platinum-iridium bar with two scratches on it exactly 1,650,763.73 wavelengths apart. The two observers could construct standard meter rules that would be identical in the following respect. The laws of physics must be the same for both observers. Applying these laws to the platinum-iridium alloy used for the standard rules, it is presumably possible to deduce theoretically the spatial arrangement of atoms in this alloy, and also to deduce the ratio of the distance between neighboring atoms to the wavelength of orange krypton 86 light. This ratio would be the same for both observers. Suppose that *A* makes a meter rule with the distance between the scratches exactly equal to 1,650,763.73 wavelengths. Then, if *B* makes an identical rule containing the same number of platinum and iridium atoms arranged in exactly the same way, the distance between his scratches must also be 1,650,763.73 wavelengths. We are assuming that each observer is concerned at this stage only to compare a meter rule stationary relative to himself with light from a krypton 86 atom stationary relative to himself. As we shall see, disagreements will not arise until an observer examines the measurements being made by the other observer. He will then discover that under certain circumstances the other observer's meter rule, even though it contains the same number of atoms as his own, appears to be shorter because the atoms appear to be closer together.

A convenient clock for our present purpose can be constructed as follows (figure 23-1). The two parallel mirrors are exactly 1 meter apart. A pulse of light is continually reflected backward and forward between these mirrors. Every time the pulse reaches the lower mirror it operates a light sensitive device and a pen makes a mark on a piece of paper moving past it. Each mark on the paper can be regarded as a "tick" of the clock. In terms of our present units, we can say that in between ticks the light travels 200 cm with a speed of 3×10^{10} cm/sec. The time between ticks is therefore

$$t_0 = \frac{200}{3 \times 10^{10}} \text{ sec} \tag{23-1}$$

$$t_0 = \tfrac{2}{3} \times 10^{-8} \text{ sec} \tag{23-2}$$

The number of ticks per second is

$$\frac{1}{t_0} = 150,000,000 \text{ ticks per sec} \tag{23-3}$$

The two space physicists, remembering how things were done back on earth, might therefore agree to take as their unit of time the time needed for the light clock to make 150,000,000 ticks. They must, of course, use the same value for the velocity of light.

It might be argued that, by designing a clock based upon the properties of light, we have prejudiced the situation by making our measurements unnecessarily sensitive to any peculiarities in the behavior of light. Suppose, then, that each observer also makes a conventional clock, using a coiled metal spring and a balance wheel as in a wrist watch, and

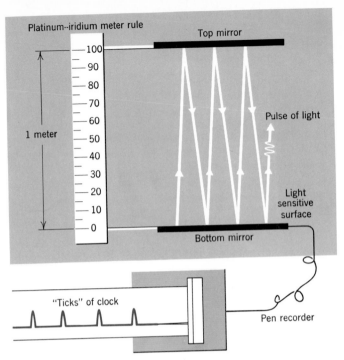

Figure 23-1 The light clock. For clarity, the successive rays of light are shown displaced sideways. Actually, the pulse of light moves up and down on the same vertical line.

compares it with the light clock. The relationship between the two types of clock depends only on the laws of physics and these are assumed to be the same for both observers. It follows that any results obtained using the light clock would also be obtained with a conventional clock.

According to present convention, the fundamental clock is an atomic clock, which utilizes the vibrations of cesium atoms (see page 20). We can readily extend our arguments and ask each observer to compare three clocks: a light clock, a wrist watch, and an atomic clock, all of which must be stationary in his frame of reference. The comparison of the rates of two clocks can be expressed as a number, such as the number of times a cesium atom vibrates while a light pulse travels forward and backward between the mirrors of the light clock. The value of this number depends only on the laws of physics and must therefore be the same for all observers. It follows that an observer may use any type of clock to make his observations.

Now consider a simple example of how the two observers A and B might make measurements *outside* their own laboratories. Suppose they wish to measure their relative velocity V_r which, for simplicity, is assumed to be along the line joining them. Suppose that observer A sends out a pulse of light at an instant which he chooses to take as the zero of time (figure 23-2a). This pulse travels to B, is reflected by a mirror and returns to A. From the number of ticks of his clock that have occurred between the sending out and return of the pulse, A deduces that the journey took t_1 secs. Assuming, as he must, that the light had a velocity c, he concludes that the reflection from B occurred at a time $\frac{1}{2}t_1$ secs and at this instant the distance between himself and B was $\frac{1}{2}ct_1$ cm. A also sends out a second pulse 1 second after the first and finds that it traveled for t_2 sec and arrived back at time $(1 + t_2)$ sec. He then concludes that the reflection of this second pulse occurred at time $(1 + \frac{1}{2}t_2)$, when the distance to B was $\frac{1}{2}ct_2$. Therefore, in between the two reflections B had traveled a distance $(\frac{1}{2}ct_2 - \frac{1}{2}ct_1)$ in a time $[(1 + \frac{1}{2}t_2) - \frac{1}{2}t_1]$. The velocity of B must therefore be

$$V_r = \frac{(\frac{1}{2}ct_2 - \frac{1}{2}ct_1)}{[(1 + \frac{1}{2}t_2) - \frac{1}{2}t_1]} \tag{23-4}$$

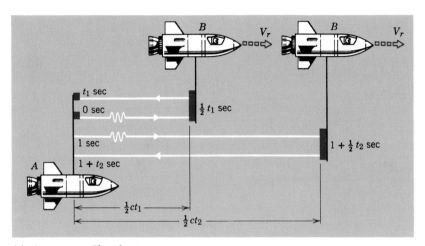

(a) A measures B's velocity.

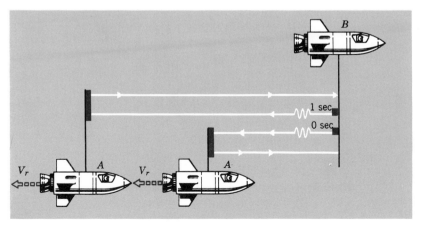

(b) B measures A's velocity.

Time and Space

Figure 23-2 A and B use light signals to measure their relative velocity.

Observer B could use the same procedure to measure the velocity of A (figure 23-2b). He must obtain a velocity having the same magnitude V_r, but in the opposite direction. To assume that one observer would obtain a larger value of the relative velocity than the other observer would be to introduce a difference between observers which is contrary to the principle of relativity.

The procedure just described is an example of what might be called the "radar technique" enabling an observer to determine distances, times, and velocities for remote objects. The observer sends out a succession of pulses of electromagnetic radiation in all directions (figure 23-3). The pulses are all slightly different so that they can be recognized. If a pulse is reflected from an object and returns to the observer t secs after it was sent out, then that object was at a distance $\frac{1}{2}ct$ at a time $\frac{1}{2}t$ after the pulse was sent out. Notice that this does not tell us where the distant object is *now*, but only where it was at some time in the past. A simple refinement of the technique enables the direction of the object to be determined. In practice our radar transmitters are not powerful enough for us to see very far in this way. Nevertheless we can see how, in principle, the location of distant objects could be determined with certainty on the basis of the assumption of the constancy of the velocity of light.

23-3   Moving Clocks

What will happen when the two physicists make observations on one another's clocks? To make a start on this problem, we shall first show that, *if the two clocks are held so that the line joining the two mirrors of each clock is perpendicular to the direction of the relative velocity, V_r, then the observers must agree that the clocks have the same length.* As the two space ships pass close to one another, the two observers hold

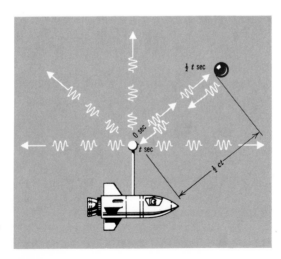

Figure 23-3   The radar technique for fixing the position of a distant object at a known instant of time.

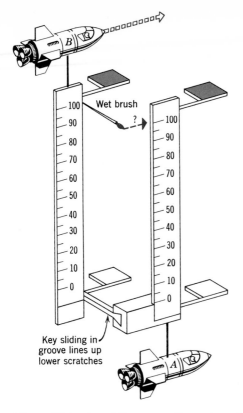

Figure 23-4   *A and B compare the meter rules attached to their clocks.*

out their clocks on poles so that they pass side by side (figure 23-4). A suitable mechanism ensures that the lower mirror of one clock passes exactly to one side of the lower mirror of the other clock. Will the two upper mirrors pass exactly side by side? Sticking out from *B*'s clock is a wet paint brush that is at exactly the same height as *B*'s upper mirror. This brush makes a mark on the meter rule supporting *A*'s mirrors. Will the mark coincide with the scratch on the rule at the same height as *A*'s upper mirror?

Suppose that the mark lies below the scratch, on the same side as the lower mirror. After examining the meter rule, *A* formulates a fundamental law of nature: a meter rule moving in a direction perpendicular to its length is shorter than a stationary meter rule. According to our original assumption this fundamental law should be equally valid for *B*. *B* would therefore apply this law to conclude that *A*'s rule is shorter than his own, and that his paint brush should therefore have made a mark *above* the scratch opposite the upper mirror. However, the evidence of the paint mark is inescapable, and is readily available to *B* if he looks at *A*'s meter rule through his telescope. The only way to avoid this logical inconsistency is to assume that the paint mark must coincide exactly with the scratch, and that the length of a meter rule moving perpendicularly to

Time
and Space

485

itself is the same as the length of a stationary rule. This is not a trivial point, because we shall see later that a meter rule does appear to shorten when it moves in the direction of its length.

Now let us consider A's point of view on the behavior of B's clock as compared with his own (figure 23-5). As far as his own clock is concerned, he says that in between ticks the light travels a distance of 200 cm with a velocity c. So A's value of the time between ticks of his own (A's) clock is

$$t_{AA} = \frac{200}{c} \tag{23-5}$$

When A looks at B's clock he does not see the pulse of light traveling backward and forward along the same path, as it does in his own clock. While B's light pulse is traveling from one mirror to the other the

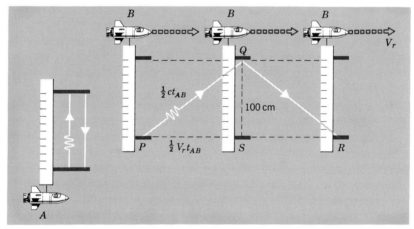

(a) A decides that B's clock is slow.

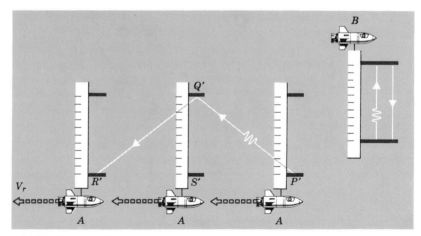

(b) B decides that A's clock is slow.

Figure 23-5 A and B compare their clocks.

mirrors themselves are moving relative to A. A therefore sees B's light pulse travel along the path PQR of figure 23-5a. According to the assumption of the constancy of the velocity of light, A must say that this path is traversed with a velocity c. Let t_{AB} be A's estimate of the time taken to traverse the path PQR, that is, A's estimate of the time interval between "ticks" of B's clock. (The first subscript of t indicates the observer and the second subscript indicates the clock he is observing.) A then says that the time from P to Q is $\frac{1}{2}t_{AB}$ and so

$$PQ = \tfrac{1}{2}ct_{AB} \qquad (23\text{-}6)$$

A says that the lower mirror traverses the path PR with a velocity V_r, which he has determined by the method described in the previous section, and so

$$PR = V_r t_{AB} \qquad (23\text{-}7)$$

Therefore

$$PS = \tfrac{1}{2}PR \qquad (23\text{-}8)$$
$$PS = \tfrac{1}{2}V_r t_{AB} \qquad (23\text{-}9)$$

SQ is the distance between the mirrors of B's clock and, as we have just shown, A agrees that this is 1 meter or 100 cm.

$$SQ = 100 \text{ cm} \qquad (23\text{-}10)$$

Applying Pythagoras' theorem to right angled triangle PQS

$$PQ^2 = PS^2 + SQ^2 \qquad (23\text{-}11)$$

Inserting the values of PQ, PS and SQ which have just been derived

$$(\tfrac{1}{2}ct_{AB})^2 = (\tfrac{1}{2}V_r t_{AB})^2 + (100)^2 \qquad (23\text{-}12)$$
$$\tfrac{1}{4}t_{AB}{}^2(c^2 - V_r{}^2) = (100)^2 \qquad (23\text{-}13)$$
$$t_{AB}{}^2 = \frac{4 \times (100)^2}{(c^2 - V_r{}^2)} \qquad (23\text{-}14)$$
$$= \frac{(200)^2}{c^2\left(1 - \dfrac{V_r{}^2}{c^2}\right)} \qquad (23\text{-}15)$$
$$t_{AB} = \frac{200}{c} \frac{1}{\sqrt{1 - \dfrac{V_r{}^2}{c^2}}} \qquad (23\text{-}16)$$

This is A's estimate of the time between ticks of B's clock and it is not the same as t_{AA} (equation 23-5), which is his estimate of the time between ticks of his own clock.
In fact

$$\frac{t_{AB}}{t_{AA}} = \frac{1}{\sqrt{1 - \dfrac{V_r{}^2}{c^2}}} \qquad (23\text{-}17)$$

Slowing down of a moving clock

$$\frac{\text{Time between ticks of a moving clock}}{\text{Time between ticks of a stationary clock}} = \frac{1}{\sqrt{1 - \dfrac{V_r{}^2}{c^2}}}$$

$$(23\text{-}18)$$

$V_r{}^2/c^2$ is a square and is necessarily positive. Moreover, if V_r is greater than c, $V_r{}^2/c^2$ is greater than 1 and $\left(1 - \dfrac{V_r{}^2}{c^2}\right)$ is negative. The square root of a negative number is imaginary, and this clearly cannot be so for a physical quantity like time. This is our first hint that velocities cannot exceed the velocity of light. At the moment let us accept the fact that $V_r{}^2/c^2$ lies between 0 and 1, and $\left(1 - \dfrac{V_r{}^2}{c^2}\right)$ therefore lies between 1 and 0. $\sqrt{1 - \dfrac{V_r{}^2}{c^2}}$ therefore also lies between 0 and 1, and so t_{AB} is necessarily larger than t_{AA}. This means that A thinks that B's clock is *slow*. If I have a clock that should tick once every second and I alter it to tick once every 2 seconds, then in 1 minute it will tick only 30 times and will register 30 secs, or half a minute, so it is running slow.

It is not permissible to say that B's clock *is* slow compared with A's. This would introduce a basic difference between the two observers which is contrary to the principle of relativity. The correct statement is that, *according to A's observations*, B's clock is slow compared with his own. What is B's point of view? Figure 23-5b shows how the situation appears to B. We can apply to this diagram exactly the same arguments that we applied to obtain A's point of view. B says that the light pulse of his own clock travels up and down the same straight line and the time interval between ticks of his own clock is

$$t_{BB} = \frac{200}{c} \tag{23-19}$$

This is the analogue of equation 23-5. B sees the light pulse of A's clock traveling along the path $P'Q'R'$ and deduces that the time interval between ticks of A's clock is

$$t_{BA} = \frac{200}{c}\frac{1}{\sqrt{1 - \dfrac{V_r{}^2}{c^2}}} \tag{23-20}$$

This is the analogue of equation 23-16.

The analogue of equation 23-17 is

$$\frac{t_{BA}}{t_{BB}} = \frac{1}{\sqrt{1 - \dfrac{V_r{}^2}{c^2}}} \tag{23-21}$$

B therefore says that *A*'s clock is slow!

The disagreement between *A* and *B* is basic. If *A* and *B* were back on earth at rest relative to one another and *A* had a well-made wrist watch, whereas *B* had a poorly made one, they could easily come to an agreement. They could synchronize their watches so that they both read 12 noon when the sun was at its highest position in the sky. They might then discover that *A*'s watch registered 1 P.M. when *B*'s watch registered 12:55 P.M. *A* would then say that *B*'s watch was slow whereas *B* might say that *A*'s watch was fast, which is logically consistent. They could come to an agreement by noticing that *A*'s watch again registered 12 noon when the sun was at its highest position the following day, and *B* could then adjust the rate of his watch to agree with *A*'s from then on. When they are moving relative to one another, however, and they both have the best watch they can make, *A* always says that *B*'s watch is slow, whereas *B* insists that *A*'s watch is slow. No adjustment of their watches can make them agree. They must agree to disagree.

Have we arrived at a conclusion which is so ridiculous that we must abandon our assumptions, as we did when we asked if the lengths of their clocks were different? In that case there was a piece of evidence common to both of them that was inconsistent with the assumptions. They could both observe the position of the paint mark on *A*'s meter rule. The important difference in the case of the time measurement is that, however hard we try, we cannot think of an observation that would convince either one that his view was wrong. Any observation that *A* makes is consistent with his view that *B*'s clock is slow. Any observation that *B* makes is consistent with his view that *A*'s clock is slow.

The reader is reminded again that these are not proofs that moving clocks *ought to* behave in this manner. They are proofs that, if the assumptions of the theory of relativity are true, moving clocks will behave in this way. The final test must be experimental. There are many experiments, such as the Michelson-Morley experiment (figure 22-4) and the Trouton-Noble experiment (figure 22-12), which can be correctly interpreted in terms of these ideas concerning moving clocks and which thereby provide a kind of indirect evidence. Fortunately, a very direct kind of evidence is provided by an experiment with π-mesons. A π-meson or pion is a fundamental particle with a mass intermediate between the mass of an electron and the mass of a proton. It is created by a certain process and a little time later suddenly disappears by changing into other particles. Not all pions live for exactly the same time, but it is possible to define a quantity called the "half-life time," τ. Its significance is that 50% of the pions live for a time less than τ, whereas the other 50% live longer than τ. If the pions are at rest relative to the laboratory, or moving with velocities so small as to be negligible, their half-life time is found to have the value

$$\tau_0 = 1.8 \times 10^{-8} \text{ sec} \tag{23-22}$$

It is easy to obtain a beam of pions moving relative to the laboratory with a very high velocity v, comparable with c. Under these circumstances the half-life time is found to be

$$\tau_v = \frac{\tau_0}{\sqrt{1 - \frac{v^2}{c^2}}} \tag{23-23}$$

This is the form equation 23-17 takes in this case. A bunch of pions can be regarded as a type of clock. The time between ticks of the clock is the time for 50% of the pions to disappear. The laboratory is frame of reference A and, when the pions are at rest, they constitute a clock stationary with respect to observer A, so τ_0 is equivalent to t_{AA}. However, when the pions move with velocity v relative to the laboratory, they constitute a moving clock. This clock would be at rest relative to an observer B moving with velocity $V_r = v$ relative to A. τ_v is therefore equivalent to t_{AB}. Experiments of this kind with pions are in complete agreement with the relativistic formula. In a typical case 50% of the moving pions were observed to be still alive after time τ_v, whereas the number would have been only 2% if the lifetime had been τ_0 as for stationary pions. Similar results have now been obtained with other fundamental particles which can also be created, live on the average for a certain time, and then disappear.

The time τ_0 clearly has a special significance for the pion. It is the half-life time from the point of view of an observer at rest relative to the pion. Presumably, even in the case of the beam of fast moving pions, if we could arrange for our laboratory to travel along with the pions, we should then measure a half-life time τ_0. Such a time is called a "proper time." It is the time interval between events occurring to a body as measured by an observer moving along with the body, so that it is stationary in his frame of reference.

Example 23-3-1

In the present state of the art rockets can be launched with speeds up to about 10^6 cm/sec. Imagine two space ships traveling through empty space with a constant relative velocity of 3×10^6 cm/sec. By how much would they disagree about the rates of their clocks.

Inserting $V_r = 3 \times 10^6$ cm/sec into equation 23-17,

$$t_{AB} = \frac{t_{AA}}{\sqrt{1 - \left(\frac{3 \times 10^6}{3 \times 10^{10}}\right)^2}} \tag{23-24}$$

$$t_{AB} = \frac{t_{AA}}{\sqrt{1 - 10^{-8}}} \tag{23-25}$$

As explained in section A7 of the Mathematical Appendix, when x is very small

$$\frac{1}{\sqrt{1 - x}} = 1 + \tfrac{1}{2}x \text{ approximately} \tag{23-26}$$

Applying this to equation 23-25

$$\frac{1}{\sqrt{1 - 10^{-8}}} \approx 1 + (0.5 \times 10^{-8}) \tag{23-27}$$

or $\qquad t_{AB} \approx t_{AA}[1 + (0.5 \times 10^{-8})]$ (23-28)

t_{AA} and t_{AB} differ by $\frac{1}{2}$ part in 10^8 parts. This means that, after A's clock has ticked out 10^8 sec, A will say that B's clock is $\frac{1}{2}$ sec slow. 10^8 sec is about 3 years. It would take about 6 years for the two clocks to disagree by one second!

It is clear from this example that relativistic effects are extremely small at the largest velocities we can give to macroscopic bodies. It is not surprising that we are not normally aware of these effects. The velocity of the earth in its orbit around the sun is also about 3×10^6 cm/sec. The fastest planet is Mercury, but its velocity is only about 4.8×10^6 cm/sec. This is why the application of classical mechanics to the solar system is so successful, but also why this success is not an argument against relativity.

23-4 Past, Present, and Future

Two observers moving relative to one another also have interesting disagreements about whether two events occurred simultaneously and about which of two events occurred first. In figure 23-6a B has set up the following experiment on a long girder held below his space ship and lined up parallel to the direction of the relative velocity. In the center of the girder at M is a small explosive charge. The flash of light from the explosion travels in both directions and enters the boxes P and Q at opposite ends of the girder at equal distances from M. When the flash of light enters P or Q it operates a light sensitive device which immediately sets off another small explosion.

B's description of what happens is as follows. The light travels with a velocity c relative to himself and has equal distances to travel to P and Q. The light therefore arrives at these two boxes at the same time and the explosions at P and Q occur simultaneously.

A's point of view is illustrated in figure 23-6b. He says that the light is traveling with a velocity c relative to himself, but the apparatus is moving with velocity V_r. Let him take the zero of time as the instant at which the flash of light leaves M. Suppose that his estimate of the length PQ is l_A. (This is not the same as B's estimate of this length, as we shall see in the next section, but this fact is irrelevant at the moment.) If his estimate of the time when the light reaches P and initiates an explosion is t_P, then he says that during this time P has moved a distance $V_r t_P$ to the right. The light has had to travel a distance $(\frac{1}{2}l_A - V_r t_P)$ in a time t_P. Therefore

$$\tfrac{1}{2}l_A - V_r t_P = ct_P \tag{23-29}$$

$$t_P = \frac{l_A}{2(c + V_r)} \tag{23-30}$$

If A's estimate of the time when the light reaches Q and initiates an

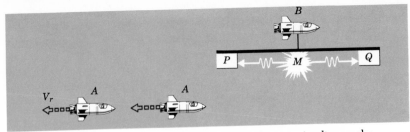

(a) From B's point of view the explosions at P and Q occur simultaneously.

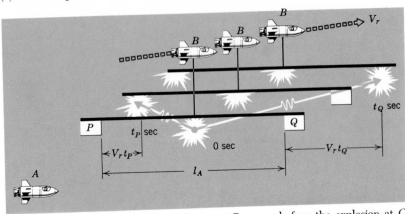

(b) From A's point of view the explosion at P occurs before the explosion at Q. The direction of V_r is really parallel to the girder, but, in order to improve the clarity, the three positions of the girder have been displaced sideways from one another.

Figure 23-6 A disagreement about simultaneity.

explosion there is t_Q, then, during this time Q has moved a distance $V_r t_Q$ to the right. The light has had to travel a distance $(\frac{1}{2}l_A + V_r t_Q)$ and it traveled this distance with speed c in time t_Q, so

$$\tfrac{1}{2}l_A + V_r t_Q = ct_Q \tag{23-31}$$

$$t_Q = \frac{l_A}{2(c - V_r)} \tag{23-32}$$

t_Q is not the same as t_P, and the difference between them is

$$t_Q - t_P = \frac{l_A}{2}\left[\frac{1}{c - V_r} - \frac{1}{c + V_r}\right] \tag{23-33}$$

$$t_Q - t_P = \frac{l_A}{2}\left[\frac{(c + V_r) - (c - V_r)}{(c - V_r)(c + V_r)}\right] \tag{23-34}$$

$$t_Q - t_P = \frac{l_A V_r}{(c^2 - V_r^2)} \tag{23-35}$$

B says that the explosions at P and Q occurred simultaneously, because the light from M had equal distances to travel to P and Q. A says that the explosion at Q occurred later than the explosion at P, because the light had farther to travel to get to Q. On the basis of our assumption concerning the constancy of the velocity of light, both observers' arguments are valid. Moreover, it is contrary to the principle of relativity to suggest that one of the observers is wrong and the other is right, because the latter has a preferred view of the universe. We are forced to conclude that there is no such thing as "absolute time" which is the same for all observers. Time is relative and is different for observers traveling with different velocities.

To illustrate the relativity of past and future, suppose that a box R is placed just to the left of Q and that it also explodes when the flash of light reaches it (figure 23-7). B says that the explosion at R occurs just before the explosion at Q, and that the explosion at P occurs at the same time as the explosion at Q. He therefore says that the explosion at R occurs *before* the explosion at P. However, A says that the explosion at R occurs just before the explosion at Q, but the explosion at P occurs much earlier than the explosion at Q. Therefore, according to A, if R is near enough to Q and $(t_Q - t_P)$ is large enough, the explosion at R occurs *after* the explosion at P.

In view of these startling conclusions, it is relevant to ask if there is any limit to the peculiar behavior of time. There are, in fact, severe restrictions on what is possible. Certain possibilities are so unreasonable that not even the theory of relativity will permit them to occur. Look again at equation 23-35, which gives A's estimate of the time difference $(t_Q - t_P)$ between two events that B says occurred simultaneously. This time difference is proportional to l_A. The disagreement about time depends upon the events occurring at different places and becomes worse the further apart they are. It is a problem that we run into when we try to determine times at *distant* places (see section 22-2 again). If $l_A = 0$, $(t_Q - t_P) = 0$ and both A and B agree that the explosions are simultaneous. They also agree about the sequence of events in time. If, for example, B talks to A over the radio, B's vocal chords are always in the same position relative to B, and so A hears the words in the correct

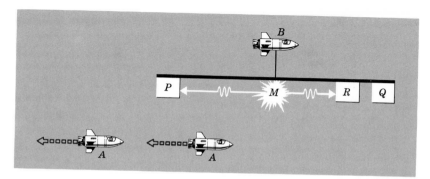

Figure 23-7 A disagreement about past and future. Box R explodes just before Q. From B's point of view, P explodes at the same time as Q and therefore explodes *after* R. From A's point of view, P explodes much earlier than Q and therefore explodes *before* R.

sequence. He does not hear B talking backwards under any circumstances!

An extraterrestrial moving observer watching events on the surface of the earth can never see time running backwards. Neither can he observe events on earth before they happen to us. In fact, anything that happens on earth must send out light signals to the observer, and he is not aware of it until some time after we are, because of the time taken for the light to reach him. However, it is possible for this observer to be aware of an event at a *distant* place before we are, so that it is in his past when it is still to come in our future. Would it then be possible for this observer to send us a radio message about this event, enabling us to predict the future? The answer is no, because the radio message has to travel through space with the speed of light and is always delayed so long that the message arrives after we ourselves have become aware of the event. Even in the theory of relativity time passes in such a way that cause always precedes effect, but a cause can only produce an immediate effect in its immediate neighborhood. Effects produced at distant places have to travel there with the speed of light.

Example 23-4-1

In section 22-2 we raised the question of the time of death of a prehistoric man relative to the instant at which light left a star in the Andromeda galaxy. We now see that this question is to be resolved by assuming that the velocity of light is measured relative to the observer on the earth. Suppose that it is done this way, with the result that the man died one year *before* the light left the star. What is the minimum speed with which an observer must move relative to the earth in order to come to the conclusion that the man died *after* the light left the star?

Suppose that the death of the man and the emission of light from the star appear to occur simultaneously to the observer when his speed relative to the earth is V_r. Then this moving observer is B and the observer on the earth is A. The explosion at P is equivalent to the death of the man and the explosion at Q is equivalent to the emission of light by the star. The time discrepancy is

$$(t_Q - t_P) = 1 \text{ year} \tag{23-36}$$

$$(t_Q - t_P) = 3.16 \times 10^7 \text{ sec} \tag{23-37}$$

The distance between the earth and the Andromeda galaxy is approximately

$$l_A = 2 \times 10^{24} \text{ cm} \tag{23-38}$$

V_r will turn out to be very small compared with c, and so the term $(c^2 - V_r^2)$ in equation 23-35 may be replaced by c^2 without introducing an appreciable error. This equation then becomes

$$(t_Q - t_P) = \frac{l_A V_r}{c^2} \tag{23-39}$$

$$\text{Whence } V_r = \frac{c^2(t_Q - t_P)}{l_A} \tag{23-40}$$

$$= \frac{(3 \times 10^{10})^2 \times (3.16 \times 10^7)}{2 \times 10^{24}} \tag{23-41}$$

$$V_r = 1.42 \times 10^4 \text{ cm/sec} \tag{23-42}$$

At this velocity the two events would occur simultaneously to the observer moving relative to the earth. At all higher velocities the death of the man would appear to occur after the light left the star. The value of V_r calculated above may easily be shown to be equivalent to about 320 mph. This speed is readily attainable by a jet plane.

23-5 Contraction of a Moving Object

In section 23-3 it was shown that, when A and B hold their rules perpendicular to the direction of their relative velocity, the rules are observed to have the same length. What happens when the rules are parallel to the direction of the relative velocity?

A has a rule that he has measured and found to have a length l_{AA}. (As in the case of the clocks the first subscript represents the observer who is measuring the length of the rule and the second subscript represents the observer relative to whom the rule is stationary.) B performs the following experiment to measure the length of A's rule (figure 23-8a). B holds out his clock so that it passes by A's rule. A probe on the bottom of B's clock touches a probe at one end of A's rule and then passes on to touch a probe at the other end. Meanwhile B's clock is ticking away and each tick results in a mark on a piece of paper as explained in section 23-2 and figure 23-1. Whenever the probe on the clock touches a probe on the rule, an electrical circuit is switched on and a large unmistakable mark is made on the same paper. Examination of the small marks between the two large marks shows that the clock made n ticks while passing from one end of the rule to the other. The evidence of the marks on the paper is inescapable and is available to both observers, so there can be no disagreement about the number n.

B's interpretation is as follows. He says that the time between ticks of his clock is t_{BB} (see section 23-3 again). The time taken for A's rule to pass by his clock was therefore nt_{BB}. But A's rule has a velocity V_r. Its length must therefore be

$$l_{BA} = V_r nt_{BB} \tag{23-43}$$

A's interpretation is different. He says that B's clock is slow and the time between its ticks is t_{AB}. The time taken for B's clock to move past his rule was therefore nt_{AB}. Since the velocity of B's clock was V_r and it took a time nt_{AB} to travel a distance l_{AA}, which is A's estimate of the length of his own rule,

$$l_{AA} = V_r nt_{AB} \tag{23-44}$$

From equations 23-43 and 23-44

$$\frac{l_{BA}}{l_{AA}} = \frac{t_{BB}}{t_{AB}} \tag{23-45}$$

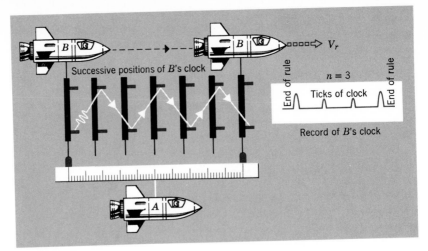

(a) B measures A's rule.

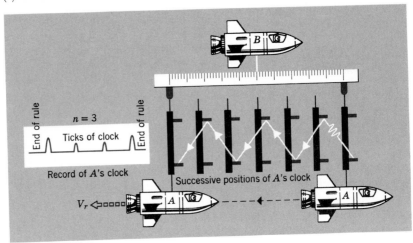

(b) A measures B's rule.

Figure 23-8 A and B measure one another's rules.

A's estimate of the time between ticks of B's clock, t_{AB}, is greater than B's estimate of this time, t_{BB}, in accordance with the equation

$$\frac{t_{BB}}{t_{AB}} = \sqrt{1 - \frac{V_r^2}{c^2}} \qquad (23\text{-}46)$$

(This is equation 23-17 turned upside down with t_{BB} replacing t_{AA}. If A and B make identical clocks, A's measurements on his own clock must give the same results as B's measurements on his own clock. Hence $t_{AA} = t_{BB}$.)

From equations 23-45 and 23-46

$$\frac{l_{BA}}{l_{AA}} = \sqrt{1 - \frac{V_r^2}{c^2}} \qquad (23\text{-}47)$$

If V_r is less than c, $\sqrt{1 - \dfrac{V_r^2}{c^2}}$ lies between 0 and 1. Therefore l_{BA} is less than l_{AA}, which means that B's estimate of the length of A's rule is less than A's estimate.

As in the case of the clocks, this fact does not represent a basic difference between A and B. If B makes a rule and estimates its length to be l_{BB}, then A will estimate its length to have a smaller value, l_{AB} (see figure 23-8b).

$$\frac{l_{AB}}{l_{BB}} = \sqrt{1 - \frac{V_r^2}{c^2}} \tag{23-48}$$

Suppose that each observer has made his rule exactly 1 meter, or 100 cm, long from his own point of view. The two rules might be identical in all respects, containing, for example, exactly the same number of platinum and iridium atoms arranged in exactly the same way. Suppose that $V_r = \frac{1}{2}c$, so that $\sqrt{1 - \dfrac{V_r^2}{c^2}} = 0.866$. Then A estimates that B's rule is only 86.6 cm long, whereas B insists that it is A's rule which is 86.6 cm long. They must again agree to disagree. A rule (or for that matter any object) which is moving relative to the observer in the direction of its length, appears to be shorter than an identical rule stationary relative to the observer. This is called the Fitzgerald-Lorentz contraction, since it is identical with the effect postulated by Fitzgerald and Lorentz to explain the Michelson-Morley experiment. It was pointed out though, in the discussion of the Kennedy-Thorndike experiment (section 22-4), that the Fitzgerald-Lorentz contraction alone is not sufficient to explain all the experiments, but must be supplemented by Einstein's ideas about moving clocks.

The Fitzgerald-Lorentz contraction

An object moving with a velocity V_r relative to the observer is contracted in the ratio $\sqrt{1 - \dfrac{V_r^2}{c^2}}$ in a direction parallel to its motion as compared with an identical object at rest relative to the observer.

It might reasonably be asked why A and B cannot compare their rules by placing them side by side as they did when the rules were perpendicular to the relative velocity (section 23-3 and figure 23-4). First note that the argument used there cannot be repeated when the rules are parallel to the direction of motion. In the latter case a paint brush on one would mark a long line on the other. Why not put the rules side by side and see if the scratches on their ends coincide? The reason why this does not work is revealing. It is necessary to decide where the two scratches at opposite ends of a rule are located *at the same instant of time*, but moving observers cannot agree on this very question of simultaneity. The contraction of a moving body is intimately related to the disagreements concerning time.

The Fitzgerald-Lorentz contraction has interesting consequences for space travel if we aspire to reach objects outside the solar system. The nearest star is Alpha Centauri, which is 4.3 light years away. This means that light takes 4.3 years to travel from it to us. Since a rocket can never exceed the speed of light, it is clear that it would take the rocket at least 4.3 years to reach Alpha Centauri. This is certainly true from the point of view of an observer left behind on earth. However, from the point of view of the pilot of the rocket, the distance between the earth and Alpha Centauri is multiplied by the factor $\sqrt{1 - \dfrac{V_r^2}{c^2}}$

and therefore appears shorter. If V_r is, say, $0.99c$, then $\sqrt{1 - \dfrac{V_r^2}{c^2}} = 0.14$

and the distance appears to be only 14% of its value as seen from the earth. The pilot would therefore deduce that light takes only $0.14 \times 4.3 = 0.60$ years to travel from the earth to Alpha Centauri. Since he sees Alpha Centauri coming toward him with 99% of the velocity of light, he would hope to get there in $0.60/0.99 = 0.606$ years, or just over seven months. By traveling at a velocity even closer to the velocity of light, the rocket might hope to arrive in an even shorter time.

In principle, therefore, it is possible to visit the stars without having to suffer a long tedious journey. In practice the power requirements to launch a rocket near the speed of light are quite prohibitive.

How does it happen that an observer on earth thinks that the rocket takes 4.3 years to reach Alpha Centauri, while the pilot thinks it takes only seven months. The answer is that, from the point of view of the observer on earth, the clock inside the rocket is slow. However, since the laws of physics are the same for the pilot as for the observer on earth, everything that happens inside the rocket keeps time with the rocket's clock. As seen by the observer on earth, inside the rocket electrons in atoms appear to be moving more slowly, molecules appear to be rotating more slowly, atoms in solids appear to be vibrating less rapidly and chemical processes appear to proceed more slowly. In particular, the metabolic processes involved in the digestion of the pilot's food proceed more slowly and he needs to eat less often. At the rate of three meals a day, he needs to eat only $3 \times 365 \times 0.606 = 662$ meals, sufficient for 0.606 years. The biochemical processes responsible for the aging of the pilot also proceed more slowly and he ages by only 0.606 years. Unfortunately he is thinking and acting more slowly, so his lengthened life is no benefit to him.

A man's span of life is a real and important thing. Have we, then, violated the principle of relativity by producing a difference between the pilot and the observer left behind on earth? No, because it is only the observer on earth who is aware of the increased span of life of the pilot. If he confines his attention to what is happening inside the rocket, the pilot can think only the same number of thoughts and do only the same number of things as in a normal span of life. The only thing that is different for him is that the universe appears contracted in the direction of his motion. Moreover, if the pilot looks back to earth, he comes to the *conclusion that it is the observer on earth who is aging less rapidly.* We shall return to this question again in the next chapter, when we shall be in a position to say more about it.

Notice that lengths are contracted by the factor $\sqrt{1 - \dfrac{V_r^2}{c^2}}$ and become smaller and smaller as V_r approaches c. In the limit when $V_r = c$, all lengths in the direction of the motion have shrunk to zero and the time needed to get from one place to another is zero. If we could travel with the speed of light, we could be in all places at the same time! From the point of view of the light (whatever this means) all parts of the universe are in instantaneous contact. However, as we shall soon see, no material body can ever quite reach the speed of light and it is not clear that there is any physical meaning in the startling statements that have just been made.

Example 23-5-1

What is the Lorentz contraction in an automobile traveling at 60 mph? 60 mph is equivalent to 2682 cm/sec.
If V_r is 2682 cm/sec.

$$\frac{V_r}{c} = \frac{2682}{3 \times 10^{10}} \tag{23-49}$$

$$= 8.94 \times 10^{-8} \tag{23-50}$$

$$\left(\frac{V_r}{c}\right)^2 = 8.0 \times 10^{-15} \tag{23-51}$$

When x is very much less than 1,

$$\sqrt{1 - x} = 1 - \tfrac{1}{2}x \text{ approximately} \tag{23-52}$$

Therefore,

$$\sqrt{1 - \left(\frac{V_r}{c}\right)^2} \approx [1 - (4.0 \times 10^{-15})] \tag{23-53}$$

This means that the change in length of a meter rule is only 4.0×10^{-15} meters, or 4.0×10^{-13} cm. Since the diameter of an atom is about 10^{-8} cm, the diameter of a nucleus is about 10^{-12} cm and the size of the electron is about 10^{-13} cm, this contraction is clearly negligible. Again we see that the difference between relativistic and classical physics is not important for the velocities we are normally concerned with.

QUESTIONS

1. Does the time between ticks of a moving light clock depend upon whether it is aligned parallel or perpendicular to the direction of its velocity?

2. Does the slowing down of a moving clock depend upon whether it is moving toward or away from the observer?

3. Do any of the phenomena in this chapter depend upon the relative *positions* of the two observers, as distinct from their relative velocity?
4. In example 23-4-1, in order to achieve the desired result will the airplane have to travel toward the Andromeda galaxy or away from it?
5. Terrestrial observations suggest that galaxies are distributed uniformly in all directions. If an astronomical observatory were launched with a velocity nearly equal to the velocity of light, would the galaxies still appear to be distributed uniformly from the point of view of this observatory? Discuss this situation in relation to the precise meaning of the modern principle of relativity.
6. It is suggested that a light wave traveling from the south toward the north has a velocity which is less than that of a light wave traveling from the north toward the south. How would you attempt to test this?

PROBLEMS

The following approximations are needed to solve some of the following problems.

When v is very small compared with c

$$\sqrt{1 - \frac{v^2}{c^2}} = 1 - \frac{1}{2}\frac{v^2}{c^2} \text{ approximately}$$

$$\frac{1}{\sqrt{1 - \frac{v^2}{c^2}}} = 1 + \frac{1}{2}\frac{v^2}{c^2} \text{ approximately}$$

If v is almost equal to c and

$$\frac{v}{c} = 1 - \alpha$$

where α is very much less than 1,

$$\sqrt{1 - \frac{v^2}{c^2}} = \sqrt{2\alpha} \text{ approximately}$$

When a material body travels with a velocity v which is almost equal to c, it is important that the small difference between v and c should be accurately known.

A

1. How many times does a cesium atom in an atomic clock vibrate between ticks of a light clock such as the one illustrated in figure 23-1?
2. A clock keeps perfect time at rest on earth. If it is given a velocity of $\frac{3}{5}c$, what time interval will an earth observer say that it records in 10 minutes of earth time?
3. A clock keeps perfect time at rest on earth. What velocity must it be given in order to record only 50 seconds in 1 minute?

4. A wrist watch keeps perfect time on earth. If it is worn by the pilot of a space ship leaving the earth with a constant velocity of 10^9 cm/sec, how many seconds does it appear to lose in one day from the point of view of an observer remaining behind on the earth?

5. A particle called a muon has a half-life time of 2.26×10^{-6} sec when at rest. What is its apparent half-life time when it has a velocity of $0.9\ c$?

6. A beam of pions all moving with the same velocity is observed to have a half-life time of 3.2×10^{-7} sec. Calculate their velocity.

7. What is the apparent length of a meter rule with a velocity of $\frac{4}{5}c$ in the direction of its length?

8. A meter rule moves past with a velocity of 2.5×10^{10} cm/sec in the direction of its length. What is its apparent length?

9. A rod has a length of 285 cm at rest. What must be its velocity in the direction of its length if its apparent length is 54 cm?

10. When at rest a box is a cube with a side of length 25 cm. What is its apparent volume when it moves with a velocity of 2×10^{10} cm/sec parallel to one of its sides?

B

11. With what constant velocity must a space ship be launched toward Alpha Centauri in order to arrive there after one day by the pilot's reckoning?

12. The "edge of the observable universe" is believed to be about 10^{10} light years away. (Light takes 10^{10} years to travel from it to us.) If the pilot of a space ship can expect to live another 50 years, with what constant velocity must he travel away from the earth in order to reach the edge of the observable universe before he dies?

13. A space ship is leaving the earth with a constant velocity of 1.8×10^{10} cm/sec. The pilot looks back at the solar system and uses his wrist watch to determine how long it takes the earth to orbit once around the sun. He takes care to make the necessary correction for the time taken for light to travel from the earth to his ship. What result does he obtain for the length of the earth's year? According to the pilot, how many rotations on its axis does the earth make in the course of orbiting once around the sun?

14. A beam of pions has a velocity of $0.99c$. How far does the beam travel in 1.8×10^{-8} sec, which is the half-life time of a pion at rest? How far does the beam actually travel before 50% of the pions have disintegrated?

15. A mass m performs simple harmonic oscillations at the end of a spring with a force constant k. (Compare figure 19-2 and the related discussion.) What is the period of the oscillation according to an observer with a constant velocity v relative to the fixed end of the spring in a direction (a) parallel to the length of the spring (b) perpendicular to the length of the spring?

16. According to an observer on the earth, a certain star is l light years from the earth. A space ship is launched toward the star and takes l

years to get there by the pilot's reckoning. What is the velocity of the space ship?

C

17. A supernova is a star that suddenly explodes and becomes much brighter. Suppose that an astronomer looking through his telescope one night sees two supernovae appear in the Andromeda galaxy at almost the same time. One is very nearly directly behind the other and the distance between them he estimates to be 5×10^{22} cm. They are stationary relative to one another, but both are moving directly toward the earth with a velocity of 3×10^7 cm sec^{-1}. What was the time interval between the two explosions (a) in the frame of reference of the observer on earth (b) in the frame of reference of an observer in the Andromeda galaxy with respect to whom the supernovae are stationary.

18. A holds his meter rule parallel to the direction of his velocity relative to B. He mounts a pistol directly above the scratch at each end of the rule. Using a light signal originating at the midpoint of the rule, he arranges to fire the two pistols simultaneously in a direction perpendicular to the rule. B holds a wooden board parallel to A's rule very close to the muzzles of the pistols, so that the bullet penetrates the board almost immediately after a pistol is fired. A says that the two bullets were simultaneously one meter apart and that the two bullet holes are therefore one meter apart. What result will B obtain if he measures the distance between the bullet holes? What is B's explanation of this result?

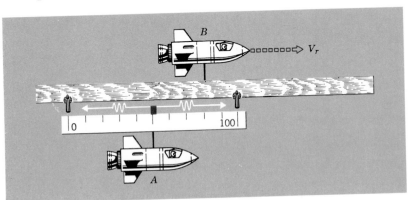

Problem 23-18

19. A holds his meter rule parallel to the direction of his velocity relative to B. He has a paint brush sticking out at right angles to the rule directly opposite the scratch at each end. He rotates the rule about an axis through its center parallel to its length. B holds a graduated bar parallel to A's rule in such a position that the paint brushes make marks on it as they sweep past. What is the distance between the two marks according to B? How does B explain this observation?

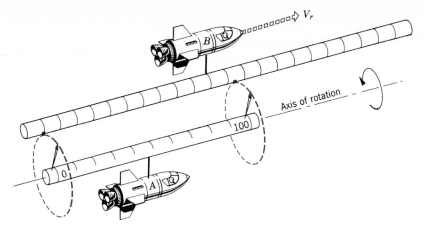

Problem 23-19

20. What is the apparent length of a meter rule moving with a velocity of $\frac{4}{5}c$ at an angle of $45°$ to the direction of its length?

21. What is the apparent length of a meter rule moving with a velocity of 2×10^{10} cm/sec at an angle of $30°$ to the direction of its length?

22. B has made a right angled triangle PQR out of sheet metal. He holds it with the side PQ parallel to the direction of his velocity relative to A and the side QR perpendicular to this direction. According to B the angle RPQ is θ_B. What is A's estimate of this angle? (Notice another unusual consequence of the theory of relativity. Different observers assign different values to angles.)

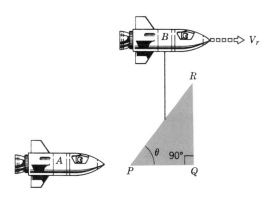

Problem 23-22

23. A and B hold their meter rules parallel to the direction of their relative velocity. A sets up a mirror at an angle of $45°$ to their relative velocity (see the diagram on the next page).
    (a) What is A's estimate of the length of the image of his own rule?
    (b) What is A's estimate of the length of the image of B's rule?
    (c) What is B's estimate of the length of the image of his own rule?
    (d) What is B's estimate of the length of the image of A's rule?

Time and Space

503

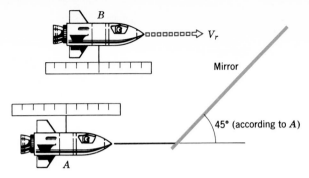

Problem 23-23

24. Repeat problem 23 for the case when *A* and *B* hold their rules *perpendicular* to the direction of their relative velocity.

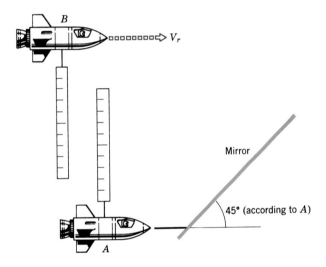

Problem 23-24

Relativistic Mechanics

24-1 Addition of Velocities

Imagine an object C, which is moving in the same direction as the relative velocity of observers A and B (figure 24-1). Using his own clock and rule, B measures the velocity of C and finds it to be v_B. A uses his clock and rule to measure the velocity of C and finds it to be v_A. A also measures the relative velocity of B and C by noting the rate at which the distance between them changes with time. The result of his measurement is the simple and obvious one:

A's estimate of the velocity of C relative to B

$$= v_A - V_r \tag{24-1}$$

The crucial point to be considered is whether this agrees with B's estimate of the velocity of C relative to himself. Classical physics assumes no difference between the measurements of A and B and therefore takes

$$C \quad v_B \text{ when measured by } B$$
$$v_A \text{ when measured by } A$$

$$B$$
$$V_r \text{ relative to } A$$

$$A$$
$$v_A = \frac{v_B + V_r}{1 + \dfrac{v_B V_r}{c^2}}$$

Figure 24-1 The relativistic law of addition of velocities.

it for granted that they agree about the relative velocity of B and C. In classical physics,

$$v_A - V_r = v_B \tag{24-2}$$

or

$$v_A = v_B + V_r \tag{24-3}$$

An important consequence of the theory of relativity is that A and B cannot agree about measurements of length and time. It is not surprising that they do not agree about the velocity of C relative to B or, for that matter, about the velocity of C relative to A. The relationship between v_A and v_B can be deduced from the results obtained in the last chapter, but the mathematics is cumbersome, so we shall be content to quote the result.

The relativistic law of addition of velocities

$$v_A = \frac{(v_B + V_r)}{\left(1 + \dfrac{v_B V_r}{c^2}\right)} \tag{24-4}$$

The proper significance of the various velocities in this discussion should be clearly understood. V_r is the relative velocity of A and B, about which they agree. v_B is B's estimate of the velocity of C relative to himself. v_A is A's estimate of the velocity of C relative to himself. A's estimate of the velocity of C relative to B is $(v_A - V_r)$, which is not the same as v_B. B's estimate of the velocity of C relative to A is $(v_B + V_r)$, which is not the same as v_A.

If v_B and V_r are both very small compared with c, $(v_B V_r)/c^2$ is very small compared with 1. The expression $\left(1 + \dfrac{v_B V_r}{c^2}\right)$ in the denominator in equation 24-4 is therefore very nearly equal to 1. The relativistic formula of equation 24-4 then differs inappreciably from the classical formula of equation 24-3. This is another illustration of the fact that relativistic mechanics is the same as Newtonian mechanics for very small velocities.

Let us apply these ideas to the case of the baseball thrown from a car, which was discussed in section 22-5. Figure 24-2 is similar to figure 22-8, except that the observer standing by the roadside has been identified as A, the observer in the car as B and the ball as C. The relative velocity, V_r, is the velocity of the car, or 20 mph.

$$V_r = 20 \text{ mph} \tag{24-5}$$

Since 1 mph is 44.7 cm/sec

$$V_r = 20 \times 44.7 \text{ cm/sec} \tag{24-6}$$

$$V_r = 894 \text{ cm/sec} \tag{24-7}$$

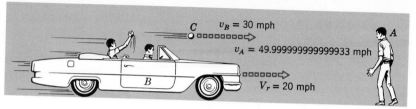

Figure 24-2 Application of the law of addition of velocities to the case of a baseball thrown from a car.

The speed with which the baseball is thrown from the car is v_B, that is, B's estimate of the speed of the ball C.

$$v_B = 30 \text{ mph.} \tag{24-8}$$

$$v_B = 30 \times 44.7 \text{ cm/sec} \tag{24-9}$$

$$v_B = 1341 \text{ cm/sec} \tag{24-10}$$

To find v_A, which is the speed of the ball as it appears to the stationary observer A, we substitute these values of v_B and V_r into equation 24-4

$$v_A = \frac{1341 + 894}{1 + \dfrac{1341 \times 894}{(3 \times 10^{10})^2}} \tag{24-11}$$

$$v_A = \frac{2235}{1 + \dfrac{1.2 \times 10^6}{9 \times 10^{20}}} \text{ cm/sec} \tag{24-12}$$

$$v_A = \frac{2235}{44.7} \frac{1}{1 + (1.33 \times 10^{-15})} \text{ mph} \tag{24-13}$$

$$= \frac{50}{1 + (1.33 \times 10^{-15})} \text{ mph} \tag{24-14}$$

When x is very small, since

$$\frac{1}{(1 + x)} = \frac{(1 - x)}{(1 - x^2)} \tag{24-15}$$

and x^2 is so small compared with 1 that it may be neglected,

$$\frac{1}{(1 + x)} = (1 - x) \text{ approximately} \tag{24-16}$$

Therefore

$$v_A = 50[1 - (1.33 \times 10^{-15})] \text{ mph} \tag{24-17}$$

$$= 50 - (6.67 \times 10^{-14}) \text{ mph} \tag{24-18}$$

$$= 50 - 0.0000000000000667 \text{ mph} \tag{24-19}$$

$$= 49.9999999999999333 \text{ mph} \tag{24-20}$$

Relativistic Mechanics

This is the value quoted previously in section 22-5 and we notice again that it is very nearly the same as the value of 50 mph which would be obtained by adding 20 mph to 30 mph, as in classical mechanics.

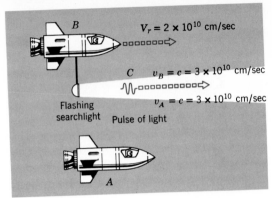

Figure 24-3   Application of the law of addition of velocities to a light beam.

Now consider the other example discussed in section 22-5, involving the speed of light relative to two space ships. In figure 24-3, which is similar to figure 22-9, it is made clear that C is now a flash of light and B's estimate of the velocity of C is

$$v_B = c \tag{24-21}$$

Substituting this value of v_B into equation 24-4, A's estimate of the velocity of C is

$$v_A = \frac{c + V_r}{1 + \dfrac{cV_r}{c^2}} \tag{24-22}$$

$$= \frac{c + V_r}{1 + \dfrac{V_r}{c}} \tag{24-23}$$

Multiply numerator and denominator by c. Then

$$v_A = \frac{c(c + V_r)}{(c + V_r)} \tag{24-24}$$

or $v_A = c$ \hfill (24-25)

The law of addition of velocities automatically ensures that the velocity

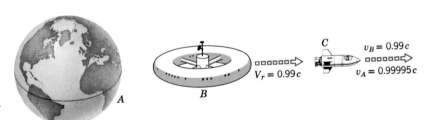

Figure 24-4   An attempt to exceed the speed of light.

of light is c for both observers! So it must, because this was the initial assumption from which everything else was derived.

Another important aspect of the law of addition of velocities is that it is not possible to exceed the speed of light by adding together two velocities less than the speed of light. Suppose, as in figure 24-4, that a large space station B is launched from earth with a velocity of

$$V_r = 0.99c \tag{24-26}$$

The space station then launches a small rocket C in the forward direction with a speed, relative to the station, of

$$v_B = 0.99c \tag{24-27}$$

According to the classical law of addition of velocities, the speed of the small rocket relative to the earth would be $1.98c$. According to the theory of relativity the velocity of the rocket C, as measured by an observer A on the earth, is

$$v_A = \frac{0.99c + 0.99c}{1 + \frac{(0.99c)(0.99c)}{c^2}} \tag{24-28}$$

$$v_A = \frac{1.98c}{1 + 0.9801} \tag{24-29}$$

$$v_A = 0.99995c \tag{24-30}$$

Even though the classical sum of the velocities is larger than c, relativity predicts that the apparent velocity of the rocket from the point of view of the observer A on earth is still slightly less than c.

It is possible for *two* objects to have a relative velocity larger than c, *from the point of view of a third observer*. For example, if *two* space stations are launched with equal velocities $0.99c$ in opposite directions, an observer on the earth estimates their relative velocity to be $1.98c$. By this he means that, according to his measurements, the distance between the two space stations is increasing at a rate $1.98c$. However, an observer on one space station estimates the velocity of the other space station to be $0.99995c$ (prove this). From the point of view of any observer, no *single* object may have a velocity greater than c.

The law of addition of velocities is confirmed by the measurements of the velocity of light in moving water which were made by Fizeau many years before the theory of relativity was proposed. The velocity of light in stationary water is c/n, where n is the refractive index of water (section 21-6). Suppose that the light is traveling through water flowing with a velocity V_r in the same direction as the light (figure 24-5). A is stationary relative to the laboratory. B moves with a velocity V_r relative to A and the water therefore appears stationary in B's frame of reference. Since the laws of physics must be the same for B as for A, B's estimate of the velocity of the light through the water, which is stationary in his frame, must be c/n.

Relativistic
Mechanics

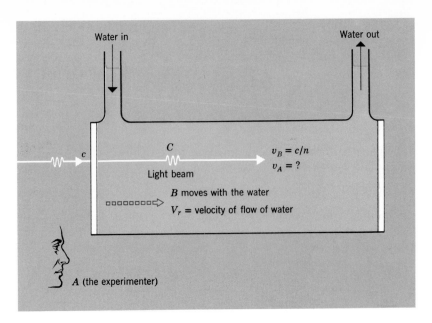

Figure 24-5 Fizeau's experiment on the velocity of light in moving water.

$$v_B = c/n \tag{24-31}$$

A's estimate of the velocity of the light in the water, which he sees flowing with velocity V_r, is obtained by inserting this value of v_B into equation 24-4.

$$v_A = \frac{\dfrac{c}{n} + V_r}{1 + \dfrac{c}{n}\dfrac{V_r}{c^2}} \tag{24-32}$$

Fizeau measured v_A for various values of V_r and obtained results in complete agreement with this formula.

24-2   Mass

In the light of these new ideas of time and space and the way in which velocities add together, the whole of Newtonian mechanics must be reconsidered. The significance of such concepts as mass, force, momentum, and energy must be carefully examined and new definitions must be provided for these quantities. This reformulation of mechanics should be guided by the following principles. The fundamental laws should have the same form for all unaccelerated observers. Relativistic mechanics should become identical with classical mechanics for velocities very small compared with the velocity of light. Moreover, if possible, it would be desirable to arrange matters so that the laws of conservation of energy and momentum are still valid. This implies, of course, that these conservation laws should be valid for all unaccelerated observers. It is not easy to justify this reformulation of mechanics without lengthy, cumbersome arguments or advanced mathematics. We shall therefore merely quote the results.

It is necessary to abandon the concept of mass as an invariable property of a body which remains constant under all circumstances. Instead, mass must be considered to vary with the velocity, v, of the body according to the formula:

Variation of mass with velocity

$$m = \frac{m_0}{\sqrt{1 - \dfrac{v^2}{c^2}}} \qquad (24\text{-}33)$$

When the velocity v is zero the mass is equal to m_0, which is called the **rest mass** of the body. When v is very small compared with c, the mass m is very little different from the rest mass m_0. As the velocity is steadily increased the mass m steadily increases, as shown in figure 24-6.

It is difficult to give a macroscopic body a velocity comparable with c, but it is now possible to produce fundamental particles, such as electrons or protons, with velocities comparable to c. Many radioactive nuclei emit electrons with velocities greater than $0.9c$. Moreover, a major part of current effort in experimental physics is devoted to accelerating charged fundamental particles up to velocities very near c. Many experiments have now been performed to measure the mass of such a particle as a

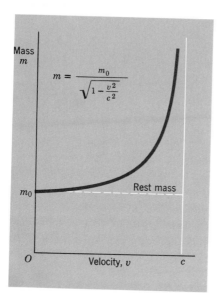

Figure 24-6   The variation of mass with velocity.

function of its velocity. The results are in complete agreement with equation 24-33 and figure 24-6.

When v is only slightly less than c, v^2/c^2 is slightly less than 1 and $\sqrt{1 - \dfrac{v^2}{c^2}}$ is small. m is then very much larger than m_0. For example, if $v = 0.99c$ the reader can easily verify that $m = 7.07m_0$. If v is equal to c, then the denominator $\sqrt{1 - \dfrac{v^2}{c^2}}$ is zero. A constant, m_0, divided by zero is infinity (see section A8 of the Mathematical Appendix). Therefore, as the velocity of the body approaches the velocity of light, its mass increases to infinity! This point is brought out quite clearly in figure 24-6.

It is now easy to see why a material body cannot have a velocity greater than the velocity of light. If we try to accelerate the body, as its velocity approaches the velocity of light its mass becomes larger and larger and it becomes increasingly more difficult to accelerate it further. In fact, since the mass becomes infinite when $v = c$, we can never quite accelerate the body up to the velocity c. We can approach closer and closer to c, by making greater and greater efforts, but we can never quite get there. This is a well known fact in connection with the machines that accelerate fundamental particles up to enormous energies and high velocities. An x-ray tube (figure 20-5) with an applied voltage of one million volts, costing about $70,000, can accelerate electrons to a velocity of $0.94c$. A synchrotron at Cornell University, which had the same effect as an accelerating voltage of 1.2 billion volts and which cost about $200,000 achieved velocities of $0.99999991c$. A linear accelerator which, at the time of this writing, is being built at Stanford University will be equivalent to about 30 billion volts and will cost at least 120 million dollars. It will achieve velocities of about $0.99999999985c$.

The conclusion that a body cannot have a velocity greater than c is basic to the whole philosophy of the theory of relativity. Otherwise the body could carry a message to a distant place and, if the velocity of the body could be increased without limit, it would be possible to establish instantaneous communication with this distant place. If observers in different places were in instantaneous communication with one another, they would have no disagreements about simultaneity because they would all be simultaneously in touch with one another. They would then have no difficulty in adjusting their clocks to agree under all circumstances.

The velocity of a body can never exceed the velocity of light.

24-3   Momentum and Force

Using the new definition of mass, the momentum of a particle can be defined as $\mathbf{p} = m\mathbf{v}$, exactly as it was in Newtonian mechanics.

Momentum

$$\mathbf{p} = m\mathbf{v} \tag{24-34}$$

$$\mathbf{p} = \frac{m_0\mathbf{v}}{\sqrt{1 - \dfrac{v^2}{c^2}}} \tag{24-35}$$

With this definition of momentum, the law of conservation of momentum is still true. In fact, if we consider the collision of two bodies (compare section 8-4), and insist that the law of conservation of momentum shall be true *from the point of view of all unaccelerated observers*, then, after a little mathematical manipulation, we can show that mass must vary with velocity in accordance with equation 24-33.

In section 6-5 force was defined as mass multiplied by acceleration. In the theory of relativity it proves preferable to use the definition of section 8-1, that force is the rate of change of momentum with time. Suppose that, from the point of view of a particular observer, a body has a momentum $\mathbf{p}_1$ at time t_1 and a momentum $\mathbf{p}_2$ at a slightly later time t_2.

Force

Force = rate of change of momentum with time $\qquad$ (24-36)

$$\mathbf{F} = \frac{(\mathbf{p}_2 - \mathbf{p}_1)}{(t_2 - t_1)} \text{ when } (t_2 - t_1) \text{ is sufficiently small} \tag{24-37}$$

If the body has velocity $\mathbf{v}_1$ at time t_1 and velocity $\mathbf{v}_2$ at time t_2, then

$$\mathbf{p}_1 = \frac{m_0\mathbf{v}_1}{\sqrt{1 - \dfrac{v_1^2}{c^2}}} \tag{24-38}$$

$$\mathbf{p}_2 = \frac{m_0\mathbf{v}_2}{\sqrt{1 - \dfrac{v_2^2}{c^2}}} \tag{24-39}$$

and

$$\mathbf{F} = \frac{1}{(t_2 - t_1)}\left[\frac{m_0\mathbf{v}_2}{\sqrt{1 - \dfrac{v_2^2}{c^2}}} - \frac{m_0\mathbf{v}_1}{\sqrt{1 - \dfrac{v_1^2}{c^2}}}\right] \tag{24-40}$$

24-4 Mass and Energy

When light falls on a surface it warms it up and exerts a pressure on it. This can be explained by assuming that electromagnetic radiation carries energy and momentum (section 20-4). If an amount of energy S is incident on unit area per second, a quantity of momentum S/c is incident

Relativistic Mechanics

on the same unit area per second (equation 20-21). A total energy E of electromagnetic radiation is associated with a momentum E/c. (Here E represents energy and *not* the electric field.) The effect is the same as if the light were a mass m traveling with a velocity c and having a momentum mc, given by

$$mc = \frac{E}{c} \qquad (24\text{-}41)$$

In this sense, therefore, an energy E of electromagnetic radiation is equivalent to a mass

$$m = \frac{E}{c^2} \qquad (24\text{-}42)$$

In the theory of relativity this equation is found to have a much deeper significance. Mass and energy can be considered as different manifestations of the same physical quantity. A quantity of energy E, whether it be kinetic energy, potential energy, chemical energy, nuclear energy or any other form of energy, can always be considered to have a mass E/c^2. Conversely, the mass m of a material body can be considered to be equivalent to an amount of energy mc^2 and, under some circumstances, can even be converted into this amount of energy.

Equivalence of mass and energy

$$E = mc^2 \qquad (24\text{-}43)$$

This is the most famous equation of modern physics. Its connection with atomic and hydrogen bombs and nuclear power is well known.

The full significance of the equation will become apparent as we apply it to various phenomena throughout the rest of this book. We shall see, for example, that a ray of light is bent in the gravitational field of the sun as though the light were a massive body. We shall see that the energy of electromagnetic radiation can be converted into the mass of fundamental particles. Conversely, fundamental particles can disappear and their mass reappear as the energy of electromagnetic radiation. The mass of a nucleus will be shown to be due, in part, to the energy of the particles inside it.

At present, let us consider only the simple case of a body of rest mass m_0 moving with a velocity v. According to equation 24-43 its total energy is

$$E = mc^2 \qquad (24\text{-}44)$$

Inserting the value of m given by equation 24-33

Total energy of a body with rest mass m_0 and velocity v

$$E = \frac{m_0 c^2}{\sqrt{1 - \dfrac{v^2}{c^2}}} \tag{24-45}$$

If the velocity v is very small compared with c, we can make use of the following approximation. When x is very small

$$\frac{1}{\sqrt{1 - x}} = 1 + \tfrac{1}{2}x \text{ approximately} \tag{24-46}$$

Therefore, if v^2/c^2 is very small

$$\frac{1}{\sqrt{1 - \dfrac{v^2}{c^2}}} = 1 + \frac{1}{2}\frac{v^2}{c^2} \text{ approximately} \tag{24-47}$$

So $E \approx m_0 c^2 \left(1 + \dfrac{1}{2}\dfrac{v^2}{c^2} \right)$ $\tag{24-48}$

or $E \approx m_0 c^2 + \tfrac{1}{2} m_0 v^2$ $\tag{24-49}$

The second term $\tfrac{1}{2}m_0 v^2$ is the classical kinetic energy! The relativistic expression for the energy therefore becomes identical with the classical expression at very small velocities, except for the addition of the **rest mass energy**, $m_0 c^2$. Since m_0 is a constant, the rest mass energy remains constant and need not be taken into account in classical physics when the conversion of energy from one form into another is being considered. However, modern physics recognizes processes in which the rest mass is annihilated and an amount of energy $m_0 c^2$ is made available.

In classical physics it was taken for granted that mass could not be created or destroyed and this was sometimes called the law of conservation of mass. Classical physics also had its law of conservation of energy. Now that mass and energy have been found to be interchangeable, these two laws must be combined into a single law of conservation of mass plus energy. It must be understood, of course, that in any equation expressing this law the mass must be multiplied by c^2 in order to enter on the same footing as energy.

Finally, let us notice an interesting feature of equation 24-45. When $v = c$, m is infinite and E is infinite. If we try to accelerate a body, as its velocity approaches c its energy becomes very large. Its velocity cannot be made equal to c without giving it an infinite amount of energy. This throws further light on the discussion in section 24-2 concerning the machines which accelerate fundamental particles up to velocities near the velocity of light. Although these machines seem to be achieving very little as far as the velocity is concerned, they are in fact giving the particles enormously high energies. In fact, the "equivalent accelerating voltage" is a measure of the energy given to the particle in electron

Relativistic
Mechanics

volts. The x-ray tube produces 1 MeV electrons, the Cornell synchrotron 1.2 BeV electrons and the Stanford linear accelerator 30 BeV electrons. (At this point the reader should review section 15-6, where the electron volt is discussed.)

Example 24-4-1

Express the rest mass energy of an electron in ergs and in electron volts.

The rest mass of an electron is

$$m_0 = 9.11 \times 10^{-28} \text{ gram} \tag{24-50}$$

Its energy is

$$m_0 c^2 = 9.11 \times 10^{-28} \times (3 \times 10^{10})^2 \text{ erg} \tag{24-51}$$
$$m_0 c^2 = 8.2 \times 10^{-7} \text{ erg} \tag{24-52}$$

Referring to equation 15-72,

$$1 \text{ electron volt} = 1.60 \times 10^{-12} \text{ erg} \tag{24-53}$$

$$\text{So } m_0 c^2 = \frac{8.2 \times 10^{-7}}{1.60 \times 10^{-12}} \text{ eV} \tag{24-54}$$

$$m_0 c^2 = 0.51 \times 10^6 \text{ eV} \tag{24-55}$$

The rest mass energy of an electron is 0.51 MeV.

Example 24-4-2

At what velocity is the mass of a particle twice its rest mass?

$$m = 2m_0 \tag{24-56}$$

$$\frac{m_0}{\sqrt{1 - \dfrac{v^2}{c^2}}} = 2m_0 \tag{24-57}$$

$$\sqrt{1 - \frac{v^2}{c^2}} = \frac{1}{2} \tag{24-58}$$

$$1 - \frac{v^2}{c^2} = \frac{1}{4} \tag{24-59}$$

$$\frac{v^2}{c^2} = \frac{3}{4} \tag{24-60}$$

$$\frac{v}{c} = \frac{\sqrt{3}}{2} \tag{24-61}$$

$$\frac{v}{c} = 0.866 \tag{24-62}$$

$$v = 0.866 \times 3 \times 10^{10} \text{ cm/sec} \tag{24-63}$$

$$v = 2.60 \times 10^{10} \text{ cm/sec} \tag{24-64}$$

Example 24-4-3

What is the velocity of an electron with an energy of 1 MeV?

It is important to realize that, when the energy of a particle is given in electron volts, the energy referred to does not include the rest mass energy m_0c^2. Consider an electron initially at rest at a point where the electric potential is 0. Its total energy is m_0c^2. If it is then accelerated by an electric field and moves to a point where the voltage is V, its potential energy decreases from 0 to $-Ve$. At the same time its velocity increases to v and its mass to m. The law of conservation of mass plus energy tells us that

$$mc^2 - Ve = m_0c^2 \tag{24-65}$$

$$\text{or} \quad Ve = mc^2 - m_0c^2 \tag{24-66}$$

The energy given in electron volts is the quantity $(mc^2 - m_0c^2)$.
Substituting the value of m from equation 24-33

$$Ve = \frac{m_0c^2}{\sqrt{1 - \dfrac{v^2}{c^2}}} - m_0c^2 \tag{24-67}$$

$$\frac{Ve}{m_0c^2} = \frac{1}{\sqrt{1 - \dfrac{v^2}{c^2}}} - 1 \tag{24-68}$$

If the rest mass energy of the electron is V_0 eV,

$$m_0c^2 = V_0e \tag{24-69}$$

V_0e was shown in example 24-4-1 to be 0.51 MeV.

Then

$$\frac{V}{V_0} = \frac{1}{\sqrt{1 - \dfrac{v^2}{c^2}}} - 1 \tag{24-70}$$

$$\frac{1}{\sqrt{1 - \dfrac{v^2}{c^2}}} = 1 + \frac{V}{V_0} \tag{24-71}$$

Putting $Ve = 1$ MeV and $V_0e = 0.51$ MeV

$$\frac{1}{\sqrt{1 - \dfrac{v^2}{c^2}}} = 1 + \frac{1}{0.51} \tag{24-72}$$

$$= 2.96 \tag{24-73}$$

$$1 - \frac{v^2}{c^2} = \left(\frac{1}{2.96}\right)^2 \tag{24-74}$$

$$= 0.114 \tag{24-75}$$

$$\frac{v^2}{c^2} = 0.886 \tag{24-76}$$

$$\frac{v}{c} = 0.94 \tag{24-77}$$

$$v = 0.94 \times 3 \times 10^{10} \text{ cm/sec} \tag{24-78}$$

$$= 2.82 \times 10^{10} \text{ cm/sec} \tag{24-79}$$

Relativistic
Mechanics

QUESTIONS

1. A body is moving along the X axis with a velocity which is not small compared to c. A force is applied to it parallel to the Y axis. What happens to the component of its velocity parallel to the X axis?
2. Why is it not strictly true to say that the number of $_6C^{12}$ atoms in exactly 12 gm of a pure diamond is exactly equal to Avogadro's number?
3. What would happen to the theory of relativity if the velocity of light were infinitely large? Illustrate your answer by inserting $c = \infty$ in all the important formulas.
4. Can you derive equation 24-4 by considering the different points of view of A and B concerning measurements of length and time?

PROBLEMS

The following approximations are needed to solve some of the following problems.

When v is very small compared with c,

$$\sqrt{1 - \frac{v^2}{c^2}} = 1 - \frac{1}{2}\frac{v^2}{c^2} \text{ approximately}$$

$$\frac{1}{\sqrt{1 - \frac{v^2}{c^2}}} = 1 + \frac{1}{2}\frac{v^2}{c^2} \text{ approximately}$$

If v is almost equal to c and

$$\frac{v}{c} = 1 - \alpha$$

where α is very much less than 1,

$$\sqrt{1 - \frac{v^2}{c^2}} = \sqrt{2\alpha} \text{ approximately}$$

When a material body travels with a velocity v which is almost equal to c, it is important that the small difference between v and c should be accurately known.

A

1. A space ship is launched from earth with a velocity of $\frac{1}{2}c$. The ship then launches a small rocket in the forward direction with a velocity of $\frac{1}{2}c$ relative to itself. What is the velocity of the rocket from the point of view of an observer on the earth?
2. A space ship is launched from earth with a velocity of 1.5×10^{10} cm/sec. The ship then launches a small rocket in the forward direction with a velocity of 2.4×10^{10} cm/sec relative to itself. What is the velocity of the rocket from the point of view of an observer on earth?

3. A body has a rest mass of 5 gm. What is its mass when its velocity is 2×10^{10} cm/sec?

4. What is the mass of a proton when its velocity is $(3/5)c$?

5. A meter rule is moving parallel to its length. What is its apparent length when its mass is twice its rest mass?

6. What mass in grams is equivalent to an energy of 1 erg?

7. What is the kinetic energy of an electron in electron volts when its mass is twice its rest mass?

8. What is the rest mass energy of a proton (a) in ergs, (b) in electron volts?

9. The "large calorie" used by dieticians is 1,000 times the "small calorie" defined by equation 12-38. What mass is equivalent to one large calorie? Comment.

10. What is the momentum of a body with a rest mass of 5 gm when its speed is 2.4×10^{10} cm/sec?

11. A body has a speed of 1.8×10^{10} cm/sec and a momentum of 2.5×10^{10} gm cm/sec. What is its rest mass?

12. A dust particle with a rest mass of 10^{-8} gm is initially at rest. What constant force must be applied to it to build up its velocity to $\frac{3}{5}c$ in 10 sec?

B

13. A space ship is launched from earth with a velocity of 2×10^{10} cm/sec. A second space ship is launched the following day and the pilot of the first ship sees it coming directly toward him with a velocity of 1.5×10^{10} cm/sec. What is the velocity of the second ship from the point of view of an observer on the earth?

14. Show that equation 24-4 may be rearranged to give

$$v_B = \frac{v_A - V_r}{1 - \dfrac{v_A V_r}{c^2}}$$

Discuss the significance of this equation.

15. A space ship is launched from earth with a velocity of 1.5×10^{10} cm/sec. The ship then fires a small rocket back to earth with a velocity of 2.5×10^{10} cm/sec relative to itself. With what velocity does an earth observer see the rocket approaching him?

16. A space ship passes by the earth with a velocity of 2.5×10^9 cm/sec. What is the pilot's estimate of the mass of the standard 1 kilogram platinum-iridium cylinder kept at Sevres?

17. At what velocity is the mass of a particle three times its rest mass?

18. The rest mass of a pion is equivalent to 140 MeV. The half-life time of a pion at rest is 1.8×10^{-8} sec. What is the half-life time of a 400 MeV pion?

19. What is the velocity of a particle when its momentum is $m_0 c$?

20. A constant force of 3×10^5 dyne is applied to a particle with a rest mass of 4×10^{-4} gm. How long will it take to accelerate the velocity from 1.8×10^{10} cm/sec to 2.4×10^{10} cm/sec?

Relativistic
Mechanics

21. What is the momentum of an electron with a velocity of 1.8×10^{10} cm/sec?

22. What is the momentum of a 1 BeV proton?

23. What is the total energy in ergs of an electron with a velocity of $0.9999\ c$?

24. Through what potential difference in volts must an electron be accelerated from rest to achieve a velocity of $\frac{1}{2}c$?

25. An electric generator supplying a small town develops 2×10^6 watt of power. How much mass would have to be destroyed to provide an equivalent amount of energy over a period of one year?

26. The cost of electrical energy supplied by a public utility company is $\frac{1}{2}$ cent for one million joules. At this rate, what is the cost of 1 gm of energy?

27. A body is initially at rest. Half of its rest mass is destroyed and given as kinetic energy to the other half. What is the resulting velocity?

28. What is the distance between two electrons when their mutual electrostatic potential energy is equivalent to the sum of their rest masses when they are infinitely far apart?

C

29. The points A and B are under water which is flowing with a velocity v from A directly toward B. The distance AB is $2l$. Two short light pulses leave A and B simultaneously (in the frame of reference in which A and B are at rest). Prove that the pulses first meet at a point which is a distance δ from the midpoint of AB, where

$$\delta = l\frac{vn}{c}\left(\frac{1 - \dfrac{1}{n^2}}{1 - \dfrac{v^2}{c^2}}\right)$$

If $l = 100$ cm, $n = 1.33$, $v = 10^3$ cm/sec, find δ and compare it with the wavelength of visible light.

30. A certain material has a density of ρ_0 gm/cm³ when at rest. What is its apparent density when its velocity is v? (Hint: consider a cube of the material with side of length a moving with velocity v parallel to one side).

31. Imagine that you have discovered a new particle, the iota, with a rest mass equal to 30 times the rest mass of an electron. What is the velocity of a 45 MeV iota?

32. A space ship and its pilot have a total rest mass of 1000 kg. How much energy must be given to the ship during launching to enable it to reach Alpha Centauri in one day by the pilot's reckoning? (Hint: this problem can be solved without first calculating the velocity).

33. A space ship and its pilot have a total rest mass of 200 kg. To launch this ship into space, a mass M is annihilated and given as kinetic energy to the ship. The pilot expects to live another fifty years and

in this time he hopes to reach the "edge of the universe," which is about 10^{10} light years away. How large must M be? Taking the average density of the rocks to be 3 gm/cm³, what would be the height in miles of a hemispherical mountain with this mass?

34. What is the increase in the mass of 1 gm of helium gas as its temperature is raised from 0°K to 300°K. How many helium atoms would have to be removed to restore the mass to the value it had at 0°K? The helium may be taken to be an ideal gas.

35. The two protons in a helium nucleus are about 2×10^{-13} cm apart. Express their mutual electrostatic potential energy in atomic mass units and decide what fractional change it produces in the atomic mass of helium.

36. What mass is equivalent to the mutual gravitational potential energy of the earth and the sun? If this amount of mass were added to the sun, how large a difference in seconds would it make to the length of the year, assuming the distance between the earth and the sun remained the same?

25

The General Theory of Relativity

25-1 Accelerating Frames of Reference

The special theory of relativity is restricted to nonaccelerating observers in inertial frames of reference and is mainly concerned with electromagnetic phenomena. The general theory of relativity considers accelerating frames of reference and thereby becomes concerned with gravitational phenomena. The special theory is supported by an overwhelming mass of experimental observations. There are only a few observations supporting the general theory, which therefore rests on a much less secure foundation. Nevertheless, the issues raised by the general theory are very profound and well worth consideration, even though there may be some doubt whether the theory has provided the ultimate answers.

Although there is no experiment that can distinguish between an observer at rest and an observer with a constant velocity, it is very easy to decide whether or not an observer is in an accelerating frame of reference. When a nonaccelerating observer looks at an isolated body, subject to no forces from other bodies, he sees it moving with a constant velocity in a straight line, which is Newton's first law of motion (section 6-1). When an accelerating observer looks at an isolated body, it appears to him to have an acceleration **a** and moreover all isolated bodies appear to have this same acceleration **a.** In reality, it is the observer himself who has an acceleration −**a** in the opposite direction. This method of detecting acceleration is very simple and unambiguous, because the nonaccelerating observer sees all isolated bodies moving along straight lines, whereas the accelerating observer sees most isolated bodies moving along curved paths.

Similar considerations may be applied to decide whether or not a frame is rotating. This is really the same question because, as was pointed out in section 5-4, uniform motion of a body in a circle inevitably implies an acceleration of the body toward the center of the circle. Imagine an observer on a turntable inside a closed room, watching the walls of the room turn around him. How can he decide whether he is rotating, or whether he is stationary and the room is rotating around him? Suppose that he places a smooth object on top of a desk which is so smooth that the frictional force it exerts on the object can be neglected. If the turntable is not rotating, when he releases the object it remains stationary in the position on the desk where he placed it (figure 25-1). However, if the turntable is rotating the observer on the turntable sees the object move across the desk away from the axis of rotation, not in a straight line but curving to one side (figure 25-2). An observer standing outside

(a) The observer is stationary and the room is rotating.

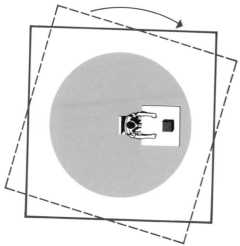

(b) The smooth object on a smooth table remains stationary relative to the table.

Figure 25-1 The observer is not rotating.

The General
Theory of
Relativity

(a) The room is stationary and the observer on the turntable is rotating.

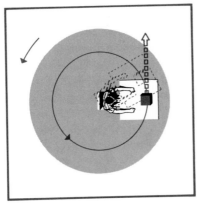

(b) When released, the smooth object on the smooth table moves in a straight line tangential to its original circular motion.

(c) Apparent motion of the smooth object in the observer's frame of reference.

Figure 25-2 The observer is rotating.

the turntable and not rotating would explain the situation as follows. Before the observer on the turntable releases his object it is moving in a circle. When the object is released it experiences no horizontal force, because the top of the desk is frictionless. In accordance with Newton's first law of motion, the object therefore continues to move in a straight line tangential to the circle (figure 25-2b). Quite apart from the explanation, the test is a simple practical matter. The observer places a smooth object on a smooth desk. If it does not move, he is not rotating. If it moves across the desk along a curved path, he is rotating.

Another method of deciding was pointed out by Newton. If a vessel of water is rapidly rotated the water is thrown outward and the surface becomes curved, with the level near the wall higher than the level in the center (figure 25-3). The observer on the turntable should therefore look at the surface of water in a vessel. If the surface is flat he is not rotating. If the surface is curved he is rotating. This phenomenon is

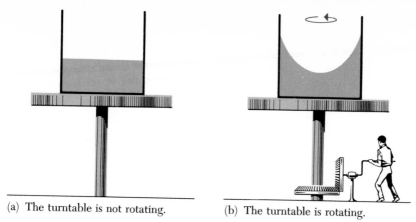

(a) The turntable is not rotating.   (b) The turntable is rotating.

Figure 25-3   When the vessel is rotating the surface of the water inside it is curved.

related to a very simple, common sense attitude to the question of how the observer decides whether he is rotating. If he is rotating he becomes dizzy! Rotation causes the fluid in the ducts of the inner ear to be thrown outward, and the pressure of this fluid is responsible for the sensation of dizziness.

The earth is rotating on its axis and the stars therefore appear to move around the earth once every day. How can we know that it is not really the earth which is stationary and the stars which move in a body around the earth? The rotation is so slow that the resulting curvature of the surface of water in a vessel is too small to detect. Neither can a desk top be made smooth enough to apply the test of figure 25-2. However, a device called a Foucault pendulum can be made to reveal the earth's rotation. It is a pendulum suspended in such a way that, as the point of suspension rotates with the earth, the couple exerted on the pendulum

(a) The pendulum oscillates in the plane of the paper and the earth rotates underneath it.

Figure 25-4   The Foucault pendulum.

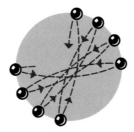

(b) As seen by an observer fixed to the earth, the plane of oscillation of the pendulum slowly rotates.

is negligibly small. Since there is nothing to twist it, the pendulum does not rotate, but continues to oscillate in a fixed direction. However, the earth rotates underneath the pendulum and the direction of oscillation of the pendulum is not fixed relative to the earth (figure 25-4). An observer standing on the earth sees the direction of oscillation of the pendulum slowly rotating, returning to its original position after approximately 24 hours. (The time taken for the earth to rotate once is not exactly 24 hours. There are complications due to the movement of the earth around the sun. Also, we are assuming that the pendulum is at the north pole. At any other latitude, the situation is more complicated and the time for one rotation is greater than 24 hours, becoming infinitely large at the equator.)

These are very real effects that we have just been discussing and the observations cannot be denied. Special relativity is based on the idea that no meaning can be attached to the concept of "motion of a body relative to space," but only to the relative motion of two observers or the relative motion of two objects. Do the considerations of the present section imply that we can nevertheless attach physical significance to the concepts of "acceleration relative to space" or "rotation relative to space"?

25-2 Mach's Principle and Relativity Reinstated

Within the limits of accuracy imposed by experimental difficulties, the direction of oscillation of a Foucault pendulum at the north pole is found to remain fixed *relative to the stars*. As an observer on earth watches the pendulum, he sees it rotating in exactly the same way that the stars in the sky above him are rotating. The "space" in which the pendulum does not rotate seems to be identical with the "astronomical rest frame" of section 22-6. When we devise a method of detecting rotation, what we are detecting is rotation *relative to the whole universe*.

Similarly, an observer moving with constant velocity is found to have no acceleration relative to the whole universe. An accelerating observer is accelerating relative to the universe, so that the stars on one side of him appear to be accelerating toward him, whereas the stars on the other side appear to be accelerating away from him. The significance of Newton's first law of motion is that an isolated body has a constant velocity relative to the universe.

The Austrian physicist and philosopher Ernst Mach interpreted all this as meaning that we cannot detect acceleration and rotation relative to "absolute space," but only relative to the matter in the universe. He suggested that there is a hitherto unrecognized interaction between a moving body and all the other matter in the universe, the most distant matter being the most important. This interaction depends on the relative acceleration of the body and the distant matter and is zero when the body is moving with a constant velocity relative to the distant matter.

The isolated body of Newton's first law is not really being left to itself. It experiences no strong force from nearby matter, but it is still subject to a Machian force from all the matter in the universe and this force

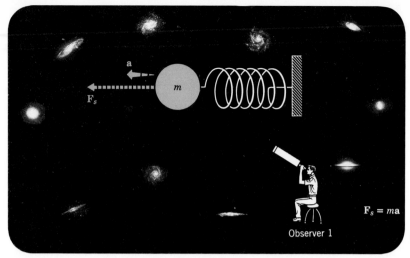

(a) Point of view of an observer at rest relative to the universe.

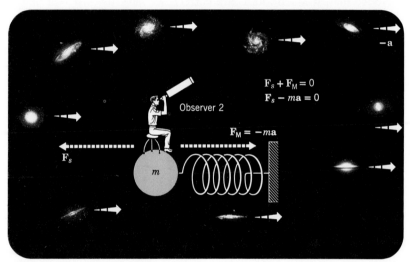

(b) Point of view of an observer moving with m.

Figure 25-5 Mach's Principle and Newton's Second Law of Motion.

constrains it to move with constant velocity (zero acceleration) relative to the universe.

Suppose that the body does experience a force $\mathbf{F}_s$ from nearby matter, such as a compressed spring (figure 25-5a). Then it acquires an acceleration **a** relative to the universe. Newton's second law tells us that, if m is the mass of the body,

$$\mathbf{F}_s = m\mathbf{a} \tag{25-1}$$

Can we adopt the underlying philosophy of relativity and describe the situation from the point of view of an observer sitting on the body and

The General
Theory of
Relativity

moving with it (figure 25-5b)? From the point of view of this observer, the body has no acceleration and so, if he can apply Newton's second law, he must conclude that there is no resultant force acting on the body. What cancels the force $\mathbf{F}_s$ exerted by the compressed spring?

Mach's answer is that the observer moving with the body sees the distant matter in the universe accelerating relative to himself with an average acceleration $-\mathbf{a}$. This accelerating matter exerts a Machian force $\mathbf{F}_M$ on the body, which is proportioned to the acceleration and to a constant m characteristic of the body

$$\mathbf{F}_M = -m\mathbf{a} \tag{25-2}$$

Since, for this observer, the net force must be zero

$$\mathbf{F}_s + \mathbf{F}_M = 0 \tag{25-3}$$

$$\text{or } \mathbf{F}_s - m\mathbf{a} = 0 \tag{25-4}$$

This is identical with the form of Newton's second law deduced by the first observer (equation 25-1).

Notice that, if Mach's ideas are correct, we have the first real hint of the true physical significance of the inertial mass m. It has something to do with the Machian force on a body when it accelerates relative to the matter in the universe. Moreover, the remarkable fact that inertial mass is equal to gravitational mass (section 6-4) provides a strong hint that the Machian force is gravitational in nature.

The same ideas can be applied to rotation. The Foucault pendulum experiences forces from distant matter that constrain it to oscillate in a direction fixed relative to this matter. In the case of the rotating bucket,

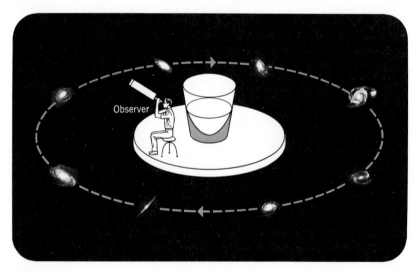

Observer

Figure 25-6 An observer rotating with the bucket of water sees the universe rotating around him. The rotating universe exerts Machian forces on the water and piles it up against the side of the bucket.

an observer rotating with the bucket sees the universe rotating around him, and the rotating distant matter exerts forces on the water that pile it up against the sides of the bucket (figure 25-6).

We shall not place Mach's principle in a colored frame because there is still much controversy amongst physicists about its validity. At the time of this writing, in the year 1974, there is no universally accepted, precise mathematical statement of Mach's principle. Einstein was very much influenced by Mach's ideas when he formulated the general theory of relativity, but the relationship between general relativity and Mach's principle is still somewhat obscure.

However, Mach's principle does show how we can still escape from the concept of absolute space and consider only the relative motion of different material objects. It encouraged Einstein to extend his principle of relativity to accelerating observers. One of the basic ideas underlying general relativity is that, when the laws of physics are properly understood, they have the same form for all accelerating observers as for all nonaccelerating observers.

> **The basic principle of general relativity**
>
> The laws of physics can be formulated in such a way that they are valid for any observer, however complicated his motion.

The laws of physics might seem to be different for rotating and nonrotating observers. When the rotating observer releases a body, it moves in a curved path. When the nonrotating observer releases a body, it moves in a straight line. However, the laws of physics should include the fact that the motion of a body is influenced by the motion of distant matter. The curved path of the body released by a rotating observer is not a consequence of rotation relative to "absolute space," but rather a consequence of the rotation of the stars relative to him. If the whole universe could be made to rotate with the observer, then a body released by him would move in a straight line. In fact, the concept of rotation of the whole universe relative to "absolute space" is meaningless, since no experiment could be performed to detect it. Only rotation relative to the matter in the universe is detectable.

Einstein's method of expressing the laws of physics in a form valid for all moving observers involves mathematical techniques too advanced to be described in detail here. Instead, we shall describe qualitatively two important aspects of his approach: the **principle of equivalence** and the need to use **non-Euclidean geometry**.

25-3   The Principle of Equivalence

Imagine an isolated observer inside a rocket in outer space, with no nearby matter to exert strong forces on him. Suppose that he cannot see outside his rocket and cannot directly measure his acceleration relative to the stars.

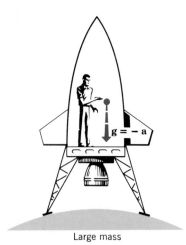

(a) An isolated rocket with an acceleration **a**.

(b) A rocket stationary in a gravitational field.

Figure 25-7 Acceleration in empty space is equivalent to being stationary in a gravitational field.

He wishes to perform experiments inside his rocket to decide whether he is accelerating. He holds up an object and releases it. Once released the object has no acceleration. If, however, the observer is in fact accelerating because his rocket engines are operating (figure 25-7a), the rocket accelerates relative to the released object, which therefore appears to the inside observer to fall in a direction opposite to the actual acceleration of the rocket. If the rocket is not accelerating, both the observer and the released object continue to move with the same velocity and so the object remains suspended in mid-air in the same position relative to the rocket where it was released.

Suppose then that the observer releases his object and observes that it falls to the floor with an acceleration $-\mathbf{a}$. Can he assume that the real situation is that the rocket has an acceleration **a**? Only if he is absolutely certain that he is far away from the gravitational attraction of nearby bodies. If he is totally ignorant of what is going on outside his rocket, the true situation (figure 25-7b) might be that he is stationary in a gravitational field in which the object falls to the floor with a gravitational acceleration $\mathbf{g} = -\mathbf{a}$!

There is a similar dilemma if the observer releases the object and it remains suspended in mid-air. This might be because the rocket is traveling through empty space with a constant velocity, its engines inoperative (figure 25-8a). But it might equally well be that the rocket is falling freely in a gravitational field (figure 25-8b). In which case, when the object is released

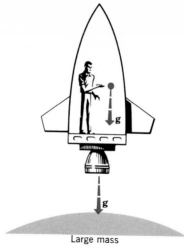

(a) An isolated rocket with
   no acceleration.

(b) A rocket falling freely in
   a gravitational field.

Figure 25-8 Unaccelerated motion in empty space is equivalent to free fall in a gravitational field.

it continues to fall freely in exactly the same way as the rocket and remains in the same position relative to the rocket.

A somewhat similar situation was considered in section 7-3. If the reader will review this section again in the light of what we are now discussing, he will appreciate the significance of the following statements. If the observer is totally unaware of the situation outside his rocket, and if he observes all bodies inside his rocket falling with an acceleration $-\mathbf{a}$, he cannot decide whether he is in a gravitational field or is accelerating relative to the stars, or both. All he can say is that, if he is in a gravitational field producing an acceleration $\mathbf{g}$ and also has an acceleration $\mathbf{a}'$ relative to the stars, then

$$\mathbf{g} - \mathbf{a}' = -\mathbf{a} \tag{25-5}$$

This conclusion rests heavily upon Galileo's famous discovery that all bodies have the same acceleration in a gravitational field (section 7-1).

It is impossible to think of any experiment on the mechanics of moving bodies that will decide whether the observer is accelerating in empty space or not accelerating in the presence of a gravitational field. Einstein generalized this conclusion to any kind of experiment whatsoever that might be performed inside the rocket, including experiments in electromagnetism or experiments involving light. He thus obtained the **Principle of Equivalence,** which is one of the foundation stones of general relativity. It says in effect that acceleration in empty space is completely equivalent to unaccelerated motion in a gravitational field.

The principle of equivalence should not be confused with the equivalence of mass and energy (equation 24-43), which is an entirely different matter.

25-4 Bending of a Beam of Light

The principle of equivalence can be used to derive a very important result concerning the bending of a beam of light in a gravitational field. Imagine a laboratory inside an isolated rocket that has an acceleration **a**. The observer in the rocket produces a beam of light traveling in a direction perpendicular to **a** and observes its path by allowing it to pass through several semi-transparent fluorescent screens, on each of which it produces a fluorescent spot (figure 25-9). An observer in an inertial frame of reference, relative to whom the laboratory is accelerating, describes the experiment in the following way (figure 25-9a). The beam of light moves along a straight line. However, while the beam is traveling from screen 1 to screen 2, screen 2 acquires an extra upward velocity and moves upward relative to the beam, which therefore strikes screen 2 at a point lower down than the point at which it struck screen 1. As the beam passes across the screens their upward velocity relative to the beam becomes progressively greater, and so the vertical displacement between adjacent spots becomes increasingly larger. The observer in the rocket, relative to whom the screens are stationary, therefore sees the beam of light bending downward as in figure 25-9b.

This is what happens when the rocket is not in a gravitational field, but has an acceleration **a**. According to the principle of equivalence, the phenomenon is exactly the same if the rocket has no acceleration, but is in the presence of a gravitational field producing a gravitational acceleration $g = -a$ (figure 25-9c).

The light "falls downward" in the field, in a similar way to a material body. There is another way of looking at this. Suppose a flash of light is sent out, carrying a total energy E. This energy is equivalent to a mass $m = E/c^2$. It is not unreasonable to expect that this mass experiences a gravitational force in the same way as any other kind of mass.

The bending of light in a gravitational field has been observed. During an eclipse of the sun by the moon, it is possible to see stars very

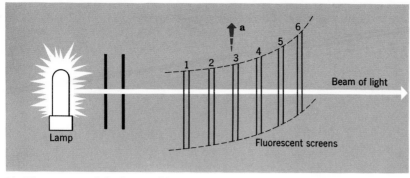

(a) The passage of the beam of light across the fluorescent screens, as seen by an observer in an inertial frame.

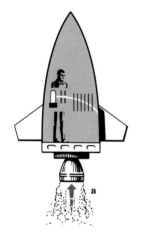

(b) The beam of light, as seen by an observer in an accelerating rocket.

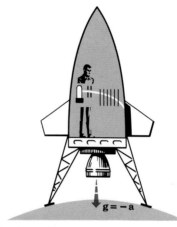

(c) The rocket is stationary in a gravitational field.

Figure 25-9 Bending of a beam of light in (b) an accelerating frame of reference, (c) a gravitational field.

nearly in line with the edge of the sun. Light from these stars has passed very near the sun and has been bent in its gravitational field. The stars therefore appear to be displaced from their normal positions in a direction away from the sun (figure 25-10). The observed magnitude of this displacement agrees with the value predicted by the general theory within the observational accuracy of about 10 per cent.

25-5 Non-Euclidean Geometry

Having shown that light is bent in a gravitational field and no longer travels along a straight line, let us pause and consider what is meant by a straight line. A good practical definition of a straight line, which might have been accepted without question before the general theory, is that it is the path followed by a beam of light in a vacuum! A farmer erecting a fence can check that it is straight by sighting along the posts to

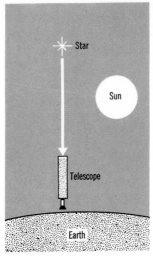

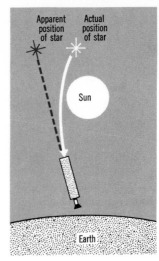

(a) The light from the star passes through a weak region of the sun's gravitational field and is not appreciably bent.

(b) When the light from the star passes near the sun the bending is detectable.

Figure 25-10 Apparent displacement of a star very nearly in line with the edge of the sun.

see that they are all "in line." Suppose, however, that the light is appreciably bent by the gravitational attraction of a nearby cow (figure 25-11a). Clearly the method breaks down. When the cow walks away the fence no longer appears straight (figure 25-11b).

Unfortunately, astronomical observations, such as observations on the solar system, are made almost exclusively with telescopes and rely heavily upon the behavior of light. Moreover, it is not possible to avoid the dif-

(a) The ray of light is bent by the gravitational attraction of the cow.

(b) The cow moves away and the posts no longer appear to be in line.

Figure 25-11 Lining up the posts of a fence by sight. (In practice the bending would be negligibly small, but it has been exaggerated to demonstrate the effect.)

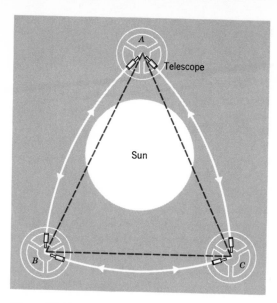

Figure 25-12 An observational proof that the sum of the three angles of a triangle is greater than 180°.

ficulty by defining a straight line in some other way, such as the shape of a string pulled taut. How do we know that a taut string is a straight line? This is obviously a complicated problem involving the forces between the atoms of the string, which are electromagnetic in character and therefore related to the behavior of light. A long string stretching from the sun to the earth and pulled as taut as possible might be bent in a way similar to a beam of light. There is a genuine problem!

Suppose that three astronomical observatories, A, B, and C, have been set up in space and are employing their rocket motors to hold themselves stationary relative to the sun (figure 25-12). Each observatory sights its telescope on the other two observatories in succession and measures the angle through which the telescope has to be swung. Because of the bending of light in the sun's gravitational field, it should be clear from the figure that the sum of the three angles measured by the three observatories is greater than 180°. The astronomers must now make a difficult choice. They can realize that light is bent in a gravitational field and try to think of a better definition of a straight line, which is not easy. Alternatively, they can decide, as a practical matter, to *define* a straight line as the path followed by a light beam. In which case they have proved from observations that the sum of the three angles of a triangle is greater than 180°, in flagrant contradiction to the precepts of Euclidean geometry, which were presented to them as inviolable logic during the years of their early education. They must therefore seriously ask whether Euclidean geometry is necessary, or whether it is not possible to use other systems of geometry. This latter course was the one followed by Einstein.

A somewhat similar situation is encountered on the surface of the

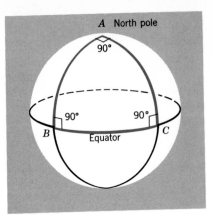

Figure 25-13 On the surface of the earth, it is possible to draw a "triangle" whose three angles add up to 270°.

earth. Imagine two pipe lines setting out from the north pole at right angles to one another and proceeding "straight" to the south pole. Straight, in this case, means following a line of longitude. A third pipe line follows the equator (figure 25-13). The first two lines cross the third line at right angles. A surveyor, who thinks that the earth is flat, measures the three angles of the triangle ABC and discovers that their sum is 270°. If the surveyor knows no way of drawing a line except on the surface of the earth, he cannot draw a "really straight" line. The best he can do is a line of longitude. The surveyor must invent a non-Euclidean geometry. Such a geometry has some peculiar features. If we confine ourselves to lines that must be drawn on the surface of the earth, the equator is a circle with its "center" at the north pole, N (figure 25-14).

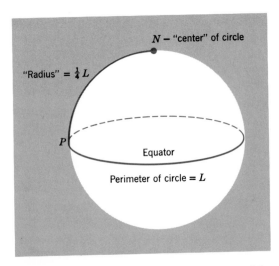

Figure 25-14 In the non-Euclidean geometry of the surface of a sphere, the ratio of the perimeter of the circle to its radius is not 2π but 4.

The "radius" of this circle is NP. If the distance around the equator is L, the length of the "radius" NP is $\frac{1}{4}L$. The ratio of the perimeter of the circle to its "radius" is not 2π but 4!

The general theory assumes that the four dimensions of space and time show the same sort of peculiar behavior as the two dimensions on the surface of the earth. Space-time is curved! This curvature is produced by the gravitational effect of nearby matter. The curvature increases as the mass of the nearby matter is increased, and also increases if this matter is brought nearer.

25-6 The Gravitational Red Shift

One important test of the general theory has already been discussed. A ray of light is bent when it passes near the sun. Consequently, during a solar eclipse the stars near the sun appear to be displaced from their normal positions. In this section and sections 25-8 and 25-9 we shall describe some other famous tests of the theory. Three of them are astronomical observations and the fourth is an experiment performed on the surface of the earth.

A gravitational field distorts time as well as space. A stationary clock runs slow in a gravitational field. Imagine a clock at rest at a distance R from a mass M (figure 25-15a). It ticks more slowly than a similar clock at rest at a great distance from M.

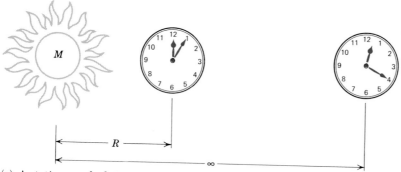

(a) A stationary clock in a gravitational field runs slow.

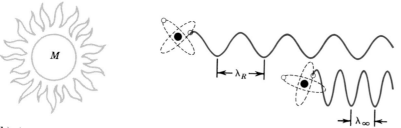

(b) An atom in a gravitational field emits light of lower frequency and longer wavelength.

Figure 25-15 The gravitational red shift.

The General
Theory of
Relativity

Slowing down of a stationary clock in a gravitational field

$$\frac{\text{Time between ticks of a clock at a distance } R \text{ from } M}{\text{Time between ticks of a clock at an infinite distance from } M}$$

$$= \frac{1}{\sqrt{1 - \dfrac{2GM}{c^2 R}}} \qquad (25\text{-}6)$$

Notice an important difference between the present situation and the behavior of *moving* clocks in the special theory (section 23-3 and equation 23-18). When two observers, A and B, are moving relative to one another, A decides that B's clock is slow, whereas B decides that A's clock is slow, and their disagreement cannot be resolved. In the present case everybody can agree that the clock in the gravitational field is slow compared with the clock outside the field. It is possible to agree that the clock outside the field is the "standard clock" and to make a suitable correction to the clock in the field to compensate for the effect of the field. The moving observers A and B cannot do this.

An atom is a kind of clock and it runs slow in a gravitational field. It emits light of longer period, smaller frequency, and longer wavelength as compared with an atom outside the field (figure 25-15*b*). When the wavelength of visible light emitted by the atom is lengthened, the light moves toward the red end of the visible spectrum. The effect is therefore called the **gravitational red shift**.

If the atom at a distance R from M emits light of wavelength λ_R and a similar atom vibrating in a similar way at infinity emits light of wavelength λ_∞, then

Gravitational red shift

$$\frac{\lambda_R}{\lambda_\infty} = \frac{1}{\sqrt{1 - \dfrac{2GM}{c^2 R}}} \qquad (25\text{-}7)$$

This effect has been observed and equation 25-7 has been verified for light emitted by atoms on the surface of the sun, but it is then a very small effect because the quantity $2GM/c^2R$ for the sun is very small compared with 1. The effect is much more pronounced for a **white dwarf star.**

White dwarf stars are believed to be formed when a star with about the same mass as our sun runs out of nuclear fuel. The blowing up effect of the thermal motion of the particles of the star then weakens and the star collapses under its own weight. As we shall explain in section 28-5, it settles down again at a diameter about one hundredth that of the sun and is then about the same size as the earth. Its density becomes about 10^6 gm/ cm^3, which means that a piece of it about the size of an acorn weighs a ton. During the collapse gravitational potential energy is converted into heat and the star warms up until it glows white hot. Hence the name white dwarf. For such a star the mass M is about the same as for the sun, but the radius R is about 100 times smaller and so the factor $\dfrac{2GM}{c^2R}$ appearing in equation 25-7 is 100 times larger than for the sun. White dwarf stars are therefore very suitable objects for checking equation 25-7.

The gravitational red shift has also been observed in a terrestrial experiment making use of the earth's gravitational field. This experiment employed γ-rays emitted by a nucleus and took advantage of a subtle effect, called the Mössbauer effect, which makes it possible to compare frequencies of these γ-rays with extraordinary accuracy. A comparison was made between the frequencies of the γ-rays emitted by two nuclei, one of which was on the surface of the earth, while the other was at a height h above it. We will prove in example 25-6-1 that when two clocks are separated by a height h near the surface of the earth

$$\frac{\text{Time between ticks of the lower clock}}{\text{Time between ticks of the upper clock}} = 1 + \frac{gh}{c^2} \qquad (25\text{-}8)$$

where g is, of course, the acceleration due to gravity. The period of the γ-rays is equivalent to the time between ticks of a clock, but the period is inversely proportional to the frequency and so

$$\frac{\text{Frequency of the } \gamma\text{-rays emitted by the } \textit{upper} \text{ nucleus}}{\text{Frequency of the } \gamma\text{-rays emitted by the } \textit{lower} \text{ nucleus}}$$

$$= 1 + \frac{gh}{c^2} \qquad (25\text{-}9)$$

The upper nucleus was in fact found to emit γ-rays of greater frequency and the magnitude of the effect agreed with equation 25-9 to within the experimental accuracy of 1%.

Example 25-6-1

Prove equation 25-8.

Suppose that the lower clock is at a distance R_e from the center of the earth and the upper clock is at a distance $R_e + h$ from the center of the earth. Then, according to equation 25-6,

$$\frac{\text{Time between ticks of the lower clock}}{\text{Time between ticks of a clock at infinity}}$$

$$= \frac{1}{\sqrt{1 - \dfrac{2GM_e}{c^2R_e}}} \qquad (25\text{-}10)$$

$$\frac{\text{Time between ticks of the upper clock}}{\text{Time between ticks of a clock at infinity}}$$

$$= \frac{1}{\sqrt{1 - \dfrac{2GM_e}{c^2(R_e + h)}}} \qquad (25\text{-}11)$$

where M_e is the mass of the earth. It follows that

$$\frac{\text{Time between ticks of the lower clock}}{\text{Time between ticks of the upper clock}} = Q$$

$$= \sqrt{\frac{1 - \dfrac{2GM_e}{c^2(R_e + h)}}{1 - \dfrac{2GM_e}{c^2 R_e}}} \qquad (25\text{-}12)$$

We will call this quantity Q and evaluate it by noting that both $\dfrac{h}{R_e}$ and

$\dfrac{2GM_e}{c^2 R_e}$ are very small compared with 1 and using the approximations in section A7 of the Mathematical Appendix. First, note that

$$\frac{2GM_e}{c^2(R_e + h)} = \frac{2GM_e}{c^2 R_e \left(1 + \dfrac{h}{R_e}\right)} \qquad (25\text{-}13)$$

From equation 95 of the Mathematical Appendix,

$$\frac{2GM_e}{c^2 R_e \left(1 + \dfrac{h}{R_e}\right)} \approx \frac{2GM_e}{c^2 R_e}\left(1 - \frac{h}{R_e}\right) \qquad (25\text{-}14)$$

$$= \frac{2GM_e}{c^2 R_e} - \frac{2GM_e h}{c^2 R_e{}^2} \qquad (25\text{-}15)$$

$$= \frac{2GM_e}{c^2 R_e} - \frac{2gh}{c^2} \qquad (25\text{-}16)$$

since equation 7-1 tells us that the acceleration due to gravity, g, is

$$g = \frac{GM_e}{R_e{}^2} \qquad (25\text{-}17)$$

The quantity Q may now be written

$$Q = \sqrt{\frac{\left(1 - \dfrac{2GM_e}{c^2 R_e}\right) + \dfrac{2gh}{c^2}}{\left(1 - \dfrac{2GM_e}{c^2 R_e}\right)}} \qquad (25\text{-}18)$$

$$= \sqrt{1 + \frac{2gh}{c^2\left(1 - \dfrac{2GM_e}{c^2 R_e}\right)}} \qquad (25\text{-}19)$$

From equation 107 of the Mathematical Appendix

$$Q \approx 1 + \frac{gh}{c^2\left(1 - \dfrac{2GM_e}{c^2 R_e}\right)} \qquad (25\text{-}20)$$

Since $\dfrac{2GM_e}{c^2 R_e}$ is very small compared with 1, it will make a negligible difference to ignore its presence inside the parentheses, and so

$$Q \approx 1 + \frac{gh}{c^2} \tag{25-21}$$

Remembering the significance of Q,

$$\frac{\text{Time between ticks of the lower clock}}{\text{Time between ticks of the upper clock}} = 1 + \frac{gh}{c^2} \tag{25-22}$$

which is the desired equation.

Example 25-6-2

In the experiment using the Mössbauer effect, the difference in height of the two nuclei was 2.2×10^3 cm. What was the fractional difference in the frequency of the γ-rays?
Inserting this value of h into equation 25-9

$$Q = 1 + \frac{gh}{c^2} \tag{25-23}$$

$$= 1 + \frac{980 \times 2.2 \times 10^3}{(3 \times 10^{10})^2} \tag{25-24}$$

$$= 1 + (2.3 \times 10^{-15}) \tag{25-25}$$

The periods, and also the frequencies, differed by only 2.3 parts in 10^{15}!

Example 25-6-3

Calculate the magnitude of the gravitational red shift for the sun.
The mass of the sun, M_s, is 1.99×10^{33} gm and its radius, R_s, is 6.97×10^{10} cm. Therefore,

$$\frac{2GM_s}{c^2 R_s} = \frac{2 \times 6.67 \times 10^{-8} \times 1.99 \times 10^{33}}{(3 \times 10^{10})^2 \times 6.97 \times 10^{10}} \tag{25-26}$$

$$= 4.2 \times 10^{-6} \tag{25-27}$$

This is very small compared with 1 and so we can use equation 108 of the Mathematical Appendix to express equation 25-7 in the approximate form

$$\frac{\lambda_R}{\lambda_\infty} \approx 1 + \frac{GM}{c^2 R} \tag{25-28}$$

$$= 1 + 2.1 \times 10^{-6} \tag{25-29}$$

for the sun. The wavelength changes by 2.1 parts in a million, which emphasizes the smallness of the effect.

Before leaving this let us remember that the observations are not made from infinity, but from the surface of the earth, and so let us check that the gravitational red shift at the surface of the earth can be neglected. The red shift produced by the gravitational field of the *sun* at the surface

The General
Theory of
Relativity

541

of the earth is clearly negligible compared with equation 25-29 because the distance of the earth from the sun is about 200 times the radius of the sun. To estimate the red shift at the surface of the earth due to the gravitational field of the earth itself, we repeat the above calculations using $M_e = 5.98 \times 10^{27}$ gm for the mass of the earth and $R_e = 6.37 \times 10^8$ cm for its radius. Then,

$$\frac{2GM_e}{c^2 R_e} = \frac{2 \times 6.67 \times 10^{-8} \times 5.98 \times 10^{27}}{(3 \times 10^{10})^2 \times 6.37 \times 10^8} \tag{25-30}$$

$$= 1.4 \times 10^{-9} \tag{25-31}$$

This clearly can be neglected in comparison with the value of the corresponding quantity for the sun quoted in equation 25-27.

25-7 Black Holes

Under most circumstances the gravitational red shift produces a very small fractional change in wavelength, but current theories of the evolution of stars suggest that there might be one situation in which the effect becomes dramatically important. Some massive stars may eventually evolve into the peculiar objects which have achieved notoriety under the name of **black holes,** but which are sometimes called "collapsars" or "frozen stars."

When a star with a mass many times that of the sun runs out of nuclear fuel, its collapse is very rapid and very violent, and it seems likely that nothing can stop it. The radius R steadily decreases and the factor $\frac{2GM}{c^2 R}$ in equations 25-6 and 25-7 steadily increases until, when this factor approaches the value 1, the gravitational red shift produces a large increase in wavelength. A crisis occurs when the radius becomes equal to the **Schwarzschild radius,** R_s, given by

$$R_s = \frac{2GM}{c^2} \tag{25-32}$$

The factor $\frac{2GM}{c^2 R_s}$ is then equal to 1; the quantity $\sqrt{1 - \frac{2GM}{c^2 R_s}}$ is equal to 0 and its reciprocal is infinite. A clock on the surface of the star runs infinitely slowly (it stops altogether!) and the surface can emit only radiation of an infinitely long wavelength.

For a star with 10 times the mass of the sun, the radius R_s at which this occurs is 3×10^6 cm or about 20 miles, as compared with 400,000 miles for the radius of our sun. The density attains the enormous value of 2×10^{14} gm/cm^3 and the star is probably composed entirely of neutrons (sections 28-5 and 32-1) squashed up against one another.

Viewed from a distance the collapsing star becomes redder as the gravitational red shift lengthens the wavelengths emitted from its surface (figure 25-16). An interesting side effect occurs when its radius shrinks below $\frac{3}{2}R_s$. Light rays from the outer edge of the star are then so strongly bent by the gravitational field that they cannot reach the viewer directly. The particles of light (photons) go into orbit around the star like a planet

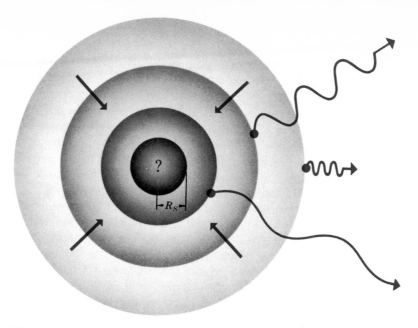

Figure 25-16 Possible formation of a "black hole." As the star collapses the wavelength of the light emitted from its surface becomes longer and longer until eventually, when its radius is R_S, no light at all can be emitted.

orbiting around the sun. Rays from the center of the star can always reach the earth, but they are effectively snuffed out as the emitted wavelengths become longer, changing from visible light through infrared to radio waves and beyond. The snuffing out stage, during which the wavelengths are multiplied by a large factor, takes less than a thousandth of a second. However, when the radius approaches very close to the Schwarzschild radius, R_s, everything slows down, including the process of collapse itself. In this final sluggish stage the radius approaches nearer and nearer to R_s but takes an infinite time to reach it.

Once the star has collapsed to the vicinity of its Schwarzschild radius, anything can fall into the strong gravitational field of the resulting "black hole," but nothing can escape from it, not even light. Classical physics had already predicted that something like this might happen. An object launched from the surface of a spherical mass M with radius R cannot completely climb out of the gravitational field of M unless its initial velocity exceeds the escape velocity given by equation 9-26.

$$\text{Escape velocity} = \sqrt{\frac{2GM}{R}} \tag{25-33}$$

As the star collapses and R becomes smaller, the required escape velocity increases and eventually becomes equal to the velocity of light when

$$c = \sqrt{\frac{2GM}{R}} \tag{25-34}$$

or $R = R_s = \dfrac{2GM}{c^2}$ (25-35)

When R is smaller than this not even light, which has the maximum permissible launching velocity, can escape from the gravitational field.

Inside the Schwarzschild radius the theory suggests some strange possibilities. There is a sense in which space and time change places. A powerful rocket is not only incapable of firing its motors in order to escape, but it cannot even fire them with sufficient force to remain hovering at a constant distance from the center. It is compelled always to fall inward. Even light cannot propagate outward, but only inward. There can be no communication with the world outside the Schwarzschild radius.

To an observer on earth watching the star collapse, it takes an infinite time to reach its Schwarzschild radius and so the question of what strange things happen inside might seem irrelevant, but there is a counterargument which really highlights the novelty of the situation. Suppose that an astronaut goes on a suicide mission to explore the collapsing star and allows his spaceship to fall freely into the black hole. An observer on earth sees that he never catches up with the falling surface of the star until both have reached the Schwarzschild radius after an infinitely long time. However, everything inside the spaceship slows down by the same amount in the strong gravitational field. The spaceship clocks run slow; the atoms of the spaceship vibrate more slowly; the astronaut's movements are slower; his metabolic processes are slower; and he thinks more slowly. This slowing down of the spaceship's clocks works in such a way that the astronaut thinks it takes him a *finite* time to reach R_s. In fact, during the final stages of breakthrough when the gravitational red shift is large, the spaceship clocks register a time interval of only about one-thousandth of a second while the earth clocks are registering an infinite time! From his own point of view, then, the astronaut can break through the Schwarzschild boundary in a very short time, and it again becomes relevant to ask whether he then enters into a strange new region of space-time completely cut off from our region of space-time. Is there any sense in saying that he continues to live his life in a region of time beyond the infinity of our time?

A possible snag is that the astronaut might not survive the journey because of the strong gravitational field he would encounter. The acceleration due to gravity on the surface of a star of mass M and radius R is given by equation 7-1.

$$g = \dfrac{GM}{R^2}$$ (25-36)

A typical black hole has a mass which is 10 times that of the sun and therefore 3×10^6 times greater than the mass of the earth and a radius 5×10^{-4} times the radius of the earth. The acceleration due to gravity near its Schwarzschild boundary is therefore about 10^{11} times greater than on the surface of the earth. The weight of the astronaut near the black hole would be about the same as the weight of a mountain on earth. The freely falling astronaut would not be directly aware of this strong gravitational field because the Principle of Equivalence tells us that there is no effective gravitational field in a freely falling frame of reference. There is a secondary effect, though, that is just as fatal. If the astronaut is falling feet first, then his feet are a little nearer to the gravitating mass than his head and so the gravitational field tries to accelerate his feet faster than his head. Near the Schwarzschild boundary this tears his body apart with forces equivalent to the weight on earth of a 100,000 tons.

The discussion is beginning to look less pragmatic, but there is another possibility. It is quite possible that a whole galaxy might collapse to form a black hole. A galaxy has a mass about 10^{11} times the mass of the sun and a Schwarzschild radius of about 3×10^{16} cm, which is about 20 times the size of the solar system out to Pluto. For such a large black hole the gravitational field is much weaker and the forces on a human body near the Schwarzschild boundary are comparable with the weight on earth of a speck of dust and therefore quite tolerable.

The dilemma therefore remains and, if the existence of galactic black holes is ever firmly established, we may be faced with two alternatives. We may be forced to entertain some strange new ideas about the nature of space and time. Alternatively, we may eventually conclude that these unusual speculations arising from the general theory of relativity merely expose the limitations of this theory and its inability to cope adequately with situations involving very strong gravitational fields.

25-8 The Radar Test

The radar technique for locating and tracking objects (figure 23-3) has been successfully applied to the nearby planets and, in particular, to the planet Mercury whose orbit is nearest to the sun. A short pulse of radio waves is sent out from the earth, reflected from the surface of Mercury, and received back again after an appropriate time delay. The time lapse between transmission and reception determines the distance to the planet at the instant of reflection. The direction from which the echo returns provides the other information needed to locate the position of the planet completely at that instant. In this way the planet can be tracked over a period of several months and its orbit determined precisely.

However, the simple method of interpreting the data given at the end of section 23-2 assumes that the radar pulse travels along a straight line with a constant velocity which is the same everywhere. According to the general theory of relativity, neither of these assumptions is valid in the strong gravitational field of the sun. We have already discussed the bending of a ray of light by the sun's gravitational field. There is another effect, more important in the present instance, which makes a light signal appear to travel more slowly the nearer it is to the sun.

The situation can be understood in principle with the help of an analogy in which the three-dimensional space around the sun is represented by a two-dimensional space on an imaginary surface. In the absence of the sun, this two-dimensional space would be a flat plain.

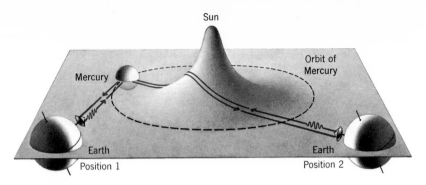

Figure 25-17 The curved space near the sun is represented by a mountain in the center of a plain. A radar pulse passing near the sun is delayed by the need to climb the mountain.

The strong gravitational field of the sun curves the space in its vicinity and so the sun can be represented by a steep mountain in the center of the plain (figure 25-17). A radar pulse traveling across the plain far from the mountain is affected very little by it and has a velocity very close to the accepted velocity of light. However, a radar pulse passing near the sun is forced to climb the mountain and this slows it down and delays its arrival time.

In terms of this analogy, the procedure in the radar experiment may be explained as follows. The orbit of Mercury is first carefully determined from a large number of observations confined to those occasions when the radar pulse traveling between the earth and Mercury does not pass near the sun (the earth is in position 1 of figure 25-17). For these observations, the effect of the sun's gravitational field is negligible because the pulses are always traveling on the flat plain far from the foot of the mountain. The accuracy of this determination of the orbit can be improved by incorporating all the suitable data on Mercury's orbit which have been accumulated over the centuries using optical telescopes. The final determination of the orbit is therefore very accurate and has avoided the complications introduced by the curvature of space near the sun.

The next step is to observe the radar echoes when Mercury is just about to disappear behind the sun. The radar pulse is then obliged to pass near the sun and climb the mountain (the earth is in position 2 of figure 25-17). From the previous determination of the orbit, it is possible to predict accurately how long the radar pulse would take to travel to Mercury and back again if it traveled with the usual velocity over a flat plain. In fact, the pulse is observed to take a longer time than this because of the extra delay produced as it travels through the curved space near the sun. Moreover, the magnitude of the additional delay agrees with the value predicted by Einstein's theory to within the experimental accuracy of about 20%

One important observational test of the theory remains to be described. It is also concerned with the orbit of the planet Mercury, but it is a very subtle effect depending upon some of the finer details of the theory. For this very reason, though, it provides a very sensitive test of the exact form of the theory. The orbit of Mercury is an ellipse, which might be described rather crudely as a slightly elongated circle. The sun is not at the center of this ellipse, but is displaced slightly to one side (figure 25-18a). The point P of the orbit at which the planet is nearest to the sun is called the perihelion. The perihelion is observed to be slowly revolving around the sun, and the orbit of Mercury is therefore as shown in figure 25-18b which, however, exaggerates the motion of the perihelion. This effect is called the advance of the perihelion of Mercury.

The advance of the perihelion is mainly a consequence of the gravitational attraction of the other planets in the solar system. However, when the influence of all the other planets was calculated with great accuracy, the result accounted for only 99.2% of the observed motion of the perihelion. There remained a small "residual" advance of the perihelion, corresponding to one complete revolution of the perihelion in about 3 million years, which defied all attempts to explain it using Newtonian mechanics. It was not possible, for example, to explain it by postulating the existence of an undiscovered planet. The general theory of relativity is able to give a completely satisfactory account of this residual advance and to derive its magnitude.

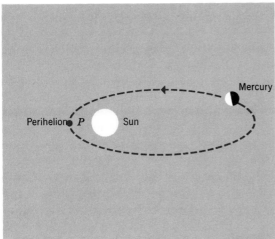

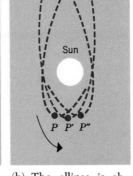

(a) An elliptical orbit. This would be the orbit of Mercury in classical mechanics in the absence of perturbations by the other planets. (The elongation has been exaggerated. The actual orbit is almost a circle.)

(b) The ellipse is observed to be slowly rotating.

The General Theory of Relativity

Figure 25-18   The advance of the perihelion of Mercury.

However, at the time of this writing there is still some controversy associated with the perihelion of Mercury. It has been suggested that a small part of the advance of the perihelion is a consequence of the fact that the rotating sun bulges out at its equator. If this effect were present, it would introduce a small distortion of the sun's gravitational field sufficient to influence the rate of advance of the perihelion and to destroy the good agreement between the observations and Einstein's theory.

25-10   The Twin Paradox

The following prediction of the theory of relativity has caused much controversial discussion among experts and non-experts. A and B are identical twins. When they are 20 years old, B starts out on a journey to a distant star and back with a certain velocity near the speed of light, while A stays at home on earth. B's clocks tell him that the journey takes 10 years, and upon his return he feels and acts like a man of 30. However, upon landing he discovers that the earth clocks have advanced 60 years and his twin brother A is an old man of 80.

This phenomenon is clearly related to the slowing down of a moving clock (sections 23-3 and 23-5). The earth-twin A, observing the space-twin B proceeding on his outward journey with a steady velocity, sees that B's clocks appear to be slow. In particular, the biochemical processes which cause B to grow older appear to A to be proceeding more slowly. On the other hand, if the space-twin B looks back at what is happening to the earth twin A, he concludes that A's clocks are slow and that it is A who is aging less rapidly. According to the special theory, there is no reason to prefer A's point of view rather than B's. Yet we have stated that, when B returns to the earth, there is a real difference in age between them, even when they are standing side by side stationary relative to one another.

This paradox might be expressed in the following way. The situation was originally described from the point of view of a frame of reference fixed relative to the earth (figure 25-19a). In this frame of reference B goes on a journey into space with a high velocity and upon his return has aged less than A. However, since all motion is relative, would it not be equally valid to use a frame of reference fixed relative to B? In this frame of reference the earth and the earth twin A make a journey into space with a high velocity (figure 25-19b). If the original postulate were true, then surely it would be A who would be younger than B when they met again.

The resolution of the paradox is that there is an important difference between the frames of reference of A and B. A's frame of reference is almost an inertial frame. The rotation of the earth and its motion around the sun produce accelerations that are too small to matter in this instance. B's frame of reference is not an inertial frame. While he is taking off, turning round, or landing, B's frame has a very large acceleration.

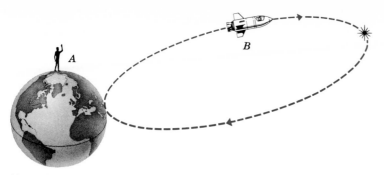

(a) Twin B goes on a journey to a star, leaving twin A behind on earth. When he returns he finds he has aged less than A.

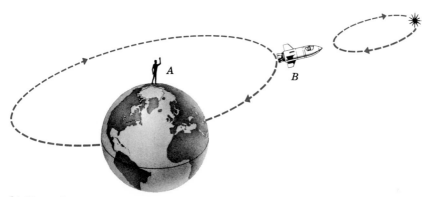

(b) From the point of view of a frame of reference fixed relative to B, the star comes on a journey to meet B. Meanwhile the earth and the earth-twin A go on a journey out into space. When the earth returns to B, will A have aged less than B?

Figure 25-19 The twin paradox.

According to the principle of equivalence an acceleration is equivalent to a gravitational field, and in a gravitational field clocks are slow.

The problem cannot therefore be satisfactorily discussed in terms of the special theory. It is necessary to apply the general theory. It can then be proved that the combined effects of B's velocities and accelerations are that, when he lands back on earth, his clocks have indeed registered a shorter period of time than A's clocks.

As a means of extending the span of human life this method has its limitations. Although B has aged 10 years compared with A's 60 years, he has eaten correspondingly fewer meals, performed fewer actions and had fewer thoughts. Nevertheless, it would be amusing to go on a year's vacation in space and return to discover what progress the earth will make over the next thousand years. (The time discrepancy can be increased by making the journey at a speed nearer to the speed of light.) Unfortunately, the necessary velocity cannot be achieved without the

The General
Theory of
Relativity

expenditure of an enormous amount of energy. The fuel required to propel one human being into space with this velocity would be sufficient to meet all the earth's power requirements for several centuries.

QUESTIONS

1. Imagine a freely falling elevator 1000 miles, or 1.61×10^8 cm, high! Why is it not possible to apply the principle of equivalence to the behavior of moving bodies inside this elevator? Use this example to demonstrate that the principle must be applied only to experiments performed inside a laboratory which is either very small or is in a uniform gravitational field.

2. Does an electric charge falling freely in a gravitational field radiate an electromagnetic wave? What reason is there for believing that it might? Why does the principle of equivalence suggest that it might not? (In the author's opinion, no completely satisfactory answer has yet been given to this question.)

3. A light clock similar to the one shown in figure 23-1 is placed in the sun's gravitational field with its meter rule aligned along a radius. An observer at a great distance from the sun sends in a light pulse along a radius so that it is reflected in turn from the two mirrors of

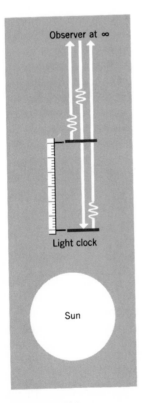

Observer at ∞

Light clock

Sun

Question 25-3

the light clock. What should the distant observer expect to be the time interval between the return of the two echoes? How might he interpret this experiment in terms of the effect of a gravitational field on the length of a meter rule and the velocity of light?

4. Twin A does not stay at home, but goes on a journey identical to B's, except that it is in exactly the opposite direction. How do the twins compare in age when they return? During the journey, what is each twin's opinion of the rate of the other twin's clock and his rate of aging?

PROBLEMS

A

1. Invent a graphical method of obtaining the shape of the curve shown in figure 25-2c.

2. In an experiment on the gravitational red shift using the Mössbauer effect, one nucleus is at the top of the Empire State Building and the other is at street level. The height of the building is 317 meters. The γ-rays emitted by the lower nucleus have a frequency of 3.47×10^{18} sec^{-1}. What is the difference in frequency between the γ-rays emitted by the two nuclei?

3. An atom in zero gravitational field emits light with a wavelength of 5.000×10^{-5} cm. What is the wavelength when the atom is at a distance of 2×10^8 cm from a collapsed star with a mass of 8×10^{33} gm?

4. Repeat problem 3 with the atom at a distance of 2×10^6 cm.

5. A collapsed object has a mass of 8.1×10^{35} gm. At what distance from its center does the gravitational red shift double the wavelength?

6. What is the Schwarzschild radius of the earth?

B

7. Derive equation 25-5. Check this equation for the particular cases (i) $g = 0$, (ii) $a' = 0$, (iii) $g = a'$.

8. The dwarf companion of the star Sirius has a mass of 1.7×10^{33} gm. The lines in its spectrum show a gravitational red shift of 1 part in 10^4. Calculate its density.

9. If the expectation of life at sea level is 70 years, how many seconds different is the expectation of life at the top of Mount Everest, which is 8.8×10^5 cm above sea level? Is it shorter or longer? Assume that the only relevant factor is the variation of the rate of a clock with gravitational potential.

10. A satellite is placed in a stable circular orbit of radius R about a star of mass M. Compare the magnitudes of the two effects which slow down a clock inside the satellite. These two effects are (a) the effect of the velocity according to the special theory, (b) the effect of the gravitational potential according to the general theory.

11. The radius of a proton is about 0.8×10^{-13} cm. Compare the rate of a "clock" at the surface of a proton with a similar clock at infinity.

12. Modify equation 25-7 by eliminating M and introducing R_s instead.
13. By what factor does the gravitational red shift multiply the wavelength at a distance $R = 2R_s$?
14. Compare the rates of two clocks at distances from the center of a black hole equal to $\frac{3}{2}R_s$ and $\frac{5}{2}R_s$.
15. The universe contains about 10^{11} galaxies each of which has a mass about 10^{11} times the mass of the sun. Calculate the Schwarzschild radius of the whole universe in light years.
16. If Newtonian Mechanics were applicable, what would be the radius of the circular orbit of a photon about a collapsed object of mass M?

C

17. The annual output of manufactured power by the whole of mankind is about 10^{27} erg. Check the statement made in section 25-10 about the power needed to send a man on a year's vacation which, by earth reckoning, would last for one thousand years of history. Take the mass of the man and his space capsule to be 500 kg.
18. A beam of light is projected horizontally near the surface of the earth. How far has it fallen after traveling a horizontal distance of 10 miles? First use Newtonian mechanics, treating the horizontal and vertical motions as entirely independent of one another. The relativistic answer is greater by a factor of 2.
19. Suppose that a mass M has collapsed to a radius very slightly larger than R_s. Find an equation relating R_s to the density ρ with the mass M eliminated. Calculate the radius (a) for $\rho = 5.5$ gm cm^{-3}, which is the density of the earth, (b) for $\rho = 2.3 \times 10^{14}$ gm cm^{-3}, which is the density inside an atomic nucleus. In each case try to find a familiar object of comparable size.
20. The acceleration due to gravity near the Schwarzschild radius of a black hole varies with what power of its mass?
21. A spaceship has an acceleration of 10^4 cm sec^{-2} in gravity free space. Compare the rates of two of its clocks which are displaced from one another by a distance of 250 cm in the direction of the acceleration. Which is slower?
22. In section 13-3 the total number of particles in the observable universe was said to be roughly equal to $\left(\dfrac{F_e}{F_G}\right)^2$. Show that this follows from the hypothesis that the positive rest mass energy of a particle is exactly cancelled by its negative mutual gravitational potential energy with all the other particles in the universe. The "radius" of an electron is

$$r_e = \frac{e^2}{m_e c^2}.$$

QUANTUM MECHANICS

26

The
Birth
of
a
Revolution

26-1   A Dilemma

At the beginning of this book, when we were introducing the concepts of classical mechanics, it was suggested that it might be possible to describe the universe in the following neat and tidy way. The universe consists of small, discrete particles moving along well defined paths through empty space. The position of each particle must be known at each instant of time. However, the paths are not arbitrary, but are determined by the accelerations which the particles produce in one another. In principle, if the positions and velocities of all the particles are known at one single instant of time, it should be possible to calculate their positions and velocities at all other past and future instants of time. Before this can be done, though, it is necessary to know the laws governing the acceleration produced in one particle by another particle. One of the major objectives of classical physics was the discovery of these laws.

So far we have considered in detail only two types of interaction, gravitational and electromagnetic. The procedure is particularly successful in the case of gravitational forces and provides a very satisfactory explanation of the behavior of falling bodies and the motion of the planets and moons of the solar system. It fails only in one small detail. The predicted magnitude of the advance of the perihelion of Mercury is only 99.2% of the observed magnitude. The remaining 0.8% requires the general theory of relativity. This theory, however, is based upon an important new concept. It postulates that the "empty" space through which the planet is moving is not featureless, but has properties that must be described by assigning appropriate numbers to each point in space. In this case the numbers describe the curvature of space—the extent to which it deviates from the "flat" space of Euclidean geometry.

The idea that properties should be associated with all points in space is particularly important in the theory of electromagnetic forces. The expression for the force between two moving charges is quite complicated, and was not correctly formulated until a late stage in the development of the theory. It proved simpler to talk instead in terms of an electric field **E** and a magnetic field **B** at each point in space. The most satisfactory formulation of electromagnetic theory is embodied in Maxwell's equations, which describe the behavior of these fields **E** and **B** in "empty" space.

Maxwell's equations imply the existence of electromagnetic waves and it is clear that light is such an electromagnetic wave. Light, however, is something of which we have very direct experience. Our most powerful way of observing the universe is to look around us and make use of the light waves falling on our eyes. On a sunny day we have a feeling of warmth caused by the energy that electromagnetic radiation transfers from the sun to our bodies. It might well be argued that we are more intimately aware of the existence of electromagnetic waves than of the existence of electrons and protons.

Suppose that a star explodes in the Andromeda galaxy, producing what is called a supernova. A large amount of energy is released in a short time and a burst of electromagnetic radiation travels outward from the star. It takes about two million years to reach the earth. Soon after it starts out the explosion dies down and everything is quiet in the Andromeda galaxy. Meanwhile, the earth is totally unaware of what has happened. Somewhere in the intervening space the energy is traveling in the form of an electromagnetic wave. It is difficult to avoid the impression that this electromagnetic wave is "something" that has just as much reality as an electron or a proton. It seems that the original picture of particles moving through empty space should be modified by the addition of electromagnetic waves traveling between the particles.

If we are to imagine the universe as consisting of both material particles and electromagnetic waves, let us compare and contrast their properties. Moving particles have energy and momentum, but so do electromagnetic waves. Particles have mass, but, according to Einstein's principle of the equivalence of mass and energy, light also has mass. The path of a moving particle is bent in a gravitational field, but so is the path of a beam of light. In one important respect, however, particles and waves are very different. A particle is a highly localized entity, which is situated at a particular point in space at a particular instant. A wave is a diffuse entity which is spread over a whole region of space.

The growing realization of the importance of electromagnetic waves was a development that lasted through most of the nineteenth century. The early twentieth century brought two startling revelations. Certain experiments were performed in which light behaved, not like a wave, but like a stream of particles! It began to look as though the universe could again be considered to be a collection of particles. However, a few years later, other experiments indicated that electrons could sometimes behave like waves! Both light and material particles sometimes exhibit the characteristics of waves and at other times exhibit the characteristics of particles. This is called the **dual nature of light** and the **dual nature of matter.**

These developments had the virtue of putting material particles and electromagnetic radiation on the same footing, but they introduced a new dilemma. How is it possible for any entity to behave sometimes like a localized particle and at other times like a diffuse wave?

In the present chapter we shall consider the origins of **quantum mechanics,** which eventually led to this complete revolution in our ideas of the nature of the physical universe. In the next chapter we shall present the decisive evidence which convinced physicists of the particle nature of light, and we shall show how it was immediately followed by decisive evidence for the wave nature of matter. Then, in chapter 28, we shall try to resolve the dilemma presented by the dual nature of light and matter.

26-2 Planck's Quantum Theory of Black Body Radiation

The first step in the direction of quantum mechanics was taken by Max Planck in 1900, five years before Einstein's first paper on relativity. Planck was trying to explain the nature of the electromagnetic radiation emitted from the surface of a hot body. Usually this radiation is characteristic of the chemical composition of the hot surface. Depending upon the particular atoms present, and their spatial arrangement relative to one another, certain frequencies are emitted preferentially and other frequencies are emitted very weakly. This is not true of a *perfect black body*. A body appears black when it completely absorbs all the light falling on it and reflects none into the eye of the observer. A perfect black body is defined as one that completely absorbs electromagnetic radiation of any frequency. The radiation which it *emits* is called **black body radiation** and is in no way characteristic of the chemical composition of the body, but depends only upon the temperature of its surface. Black body radiation was briefly discussed in section 20-3.

Figure 26-1 shows how the energy radiated by the black body is distributed over the various possible frequencies of the emitted electromagnetic waves. At any temperature all possible frequencies from zero to infinity are emitted. At a particular temperature, T°K, a maximum intensity of radiation is emitted at a particular frequency, ν_m. Most of the energy is radiated at frequencies not very different from ν_m. For example, about three quarters of the energy is contained within the frequency range between $\frac{1}{2}\nu_m$ and $2\nu_m$. As the temperature is raised, the energy radiated at any frequency rapidly increases. The total energy radiated at all frequencies is proportional to T^4 (Stefan's law). At the same time the frequency of maximum radiation, ν_m, increases in direct proportion to the absolute temperature (Wien's law)

$$\nu_m = 5.9 \times 10^{10} T \tag{26-1}$$

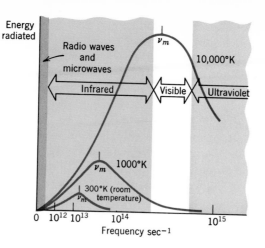

Figure 26-1 The spectrum of black body radiation. In order to bring out the main qualitative features, some distortion has been introduced. In reality, the curve at 10,000°K is about 37,000 times higher than the curve at 300°K and the radio and microwave region occupies only $\frac{1}{3000}$th of the horizontal axis.

An attempt to explain these facts on the basis of classical physics would proceed somewhat as follows. All bodies contain electrons. At any temperature above 0°K these electrons are in rapid vibration and the vibration becomes more energetic as the temperature is raised. An oscillating electron radiates an electromagnetic wave. Since a black body radiates at all frequencies, it must contain electrons capable of oscillating at all frequencies. The theory must therefore decide how much energy is associated with each frequency of oscillation at a temperature T and how much radiation is consequently emitted. However, all attempts to proceed in this way failed to explain the observed facts. In particular, the classical theory predicted that an infinite amount of energy should be radiated at the higher frequencies!

Planck pointed out that the difficulty could be overcome by making the following rather startling assumption. An electron oscillating with a frequency v radiates electromagnetic waves of frequency v, but the energy is not emitted continuously. Instead it is emitted in "bursts" of well-defined amount, ε. The quantity of energy in a burst is called a **quantum** of energy. It is proportional to the frequency of the radiation and the constant of proportionality is called Planck's constant.

Planck's quantum of energy

$$\varepsilon = hv \tag{26-2}$$

Planck's constant

$$h = 6.625 \times 10^{-27} \text{ erg sec} \tag{26-3}$$

The Birth
of a
Revolution

Planck's idea was that the oscillating electron cannot emit as small an amount of energy as it wishes. It must either emit a quantum of amount $h\nu$ or nothing. It can then go on to emit subsequent quanta, each of amount $h\nu$, but it cannot, for example, emit a total amount $(3/2)\,h\nu$. On a fine scale the emission of radiation is discontinuous.

Notice, however, how fine the scale is. For visible light of frequency about 5×10^{14} sec^{-1}, the energy in a quantum is 3.31×10^{-12} erg. To acquire a kinetic energy of this magnitude, a grain of sand would have to fall through a height of only about 3×10^{-10} cm, which is much less than the diameter of a single atom (see example 26-2-1)! In our everyday experience of macroscopic bodies the energies involved are so large compared with Planck's quanta that we are not normally aware of the discontinuous nature of energy. This is an important point because, whenever we encounter a startling new idea, it is always relevant to ask why it was not apparent to the many generations of intelligent men who preceded us. It is only when we deal with very small entities of the size of atoms, or fundamental particles, that the quantum aspects of nature become apparent. If we ask how far a single proton must fall before it acquires enough kinetic energy to be able to create a quantum of visible light, the answer is interesting. A proton starting at rest a long way from the earth, at an effectively infinite distance, and falling freely all the way to the earth's surface, would still not have quite enough kinetic energy to create a quantum of *visible* light (see example 26-2-2)!

Planck's constant, h, is a very important fundamental constant of nature, which can be justifiably compared with the velocity of light, c. In the discussion of the theory of relativity we realized that some of our everyday prejudices about time, space, and motion are a consequence of the fact that we normally deal with velocities small compared with c. As soon as we encounter velocities comparable with c, unexpected new features appear. Moreover, c plays the role of an upper limit to all possible velocities. Similarly, as long as we are concerned with large objects we do not have to worry about the strange new ideas introduced by quantum mechanics. However, when we deal with very small objects such as atoms and fundamental particles, the unexpected again turns up. We have just encountered the first instance of this in the discontinuous nature of energy. The constant h then serves to set a *lower* limit on the smallness of such things as quantity of energy.

Example 26-2-1

A grain of sand has a mass m of 10^{-5} gram. Through what height must it fall from rest in order to have a kinetic energy equal to the energy of a quantum of visible light with a frequency of 5×10^{14} sec^{-1}?

The energy of the quantum is

$$\varepsilon = h\nu \tag{26-4}$$

$$= 6.625 \times 10^{-27} \times 5 \times 10^{14} \tag{26-5}$$

$$= 3.31 \times 10^{-12} \text{ erg} \tag{26-6}$$

When the grain of sand falls through a height H from rest, its final kinetic energy, $\frac{1}{2}mv^2$, is equal to its decrease in gravitational potential energy, mgH.

$$\frac{1}{2}mv^2 = mgH \qquad (26\text{-}7)$$

Therefore $mgH = h\nu \qquad (26\text{-}8)$

$$H = \frac{h\nu}{mg} \qquad (26\text{-}9)$$

$$= \frac{3.31 \times 10^{-12}}{10^{-5} \times 980} \qquad (26\text{-}10)$$

$$= 3.38 \times 10^{-10} \text{ cm} \qquad (26\text{-}11)$$

The grain of sand would have to fall through a height of only 3.38×10^{-10} cm. The diameter of an atom is about 10^{-8} cm.

Example 26-2-2

A proton is at rest a long way away from the earth. It falls toward the earth under the influence of the earth's gravitational attraction. When it reaches the earth, all its kinetic energy is converted into the energy of a single quantum of electromagnetic radiation. What is the frequency of this quantum?

When the proton is at an infinite distance from the earth, their mutual gravitational potential energy is zero. When the proton reaches the surface of the earth, their mutual potential energy is

$$\Phi_G = -\frac{GM_e m}{R_e} \qquad (26\text{-}12)$$

where m is the mass of the proton, M_e is the mass of the earth, and R_e is the radius of the earth. The kinetic energy gained by the proton is equal to the potential energy lost, and so, if v is the velocity with which the proton strikes the earth,

$$\frac{1}{2}mv^2 = \frac{GM_e m}{R_e} \qquad (26\text{-}13)$$

All of this energy is converted into the energy of the quantum of electromagnetic radiation.
Therefore

$$h\nu = \frac{GM_e m}{R_e} \qquad (26\text{-}14)$$

$$\nu = \frac{GM_e m}{hR_e} \qquad (26\text{-}15)$$

$$= \frac{6.67 \times 10^{-8} \times 6 \times 10^{27} \times 1.67 \times 10^{-24}}{6.625 \times 10^{-27} \times 6.37 \times 10^{8}} \qquad (26\text{-}16)$$

$$= 1.6 \times 10^{14} \text{ sec}^{-1} \qquad (26\text{-}17)$$

Referring to table 20-1, this frequency is seen to be in the infrared.

Let us now go farther than Planck himself was prepared to go and assume that the emission of a quantum of energy corresponds to the emission of a "particle" of electromagnetic radiation. A particle of electromagnetic radiation is called a **photon**. Its velocity is the velocity of light, c, and its total energy is $\varepsilon = h\nu$. Now consider the electromagnetic radiation inside a closed box with its walls at a uniform temperature T°K. The hot walls of the box continuously emit radiation that travels across the box until it strikes the wall at another place where it is partially absorbed and partially reflected. After several such reflections, all of the radiation is eventually reabsorbed. In equilibrium the amount of radiation emitted is equal to the amount absorbed and a constant amount is "in transit" inside the box. This radiation is sometimes called **cavity radiation** and it can be shown to have the same characteristics as black body radiation. For example, the energy of cavity radiation is distributed among the various frequencies in the manner illustrated in figure 26-1.

From the particle point of view we can say that the walls emit photons. A photon bounces backward and forward inside the box until, at one of its collisions with a wall, it is absorbed and disappears. The box can be visualized as filled with a "gas of photons," which is very similar in many ways to the ideal monatomic gas discussed in chapter 11 (see figure 26-2). The atoms of an ideal gas move in all

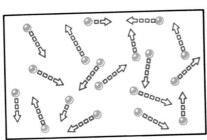

(a) An ideal monatomic gas.

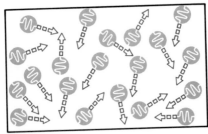

(b) Cavity radiation. A gas of photons.

Figure 26-2   Electromagnetic radiation inside an enclosure with hot walls (cavity radiation) can be visualized as a gas of photons. It is similar in some ways to an ideal monatomic gas of atoms.

directions with velocities which range from zero up to c and kinetic energies which range from zero to infinity. The photons of cavity radiation move in all directions, but they all have the same velocity, c. However, their frequencies, ν, range from zero to infinity and so their energies, $h\nu$, also range from zero to infinity. Most of the atoms of an ideal gas have velocities close to an average velocity v_a determined by the fact that the average kinetic energy is

$$\tfrac{1}{2}mv_a^2 = \tfrac{3}{2}kT \tag{26-18}$$

Most of the photons of cavity radiation have frequencies close to a most probable frequency ν_m. According to Planck's theory, the energy of the photon with this most probable frequency is

$$h\nu_m = 2.822\,kT \tag{26-19}$$

Apart from the numerical factor, there is a strong analogy between equations 26-18 and 26-19.

Equation 26-19 may be rewritten

$$\nu_m = \frac{2.822k}{h}\,T \tag{26-20}$$

Inserting the values of Planck's constant, h, and Boltzmann's constant, k, this equation is found to be the same as equation 26-1, which was discovered experimentally and is known as Wien's law.

In figure 26-3 a comparison is made between the distribution of the atoms of a gas over their possible kinetic energies, $\tfrac{1}{2}mv^2$, and the distribution of the photons of cavity radiation over their possible energies, $h\nu$. Although the details are different, there is a strong qualitative similarity.

26-4 The Photoelectric Effect

For five years Planck's theory remained a bold speculation which created more problems than it solved. Then, in 1905, the same year that he

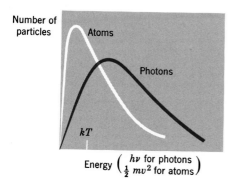

Figure 26-3 Distribution of the atoms of a gas over their possible kinetic energies compared with the distribution of the photons of cavity radiation over their possible energies (and frequencies).

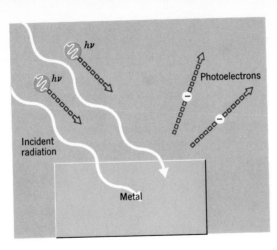

Figure 26-4 The photoelectric effect.

published his first paper on special relativity, Einstein pointed out that the nature of the photoelectric effect is readily explainable only in terms of Planck's quanta.

When electromagnetic radiation of a sufficiently high frequency falls on the surface of a metal, electrons are ejected from it (figure 26-4). This is called the **photoelectric effect** and the ejected electrons are called **photoelectrons**. The classical explanation is fairly obvious. The oscillating electric vector of the incident electromagnetic wave exerts an oscillating force on the free electrons inside the metal. These electrons are thereby set into oscillation with a steadily increasing amplitude until they are oscillating so violently that they are able to escape from the surface of the metal.

This classical theory is completely at variance with the experimental facts. The theory suggests that, if the incident radiation is very weak, there should be an appreciable time lag while the electron builds up a sufficient amplitude of oscillation to be able to escape. In fact, however weak the radiation, the first photoelectron is ejected within 10^{-8} sec after the radiation is switched on. The theory also suggests that, if the intensity of the radiation is increased, thereby increasing the size of the oscillating electric vector, the photoelectrons should be ejected with larger velocities. Experimentally it is found that the *number* of photoelectrons ejected per second is directly proportional to the energy of the radiation falling upon the surface per second, but that the *velocities* with which they are ejected are independent of this energy. Finally, the theory suggests that radiation of very low frequency should be capable of ejecting photoelectrons, provided it is given sufficient time to build up a large oscillation in the electrons. Experimentally, it is found that no photoelectrons are produced until the frequency exceeds a certain threshold frequency, ν_c. This threshold frequency is often in the ultraviolet, but is sometimes in the visible region of the spectrum and, for certain special materials, can be in the infrared.

Einstein's explanation, put into modern terms, is as follows. A photon strikes the metal and is completely absorbed. It disappears and gives up a definite amount of energy, $h\nu$, to one of the free electrons. Before it can escape from the surface of the metal, this electron loses part of its energy in two ways. Those photoelectrons originating at some depth below the surface are not able to escape without suffering several collisions with the atoms of the metal and losing a considerable fraction of their energy in the process. The energy lost in this way is a variable quantity and the electrons therefore emerge with a wide spread of velocities. This complication can be avoided by concentrating on those photoelectrons emerging with maximum velocity, v_m, because these presumably originated very near the surface and escaped before experiencing a collision.

Even these maximum velocity electrons do not emerge with the full energy $h\nu$ of the photon. An electron inside a metal is firmly bound to the metal in much the same way that an electron in an atom is firmly bound to the atom. In order to pluck an electron out of a metal it is necessary to exert a force on it to counteract the attraction of the rest of the metal. In the process of pulling the electron out and depositing it at rest at some point well outside the surface, this force does an amount of work ϕ ergs. ϕ is called the **work function** of the metal and its value is characteristic of the metal in question. A photoelectron breaking through the surface of the metal does not receive the assistance of an external force, and so the work ϕ is done at the expense of the electron's kinetic energy. An electron originating near the surface is given an amount of energy $h\nu$ by the photon, but loses an amount ϕ while escaping. The maximum kinetic energy of the photoelectrons is therefore

$$\tfrac{1}{2}mv_m{}^2 = h\nu - \phi \tag{26-21}$$

The following analogy may prove helpful. A soccer ball of mass m at rest at the bottom of a pit of depth H is to be kicked out of the pit up a ramp (figure 26-5). The kick delivers a definite amount of energy, ε, which is analogous to the energy $h\nu$ given to the photoelectron by the photon. The ball therefore starts out with a velocity V and a kinetic energy

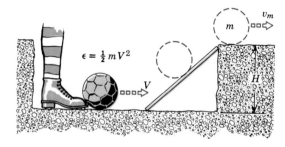

Figure 26-5   An analogy to illustrate Einstein's equation for the photoelectric effect.

$$\tfrac{1}{2}mV^2 = \varepsilon \qquad\qquad (26\text{-}22)$$

While climbing the ramp, the ball gains potential energy mgH, which is analogous to the work function, ϕ. The velocity, v_m, of the ball as it emerges from the pit is therefore given by

$$\tfrac{1}{2}mv_m{}^2 = \varepsilon - mgH \qquad\qquad (26\text{-}23)$$

which is analogous to

$$\tfrac{1}{2}mv_m{}^2 = h\nu - \phi \qquad\qquad (26\text{-}24)$$

Einstein's theory is completely in accord with the experimental facts. The first photoelectron can be ejected as soon as the first photon strikes the metal, explaining why there is no time lag. The velocity of ejection depends only on the energy of a single photon, which depends only on the frequency of the radiation. Increasing the intensity of the radiation increases the number of photons arriving per second and therefore increases the number of photoelectrons ejected per second, but in no way affects the energy given to an electron when it absorbs a single photon. The significance of the threshold frequency, ν_c, is also clear. The electron can escape only if the energy given to it by the photon is greater than the energy ϕ which it must have to break through the surface. The minimum photon energy which will produce photoelectrons is therefore given by

$$h\nu_c = \phi \qquad\qquad (26\text{-}25)$$

Equation 26-21 may therefore be rewritten as

$$\tfrac{1}{2}mv_m{}^2 = h(\nu - \nu_c) \qquad\qquad (26\text{-}26)$$

Note that, since the kinetic energy $\tfrac{1}{2}mv_m{}^2$ is necessarily positive, this clearly implies that ν must be greater than ν_c. If the maximum velocity of the photoelectrons is measured for various values of the frequency of the incident radiation, equation 26-26 is verified in detail. The plot of $\tfrac{1}{2}mv_m{}^2$ against ν is a straight line (figure 26-6), which cuts the axis of ν at the threshold frequency ν_c. The slope of this straight line enables h to be measured and the value obtained is found to be in good agreement with the value needed to explain black body radiation.

Modern detection techniques are so sensitive that it is possible to magnify the effect of each photoelectron so that it produces an audible click in a loudspeaker. With a sufficiently low intensity of illumination, it is then possible to "hear" the arrival of a photon! The human eye is also extremely sensitive and can see a very faint light when it sends only about 100 photons into the eye every second.

Example 26-4-1

When ultraviolet light of frequency 1.3×10^{15} sec^{-1} is shone on a metal, photoelectrons are ejected with a maximum energy of 1.8 electron volts. Calculate the work function of the metal in ergs and electron volts. What is the threshold frequency of this metal?

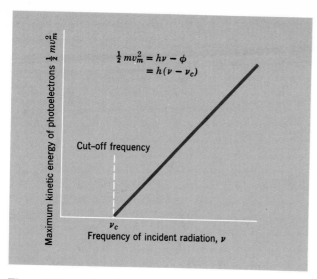

$$\tfrac{1}{2}mv_m^2 = h\nu - \phi$$
$$= h(\nu - \nu_c)$$

Cut-off frequency

ν_c
Frequency of incident radiation, ν

Maximum kinetic energy of photoelectrons $\tfrac{1}{2}mv_m^2$

Figure 26-6   A plot of Einstein's equation for the photoelectric effect.

The energy of a photon of the ultra-violet light is

$$h\nu = 6.625 \times 10^{-27} \times 1.3 \times 10^{15} \tag{26-27}$$
$$= 8.61 \times 10^{-12} \text{ erg} \tag{26-28}$$

Since 1 electron volt $= 1.60 \times 10^{-12}$ erg, $\tag{26-29}$

the maximum kinetic energy of the photoelectrons is

$$\tfrac{1}{2}mv_m^2 = 1.8 \times 1.6 \times 10^{-12} \tag{26-30}$$
$$= 2.88 \times 10^{-12} \text{ erg} \tag{26-31}$$

From equation 26-21, the work function is

$$\phi = h\nu - \tfrac{1}{2}mv_m^2 \tag{26-32}$$
$$= 8.61 \times 10^{-12} - 2.88 \times 10^{-12} \text{ erg} \tag{26-33}$$
$$= 5.73 \times 10^{-12} \text{ erg} \tag{26-34}$$
$$= \frac{5.73 \times 10^{-12}}{1.60 \times 10^{-12}} \text{ electron volt} \tag{26-35}$$
$$= 3.58 \text{ eV} \tag{26-36}$$

The threshold frequency is obtained from equation 26-25

$$\nu_c = \frac{\phi}{h} \tag{26-37}$$

$$= \frac{5.73 \times 10^{-12}}{6.625 \times 10^{-27}} \tag{26-38}$$
$$= 8.65 \times 10^{14} \text{ sec}^{-1} \tag{26-39}$$

This frequency is in the ultra-violet, only just beyond the blue end of
the visible spectrum.

26-5 The Continuous X-ray Spectrum

An x-ray tube is illustrated in figure 20-5. Electrons are accelerated through a potential difference V and strike a metal target, where they create x-rays. These x-rays consist in part of several discrete frequencies characteristic of the metal of the target. This part is not our present concern. In addition there is a continuous part consisting of all frequencies from zero up to a maximum frequency ν_{max}. This maximum frequency provides good evidence for the quantum nature of x-rays.

Classically, the x-rays are the radiation from a rapidly decelerating electron and classical theory suggests that they should contain all frequencies from zero to infinity. The modern picture is that, when the electron collides with the atoms of the target, its energy is dissipated in the creation of one or more photons. Clearly, the maximum energy that a photon can have corresponds to all the kinetic energy of the electron being used up in the creation of this one photon. When an electron strikes the target its kinetic energy is Ve, where V is the potential difference across the x-ray tube. The maximum frequency, ν_{max}, that an x-ray photon can have is therefore given by

$$h\nu_{max} = Ve \qquad (26\text{-}40)$$

This equation is in complete agreement with the observed values of ν_{max}.

Example 26-5-1

Establish the numerical relationship between ν_{max} in $\sec^{-1}$ and V in volts. What is the minimum voltage which must be applied to the x-ray tube to produce x-rays of wavelength 10^{-8} cm?

If V is in volts and e in coulombs, the kinetic energy Ve is in joules. Kinetic energy of an electron striking the target

$$= Ve \qquad (26\text{-}41)$$

$$= V \text{ (volts)} \times 1.60 \times 10^{-19} \text{ joule} \qquad (26\text{-}42)$$

$$= 1.60 \times 10^{-12} \, V \text{ erg} \qquad (26\text{-}43)$$

(This could also have been obtained directly from the knowledge that 1 eV is 1.60×10^{-12} erg.) Therefore

$$h\nu_{max} = 1.60 \times 10^{-12} V \qquad (26\text{-}44)$$

$$\nu_{max} = \frac{1.60 \times 10^{-12}}{6.625 \times 10^{-27}} V \qquad (26\text{-}45)$$

$$\nu_{max} = 2.41 \times 10^{14} V \qquad (26\text{-}46)$$

This is the required relationship between ν_{max} and V.

A wavelength, λ, of 10^{-8} cm corresponds to a frequency

$$\nu = \frac{c}{\lambda} \qquad (26\text{-}47)$$

$$= \frac{3 \times 10^{10}}{10^{-8}} \qquad (26\text{-}48)$$

$$= 3 \times 10^{18} \sec^{-1} \qquad (26\text{-}49)$$

In order to produce this frequency, the voltage across the tube must be at least

$$V_{min} = \frac{\nu}{2.41 \times 10^{14}} \qquad (26\text{-}50)$$

$$= \frac{3 \times 10^{18}}{2.41 \times 10^{14}} \qquad (26\text{-}51)$$

$$= 12{,}500 \text{ volts} \qquad (26\text{-}52)$$

QUESTIONS

1. Give examples from everyday experience of hot bodies which are not black bodies, but emit radiation of certain frequencies preferentially.
2. In section 26-3 it was stated that the atoms of an ideal gas have velocities ranging from 0 to c, but energies ranging from 0 to ∞. Explain this statement.
3. In section 26-2 it was stated that a black body emits radiation with all frequencies from zero up to *infinity*. This is not quite true. Can you think of any obvious upper limit to the frequency of a photon emitted by a body of finite size? Show that, in any case of practical interest, this highest frequency is so high that it is effectively infinite. (Hint: for any hot body the order of magnitude of the thermal energy is kT per atom.)
4. If the container of figure 26-2a is expanded at constant temperature the pressure of the gas decreases in accordance with Boyle's law. If the container of figure 26-2b is expanded at constant temperature, what happens to the pressure of the radiation on the walls? What essential difference between the properties of photons and atoms causes this difference in behavior?
5. Consider the question of why it is permissible to use the classical formula for kinetic energy in Einstein's equation for the photoelectric effect.
6. Invent a precise definition of a particle. Suggest experiments to discover if photons of light satisfy this definition.
7. Invent a precise definition of a wave. Suggest experiments to discover if light satisfies this definition.

PROBLEMS

A

1. The temperature of the surface of the sun is $6000°K$. Assuming that it is a black body, at what frequency does it emit most radiation? Where is this frequency relative to the visible spectrum?
2. At what temperature does the radiation emitted by a black body consist mainly of:
   (a) Radio waves with frequencies about 10^6 sec^{-1}.

The Birth
of a
Revolution

(b) Infrared radiation with wavelengths about 10^{-3} cm.

(c) Blue light with a wavelength of 4.5×10^{-5} cm.

(d) X-rays with wavelengths about 10^{-8} cm.

(e) γ-rays with wavelengths about 10^{-12} cm.

3. If a photon has an energy of 2×10^{-14} erg, what is its frequency?

4. What is the energy of a photon of a radio wave with a wavelength of 200 m?

5. What is the wavelength of a photon if its energy is 1 erg?

6. A beam of infrared radiation has a wavelength of 10^{-4} cm. What is the energy of one of its photons in electron volts?

7. A photon has an energy of 10^{-20} erg. What type of electromagnetic radiation is it?

8. What is the most probable energy of a photon radiated from the surface of a black body at a temperature of $10^{5}\,^\circ$K?

9. What must be the temperature of a black body if most of the photons it emits have energies in the vicinity of (a) 1 eV, (b) 10^6 eV, (c) 10^9 eV?

10. The universe is believed to be full of black body radiation at a temperature of $3°$K. About what is the wavelength of most of the photons?

11. A metal has a work function of 4.5 eV. What is its photoelectric threshold frequency?

12. What is the minimum voltage which must be applied across an x-ray tube to produce x-rays with a frequency of 10^{20} $\sec^{-1}$?

B

13. What is the wavelength of a 1 MeV photon?

14. What is the equivalent mass of a photon of red light with a wavelength of 6.75×10^{-5} cm?

15. What is the frequency of a photon if its energy is equal to the rest mass energy of an electron? What type of electromagnetic radiation is it?

16. When ultraviolet light with a wavelength of 1.2×10^{-5} cm is shone on a metal, it ejects photoelectrons with a maximum energy of 10^{-11} erg. What is the photoelectric threshold frequency of the metal?

17. The photoelectric work function of a metal is 5.2 eV. What frequency of radiation must be shone on it to give photoelectrons with velocities up to 8×10^7 cm/sec?

18. What is the minimum voltage which must be applied across an x-ray tube to produce photons each of which has an energy equal to the rest mass energy of an electron?

19. What is the minimum voltage which must be applied across an x-ray tube to produce photons with a wavelength equal to the radius of a proton $(0.8 \times 10^{-13}$ cm)?

20. Calculate the period of a photon whose energy is equal to the mutual *gravitational* potential energy of two electrons 10^{-13} cm apart. Comment on the order of magnitude of the answer.

21. An electron starts from rest at a great distance from a certain star, which has a mass of 10^{34} gm and a radius of 10^{10} cm. It then falls toward the star under the influence of its gravitational attraction. When it strikes the surface of the star, all its kinetic energy is converted into the energy of a single photon. What is the frequency of this photon and what type of electromagnetic radiation is it? You may assume that the electromagnetic energy radiated by the electron before it strikes the surface is negligibly small. Is it necessary to consider the relativistic variation of mass with velocity?

22. The flux of solar radiation flowing perpendicularly across unit area in the vicinity of the earth is 1.4×10^6 erg cm^{-2} sec^{-1}. Approximately how many photons does the sun emit per second? Take the temperature of the surface of the sun to be $6000\,°$K and assume that you can ignore photons with frequencies appreciably different from ν_m.

23. If the sun were removed to a distance of about 5×10^{19} cm, it would be a sixth magnitude star just barely visible to the unaided eye. Under these circumstances, how many photons from the sun would enter the eye in one second. Compare the answer with the statement made in the text about the minimum number of photons per second which the eye can detect. Comment.

24. Approximately how many photons are there per cm^3 in the vicinity of the earth? (Hint: review the proof of equation 15-7).

The Birth
of a
Revolution

27

Particles
and
Waves

27-1 The Compton Effect

The phenomena discussed in the previous chapter strongly indicate that the energy of electromagnetic radiation exists in bundles of amount $h\nu$, but they give no *direct* evidence for *particles* of electromagnetic radiation. The Compton effect, discovered by A. H. Compton in 1922, makes the existence of such particles very plausible.

When x-radiation passes through a solid, typically graphite, part of it is scattered in all directions. The scattered radiation has two components. One component has exactly the same frequency as the incident radiation. The other component, which is the one that interests us, has a frequency slightly smaller than that of the incident radiation (see figure 27-1a). Moreover, the decrease in frequency becomes greater as the angle of scattering, α, increases.

This effect is easily understood if we assume that it is a consequence of a collision between a photon of the incident x-rays and an electron in the solid. This collision may be discussed in exactly the same way as the collision of two billiard balls, except that we must use the relativistic formulas for the energy and momentum of the electron, which acquires a velocity comparable with c.

The situations before and after the collision are shown in figure 27-1b. The law of conservation of energy may be applied in the following way. Before the collision the incident photon has a frequency ν and an energy $h\nu$. Without introducing any appreciable error the electron may be considered to be initially at rest, and its total energy is therefore m_0c^2, where m_0 is its rest mass. Therefore,

$$\text{Total energy before collision} = h\nu + m_0c^2 \tag{27-1}$$

After the collision the photon bounces off at an angle with a diminished

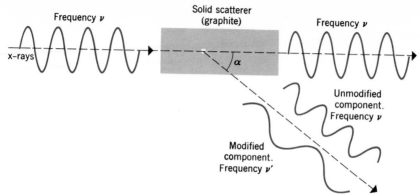

(a) The wave point of view.

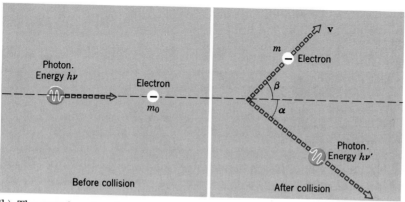

(b) The particle point of view.

Figure 27-1   The Compton effect.

frequency, ν', and a diminished energy $h\nu'$. The energy lost by the photon is given to the electron, which acquires a velocity v and an energy

$$mc^2 = \frac{m_0 c^2}{\sqrt{1 - \dfrac{v^2}{c^2}}} \tag{27-2}$$

Therefore,

Total energy after collision $= h\nu' + mc^2$ $\qquad$ (27-3)

Equating the total energies before and after the collision

$h\nu + m_0 c^2 = h\nu' + mc^2$ $\qquad$ (27-4)

or $h(\nu - \nu') = mc^2 - m_0 c^2$ $\qquad$ (27-5)

$$h(\nu - \nu') = \frac{m_0 c^2}{\sqrt{1 - \dfrac{v^2}{c^2}}} - m_0 c^2 \tag{27-6}$$

Particles
and Waves

571

The decrease in frequency of the scattered radiation is clearly seen to be a consequence of the fact that a single photon gives up part of its energy to the electron. (The scattered radiation that has the same frequency as the incident radiation is produced by a different process (see question 27-1).)

Notice an important difference between the photoelectric effect and the Compton effect. In the photoelectric effect the photon completely disappears and *all* of its energy is given to the photoelectron. In the Compton effect there is still a photon after the collision, but its frequency is less than that of the incident photon and *part* of the energy of the incident photon has been given to the Compton electron.

27-2 The Momentum of a Photon

As explained in section 20-4, a quantity of electromagnetic radiation with energy ε has momentum ε/c. A photon with energy $h\nu$ must therefore have momentum $h\nu/c$. Since $c = \nu\lambda$, where λ is the wavelength, this might also be expressed as h/λ.

Momentum of a photon

$$p = \frac{h\nu}{c} \tag{27-7}$$

$$= \frac{h}{\lambda} \tag{27-8}$$

We can now apply the law of conservation of momentum to the collision of a photon and an electron in the Compton effect. Momentum is a vector and the component of the total momentum resolved in any direction is conserved. Referring to figure 27-2 and considering

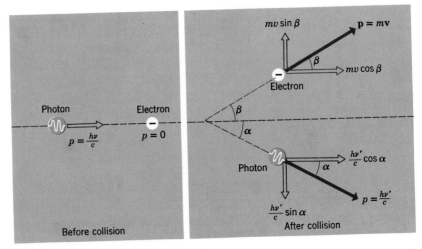

Figure 27-2   Conservation of momentum in the Compton effect.

the components of momentum in the direction of travel of the incident photon, this photon initially has a momentum $h\nu/c$ in this direction. The electron is initially at rest and has no momentum. After the collision the component of the momentum of the scattered photon in this direction is $(h\nu'/c) \cos \alpha$, and the component of the momentum of the electron is $m\upsilon \cos \beta$. The law of conservation of momentum therefore tells us that

$$\frac{h\nu}{c} = \frac{h\nu'}{c} \cos \alpha + m\upsilon \cos \beta \qquad (27\text{-}9)$$

Before the collision there is no component of the momentum in a direction perpendicular to the direction of travel of the incident photon. After the collision the component of the momentum of the scattered photon in this perpendicular direction is $(h\nu'/c) \sin \alpha$. The electron has a component of momentum $m\upsilon \sin \beta$ in exactly the opposite direction. Since the resultant component in this direction must remain zero,

$$\frac{h\nu'}{c} \sin \alpha = m\upsilon \sin \beta \qquad (27\text{-}10)$$

Given the frequency, ν, of the incident photon and the angle of scattering, α, equations 27-6, 27-9 and 27-10 enable all the other relevant quantities to be calculated. The necessary algebraic manipulation is moderately complicated, but should not be beyond the capabilities of many students, who are advised to attempt it as an exercise. The result of this algebraic manipulation is

$$\frac{1}{\nu'} = \frac{1}{\nu} + \frac{h}{m_0 c^2}(1 - \cos \alpha) \qquad (27\text{-}11)$$

This formula is in satisfactory agreement with the experimental results on the change in frequency as a function of the angle of scattering, α.

The concept of photons with energy and momentum enables us to form a vivid picture of the pressure of light. It can now be visualized as due to a hail of photons striking the surface on which the pressure is exerted. It is therefore very similar to the pressure exerted by the atoms of a gas on the walls of their container (section 11-6).

We can also obtain a clearer understanding of the laws of conservation of energy and momentum when applied to electromagnetic phenomena. In section 20-4 we discussed the fact that the forces between two moving charges are not equal and opposite. The laws of conservation of energy and momentum do not therefore apply if we consider only the energy and momentum of the moving charges. We pointed out that the apparently lost energy and momentum is really carried away by electromagnetic radiation. We can now describe the situation as in figure 27-3. The two charges are continually emitting or absorbing photons, which carry away or bring up energy and momentum. The laws of conservation of energy and momentum are applicable if we include the photons as well as the charged particles.

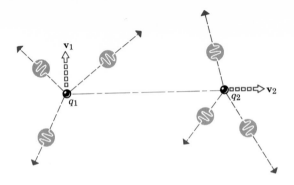

Figure 27-3 Two moving charges, producing accelerations in one another, radiate photons. The laws of conservation of energy and momentum cannot be applied unless the energy and momentum of these photons is included. (Compare figure 20-7.)

27-3 The Wave Nature of Matter

In 1922 the Compton effect provided conclusive evidence that electromagnetic radiation sometimes exhibits the properties of particles. Yet interference and diffraction experiments clearly indicate that electromagnetic radiation also exhibits the properties of a wave. In some sense (which we shall discuss in chapter 28), it must combine the properties of both particles and waves. In 1924 Louis de Broglie made the bold suggestion that electrons and protons, which had been shown in many experiments to behave like particles, might also behave like waves. By 1927 Davisson and Germer in the United States, and G. P. Thomson in Scotland, had demonstrated interference and diffraction effects with electrons.

In these experiments the diffraction was not produced by slits, as in the experiments of Young and Fresnel with light (chapter 21), but by crystalline solids. A crystal is an ordered array of atoms. When a wave falls on this array, each atom is the source of a scattered wavelet and the wavelets interfere constructively in some directions and destructively in other directions. The result is a complicated diffraction pattern, the details of which need not concern us here. Diffraction by crystals was first observed for x-rays and was successfully used both to measure the wavelengths of the x-rays and to study the spatial arrangement of the atoms in the crystals.

Part *a* of figure 27-4 shows a set of circular diffraction fringes obtained by passing x-rays through an aluminum foil. Part *b* shows a set of fringes produced by passing a beam of electrons through the same foil. The velocity of the electrons was adjusted to ensure that the wavelength associated with the electron beam was the same as the wavelength of the x-rays. The similarity of the two sets of fringes is striking evidence that electrons can be diffracted in exactly the same way as electromagnetic waves.

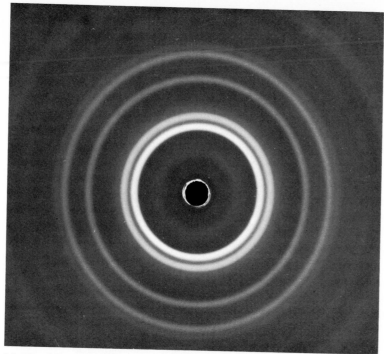

(a) The diffraction pattern obtained by passing x-rays through an aluminum foil.

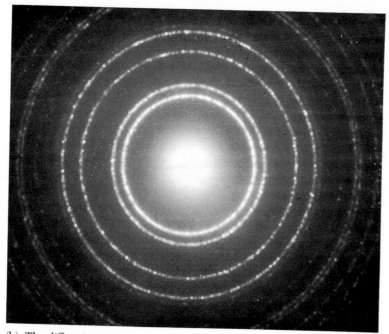

(b) The diffraction pattern obtained by passing a beam of electrons through the same foil.

Figure 27-4   A direct comparison of electron diffraction and x-ray diffraction. The wavelengths are the same in both cases. (From the P.S.S.C. film *Matter Waves*.)

Figure 27-5 Interference fringes obtained by passing electrons through a double slit. (Photograph by G. Mollenstedt.)

With modern improvements in technique, it is possible to observe diffraction of electrons by slits and straight edges. Figure 27-5 shows the interference fringes obtained when a beam of electrons passes through two parallel slits. It should be compared with the analogous fringes produced by light, as shown in figure 21-10. Figure 27-6a shows

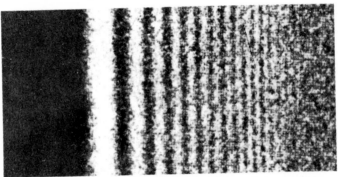

(a) Diffraction fringes produced when a beam of electrons passes near the straight edge of a small cubic crystal. (Courtesy of *Physics*, P.S.S.C., D. C. Heath & Co., 1960.)

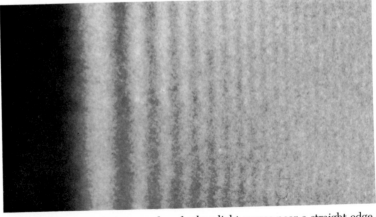

(b) The diffraction pattern produced when light passes near a straight edge. (Photograph by Brian Thompson.)

Figure 27-6 Comparison of the diffraction patterns produced by electrons and by light passing near a straight edge.

the fringes near the edge of the shadow when a beam of electrons passes near one of the straight edges of a small cubic crystal. Part *b* of the same figure shows similar diffraction fringes near the edge of a shadow produced by light.

Subsequent investigations have revealed the existence of diffraction effects with other fundamental particles, such as protons and neutrons, and even with helium, neon and argon atoms and hydrogen molecules. Figure 27-7 compares the diffraction patterns produced when (*a*) neutrons and (*b*) photons (x-rays) are incident on a single crystal of common salt, NaCl.

The wave associated with a material particle is called a **de Broglie wave.** The quantity that oscillates in a wave-like fashion is normally denoted by the symbol ψ, and is called the **wave function.** It is analogous to the vertical displacement of the surface of the water in the case of an ocean wave, or to the oscillating electric vector **E** in the case of an electromagnetic wave. The physical significance of ψ will be discussed in the next chapter.

27-4 The Equations Relating Wave Properties and Particle Properties

From the details of the diffraction patterns it was possible to deduce the wavelengths of the de Broglie waves. (See, for example, equation 21-27 and equations 21-43, 21-44, and 21-45.) These wavelengths were found to agree with de Broglie's theory. In order to satisfy certain formal requirements of the special theory of relativity, he had suggested that the equations for the frequency and wavelength of matter waves should be analogous to equations 26-2 and 27-8 for photons.

Formulas relating wave properties and particle properties.

Total energy of a particle,

$$\varepsilon = h\nu \tag{27-12}$$

$$= mc^2 \tag{27-13}$$

$$= \frac{m_0 c^2}{\sqrt{1 - \dfrac{v^2}{c^2}}} \tag{27-14}$$

Momentum of a particle,

$$p = \frac{h}{\lambda} \tag{27-15}$$

$$= mv \tag{27-16}$$

$$= \frac{m_0 v}{\sqrt{1 - \dfrac{v^2}{c^2}}} \tag{27-17}$$

Particles and Waves

(a) Neutron diffraction

(b) X-ray diffraction

Figure 27-7 Diffraction patterns produced when (*a*) neutrons and (*b*) photons (x-rays) are incident on a single crystal of NaCl. The difference between these spot patterns and the ring patterns of figure 27-4 arises because the aluminum foil used to obtain the ring patterns consisted of many small crystals aligned in all possible directions. A ring pattern is therefore a superposition of a large number of spot patterns with a random distribution of orientations. (Photograph (*a*) by E. O. Wollan, (*b*) from Lapp and Andrews, *Nuclear Radiation Physics*, 3rd ed., 1963, Prentice-Hall, Inc.)

Equations 27-15 and 27-16 are equivalent to

$$\lambda = \frac{h}{mv} \tag{27-18}$$

This is the equation that was verified by the diffraction experiments with electrons and subsequent experiments with protons, neutrons, atoms and molecules.

Notice that the relationship between momentum and wavelength, equation 27-15, is the same as equation 27-8 for photons, but that equation 27-7 is not true for material particles. Since the velocity of an electromagnetic wave is c, $c = \lambda v$ for a photon. Therefore, $p = h/\lambda$ is equivalent to $p = hv/c$ *for a photon.* However, the velocity of a de Broglie wave is not c. From equations 27-12 and 27-13, the frequency of the de Broglie wave is

$$\nu = \frac{mc^2}{h} \tag{27-19}$$

Combining this with equation 27-18, the velocity of a de Broglie wave is seen to be

$$V = \lambda \nu \tag{27-20}$$

$$= \frac{h}{mv} \frac{mc^2}{h} \tag{27-21}$$

$$V = \frac{c^2}{v} \tag{27-22}$$

This is not equal to c, except in the special case where the velocity of the particle v is equal to c. A material particle can never reach the speed of light, but a photon always has the speed of light. In the case of the photon, the general equation $p = h/\lambda$ also implies $p = hv/c$, but this is not so for particles moving with velocities less than c.

For a material particle, equation 27-22 has two disturbing features. First, the velocity of the de Broglie wave, c^2/v, is not the same as the velocity of the particle, v. Second, since v must be less than c, the velocity of the wave, c^2/v, must be greater than c. This appears to violate a fundamental requirement of the theory of relativity. We shall return to this point in the next chapter.

The set of equations 27-12 to 27-17 is universally true for all particles, including photons. However, if we insert $v = c$ into equation 27-14, the denominator becomes zero and it appears that the energy might become infinite. We have already stated that this is the reason why material particles can never attain the speed of light. This difficulty is easily avoided in the case of a photon, by assuming that its rest mass is zero.

$$m_0 = 0 \text{ for a photon} \tag{27-23}$$

The expression for the energy then becomes $0/0$, which is indeterminate and may have any value between 0 and infinity. It has, in fact, the finite value $h\nu$.

Conversely, if the rest mass of a particle is zero and its speed is less than c, then the numerator is zero and the denominator is not zero, so that the total energy is zero. A zero rest mass particle can possess energy only if it has the speed of light, because the numerator and denominator are then both zero. Unless it possesses energy it cannot produce any physical effects to make us aware of its existence.

A photon is able to travel with the speed of light because its rest mass is zero. If a particle has rest mass zero, it must travel with the speed of light in order to have any energy.

Example 27-4-1

What is the de Broglie wavelength of an electron with a kinetic energy of 1 eV?

When calculating the de Broglie wavelength, it is important to know whether or not to use relativistic formulas. One way of deciding this is to realize that, when the velocity v is very small compared with c, the kinetic energy is small compared with the rest mass energy, m_0c^2. Referring to example 24-4-1, the rest mass energy of an electron is 0.51×10^6 eV which *is* large compared with 1 eV. We need not use relativistic formulas in the present example.

In the formula for the de Broglie wavelength,

$$\lambda = \frac{h}{mv} \tag{27-24}$$

it is permissible to insert the rest mass, m_0, for m.

$$m \approx m_0 = 9.11 \times 10^{-28} \text{ gram} \tag{27-25}$$

The kinetic energy is

$$\tfrac{1}{2}m_0v^2 = 1 \text{ eV} \tag{27-26}$$
$$= 1.6 \times 10^{-12} \text{ erg} \tag{27-27}$$

Therefore

$$v = \sqrt{\frac{2 \times 1.6 \times 10^{-12}}{m_0}} \tag{27-28}$$

$$= \sqrt{\frac{2 \times 1.6 \times 10^{-12}}{9.11 \times 10^{-28}}} \tag{27-29}$$

$$= 5.92 \times 10^7 \text{ cm/sec} \tag{27-30}$$

Notice that this is, in fact, very much less than the velocity of light, 3×10^{10} cm/sec.

Inserting these values of m and v into equation 27-24,

$$\lambda = \frac{6.625 \times 10^{-27}}{9.11 \times 10^{-28} \times 5.92 \times 10^{7}} \tag{27-31}$$

$$= 1.23 \times 10^{-7} \text{ cm} \tag{27-32}$$

This is about ten times larger than the diameter of an atom.

Example 27-4-2

What is the de Broglie wavelength of an electron with a kinetic energy of 1 MeV?

The kinetic energy is no longer small compared with the rest mass energy and so we must use relativistic equations. The total energy is the sum of the kinetic energy, 1.0×10^{6} eV, and the rest mass energy, 0.51×10^{6} eV, so

$$mc^2 = 1.51 \times 10^6 \text{ eV} \tag{27-33}$$

$$= 1.51 \times 10^6 \times 1.6 \times 10^{-12} \text{ erg} \tag{27-34}$$

$$m = \frac{1.51 \times 10^6 \times 1.6 \times 10^{-12}}{(3 \times 10^{10})^2} \tag{27-35}$$

$$= 2.68 \times 10^{-27} \text{ gm} \tag{27-36}$$

To find the velocity v, remember that

$$m = \frac{m_0}{\sqrt{1 - \dfrac{v^2}{c^2}}} \tag{27-37}$$

$$\sqrt{1 - \frac{v^2}{c^2}} = \frac{m_0}{m} \tag{27-38}$$

$$= \frac{9.11 \times 10^{-28}}{2.68 \times 10^{-27}} \tag{27-39}$$

$$= 0.340 \tag{27-40}$$

Squaring, $1 - \dfrac{v^2}{c^2} = 0.116$ (27-41)

$$\frac{v^2}{c^2} = 0.884 \tag{27-42}$$

Taking the square root,

$$\frac{v}{c} = 0.94 \tag{27-43}$$

$$v = 0.94 \times 3 \times 10^{10} \tag{27-44}$$

$$= 2.82 \times 10^{10} \text{ cm/sec} \tag{27-45}$$

The de Broglie wavelength is therefore

$$\lambda = \frac{h}{mv} \tag{27-46}$$

$$= \frac{6.625 \times 10^{-27}}{2.68 \times 10^{-27} \times 2.82 \times 10^{10}} \tag{27-47}$$

$$= 8.8 \times 10^{-11} \text{ cm} \tag{27-48}$$

Particles and Waves

QUESTIONS

1. In the Compton experiment, the radiation scattered without change in frequency can be considered to be the result of a collision between an incident photon and the whole block of scattering material. Explain carefully how such a collision can change the direction of the momentum of the photon without producing any appreciable change in its energy.

2. Why is it necessary to use relativistic equations for the electron in the Compton effect, when it was not necessary in the photoelectric effect?

3. How would the Compton effect differ if protons were involved rather than electrons?

4. Which of the following types of radiation has a photon with (a) the least, (b) the greatest momentum? (i) Radio waves, (ii) ultraviolet light, (iii) red light, (iv) violet light, and (v) x-rays.

5. Which of the following equations are not applicable to all particles? (a) $\frac{1}{2}mv^2 = h\nu$ (b) $\varepsilon = mc^2$ (c) $\varepsilon = h\nu$ (d) $p = h/\lambda$ (e) $p = h\nu/c$ (f) $p = mv$.

6. Consider a process in which a photon of visible light strikes an isolated electron at rest in otherwise empty space. The photon disappears and its energy is given as kinetic energy to the electron. Show that this process cannot simultaneously conserve both energy and momentum. Why is this objection not valid for the photoelectric effect?

PROBLEMS

A

1. What is the momentum of a photon with an energy of 6×10^{-10} erg?

2. What is the momentum of a photon with a frequency of 4.5×10^{17} sec^{-1}?

3. If a photon has a momentum of 3×10^{-21} gm cm sec^{-1}, what is its wavelength?

4. What is the momentum of a 1 MeV photon?

5. What is the momentum of a mass of 15 gm with a de Broglie wavelength of 10^{-35} cm?

6. What is the de Broglie wavelength of a mass of 1.5×10^{-5} gm with a velocity of 3×10^2 cm/sec?

7. What is the de Broglie wavelength of a body with a rest mass of 4×10^{-3} gm and a velocity of $\frac{3}{5} c$?

8. To meet the requirements of the United States Golf Association, a golf ball must have a mass of 45.6 gm and acquire a velocity of 7600 cm/sec when tested on a certain machine. What is its de Broglie wavelength under these circumstances?

9. What is the de Broglie wavelength of an electron with a velocity of 2×10^6 cm/sec?

10. What is the de Broglie wavelength of an electron with a velocity of $\frac{3}{5} c$?

11. In a Compton scattering process the incident radiation had a frequency of 1.000×10^{19} sec^{-1} and the scattered radiation a frequency of 0.990×10^{19} sec^{-1}. What energy was given to the scattered electron? Express the answer in ergs and in electron volts.

12. Show that equation 27-11 may be put in the form

$$\lambda' = \lambda + \frac{h}{m_0 c}(1 - \cos \alpha)$$

Find the magnitude of the "Compton wavelength"

$$\lambda_C = \frac{h}{m_0 c}$$

B

13. In a Compton scattering process the incident radiation had a frequency of 1.200×10^{20} sec^{-1} and the scattered electron acquired a velocity of 1.5×10^{10} cm/sec. What was the frequency of the scattered photon?

14. A beam of blue light with a wavelength of 4.5×10^{-5} cm shines perpendicularly on the surface of a polished metal mirror and is totally reflected back in the direction from which it came. If 10^{20} photons strike the mirror in one second, what is the force exerted on the mirror by the pressure of the light?

15. What is the velocity of a particle if its de Broglie wavelength is $h/m_0 c$?

16. What is the de Broglie wavelength of a 5 eV electron?

17. What is the de Broglie wavelength of a 1 BeV electron?

18. Calculate the de Broglie wavelength of a proton under the following circumstances:
(a) Its velocity is 1 cm/sec, (b) Its velocity is 2.9×10^{10} cm/sec, (c) Its kinetic energy is 10 eV, (d) Its kinetic energy is 10^9 eV, (e) Its mass is twice its rest mass.

19. If the de Broglie wavelength of a neutron is 10^{-8} cm, what are its velocity and its kinetic energy?

20. Through what voltage must an electron be accelerated from rest to give it a de Broglie wavelength of 4.5×10^{-5} cm, which is the same as the wavelength of blue light.

21. What is the de Broglie wavelength of a diatomic oxygen molecule when its velocity is 10^6 cm/sec?

22. Find the frequency and wavelength of the wave associated with (a) a 100 eV photon (b) a 100 eV electron.

23. An electron and a photon both have a wavelength of 10^{-8} cm. Find the momentum, the total energy and the kinetic energy in each case. (Since the rest mass of a photon is zero, its kinetic energy is the same as its total energy).

C

24. Make a rough estimate of the de Broglie wavelength of a baseball hit for a home run.

25. What is the kinetic energy of an electron in electron volts if its de Broglie wavelength is 10^{-13} cm?

26. Find the order of magnitude of the de Broglie wavelength of most of the atoms in helium gas at room temperature.

27. If its velocity is very small compared with the velocity of light, show that the de Broglie wavelength of a V electron volt electron is given by the formula

$$\lambda = \frac{1.23 \times 10^{-7}}{\sqrt{V}} \text{ cm}$$

28. If its velocity is very nearly equal to the velocity of light, show that the de Broglie wavelength of a V electron volt electron is given by the formula

$$\lambda = \frac{1.24 \times 10^{-4}}{V} \text{ cm}$$

29. If E is the kinetic energy of a particle with a rest mass m_0, prove that its de Broglie wavelength is

$$\lambda = \frac{hc}{\sqrt{E(E + 2m_0c^2)}}$$

Consider the limiting cases (a) E very small compared with m_0c^2, (b) E very large compared with m_0c^2. Show that the formula gives the right answer in these two cases.

30. An electron is moving in a circular orbit in a magnetic field. Consider the possibility that the radius of the orbit might be smaller than the de Broglie wavelength of the electron. Assume that it is possible to use any strength of magnetic field between 0 and 10^6 gauss, and to observe any radius of orbit between 10^{-2} cm and 10^4 cm. Find the range of electron velocities and electron energies (in electron volts) over which the above possibility might occur.

28

Probability
and
Uncertainty

28-1 Reconciliation of the Wave and Particle Viewpoints

How can an electron be both a particle and a wave? Is it "really" a particle or "really" a wave? What is it that oscillates in the wave? The attempt to answer these questions led to a revolutionary new conception of the nature of the physical universe and raised some profound philosophical issues.

Consider the diffraction of light by a single slit, as already discussed in section 21-4. After passing through the slit, the electromagnetic wave spreads out and forms a pattern of diffraction fringes on a screen. The intensity of illumination (brightness) at a point on the screen is related to the rate of arrival of energy per unit area in the vicinity of the point. According to equation 20-23, this is:

Instantaneous rate of arrival of energy per unit area of the screen

$$= \frac{cE^2}{4\pi} \tag{28-1}$$

E is the magnitude of the oscillating electric vector of the electromagnetic wave in the immediate vicinity of the point on the screen. This electric vector performs simple harmonic motion and, as explained in section 19-1 and equation 19-5, its variation with time is represented by

$$E = E_0 \cos 2\pi\nu t \tag{28-2}$$

Averaging over a time long compared with the period of the wave, it can be shown that:

Average energy arriving per sec per unit area of the screen

$$= \frac{cE_0{}^2}{8\pi} \tag{28-3}$$

Now look at this from the photon point of view. If photons behaved like machine gun bullets, they would pass through the slit along straight lines and strike a region of the screen having the same shape as the slit (figure 28-1a). In fact, the behavior of the photons is much more subtle. They emerge from the slit as a set of diverging beams that strike the screen to form the bright fringes of the diffraction pattern (figure 28-1b). A large number of photons strike the screen in the center of a bright fringe, whereas no photons strike the screen in the center of a dark fringe.

Since each photon carries the same amount of energy, $h\nu$, the energy arriving at a small area of the screen per second must be directly proportional to the number of photons striking that area per second. Let n be the number of photons per unit volume (the density of photons) in a small region of space just in front of the point on the screen we are considering. These photons are streaming into the screen with a velocity c in a direction which is very nearly perpendicular to the screen. The number striking unit area of the screen per second is therefore nc (compare the proof of equation 15-7). It follows that:

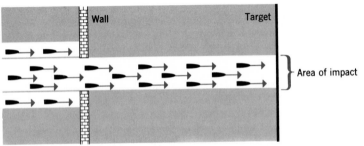

(a) Passage of a stream of bullets through a gap in a wall.

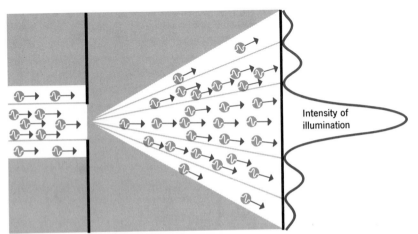

(b) Diffraction of a stream of photons passing through a slit.

Figure 28-1 Photons do not behave like machine gun bullets.

Energy arriving per sec per unit area of the screen $= nch\nu$ \hfill (28-4)

Since equations 28-3 and 28-4 represent the same quantity,

$$nch\nu = \frac{cE_0^2}{8\pi}$$ \hfill (28-5)

> **The density of photons in the immediate vicinity of a point is directly related to the square of the oscillating electric vector at that point:**
>
> $$n = \frac{E_0^2}{8\pi h\nu}$$ \hfill (28-6)

Similar reasoning can be applied to electrons. Imagine a beam of electrons of suitable de Broglie wavelength passing through a slit of suitable width. They emerge as diverging beams and form a pattern of diffraction fringes on a fluorescent screen or a photographic plate. Instead of the electric vector, E, the de Broglie wave is represented by the wave-function, ψ, which has an oscillating value at any point in space,

$$\psi = \psi_0 \cos 2\pi\nu t$$ \hfill (28-7)

The brightness of the fluorescent screen, or the degree of blackening of the photographic plate, is proportional to ψ_0^2. It is also proportional to the number of electrons striking unit area of the screen or plate per second. This, in its turn, is proportional to the number of electrons per unit volume, n_e, in a small region of space just in front of the screen or plate.

> **The density of electrons in the immediate vicinity of a point in space is directly related to the square of the wave function at that point:**
>
> $$n_e = \alpha\psi_0^2$$ \hfill (28-8)

The constant of proportionality, α, must be chosen so that the total number of electrons, added up for all regions of space, is equal to the number of electrons known to be present.

28-2   The Role of Probability

We have still not achieved a complete understanding of the relationship between particles and waves, because we can ask the following penetrating question. What would happen if we were to send a *single* electron through the slit on its own? Would it spread itself out over the whole diffraction pattern, or would it strike a definite point on the screen? If the latter, which point would it strike? It is convenient to replace the screen by a row of counters, as in figure 28-2. If the electron passes

Probability and Uncertainty

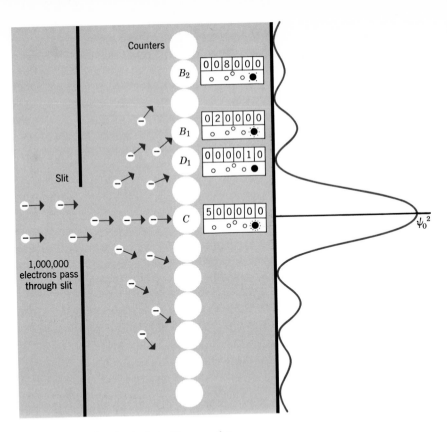

Figure 28-2 The physical significance of ψ.

through a counter, its presence there is heralded by a click in a loud speaker, and at the same time the number recorded on an instrument panel attached to this particular counter advances by one unit. Will all the loudspeakers click and will each counter advance by one unit? If not, how can we decide which one of the counters will register the arrival of the electron?

The answer, which is supported by an impressive amount of experimental evidence, is as follows. Only one counter will respond, as might be expected if the electron is a particle which can be in only one place at a time. However, it is quite impossible to predict in advance which counter this will be. The best that can be done is to assess the **probability** that the electron will strike a particular counter. This probability is proportional to the square of the wave function, $\psi_0{}^2$, in the vicinity of the counter. The important role played by probability in quantum mechanics was first realized by Max Born.

Let us first pass a beam of electrons through the slit of figure 28-2, and observe the number of particles recorded by each counter after 1,000,000 electrons have passed through. The number of electrons passing through a counter is related to the value of $\psi_0{}^2$ in the vicinity of this counter. A graph of $\psi_0{}^2$ is shown to the right of the figure. Most of the electrons arrive inside the broad central fringe and the counter C at the center might register 500,000 arrivals. (The figures to be quoted have the right order of magnitude, but no precise significance, since they depend, amongst other things, on the width of a counter.) No electrons arrive at the center of the first dark fringe where $\psi_0{}^2 = 0$. The counter D_1 opposite this point will, however, receive a few particles, say 10, since it has a finite width and extends into regions where $\psi_0{}^2$ is slightly larger than 0. The counter B_1 opposite the center of the first outer bright fringe might count 20,000 particles. This is much smaller than the number counted by C, because the value of $\psi_0{}^2$ at B_1 is only about one twentieth of its value at C. Similarly the counter B_2 opposite the center of the second outer bright fringe might count 8,000 particles.

Now pass a single electron through the slit, on its own. According to the contemporary interpretation of quantum mechanics, it is quite impossible to predict which counter this electron will strike. Suppose that, in fact, it strikes B_1. Now pass through a second electron. It might strike C. A third electron might strike B_2 and a fourth electron again strike C. Repeat this experiment 1,000,000 times, passing through 1,000,000 electrons one at a time. The numbers registered by each counter will be the same as in the original experiment in which the 1,000,000 electrons were passed through together in a beam. (There will really be small differences, but these are secondary and will be ignored for the sake of exposition.) In 500,000 of the experiments the electron strikes C. In 10 of the experiments it strikes D_1. Notice, though, that it is impossible to predict which 10 experiments these will be. The result of any single experiment is unpredictable. The net effect of a large number of experiments *is* predictable.

In 500,000 out of 1,000,000 experiments the electron strikes C. The probability that the electron strikes C is defined as 500,000/1,000,000 or 0.5. Clearly a probability of 1.0 means complete certainty, whereas a probability of 0 means that the event never occurs. The probability that the electron strikes B_1 is 20,000/1,000,000 or 0.02. The probability that it strikes B_2 is 8,000/1,000,000 or 0.008. The probability that it strikes D_1 is only 10/1,000,000 or 0.00001.

The number of particles striking a counter is proportional to $\psi_0{}^2$. The probability that a single electron strikes a particular counter is consequently proportional to the value of $\psi_0{}^2$ in the immediate vicinity of this counter. The significance of the wave function ψ_0 for a single particle may therefore be expressed as follows:

Interpretation of ψ in terms of probability

If the wave function of a single particle has the value ψ_P at a point P and a small region of space of volume v surrounds P, then the probability of finding the particle inside this region of space is $\psi_P{}^2 v$.

(In this context the symbol v represents a small volume of space and not a velocity.) The de Broglie wave is a wave of probability! It is described by giving the value of ψ at all points in space. At a particular point P suppose that ψ has the particular value ψ_P. Design an experiment to locate the position of the particle accurately and repeat this experiment N times, N being a very large number. Then, in a number $N\psi_P^2 v$ of these experiments the particle will be found inside a small volume v surrounding P. It should be obvious that the probability of finding the particle inside a small volume increases in proportion with the size of the volume.

These considerations are readily extended to photons. If E_0 is the amplitude of an oscillating electric vector at a point P, the probability of finding a photon in a volume v surrounding P is proportional to $E_0^2 v$.

28-3   The Philosophical Implications

According to the contemporary interpretation of quantum mechanics, it is quite impossible to predict in advance which counter a particular electron will strike. This is *not* believed to be a consequence of the presence of unknown factors that guide the electron along its path by means that we have not yet discovered. The idea is rather that the future behavior of an electron is not completely determined by its past history. Several possibilities are available for its future behavior and one of these is chosen *purely by chance*, for no reason that can ever be determined. This applies to all aspects of the behavior of an electron, not only to its diffraction by a slit. It also applies to all moving bodies although, as we shall see, its consequences are more important for bodies of small mass, such as fundamental particles, atoms and molecules.

This indeterminism of quantum mechanics is in sharp contrast to the determinism of classical mechanics, as discussed in section 6-2. There it was suggested that the past, present, and future behavior of all the particles in the universe is completely determined. In principle, by applying the laws of classical physics to these particles, I could predict their future behavior and discover that I am predestined to be killed by a car as I cross the road in search of my lunch. Moreover, it would be beyond my power to do anything to prevent this. Quantum mechanics suggests a very different state of affairs. Whatever the past behavior of the particles in the universe, several possibilities are available for their future behavior. The possibility that is actually realized is purely a matter of chance. The universe may not have predetermined my death in an automobile accident, but it may well be "playing roulette" to decide whether I shall be killed by the automobile or not.

The behavior of electrons is very remote from our everyday experience of human bodies and automobiles. Perhaps these considerations about probability and chance are important to the physicist in his calculations of the behavior of small particles, but they have no relevance to the macroscopic bodies of normal experience. The following experiment is designed to demonstrate that the indeterminism of the behavior of electrons may have serious macroscopic consequences.

The electron gun at the extreme left of figure 28-3 ejects single electrons at a rate slow enough to ensure that only one electron is in the apparatus at any time. Consider the very first electron ejected after the gun is switched on. This electron passes through a very narrow slit which produces a broad diffraction pattern. If it should happen to be diffracted to the left, it strikes Counter 1. The arrival of the electron at Counter 1 results in a pulse of electric current, which travels to the component labeled "Trigger." This in its turn explodes a hydrogen bomb. If the electron should happen to be diffracted to the right, it strikes Counter 2, which delivers its pulse of current to the component labeled "Deactivator." This component operates a mechanism inside the hydrogen bomb which renders it incapable of exploding at any future time.

The student sits on the hydrogen bomb contemplating the philosophical implications of quantum mechanics. The outcome of the experiment is a matter of great consequence to him. Yet, according to the contemporary point of view, there is no way of predicting whether the first electron will be diffracted to the right or to the left. The best team of scientists available, having at their disposal all that they might require in the way of technical or computational facilities, could still not predict the outcome. Moreover, rightly or wrongly, the current belief is that it will never be possible to predict the outcome of such an experiment, however much our understanding of the physical nature of the universe progresses. The element of chance is fundamental to the behavior of nature. All that can be said is that the chances are 1 in 2 that the student survives.

In spite of the importance of this element of chance, the behavior of the universe is not completely chaotic. Although we cannot completely predict the outcome of an experiment, we can certainly "lay odds" on the various possibilities. Some consequences are much more likely than others. If we had to bet on the outcome of the experiment of figure 28-2, we would be well advised to place our money on the central counter C. The probability that the electron hits this counter is 1 in 2, whereas the probability that it hits D_1 is only 10 in 1,000,000. Moreover, if we pass

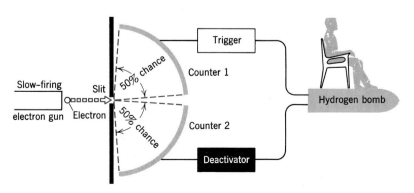

Figure 28-3 An experiment in philosophy.

through a large number of electrons, either together or one at a time, we can be quite certain that approximately one half of them will hit C. If the experiment of figure 28-3 were repeated with 1,000,000 students, we could be quite certain that approximately 500,000 of them would survive. The larger the number of experiments performed, the more nearly would the survival rate approach 50%. The laws of probability operate in such a way that, even when it is not possible to make any prediction about a single event, it is possible to predict with some accuracy the outcome of a large number of events.

The macroscopic bodies of everyday experience contain an enormous number of electrons, protons, and neutrons. Although it is not possible to predict the behavior of a single one of these particles, it is possible to predict with good accuracy the combined behavior of the large number of particles. The indeterminacy of quantum mechanics is therefore not obvious in our everyday experience.

28-4   Heisenberg's Uncertainty Principle

In 1927 Heisenberg tried to resolve the problem of exactly what is meant by saying that the electron is a particle. He decided that the essential characteristic of a particle is that, at a fixed instant of time it must have a definite location at a definite point in space and must have a well-defined velocity. He then asked how it might be possible to determine experimentally this definite position and definite velocity. He came to the startling conclusion that, in the case of a small particle like an electron, it is *not* possible to measure precisely both its position and its velocity. In any conceivable experiment there is always an uncertainty in the measured value of the position, or the velocity, or both. Moreover, this is not due to imperfections in the design or construction of the apparatus used in the experiment. With the best possible apparatus that could ever be constructed, the uncertainty would still be present. It is an unavoidable consequence of the way in which nature behaves.

As explained in section 2-3, one way of fixing the position of a particle is to determine its Cartesian coordinates, x, y and z. Consider the following attempt to determine precisely the x coordinate of an electron (figure 28-4). A beam of electrons is passed through a narrow slit of width w. The beam is traveling parallel to the z axis and the long side of the slit is parallel to the y axis. All points inside the slit have x coordinates lying between x_1 and x_2, where

$$x_2 - x_1 = w \tag{28-9}$$

If an electron passes through the slit we can be certain that, at the instant it passed through, its x coordinate had a value between x_1 and x_2. However, the apparatus does not enable us to determine exactly where the electron was inside the slit and so there is an uncertainty, Δx, in its x coordinate defined by

$$\Delta x = x_2 - x_1 \tag{28-10}$$

or

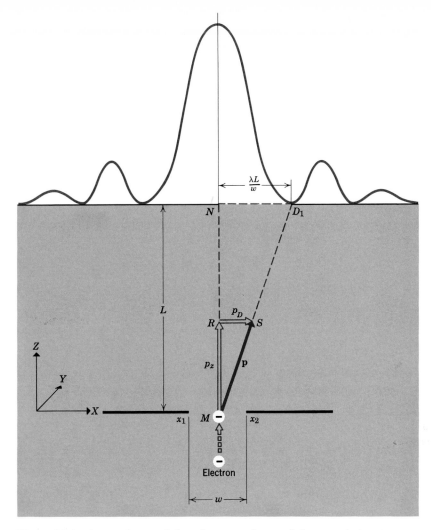

Figure 28-4 An attempt to define the x coordinate of the particle by passing it through a narrow slit inevitably introduces an uncertainty in the x component of its momentum.

$$\Delta x = w \tag{28-11}$$

This uncertainty can be made as small as desired by making the slit sufficiently narrow. There is no limit to the accuracy with which the x coordinate can be fixed in this way.

Consider, however, what happens to the velocity and momentum of the electron as it emerges from the slit. It is diffracted in an arbitrary and unpredictable direction and therefore acquires an indeterminate component of velocity and momentum in the x direction. The process of determining the x coordinate inevitably introduces an indeterminate component of velocity and momentum in the x direction.

In analyzing the situation, it is convenient to talk first about momentum. There is a small probability that an electron will be diffracted through a large angle and acquire a large extra component of momentum in the x direction. However, most electrons fall inside the broad central fringe. Let p_D be the x component of momentum acquired by an electron which falls slightly to the left of D_1. (No electron actually strikes D_1 which is the center of the first dark fringe where $\psi_0{}^2 = 0$.) Then most of the electrons acquire an extra x component of momentum in the range between $-p_D$ and $+p_D$. The uncertainty in the x component of momentum is therefore conveniently defined as

$$\Delta p_x = p_D \tag{28-12}$$

In figure 28-4 the triangle of vectors MRS shows how the extra momentum p_D is added to the original momentum p_z to produce the resultant momentum $\mathbf{p}$ directed toward D_1. This triangle is similar to triangle MND_1, so

$$\frac{RS}{MR} = \frac{ND_1}{MN} \tag{28-13}$$

or $\quad \dfrac{p_D}{p_z} = \dfrac{ND_1}{L} \tag{28-14}$

Referring to section 21-4

$$ND_1 = \frac{\lambda L}{w} \tag{28-15}$$

Whence

$$\frac{p_D}{p_z} = \frac{\lambda}{w} \tag{28-16}$$

$$p_D w = p_z \lambda \tag{28-17}$$

But λ is the de Broglie wavelength and is related to the original momentum, p_z, of the electron by equation 27-15

$$p_z = \frac{h}{\lambda} \tag{28-18}$$

Therefore,

$$p_D w = h \tag{28-19}$$

Moreover, p_D is the uncertainty in the x component of momentum (equation 28-12) and w is the uncertainty in the x coordinate (equation 28-11). Therefore,

Heisenberg's uncertainty principle

$$\Delta p_x \, \Delta x \approx h \tag{28-20}$$

We have chosen to use the symbol $\simeq$, meaning "has a value not very different from," rather than an equality sign, $=$, because we have not developed the mathematical analysis carefully enough to give Δp_x a precise definition.

We have concentrated on the x coordinate, but there are also uncertainties Δy and Δz in the y and z coordinates, with corresponding uncertainties Δp_y and Δp_z in the y and z components of the momentum.

$$\Delta p_y \, \Delta y \simeq h \tag{28-21}$$
$$\Delta p_z \, \Delta z \simeq h \tag{28-22}$$

If the velocity of the electron is small compared with c and its rest mass is m_0,

$$\Delta p_x = m_0 \, \Delta v_x \tag{28-23}$$

where Δv_x is the uncertainty in the x component of its velocity. The uncertainty principle may then be written

$$\Delta v_x \, \Delta x \simeq \frac{h}{m_0} \tag{28-24}$$

This equation is not true when the velocity is comparable with c, whereas equation 28-20 is always true.

Suppose that we attempt to define the position of the electron precisely by making the width of the slit, and consequently Δx, very small. Since

$$\Delta v_x \simeq \frac{h}{m_0 \, \Delta x} \tag{28-25}$$

the small value of Δx in the denominator makes Δv_x very large. The physical reason is that, when the slit is made narrower, the diffraction pattern becomes broader (section 21-4). An increase in the precision of determining x can be achieved only at the expense of producing a greater uncertainty in the x component of velocity (or, to be more precise, momentum). Conversely, we can reduce the uncertainty in velocity by making the slit wider, but there is then an increased uncertainty in x.

Equations 28-20 and 28-24 apply to all particles, including macroscopic bodies. Suppose that we are playing miniature golf and attempting to putt the ball through a gap in a wall so that it falls into the hole on the other side (figure 28-5). After passing through the gap in the wall, the golf ball acquires an uncertainty in its transverse velocity and its subsequent motion is determined by a diffraction pattern and the laws of probability. Why, then, are we confident that, with sufficient skill, we can be certain of putting the ball into the hole? The reason is that, in equation 28-25, the mass of the moving body appears in the denominator. The mass of a golf ball is overwhelmingly larger than the mass of an electron. The uncertainty Δv_x in the transverse velocity of the golf ball is therefore negligibly small. Another way to look at this is that the de Broglie wavelength of the golf ball is

$$\lambda = \frac{h}{m_0 v} \tag{28-26}$$

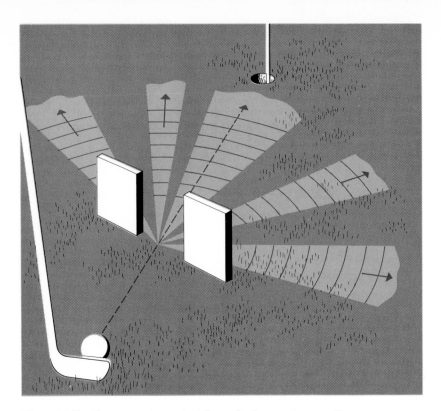

Figure 28-5   The uncertainty principle applied to miniature golf.

Since m_0 is very large, λ is very small and the diffraction fringes are very close together (section 21-4). The apparent straight line path of the golf ball is therefore a consequence of its small de Broglie wavelength, in the same way that the apparent rectilinear propagation of light is due to its small wavelength (section 21-5).

28-5   Solids, Gases and Stars

In classical physics a solid at $0°K$ was visualized as an array of atoms at rest on the points of a periodically repeating lattice (figure 28-6*a*). This is clearly inconsistent with the uncertainty principle. To require that the atoms shall be located exactly on the lattice points implies no uncertainty in their Cartesian coordinates. Inserting $\Delta x = 0$ into equation 28-20 would make Δp_x infinite. The momentum of the atom would be infinitely uncertain, which means that the atom might have any velocity between zero and c. Conversely, to require that an atom shall be at rest means that its velocity is always exactly zero and $\Delta v_x = 0$. This would make Δx infinite and the atom might be located anywhere in the whole universe!

The atoms of an actual solid make the following compromise. An atom is not permanently located at a lattice point, but roams around inside a region of diameter d surrounding the lattice point, where d is less than the distance to a neighboring atom (figure 28-6*b*). During its random

(a) Classical picture of a solid at 0°K.

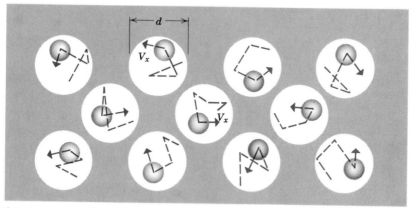

(b) Quantum mechanical picture of a solid at 0°K.

Figure 28-6 The uncertainty principle requires that the atoms of a solid shall be in rapid motion at 0°K.

motion inside this region, the x component of the velocity varies from 0 upward, frequently having a value near a certain average velocity V_x but rarely having a value greatly in excess of V_x. The uncertainty in the x coordinate, Δx, then has the order of magnitude of d and the uncertainty in velocity, Δv_x, has the order of magnitude of V_x. The uncertainty principle, in the form of equation 28-24, is not violated if

$$V_x d \approx \frac{h}{m_0} \tag{28-27}$$

where m_0 is now the mass of the atom.

At 0°K, therefore, the atoms of a solid are not at rest but are in a state of rapid chaotic motion. Direct measurement of the energy of a solid near 0°K confirms that there is in fact a large kinetic energy and, as might be expected from equation 28-27, the average velocity of the atoms is still about 1000 mph! It might be argued that, since the atoms have these high velocities at 0°K, why can we not obtain a temperature below 0°K by slowing them down still further. The answer is that we cannot do this without violating the inviolable uncertainty principle. As we cool a solid, its atoms slow down. At 0°K they have the lowest possible average velocity permitted by the uncertainty principle.

The atoms of an ideal *gas* can be at rest at $0°K$, for the following reason. According to equation 11-20, the average velocity v_a of an atom of an ideal gas is given by

$$\tfrac{1}{2}mv_a{}^2 = \tfrac{3}{2}kT \qquad (28\text{-}28)$$

Consequently, when $T = 0$, $v_a = 0$. If we cool down a gas in a container of fixed volume, before it reaches $0°K$ it either liquefies or solidifies. To prevent it from doing so, we must continually increase the volume of the container as the temperature is lowered. In order to approach very close to $0°K$ without condensing the gas, it is necessary to make the volume of the container infinitely large. Under these circumstances it *is* possible for the gas atom to be at rest with $\Delta v_x = 0$, because it can be located anywhere inside the infinitely large container and the uncertainty in its position, Δx, can then be infinite.

The Heisenberg Uncertainty Principle is also important for the stability of **white dwarf stars.** It will probably be the fate of our own sun to become a white dwarf star about five billion years from now. All stars tend to collapse under their own weight because the inner layers exert strong attractive gravitational forces on the outer layers, pulling them downward in the gravitational field, inward toward the center of the star. In stars similar to the present sun this is prevented by the thermal motion of the atoms, ions, electrons and bare nuclei of which the star is composed. Left to itself one of these particles would fall downward toward the center, but it is actually subjected to a continual bombardment by the surrounding particles which knock it in all directions. The temperature and density increase toward the center of the star and so the particle receives more knocks from below than above, because below it there are more particles moving with greater speeds. The particle is therefore knocked outward more often than inward and the material of the star settles down to a condition in which the excess outward knocks exactly compensate the inward gravitational pull.

The temperature at the surface of the star assumes exactly the right value to allow it to radiate away energy at the same rate that energy is generated by nuclear reactions near its center. Eventually, about five billion years from now in the case of the sun, the star exhausts its nuclear fuel, but the loss of energy by radiation must continue and so the star cools off. As the temperature falls the particles move less rapidly and collisions are no longer able to prevent the inward fall under gravity. Gravitational collapse sets in, the radius shrinks and the density of the material of the star steadily increases.

For a star with a mass not greater than 1.2 times the mass of the sun, the collapse is halted when the diameter becomes about the same as the diameter of the earth, and the average density is about 10^6 gm/cm³. At this high density the average distance between nuclei, d_n, is about 3×10^{-10} cm. Since the diameter of an atom is about 10^{-8} cm, some 30 times greater than the distance between nuclei in the white dwarf star, it is clear that the star cannot be composed of ordinary atoms. The electrons, which in an ordinary atom roam freely throughout a region of size 10^{-8} cm, are squashed into a confined space of size $d_n \approx 3 \times 10^{-10}$ cm bounded by nuclei on all sides. Since an electron is obliged to be somewhere inside this space, the uncertainty in its position cannot exceed d_n. The uncertainty in its momentum must be at least h/d_n and it must have an average velocity of about $h/m_e d_n$. The relevant mass m_e is here the mass of an *electron*, which is much smaller than the mass of an atom, and the average velocity is therefore very large, almost as large as the velocity of light, c.

A particle falling inward in a white dwarf star is knocked outward again by the bombardment of these fast moving electrons. The important difference is that the high velocity of the electrons arises from the Uncertainty Principle and is independent of the temperature. The electrons would still move with velocities near c even if the star cooled off to temperatures close to 0°K. The future of a white dwarf star is in fact to remain stable at about the size of the earth and slowly cool off over a period of tens of billions of years. Astronomers have discovered **red dwarf stars** which have already cooled off to a surface temperature at which they glow a dull red. The ultimate fate of our own sun, about a hundred billion years from now, may well be to become a **black dwarf star,** a dark earth-sized cinder probably composed largely of carbon with a density about a million times greater than that of coal and a temperature near the absolute zero of temperature.

A star with a mass greater than 1.2 times the mass of the sun collapses too strongly to be halted at the white dwarf stage by the pressure of the electrons. When its density becomes about 10^{11} gm/cm^3 its electrons combine with its protons to form neutrons (section 32-1). It is then composed almost entirely of neutrons packed closely together and is called a **neutron star.** What happens next probably depends upon its mass. If it is very massive (more than 3 times the mass of the sun, perhaps, but the theories are not too precise about this) it never stops collapsing and becomes a **black hole** (section 25-7). If it has an intermediate mass, it probably stops collapsing when the neutrons touch one another and repel strongly. The collapse then turns into a **supernova** explosion making the star temporarily as bright as 100 billion ordinary stars. The explosion blows off most of the material of the star, leaving behind at the core a small neutron star with a mass about the same as the mass of the sun and a stable diameter of about 10 miles. The **pulsars** discovered by radio astronomers are believed to be such remnant neutron stars (section 10-6). In the stable neutron star it is a neutron which is boxed into a small confined space by its neighbors and therefore has a large uncertainty in its velocity. The fast moving neutrons stabilize the star in the same way as the electrons in a white dwarf star.

28-6   Locating an Electron with a Microscope

As a further illustration of Heisenberg's uncertainty principle, suppose that we attempt to fix the position of an electron (or any particle) by looking at it in a microscope (figure 28-7). This involves shining light on the electron, which scatters it in all directions. Part of the scattered

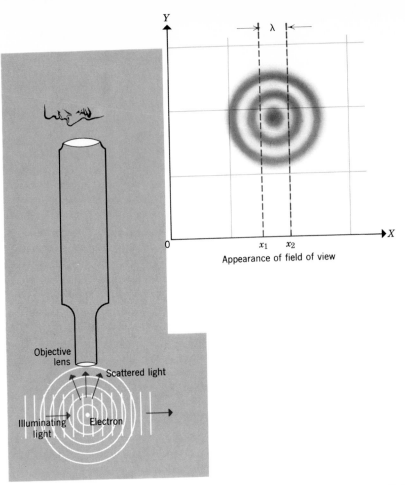

Figure 28-7 Looking at an electron in a microscope. Here the illuminating light is visualized to be an electromagnetic wave. The electron is not seen as a sharp point, but as a diffraction pattern of concentric rings.

light enters the microscope, which is an arrangement of glass lenses designed to form a magnified image of the object being viewed. A ruled grating can be incorporated in the microscope in such a way that the field of view looks like graph paper, providing convenient Cartesian axes relative to which the x and y coordinates of the electron can be determined. A second microscope, at right angles to the first one, might be used to determine the z coordinate.

However, as already mentioned in section 21-5, diffraction effects result in a blurred image. The diameter of an electron is about 10^{-12} cm and is small compared with the wavelength of visible light, which is about 5×10^{-5} cm. If the microscope is properly designed and the electron is very near the objective lens, the diameter of the blurred image can be shown to be approximately equal to the wavelength of the light, λ. It is therefore much greater than the actual diameter of the electron.

(Note that λ is now being used for the wavelength of the *light*, and not the de Broglie wavelength of the electron.) Thus the image of the electron is not seen at a point with a precise x coordinate, but spreads out over a region whose x coordinates vary from x_1 to x_2, where $(x_2 - x_1) \approx \lambda$.

It would seem that the way out of this difficulty is to make a careful study of the diffraction pattern and locate its exact center, which is the precise position of the electron. To see why this is not possible, let us treat the illuminating light as a stream of photons (figure 28-8). The scattering of light into the microscope now involves Compton effect collisions (sections 27-1 and 27-2), in each of which a photon collides with the electron and bounces off into the objective lens. A single photon does not, of course, form a complete diffraction pattern, but strikes a particular point somewhere inside this pattern. In the same way that the majority of electrons passing through a slit remain within the broad central fringe, most of the photons strike within a distance λ from the center of the diffraction pattern. If we were to observe only the point of arrival of a single photon, we could still say that the actual position of the electron was very likely to be within a distance λ from this point. The observation of the scattering of a single photon would therefore fix the x coordinate of the electron with an uncertainty

$$\Delta x \approx \lambda \tag{28-29}$$

After colliding with the photon, the electron changes its velocity and momentum. The very act of using the photon to observe the electron automatically changes the momentum of the electron, so that any subsequent observation of the path of the electron cannot possibly tell us precisely what was the velocity and momentum before the observations commenced. At first sight, it might seem possible to use the mathematical analysis of the Compton effect, as given in sections 27-1 and 27-2, to calculate the change in momentum of the electron produced by the collision. However, the apparatus provides us with no means of knowing where the photon entered the objective lens. The focusing action of the microscope is such that, wherever the photon enters the lens, its subsequent path makes it strike the field of view somewhere within the diffraction pattern. If the photon experiences a glancing collision and enters the objective at the far right (figure 28-8b), very little momentum is transferred to the electron. However, if the photon collides head on and enters the objective at the far left (figure 28-8c), it can be shown that the momentum transferred to the electron has a value not very different from h/λ (see problem 28-18). λ is still the wavelength of the photon. The uncertainty in the momentum transferred to the electron is therefore

$$\Delta p_x \approx \frac{h}{\lambda} \tag{28-30}$$

Combining this with equation 28-29

$$\Delta p_x \, \Delta x \approx h \tag{28-31}$$

which is Heisenberg's uncertainty principle again.

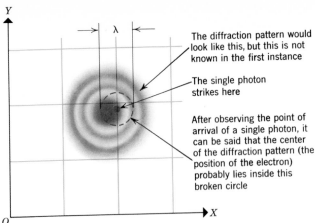

The diffraction pattern would look like this, but this is not known in the first instance

The single photon strikes here

After observing the point of arrival of a single photon, it can be said that the center of the diffraction pattern (the position of the electron) probably lies inside this broken circle

(a) Conclusion to be drawn from observation of the point of arrival of a single photon.

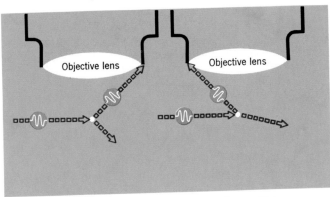

Objective lens   Objective lens

(b) A glancing collision.   (c) A head-on collision.

Figure 28-8 Looking at an electron in a microscope. The photon point of view.

Let us consider various ways in which we might try to beat the uncertainty principle. We could locate the particle more accurately by using electromagnetic radiation of small wavelength, λ, thereby reducing the spread of the diffraction pattern. This would involve designing a "microscope" using x-rays or γ-rays. The momentum, h/λ, of the photons would then be increased and each collision would give the electron a harder kick, thereby increasing the uncertainty in its momentum. It *is* possible to locate the electron very precisely, but only at the expense of producing a large uncertainty in its momentum.

Suppose that we design a microscope with a small objective lens which is far away from the electron (figure 28-9). Then the angle of rebound of the photon is much better defined and it is possible to make a better calculation of the momentum transferred to the electron, thereby reducing Δp_x. However, the theory of the microscope tells us that a microscope designed in this way produces a much broader diffraction pattern, with a corresponding increase in Δx. A detailed analysis shows that the gain in Δp_x is completely offset by the loss in Δx, and equation 28-31 still holds.

So far we have discussed a rather unconventional use of the microscope, in which we observe the point of arrival of a single photon on the field of view. Let us now return to our earlier suggestion that we should make a detailed study of the diffraction pattern and locate its exact center. This would involve scattering a large number, N, of photons from the electron, preferably all in a very short interval of time to give the electron no chance to change its position during the procedure. These photons would then distribute themselves over the diffraction pattern, clearly enabling a better decision to be made about the location of its center (figure 28-10). The degree of improvement achieved can be found by applying a branch of mathematics known as statistics. The uncertainty in the x coordinate can be shown to be reduced to

$$\Delta x \approx \frac{\lambda}{\sqrt{N}} \tag{28-32}$$

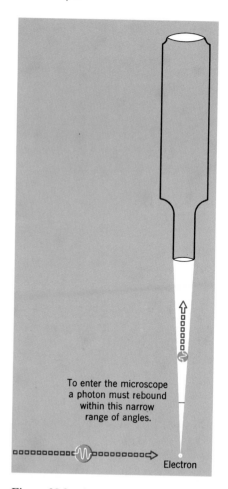

To enter the microscope a photon must rebound within this narrow range of angles.

Electron

Probability and Uncertainty

Figure 28-9 An attempt to define more precisely the momentum transferred to the electron.

However, instead of receiving one kick from one photon, the electron receives N kicks in various directions. There is a correspondingly greater uncertainty produced in its momentum, which can be shown to be

$$\Delta p_x \approx \sqrt{N}\frac{h}{\lambda} \tag{28-33}$$

Therefore, as before

$$\Delta p_x \, \Delta x \approx h$$

and nothing has been gained.

28-7   Wave-Packets

One way of making a wave look something like a particle is to consider a **wave-packet.** This is a wave confined to a region of space of length l, as shown in figure 28-11a. Since the square of the wave amplitude at a point is related to the probability of finding the particle at that point, there is a high probability of finding the particle anywhere inside the region occupied by the wave-packet, but no probability of finding it

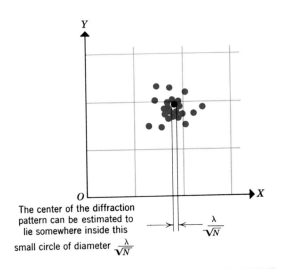

The center of the diffraction pattern can be estimated to lie somewhere inside this

small circle of diameter $\dfrac{\lambda}{\sqrt{N}}$

$\dfrac{\lambda}{\sqrt{N}}$

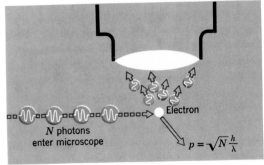

N photons
enter microscope

Electron

$p = \sqrt{N}\dfrac{h}{\lambda}$

Figure 28-10   Rapid bombardment of the electron by a large number of photons in an attempt to determine the center of the diffraction pattern.

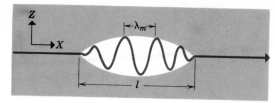

(a) This wave-packet can be synthesized by adding together a large number of infinitely long sine waves similar to the ones shown below.

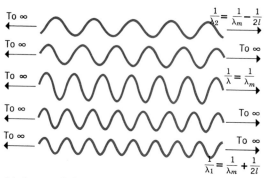

(b) Some of the Fourier components of the wave-packet.

Figure 28-11 A wave-packet and its Fourier components.

outside. The possible values of the x coordinate of the particle therefore extend over a range of values of length l, and

$$\Delta x = l \tag{28-34}$$

It might seem that we should take the wavelength to be the length λ_m shown on the figure, and assign to the momentum the precise value

$$p_m = \frac{h}{\lambda_m} \tag{28-35}$$

However, the strict mathematical formulation of quantum mechanics allows us to apply a relationship of this kind only to an *infinitely long* sine wave. Fortunately, the wave-packet of finite length can be replaced by a large number of infinitely long sine waves of various wavelengths, similar to the ones shown in figure 28-11b. These are called the *Fourier components* of the wave-packet. If their wavelengths, amplitudes, and phases are properly chosen, when they are all added together they completely cancel one another outside the wave-packet, but reproduce the wave of figure 28-11a inside the wave-packet.

Nearly all of the Fourier components have wavelengths lying between λ_1 and λ_2, where

$$\frac{1}{\lambda_1} = \frac{1}{\lambda_m} + \frac{1}{2l} \tag{28-36}$$

Probability and Uncertainty

605

$$\frac{1}{\lambda_2} = \frac{1}{\lambda_m} - \frac{1}{2l} \tag{28-37}$$

The value of λ_m is in between the values of λ_1 and λ_2. The corresponding momenta are

$$p_1 = \frac{h}{\lambda_1} \tag{28-38}$$

$$= \frac{h}{\lambda_m} + \frac{h}{2l} \tag{28-39}$$

$$= p_m + \frac{h}{2l} \tag{28-40}$$

$$\text{and} \quad p_2 = \frac{h}{\lambda_2} \tag{28-41}$$

$$= \frac{h}{\lambda_m} - \frac{h}{2l} \tag{28-42}$$

$$= p_m - \frac{h}{2l} \tag{28-43}$$

The momentum to be assigned to the particle is therefore not a well defined quantity, but extends over a range

$$\Delta p_x \approx p_1 - p_2 \tag{28-44}$$

$$\text{or} \quad \Delta p_x \approx \frac{h}{l} \tag{28-45}$$

Combining equations 28-34 and 28-45

$$\Delta p_x \, \Delta x \approx h \tag{28-46}$$

Heisenberg's uncertainty principle is an intrinsic part of the nature of such a wave-packet.

In the case of an electron (or any other particle with non-zero rest mass), each Fourier component is a de Broglie wave. According to equation 27-22, the velocity of a de Broglie wave is

$$V = \frac{c^2}{v} \tag{28-47}$$

Equations 27-15 and 27-17 imply that, for each value of the wavelength λ of a Fourier component there is a corresponding value of the associated particle velocity v. The velocity V of a Fourier component is therefore different for components of different wavelengths. This is the phenomenon of dispersion and is similar to the behavior of light passing through glass (section 21-6). As the wave-packet moves forward, the various Fourier components get "out of step." Although they start by adding up to give the wave-packet of figure 28-11a, they do not continue to do so as time passes. What happens is illustrated in figure 28-12.

As the wave-packet travels onward, it grows broader. The velocity of its left hand edge L is less than the velocity of its right hand edge R. This is related to the fact that the velocity of the particle is not a well-defined quantity, but has an uncertainty Δv_x.

The *envelope* of the wave-packet is the upper boundary of the white area in the figure and P is its peak. The velocity of P is called the **group velocity** and it can be shown to be equal to

$$v_m = \frac{p_m}{m} \tag{28-48}$$

$$= \frac{h}{m\lambda_m} \tag{28-49}$$

The velocity of P is the same as the average velocity of the particle, which is a reasonable result.

We can now resolve the difficulty that the velocity of a de Broglie wave is c^2/v, which is not v, and which is even greater than c. Concentrate on the wave inside the envelope and the peak p which is marked by a heavy dot on the figure. This peak p does not travel with the same velocity as the peak P of the envelope, but with the greater velocity

$$V_m = \frac{c^2}{v_m} \tag{28-50}$$

This is the velocity of the Fourier component with the average wavelength, λ_m. It is called the **phase velocity** of the wave-packet. The peak p therefore travels faster than the envelope and moves from the left hand edge L toward the right hand edge R. Waves are continually emerging from L and disappearing into R.

We need not worry that V_m is greater than c. The electron, which is

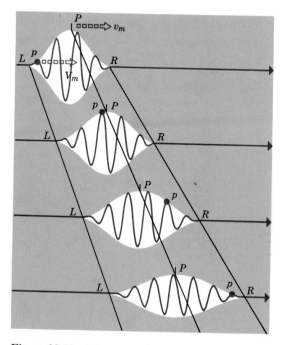

Figure 28-12 Distortion of a de Broglie wave-packet as it travels through space.

Probability and Uncertainty

the important physical entity, is obliged to remain inside the slowly moving envelope and so can never have a velocity in excess of c.

28-8   Heisenberg's Uncertainty Principle for Energy and Time

There is another important form of the uncertainty principle in which energy, ε, replaces momentum, p_x, and time, t, replaces the x coordinate.

$$\Delta\varepsilon\,\Delta t \approx h \tag{28-51}$$

Imagine a beam of light of a single frequency passing through a hole that can be closed by a shutter (figure 28-13). We are trying to determine the exact instant at which a photon passes through the hole. The shutter is initially closed, is opened at time t_1 secs and closed again at a slightly later time t_2 secs. We can then be certain that any photon emerging on the far side was in the hole at a time between t_1 and t_2 secs. The uncertainty in the time is therefore

$$\Delta t = (t_2 - t_1) \tag{28-52}$$

The electromagnetic wave which passes through the hole is no longer an infinitely long sine wave, but is a wave-packet which takes a time $(t_2 - t_1)$ to pass by any point. It can be decomposed into Fourier components with a range of wavelengths and a range of frequencies. It can be shown that most of these frequencies lie between ν_1 and ν_2, where

$$(\nu_2 - \nu_1) = \frac{1}{(t_2 - t_1)} \tag{28-53}$$

The photons emerging from the hole consequently do not have a well defined energy, but an energy somewhere in the range between $h\nu_1$ and $h\nu_2$. The spread in energy is

$$\Delta\varepsilon \approx (h\nu_2 - h\nu_1) \tag{28-54}$$

or, using equation 28-53

$$\Delta\varepsilon \approx \frac{h}{(t_2 - t_1)} \tag{28-55}$$

Combining equations 28-52 and 28-55

$$\Delta\varepsilon\,\Delta t \approx h \tag{28-56}$$

The apparatus was designed to fix as precisely as possible the instant of time at which the photon is in the hole. It does so only at the expense of producing a large uncertainty in the energy to be assigned to the photon after it emerges from the hole. If the shutter is left open for a shorter time, in order to fix the instant more precisely, the wave-packet emerging is shorter. A shorter wave-packet has Fourier components spreading over a wider range of frequencies. There is consequently a wider range of possibilities for the energy of the photon emerging.

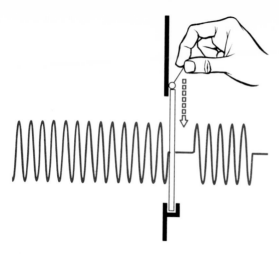

Figure 28-13 An attempt to fix the exact instant at which a photon passes through an opening.

QUESTIONS

1. The classic example of determinism in Newtonian mechanics is the accuracy with which it is possible to predict the time of a future eclipse of the sun. Imagine that you have at your disposal all technical facilities that the human race might be expected to develop in the foreseeable future. Devise a method of perturbing the moon's orbit in order to change the times of future eclipses. Arrange for the exact result of the procedure to depend upon an unpredictable factor, as in the experiment of figure 28-3.

2. Since the velocity of a photon in a vacuum is always exactly c, is its position therefore infinitely uncertain ($\Delta x = \infty$)?

3. When a particle has a total energy which is enormously large compared with its rest mass energy, we can be quite certain that its velocity differs from c by a very small amount. If we determine its position very accurately, so that Δx is very small, how is it possible for the uncertainty in its momentum, Δp_x, to be large?

4. I wish to determine the position of an electron with an uncertainty of about 10^{-4} cm, and at the same time determine its momentum as accurately as possible. Which of the following types of electromagnetic radiation would be most suitable for my purpose? (a) Radio waves, (b) microwaves, (c) visible light, (d) x-rays, (e) γ-rays.

5. Is the total energy of a solid very near $0°$K necessarily greater than the total energy of the same mass of gas at the same temperature?

6. A single electron passes through a narrow slit and is diffracted to one side. The angle of diffraction is determined as accurately as possible by placing a dense array of very small counters on the other side of the slit and observing which counter responds to the arrival of the electron. Is there any limit to the accuracy with which this

Probability
and
Uncertainty

can be done? Use this example to discuss the precise meaning of the uncertainty principle.

7. For the situation described in the previous question, is it permissible to apply the law of conservation of momentum to the passage of the electron through the first slit? (This is a very profound question. Think about it carefully. Discuss it exhaustively.)

8. Refer back to question 1 of Chapter 14. Revise your answer in the light of any additional knowledge you now have.

PROBLEMS

A

1. About what is the uncertainty in the momentum of a particle if the uncertainty in its position is 10^{-7} cm?

2. About what is the uncertainty in the velocity of a mass of 2×10^{-22} gm if the uncertainty in its position is 10^{-5} cm?

3. Check the statement in section 28-5 that the velocity of an electron in a white dwarf star is comparable with c.

4. The x component of the velocity of an electron is known to lie somewhere between 2.16×10^4 cm/sec and 2.35×10^4 cm/sec. Approximately what is the minimum uncertainty in its x coordinate?

5. The x coordinate of a proton is known to be somewhere between $+9.16534$ cm and $+9.16567$ cm. Approximately what is the minimum uncertainty in the x component of its momentum?

6. The x coordinate of a 200 eV proton is known to lie somewhere between $+0.00123$ cm and $+0.00127$ cm. Approximately what is the minimum uncertainty in the x component of its velocity? How did you use the information given about the kinetic energy of the proton?

7. A radar pulse lasts for 2.5×10^{-7} sec. Approximately what is the uncertainty in the energy of its photons?

8. It is known that an atom will probably emit a photon some time within the next 10^{-8} sec. About what is the uncertainty in the frequency of this photon?

B

9. A beam of electrons with a velocity of 10^7 cm/sec passes through a slit 10^{-3} cm wide. What is the width of the central fringe of a diffraction pattern formed on a screen 2 meters away?

10. A beam of 10 eV electrons passes through a slit 10^{-4} cm wide. What is the width of the central fringe of a diffraction pattern formed on a screen 1 meter away?

11. The electron in a hydrogen atom is confined within a region of space with a diameter of about 10^{-8} cm. Approximately what is the minimum value of the uncertainty in its velocity? What then is the order of magnitude of its minimum permissible average velocity? Need you use relativistic formulas?

12. It was once thought that the nucleus of an atom contains protons and electrons. The diameter of a nucleus is about 10^{-12} cm. Set a lower limit on the order of magnitude of the average momentum of an electron inside a nucleus. Then calculate the corresponding average velocity and average energy (in ergs and electron volts). Need you use relativistic formulas?

13. Assuming that the radius of the observable universe is 10^{10} light years and that it has the mass of 10^{11} galaxies each containing 10^{11} suns, what ultimate limit does this place on the smallest velocity that has any cosmological significance?

14. Given that the average velocity of the neutrons in a neutron star is about $\frac{1}{3}c$, make a rough estimate of the average distance between neutrons and the average density.

15. A hot gas has a prominent line in the blue region of its spectrum. The line is spread over a range of frequencies with a width equal to about 10^{-5} of the average frequency. What is the best accuracy with which one could pinpoint the instant of emission of a photon of this line by an atom of the gas?

C

16. The density of diamond is 3.5 gm cm^{-3}. Calculate the order of magnitude of the average velocity and the average energy of a carbon atom in the diamond very near 0°K. At about what temperature would the atoms of carbon vapor have the same average energy?

17. A beam of 1 BeV electrons is passed through a shutter which is kept open for 10^{-20} sec. What is the uncertainty in the mass of the electrons which emerge? What is the uncertainty in their velocity?

18. In connection with figure 28-8c, if the photon strikes the electron head on and is deflected through 180°, show that the change in momentum of the electron lies between h/λ and $2h/\lambda$, where λ is the initial wavelength of the photon.

19. An experiment is being designed to measure the mutual *gravitational* potential energy of two protons in a nucleus. The average distance between the two protons is 3×10^{-13} cm. If an accuracy of 1% is required, about how long will the experiment last?

Probability
and
Uncertainty

29

The Nature of an Atom

29-1 Bohr's Theory of the Atom

Niels Bohr's famous theory of the atom was proposed in 1913. It came several years after Planck's discussion of black body radiation and Einstein's explanation of the photoelectric effect. It preceded by more than a decade the discovery of the wave nature of the electron. The rapid developments following this latter discovery revealed certain flaws in Bohr's theory and we do not now look upon it as a completely adequate description of an atom. Nevertheless, it played a very important role in the development and understanding of quantum mechanics. It may also prove helpful to the reader as an intermediate link between the classical approach and the quantum approach.

The primary objective of the theory was to explain the line spectrum of a luminous gas of atoms. If the light emitted by the atoms is passed through a spectroscope, the spectrum is observed to consist of several narrow bright lines separated by dark regions (see figure 29-1 and compare the discussion in sections 1-1 and 21-6). The atoms do not emit all frequencies of electromagnetic radiation, but only a few distinct frequencies. This is difficult to explain in terms of classical physics but can be given a simple explanation within the spirit of the quantum theory. The energy of an atom cannot have any conceivable value, but only one of certain particular values. Suppose that the atom initially has one of these permissible energies, ε, and then suddenly changes to another state in which it has a different permissible energy, ε'. Assuming that ε' is less than ε, an amount of energy $(\varepsilon - \varepsilon')$ is released, and this can be used up in the emission of a photon with energy $h\nu$, where

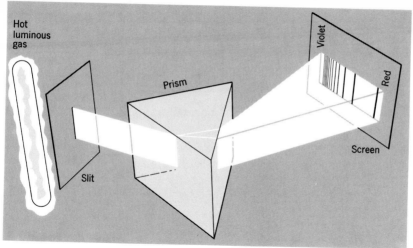

(a) Line spectrum of a luminous gas.

(b) The spectrum of monatomic hydrogen in the visible region.

Figure 29-1   An atom can emit only certain special frequencies of light.

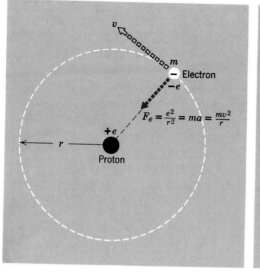

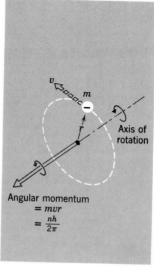

Figure 29-2   Bohr's theory of the hydrogen atom.

$$hv = (\varepsilon - \varepsilon') \tag{29-1}$$

Since there are only a limited number of permissible values for ε and ε', there are only a limited number of possible frequencies that can be emitted in this way. The problem, then, is to explain why the energy of an atom is restricted to these particular values and to calculate their magnitudes.

We shall discuss the simplest atom, hydrogen, although, with a few refinements, the theory can readily be applied to more complicated atoms. Mainly as a result of the research of Ernest Rutherford, Bohr knew that the hydrogen atom consisted of a small, positively charged nucleus (a proton) with a negatively charged electron moving around it. He imagined the electron was moving in a circle with the proton at its center (figure 29-2), in much the same way that the earth moves around the sun. The classical treatment of the electron's motion is very similar to the treatment of the orbit of a satellite (section 7-5), except that the force on the electron is electrostatic rather than gravitational. This electrostatic force of attraction, F_e, exerted by the proton on the electron, is obtained from Coulomb's law (equation 13-3) by putting the two charges q_1 and q_2 both equal to e and the distance between the charges equal to the radius of the circle, r.

$$F_e = \frac{e^2}{r^2} \tag{29-2}$$

If the velocity of the electron is v, its acceleration toward the center of the circle is

$$a = \frac{v^2}{r} \tag{29-3}$$

If m is the mass of the electron, the force may be equated to the product of mass and acceleration, to give

$$\frac{e^2}{r^2} = m\frac{v^2}{r} \tag{29-4}$$

$$\text{or} \quad mv^2r = e^2 \tag{29-5}$$

There is nothing in this classical approach to impose any restriction on the value of the radius r provided that the velocity has the corresponding value to satisfy equation 29-5. The crucial assumption of Bohr's theory is that only certain values of r are permissible, and that an allowed value of r is such that the angular momentum of the electron is an integral multiple of a fundamental unit of angular momentum, $h/2\pi$. Angular momentum was defined in Chapter 10 as

$$A = I\omega \tag{29-6}$$

The electron in the hydrogen atom is rotating about an axis through the proton perpendicular to the plane of the orbit. Its moment of inertia about this axis is

$$I = mr^2 \tag{29-7}$$

Its angular velocity is

$$\omega = \frac{v}{r} \tag{29-8}$$

Its angular momentum is therefore

$$I\omega = mvr \tag{29-9}$$

Bohr's basic assumption is

$$mvr = n\frac{h}{2\pi} \tag{29-10}$$

where n is an integer, $(1, 2, 3, 4$ etc.)
If we square equation 29-10

$$m^2v^2r^2 = n^2\frac{h^2}{4\pi^2} \tag{29-11}$$

and then divide by equation 29-5

$$\frac{m^2v^2r^2}{mv^2r} = n^2\frac{h^2}{4\pi^2e^2} \tag{29-12}$$

$$mr = n^2\frac{h^2}{4\pi^2e^2} \tag{29-13}$$

$$r_n = \frac{h^2}{4\pi^2e^2m}n^2 \tag{29-14}$$

The radius has been written with a subscript n to indicate that it corresponds to a particular value of the integer n. If we insert known values of the fundamental constants h, e, and m,

$$r_n = 0.53 \times 10^{-8}\, n^2 \text{ cm} \tag{29-15}$$

In the normal state of the hydrogen atom $n = 1$, and the radius has its smallest value, which is 0.53×10^{-8} cm. This is in very satisfactory agreement with the known size of the atom. As n takes successive values of 1, 2, 3, 4, etc., the radii are in the ratio $1:4:9:16$ etc. (figure 29-3).

The *energy* of the atom has two parts, a negative electrostatic potential energy $-e^2/r_n$, and the positive kinetic energy of the electron, $\frac{1}{2}mv^2$. The rest mass energy need not be included because v is much less than c, and a classical, non-relativistic treatment is adequate. The total energy of the atom in its nth orbit is therefore

$$\varepsilon_n = -\frac{e^2}{r_n} + \tfrac{1}{2}mv^2 \tag{29-16}$$

Notice, however, that equation 29-5 may be written in the form

$$mv^2 = \frac{e^2}{r_n} \tag{29-17}$$

whence

$$\varepsilon_n = -\frac{e^2}{2r_n} \tag{29-18}$$

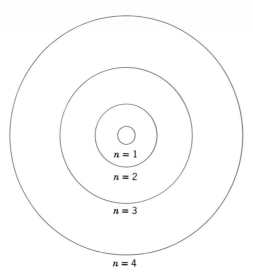

$n = 1$

$n = 2$

$n = 3$

$n = 4$

Figure 29-3 The first four Bohr orbits drawn to scale.

Substituting the value of r_n from equation 29-14

$$\varepsilon_n = - \frac{2\pi^2 e^4 m}{h^2} \frac{1}{n^2} \qquad (29\text{-}19)$$

This energy is negative, because the negative potential energy pre-dominates. If the electron and proton were both at rest at an infinite distance apart, the energy would be zero. ε_n is therefore the energy which must be supplied to the atom to remove the electron from the nth orbit and place it at rest at an infinite distance from the proton. ε_1 is commonly called the *ionization energy*.

Suppose that the atom is in an orbit described by the integer n and then falls inward to an orbit nearer the center, described by a smaller integer n'. It is clear from equation 29-19 that, since n' is smaller than n, the energy of the inner orbit, $\varepsilon_{n'}$, is *more negative* than the energy of the outer orbit, ε_n. Therefore, when the electron falls inward, a positive amount of energy $(\varepsilon_n - \varepsilon_{n'})$ is made available. If this energy appears in the form of a photon with frequency $\nu_{nn'}$

$$h\nu_{nn'} = \varepsilon_n - \varepsilon_{n'} \qquad (29\text{-}20)$$

$$= - \frac{2\pi^2 e^4 m}{h^2} \frac{1}{n^2} - \left(- \frac{2\pi^2 e^4 m}{h^2} \frac{1}{(n')^2} \right) \qquad (29\text{-}21)$$

Rearranging this expression one obtains:

The frequencies of the hydrogen spectrum according to the Bohr theory

$$\nu_{nn'} = \frac{2\pi^2 e^4 m}{h^3} \left(\frac{1}{(n')^2} - \frac{1}{n^2} \right) \qquad (29\text{-}22)$$

Substituting all possible values of the two integers n and n' into this equation, it should be possible to deduce all the frequencies in the hydrogen spectrum. The equation is found to be in almost complete agreement with the observed spectrum. A more careful investigation reveals, however, that a line predicted by the theory often turns out to be a group of several lines close together, but with all their frequencies very close to the predicted value. The Bohr theory is clearly a first approximation to the truth, and there must be some further complications.

29-2 Clouds of Probability

Bohr's circular orbits are clearly inconsistent with the uncertainty principle. It is not permissible to require that the electron shall always be exactly at a distance r from the proton and that its velocity shall always be exactly v. The correct way to describe a hydrogen atom is to state the value of the wave function, ψ_P, of the electron at each point P in the vicinity of the proton. $\psi_P^2 v$ is then the probability that the electron will be found inside a small volume v surrounding this point P.

Suppose that we devise a method of locating exactly the position of the electron in the atom at a particular instant of time. This does not violate the uncertainty principle as long as we do not demand any knowledge of the simultaneous velocity of the electron. It is not possible to predict where the electron will be found, although it is more likely to be found in those regions where ψ^2 is large. Having found the electron, let us mark its position with a dot. If we repeat the experiment a second time, the electron will be found in an entirely different position, which we again mark with a dot. After repeating the experiment a very large number of times, we obtain a cloud of dots occupying a region of space all around the proton. The density of this cloud in the vicinity of a point is proportional to the value of ψ^2 at that point.

The atom may therefore be visualized as a "cloud of negative electricity." We must remember, though, that different parts of the cloud do not co-exist simultaneously. A slightly better description of this picture of an atom is that it is a "cloud of probability"! Some typical clouds are shown in figure 29-5.

For an electron in an atom, the wave function ψ must not be thought to describe a wave traveling through space and transferring energy from place to place. It is more analogous to a **standing wave** on a plucked string. Figure 29-4 shows a few possible modes of vibration of a plucked string which is firmly held at each end. In this situation, one receives no impression that the wave is traveling to the right or to the left. Instead of the wave passing over each point of the string in succession, each point continually oscillates up and down with a fixed amplitude, which varies along the string. At points such as N, which are called *nodes*, there is never any oscillation. At points such as A, called *antinodes*, the amplitude always has its maximum value. This state of "unvarying oscillation" is the way we should visualize a hydrogen atom in a particular state with a particular energy. The amplitude of oscillation of the string is independent of time, but varies from point to point along the string.

The Nature
of an
Atom

(a) A traveling wave on a taut string.

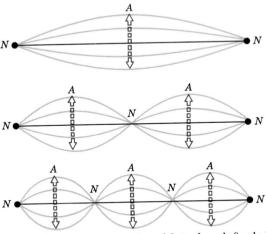

(b) Standing waves on a string of finite length firmly held at each end.

Figure 29-4 The distinction between traveling waves and standing waves.

Similarly, the oscillating wave function ψ of the electron has an amplitude that is independent of time at a particular point, but varies from point to point. The variable density of a cloud in figure 29-5 is an indication of how ψ^2 varies from point to point.

Notice that there are several possible modes of vibration of a plucked string, only a few of which are shown in figure 29-4. To distinguish between them, each mode could be designated by the number of its antinodes. Similarly, there are many different **states** of a hydrogen atom, corresponding to different clouds of various sizes and shapes, a few of which are shown in figure 29-5. To designate a particular cloud, it is necessary to give the value of each of three integers, n, l and m_l. These integers are called the **quantum numbers** of the electron in the atom.

To find the various states of the hydrogen atom, it is necessary to solve a very famous equation discovered by Erwin Schrodinger in 1925. **Schrodinger's equation** enables us to determine ψ if we know the potential energy of the electron at all points in space. It plays a similar role in the theory of material particles to the role played by Maxwell's equations in the theory of electromagnetism and photons. Unfortunately, like Maxwell's equations, it involves advanced mathematics and cannot be discussed in detail here. Schrodinger's equation is found to have physically meaningful solutions only for certain particular values of the energy and angular momentum of the atom. These particular values are determined by the quantum numbers n, l and m_l. Once three integral values have been assigned to these three quantum numbers, the equation can be solved to give ψ at all points in space and hence the size and shape of the "cloud of probability." Corresponding to the various possible values of n, l and m_l, the cloud can have many different sizes and shapes, but it cannot have every conceivable size or shape. This is analogous to the fact that the Bohr circular orbit can have many different radii, but not every conceivable radius.

29-3   The Physical Significance of the Quantum Numbers

The **principal quantum number,** n, may be any integer from 1 to infinity, but may not be 0. It is very similar to the integer n appearing in Bohr's theory. The "radius" of the cloud, r_n, may be defined by an equation which is identical with equation 29-14 for the radius of a Bohr orbit.

The radius of the cloud (the size of the atom)

$$r_n = \frac{h^2}{4\pi^2 e^2 m} n^2 \qquad (29\text{-}23)$$

n is the **principal quantum number** which may have any positive integral value, but may not be zero.

The cloud does not have a well-defined boundary, but rapidly becomes more tenuous as we move away from the atom. The significance of r_n is that most of the cloud lies at a distance of about r_n from the nucleus, and very little of the cloud lies at distances greater than $2r_n$. Referring to the procedure of locating the electron and marking it with a dot, we can say that on most occasions the distance of the electron from the nucleus is found to be not very different from r_n, and very rarely is it found to be greater than $2r_n$. In figure 29-5 the radii of the corresponding Bohr orbits are shown as the lengths r_1 and r_2. Notice that the clouds with $n = 2$ are approximately 4 times as wide as the cloud with $n = 1$.

The energy associated with a cloud is also determined mainly by the principal quantum number n, and it has a value which is very nearly equal to its value in the Bohr theory (equation 29-19). However, there are small extra contributions to the energy which depend upon the other quantum numbers. For a fixed value of n, these other quantum numbers may have several possible values, and so there are several possible values of the energy, which are very close together. This explains the fact that, where the Bohr theory predicts one spectral line, we frequently find a group of lines close together.

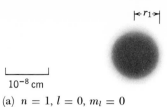

10^{-8} cm

(a) $n = 1, l = 0, m_l = 0$

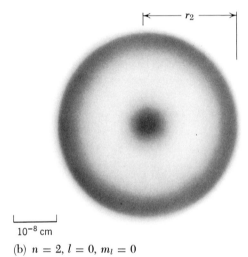

10^{-8} cm

(b) $n = 2, l = 0, m_l = 0$

The energy associated with a cloud

$$\varepsilon = -\frac{2\pi^2 e^4 m}{h^2} \frac{1}{n^2}$$

+ a small contribution depending on the other quantum
numbers   (29-24)

The basic postulate of the Bohr theory is that the angular momentum
of the atom is $n\, h/2\pi$. In the more refined theory the angular momentum is not determined by n, but by the **orbital quantum number,** l.

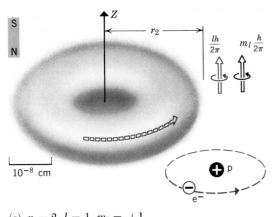

(c) $n = 2, l = 1, m_l = +1$

Figure 29-5 Probability clouds for the first few states of the hydrogen atom. In parts c, d, and e the corresponding circular Bohr orbits are also shown.

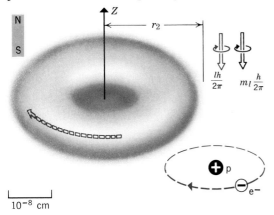

(d) $n = 2, l = 1, m_l = -1$

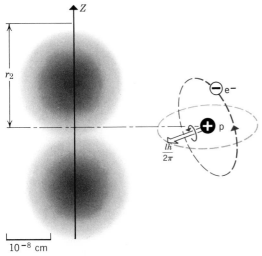

(e) $n = 2, l = 1, m_l = 0$

Figure 29-5 (*Continued*)

This is also an integer, with values, 0, 1 etc. up to $(n-1)$. For example, when n is 5, l can be 0, 1, 2, 3, or 4. The angular momentum due to the motion of the electron around the nucleus is called the **orbital angular momentum** and its magnitude is $l(h/2\pi)$. (Strictly speaking its magnitude is $\sqrt{l(l+1)}\, h/2\pi$, but this is a subtle point depending upon the exact definition of the angular momentum. We shall not deviate very far from the truth, and we shall certainly avoid some unnecessary complications, if we take the simpler value $l(h/2\pi)$.)

Angular momentum is a vector (section 10-5). The direction of this vector is the axis about which the system is rotating. If we look in the direction of the arrowhead on the vector, the system is seen to be rotating clockwise. The orientation of the axis of rotation of the atom is influenced by the direction of the magnetic field (or sometimes the electric field) in the vicinity of the atom. Such a field is always present and, even though it may be very small, it serves to mark out a preferred direction in space, which we shall take to be the direction of the Z axis.

The orbital angular momentum may not point in any direction whatsoever, but must have one of $(2l+1)$ different orientations relative to the Z axis. The component of angular momentum resolved along the Z axis is $m_l(h/2\pi)$, where m_l is the **magnetic quantum number**. It may have any one of the $(2l+1)$ integral values in the range $-l$, $(-l+1), \ldots, 0, \ldots, (l-1), +l$. For example, if $l=2$, m_l can be $-2, -1, 0, +1$ or $+2$. The case $l=2$ is illustrated in figure 29-6.

Orbital angular momentum of an electron in an atom.

$$\text{Orbital angular momentum} = l\frac{h}{2\pi} \qquad (29\text{-}25)$$

l is the **orbital quantum number** and may have any one of the integral values 0, 1, 2 up to $(n-1)$.

Component of angular momentum along the Z axis

$$= m_l \frac{h}{2\pi} \qquad (29\text{-}26)$$

m_l is the **magnetic quantum number** and may have any one of the $(2l+1)$ integral values $-l$, $(-l+1), \ldots, 0, \ldots$ $(+l-1), +l$.

Besides determining the angular momentum, the quantum numbers l and m_l also determine the shape of the cloud, its size having already been determined by n. Figure 29-5 shows the shapes of the clouds for the first few states of the hydrogen atom. The smallest cloud with $n=1$ is the normal state of the atom. Since $(n-1)$ is then 0, the only possible value of l is 0 and the only possible value of m_l is 0. This state has no

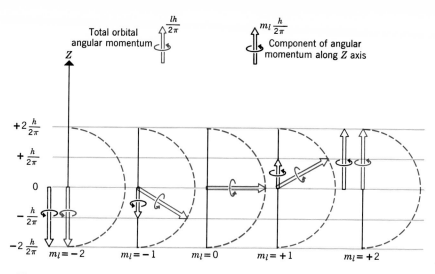

Figure 29-6 The 5 orientations of the orbital angular momentum relative to the Z axis when $l = 2$.

orbital angular momentum. The corresponding cloud is a spherical ball (figure 29-5a).

If $n = 2$, l can be 0 or 1. If $l = 0$, m_l is also zero and the atom has no orbital angular momentum. The corresponding cloud is a spherical ball surrounded by a spherical shell (figure 29-5b).

If $n = 2$ and $l = 1$, m_l can be -1, 0 or $+1$. The orbital angular momentum is then $h/2\pi$ and its component along the Z axis can be $-h/2\pi$, 0 or $+h/2\pi$. If $m_l = +1$, the cloud has the shape of a dough-nut encircling the Z axis (figure 29-5c). If we were to invent an experi-ment to measure the velocity of the electron at a fixed instant of time, we would not always obtain the same result. More often than not, though, we would find the electron moving around the doughnut in a clockwise direction looking from below. We may therefore imagine the cloud to be streaming around the doughnut in this direction, thereby giving the atom an angular momentum about the Z axis. If $m_l = -1$, the cloud has exactly the same shape but is streaming around in a counterclockwise direction looking from below, thereby producing an angular momentum vector pointing in the negative Z direction (figure 29-5d).

If $n = 2$, $l = 1$ and $m_l = 0$, the orbital angular momentum is $h/2\pi$ but its component along the Z axis is zero. The cloud consists of two balls with their centers on the Z axis at equal distances on opposite sides of the origin (figure 29-5e). If we were to measure the velocity of the electron, it would be as likely to be going round clockwise as counter-clockwise. There would be no angular momentum on the average. This is consistent with the fact that $m_l(h/2\pi)$ is zero, but what about $l(h/2\pi)$, which is not zero? We should imagine the atom to be like a top spinning about an axis perpendicular to the Z axis. The axis of rotation, in its turn, precesses rapidly about the Z-axis as shown in the inset to figure 29-5e. At a given instant, the top has a well defined angular momentum represented by a vector perpendicular to the Z-axis. However, if we average over a long interval of time, every occasion when the vector points in a particular direction is cancelled by another occasion on which it points in exactly the opposite direction. The average angular momentum is therefore zero, even though the instantaneous angular momentum is $l(h/2\pi)$.

Let us form a picture of what happens when the atom emits a photon of light. It is initially in one state with particular values of n, l and m_l, represented by a cloud of a certain size and shape. It suddenly collapses into a smaller cloud with a different shape. Since the smaller cloud has less energy, energy is released in the form of a photon. The energy $h\nu$, and hence the frequency ν, of this photon is determined mainly by the change in n, as in equation 29-22. However, there are also small differences in energy depending upon the shapes of the clouds. For fixed values of n and n' for the initial and final states, transitions between clouds of various shapes produce a group of spectral lines lying close together.

Students who have encountered **molecular orbitals** in chemistry are sometimes confused by the shapes of the clouds in parts c, d, and e of figure 29-5. For their benefit it should be emphasized that these cloud shapes are relevant only when the z axis has some particular significance, such as when there is a magnetic field pointing in the z direction. When the hydrogen atom is part of a molecule, it is sometimes more relevant to deal with three different cloud shapes known as the p_x, p_y, and p_z orbitals. In the intricate terminology of atomic spectroscopy the letter p implies that $l = 1$. The p_z orbital is identical with the cloud of figure 29-5e. The p_x and p_y orbitals have similar shapes except that the line joining the centers of the two balls lies along the x axis for the p_x orbital and along the y axis for the p_y orbital.

29-4   Spin

Our description of the atom is still not complete. The analysis of the last section, involving the quantum numbers n, l and m_l still does not explain all the lines in the spectrum of an atom! A line that would be expected to be single, according to this analysis, frequently turns out to be double. The additional concept needed was supplied by Uhlenbeck and Goudsmit in 1925. The electron has an intrinsic angular momentum.

The *orbital* angular momentum of the electron is due to its bodily motion around the nucleus. It is analogous to the angular momentum of the earth about an axis through the sun due to the annual orbital motion of the earth around the sun. It is well known that the earth also performs a daily rotation about an axis through its north and south poles, which results in an angular momentum about this axis. The electron is also rotating about an axis through itself and consequently has an *intrinsic* angular momentum, which is always present, even when the electron has zero velocity. In terms of the fundamental unit of angular momentum, $h/2\pi$, the intrinsic angular momentum of the electron is rather surprisingly found to be one half of a unit or $\frac{1}{2}h/2\pi$. This is frequently referred to as the **spin** of the electron.

The orbital angular momentum has l units of $h/2\pi$ and $(2l + 1)$ possible orientations, corresponding to the $(2l + 1)$ possible values of m_l (figure 29-6). The intrinsic angular momentum has $s(= \frac{1}{2})$ units of $h/2\pi$ and $(2s + 1)$, that is 2, possible orientations, corresponding to 2 possible values of the **spin magnetic quantum number**, m_s. If $m_s = +\frac{1}{2}$, the intrinsic angular momentum vector of the spinning electron points in the positive Z direction. If we look in the positive Z direction, the electron is then seen to be spinning clockwise. If $m_s = -\frac{1}{2}$, the intrinsic angular momentum vector points in the negative Z direction. If we look in the *positive* Z direction still, the electron is then seen to be spinning counterclockwise (figure 29-7). The two orientations of the spin result in slightly different energies, explaining the fact that some spectral lines are found to consist of two lines very close together.

Spin of an electron

Intrinsic angular momentum of an electron

$$= \frac{1}{2}\frac{h}{2\pi} \tag{29-27}$$

Component of this angular momentum in a preferred direction

$$= m_s\frac{h}{2\pi} \tag{29-28}$$

m_s is the **spin magnetic quantum number** and its value is either $-\frac{1}{2}$ or $+\frac{1}{2}$

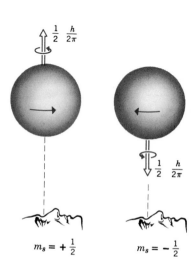

$m_s = +\frac{1}{2}$ $m_s = -\frac{1}{2}$

Figure 29-7 The spinning electron and its two possible orientations relative to a preferred direction.

The spinning electron, with its one half a unit of angular momentum and two possible orientations in space, seems a strange concept, even though it is needed to explain the fine structure of spectra. Fortunately, when P.A.M. Dirac modified Schrodinger's equation in 1928 to make it consistent with the theory of relativity, these properties of the electron were found to be a natural consequence of his modified equation, which is known as **Dirac's relativistic wave equation.**

The information to be obtained from atomic spectra has still not been exhausted. If extremely refined techniques are used to examine a spectral line that would be expected to be single according to the above analysis, including n, l, m_l and m_s, this line is often found to consist of a group of lines, which are *very*, *very* close together. This is called **hyperfine structure.** It has several causes, but the one that concerns us at the moment is that the nucleus, like the electron, may have an intrinsic angular momentum. This angular momentum may have any one of several different orientations, each of which has a *slightly* different energy, thus giving rise to hyperfine structure in the spectrum.

A nucleus is believed to consist of protons and neutrons. The intrinsic angular momentum of either a proton or a neutron is exactly the same as that of an electron.

Spin of a proton or a neutron

Intrinsic angular momentum of a proton or a

$$\text{neutron} = \frac{1}{2}\frac{h}{2\pi} \qquad (29\text{-}29)$$

Component of this angular momentum in a preferred

$$\text{direction} = -\frac{1}{2}\frac{h}{2\pi} \text{ or } +\frac{1}{2}\frac{h}{2\pi} \qquad (29\text{-}30)$$

The way in which the spins of the protons and neutrons combine to form the total angular momentum of the nucleus is complicated and not yet understood in all its ramifications. Considering the first of two simple cases, a deuterium nucleus is called a deuteron and consists of one proton and one neutron. Their spins point in the same direction and the total angular momentum of the deuteron is $h/2\pi$. A helium nucleus contains two protons and two neutrons. The spin of each proton is opposed by the spin of a neutron and the total angular momentum is zero.

A very interesting and important case is a hydrogen atom in its normal state with $n = 1$, $l = 0$ and $m_l = 0$. There is a single electron with a spin of $\frac{1}{2}$ and two possible orientations. The nucleus is a proton with a spin of $\frac{1}{2}$ and two possible orientations. (Notice the phraseology, in which it is inherently assumed that the unit of angular momentum is

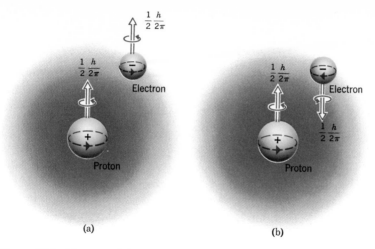

Figure 29-8 Two possibilities for the normal state of a hydrogen atom ($n = 1$, $l = 0$, $m_l = 0$). (a) has a slightly greater energy than (b). A transition from (a) to (b) produces a photon of "21 centimeter radiation."

$h/2\pi$.) There are two values of the energy, depending upon whether the electron and the proton are spinning in the same direction or opposite directions (figure 29-8). The state in which the two spins are in the same direction has the higher energy. If the atom is initially in this state and one of the spins then "flips over," so that the two are in opposite directions, a small amount of energy is released as a photon. The energy is very small indeed and the photon is in the microwave region (table 20-1), with a wavelength of 21.11 cm.

This flipping over procedure is a rather rare occurrence and the hydrogen atom needs to be left alone for rather a long time (about 10 million years!) to give the process a chance to occur. Under conditions normally encountered on earth, the atom is likely to experience some more drastic event, such as a collision with another atom, before the flip can occur. However, in regions of interstellar space, where the density of matter is very low, there are hydrogen atoms which do stay unmolested for sufficiently long periods of time. Radio astronomers have observed this "21 centimeter radiation" coming from apparently empty regions of space. It is proving to be a very valuable tool in their investigations of the universe.

It will now be realized that the total angular momentum of an atom is the vector sum of the orbital angular momenta of the electrons, plus the spin angular momenta of the electrons, plus the nuclear spin angular momentum. If we apply the law of conservation of angular momentum (section 10-5) to processes in which the atom emits a photon and changes its state and its total angular momentum, we are forced to the conclusion that the photon must have an intrinsic angular momentum of one whole unit.

The Nature
of an
Atom

627

Spin of a photon

Intrinsic angular momentum of a photon

$$= \frac{h}{2\pi} \qquad (29\text{-}31)$$

We are not normally aware of this because ordinary light contains equal numbers of photons spinning in both directions. However, circularly polarized light, in which the electric vector does not oscillate up and down but swings around in a circle, consists of photons all spinning the same way. If these photons fall on a blackened disk and are absorbed by it, their angular momentum is transferred to the disk, which is observed to rotate (figure 29-9).

29-5   Magnetic Properties of an Atom

The magnetic properties of an atom are partly due to the orbital motion of the electrons, and partly due to their intrinsic spin (compare section 17-4). The nucleus makes a very small contribution which can often be neglected. The atom can be visualized as a small bar magnet and the strength of this bar magnet is conveniently measured by its **magnetic moment,** which was defined on page 345.

The magnetism of the orbital motion can be most simply discussed in terms of Bohr's theory. An electron moving with velocity v in a circular orbit of radius r is similar to an electric current flowing in a circular loop of wire. The current i is defined as the charge passing a point P on the circle in 1 sec. The distance around the orbit is $2\pi r$ and so the electron goes round $v/2\pi r$ times each second. A charge e therefore passes the point $v/2\pi r$ times per second, and the total charge passing per sec is therefore

$$i = \frac{ev}{2\pi r} \qquad (29\text{-}32)$$

The magnetic moment was defined by equation 17-25 as

$$M = \frac{i\mathcal{A}}{c} \qquad (29\text{-}33)$$

where $\mathcal{A}$ is the area of the loop, which is πr^2. Therefore, the magnetic moment of the hydrogen atom is

$$M = \frac{ev}{2\pi r} \frac{\pi r^2}{c} \qquad (29\text{-}34)$$

or $\quad M = \dfrac{evr}{2c} \qquad (29\text{-}35)$

If we multiply numerator and denominator by m, the mass of the

electron,

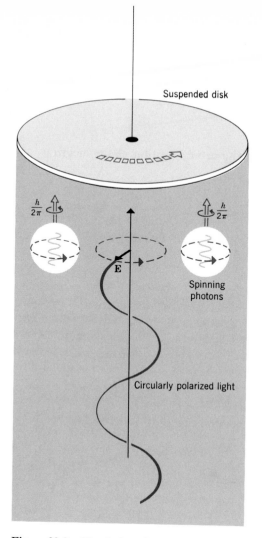

Suspended disk

$\frac{h}{2\pi}$

E

$\frac{h}{2\pi}$

Spinning
photons

Circularly polarized light

Figure 29-9 Circularly polarized light has all its photons spinning the same way. When they fall upon a suspended disk, they transfer their angular momentum to it and cause it to rotate.

$$M = \frac{emvr}{2mc} \tag{29-36}$$

and notice from equation 29-10 that

$$mvr = n\frac{h}{2\pi} \tag{29-37}$$

then $\quad M = \dfrac{e}{2mc} n \dfrac{h}{2\pi}$ $\tag{29-38}$

This is usually written

$$M = n\beta \qquad \text{(29-39)}$$

$$\text{where} \quad \beta = \frac{eh}{4\pi mc} \qquad \text{(29-40)}$$

is a fundamental unit of magnetic moment called the **Bohr magneton**.

The more sophisticated theory gives a very similar result, except that n is replaced by the orbital quantum number, l.

Magnetic moment due to the orbital motion of the electron

$$= l\frac{eh}{4\pi mc} \qquad \text{(29-41)}$$

$$= l\beta \qquad \text{(29-42)}$$

β is the fundamental unit of magnetic moment and is called the **Bohr magneton**.

Since the intrinsic angular momentum of an electron is $\frac{1}{2}h/2\pi$, it might be expected that its magnetic moment would be $\frac{1}{2}\beta$. In fact its intrinsic magnetic moment is one whole Bohr magneton. Dirac's relativistic wave equation is able to explain why, but the mathematics is too complicated to be presented here.

Magnetic moment of an electron due to its intrinsic spin

$$= \frac{eh}{4\pi mc} \qquad \text{(29-43)}$$

$$= \beta \qquad \text{(29-44)}$$

Since the proton has a charge with the same magnitude as the electron and it also has a spin of $\frac{1}{2}h/2\pi$, we might guess that its magnetic moment would be obtained by replacing the mass of the electron, m, by the mass of the proton, m_p, in equation 29-43, giving

$$\beta_n = \frac{eh}{4\pi m_p c} \qquad \text{(29-45)}$$

β_n is called the **nuclear magneton**. Since m_p is 1836 times bigger than m, the nuclear magneton is much smaller than the Bohr magneton. Nuclear magnetism is therefore much weaker than electronic magnetism.

Experiment reveals that the magnetic moment of the proton is not β_n, but $2.79\beta_n$. Moreover, the neutron, which is uncharged and therefore might not be expected to have a magnetic moment at all, is found to have a magnetic moment of $1.91\beta_n$. A qualitative explanation of these complications is given in section 33-3.

Magnetic moment of a proton

$$= 2.79 \, \beta_n \tag{29-46}$$

Magnetic moment of a neutron

$$= 1.91 \, \beta_n \tag{29-47}$$

$$\beta_n = \frac{eh}{4\pi m_p c} \tag{29-48}$$

and is called the **nuclear magneton.** m_p is the rest mass of a proton.

The total magnetic moment of an atom is a suitable vector sum of contributions from the orbital motion of the electrons, the spin of the electrons and a small contribution from the nucleus.

29-6 The Pauli Exclusion Principle and Indistinguishability

In an atom containing several electrons, the electrons exert forces on one another and perturb one another's motion. Nevertheless, it is found to be a good approximation to treat the state of a single electron as though it were alone in a hydrogen atom and to describe it by the set of quantum numbers n, l, m_l and m_s. From a consideration of atomic spectral lines and the set of energy values needed to explain them, Wolfgang Pauli discovered in 1925 that there is an important restriction on the quantum numbers of electrons in an atom.

Pauli's exclusion principle

No two electrons in the same atom may have all their quantum numbers the same.

This means, for example, that if the two electrons have the same values of n, l, and m_l, they must have different values of m_s. They must be spinning in opposite directions so that one has $m_s = -\frac{1}{2}$ and the other has $m_s = +\frac{1}{2}$.

Imagine a special type of piano with its mechanism designed in such a way that, when a key is depressed, the corresponding string is struck a blow of invariable strength and duration, quite independently of the force applied to the key or the length of time for which it is depressed. Adding electrons to an atom is somewhat similar to striking a chord of several simultaneous notes on such a piano. The number of different notes which can be struck simultaneously is limited only by the size of the keyboard, or by the number of players available and the number of fingers they possess. It is quite impossible, though, to strike any one note twice over at a single instant of time, to give it twice its normal loudness.

The Nature
of an
Atom

This analogy may also help to illustrate another important property of electrons, their **indistinguishability.** The sound emitted by the piano is independent of which fingers are used to strike the various notes. If two notes are struck, the first with the index finger of the right hand and the second with the index finger of the left hand, the sound is exactly the same as if the hands were crossed and the first note were struck by the index finger of the *left* hand. Similarly, all we can profitably say about an atom is that certain states are occupied by electrons, and it is not relevant to ask which electron is in which state. We cannot take two electrons, call them Tom and Bill, and put Tom into state 1 and Bill into state 2. Putting Tom into state 2 and Bill into state 1 would then clearly produce a different atom, but the fact is that no way has ever been found to distinguish between these two cases experimentally. An electron does not have a name and serial number, although it does have a rank in the sense that it can be distinguished from a proton.

The principle of indistinguishability of similar particles applies to all fundamental particles, for example protons, neutrons, photons, and even to atoms of the same element in the same state. A photon cannot be distinguished from any other photon with the same frequency, though it can be clearly distinguished from an electron. A helium atom in its normal state cannot be distinguished from any other helium atom in its normal state, though it can clearly be distinguished from a neon atom.

Indistinguishability of similar particles

It is sufficient to say that a state is, or is not, occupied by a particle of a particular kind. It is not permissible to try to establish the identity of the particle occupying the state.

29-7   The Periodic Table

The various chemical elements may be reproduced by adding electrons to the atom one at a time, each time also adding one proton and an appropriate number of neutrons to the nucleus. This places the elements in a sequence such that the Zth element in the sequence has Z electrons. Z is called the **atomic number.** Mendeleev showed, during the latter part of the nineteenth century, that the sequence can be tabulated in such a way that elements with similar chemical properties all lie in a vertical column and are called a *Group.* A modern form of this **periodic table** is shown in table 29-1. Notice, for example, Group 1 on the extreme left, consisting of the alkali metals hydrogen (H), lithium (Li), sodium (Na), potassium (K), rubidium (Rb), caesium (Cs) and francium (Fr). On the extreme right is Group 0, consisting of the inert gases helium (He), neon (Ne), argon (Ar), krypton (Kr), xenon (Xe) and radon (Rn). In the center are the *transition elements,* which do not readily fit into any of the main eight groups.

The periodic table is clearly a mixture of order and complexity. Both

Table 29-1 Periodic Table of the Elements

	Group I	Group II	Transition elements														Group III	Group IV	Group V	Group VI	Group VII	Group O
Period 1	H 1																					He 2
Period 2	Li 3	Be 4															B 5	C 6	N 7	O 8	F 9	Ne 10
Period 3	Na 11	Mg 12															Al 13	Si 14	P 15	S 16	Cl 17	Ar 18
Period 4	K 19	Ca 20	Sc 21	Ti 22	V 23	Cr 24	Mn 25	Fe 26	Co 27	Ni 28	Cu 29	Zn 30					Ga 31	Ge 32	As 33	Se 34	Br 35	Kr 36
Period 5	Rb 37	Sr 38	Y 39	Zr 40	Nb 41	Mo 42	Tc 43	Ru 44	Rh 45	Pd 46	Ag 47	Cd 48					In 49	Sn 50	Sb 51	Te 52	I 53	Xe 54
Period 6	Cs 55	Ba 56	° 57–71	Hf 72	Ta 73	W 74	Re 75	Os 76	Ir 77	Pt 78	Au 79	Hg 80					Tl 81	Pb 82	Bi 83	Po 84	At 85	Rn 86
Period 7	Fr 87	Ra 88	† 89–?																			

° Rare earth metals	La 57	Ce 58	Pr 59	Nd 60	Pm 61	Sm 62	Eu 63	Gd 64	Tb 65	Dy 66	Ho 67	Er 68	Tm 69	Yb 70	Lu 71		
† Actinide metals	Ac 89	Th 90	Pa 91	U 92	Np 93	Pu 94	Am 95	Cm 96	Bk 97	Cf 98	Es 99	Fm 100	Md 101	No 102	Lw 103		?

The Nature
of an
Atom

these aspects can be explained in terms of the various concepts presented earlier in this chapter. In fact the whole of chemistry can, in principle, be derived from the quantum mechanical theory of the atom. To explain even a small part of how this is possible would take us too far afield. We shall be content to indicate how the periodic table is a consequence of Pauli's exclusion principle and the quantum numbers of an electron in an atom.

When an atom is at a low temperature, and the temperature of the room is low enough for this purpose, its electrons settle down into the allowed states of lowest energy, no two of which can be identical. Table 29-2 shows these states for the first eleven elements. In order to circumvent some non-essential complications and to allow the use of the quantum numbers n, l, m_l and m_s, it has been assumed that each atom is in a very large magnetic field. A more complete analysis shows that the important qualitative conclusions are independent of the presence of this large field.

The first electron, going in to form the lowest energy state of the hydrogen atom, has $n = 1$. The only possible values of l and m_l are then 0. m_s can be either $-\frac{1}{2}$ or $+\frac{1}{2}$, but $-\frac{1}{2}$ is preferred, since it can be shown to result in a lower energy for the magnetic moment of the electron in the large magnetic field. The second electron, which forms a helium atom, is not allowed by Pauli's exclusion principle to go into the same low energy state as the first electron. Nevertheless, it can still take advantage of the low energy associated with $n = 1$, $l = 0$, $m_l = 0$ if it has $m_s = +\frac{1}{2}$. It then differs from the first electron because it is spinning the opposite way round. This exhausts the possibilities for $n = 1$ and any further electrons must have n different from 1.

Table 29-2   Quantum numbers of the electrons added to an atom to build up the first eleven elements. A very large magnetic field is assumed to be present.

Element	Atomic number Z	Quantum numbers of last electron added				
		n	l	m_l	m_s	
Hydrogen (H)	1	1	0	0	$-\frac{1}{2}$	K shell
Helium (He)	2	1	0	0	$+\frac{1}{2}$	
Lithium (Li)	3	2	0	0	$-\frac{1}{2}$	
Beryllium (Be)	4	2	0	0	$+\frac{1}{2}$	
Boron (B)	5	2	1	-1	$-\frac{1}{2}$	
Carbon (C)	6	2	1	0	$-\frac{1}{2}$	L shell
Nitrogen (N)	7	2	1	$+1$	$-\frac{1}{2}$	
Oxygen (O)	8	2	1	-1	$+\frac{1}{2}$	
Fluorine (F)	9	2	1	0	$+\frac{1}{2}$	
Neon (Ne)	10	2	1	$+1$	$+\frac{1}{2}$	
Sodium (Na)	11	3	0	0	$-\frac{1}{2}$	M shell
etc.						

The two states for which $n = 1$ are called the K shell. They are both represented by the spherical cloud of figure 29-5a. In general, a set of clouds all having the same value of n and consequently all having the same size, r_n, (equation 29-23) are said to constitute a **shell.** The K shell has $n = 1$, the L shell has $n = 2$, the M shell has $n = 3$, the N shell has $n = 4$, and so on. Since r_n is proportional to n^2, it is clear that the K shell is nearest to the nucleus, and the L, M, and N shells are at successively greater distances from the nucleus.

Hydrogen and helium make up the first period of the periodic table, and in helium the K shell is full. A full shell can be shown to be particularly stable and difficult to break up. This means that atoms with all their electrons in full shells do not readily react chemically with other atoms. They are the inert gases and helium is the first of them.

The third electron, forming lithium, is obliged to have $n = 2$ and so goes into the L shell. It then has a choice of $l = 0$ or 1, but $l = 0$ can be shown to have the lowest energy. m_l must then be 0 and $m_s = -\frac{1}{2}$, as in hydrogen. The chemical properties of an atom depend upon the number of electrons in the outermost shell. Atoms with only one electron in their outermost shell are alkali metals and lithium is the first of them. (Hydrogen is a special case, although it is possible that hydrogen might be an alkali metal at very high pressures.) Since this one electron is very loosely held, the alkali metals are very reactive chemically.

The fourth electron has $n = 2$, $l = 0$, $m_l = 0$ and $m_s = +\frac{1}{2}$, forming beryllium, the first of the alkaline earths. This then exhausts the possibilities for $n = 2$, $l = 0$, $m_l = 0$. The fifth electron must have $n = 2$ and $l = 1$. There are then three possibilities for m_l (-1, 0 and $+1$), and for each of these possibilities two possibilities for m_s ($-\frac{1}{2}$ and $+\frac{1}{2}$), making six possibilities in all. Table 29-2 shows how these possibilities are realized as we pass through the sequence boron, carbon, nitrogen, oxygen, fluorine, and neon. Altogether, 8 electrons can go into the L shell and there are eight elements in the second period of the periodic table.

The eighth electron fills the L shell and produces the inert gas neon, which has both its K and L shells full. The ninth electron is obliged to go into the M shell, with $n = 3$. It has $l = 0$, $m_l = 0$ and $m_s = -\frac{1}{2}$. It is very similar to the third electron which produced lithium, except that it goes into the M shell rather than the L shell. This ninth electron therefore produces an alkali metal, sodium.

Using similar arguments, we can show that the M shell, with $n = 3$, holds up to 18 electrons. However, after 8 electrons have gone into the M shell, the orderly sequence of events breaks down. The next electron can realize a lower energy by avoiding the M shell and going into the N shell, where $n = 4$. From this point on, there are many complications, as electrons go into outer shells before inner shells are filled up. Nevertheless, with sufficient painstaking, all the details can be explained. In particular, the transition elements are seen to be the result of going back to fill up some of the incomplete inner shells before proceeding with the outer shells.

QUESTIONS

1. Do the Bohr theory and the uncertainty principle both give the same order of magnitude for the velocity of an electron in an atom of given radius?
2. Why is the hydrogen atom normally considered to be a stationary proton with an electron moving around it, rather than a stationary electron with a proton moving around it? Is either picture strictly true?
3. What are the possible values of l for an electron in the N shell?
4. If $l = 5$, what are the possible values of m_l?
5. Write down all the possible combinations of n, l, m_l and m_s for the M shell. How many electrons are there in a full M shell?
6. Write down all the possible combinations of n, l, m_l and m_s for the N shell. How many electrons are there in a full N shell?
7. When an electron is in a magnetic field, in which direction does the torque on it tend to align the angular momentum vector?
8. Very large molecules containing many atoms can be seen in a microscope. Imagine two such molecules with identical composition and structure in the field of view of a microscope. They can clearly be distinguished as "the left hand molecule" and "the right hand molecule". Is this consistent with the idea of indistinguishability?

PROBLEMS

A

1. Show that the energy of the nth Bohr orbit is

$$\varepsilon_n = -\frac{2.18 \times 10^{-11}}{n^2}\,\text{erg}$$

$$= -\frac{13.6}{n^2}\,\text{eV}$$

2. What is the minimum frequency a photon must have to be able to lift the electron completely out of a hydrogen atom in its state of lowest energy? What type of electromagnetic radiation would it be?
3. What is the maximum frequency that appears in the line spectrum of hydrogen?
4. What frequency is emitted when the electron in a hydrogen atom falls from the N shell to the L shell?
5. What is the "radius" of the M shell in a hydrogen atom?
6. An electron in a hydrogen atom has $n = 4$, $l = 2$ and $m_l = +1$. What is the numerical value of its orbital angular momentum?
7. An electron in a hydrogen atom has $n = 5$, $l = 3$ and $m_l = -3$. What is the numerical value of the component of its orbital angular momentum along the Z axis?
8. If an electron is in the M shell, what is the maximum possible value of the component of its orbital angular momentum along the Z-axis.

9. What is the velocity of an electron in the first Bohr orbit of a hydrogen atom? What is the ratio of its mass to its rest mass?

10. Find an expression for the time to make one revolution in the nth Bohr orbit. With what power of n does it vary? An electron excited into the L shell usually waits about 10^{-8} sec before falling back into the K shell. How many revolutions does it make in this time?

11. What is the minimum frequency of a photon that can be absorbed by a hydrogen atom in its ground state?

12. Imagine the electron to be a rigid uniform solid sphere with a radius of 2.8×10^{-13} cm. Find the equatorial velocity corresponding to its intrinsic angular momentum. Comment. (The moment of inertia of a sphere is $(2/5) MR^2$.)

13. Imagine the electron to be a rigid uniform solid sphere with a radius of 2.8×10^{-13} cm. Using non-relativistic formulas, deduce its rotational kinetic energy from its known angular momentum. Compare the result with the rest mass energy of the electron.

C

14. Adapt the Bohr theory to a He^+ ion, which has two protons and two neutrons in its nucleus and one extranuclear electron. What is the ratio of the radii of the nth orbits in He^+ and H? What is the ratio of ε_n for He^+ and H?

15. The electron in a hydrogen atom can be replaced by another fundamental particle called a μ-meson, which is very similar to an electron except that its rest mass is 207 times larger. What is the radius of the K shell for μ-mesonic hydrogen? Compare its energy levels with those of ordinary hydrogen. What would be the frequency of a photon resulting from the transition $n = 2$ to $n = 1$? What type of electromagnetic radiation would it be?

16. What fractional contribution does the ionization energy make to the atomic mass of hydrogen?

17. At what temperature would the average kinetic energy of the atoms of hydrogen gas be just sufficient to ionize a hydrogen atom during a collision?

18. Show that the number of electrons in a full shell with principal quantum number n is $2n^2$.

19. A current of 10^{-15} amp flowing in a single circular loop produces a magnetic moment of one Bohr magneton. What is the radius of the loop?

20. An electron with $l = 5$ and $m_l = +3$ is in a magnetic field of 10^4 gauss. What is the torque on it?

21. A positron is identical with an electron except that it is positively charged. Positronium consists of a positron and an electron revolving around one another. Modify the Bohr theory to apply to positronium, and derive equations analogous to equations 29-14 and 29-19.

The Nature
of an
Atom

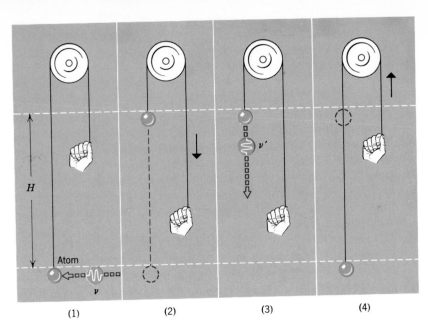

H

Atom

v

v'

(1) (2) (3) (4)

Problem 29-22

22. With the help of the diagram, consider the following sequence of operations. (1) The atom absorbs a photon of frequency v, which raises one of its electrons to a higher energy level. (2) The atom is then raised through a height H in the earth's gravitational field. (3) The electron falls back to the lower energy level, emitting a photon of frequency v'. (4) The atom is lowered through a distance H back to its original position. Assuming that the law of conservation of energy is applicable, find the difference between v and v'. Compare your answer with equation 25-9.

H

THE SEARCH FOR THE ULTIMATE CONSTITUENTS

30

The Nucleus

30-1 Some Basic Properties of Nuclei

Chemical properties of atoms depend almost entirely upon the behavior of the electrons moving in various shells around the nucleus. Interatomic forces, which are the chief determining factor for the physical properties of solids and liquids, also depend almost entirely on the behavior of the electrons. The theory outlined in the previous chapter therefore leads to an almost complete understanding of the chemical and physical properties of macroscopic matter. In this connection, the principal importance of the nucleus is that it contains 99.9% of the mass of the atom. It also provides a positive charge which neutralizes the negative charge of the electrons and acts as an almost stationary center of attraction for the electrons. Study of the nucleus has served to elucidate only some of the finer details of the properties of macroscopic matter. On the other hand, it has opened up new, and still only partially explored, avenues in our search for the basic nature of matter and its ultimate constituents.

The nucleus is believed to be composed of protons and neutrons. A proton has a positive electric charge which is equal in magnitude to the negative charge on the electron. Its rest mass is 1836.1 times the rest mass of the electron. The neutron has no charge and its rest mass is 1838.6 times the rest mass of the electron, which means that it is slightly more massive than the proton, by about 0.14%. Both the proton and the neutron have an intrinsic angular momentum of $\frac{1}{2}h/2\pi$ and they also have small magnetic moments (section 29-5). In addition to this intrinsic angular momentum, their motion within the small confines of the nucleus also results in a kind of orbital angular momentum, quantized in units of $h/2\pi$, but this is not as well understood as in the case of electrons in an atom. It is common practice to refer to the protons and neutrons in the nucleus collectively as **nucleons**. The justification for this is that the properties of the proton and the neutron are almost identical except for their electric charges, and electromagnetic forces play only a subsidiary role inside the nucleus.

Since an atom is usually electrically neutral, the number of positively charged protons in its nucleus must equal the number of negatively charged electrons moving around the nucleus. The Zth element in the periodic table has Z electrons and therefore has Z protons in its nucleus. As already mentioned, Z is called the **atomic number.** If there are also N neutrons in the nucleus, the total number of protons and neutrons is

$$A = Z + N \tag{30-1}$$

A is called the **mass number.** It is an exact integer and must not be confused with the **nuclear mass,** which is never an exact integer. Actually, when measured in atomic mass units (section 11-3 and equation 11-7), the nuclear mass always differs only very slightly from an integer, and the mass number, A, is this nearest integer. If the chemical symbol for an element is X, it is customary to denote its nucleus by $_ZX^A$. The subscript in front of the chemical symbol gives the number of protons and the superscript following it gives the total number of nucleons (protons and neutrons).

> **Atomic number,** Z = number of protons in nucleus $\qquad$ (30-2)
> **Mass number,** A = total number of protons and
> neutrons in nucleus $\qquad$ (30-3)
> Symbol denoting the nucleus of a chemical element
> X is $_ZX^A$ $\qquad$ (30-4)

Although, for a particular element, the number of electrons and hence the number of protons is fixed, the number of neutrons in the nucleus may be varied. The resulting nuclei, with various mass numbers and various atomic masses are called **isotopes** of the element. For example, ordinary uranium is $_{92}U^{238}$ and its nucleus contains 92 protons and $238 - 92 = 146$ neutrons. Its atomic mass is 238.051. The famous isotope "uranium 235" is $_{92}U^{235}$, with 92 protons and $235 - 92 = 143$ neutrons. Its atomic mass is 235.044. A few of the lighter nuclei are illustrated in figure 30-1, including the three isotopes of hydrogen (hydrogen, deuterium and tritium) and the two isotopes of helium.

Although most of the mass of the atom resides in the nucleus, its volume is only about one trillionth of the total volume of the atom. If the atom were magnified to the size of Mount Everest, the nucleus would still be only about as large as a football. It follows that the density of matter inside the nucleus is very high. It is, in fact, about 2.3×10^{14} times greater than the density of water. This density is almost constant for all nuclei. Since protons and neutrons have approximately the same mass, this must mean that the average volume available per nucleon is approximately constant for all nuclei. Another way of expressing this is that the average distance between adjacent nucleons is approximately constant. Its value is in fact very near to 1.9×10^{-13} cm for most nuclei. The radius of a proton or neutron is believed to be about 0.8×10^{-13} cm.

The Nucleus

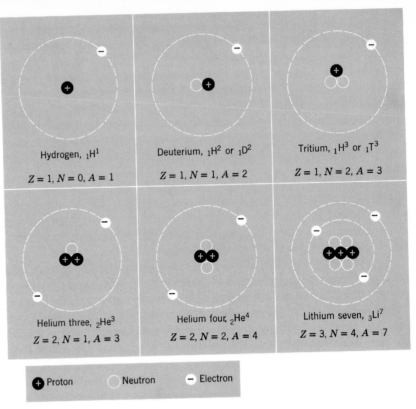

Figure 30-1 Composition of some of the lighter atoms.

30-2 Nuclear Forces

The gravitational attraction between the nucleons is far too weak to hold the nucleus together and the electrostatic repulsion between the protons would only blow it apart. The forces that hold the nucleons together are called **nuclear forces** and they are very different in character from either gravitational or electromagnetic forces. Nuclear forces are not yet completely understood and so we shall confine ourselves to a few qualitative remarks.

It is not even known exactly how the nuclear force varies with the distance between the nucleons, although it is quite certain that it does not vary inversely as the square of this distance, as in the case of gravitational forces (equation 6-20) and electrostatic forces (equation 6-26). When the two nucleons are further apart than 10^{-12} cm, the nuclear force is negligible. As they are brought closer together, to distances less than 10^{-12} cm, the force is attractive and increases very rapidly (more rapidly than $1/R^2$). At a distance of about 2×10^{-13} cm, the force becomes much stronger than the electrostatic repulsion between two protons at the same distance apart. At distances below about 0.5×10^{-13} cm, the attraction probably changes into a strong repulsion. These qualitative aspects of the force are illustrated in figure 30-2.

THE SEARCH
FOR THE
ULTIMATE
CONSTITUENTS

The constant density of nuclear matter can be understood in terms of these characteristics of the nuclear force. It might be thought that two neighboring nucleons would settle down at that separation, r_c, where the force changes from an attraction to a repulsion. However, there is another consideration, reminiscent of the phenomena discussed in section 28-5. If the average distance between nucleons is d, the uncertainty in the coordinate of a nucleon cannot be greater than d. There is then an uncertainty in the momentum of at least h/d. The nucleons are consequently in rapid motion and their average velocity increases as d decreases. If d were as small as r_c the motion would be so rapid that the nucleus would fly apart. As a compromise, the two neighboring nucleons settle down at a distance of 1.9×10^{-13} cm apart, where the attractive force is still large. The distance from a nucleon to its second nearest neighbors is then about 3.8×10^{-13} cm, but at this distance the attractive force is very much smaller. The force on a nucleon therefore depends primarily on its nearest neighbors and very little on the rest of the nucleus. The distance between nearest neighbors adjusts itself to its most favorable value (1.9×10^{-13} cm) independently of how many other nucleons are also in the nucleus.

The force between two nucleons seems to be almost independent of whether they are protons or neutrons. To an accuracy of about 1%, the force between two protons is the same as the force between two neutrons or the force between a proton and a neutron. This is called **charge independence** of the nuclear forces. It is one justification for the use of the word nucleon, which de-emphasizes the difference between protons and neutrons. One obvious difference, of course, is that in the case of two protons there is an additional electrostatic repulsion, but at a distance of 1.9×10^{-13} cm this is in fact less than 1% of the nuclear force.

The nuclear force does seem to depend upon whether the two nucleons are spinning in the same or opposite directions. It is much

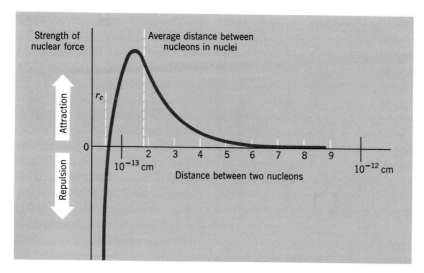

Figure 30-2 Variation of nuclear force with distance (qualitative).

The Nucleus

643

weaker when they spin in opposite directions. Clearly, the complications which were encountered when we tried to understand electromagnetic forces are trivial compared with those that will have to be overcome before we properly understand nuclear forces.

30-3 Mass Defects

Since the mass of a neutron is slightly greater than the mass of a proton, nuclear masses would not be expected to be integral multiples of the mass of a proton. It is significant, however, that the rest mass of a nucleus is always less than the sum of the rest masses of the protons and neutrons it contains. This is most easily discussed in the case of the deuteron, which is the nucleus of deuterium and contains one proton and one neutron (figure 30-1). The relevant masses are listed below.

$$\text{Rest mass of proton} \quad = 1.67239 \times 10^{-24} \text{ gm} \tag{30-5}$$
$$\text{Rest mass of neutron} \quad = 1.67470 \times 10^{-24} \text{ gm} \tag{30-6}$$

Rest mass of proton

$+$

$$\text{Rest mass of neutron} \quad = 3.34709 \times 10^{-24} \text{ gm} \tag{30-7}$$
$$\text{Rest mass of deuteron} = 3.34313 \times 10^{-24} \text{ gm} \tag{30-8}$$
$$\text{Mass defect} \qquad\qquad = 0.00396 \times 10^{-24} \text{ gm} \tag{30-9}$$

The difference between the mass of a nucleus and the sum of the masses of its constituent nucleons is called its **mass defect.** For the deuteron, this is seen to be 3.96×10^{-27} gm.

Imagine a proton and a neutron initially at rest at an infinite distance apart. Then allow them to come together to form a deuteron. At distances less than 10^{-12} cm they have a strong attraction for one another, and this results in a negative potential energy, in exactly the same way that the attractive gravitational force produces a negative potential energy (section 9-3). When the deuteron has formed, this negative potential energy is only partially offset by the kinetic energy of the two nucleons. The energy of the deuteron is consequently less than the energy of the proton and neutron at rest at infinity. The difference is called the **binding energy.**

Now let us make use of Einstein's principle of the equivalence of mass and energy (section 24-4). If m_p is the rest mass of the proton and m_n the rest mass of the neutron,

Total energy of neutron and proton
$$\text{at rest at infinity} = m_p c^2 + m_n c^2 \tag{30-10}$$

If m_d is the rest mass of the deuteron,

$$\text{Total energy of deuteron} = m_d c^2 \tag{30-11}$$

Consequently,

$$\text{Binding energy of deuteron} = (m_p c^2 + m_n c^2) - m_d c^2 \tag{30-12}$$
$$= (m_p + m_n - m_d)c^2 \tag{30-13}$$

The binding energy is the mass defect multiplied by c^2.

These ideas are confirmed by observations of the **photodisintegration of the deuteron.** When deuterium is bombarded by γ-rays, a deuteron can absorb a photon and make use of its energy to break apart into a proton and a neutron (figure 30-3). A part of the photon's energy equal to the binding energy is required to separate the neutron and proton and place them at rest a large distance apart. Any energy in excess of this appears as kinetic energy of the nucleons after they have been separated. Clearly the process cannot occur at all unless the energy of the photon is at least as great as the binding energy. There is therefore a "threshold frequency," ν_c, given by

$$h\nu_c = (m_p + m_n - m_d)c^2 \tag{30-14}$$

Everything in this equation can be measured and its validity has been established with good accuracy. Inasmuch as it relates the energy of a photon to the rest masses of particles, it provides an excellent verification of the equivalence of mass and energy.

Actually, a photon with frequency ν_c will not disintegrate the deuteron and the reason is instructive. If it were to do so, the proton and neutron would be at rest after separation and the total momentum at the end of the process would be zero. Initially the photon had a momentum $h\nu_c/c$ (equation 27-7) and the deuteron was at rest. The process would therefore violate the law of conservation of momentum. The photon frequency must exceed a value which is slightly larger than ν_c. The proton and neutron are then liberated with non-zero velocities and may have a combined momentum equal to the initial momentum of the photon. This is discussed in more detail in example 30-3-1, where it is shown that the actual threshold frequency is about 0.1% larger than the frequency given by equation 30-14.

In general, the binding energy of a nucleus is defined as the energy

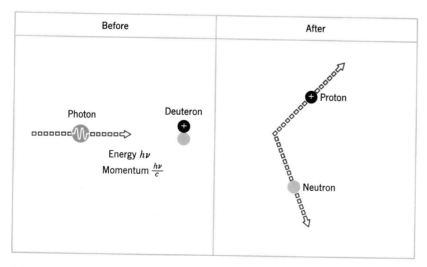

Before	After

Photon　　　　　　　Deuteron

Energy $h\nu$

Momentum $\dfrac{h\nu}{c}$

Proton

Neutron

The Nucleus

645

Figure 30-3 The photodisintegration of the deuteron.

required to break up a nucleus into its constituent protons and neutrons and place them at rest at infinite distances from one another. The mass defect is defined as the difference between the rest mass of the nucleus and the sum of the rest masses of its constituent protons and neutrons. The binding energy is always equal to the mass defect multiplied by c^2. The mass defect divided by A, which is the number of nucleons in the nucleus, is a measure of the average binding energy per nucleon. This quantity indicates how strongly the nucleus is held together and how low is its energy as compared with separate neutrons and protons. In figure 30-4 it is plotted against the mass number A for naturally occurring nuclei. The curve has two features which will be important in our subsequent discussions. The helium nucleus, $_2\text{He}^4$, stands out as being very strongly bound. The curve has a maximum at $A = 56$, and the most strongly bound nucleus is iron, $_{26}\text{Fe}^{56}$.

Example 30-3-1

Calculate the energy, in ergs and eV, of the photon which has just sufficient energy to disintegrate a deuteron, according to equation 30-14. What are its frequency and wavelength? Make a rough estimate of the magnitude of the correction introduced by the considerations involving conservation of momentum.

According to equation 30-9,

$$\text{Mass defect of deuteron} = 3.96 \times 10^{-27} \text{ gm} \tag{30-15}$$

The binding energy, which is also equal to the required photon energy, is therefore

$$h\nu_c = 3.96 \times 10^{-27} \times (3 \times 10^{10})^2 \tag{30-16}$$

$$= 3.564 \times 10^{-6} \text{ erg} \tag{30-17}$$

$$= \frac{3.564 \times 10^{-6}}{1.602 \times 10^{-12}} \text{ eV} \tag{30-18}$$

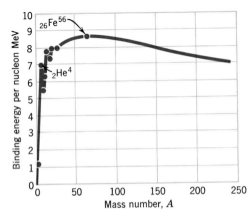

Figure 30-4 The binding energy per nucleon and its variation with mass number for naturally occurring nuclei.

$$= 2.225 \times 10^6 \text{ eV} \tag{30-19}$$

$$= 2.225 \text{ MeV.} \tag{30-20}$$

Dividing the energy *in ergs* by h, the threshold frequency is

$$\nu_c = \frac{3.564 \times 10^{-6}}{6.625 \times 10^{-27}} \tag{30-21}$$

$$= 5.38 \times 10^{20} \text{ sec}^{-1} \tag{30-22}$$

The corresponding wavelength is

$$\lambda_c = \frac{c}{\nu_c} \tag{30-23}$$

$$= \frac{3 \times 10^{10}}{5.38 \times 10^{20}} \tag{30-24}$$

$$= 5.57 \times 10^{-11} \text{ cm} \tag{30-25}$$

This is clearly a very energetic γ-ray (see table 20-1).

Suppose that a photon with a slightly larger energy, $h\nu_m$, is just able to give the neutron and proton high enough velocities to conserve momentum. In the absence of a detailed theory of the phenomenon, we do not know how the available kinetic energy is shared between the proton and the neutron. To obtain a rough order of magnitude estimate of the effect, let us assume that they both have the same velocity v in the direction of motion of the photon. The equation expressing conservation of momentum is then

$$\frac{h\nu_m}{c} = m_p v + m_n v \tag{30-26}$$

The energies involved in this problem are much too small to give the neutron or proton a velocity anywhere near the velocity of light. The masses in these formulas are therefore still the rest masses. We may also use the non-relativistic formula for kinetic energy.

Kinetic energy of proton and neutron $= \frac{1}{2}m_p v^2 + \frac{1}{2}m_n v^2$ (30-27)

$$= \frac{(m_p + m_n)^2 v^2}{2(m_p + m_n)} \tag{30-28}$$

Using equation 30-26,

Combined kinetic energy $= \dfrac{h^2 \nu_m{}^2}{2(m_p + m_n)c^2}$ (30-29)

Assuming that ν_m is not very different from ν_c, let us estimate the order of magnitude of this quantity by using equation 30-17 for the magnitude of $h\nu_m$ and equation 30-7 for $(m_p + m_n)$. Then,

Combined kinetic energy

$$= \frac{(3.564 \times 10^{-6})^2}{2 \times (3.347 \times 10^{-24}) \times (3 \times 10^{10})^2} \tag{30-30}$$

$$= 2.12 \times 10^{-9} \text{ erg} \tag{30-31}$$

This is less than one thousandth of the threshold photon energy given by equation 30-17. Thus, increasing the threshold frequency by about one part in a thousand is sufficient to deal with the difficulty concerning conservation of momentum. This difference is only slightly larger than the experimental error.

Example 30-3-2

The photoneutron effect is the absorption of a photon by a heavy nucleus with subsequent emission of a neutron. The threshold energy of the photon is found to be about 10 MeV. Assume that the neutron is removed from the nucleus by applying to it a constant force which has to move it through a distance of 10^{-13} cm outward. Calculate the order of magnitude of this "average nuclear force" and compare it with the electrostatic force between two protons separated by 1.9×10^{-13} cm.

The minimum energy needed to remove the neutron from the nucleus is 10 MeV. This is the work which must be done by the constant force to "pluck out" the neutron.

$$\text{Work done by force} = 10 \text{ MeV} \tag{30-32}$$
$$= 10^7 \times 1.602 \times 10^{-12} \text{ erg} \tag{30-33}$$
$$= 1.602 \times 10^{-5} \text{ erg} \tag{30-34}$$
But, $$\text{Work} = \text{Force} \times \text{distance} \tag{30-35}$$
So, $$\text{Force} \times (10^{-13} \text{ cm}) = 1.602 \times 10^{-5} \text{ erg} \tag{30-36}$$
$$\text{Force} = 1.602 \times 10^8 \text{ dyne} \tag{30-37}$$

Notice, first, that this force, which is present inside a single nucleus, is equivalent to the weight of about 160 kilograms. It could support the weight of two men.

The electrostatic force between the two protons is

$$\text{Electrostatic force} = \frac{e^2}{R^2} \tag{30-38}$$

$$= \left(\frac{4.8 \times 10^{-10}}{1.9 \times 10^{-13}}\right)^2 \tag{30-39}$$

$$= 6.4 \times 10^6 \text{ dyne} \tag{30-40}$$

The "average nuclear force" is about 25 times larger than the electrostatic force. Since, in reality, the nuclear force falls off very rapidly as the neutron is pulled outward, the force in the initial position is considerably larger than the average.

Example 30-3-3

What is the difference in mass between a hydrogen atom in its normal state with the principal quantum number $n = 1$ and a proton and electron at rest an infinite distance apart?

This is very similar to the nuclear case. The binding energy of the hydrogen atom is the energy needed to remove the proton and electron from the atom and place them at rest an infinite distance apart. This is usually called the *ionization energy* of the atom and its value is given by equation 29-24, with $n = 1$.

$$\text{Binding energy of hydrogen atom} = -\frac{2\pi^2 me^4}{h^2} \tag{30-41}$$

The "mass defect" is obtained by dividing this by c^2

$$\text{Mass defect} = -\frac{2\pi^2 me^4}{c^2 h^2} \tag{30-42}$$

$$= \frac{2\pi \times (9.11 \times 10^{-28}) \times (4.8 \times 10^{-10})^4}{(3 \times 10^{10})^2 \times (6.625 \times 10^{-27})^2} \tag{30-43}$$

$$= 2.42 \times 10^{-32} \text{ gm} \tag{30-44}$$

The mass of a hydrogen atom is 1.67×10^{-24} gm. In 1 gm of hydrogen the binding energy makes a difference of $(2.42 \times 10^{-32})/(1.67 \times 10^{-24})$ gm or 1.45×10^{-8} gm. This is just about the limit of the accuracy with which 1 gm of hydrogen can be weighed, and so the "mass defect" of atoms is not a matter of much practical importance.

30-4 The Stability of Nuclei

If the number of protons and the number of neutrons could be varied freely, the number of possible nuclei would be extremely large. Only a few of these possibilities have been found to exist and they are all shown in figure 30-5. Each nucleus is represented by a square symbol, ■, □ or ⊡. The center of this square has an x coordinate equal to the number of protons, Z, in the nucleus and a y coordinate equal to the number of neutrons, N. All known nuclei lie within a narrow band, which should be compared with the straight line $Z = N$. For the lighter nuclei, the number of neutrons is never very different from the number of protons. For the heavier nuclei, there is a tendency for the number of neutrons to exceed the number of protons, but never by more than 60%.

The nuclei of figure 30-5 have been divided into three categories. Those denoted by the symbol ■ are stable nuclei which are found to occur naturally. They tend to lie near the center of the band. The nuclei denoted by the symbol □ are unstable and have to be produced artificially, as will be explained in the next chapter. They are very loosely bound and have an excessive amount of energy. They break up to form stable nuclei, which have less energy, and the excess energy is converted into kinetic energy of the fragments, enabling them to fly apart. The various ways in which this disintegration can occur will be discussed in the next chapter. If these unstable nuclei ever existed in nature, they would long ago have disintegrated to form stable nuclei.

Some unstable nuclei do occur naturally. They are the **naturally radioactive nuclei** denoted by the symbol ⊡. There are two reasons for their natural occurrence. Some of them disintegrate very slowly, taking several billion years to disappear. Since there is reason to believe that the earth and its elements were formed only a few billion years ago, these long-lived unstable nuclei have not yet had time to disappear

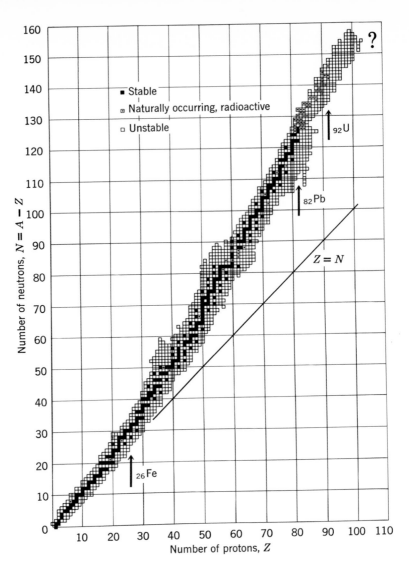

Figure 30-5 Chart of the known nuclei.

completely. Others of the naturally occurring radioactive nuclei are short-lived, but are produced during the disintegration of the long-lived radioactive nuclei. Thus, although they disappear soon after being formed, they are continually being replenished by disintegration of the long-lived nuclei. Notice that there are no stable nuclei beyond lead, which has $Z = 82$, and there are no naturally occurring nuclei at all beyond uranium, which has $Z = 92$.

The stable nuclei are not all equally strongly bound. As figure 30-4 shows, the binding energy per nucleon is greatest in the vicinity of $Z = 26$ and $N = 30$ ($_{26}Fe^{56}$). This means that, if we could take all the naturally occurring nuclei and shuffle their nucleons, we would achieve the lowest possible energy by converting everything into iron, $_{26}Fe^{56}$.

Why, then, have all the chemical elements on earth not long ago degenerated into iron? The answer is that the shuffling process is not easy to achieve in nature. It requires bringing two nuclei closer together than 10^{-12} cm, so that the nuclear forces can come into play and mix the nucleons. However, before they come this close, the electrostatic repulsion between the two positively charged nuclei becomes enormously large. Unless the two nuclei start out with very high velocities, they turn back before approaching closely enough. Sufficiently high velocities are rarely encountered on the surface of the earth. If they were due to thermal agitation (equation 11-20), the temperature would have to be as high as $10^9°$ K. We shall return to this point in section 31-3.

It is possible to understand qualitatively why the stable nuclei have the compositions shown in figure 30-5. The first thing to appreciate is that a nucleus is stable when it has the lowest possible energy. Otherwise it would be able to undergo some change to a state of lower energy, giving off its excess energy in some way such as the emission of a photon or violent ejection of one of its particles. Suppose, then, that we try to build up nuclei by adding neutrons or protons one at a time, attempting in each case to realize the state of lowest energy, as we did when we were building up atoms with electrons (section 29-7). As far as the nuclear forces and their potential energy are concerned, there is very little difference between a neutron and proton. However, a proton has the disadvantage that the electrostatic repulsion between the protons in the nucleus introduces a positive electrostatic potential energy. At first sight, it would therefore seem advantageous to construct a nucleus entirely from neutrons. In fact, there are no stable nuclei composed entirely of neutrons. Not even the di-neutron, composed of two neutrons bound together, is found to exist in a stable state. The explanation depends upon the fact that protons and neutrons obey the Pauli exclusion principle, which plays as important a part here as it did in the case of electrons in atoms.

The neutrons in a nucleus must all be in different states of motion. The first neutron to go in may be put into the state of lowest energy, but succeeding neutrons must go into states of successively higher energy. However, a proton and a neutron, being dissimilar particles, may be in the same state of motion. The first proton which goes in may go into the state of lowest energy, even though there is a neutron already there. The second proton cannot go into the state of lowest energy, because there is a proton already in it, but it can go into the next highest state, even though this is already occupied by a neutron. As far as this aspect of the situation is concerned, the lowest possible energy would be achieved by having equal numbers of protons and neutrons.

Consider the following analogy, which is illustrated in figure 30-6. A set of spherical 1 kilogram weights can be placed in the circular holders which are arranged in a vertical column at equal intervals of height above one another. A set of cubical 1 kilogram weights can be placed in square holders arranged in the same way at the same heights. The object of the game is to place a fixed number, A, of weights on to the rack with the minimum expenditure of work in lifting the weights, pro-

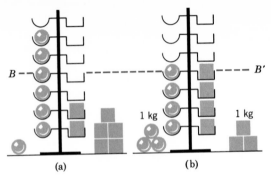

Figure 30-6 An analogy to illustrate the effect of the Pauli exclusion principle on the building up of a nucleus with the lowest possible energy. A fixed number of weights (8 in the case illustrated) must be placed on the rack with the expenditure of a minimum amount of work to lift the weights. The answer is to place an equal number of weights in the circular and square holders, as in (*b*). The advantage of (*b*) over (*a*) is that, instead of placing two spherical weights *above* the line BB', we place two cubical weights *below* this line.

ducing the arrangement with the minimum gravitational potential energy. The spherical weights represent neutrons and the cubical weights represent protons. There is a complete freedom of choice in selecting spherical or cubical weights, but not more than one weight may be placed in each holder. The answer is to choose an equal number of cubical and spherical weights, filling both sides of the rack up to the line BB', as in *b*. In any other arrangement, such as *a*, one side of the rack must be filled above BB', which requires a greater amount of work in lifting weights to a greater height than if the other side of the rack were first filled below BB'.

For the lighter nuclei the above consideration is the dominant one, and lighter nuclei do have approximately equal numbers of protons and neutrons (see figure 30-5). The other important feature of light nuclei is that the binding energy per nucleon increases (the energy per nucleon becomes *more negative*) as the nucleus becomes bigger (figure 30-4). The reason for this is that a nucleon in the interior of the nucleus is surrounded on all sides by other nucleons and has a negative mutual potential energy with each one. On the other hand, a nucleon at the surface of the nucleus has fewer nearest neighbors and a less negative total potential energy (figure 30-7a). As the nucleus becomes larger, the fraction of nucleons in the interior becomes larger and so the average potential energy per nucleon becomes more negative (figure 30-7b and c).

The positive electrostatic potential energy of the protons is not very important for lighter nuclei and has little effect on the arguments just advanced. However, it becomes progressively more important as the number of protons in the nucleus is increased. If we try to add another proton to a nucleus which already contains Z protons, the electrostatic repulsive force on it as it is brought up is Ze^2/r^2 (figure 30-8). The work done in overcoming this electrostatic repulsion therefore increases in

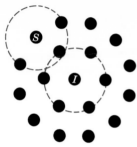

(a) The nucleon I in the interior of the nucleus has 6 nearest neighbors. The nucleon S on the surface has only 3 nearest neighbors.

(b) Fraction of surface nucleons = $\frac{6}{7} = 0.857$.

(c) Fraction of surface nucleons = $\frac{12}{19} = 0.632$.

Figure 30-7 Surface energy of a nucleus. These patterns are 2-dimensional, of course, but a similar situation pertains in 3 dimensions.

proportion to Z. The contribution of the positive electrostatic potential energy in partially counterbalancing the negative binding energy per particle therefore becomes more important for the heavier nuclei. In fact, it eventually offsets the surface effect and, in the vicinity of $Z = 26$, $A = 56$, the binding energy per nucleon starts to decrease (figure 30-4). For the heavier nuclei it becomes a real advantage to add an uncharged neutron rather than a proton, even though the exclusion principle requires the neutron to go into a state of higher energy. This is why the

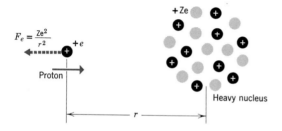

Figure 30-8 If we add an extra proton to a nucleus, the work done in bringing it up against the electrostatic repulsive force is proportional to the charge already on the nucleus.

heavier nuclei contain appreciably more neutrons than protons (figure 30-5).

The existence of no stable nuclei above $Z = 82$ and no naturally occurring nuclei above $Z = 92$ is also related to the increasing importance of the electrostatic potential energy. It will be better understood after we have discussed α-decay (section 31-4).

QUESTIONS

1. How many neutrons are there in the nucleus of $_{43}Tc^{97}$?
2. Write down the symbol for the nucleus of that isotope of Einsteinium which contains 148 neutrons. You may find the tables at the back of the book helpful.
3. Using figure 30-5 and the tables at the back of the book, write down the symbol for a stable nucleus with 50% more neutrons than protons.
4. Comment on the following statements. (a) The atomic mass of $_{41}Nb^{93}$ is 41.006. (b) The mass number of $_{76}Os^{192}$ is 191.96. (c) $_6C^{12}$ is the twelfth element in the Periodic Table. (d) The nucleus of $_{22}Ti^{48}$ contains 48 neutrons. (e) $_{235}U^{92}$ is used in nuclear reactors. (f) $_{21}Sc^{114}$ is a stable nucleus.
5. Is the mass defect proportional to the difference between the mass number and the atomic mass? Explain your answer.
6. "In the photodisintegration of a nucleus, infinity means 10^{-12} cm or greater." Explain this statement.
7. Referring to figure 30-5, in which region of the periodic table do the elements have most isotopes? In which region do the elements have fewest isotopes?
8. Do you think a nucleus is more likely to be shaped like a sphere or a long thin cigar? Why?
9. Would you expect $_2He^3$ to have a greater or smaller binding energy than $_1T^3$? Why?
10. The orbital and intrinsic angular momenta of nucleons obey rules similar to those applicable to extranuclear electrons. Is it possible for the nucleus of $_2He^3$ to have zero total angular momentum?
11. The total angular momentum of the deuteron is $\dfrac{h}{2\pi}$. Is this consistent with the possibility that a deuteron might be composed of two protons and an electron?

PROBLEMS

A

1. How many electrons are there in an uncharged atom of (a) $_{92}U^{235}$, (b) $_{92}U^{238}$?
2. How many protons are there in 12 gm of $_6C^{12}$?
3. How many nucleons are there in 12 gm of $_6C^{12}$?

4. How many fundamental particles are there in 12 gm of $_6C^{12}$ (with no net electric charge)?
5. How many neutrons are there in 1 gm of (a) $_1H^1$, (b) $_6C^{12}$, (c) $_7N^{14}$, (d) $_{92}U^{238}$?
6. How many protons are there in one mole of diatomic $_8O^{16}$?
7. How many protons are there in 15.9949 gm of diatomic $_8O^{16}$?
8. Osmium has an atomic mass of 191.96 and its nucleus contains 116 neutrons. How many electrons are there in the neutral atom?
9. How many electron volts are equivalent to one atomic mass unit?

B

10. Calculate the density of nuclear matter if the average distance between nucleons is 1.9×10^{-13} cm?
11. With the help of Heisenberg's Uncertainty Principle, make a rough estimate of the average velocity of a nucleon in a large nucleus in order to decide whether or not it is small compared with the velocity of light.
12. Repeat the last part of example 30-3-1 assuming that the proton remains at rest and all the available kinetic energy is given to the neutron.
13. If photodisintegration of a deuteron is caused by absorption of a photon with a wavelength of 5×10^{-11} cm, how much kinetic energy is given to the escaping neutron and proton?
14. If the kinetic energy given to the proton and neutron after photo-disintegration of a deuteron is 0.85 MeV, what was the frequency of the incident photon?
15. Approximately how much of the mass of 1 kg of iron is the mass equivalent of kinetic energy and potential energy, rather than rest mass of protons, neutrons, and electrons?

C

16. Assuming that air contains 24% $_7N^{14}$ and 76% $_8O^{16}$ by weight, how many protons and how many neutrons are there in 1 liter of air at $300°K$ and a pressure of 10^6 dyne cm^{-2}? (Assume that air obeys the ideal gas laws under these conditions).
17. What is the "mass defect" of a hydrogen atom when its electron is in the N shell?
18. Calculate the electrostatic potential energy of two $_{10}Ne^{20}$ nuclei when the distance between them is 10^{-12} cm. Suppose that the two nuclei belong to an ideal monatomic gas at a temperature of $T°K$ and their kinetic energies are equal to the average kinetic energy of the atoms of this gas. How large must T be in order that the two nuclei might approach as close as 10^{-12} cm? (The extranuclear electrons may be ignored.)
19. Calculate the classical escape velocity from the earth if it had its present radius, but the density of nuclear matter. Comment.

20. Imagine a universe in which all the matter is in the form of an ideal monatomic gas of hydrogen at a temperature of about $10°K$. Suppose that the hydrogen atoms are able to come together to form atoms of iron $_{26}Fe^{56}$ and that all the energy released is converted into thermal kinetic energy of these atoms. Calculate the final temperature of the ideal monatomic gas of iron vapor. At this temperature, would the iron remain in the form of neutral atoms?

31

Nuclear
Transformations

31-1   Nuclear Reactions

A chemical reaction is the reshuffling of the atoms of two reacting molecules to form different product molecules. A nuclear reaction is the reshuffling of the nucleons of two colliding nuclei to form different nuclei. In order to bring the two reacting nuclei close enough together, one of them must be accelerated to a high velocity, so that it will not be turned back too soon by the strong electrostatic repulsion between the two positively charged nuclei. The bombarding particle is usually a very light nucleus, such as a proton ($_1H^1$), a deuteron ($_1H^2$) or an α-particle (a helium nucleus, $_2He^4$). Occasionally heavier nuclei are used. A simple method of achieving high impact velocities is to accelerate the particle through an electric potential difference of several million volts. One of the first "atom smashing" machines of this kind is shown in figure 31-2, and figure 31-3 shows a more modern version.

Let us consider the first nuclear reaction to be discovered (by Rutherford in 1919). A helium nucleus ($_2He^4$) strikes a nitrogen nucleus ($_7N^{14}$) and merges with it. A proton ($_1H^1$) is subsequently ejected, leaving behind an oxygen nucleus ($_8O^{17}$). This reaction is illustrated in figure 31-1 and is represented by the equation

$$_2He^4 + _7N^{14} \rightarrow _8O^{17} + _1H^1 \tag{31-1}$$

The two protons and two neutrons of $_2He^4$ mix with the seven protons and seven neutrons of $_7N^{14}$ to form a **compound nucleus** containing 9 protons and 9 neutrons. Since this compound nucleus has $Z = 9$ and $A = 18$, it must be an isotope of fluorine, $_9F^{18}$. This fluorine nucleus is not formed in its normal state, but in a state of much higher energy. This makes it very unstable and it gets rid of its excess energy by ejecting a proton. The compound nucleus exists for only a very short time (about 10^{-19} sec!) but, by nuclear time standards, this is long enough to give it a separate existence. The reaction should therefore be considered to proceed in two separate stages

657

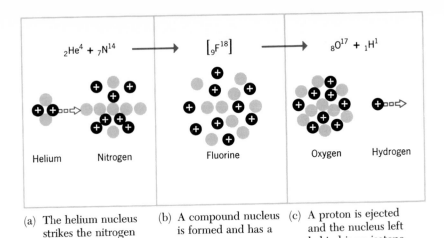

$_2He^4 + _7N^{14}$ ⟶ $[_9F^{18}]$ ⟶ $_8O^{17} + _1H^1$

Helium   Nitrogen   Fluorine   Oxygen   Hydrogen

(a) The helium nucleus strikes the nitrogen nucleus.

(b) A compound nucleus is formed and has a transitory existence.

(c) A proton is ejected and the nucleus left behind is an isotope of oxygen.

Figure 31-1 A nuclear reaction.

Figure 31-2 The first atom-smashing machine, built by Cockcroft and Walton in 1932. Protons were accelerated up to about 0.5 MeV. (Courtesy of J. Cockcroft.)

Figure 31-3 A modern atom-smashing machine. The tandem accelerator at the University of Pennsylvania. It accelerates light nuclei up to about 15 MeV.

$$_2He^4 + {}_7N^{14} \rightarrow [{}_9F^{18}] \tag{31-2}$$

$$[{}_9F^{18}] \rightarrow {}_8O^{17} + {}_1H^1 \tag{31-3}$$

The concept of the compound nucleus, which was introduced by Niels Bohr, is useful in many nuclear reactions. It breaks down when the time interval, Δt, between the arrival of the bombarding particle and the ejection of the products is very short. Under these circumstances it would be possible to fix the time of formation and the time of disintegration of the compound nucleus with high precision, the maximum uncertainty being Δt. The Heisenberg uncertainty principle, in the form of equation 28-51, then requires that there shall be a large uncertainty in the energy of the compound nucleus given by

$$\Delta E \approx \frac{h}{\Delta t} \tag{31-4}$$

When Δt is less than 10^{-21} sec, ΔE is so large that the compound nucleus becomes too indefinite a concept to be useful.

Nuclear reactions may proceed in many different ways. Except in the case of fission, which will be discussed in the next section, the compound nucleus usually ejects a proton or a neutron, or a small nucleus such as a deuteron, $_1H^2$, or an α-particle, $_2He^4$. Frequently the product nucleus is an unstable nucleus which does not occur in nature, and which is radioactive. It disintegrates over a period of time which is long compared with the typical life-time of a compound nucleus and it can therefore

Nuclear Transformations

659

be separated out and studied at leisure. Such a nucleus is called a **radio-isotope**. A few typical reactions are given below.

$$_1H^1 + _{13}Al^{27} \rightarrow [_{14}Si^{28}] \rightarrow _{12}Mg^{24} + _2He^4 \tag{31-5}$$

$$_1H^2 + _{12}Mg^{26} \rightarrow [_{13}Al^{28}] \rightarrow _{11}Na^{24} + _2He^4 \tag{31-6}$$

$$_0n^1 + _{13}Al^{27} \rightarrow [_{13}Al^{28}] \rightarrow _{12}Mg^{27} + _1H^1 \tag{31-7}$$

In the last equation the symbol $_0n^1$ represents a neutron. The use of a neutron as a bombarding particle has one particular advantage. Since the neutron is uncharged, it experiences no electrostatic repulsion as it approaches the target nucleus. A neutron with a very small velocity is therefore able to "trickle" into a nucleus and cause a reaction, whereas charged bombarding particles have to be accelerated to high velocities.

Since a nuclear reaction is a reshuffling process, the number of protons or neutrons remains unchanged. In the equation representing the reaction a subscript represents the number of protons in a nucleus and so the sum of the subscripts must be the same on both sides of the equation. In equation 31-5, for example, we start with $1 + 13$ protons, form a compound nucleus with 14 protons, and end up with $12 + 2$ protons. A superscript represents the total number of nucleons (protons and neutrons) in a nucleus and the sum of the superscripts must also be the same on both sides. In equation 31-5, we start with $1 + 27$ nucleons, form a compound nucleus with 28 nucleons, and end up with $24 + 4$ nucleons.

Nuclear reactions provide yet another means of checking the equivalence of mass and energy. The sum of the rest masses of the reacting nuclei is never equal to the sum of the rest masses of the product nuclei. The difference is compensated by a difference in the kinetic energy of the reacting nuclei and the product nuclei. The sum of the rest mass energies of the reacting nuclei plus their kinetic energies *is* always equal to the sum of the rest mass energies of the product nuclei plus their kinetic energies.

Notice that the compound nucleus is the same in equations 31-6 and 31-7, but the product nuclei are different. The reason is that the compound nucleus has different amounts of excess energy in the two cases. The way in which a nucleus prefers to disintegrate depends upon the magnitude of its excess energy.

31-2 Nuclear Fission

One of the important features of figure 30-4 is that the binding energy per nucleon of a heavy nucleus, with A in the vicinity of 200, is appreciably less than the binding energy per nucleon of a nucleus with A in the vicinity of 100. If the heavy nucleus were to split up into approximately equal fragments, the average energy per nucleon would become more negative and kinetic energy would be made available for the fragments flying apart. Another way to look at this is that the combined rest mass of the two fragments would be less than the rest mass of the original nucleus and the difference would be converted into kinetic energy. This type of disintegration, in which a very heavy nucleus splits up into two parts of approximately equal size, is called **fission**. The nature of the fission fragments is not unique. A particular kind of heavy nucleus may split up in many different ways into a wide choice of product nuclei of intermediate masses.

Although fission is energetically possible in naturally occurring heavy nuclei, it is a very rare event. For example, a nucleus of uranium 238, $_{92}U^{238}$, waits on the average for 8×10^{15} years before undergoing fission. It is more likely to emit an α-particle ($_2He^4$), which it does on the average after 4.5×10^9 years. The reason why fission does not readily occur can be understood with the help of figure 31-4. Part a shows the original nucleus and E_a is its energy, which is, of course, its rest mass multiplied by c^2. Part c shows the fission fragments when they are well separated. E_c is the sum of their rest mass energies, excluding their kinetic energies. E_c is less than E_a and the difference $(E_a - E_c)$ is the combined kinetic energy of the fragments. Part b shows the instant at which the two fragments are just about to separate and E_b is the energy associated with this situation, again excluding the kinetic energies. E_b is greater than E_c, because at b the two fragments have a large positive electrostatic potential energy. The important point, though, is that E_b is also larger than E_a. The reason is that the two fragments at b have a larger total surface area than the original nucleus at a. We explained in section 30-4 that a nucleon near the surface of a nucleus has a greater energy than a nucleon in the interior.

Thus, if situation a could be transformed into situation c, there would be energy to spare for kinetic energy, but it is not possible to go from a to the intermediate situation b without supplying energy from outside. A simple analogy is a ball at rest in a hollow at the top of a hill, as in figure 31-4d. The gravitational potential energy at A is less than at B, but greater than at C. If the ball could get from A to A', it could easily roll down the hill. However, it cannot be lifted over the hump at B unless energy is supplied from outside. The part of the hill between A

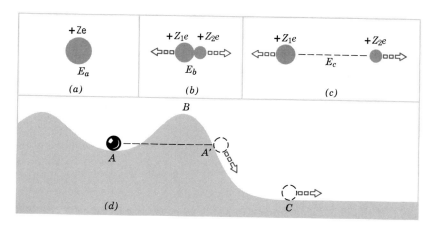

(a)   (b)   (c)   (d)

Nuclear Transformations

Figure 31-4   The potential barrier opposing spontaneous fission.

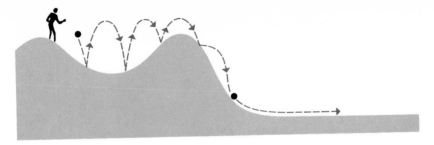

Figure 31-5 The analog of induced fission.

and A' is called **a potential barrier.** The high energy of situation b produces a potential barrier opposing fission.

The problem has now reversed itself. The ball would not escape from the hollow even after 8×10^{15} years, so why does the uranium nucleus eventually succeed in undergoing fission? We shall encounter a similar problem in the case of α-emission and we will defer its discussion until then.

The way to get the ball out of the hollow is to give it some extra energy by, for example, bouncing it up to a sufficient height (figure 31-5). A heavy nucleus may be induced to undergo fission by supplying it with extra energy. This might be simply the result of the nucleus absorbing a photon (photofission). Alternatively, a heavy nucleus might be bombarded by a light particle to form a compound nucleus with sufficient excess energy to be able to surmount the potential barrier. The most important type of bombarding particle is a neutron which, we recall, can enter a nucleus without having to be accelerated to a high velocity to overcome the electrostatic repulsion of the nuclear charge. A typical fission reaction, induced by the addition of a neutron to uranium 235, is illustrated in figure 31-6.

Since very heavy nuclei contain a higher proportion of neutrons than stable nuclei of intermediate mass (figure 30-5), the fission fragments are

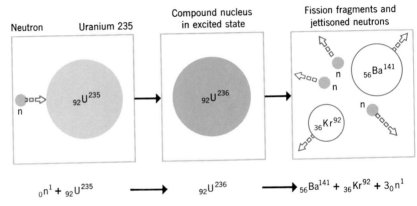

$$_0n^1 + {}_{92}U^{235} \longrightarrow {}_{92}U^{236} \longrightarrow {}_{56}Ba^{141} + {}_{36}Kr^{92} + 3{}_0n^1$$

Figure 31-6 Fission induced in uranium 235 by the addition of a neutron.

initially formed with too many neutrons to be stable. They rapidly jettison these excess neutrons, in about 10^{-14} sec, and each fission usually results in the liberation of two or three free neutrons. Some of these neutrons may enter other fissionable nuclei and produce further fissions and still more neutrons. If, on the average, more than one neutron from each fission goes on to produce further fissions, the number of fissions taking place at each succeeding stage increases very rapidly (figure 31-7a). This is called a **chain reaction.** Within a small fraction of a second, the number of nuclei having undergone fission corresponds to many grams of material and the energy released, initially in the form of kinetic energy of the fission fragments, is equivalent to the explosion of many thousands of tons of TNT. This is the principle underlying the atomic bomb.

If, on the average, exactly one neutron from each fission goes on to produce a further fission, the number of fissions occurring per second remains constant and the situation does not get out of hand (figure 31-7b). This is the principle of a **nuclear reactor,** which is a controllable source of useful nuclear energy. A careful balance is achieved in its design, so that each fission produces more than one neutron which *could* produce a further fission, but a sufficient number of these neutrons are absorbed in non-fission processes to ensure that *exactly one* does go on to produce a further fission. The following example should make it clear why a nuclear fuel or explosive is much more effective than an equal weight of a conventional fuel or explosive.

Example 31-2-1

Given the rest masses of the various particles and nuclei involved in the fission shown in figure 31-6, calculate the energy released in ergs and electron volts.

The nuclear masses will be quoted in atomic mass units or u. The atomic mass unit is $\frac{1}{12}$ of the mass of a $_6C^{12}$ atom (section 11-3). The reaction is

$$_0n^1 + _{92}U^{235} \rightarrow _{56}Ba^{141} + _{36}Kr^{92} + 3_0n^1 \tag{31-8}$$

The masses of the reacting particles are

Mass of neutron, $_0n^1$	$= 1.0087$ u	(31-9)
Mass of $_{92}U^{235}$	$= 235.0439$ u	(31-10)
Combined mass of reactants	$= 236.0526$ u	(31-11)

The masses of the products are

Mass of $_{56}Ba^{141}$	$= 140.9139$ u	(31-12)
Mass of $_{36}Kr^{92}$	$= 91.8973$ u	(31-13)
Mass of three neutrons, 3_0n^1	$= 3.0261$ u	(31-14)
Combined mass of products	$= 235.8373$ u	(31-15)

Nuclear Trans-
formations

Subtracting the mass of the products from the mass of the reactants,

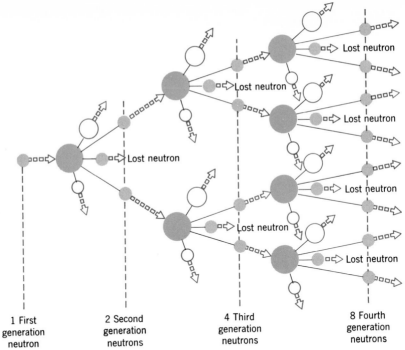

1 First generation neutron
2 Second generation neutrons
4 Third generation neutrons
8 Fourth generation neutrons

(a) An uncontrolled chain reaction. The principle of the atomic bomb.

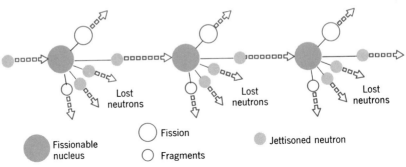

Lost neutrons

Lost neutrons

Lost neutrons

Fissionable nucleus

Fission Fragments

Jettisoned neutron

(b) A controlled chain reaction. The principle of a nuclear reactor.

Figure 31-7 Fission chain reactions.

Mass converted into energy = 0.2153 u (31-16)

But, 1 atomic mass unit $= 1.66 \times 10^{-24}$ gm (31-17)

Therefore, mass converted $= 3.57 \times 10^{-25}$ gm (31-18)

Multiplying by c^2

Energy released $= 3.57 \times 10^{-25} \times (3 \times 10^{10})^2$ (31-19)

$= 3.21 \times 10^{-4}$ erg (31-20)

$= \dfrac{3.21 \times 10^{-4}}{1.60 \times 10^{-12}}$ eV (31-21)

$= 2.01 \times 10^8$ eV (31-22)

The energy released in the fission of one uranium 235 nucleus is 201 MeV.

Conventional fuels and conventional explosives release their energy by means of chemical reactions. The energy liberated in a chemical reaction is due to a readjustment of the behavior of electrons in atoms and molecules. A good idea of its order of magnitude may be obtained by using the Bohr theory to calculate the energy needed to remove an electron from a hydrogen atom (the ionization energy of hydrogen). Inserting numbers into equation 29-24 (or multiplying equation 30-44 by c^2), we find that,

$$\text{Ionization energy of hydrogen} = 2.18 \times 10^{-11} \text{ erg} \qquad (31\text{-}23)$$

$$= \frac{2.18 \times 10^{-11}}{1.60 \times 10^{-12}} \text{ eV} \qquad (31\text{-}24)$$

$$= 13.6 \text{ eV} \qquad (31\text{-}25)$$

Comparing equations 31-22 and 31-25, we see that the fission of one nucleus of uranium 235 releases about ten million times as much energy as a chemical reaction between two or three atoms. One *pound* of uranium releases as much energy as several thousand *tons* of a conventional fuel.

31-3 Nuclear Fusion

Another important feature of figure 30-4 is a strong tendency among the lighter nuclei for the binding energy per nucleon to increase with increasing mass number. This means that, if two light nuclei are fused together, the average energy per nucleon decreases and energy is released. This is called **nuclear fusion** and is the principle underlying the hydrogen bomb and thermonuclear power.

Before fusion can take place, the nuclei must be brought together against the opposition of their mutual electrostatic repulsion. This problem is the reverse of the escape velocity of a rocket (section 9-5). The rocket must be given sufficient initial kinetic energy to allow for the increase in its gravitational potential energy as it moves away from the earth. As the two nuclei move closer together, their mutual electrostatic potential energy becomes larger and more positive. Since energy is conserved, their kinetic energy decreases as their potential energy increases. They turn back when all their kinetic energy has been converted into potential energy. For fusion to take place, this must not happen until they are within the range of one another's *nuclear* forces.

Sufficiently high kinetic energies are readily achieved by accelerating one of the nuclei in an "atom smashing machine." In fact nuclear fusion is merely a special case of the nuclear reactions discussed in section 31-1. However, such experiments deal with a small number of nuclei and yield only small amounts of energy. To produce fusion in a large mass of material, the kinetic energies must be due to the thermal motion of the nuclei. If the material is in the form of a gas at a high temperature, the average kinetic energy of a nucleus is given, at least approximately, by the equation for an ideal gas (equation 11-20)

Nuclear Transformations

665

$$\tfrac{1}{2}mv_a{}^2 = \frac{3}{2}\,kT \qquad\qquad (31\text{-}26)$$

The two nuclei start out with a combined kinetic energy of $mv_a{}^2$ or $3kT$. If their atomic numbers are Z_1 and Z_2 and they approach to within a distance R, their potential energy is then $Z_1Z_2e^2/R$. So

$$3kT = \frac{Z_1Z_2e^2}{R} \qquad\qquad (31\text{-}27)$$

$$T = \frac{Z_1Z_2e^2}{3kR} \qquad\qquad (31\text{-}28)$$

If $\quad R = 10^{-12}$ cm

$$T = \frac{Z_1Z_2(4.8 \times 10^{-10})^2}{3 \times 1.38 \times 10^{-16} \times 10^{-12}} \qquad\qquad (31\text{-}29)$$

$$T = 5.5 \times 10^8 Z_1 Z_2 \ {}^\circ\text{K} \qquad\qquad (31\text{-}30)$$

Since Z_1 and Z_2 are small integers, T has the order of magnitude of one billion degrees.

Actually, for reasons that will be explained in more detail in section 31-4, we can get by with a temperature of about ten million degrees. Even so, this is a very high temperature, at which all atoms are broken up into separate electrons and nuclei. The material is therefore not a gas of atoms but a **plasma**, which is a mixture of free negatively and positively charged particles. Nuclear fusion produced by a very high temperature is called a **thermonuclear reaction.**

Temperatures of $10^7 \,^\circ$K are not easy to achieve on earth, but they occur naturally near the center of a star, such as the sun. Thermonuclear reactions therefore proceed readily near the center of the sun and they are the immediate source of most of the energy poured out by the sun, or any similar star. The reactions occurring inside stars may not yet be fully understood, but the following two sequences are probably important. The **proton-proton cycle** is

$$_1\text{H}^1 + {}_1\text{H}^1 \rightarrow {}_1\text{H}^2 + e^+ + \nu \qquad\qquad (31\text{-}31)$$

$$_1\text{H}^1 + {}_1\text{H}^2 \rightarrow {}_2\text{He}^3 \qquad\qquad (31\text{-}32)$$

$$_2\text{He}^3 + {}_2\text{He}^3 \rightarrow {}_2\text{He}^4 + {}_1\text{H}^1 + {}_1\text{H}^1 \qquad\qquad (31\text{-}33)$$

The **carbon cycle** is

$$_1\text{H}^1 + {}_6\text{C}^{12} \rightarrow {}_7\text{N}^{13} \qquad\qquad (31\text{-}34)$$

$$_7\text{N}^{13} \rightarrow {}_6\text{C}^{13} + e^+ + \nu \qquad\qquad (31\text{-}35)$$

$$_6\text{C}^{13} + {}_1\text{H}^1 \rightarrow {}_7\text{N}^{14} \qquad\qquad (31\text{-}36)$$

$$_7\text{N}^{14} + {}_1\text{H}^1 \rightarrow {}_8\text{O}^{15} \qquad\qquad (31\text{-}37)$$

$$_8\text{O}^{15} \rightarrow {}_7\text{N}^{15} + e^+ + \nu \qquad\qquad (31\text{-}38)$$

$$_7\text{N}^{15} + {}_1\text{H}^1 \rightarrow {}_6\text{C}^{12} + {}_2\text{He}^4 \qquad\qquad (31\text{-}39)$$

The details need not concern us unduly. In fact some of these reactions involve the emission of a positron (e^+) and a neutrino (ν), which will not be discussed until the next chapter. The important thing is that the net result of either cycle is the fusing of four protons to produce a helium nucleus.

Source of stellar and solar energy

$$_1H^1 + {}_1H^1 + {}_1H^1 + {}_1H^1 \rightarrow$$
$$_2He^4 + e^+ + e^+ + \nu + \nu \quad (31\text{-}40)$$

The binding energy of $_2He^4$ is particularly large (figure 30-4) and the decrease in mass in this fusion reaction is therefore large. As a result, 26.7 MeV of energy is produced for every helium nucleus formed. This is smaller than the 201 MeV resulting from the fission of $_{92}U^{235}$, but we must remember that only 4 gm of hydrogen are needed as compared with 235 gm of uranium. The energy yield per gram is therefore greater in the fusion process by a factor of $(26.7/201) \times (235/4) = 7.8$.

The explosion of an atomic bomb produces a temperature of about $5 \times 10^{7} {}^\circ K$, which is high enough to make fusion possible. A hydrogen bomb consists of a mixture of light nuclei, such as deuterium, tritium, and lithium, with an atomic bomb to act as a fuse and initiate the fusion of these light elements. The size of an *atomic* bomb is limited by the following design problem. Before firing there must be a situation in which most neutrons are lost and less than one per fission produces a further fission. The firing must produce a sudden change to a situation in which fewer neutrons are lost and more than one per fission produces a further fission. The size of a *hydrogen* bomb is not limited by these considerations and its destructive power depends only upon the total quantity of light nuclei which can be incorporated into its design.

Vigorous attempts are being made to produce controlled thermo-nuclear reactions in the laboratory, with a view to using them as a possible future source of commercial power. At the time of this writing the technical problems have not yet been solved.

31-4 Availability of Energy in the Universe

Now that mankind is acutely aware of the importance of the various sources of energy available to him, it might be interesting to take a broad view of the situation and look for the ultimate sources of energy in the universe. Let us first examine the various sources of energy directly available to us on earth. Fuels such as gasoline, oil, coal, peat, wood and natural gas are all the remains of living organisms which, during their lifetime, absorbed energy from sunlight and stored it in the form of **chemical energy** by manufacturing complicated molecules. The energy is present in these molecules as kinetic energy of the electrons and mutual electrostatic potential energy of electrons and nuclei. When the fuel is burnt, energy is released because the molecules react with oxygen molecules to produce new molecules having less energy. **Hydroelectric power** is possible because the energy of sunlight evaporates water from the oceans so that fast moving steam molecules can rise in the earth's atmosphere to form clouds with the increased gravitational potential energy appropriate to their height. The water of the clouds falls as rain on the mountains and flows downhill, converting its gravitational potential energy into the kinetic energy of flow of rivers. This energy can then be used to generate electricity if a river is diverted through the turbines of a hydroelectric power generating station. Energy abstracted from the **wind** to drive windmills also comes from sunlight when different regions of the atmosphere with different climatic conditions are warmed unequally, creating temperature differences which set the air into motion. Various **solar energy** projects hope to capture the energy of sunlight directly and convert it into economically useful power.

All the energy sources so far discussed therefore derive their energy from sunlight, but we have already learnt in the previous section that the sun's energy is **nuclear energy** liberated by nuclear reactions at the center of the sun. What are rather loosely called **atomic energy** projects are attempts to generate **nuclear energy** on earth directly, without relying upon the sun. **Thermonuclear power,** when it is fully developed, will use nuclear fusion processes similar to those taking place in the sun. **Nuclear reactors** produce energy from the fission of very large nuclei like uranium. These large nuclei are available in ores on earth, but we shall have to explain later where they came from and where they got their excess energy. A promising source of energy for the future is **geothermal energy,** which is the heat energy of the hot interior of the earth. The interior of the earth is made hot by the energy released during the decay of radioactive elements in the earth. As we shall explain more fully in subsequent sections, radioactivity is the ejection of a small particle by a large nucleus, raising again the question of the origin of these large nuclei. The conclusion so far, then, is that all the forms of energy used by mankind originate as nuclear energy, but we shall have to qualify this statement somewhat as we proceed with the analysis.

To understand why there is any available energy at all in the universe we ought to consider the state of the universe at the very beginning (if there was a beginning—the popular current view amongst cosmologists is that there was). In recent years some interesting ideas about the very early universe have emerged, but they are still too tentative for our purpose, so we will pick up the story at a time in the evolution of the universe when the theory can be accepted with some confidence. When the universe was only a few million years old, as compared with its present age of more than 10 billion years, matter was mainly in the form of single hydrogen atoms at a temperature not very different from the temperature on the surface of the earth. The distance between the atoms was more than 10^7 times the diameter of an atom, which corresponds to a density far less than the density of the small amount of gas remaining in the best "vacuum" which can be produced in the laboratory. The dominant contribution to the energy of such a diffuse hydrogen gas is the **rest mass energy** m_0c^2 of the hydrogen atoms. Throughout the following discussion we shall always express amounts of energy as a fraction of the maximum obtainable by completely converting the rest mass of the primeval hydrogen atoms into energy.

The mutual electrostatic potential energy of the proton and electron in one of these hydrogen atoms is only about 10^{-8} of the rest mass energy. The kinetic energies of the particles have a similar small magnitude. Later in the history of the universe, when more complicated chemical elements have been formed, energy can be released in chemical reactions, but these merely involve a rearrangement of the electrons and nuclei into new structural patterns with different electrostatic potential energies and different kinetic energies. Whenever molecules react to produce new molecules with less net energy, the energy released never exceeds 10^{-8} of the rest mass energy. The combustion of conventional fuels is very important to the economy of mankind, but it utilizes a trivial fraction of the total available energy.

There is no direct method of destroying the total rest mass of the protons and electrons (except in conjunction with their antiparticles—see sections 32-3 and 32-4) and so the energy must be released in more subtle and devious ways. There are two important possibilities based on the **negative potential energy** associated with either **nuclear forces** or **gravitational forces**. If two particles with a negative potential energy can be brought closer together, the potential energy becomes more negative and a positive amount of energy is released. If some of the primeval protons can be brought to within distances of 10^{-12} cm of one another and some of them converted into neutrons, they form a nucleus with a negative binding energy equivalent to a mass defect and the lost mass is converted into energy. This is nuclear energy and the thermonuclear reactions which bring the protons together can take place only at sufficiently high temperatures. The release of nuclear energy therefore requires that very high temperatures should occur in suitable places during the subsequent history of the universe.

The second possibility is that a large number of hydrogen atoms should come close together and acquire a large negative *gravitational* potential energy. In fact this is the first thing that happens and it leads to the birth of stars. For some reason, perhaps merely as a result of a chance fluctuation, a region of the diffuse hydrogen gas acquires a slightly larger density than the average. If this increase in density is large enough, gravitational collapse sets in as the inner parts of the region exert attractive gravitational forces on the outer parts. If the shrinking cloud of hydrogen has a mass M, when its radius is R it has a negative gravitational potential energy of the order of magnitude of $-GM^2/R$. (It is pointless to be more precise because the exact value depends on how the density varies with distance from the center.) As the numerical value of R to be inserted in the denominator of this expression becomes smaller, the potential energy becomes more negative and positive energy is continually made available. It might even seem that an infinite amount of energy could be obtained by allowing the mass to collapse to an infinitesimally small radius, but we shall see later why this is not possible. In the first instance, at least, the shrinking hydrogen cloud warms up and then settles down as a star with a stable diameter. When the surface temperature reaches about $5000°$K, the star radiates away energy as visible light and starts to shine. The gravitational potential energy released in the interior cannot escape so readily and the central temperature rises to about $10^7°$K. Nuclear reactions can then take place and the star acquires a new source of energy. It becomes a star like our own sun and can remain in equilibrium with a fixed radius for the reasons described in some detail in section 28-5. For the next several billion years it no longer taps its source of gravitational potential energy but relies entirely upon nuclear energy. The total gravitational potential energy released during this embryonic period of collapse from a tenuous cloud to about the diameter of the sun amounts to about 10^{-6} of the rest mass of the star.

The nuclear reactions taking place at the center of the star convert hydrogen into helium, as explained in the previous section.

$$_1H^1 + {}_1H^1 + {}_1H^1 + {}_1H^1 \Longrightarrow {}_2He^4 + e^+ + e^+ + \nu + \nu \qquad (31\text{-}41)$$

The energy released is about 0.7% of the rest mass energy of the hydrogen, but not all of the hydrogen in the star is near enough to the high temperatures at the center to participate in the reactions. Eventually the central temperature rises to a high enough value to permit the conversion of helium into carbon.

$$_2He^4 + {}_2He^4 + {}_2He^4 \Longrightarrow {}_6C^{12} \qquad (31\text{-}42)$$

A higher temperature is required because a helium nucleus has twice the charge of a hydrogen nucleus. When two helium nuclei try to approach to within the range of one another's nuclear forces, the repulsive electrostatic force between them is four times as large as for two hydrogen nuclei. The energy released during this conversion of helium into carbon is only 0.06% of the rest mass energy. This is the end of the matter for a star like the sun and the net result of all the nuclear reactions is to release less than 1% of the total rest mass energy of the original hydrogen.

Stars more massive than the sun liberate energy faster and develop higher central temperatures. This permits nuclear reactions involving still larger nuclei. The ultimate achievement is to convert hydrogen into iron, $_{26}Fe^{56}$, which has the nucleus with the largest mass defect per nucleon and therefore releases the largest fraction of the rest mass energy. This happens in the most massive stars, but it still yields less than 1% of the rest mass energy of the participating hydrogen. However, it does manufacture the chemical elements from $_1H^1$ up to $_{26}Fe^{56}$ and slightly beyond.

Recent astronomical discoveries have indicated that the really dramatic processes of energy production involve gravitational potential energy. When a star exhausts its nuclear fuel, gravitational collapse starts again and more gravitational potential energy is released. A star with a mass less than 1.2 times the mass of the sun finally settles down as a white dwarf star (section 28-5). It has then released a further amount of gravitational potential energy equal to about 10^{-4} of its rest mass energy, which still compares unfavorably with the nuclear energy it has previously released. However, the collapse of more massive stars cannot be halted at the white dwarf stage and they turn into rapidly collapsing neutron stars (sections 28-5 and 32-1). If the mass is no more than 2 or 3 times the mass of the sun, this may lead to a supernova explosion leaving behind at its core a small stable neutron star observable on earth as a pulsar. The mass of this remnant neutron star is about the same as the mass of the sun but its diameter is only about 10 miles, so it has a very large negative gravitational potential energy. The details are not yet fully understood, but it is quite likely that the total gravitational potential energy released amounts to 2 or 3% of the rest mass energy of the original star and is therefore larger than the nuclear energy previously released.

The supernova explosion creates temperatures in the vicinity of $10^{10\,\circ}\text{K}$ for a short period of time. At these extremely high temperatures it is possible to have thermonuclear reactions which synthesize the very largest nuclei, and this probably accounts for the presence in the universe of the heavier elements from iron up to uranium. These elements have somehow found their way into the earth, and the presence on earth of ores containing elements such as uranium is a prerequisite for the generation of atomic energy by nuclear reactors. Moreover, the occurrence inside the earth of heavy radioactive elements warms up the earth's interior and leads to the possibility of geothermal power. It is interesting that these two promising methods of solving our energy crisis can be traced back to the supernova explosions of collapsing neutron stars and the original source of the energy was the gravitational potential energy of the star.

During the collapse leading to the stable neutron star, conservation of angular momentum results in a very rapid rotation (section 10-6). With the angular momentum $I\omega$ constant, the kinetic energy $\frac{1}{2}I\omega^2 = \frac{1}{2}(I\omega)\omega$ is proportional to the angular velocity ω, and so some of the lost gravitational potential energy reappears as rotational kinetic energy. Complicated processes not yet fully understood gradually slow down the rotation and convert the rotational kinetic energy into the energy of the various radiations emitted by the pulsar. On a smaller scale, rotational kinetic energy would be involved if we ever succeeded in harnessing the energy of the **tides.** The ocean is humped up by the gravitational pull of the moon and the earth rotates under the hump. The energy producing device would increase the frictional drag between the ocean and the earth's solid crust and would produce energy at the expense of slowing down the earth's rotation and gradually increasing the length of the day. Presumably the earth obtained its rotational kinetic energy from gravitational potential energy during the application of the law of conservation of angular momentum to whatever processes of gravitational collapse took place during the formation of the solar system.

Finally, in the collapse of a very massive star into a black hole (section 25-7) there is a distinct possibility that all or most of the rest mass energy might become available. The negative gravitational potential energy of a star just about to cross its Schwarzschild boundary has a magnitude comparable with the total rest mass energy of the material of the star in its original uncollapsed state. We can understand, though, why it is not possible to extract an infinite amount of gravitational potential energy by allowing the radius to shrink to smaller and smaller values. We have reason to believe that it takes an infinitely long time for a collapsing star to cross its Schwarzschild boundary. Moreover, once it has crossed and the radius has fallen below the Schwarzschild radius, R_s, nothing can escape from the star.

Table 31-1 summarizes this discussion by giving a rough order of magnitude for the fraction of the rest mass energy released in each of the basic processes we have mentioned.

Table 31-1   Order of Magnitude of the Fraction of the Rest Mass Energy Released in Various Processes.

Chemical Energy
   Chemical reactions 10^{-9}

Nuclear Energy
   Radioactive decay 10^{-4}
   Nuclear fission processes 10^{-3}
   Thermonuclear fusion reactions 10^{-2}

Gravitational Potential Energy
   Birth of a star 10^{-6}
   Collapse to a white dwarf 10^{-4}
   Collapse to a stable neutron star 10^{-2}
   Collapse to a black hole   1

31-5   Alpha-Decay

The first nuclear transformations to be discovered were a consequence of natural radioactivity. In 1896 Henri Becquerel observed that uranium-bearing salts blackened a photographic plate in their vicinity, even if the plate was wrapped in several layers of paper opaque to visible light. During the next few years this was shown to be due to the emission of three kinds of rays. The α-rays were found to be positively charged and were identified as helium nuclei, $_2He^4$. The β-rays were found to be negatively charged and were identified as electrons. β-emission leads to some important new ideas and will not be discussed until the next chapter. The γ-rays were found to be photons, and they will be discussed in section 31-7.

Alpha-emission is the ejection from the nucleus of an α-**particle**, which is a helium nucleus, $_2He^4$, containing two protons and two neutrons. The disintegrating nucleus therefore loses two protons and its atomic number Z decreases by 2. It becomes the nucleus of a different element two places earlier in the periodic table. Altogether it loses 4 nucleons and its mass number is therefore decreased by 4. Its atomic weight decreases by approximately 4 atomic mass units. The general equation describing α-decay is

$$_ZX^A \rightarrow _{Z-2}Y^{A-4} + _2He^4 \tag{31-43}$$

It is illustrated in figure 31-8. Although the possibility of one chemical element being transmuted into another is now a matter of common knowledge, its discovery had a dramatic impact upon a scientific world that had come to regard the elements as distinct and immutable.

Alpha-emission occurs only for the heavier nuclei, in which the electrostatic repulsion between the protons has become inconveniently large.

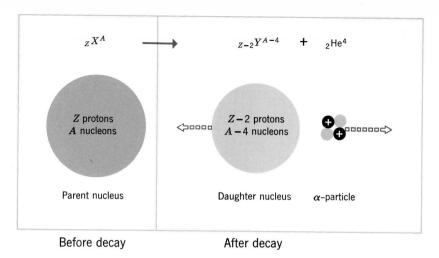

$_ZX^A$ $\longrightarrow$ $_{Z-2}Y^{A-4}$ $+$ $_2He^4$

Z protons
A nucleons

Z − 2 protons
A − 4 nucleons

Parent nucleus   Daughter nucleus   α-particle

Before decay   After decay

Figure 31-8   Alpha-emission.

These are the nuclei which attempt to lower their energies by undergoing fission. Figure 30-4 reveals that the helium nucleus has a particularly large binding energy, which is not very much less than the binding energies of nuclei of intermediate mass. Suppose that a heavy nucleus with mass number A and mass M_A emits an α-particle of mass M_α and changes into a nucleus with mass number $A − 4$ and mass M_{A-4}. This is energetically possible, with energy left over for kinetic energy, if M_A is greater than $(M_{A-4} + M_\alpha)$. This is found to be so for nuclei with Z greater than 82 and is the main reason why there are no permanently stable nuclei above lead, $_{82}Pb^{208}$.

Alpha-emission resembles fission inasmuch as it does not always occur readily, even when it is energetically possible. For example, a nucleus of bismuth, $_{83}Bi^{209}$, waits on the average 3×10^{17} years before emitting an α-particle. This is about one hundred million times the age of the earth! The reason is the same as in the case of fission. It is only a slight distortion of the truth to assume that the α-particle has a separate, individual existence inside the nucleus. Let us consider how its potential energy changes as it escapes (figure 31-9). Inside the nucleus the α-particle is within range of the *attractive* nuclear forces of the other nucleons and this gives it a *negative* potential energy $-\phi_a$. At B, where the α-particle has just left the range of the nuclear forces, it experiences a large electrostatic *repulsion*, giving it a large *positive* potential energy $+\phi_b$. At C, where the α-particle has escaped to a large (effectively infinite) distance from the nucleus, it experiences a negligible force and its potential is effectively zero. The potential therefore looks like a volcano with a deep crater.

Inside the nucleus the α-particle also has kinetic energy E_k, giving it a total energy

$$E_0 = E_k - \phi_a \tag{31-44}$$

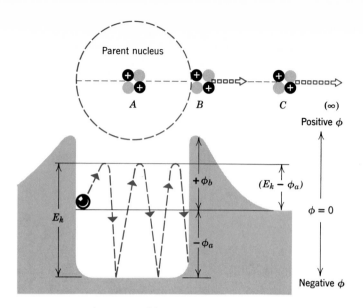

Figure 31-9   The potential barrier opposing α-emission.

This is positive and greater than its potential energy at C, so that it is energetically possible for it to escape to infinity, with energy $(E_k - \phi_a)$ left over for kinetic energy there. However, its total energy is less than the energy ϕ_b of the top of the potential barrier and it does not have enough energy to reach B. The analog is a ball bouncing inside the crater to a height which is below the height of the rim, but above the level of the surrounding plain.

Classically the ball could never escape from the crater, but quantum mechanics offers the α-particle a means of escaping from the nucleus. Imagine the α-particle as a de Broglie wave inside the nucleus (figure 31-10). When this wave strikes the potential barrier, it is not completely reflected but penetrates into the barrier as a disturbance whose amplitude decreases *very rapidly* with distance. However, when it reaches the other side of the barrier, it has not quite died down and it emerges as an oscillating wave of *very small* amplitude. The fact that the

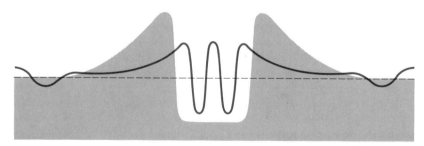

Figure 31-10   The "tunnel" effect in α-emission. The black area represents the potential barrier and the green line is the de Broglie wave associated with the α-particle.

de Broglie wave has any amplitude at all outside the nucleus means that there is a small, but non-zero, probability that the α-particle will be found outside the nucleus. If we invent a method of locating the position of the α-particle, it is overwhelmingly more probable that we shall find it inside the nucleus, but if we keep on looking we shall eventually find it outside the nucleus. This is frequently called the **tunnel effect,** because the α-particle appears to have tunneled its way through the potential barrier, even though it is classically impossible for it to escape over the top.

A deeper insight into what is going on may be obtained in the following way. Since the de Broglie wave has an amplitude inside the barrier, there is a non-zero probability of finding the α-particle there. This does not seem possible, since the potential energy inside the barrier is greater than the total energy of the particle and the kinetic energy cannot be negative. It would seem, though, that as the particle escapes it must spend some time inside the barrier. If we were to find it inside the nucleus at one instant and completely outside the very next instant, this would be equivalent to an infinite velocity and an infinite kinetic energy, clearly violating the law of conservation of energy.

The answer is that we must not push conservation of energy too far. According to Heisenberg's uncertainty principle, in the form of equation 28-51, if we make measurements of the energy of a system in experiments each of which takes a time Δt, the resulting values spread out over a range of width ΔE, where

$$\Delta E \approx \frac{h}{\Delta t} \qquad (31\text{-}45)$$

Energy is a well-defined quantity only if each experiment lasts a long time or if we average over a large number of experiments. In a single determination of energy taking a short time, Δt, the energy is too indefinite for conservation of energy to be applied as a precise principle.

Now let us consider exactly what happens if we measure the energy of the α-particle on a large number of successive occasions, on each occasion devoting a time Δt to the measurement. The average of all the measurements is E_0. Most of the measurements lie in the range between $E_0 - \Delta E$ and $E_0 + \Delta E$. Nevertheless there is a significant chance of obtaining a value $E_0 - 2\Delta E$ or $E_0 + 2\Delta E$. There is even a *very* small chance of obtaining say $E_0 + 100\Delta E$. We can therefore imagine the energy of the α-particle to be fluctuating rapidly over a wide range of values around E_0. Since $\Delta E \approx h/\Delta t$, the extent of a fluctuation increases as its duration becomes shorter.

For a very short interval of time the energy of the α-particle may therefore fluctuate to a high enough value to take it over the top of the potential barrier. This does not necessarily ensure that it will escape. In addition, it must have a high enough velocity v so that, in the time Δt, it can travel a distance $v\Delta t$ great enough to take it right through the barrier. Thus, the width of the barrier is important as well as its height. When this idea is worked out in detail, a suitable fluctuation is found to

Nuclear Trans-
formations

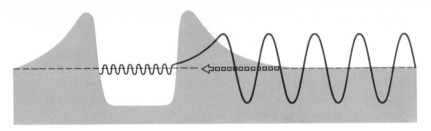

Figure 31-11 Tunneling of a charged bombarding particle into a target nucleus.

be a very rare event. For uranium 238 a suitable fluctuation would last for about 10^{-21} sec, but there is only one chance in about 10^{38} that the energy would fluctuate widely enough to enable the α-particle to escape. Consequently, if we were to observe the α-particle during successive intervals of 10^{-21} sec, we would have to look at it about 10^{38} times before there was a good chance of it escaping. On the average, the time taken to escape would be about $10^{-21} \times 10^{38} = 10^{17}$ sec (a few billion years).

The passage of a particle across a potential barrier may therefore be described in two ways. We may say that the de Broglie wave associated with the particle can propagate through the barrier and emerge on the other side. Alternatively, we may say that the uncertainty principle allows the particle to undergo an occasional short-lived fluctuation in energy which carries it over the top of the barrier.

These arguments may obviously be applied to the case of fission. They enable us to understand how spontaneous fission can eventually occur in spite of the potential barrier of figure 31-4. Similar considerations are also relevant to nuclear fusion and the bombardment of a nucleus by a charged particle. Imagine that we try to reverse α-emission by throwing the α-particle back at the product nucleus with the same total energy E_0 (figure 31-11). Classically, the bombarding particle does not have enough energy to surmount the potential barrier and enter the nucleus. However, it is able to take advantage of the quantum effects which we have just described, and there is a definite probability that it will enter the nucleus. The α-particle does not have to be given enough energy to surmount the barrier, but it still has to be given a fairly large energy, otherwise the probability of its entering will be negligible. Similar reasoning applies to the bombardment of any nucleus by any charged particle and to the approach of two light nuclei in fusion. This is the reason why thermonuclear reactions can proceed at rather lower temperatures than we might first expect.

31-6 The Half-Life and the Nature of Radioactive Decay

So far we have talked about "the average time for a nucleus to disintegrate." We can now be a little more explicit. Suppose that we look at the α-particle in a uranium nucleus at successive intervals of 10^{-21} sec. Then we shall have to observe it about 10^{38} times before there is an appreciable chance of its escaping. However, the occurrence of a large enough fluctuation to enable the α-particle to escape is purely a matter of chance, determined by the laws of probability. There is a very small, but definite chance that it will escape the very first time we look at it, or the one hundred and second time, or the one millionth and tenth time, or any time.

If we look at 10^{38} nuclei during each interval of 10^{-21} sec, since there is one chance in 10^{38} that an α-particle will escape from each nucleus each time we look, there is a very good chance that one of the nuclei will emit an α-particle the very first time we look. To be more realistic, suppose we have 10^{22} nuclei, weighing about four gm. Then, after 10^{16} intervals of 10^{-21} sec, we shall have given the nuclei $10^{22} \times 10^{16} = 10^{38}$ opportunities to fluctuate and there is then a good chance that an α-particle will escape from one of them. This would take $10^{16} \times 10^{-21} = 10^{-5}$ sec. We would therefore expect about 10^5 α-particles to be emitted each second.

Obviously, the chance of an α-particle being emitted is proportional to the number of nuclei present. The number of nuclei disintegrating in a short time interval is therefore proportional to the number of nuclei present at the beginning of that interval. The *fraction* of nuclei disintegrating in a given time interval is independent of the number of nuclei. Suppose that the fraction of nuclei *remaining* at the end of one second is f. If there are initially N nuclei, then after one second Nf of them remain and $N(1 - f)$ have disintegrated. At the beginning of the next second there are only Nf nuclei to start with and so only Nf^2 remain at the end, $Nf(1 - f)$ having disintegrated. At the beginning of the third second there are Nf^2 nuclei of which $Nf^2(1 - f)$ disintegrate, leaving Nf^3 at the end of the third second, and so on. As the material decays the number of α-particles emitted per second decreases as the number of nuclei which have not yet disintegrated decreases. This is shown in figure 31-12.

The time which elapses before exactly one half of the nuclei have disintegrated is called the **half-life,** τ_h. The nucleus has an even chance of distintegrating before a time τ_h has elapsed or after this time. If we watch a single nucleus for a time τ_h, there is a 50% chance of a fluctuation which will enable the α-particle to escape.

Suppose that we start with N nuclei. After a time τ_h, $\frac{1}{2}N$ have disintegrated and $\frac{1}{2}N$ remain. It is one of the important aspects of a situation obeying the laws of probability that the chance of an event occurring in the future is quite independent of anything that happened in the past. Thus, the $\frac{1}{2}N$ nuclei present at time τ_h still have a 50% chance of surviving during the next interval of τ_h secs. During this interval $\frac{1}{2}(\frac{1}{2}N) = \frac{1}{4}N$ nuclei disintegrate and $\frac{1}{4}N$ nuclei remain at time $2\tau_h$. At time $3\tau_h$, $N/8$ nuclei remain, and so on. There will still be a few nuclei remaining after a time very long compared with τ_h.

These ideas are applicable to any decay process, such as fission, the ejection of a particle from a compound nucleus during a nuclear reaction, or the radioactive processes of γ emission and electron or positron emission, which will be explained later. They also apply to the emission

Nuclear Transformations

677

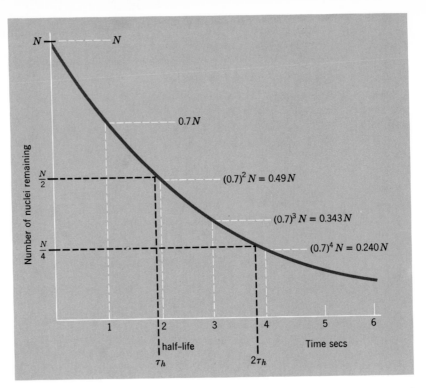

$N \cdots \mid N$

$0.7 N$

$(0.7)^2 N = 0.49 N$

$(0.7)^3 N = 0.343 N$

$(0.7)^4 N = 0.240 N$

half-life

τ_h

$2\tau_h$

Time secs

Number of nuclei remaining

Figure 31-12 The nature of radioactive decay. It has been assumed that 30% of the nuclei decay in one second, leaving 70% remaining.

of a photon by an excited atom (see Chapter 29). An excited state of an atom has a half-life which, in a typical case, may be about 10^{-9} sec.

The element of chance in radioactive decay is another instance of the basic indeterminacy of natural processes. The instant at which a particular nucleus disintegrates is entirely a matter of chance and cannot be predicted on the basis of any information about the past behavior of the nucleus, however complete. Suppose that a speck of uranium emits an average of 10 α-particles per second. Then, during one particular second, because of the chance nature of the process, it might well emit as few as 8 or as many as 12 α-particles. It is easily possible to design an apparatus to count these particles and to arrange that, after counting 10 or more particles, it triggers a chain of events which result in the explosion of a hydrogen bomb. If this apparatus were switched on for exactly one second, there would be approximately even chances of exploding or not exploding the bomb. The situation is analogous to the one illustrated in figure 28-3 and, as in that case, the outcome is quite unpredictable.

Example 31-6-1

A sample of a radioactive material contains one million radioactive nuclei. The half-life is 20 sec. How many nuclei remain after 10 sec?

Suppose that a fraction f remains after 10 sec. Then, at time $t = 0$ sec there are 10^6 nuclei. At time $t = 10$ sec there are $10^6 f$ nuclei. At the end of the next 10 sec a fraction f of these remain. Therefore, at time $t = 20$ sec there are $10^6 f^2$ nuclei. But 20 sec is the half-life, during which half the nuclei disintegrate.

Therefore

$$f^2 = \tfrac{1}{2} \tag{31-46}$$
$$f = 0.707 \tag{31-47}$$

The number of nuclei remaining after 10 sec is $10^6 \times 0.707$ or 707,000.

31-7 Gamma-Rays

After one of the many kinds of nuclear transformation we have just discussed, the product nucleus is sometimes formed in an excited state with excess energy. This excess energy is usually jettisoned in the form of one or more photons as the nucleus is de-excited to its normal state of lowest energy. Since nuclear energies are many times larger than the energies needed to excite electrons in atoms, the emitted photons have frequencies much larger than visible light and wavelengths much shorter. Their frequencies are comparable with those of x-rays or even higher. They are the γ-rays and have energies up to about 20 MeV, as compared with about 5 eV for visible light.

The term γ-rays is now generally used to denote all high energy photons, whatever their origin. The highest energy γ-rays are now produced by accelerating charged particles to high energies and then decelerating them rapidly by allowing them to strike a target. This is exactly the same principle that is used in an x-ray tube (figure 20-5). Photons with energies of several billion electron volts can be produced in this way. Their wavelengths are about 10^{-14} cm, which is less than the diameter of a proton or an electron.

QUESTIONS

1. Which of the following are possible reactions?

(a) $_1H^1 + {}_2He^3 \rightarrow {}_2He^4$

(b) $_1H^1 + {}_3Li^7 \rightarrow {}_2He^4 + {}_2He^4$

(c) $_{88}Ra^{224} \rightarrow {}_{86}Rn^{219} + {}_2He^4$

(d) $_{94}Pu^{238} + {}_0n^1 \rightarrow {}_{54}Xe^{141} + {}_{40}Zr^{97} + 2{}_0n^1$

(e) $_5B^{11} + {}_1H^1 \rightarrow {}_4Be^8 + {}_2He^4$

(f) $_2He^4 + {}_{13}Al^{27} \rightarrow {}_{15}P^{30} + {}_0n^1$

(g) $_1H^2 + {}_{15}P^{31} \rightarrow {}_{13}Al^{29} + {}_2He^4$

2. Insert the missing symbol.

(a) $_2He^4 + {}_6C^{12} \rightarrow {}_7N^{15} + ?$

Nuclear Transformations

(b) $_{84}Po^{210} \rightarrow {}_{82}Pb^{206} + ?$

(c) $? + {}_6C^{12} \rightarrow {}_6C^{13} + {}_1H^1$

(d) $_1H^1 + ? \rightarrow {}_2He^3 + {}_0n^1$

(e) $_{96}Cm^{240} + {}_0n^1 \rightarrow ?$

3. What is wrong with the following sentence?
"The half-life for α-decay of deuterium is 10.2 sec and so 25% of the original sample remains after 20.4 sec."

4. The masses quoted in example 31-2-1 are really atomic masses, not nuclear masses. Does this necessitate (a) a correction which makes a significant difference to the answer, or (b) no correction at all, or (c) a correction which is very small compared with the accuracy of the calculation?

5. Explain carefully why fusion can occur at lower temperatures than would be expected from equation 31-30.

6. A 10 MeV proton strikes a stationary uranium nucleus, $_{92}U^{238}$. According to the theory of relativity, it is permissible to describe the collision in the frame of reference in which the incident proton is initially at rest. Find the kinetic energy of the approaching uranium nucleus in this new frame, and show that it is much greater than 10 MeV. This suggests the possibility of new reactions which cannot take place when the available kinetic energy is only 10 MeV, but which become possible when more initial kinetic energy is available. What is wrong with this argument?

PROBLEMS

A

1. The α-particle emitted by $_{92}U^{235}$ has a kinetic energy of 4.58 MeV. Deduce the atomic mass of the $_{90}Th^{231}$ nucleus left behind after its emission. (Use the atomic masses given on the inside of the cover.)

2. Find the kinetic energy in MeV of the α-particle emitted by $_{88}Ra^{226}$ given that its atomic mass is 226.0254 u and that the atomic mass of $_{86}Em^{222}$ is 222.0175 u.

3. $_{94}Pu^{239}$ emits a 5.2 MeV α-particle. Find its atomic mass as accurately as you can.

4. A radioactive element has a half-life of 0.15 sec. If there are initially 0.002 gm of it, what mass of this element remains after (a) 0.15 sec, (b) 0.30 sec, (c) 0.60 sec?

5. A radioactive element has a half-life of 10^9 years. At the zero of time it emits 10^6 α-particles per second. How many α-particles per second will it emit after a lapse of (a) 10^9 years, (b) 2×10^9 years, (c) 10^{10} years? Assume that any products of its decay which also emit α-rays are immediately removed. Why is this last assumption necessary?

6. The mass of a radioactive element is initially 0.008 gm, but 15 sec later only 0.001 gm remains. What is its half-life?

7. If a nucleus emits a 20 MeV γ-ray, calculate the mass difference between its excited state and its ground state in atomic mass units.

B

8. A radioactive material with a half-life of 2×10^{-3} sec initially emits 4×10^8 α-particles per second. How many α-particles per second does it emit after a lapse of (a) 10^{-3} sec, (b) 5×10^{-4} sec? Assume that all the products of the decay are immediately removed.

9. If a radioactive material has a half-life of 8 sec, what fraction of it decays in 1 sec?

10. In a fission process a nucleus with an atomic mass of 205.9752 u divides into two smaller nuclei with atomic masses of 116.9041 u and 88.9108 u. Calculate the energy released during the fission of 1 gm of the original nuclei.

11. The energy calculated in question 10 is to be used to raise the temperature of helium gas from 300°K to 1000°K without change of volume. What mass of helium gas can it warm up?

12. If a photon is capable of disintegrating a $_6C^{12}$ nucleus into three α-particles, set a lower limit on its frequency.

13. $_{92}U^{235}$ emits a 4.58 MeV α-particle. Find the velocity of the recoiling $_{90}Th^{231}$ nucleus.

14. $_{94}Pu^{239}$ emits a 5.2 MeV α-particle. Find the kinetic energy in MeV of the recoiling daughter nucleus.

15. If the nucleus $_ZX^A$ emits an α-particle with a kinetic energy of E_α MeV, find an expression for the kinetic energy of the recoiling daughter nucleus.

16. Imagine that a 1 MeV proton strikes a stationary deuteron to produce the reaction

$$_1H^1 + {_1H^2} \rightarrow {_2He^3}$$

If the $_2He^3$ nucleus is produced in its state of lowest energy and no energy is lost by unknown processes, calculate the velocity of the $_2He^3$ nucleus. Consider the question of whether momentum can be conserved.

C

17. Imagine that a 2 MeV proton strikes a stationary tritium nucleus to produce the reaction

$$_1H^1 + {_1T^3} \rightarrow {_2He^3} + {_0n^1}$$

Calculate the sum of the kinetic energies of the products. Assume that the $_2He^3$ nucleus is formed in its ground state and that no energy is lost by unknown processes. Consider the question of whether momentum can be conserved.

18. Nucleus A decays into nucleus B with a half-life of 10^8 years. B decays with a half-life of 10^6 years. After a sufficient time has elapsed, the ratio of the number of nuclei of A to the number of nuclei of B settles down to a constant value. What is this constant value? Why?

19. Calculate the height in electron volts of the barrier for α-decay of $_{92}U^{238}$ (ϕ_b of figure 31-9). Make the following assumptions. The point B corresponds to an undistorted α-particle just touching an undistorted $_{90}Th^{234}$ nucleus. At B the *nuclear* force between the α-particle and the thorium nucleus may be neglected. Both nuclei are spherical and their density has the average value 2.3×10^{14} gm cm^{-3}.

20. The α-particle emitted by $_{92}U^{238}$ has a kinetic energy of 4.20 MeV. What must be its minimum separation from the $_{90}Th^{234}$ nucleus it leaves behind before it can be considered to have escaped? Compare this distance with the radius of the $_{90}Th^{234}$ nucleus.

21. Calculate the age of the sun from the following data. Assume that the solar radiation of 1.4×10^6 erg cm^{-2} sec^{-1} in the vicinity of the earth is a reliable guide to the total energy output of the sun, and that it has maintained this output at this level all its life. This energy is derived from the reaction of equation 31-40 and all other processes may be ignored. 5% of the mass of the sun is now $_2He^4$, but it originally contained no $_2He^4$.

22. Show that the ratio of the gravitational potential energy of a star to its rest mass energy is $\dfrac{\alpha R_s}{R}$, where R_s is the Schwarzschild radius and α is not very different from 1.

23. The annual consumption of energy by the whole of mankind is about 10^{27} ergs. Compare this with the rate at which the whole earth receives energy from sunlight.

24. If the rest mass of the moon could be completely converted into energy, for how long could it supply mankind with energy at the present rate of consumption?

25. Suppose that geologists could devise a method of making the earth's radius shrink by 1 meter and utilizing all the gravitational potential energy released. For about how long could this meet the energy needs of mankind at the present rate of consumption?

A Profusion of Particles and Processes

32-1 Beta-Decay

Beta-decay is the emission of an electron by a nucleus. All other properties of nuclei firmly indicate that they are composed of protons and neutrons, but contain no electrons. Where, then, does the electron come from? The only satisfactory explanation is that beta-decay occurs when a neutron inside the nucleus changes into a proton by emitting two small particles, an electron and an antineutrino. (The nature of neutrinos and antineutrinos will be explained in due course.) This differs fundamentally from the processes discussed in the previous chapter, which involved the reshuffling of neutrons and protons. We are now introducing a new and very important type of process in which a "fundamental" particle changes into other fundamental particles. This is often called a **fundamental process.**

Experiments on beams of free neutrons have shown that a neutron can indeed change into a proton with simultaneous emission of an electron. Like α-decay and fission, the process occurs at an unpredictable time. Its half-life is 11 minutes. A free neutron is therefore an unstable particle which does not live very long by human standards. The decay is depicted in figure 32-1 and is represented by the equation:—

$$n \longrightarrow p + e^- + \bar{\nu} \qquad (32\text{-}1)$$

neutron $\longrightarrow$ proton + electron + antineutrino

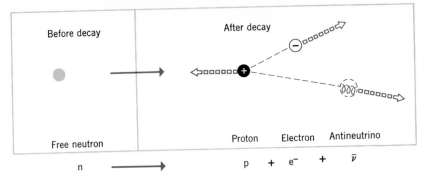

Figure 32-1 Decay of a free neutron.

It is conventional to represent the neutron by the symbol n, the proton by p, the electron by e⁻ and the neutrino by ν. The superscript associated with e indicates that the charge on the electron is negative. The proton might therefore be represented by p^+, but this is not normally done. The bar *above* the ν indicates that this particle is an *anti*-neutrino rather than a neutrino. The distinction will be explained later.

When a neutron inside a nucleus changes into a proton, the electron and the antineutrino are violently ejected from the nucleus. The emitted electrons are the β-rays, which have long been known to constitute part of the radiation coming from radioactive materials. The nucleus loses one neutron and gains an extra proton. Its atomic number therefore increases from Z to Z + 1 and it becomes the nucleus of an element one position higher in the periodic table. The total number of neutrons plus protons remains the same and the mass number A is consequently unaltered. The process is depicted in figure 32-2 and is represented by the equation

$$_Z X^A \rightarrow \ _{Z+1} Y^A + e^- + \bar{\nu} \tag{32-2}$$

Why, then, do not all nuclei containing neutrons emit electrons? Why

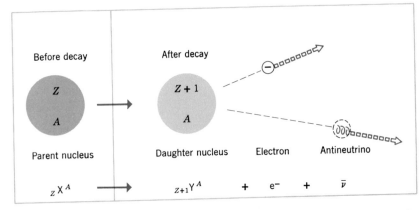

Figure 32-2 Beta-decay. A neutron in the parent nucleus has become a proton in the daughter nucleus.

do not all neutrons decay into protons so that, after a time not very much longer than 11 minutes, very few neutrons are left and the universe is composed almost entirely of protons and electrons? The answer is that a neutron in a nucleus will spontaneously beta-decay into a proton if energy is released in the process, but not if additional energy has to be supplied to the nucleus before the decay can take place. The rest mass of a *free* neutron is greater than the sum of the rest masses of its decay products. Some mass is therefore available to be converted into the kinetic energies of the products, enabling them to fly apart. Inside a nucleus the situation is more complicated. The replacement of a neutron by a proton sometimes results in a sizable *increase* in the energy of the nucleus. In a large nucleus this might be due to the positive electrostatic potential energy of the newly created proton and the other protons already in the nucleus. In addition, the Pauli exclusion principle does not allow the proton to be created in a state which is already occupied by an existing proton, and the available unoccupied states might all have energies in excess of the energy of the original neutron. In figure 30-6b, it is not possible to remove a spherical weight (neutron) from the left hand rack and add a cubical weight (proton) to the right hand rack without increasing the energy of the system. Thus, in many nuclei it is not possible for a neutron to change into a proton unless a large amount of energy is added from outside, which rarely happens in nature. In section 30-4 there was a discussion of the factors determining the most favorable ratio of neutrons to protons. Electron emission is most likely to occur in a nucleus with too many neutrons. The replacement of a neutron by a proton then brings the nucleus closer to the region of stable nuclei shown in figure 30-5.

Neutron stars are composed almost entirely of stable neutrons (section 28-5). We are now in a position to understand how they are formed and why their neutrons do not beta-decay. When a star exhausts its nuclear fuel it goes into gravitational collapse and its electrons are squeezed into restricted spaces between its nuclei. The Heisenberg Uncertainty Principle then compels the electrons to acquire large average velocities with an order of magnitude of about $\dfrac{h}{m_e d_n}$, where m_e is the mass of an electron and d_n is the average distance between nuclei. If the mass of the star exceeds the mass of the sun by a factor of more than 1.2, the star does not settle down as a white dwarf with a density of 10^6 gm/cm^3 but continues to collapse to higher densities. As the distance between nuclei, d_n, decreases the average velocity of the electrons increases and their average kinetic energy also increases. When the density reaches 10^{11} gm/cm^3 the kinetic energy of the electrons is large enough to enable them to enter the protons inside the nuclei by a process of "inverse beta-decay."

$$e^- + p \rightarrow n + \nu \tag{32-3}$$

Because the rest mass of the neutron is greater than the sum of the rest masses of the proton and electron, this process cannot occur spontaneously but requires the addition of extra energy. The large kinetic energy of the electrons inside the high density star provides this extra energy. Since the star is electrically neutral, there is one proton for each electron and so almost all of the protons are converted into neutrons. (1% of the matter in the star remains in the form of electrons and protons).

A very massive neutron star probably continues to collapse into a black hole, whereas a star of intermediate mass probably experiences a supernova explosion, blowing off a large fraction of its mass but leaving a small stable neutron star at its core. This has about the same mass as the sun, a diameter of about 10 miles and a density of about 10^{14} gm/cm^3. This is not very different from the density of an atomic nucleus and the star could be looked upon as a giant nucleus composed almost entirely of neutrons very nearly touching one another. It is held together by attractive gravitational forces with some help from the attractive nuclear forces between neighboring neutrons. It is prevented from collapsing by the blowing up effect of the Heisenberg Uncertainty Principle (section 28-5) acting in this case upon the neutrons.

A neutron inside the star cannot beta-decay because if it did it would have only enough energy to emit an electron with a small velocity. However, the electron would be emitted into a small confined space between nuclei, which would set an upper limit on the uncertainty in its position and therefore necessitate a large uncertainty in its velocity. Energy considerations therefore imply a small average velocity for the electron, whereas the Uncertainty Principle requires a large average velocity. The two requirements are incompatible and the beta-decay is not allowed to take place.

32-2 The Neutrino

The neutrino is an elusive particle which does not make its presence felt in any obvious way. It was therefore originally thought that β-decay results in the emission of an electron only.

$$_zX^A \rightarrow _{z+1}Y^A + e^- \tag{32-4}$$

If this process is to take place, the rest mass of the parent nucleus, M_X, must be greater than the rest mass of the daughter nucleus, M_Y, so that an amount of energy $(M_X - M_Y)c^2$ is released. Part of this is needed for the rest mass energy, m_0c^2, of the electron and the rest, $(M_X - M_Y - m_0)c^2$, is available for kinetic energy. Momentum must be conserved and the daughter nucleus $_{z+1}Y^A$ must recoil in the opposite direction to the electron in much the same way that the earth recoils when a rocket is fired (section 8-5). The earth is much more massive than the rocket and the recoiling nucleus is also much more massive than the electron. Therefore, as explained in section 9-5, only a negligible fraction of the kinetic energy is given to the nucleus. We would therefore expect that the electron would always be emitted with the same velocity, v_m, corresponding to a kinetic energy $(M_X - M_Y - m_0)c^2$.

The electrons leaving the nucleus are actually observed to have all possible velocities from 0 up to v_m. A few of them do have a velocity v_m and the expected kinetic energy, but most of them are emitted with too small a velocity and some kinetic energy appears to be missing. Pauli suggested in 1930 that the missing kinetic energy is given to a new kind of particle emitted simultaneously with the electron. Fermi developed this idea and called the new particle a **neutrino** which is Italian for "little neutral one." The particle accompanying an electron in β-decay is now known to be an *anti*-neutrino, but we shall ignore this distinction at the moment. The energy $(M_X - M_Y)c^2$ lost by the nucleus is shared between the electron and the neutrino in all possible proportions. Usually the electron is emitted with less than the total available energy. Occasionally, however, it receives all of the available energy and the neutrino receives none. This must mean that the rest mass of the neutrino is zero. Otherwise, a minimum amount of energy would be taken by the neutrino to provide its rest mass energy, even if it had zero velocity.

The only other known particle with zero rest mass is the photon. As explained in section 27-4, the photon, which has a velocity c, avoids having infinite energy by having zero rest mass. In the expression for its energy,

$$\varepsilon = \frac{m_0 c^2}{\sqrt{1 - \dfrac{v^2}{c^2}}} \tag{32-5}$$

both the numerator and denominator are zero and the energy is not necessarily zero or infinite. The neutrino is similar. It has zero rest mass, but avoids having zero energy by always moving with the velocity of light.

The neutrino salvages not only the law of conservation of energy, but also the laws of conservation of linear momentum and angular momentum. If only an electron were emitted, the daughter nucleus would recoil in a direction exactly opposite to the direction of emission of the electron, with a linear momentum equal, but opposite, to that of the electron (figure 32-3). It is found, however, that the electron and the nucleus do not usually fly apart in exactly opposite directions. Their two vector momenta therefore cannot possibly be added together vectorially to give zero, whereas the parent nucleus is usually moving so slowly that its momentum is very nearly zero. The assumption that a neutrino is simultaneously emitted remedies the situation. The vector sum of the three vectors representing the momenta of the daughter nucleus, the electron and the neutrino is always found to be zero.

The intrinsic angular momentum of a neutron, a proton, or an electron is $\frac{1}{2}$ of the fundamental unit of angular momentum, $h/2\pi$ (section 29-4). Moreover, the component of this angular momentum in a preferred direction is either $+\frac{1}{2}$ or $-\frac{1}{2}$ of a unit. If the neutron were to decay into a proton and an electron only, there would be no way of conserving angular momentum. The initial value of the angular momentum in the preferred direction would be $+\frac{1}{2}$ or $-\frac{1}{2}$, whereas its final value would be restricted to the following four possibilities:

(1) $+\frac{1}{2} + \frac{1}{2} = +1$ (2) $+\frac{1}{2} - \frac{1}{2} = 0$

(3) $-\frac{1}{2} + \frac{1}{2} = 0$ (4) $-\frac{1}{2} - \frac{1}{2} = -1$

(a) The parent nucleus before β-decay. The nucleus is at rest and the total momentum is zero.

(b) Emission of an electron only. The two momenta are equal in magnitude, but in exactly opposite directions, ensuring that the total momentum remains zero.

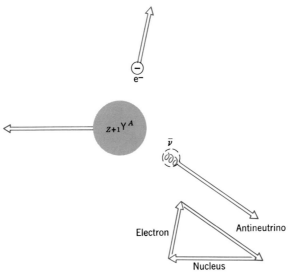

(c) Emission of an electron and an antineutrino. The vector sum of the three momenta must be zero, but it is no longer necessary for the recoiling nucleus and the electron to move in exactly opposite directions.

Figure 32-3   The law of conservation of momentum applied to β-decay.

The law of conservation of angular momentum can be satisfied if the angular momentum of the neutrino (or antineutrino) is also $\frac{1}{2}(h/2\pi)$. The three $\frac{1}{2}$ unit spins of the products of the decay can then easily be combined to give the $\frac{1}{2}$ unit of spin of the original neutron (for example: $+\frac{1}{2} - \frac{1}{2} + \frac{1}{2} = +\frac{1}{2}$).

We can now see how a neutrino differs from a photon. Although they both have zero rest mass and the velocity of light, a neutrino has a spin of $\frac{1}{2}$ a unit, whereas a photon has a spin of one whole unit, $h/2\pi$ (section 29-4). Experimentally, it is very clear that a neutrino is not a photon.

(a) The parent nucleus before β-decay. The nucleus is at rest and the total momentum is zero.

(b) Emission of an electron only. The two momenta are equal in magnitude, but in exactly opposite directions, ensuring that the total momentum remains zero.

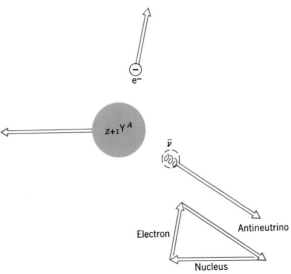

(c) Emission of an electron and an antineutrino. The vector sum of the three momenta must be zero, but it is no longer necessary for the recoiling nucleus and the electron to move in exactly opposite directions.

Figure 32-3 The law of conservation of momentum applied to β-decay.

The law of conservation of angular momentum can be satisfied if the angular momentum of the neutrino (or antineutrino) is also $\frac{1}{2}(h/2\pi)$. The three $\frac{1}{2}$ unit spins of the products of the decay can then easily be combined to give the $\frac{1}{2}$ unit of spin of the original neutron (for example: $+\frac{1}{2} - \frac{1}{2} + \frac{1}{2} = +\frac{1}{2}$).

We can now see how a neutrino differs from a photon. Although they both have zero rest mass and the velocity of light, a neutrino has a spin of $\frac{1}{2}$ a unit, whereas a photon has a spin of one whole unit, $h/2\pi$ (section 29-4). Experimentally, it is very clear that a neutrino is not a photon.

small a velocity and some kinetic energy appears to be missing. Pauli suggested in 1930 that the missing kinetic energy is given to a new kind of particle emitted simultaneously with the electron. Fermi developed this idea and called the new particle a **neutrino** which is Italian for "little neutral one." The particle accompanying an electron in β-decay is now known to be an *anti*-neutrino, but we shall ignore this distinction at the moment. The energy $(M_X - M_Y)c^2$ lost by the nucleus is shared between the electron and the neutrino in all possible proportions. Usually the electron is emitted with less than the total available energy. Occasionally, however, it receives all of the available energy and the neutrino receives none. This must mean that the rest mass of the neutrino is zero. Otherwise, a minimum amount of energy would be taken by the neutrino to provide its rest mass energy, even if it had zero velocity.

The only other known particle with zero rest mass is the photon. As explained in section 27-4, the photon, which has a velocity c, avoids having infinite energy by having zero rest mass. In the expression for its energy,

$$\varepsilon = \frac{m_0 c^2}{\sqrt{1 - \dfrac{v^2}{c^2}}} \tag{32-5}$$

both the numerator and denominator are zero and the energy is not necessarily zero or infinite. The neutrino is similar. It has zero rest mass, but avoids having zero energy by always moving with the velocity of light.

The neutrino salvages not only the law of conservation of energy, but also the laws of conservation of linear momentum and angular momentum. If only an electron were emitted, the daughter nucleus would recoil in a direction exactly opposite to the direction of emission of the electron, with a linear momentum equal, but opposite, to that of the electron (figure 32-3). It is found, however, that the electron and the nucleus do not usually fly apart in exactly opposite directions. Their two vector momenta therefore cannot possibly be added together vectorially to give zero, whereas the parent nucleus is usually moving so slowly that its momentum is very nearly zero. The assumption that a neutrino is simultaneously emitted remedies the situation. The vector sum of the three vectors representing the momenta of the daughter nucleus, the electron and the neutrino is always found to be zero.

The intrinsic angular momentum of a neutron, a proton, or an electron is $\frac{1}{2}$ of the fundamental unit of angular momentum, $h/2\pi$ (section 29-4). Moreover, the component of this angular momentum in a preferred direction is either $+\frac{1}{2}$ or $-\frac{1}{2}$ of a unit. If the neutron were to decay into a proton and an electron only, there would be no way of conserving angular momentum. The initial value of the angular momentum in the preferred direction would be $+\frac{1}{2}$ or $-\frac{1}{2}$, whereas its final value would be restricted to the following four possibilities:

(1) $+\frac{1}{2} + \frac{1}{2} = +1$ (2) $+\frac{1}{2} - \frac{1}{2} = 0$
(3) $-\frac{1}{2} + \frac{1}{2} = 0$ (4) $-\frac{1}{2} - \frac{1}{2} = -1$

A Profusion
of Particles
and Processes

687

A γ-ray with the same energy as a neutrino would produce some easily observable effects such as the photoelectric effect (section 26-4) or the Compton effect (section 27-1). Interaction between a neutrino and any other form of matter is an extremely rare event. It was not until 1956, sixty years after the discovery of β-decay, that any such interaction was observed. This interaction provides the most direct evidence for the existence of the neutrino, but we shall defer its discussion until after the positron has been introduced. The rarity of this interaction can be emphasized in the following way. A photon of appropriate energy passing through solid lead would travel on the average for about ten billionths of a second and cover a distance of about 3 meters before interacting with an atom. A neutrino with the same energy would travel for about fifty years before interacting! In order to stop the neutrino, we would have to use a lead shield with a thickness more than ten times the distance between the sun and the nearest star!

The weakness of its interaction with matter is conclusive proof that the neutrino carries no electric charge. (It is the "little neutral one"). Any charged particle moving rapidly through matter exerts strong electric fields on the atoms as it passes by and very soon pries loose an electron. The absence of charge on the neutrino may also be deduced from the **law of conservation of charge,** which says that there is no known process which creates or destroys electric charge, (equal amounts of positive and negative charge being assumed to cancel one another to give zero net charge). The charge on a neutron is zero. When it decays, the positive charge of the proton is canceled by the negative charge of the electron. If the net charge is to remain zero, the neutrino must be uncharged.

Neutrinos are probably very abundant throughout the universe, although their presence is not very obvious because they interact so rarely with matter. The thermonuclear reactions taking place at the center of the sun produce neutrinos (section 31-3 and equation 31-40). The sun is therefore continually bombarding the earth with a strong flux of neutrinos. As you read this, there are probably about one hundred thousand neutrinos inside your body passing through with the speed of light. This statement is independent of whether it is day or night, because, even if the sun is on the other side of the earth, the neutrinos pass through the earth with negligible interference and then enter your body. As they pass through the earth, only about one in every trillion

The neutrino, ν

Charge $= 0$ (32-6)

Rest mass $= 0$ (32-7)

Velocity $= c$ always   (32-8)

Intrinsic angular momentum $= \frac{1}{2}(h/2\pi)$ (32-9)

(10^{12}) is likely to interact with an atom of the earth. As far as the neutrinos passing through your body are concerned, you will probably have to wait several hours for the next interaction between one of these neutrinos and an atom of your body.

32-3   The Positron

The positron is identical with the electron in all respects, except that it is positively charged. It has exactly the same rest mass as the electron, its spin angular momentum is $\frac{1}{2}(h/2\pi)$, but its charge is $+e$ as compared with $-e$ for the electron. It is called the **antiparticle** of the electron and is represented by the symbol e^+, whereas the electron is represented by e^-.

Some nuclei decay by emitting a positron. This is called **positron-emission** and is very similar to electron-emission. The term β-decay is frequently used to denote both processes. Positron-emission may be looked upon as the result of a proton in the nucleus changing into a neutron.

$$p \rightarrow n + e^+ + \nu \tag{32-10}$$

Notice that charge is conserved, since there is one unit of positive charge on the proton before the reaction, and the only charged product is the positron with one unit of positive charge. As in the case of electron-emission, a neutrino is needed, but this is now really a neutrino rather than its antiparticle, the antineutrino. (The reason for needing to distinguish between them and the difference between them will become clear later).

A free proton does not decay into a neutron. This is energetically impossible, since the rest mass of a neutron is greater than the rest mass of a proton. However, in a nucleus which has too many protons, the substitution of a neutron for a proton often decreases the total energy. Positron-emission may be represented by the equation:

$$_Z X^A \rightarrow _{Z-1} Y^A + e^+ + \nu \tag{32-11}$$

The result is to lower the nucleus by one position in the periodic table, without changing its mass number. This is possible if the mass of the daughter nucleus Y is less than the mass of the parent nucleus X by at least the mass of the positron.

In some sense, not yet fully understood, a positron is the exact opposite of an electron. If a positron encounters an electron, they cancel one another out and both disappear. Their energy, momentum and angular momentum reappear usually in the form of two photons (although occasionally three or even more photons may be formed). This is called the **annihilation of matter** and is illustrated in figure 32-4. Using the conventional symbol γ to represent a photon, the equation describing the process is

$$e^+ + e^- \rightarrow 2\gamma \tag{32-12}$$

Before	After

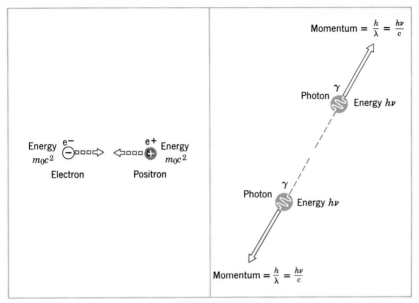

(a) A positron and an electron approach one another. We are considering the important special case when their velocities are small compared with c.

(b) The electron and positron change into two photons with equal frequencies, moving in exactly opposite directions.

Figure 32-4 The annihilation of matter.

This process is different from any other that we have so far discussed inasmuch as the rest mass is completely destroyed and completely converted into the energy of electromagnetic radiation. It therefore provides the final vindication of Einstein's principle of the equivalence of mass and energy (section 24-4). Consider the very common case when the electron and positron are moving so slowly that their kinetic energies are negligibly small compared with their rest mass energies. Then the total energy available before the collision is approximately $2m_0c^2$, where m_0 is the rest mass of either the electron or the positron. Since the total momentum before the collision is negligibly small, the two photons must be emitted in exactly opposite directions with equal, but opposite, momenta. The momentum of a photon is $h\nu/c$, and so the two photons must have equal frequencies and equal energies. The Einstein equation

$$E = mc^2 \tag{32-13}$$

takes the form

$$2h\nu = 2m_0c^2 \tag{32-14}$$

The energy of each photon is equal to the rest mass energy of an electron, which is 0.511 MeV.

The reverse process of conversion of energy into rest mass occurs in **pair production**, which is sometimes called **the materialization of light.**

If a photon strikes a nucleus, the strong electric field in the vicinity of the nucleus induces the photon to change into an electron-positron pair.

$$\gamma \rightarrow e^+ + e^- \tag{32-15}$$

This is illustrated in figure 32-5. The initial energy of the photon is $h\nu$ and the energies of the electron and positron are $m_e c^2$ and $m_p c^2$. Here m_e and m_p are the "total" masses of the electron and positron, including both rest mass and kinetic energy. The law of conservation of energy is

$$h\nu = m_e c^2 + m_p c^2 \tag{32-16}$$

Since the rest mass energy of either particle is 0.511 MeV, the photon must have an energy of at least twice this or 1.022 MeV. It is not visible light that is being materialized, but γ-rays.

We are now in a position to discuss the experiment which gave very direct evidence for the existence of neutrinos. In the vicinity of a nuclear reactor there are many antineutrinos arising from β-decay of unstable nuclei inside the reactor. A tank of water was placed near such a reactor, and occasionally an antineutrino was absorbed by one of the protons in the water to form a neutron and a positron.

$$\bar{\nu} + p \rightarrow n + e^+ \tag{32-17}$$

The positron was almost immediately captured by an electron, and their annihilation produced two photons traveling in opposite directions, each with an energy of about 0.5 MeV. The neutron wandered through the water for a few millionths of a second and was then absorbed by the nucleus of a cadmium atom dissolved in the water. The resulting excited cadmium nucleus emitted one or more γ-ray photons of total energy about 9 MeV.

Above and below the tank of water was a tank of a special liquid

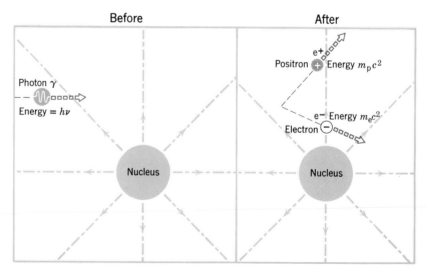

Figure 32-5   The materialization of light (positron-electron pair production).

"scintillator," which emits a flash of light when a photon enters it. The intensity of the flash is an indication of the energy of the photon. The apparatus was specially designed to recognize the simultaneous occurrence of a 0.5 MeV flash in both tanks, followed a few millionths of a second later by one or more flashes corresponding to a total energy of 9 MeV.

The occurrence of all these events was a reasonable guarantee that the process responsible was indeed the one shown in equation 32-17. This combination of events was found to occur with a frequency which could be predicted from the expected density of antineutrinos in the vicinity of the nuclear reactor. Moreover, the frequency of occurrence decreased markedly when the reactor was turned off.

32-4   Antimatter

Nearly all the fundamental particles have clearly distinguishable antiparticles. The antiparticle of the proton is the **antiproton,** which is represented by the symbol $\bar{p}$. A bar over the symbol for a particle always denotes its antiparticle. The antiproton is identical with the proton in all respects, except that it is *negatively* charged. However, the difference between a particle and its antiparticle is more profound than a mere reversal of the sign of the charge. The uncharged neutron has an antiparticle, the **antineutron,** $\bar{n}$, which is clearly distinct from the neutron, as we shall see. We have already mentioned that there is an **antineutrino,** $\bar{\nu}$.

Since the rest mass of a proton is much larger than that of an electron, much more energy is needed to create a proton-antiproton pair than to create an electron-positron pair. The procedure is to accelerate a proton to an enormously high energy and then allow it to strike a stationary proton. Both protons survive the collision, but part of the energy is used to create another proton and an antiproton.

$$(p + p) \rightarrow (p + p) + (p + \bar{p}) \qquad (32\text{-}18)$$

Sometimes a neutron and an antineutron are formed

$$(p + p) \rightarrow (p + p) + (n + \bar{n}) \qquad (32\text{-}19)$$

The combined rest mass energy of a proton and antiproton is 1.872 BeV (1 BeV $= 10^9$ electron volts), but a much higher energy is needed before the pair creation processes become reasonably probable. The first particle accelerator to achieve the feat was the Bevatron, which accelerates protons up to energies of 6.2 BeV. It is shown in figure 32-6, which gives an impression of the enormous amount of effort and money devoted to research in this field.

The antiproton is a rare particle in nature. One that is produced in any way very soon encounters a proton and annihilates with it. The energy is usually converted into pions

$$p + \bar{p} \rightarrow \pi^+ + \pi^0 + \pi^- \qquad (32\text{-}20)$$

The pion, or π-meson, is yet another fundamental particle that we shall discuss a little later. In the particular reaction just quoted all three kinds

A Profusion
of Particles
and Processes

693

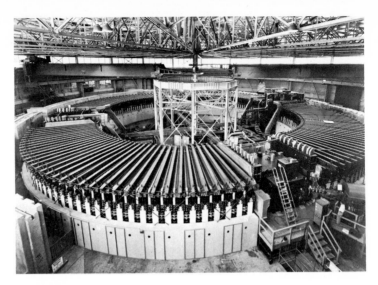

Figure 32-6 This photograph shows the "Bevatron," a giant atom-smasher at the University of California Lawrence Radiation Laboratory, Berkeley. The machine accelerates protons to 6.2 billion electron volts. At the lower right is a linear accelerator that injects protons into a chamber in the giant circular 10,000 ton magnet (diameter, 120 feet). The protons circle this chamber 4 million times in 1.8 seconds, traveling over 300,000 miles, farther than from the earth to the moon. With this machine, the antiproton and numerous other particles have been discovered. Compare the size of the machine with the man in the bottom right-hand corner. (Courtesy of the Lawrence Radiation Laboratory, University of California, Berkeley.)

of pion are produced, with positive charge, negative charge and no charge. A different number of pions may result, although conservation of charge requires that the number of positive and negative pions shall always be equal. Figure 32-7 is a bubble chamber photograph of an annihilation in which two positive and two negative pions are produced.

In a similar way, an antineutron can annihilate with a neutron. The reaction might be

$$n + \bar{n} \rightarrow \pi^+ + \pi^+ + \pi^0 + \pi^- + \pi^- \tag{32-21}$$

The existence of this process is a clear indication that an antineutron is different from a neutron, being in some way its opposite. Although they cannot be distinguished by the sign of their electric charge, there is at least one way in which they are clearly different. They are both spinning like tops and each has an angular momentum of $\frac{1}{2}h/2\pi$. Also, they both behave like small bar magnets (section 29-5) and have magnetic moments of equal strength. However, if a neutron and an antineutron are both spinning in the same direction, their bar magnets are oriented in opposite directions (figure 32-8).

The atoms of ordinary matter have positively charged nuclei composed of protons and neutrons, surrounded by negatively charged elec-

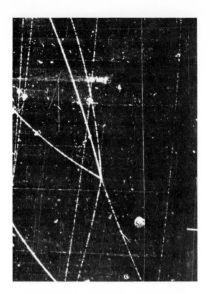 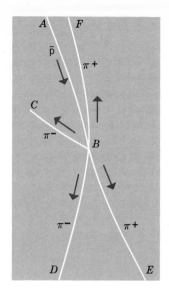

Figure 32-7 A bubble chamber photograph of a proton annihilating with an antiproton. A white track is a chain of bubbles left behind by a charged particle as it passes through a superheated liquid. The incoming antiproton enters at A and collides with a proton at B. Since this proton is at rest it does not produce a track of bubbles. The four pions resulting from the annihilation produce the tracks BC, BD, BE, and BF. (Courtesy of the Lawrence Radiation Laboratory, University of California, Berkeley.)

trons. **Antimatter** would have atoms with negatively charged nuclei, composed of antiprotons and antineutrons, surrounded by positively charged positrons (figure 32-9). Its properties would be identical with those of ordinary matter in almost all respects, and it would be difficult to distinguish between them. Of course, if antimatter came into contact with ordinary matter, they would annihilate one another and there would be an explosive release of large amounts of energy in the form of photons and pions. However, it is conceivable that some isolated regions of the universe are composed exclusively of antimatter. We may well be

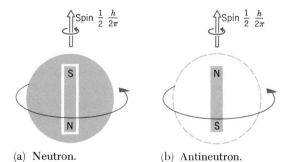

(a) Neutron.   (b) Antineutron.

Figure 32-8 If a neutron and an antineutron spin in the same direction, their magnetic moments are in opposite directions.

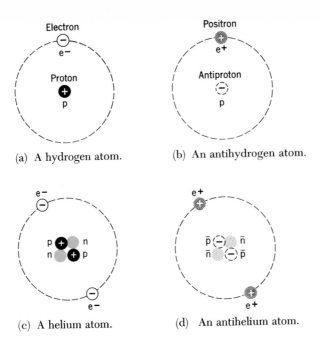

(a) A hydrogen atom.   (b) An antihydrogen atom.

(c) A helium atom.   (d) An antihelium atom.

Figure 32-9   Matter and antimatter.

able to see some of these regions, but their visual appearance would be exactly the same as if they were composed of ordinary matter.

32-5   Classification of the Known Fundamental Particles

The proliferation of the fundamental particles has proceeded to the point where, at the time this is being written, there are more than one hundred of them! It seems unreasonable to regard them all as fundamental, but there is no reason as yet to believe that any one of them is necessarily less fundamental than the original two, the electron and the proton. It is conceivable that a few of them are truly fundamental and the rest are really composite structures made from these few. It is also conceivable that they are all composite structures made from fundamental constituents which have not yet been discovered. On the other hand, it may be a mistake to picture them as composed of anything, in the sense that a house is composed of bricks. They may be different "forms" of matter, which are to be described and distinguished from one another only by quoting for each one such characteristic properties as electric charge, intrinsic angular momentum, and any other properties which turn out to be basic to their nature. A somewhat analogous case in plane geometry might be the sequence of regular polygons, starting with the equilateral triangle, the square, the pentagon and the hexagon. The equilateral triangle is simpler, but not necessarily more "fundamental" than the others. Moreover, although the hexagon can be built up out of equilateral triangles, the square and the pentagon cannot. Although we have a very incomplete understanding of these particles,

THE SEARCH
FOR THE
ULTIMATE
CONSTITUENTS

some progress has been made in classifying them and introducing order into their properties. This can be explained most easily by restricting the discussion to the "metastable" particles. In this context, the word "metastable" implies that the half-life of the particle is much longer than a certain characteristic time, which is actually the time taken for light to travel across the diameter of a fundamental particle. Since this diameter is about 10^{-13} cm, the characteristic time is $10^{-13}/(3 \times 10^{10}) = 3.3 \times 10^{-24}$ sec. Table 32-1 lists all the known particles that have half-lives at least one hundred times longer than this characteristic time. The table also includes some of their more important properties and shows some of the categories into which they can be classified. Notice that the rest mass steadily increases from the top of the list downward.

The first broad division is into **particles** and **antiparticles.** Every particle has a corresponding antiparticle. In the case of the three particles shown surrounded by a circle, the photon, the neutral pion and the eta-meson, their properties are such that the particle and antiparticle are identical in all respects and cannot be distinguished from one another. (If -1, -2, -3 etc. are called the *antinumbers* of $+1$, $+2$, $+3$, etc., then 0 is indistinguishable from its antinumber.) The electric charge on a metastable fundamental particle is always $+e$, 0 or $-e$. If the particle is charged, the antiparticle has a charge of the opposite sign. Notice the interesting case of the sigma-particle, with three particles, Σ^+, Σ^0 and Σ^- and three distinguishably different antiparticles $\overline{\Sigma}^-$, $\overline{\Sigma}^0$ and $\overline{\Sigma}^+$. To obtain a feeling for how this might occur, consider a neutral helium atom with two electrons, a positive helium ion with one electron and a negative helium ion with three electrons. The corresponding structures of antimatter would be an antihelium atom with two positrons, a negative antihelium ion with one positron and a positive antihelium ion with three positrons.

The second broad division is into **fermions** and **bosons.** A fermion has a half-integral spin ($\frac{1}{2}$, $\frac{3}{2}$, $\frac{5}{2}$ etc) and obeys the Pauli exclusion principle. Two fermions of the same kind cannot be in the same state. A boson has zero or integral spin (0, 1, 2 etc), and is not restricted by the Pauli exclusion principle. Two bosons of the same kind are allowed to be in the same state.

The fundamental particles are conveniently divided into four groups: (1) the Photon, (2) Leptons, (3) Mesons, and (4) Baryons. Mesons and baryons are sometimes grouped together and called **hadrons.** We are already familiar with the photon. It is the particle associated with the electromagnetic field. It has no charge and no rest mass and moves with the velocity of light. Its angular momentum is 1 unit of $h/2\pi$, so it is a boson and does not obey the Pauli exclusion principle. A separate section will be devoted to each of the other three groups.

32-6 Leptons

Leptons all have a spin of $\frac{1}{2}$ a unit and are therefore fermions, obeying the Pauli exclusion principle. The best-known members of the group are

Table 32-1 The metastable particles

	Name	Particles Charge +e	0	−e	Antiparticles Charge +e	0	−e	Rest mass (Units such that rest mass of electron = 1)	Spin. Units of $\frac{h}{2\pi}$
Photon	photon		γ			γ		0	1 (Boson)
Leptons	neutrino, antineutrino		ν			$\bar{\nu}$		0	½ (Fermions)
	μ-neutrino, anti-μ-neutrino		ν_μ			$\bar{\nu}_\mu$		0	
	electron, positron			e^-	e^+			1	
	muons			μ^-	μ^+			207	
Mesons	π-mesons (pions)	π^+	π^0			π^0	π^-	264 273	0 (Bosons)
	K-mesons (Kaons)	K^+	K^0			$\overline{K^0}$	K^-	966 974	
	η-meson (eta)		η^0			η^0		1074	
Hadrons — Baryons	proton, antiproton	p					$\bar{p}$	1836.1	½ (Fermions)
	neutron, antineutron		n			$\bar{n}$		1838.6	
	Lambda		Λ^0			$\overline{\Lambda^0}$		2183	
	Sigma	Σ^+					$\overline{\Sigma}^-$	2328	
			Σ^0			$\overline{\Sigma}^0$		2334	
				Σ^-	$\overline{\Sigma}^+$			2343	
	Xi		Ξ^0			$\overline{\Xi}^0$		2573	
				Ξ^-	$\overline{\Xi}^+$			2586	
	Omega			Ω^-	$\overline{\Omega}^+$			3276	3/2

the electron, the positron, the neutrino, and the antineutrino. In addition there are two particles μ^- and μ^+, which were originally called μ-mesons, but are more appropriately called **muons,** since they are not really members of the group of mesons. Associated with these muons are two very recently discovered particles, ν_μ and $\bar{\nu}_\mu$, which are very similar to neutrinos and which will be called μ-**neutrinos.**

The properties of the muons, μ^- and μ^+, are identical with the properties of the electron and positron, e^- and e^+, in all except two respects. A muon has a rest mass which is equal to 207 electron rest masses, and it is therefore sometimes described as a "heavy electron." In addition, a muon has a half-life of only 2.2×10^{-6} sec, for it very soon decays into an electron, a μ-neutrino, and an ordinary neutrino.

$$\mu^- \to e^- + \bar{\nu} + \nu_\mu \tag{32-22}$$

$$\mu^+ \to e^+ + \nu + \bar{\nu}_\mu \tag{32-23}$$

One reason for grouping together electrons, muons and neutrinos is the **law of conservation of leptons.** For the purpose of this law, each of the particles ν, ν_μ, e^- and μ^- is to be counted as $+1$ lepton, whereas each of the antiparticles $\bar{\nu}$, $\bar{\nu}_\mu$, e^+ and μ^+ is to be counted as -1 lepton. The law then states that a process in which fundamental particles are transformed into other fundamental particles can occur only if it leaves the total number of leptons unaltered. The number of leptons in the universe remains constant, provided that the sum of a particle and an antiparticle is counted as zero.

For example, in equation 32-22 the original negative muon, μ^-, counts as $+1$ lepton. Among its products, e^- counts as $+1$, $\bar{\nu}$ as -1, ν_μ as $+1$, and the law is satisfied because $+1 = +1 - 1 + 1$. Similarly, in equation 32-23 μ^+ counts as -1, e^+ as -1, ν as $+1$, and $\bar{\nu}_\mu$ as -1, so $-1 = -1 + 1 - 1$. In the fundamental equation of β-decay,

$$n \to p + e^- + \bar{\nu} \tag{32-24}$$

the original neutron, n, is not a lepton and counts as 0. The proton, p, also counts as 0, while the electron, e^-, counts as $+1$ and the antineutrino, $\bar{\nu}$ as -1. The law is then satisfied because $0 = 0 + 1 - 1$. This is the reason for establishing the convention that the neutrino accompanying electron-emission is the *anti*-neutrino. For positron-emission,

$$p \to n + e^+ + \nu \tag{32-25}$$

the count of leptons gives $0 = 0 - 1 + 1$.

It seems probable that there is also a **law of conservation of μ-leptons,** in which μ^- and ν_μ are counted as $+1$ and μ^+ and $\bar{\nu}_\mu$ as -1. For example, in equation 32-22 we start with $+1$ μ-lepton, μ^-, and the only μ-lepton among the products is ν_μ, which also counts as $+1$. This law excludes, for example, the process

$$\mu^- \to e^- + \bar{\nu} + \nu \tag{32-26}$$

which conserves leptons of both types, but does not conserve μ-leptons by themselves. There is initially $+1$ μ-lepton, but there are no μ-leptons among the products.

32-7   Mesons

Mesons have no spin, are bosons and are not restricted by the Pauli exclusion principle. Any number of mesons may be in the same state. The word "meson" was originally intended to apply to all particles with rest masses intermediate between the rest mass of the electron and the rest mass of the proton, but the μ-meson is now known to be a fermion, is classified with the leptons and is preferably called the muon. This leaves three categories of metastable mesons which are bosons. They are π-mesons or **pions**, K-mesons or **kaons**, and the η-meson or **eta**.

There are three pions. The neutral pion, π^0, has a rest mass 264 times the electron rest mass and is its own antiparticle. The negative pion, π^-, is the antiparticle of the positive pion, π^+, and each has a rest mass of 273 electron rest masses, slightly more than π^0. Pions are produced by bombarding nucleons with high energy protons, when the following reactions occur:

$$p + p \rightarrow p + p + \pi^0 \qquad (32\text{-}27)$$

$$p + p \rightarrow p + n + \pi^+ \qquad (32\text{-}28)$$

$$p + n \rightarrow p + p + \pi^- \qquad (32\text{-}29)$$

Since the rest mass energy of a pion is about 140 MeV, the incident proton must have a kinetic energy somewhat in excess of this. Pions are also produced when an antiproton annihilates with a proton, or an antineutron annihilates with a neutron. For example,

$$p + \bar{p} \rightarrow \pi^+ + \pi^0 + \pi^- \qquad (32\text{-}30)$$

$$n + \bar{n} \rightarrow \pi^+ + \pi^0 + \pi^0 + \pi^- \qquad (32\text{-}31)$$

Any number of pions may be produced in these reactions, as long as charge is conserved.

The neutral pion has a half-life of only about 10^{-16} sec and decays into two photons

$$\pi^0 \rightarrow \gamma + \gamma \qquad (32\text{-}32)$$

A charged pion has a longer half-life of 2.6×10^{-8} sec and usually decays into a muon and a μ-neutrino

$$\pi^+ \rightarrow \mu^+ + \nu_\mu \qquad (32\text{-}33)$$

$$\pi^- \rightarrow \mu^- + \bar{\nu}_\mu \qquad (32\text{-}34)$$

Notice that these processes conserve μ-leptons. Figure 32-10 is a photograph of a pion decaying into a muon, which subsequently decays into a positron (equation 32-23). Direct decay of a charged pion into an electron and an ordinary neutrino is also possible, but less likely

$$\pi^+ \rightarrow e^+ + \nu \qquad (32\text{-}35)$$

$$\pi^- \rightarrow e^- + \bar{\nu} \qquad (32\text{-}36)$$

These processes clearly conserve leptons.

Pions interact strongly with nucleons and are responsible for nuclear forces, as will be explained in section 33-1. They have been called "the glue which holds the nucleus together."

The positive kaon, K^+, and its negative antiparticle, K^-, have rest masses 966 times the electron rest mass. The neutral kaon, K^0, is dis-

Figure 32-10 The decay of a pion into a muon, which subsequently decays into a positron.

$$\pi^+ \rightarrow \mu^+ + \nu_\mu$$

$$\mu^+ \rightarrow e^+ + \nu + \bar{\nu}_\mu$$

When a charged particle passes through a photographic emulsion the grains along its path are activated and turn black after development. The pion enters at A and disintegrates at B. The resulting muon travels along BC and changes into a positron, which leaves the track CD. The neutrinos are uncharged and react very weakly with matter, so they do not leave tracks. However, it is very obvious at B and C that the charged decay product has a velocity in a different direction from the decaying particle. If momentum is to be conserved, an unseen particle must also be produced by the decay. (Courtesy of C. F. Powell.)

tinct from its antiparticle, $\overline{K^0}$, and each has a rest mass of 974 electron rest masses, being slightly heavier than the charged kaons. A typical reaction producing a kaon is

$$\pi^- + p \rightarrow \Lambda^0 + K^0 \qquad (32\text{-}37)$$

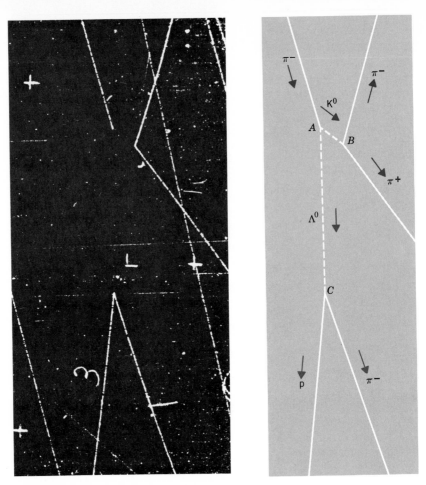

Figure 32-11 Bubble chamber photograph of the reactions

(A) $\pi^- + p \rightarrow \Lambda^0 + K^0$

(B) $K^0 \rightarrow \pi^+ + \pi^-$

(C) $\Lambda^0 \rightarrow \pi^- + p$

K^0 and Λ^0 are uncharged and leave no tracks. (Courtesy of Lawrence Radiation Laboratory, University of California, Berkeley.)

In the bubble chamber photograph of figure 32-11, this process is shown occurring at the point A.

Charged kaons have a half-life of 1.2×10^{-8} sec and can decay in many different ways

$$K^+ \rightarrow \mu^+ + \nu_\mu \tag{32-38}$$

$$K^+ \rightarrow \pi^+ + \pi^0 \tag{32-39}$$

$$K^+ \rightarrow \pi^+ + \pi^+ + \pi^- \tag{32-40}$$

$$K^+ \rightarrow \pi^+ + \pi^0 + \pi^0 \tag{32-41}$$

$$K^+ \rightarrow e^+ + \pi^0 + \nu \tag{32-42}$$

$$K^+ \rightarrow \mu^+ + \pi^0 + \nu_\mu \tag{32-43}$$

The decay equations for K⁻ may be obtained from the above by changing each particle into its antiparticle. A photograph of the decay mode given in equation 32-40 is shown in figure 32-12. The decay of the neutral kaon will not be treated in full detail since it involves certain complications, the discussion of which would divert us too far from our main theme. One of the possible decay modes

Figure 32-12   The decay of a kaon into three pions

$$K^+ \rightarrow \pi^+ + \pi^- + \pi^+$$

The kaon enters at A and decays at B into three pions producing the tracks BC, BD and BE. At C one of the pions interacts strongly with a nucleus and disintegrates it into two fragments. (R. Brown, U. Camerini, P. H. Fowler, H. Muirhead, C. F. Powell, and D. M. Ritson.)

$$K^0 \to \pi^+ + \pi^- \tag{32-44}$$

is shown occurring at point B in figure 32-11.

The η-meson is the most recently discovered of the group. It is uncharged and is its own antiparticle. Its rest mass is 1074 times the rest mass of an electron. Its dominant decay mode is into two photons with a lifetime in the vicinity of 10^{-18} sec. It can also decay into three pions.

A careful perusal of the equations in this section will reveal that there is no conservation law for mesons analogous to the law of conservation of leptons. In equation 32-27, for example, we start with no mesons and end up with one meson. In equation 32-33 a meson disappears and only leptons appear in its place. This is also true of the other boson, the photon. There are many processes in which photons are created or destroyed. When an atom emits light, a photon is created. In the photoelectric effect a photon disappears. During pair production (equation 32-15) a photon disappears and during an annihilation process (equation 32-12) photons are created.

32-8 Baryons

All baryons have half-integral spin, are fermions and obey the Pauli exclusion principle. The best-known baryons are the nucleons—the proton, the antiproton, the neutron, and the antineutron. The other members of the group are the lambdas, the sigmas, the xis, and the omegas, all of which are sometimes called **hyperons,** because their rest masses are greater than the rest masses of the nucleons.

As in the case of leptons, there is a **law of conservation of baryons.** Particles count as $+1$ baryon, antiparticles as -1 baryon and particles outside the group as 0. In the now familiar equation for β-decay,

$$n \to p + e^- + \bar{\nu} \tag{32-45}$$

the original neutron counts as $+1$ baryon. Among the products, the proton is $+1$ baryon and the electron and neutrino are both leptons and count as 0 baryons. In the equation describing the method of producing antineutrons,

$$p + p \to p + p + (n + \bar{n}) \tag{32-46}$$

p and n each count as $+1$, while $\bar{n}$ counts as -1. So we start with $+2$ baryons and end up with $3 - 1$ baryons.

The only two **lambda-particles** are the neutral Λ^0 and its antiparticle, $\overline{\Lambda^0}$, each with a rest mass of 2183 electron rest masses. An example of a reaction producing lambdas is an annihilation mode of a proton and an antiproton,

$$p + \bar{p} \to \Lambda^0 + \overline{\Lambda^0} \tag{32-47}$$

On each side of this equation there is $+1$ -1 baryons. The lambda has a half-life of 2.5×10^{-10} sec and decays in two ways

$$\Lambda^0 \to p + \pi^- \tag{32-48}$$
$$\Lambda^0 \to n + \pi^0 \tag{32-49}$$

On each side of these equations there is $+1$ baryon. The half-life of the antilambda, $\overline{\Lambda^0}$, is, of course, also 2.5×10^{-10} sec. Its decay modes are obtained by changing particles into antiparticles

$$\overline{\Lambda^0} \to \bar{p} + \pi^+ \tag{32-50}$$
$$\overline{\Lambda^0} \to \bar{n} + \pi^0 \tag{32-51}$$

The processes of equations 32-47, 32-48, and 32-50 are shown in figure 32-13.

Λ^0 is similar in many ways to a neutron and may replace a neutron in a nucleus, producing what is called a **hypernucleus.** Such a nucleus lasts for only about 10^{-10} sec before the Λ^0 decays, but this is a long time by nuclear standards.

The **sigma-particles** are interesting because there are six of them, all different. Since there can be some confusion about the notation, it should be made clear that $\overline{\Sigma^-}$ means the antiparticle of Σ^+, and is sometimes written $\overline{\Sigma^+}$, which tends to obscure the fact that it is negatively charged. We shall discuss only the particles. The reader should have no difficulty in deducing the corresponding properties of the antiparticles. The lightest is Σ^+, with a rest mass of 2328 electron rest masses, and a half-life of 0.8×10^{-10} sec, the decay modes being

$$\Sigma^+ \to p + \pi^0 \tag{32-52}$$
$$\Sigma^+ \to n + \pi^+ \tag{32-53}$$

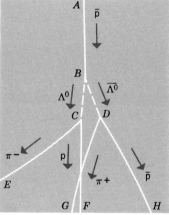

Figure 32-13 An antiproton enters at A and annihilates a proton at rest at B, producing two lambdas

$$p + \bar{p} \to \Lambda^0 + \overline{\Lambda^0}$$

The Λ^0 decays at C to produce a negative pion CE and a proton CF

$$\Lambda^0 \to \pi^- + p$$

The $\overline{\Lambda^0}$ decays at D to produce a positive pion DG and an antiproton DH

$$\overline{\Lambda^0} \to \pi^+ + \bar{p}$$

(Courtesy of Lawrence Radiation Laboratory, University of California, Berkeley.)

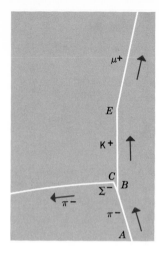

Figure 32-14   A negative pion enters at A and hits a proton at rest at B, producing a negative sigma and a positive kaon

$$\pi^- + p \rightarrow K^+ + \Sigma^-$$

At C the sigma decays into a negative pion and a neutron, which does not leave a track,

$$\Sigma^- \rightarrow \pi^- + n$$

At E the kaon decays into a muon and a μ-neutrino, which leaves no track,

$$K^+ \rightarrow \mu^+ + \nu_\mu$$

(Courtesy of Lawrence Radiation Laboratory, University of California, Berkeley.)

Σ^0 has a rest mass of 2334 electron rest masses and a half-life of about 10^{-20} sec, decaying very rapidly in the following way

$$\Sigma^0 \rightarrow \Lambda^0 + \gamma \tag{32-54}$$

Σ^- has a rest mass of 2343 electron rest masses and a half-life of 1.6×10^{-10} sec, decaying in the following way

$$\Sigma^- \rightarrow n + \pi^- \tag{32-55}$$

This last process is shown in figure 32-14.

There are four **xi-particles**, all with rest masses about 2580 times the rest mass of an electron. They are Ξ^0 and its antiparticle $\overline{\Xi^0}$, and Ξ^- and its antiparticle Ξ^+. They have lifetimes of about 10^{-10} sec and decay into a lambda and a pion,

$$\Xi^0 \rightarrow \Lambda^0 + \pi^0 \tag{32-56}$$

$$\Xi^- \rightarrow \Lambda^0 + \pi^- \tag{32-57}$$

with corresponding equations for the antiparticles.

The **omega-particle** has a very special significance, because its existence was predicted before its discovery. The theory which made this successful prediction is known as "the eightfold way." It is a method of dividing baryons and mesons into groups according to the values of their

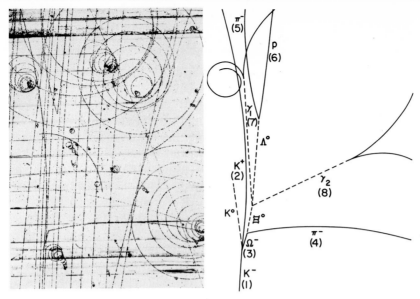

Figure 32-15 A K⁻ enters from the bottom (track 1) and reacts with a stationary proton (no track) to form a K⁺ (track 2), a K⁰ (no track), and a Ω⁻ (track 3). The Ω⁻ travels for a short distance in about 10^{-11} sec and then decays into a Ξ⁰ (no track) and a π⁻ (track 4). The Ξ⁰ decays into a Λ⁰ (no track) and a π⁰ (no track). The π⁰ decays almost immediately into 2 photons, γ₁ and γ₂, which eventually create electron-positron pairs. The Λ⁰ eventually decays into a p (track 6) and a π⁻ (track 5). The uncharged particles which leave no track in the bubble chamber are represented by broken lines in the explanatory diagram to the right of the photograph. (Courtesy of Brookhaven National Laboratories.)

spin. The metastable baryons discussed so far (p, n, Λ, Σ and Ξ) all have a spin of ½ and constitute one of these groups. The years from 1961 to 1964 saw the discovery of many new unstable particles with short lifetimes of about 10^{-23} sec. Some of these particles had a spin of ³⁄₂ and therefore formed another group, which however lacked one member. This missing baryon was expected to exist only as a negatively charged particle, Ω⁻, and a positively charged antiparticle, $\overline{\Omega}^+$. Its mass was predicted to be in the range between 3280 and 3288 electron masses. When it was discovered in 1964, its measured mass was found to be somewhere in the range between 3276 and 3325 electron masses, which includes the predicted range. The first bubble chamber photograph of the omega minus is shown in figure 32-15. Its half-life is 1.5×10^{-10} sec and it decays into a Ξ⁰ and a π⁻

$$\Omega^- \rightarrow \Xi^0 + \pi^- \tag{32-58}$$

A Profusion of Particles and Processes

QUESTIONS

1. "Just before falling under the death-rays of the Martians' guns, the scientist has time to scribble down part of his secret formula

$$_{151}X^{362} \rightarrow {}_{150}Y^{362} + \cdots"$$

Can you guess what type of process he is describing and fill in the missing part?

2. Refer to figure 30-5 and decide whether the following nuclei are more likely to emit electrons or positrons.

(a) $_{55}Cs^{145}$, (b) $_{80}Hg^{192}$, (c) $_{38}Sr^{81}$, (d) $_{36}Kr^{94}$, (e) $_{20}Ca^{49}$.

3. From the atomic masses quoted on the inside of the cover, can you decide whether tritium, $_{1}T^{3}$, is able to emit an electron? It does in fact emit an electron with a maximum kinetic energy of 0.019 MeV. Is this consistent with the atomic masses quoted? What can you deduce from the maximum energy of the emitted electron? (Hint: The figures quoted are the *atomic* masses, not the *nuclear* masses.)

4. Discuss conservation of angular momentum for the following reactions

(a) $e^+ + e^- \rightarrow 2\gamma$

(b) $\gamma \rightarrow e^+ + e^-$

(c) $\bar{\nu} + p \rightarrow n + e^+$

5. The particles e^+, e^-, p and $\bar{p}$ all have magnetic moments and are equivalent to small bar magnets. In each case consider whether the angular momentum vector points from the south pole toward the north pole, or in the opposite direction.

6. A new particle, the subiota, and its antiparticle, the antisubiota, are produced in the following way. X-rays from a tube with 10,000 volts across it travel an average distance of 10^5 cm along a partially evacuated tube and then create subiota pairs. On the average the subiotas return to the x-ray tube 2×10^{-4} sec after the x-ray pulse is sent out. How much can you deduce about the properties of the subiota from this information?

7. Rewrite equations 32-38 to 32-43 with all particles replaced by their antiparticles.

8. For each of the following equations, state which conservation laws are violated.

(a) $\mu^+ \rightarrow e^+ + \nu + \gamma$

(b) $\pi^+ \rightarrow \mu^+ + \nu_\mu + \bar{\nu}$

(c) $n \rightarrow p + \bar{p} + \pi^0$

(d) $\Sigma^+ \rightarrow \Lambda^0 + \bar{p} + \pi^0$

(e) $p \rightarrow \Sigma^+ + K^0$

9. In inverse beta-decay an electron is captured by a nucleus (equation 32-3). If a negative muon were captured by a nucleus in a similar way, what would the process be?

10. Since beta-decay produces an antineutrino (equation 32-1), why does inverse beta-decay produce a neutrino (equation 32-3)? Formulate a rule for "transferring a particle to the other side of the equation." Does

your rule apply equally to bosons and fermions? Does your rule ever conflict with any of the conservation laws?

11. For each of the following descriptions make a list of all the particles you know which satisfy this description:
    (a) The antiparticle of the photon.
    (b) A lepton moving with the speed of light.
    (c) The meson of smallest rest mass.
    (d) A "heavy electron"
    (e) A baryon with a rest mass less than 1,200 MeV.

12. Well up beyond the tropostrata
    There is a region stark and stellar
    Where, on a streak of anti-matter,
    Lived Dr. Edward Anti-Teller.

    Remote from Fusion's origin,
    He lived unguessed and unawares
    With all his anti-kith and kin
    And kept macassars on his chairs.

    One morning, idling by the sea,
    He spied a tin of monstrous girth
    That bore three letters: A.E.C.
    Out stepped a visitor from Earth.

    Then, shouting gladly o'er the sands,
    Met two who in their alien ways
    Were like as lentils. Their right hands
    Clasped, and the rest was gamma rays.

    H. P. F.
    (Copr. © 1956 The New Yorker Magazine, Inc.)

    Do the last two words of this poem provide a complete description of the ultimate remnants of the encounter?

PROBLEMS

A

1. What is the wavelength of the lowest energy photon capable of producing a proton-antiproton pair? (Ignore the question of conservation of linear momentum, and assume that the proton and antiproton can be created with zero velocities.)

2. A positron and electron with negligible kinetic energy annihilate one another to produce two photons. What are their frequencies?

3. A 1 MeV electron encounters a 1 MeV positron traveling in exactly the opposite direction. What are the wavelengths of the two photons produced?

4. An iota has a rest mass equal to 50 electron rest masses. What is the frequency of a photon which is just able to produce an iota anti-iota pair? (Ignore the correction caused by the need to conserve momentum.)

A Profusion
of Particles
and Processes

709

5. Calculate the energy in ergs and electron volts which is released when a helium atom encounters an antihelium atom. Include the energy of all the products.

6. If an η-meson at rest decays into two photons, calculate their wavelengths.

7. Is it energetically possible for an η-meson to decay into (a) four pions (which four?), (b) five pions?

B

8. 10^{-12} gm of antihelium is introduced into 1 mole of helium gas. Put an upper limit on the resulting rise in temperature. Why will the actual rise in temperature be less than this?

9. What is the momentum of a photon which has just sufficient energy to create an electron-positron pair? Calculate the velocity of an electron with an equal momentum. What is the kinetic energy of this electron in electron volts?

10. What is the maximum kinetic energy in ergs and in electron volts of an electron produced by the β-decay of a free neutron at rest?

11. What is the minimum energy which the antineutrino must have in equation 32-17, if the proton is free and at rest?

12. A newly discovered particle and its antiparticle come slowly together and produce two photons. One of these photons creates an electron-positron pair. Both the electron and the positron have velocities of $0.99c$. Calculate the rest mass of the new particle.

13. Calculate the total kinetic energy in ergs and in electron volts of the products of the following decays, assuming that the decaying particle is at rest

(a) $\mu^+ \rightarrow e^+ + \nu + \bar{\nu}_\mu$

(b) $\pi^+ \rightarrow \mu^+ + \nu_\mu$

(c) $K^+ \rightarrow \mu^+ + \pi^0 + \nu_\mu$

(d) $\Lambda^0 \rightarrow p + \pi^-$

(e) $\Sigma^0 \rightarrow \Lambda^0 + \gamma$

(f) $\Omega^- \rightarrow \Xi^0 + \pi^-$

C

14. Show that, in the absence of a nearby nucleus, the reaction

$$\gamma \rightarrow e^+ + e^-$$

cannot possibly conserve both energy and momentum. How does the nearby nucleus save the situation?

15. Calculate as precisely as you can the energy in electron volts released in the fusion process of equation 31-40 when four protons fuse to form a helium nucleus. Use the atomic masses given on the inside of the cover and remember that they are not nuclear masses. Include any energy that might result from the subsequent fate of the positrons. Explain why all of the energy released is not necessarily

available to convert into useful power? Calculate the maximum amount that might not be available. What further information would you need in order to calculate more precisely the actual available energy in a macroscopic process.

16. Light takes about 10^{10} years to travel across the observable universe. The average density of matter in the whole of the universe probably lies somewhere between 10^{-30} gm/cm³ and 10^{-28} gm/cm³. Does a neutrino have a good chance of traveling right across the observable universe without being absorbed?

17. A neutron at rest decays in such a way that the electron and neutrino are emitted at right angles to one another and the proton recoils at an angle of 135° to either. Calculate the recoil velocity of the proton. Proceed by ignoring the kinetic energy of the proton, but not its momentum. When you have calculated the velocity of the proton in this way, show that its kinetic energy is in fact negligible.

18. What must be the temperature of a hot body if most of the photons it emits are just capable of creating a particle of rest mass m_0 in conjunction with its antiparticle? Calculate this temperature for (a) electron-positron pairs, (b) proton-antiproton pairs.

A Profusion
of Particles
and Processes

33

Inklings

33-1 Forces, Interactions, and Fundamental Processes

The long list of "fundamental" particles in the previous chapter is a clear indication that physics has not yet arrived at a simple, elegant description of the universe. Nevertheless, some orderliness is beginning to emerge, and the outlines of a new conception of the basic nature of the universe are beginning to take shape.

The universe of Newtonian mechanics consisted of isolated bodies moving through empty space and exerting instantaneous forces on one another. The classical theory of electromagnetism, achieving its final expression in Maxwell's equations, shifted the emphasis to a field existing at all points in space. Interactions between charged bodies took place through the intermediary of this field and traveled from one body to another with the speed of light. The picture of energy, in the form of an electromagnetic wave, traveling through the field from one body to a distant body created the impression that the field was a very "real" entity. This impression was strengthened by the quantum mechanical description of an electromagnetic wave as a stream of photons. In the list of fundamental particles, the photon has a valid claim to inclusion on an equal footing with the charged particles which exert electromagnetic forces on one another. In fact, the electromagnetic interaction between two charged particles, instead of being described in terms of these forces, may now be visualized somewhat differently in the following way.

In figure 33-1 we consider two interacting electrons, but the argument is applicable to any two charged fundamental particles. The electron e_1 first emits a photon by means of the fundamental process

$$e_1 \rightarrow e_1' + \gamma \tag{33-1}$$

The prime on the symbol e_1' is intended to indicate that this electron has transferred to a different state of motion, giving up energy and momentum to the photon. The photon then travels through space with a velocity c until it reaches the other electron e_2, which absorbs it by means of the fundamental process

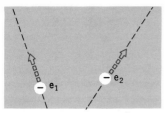

(a) Two electrons, e_1 and e_2.

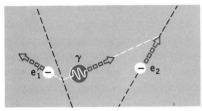

(b) e_1 emits a photon.

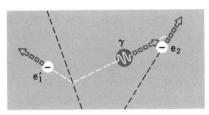

(c) The photon in flight.

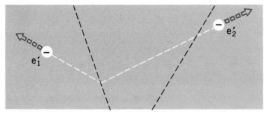

(d) The photon is absorbed by e_2.

Figure 33-1   The electromagnetic interaction between two electrons can be described in terms of the interchange of photons between them.

$$\gamma + e_2 \rightarrow e_2' \tag{33-2}$$

The net result is that the two electrons have exchanged energy and momentum and each has changed its state of motion. In classical terms, the two electrons have "accelerated" one another. As the electrons proceed on their way, a continual stream of photons passes backward and forward between them, maintaining a continuous interaction between them.

The electromagnetic interaction is consequently seen to depend upon the two fundamental processes of equations 33-1 and 33-2. However, there are many different fundamental processes and they are all presum-

Inklings

ably capable of producing interactions in a similar way. We shall now turn to the fundamental processes responsible for nuclear forces.

33-2   Nuclear Forces

Electromagnetic forces are caused by the passing of photons backward and forward between charged particles. Nuclear forces are caused by the passing of *pions* backward and forward between nucleons. Consider a proton and a neutron inside a nucleus (figure 33-2). The proton can emit a positive pion and change into a neutron, the fundamental process being

$$p \rightarrow n + \pi^+ \tag{33-3}$$

This pion then travels to the original neutron, which absorbs it and changes into a proton. This second fundamental process is the reverse of the previous one:

$$\pi^+ + n \rightarrow p \tag{33-4}$$

As a result, the two nucleons have not only exchanged energy and momentum, but the proton and neutron have exchanged places. The neutron which caught the pion, and changed into a proton, then throws it back again and restores the situation to its initial state. The pion is continually thrown backward and forward, producing a continuously acting nuclear force between the two nucleons.

A negative pion, π^-, may also act as carrier of the interaction. The neutron first emits the negative pion and changes into a proton:

$$n \rightarrow p + \pi^- \tag{33-5}$$

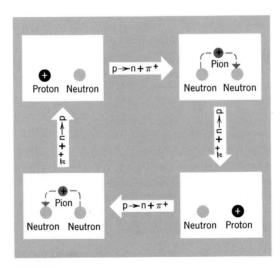

Figure 33-2   The nuclear force between a proton and a neutron is partly due to the passage of a positive pion, π^+, backward and forward between them. Other contributions to the force arise from a similar interchange of negative pions, π^-, or neutral pions, π^0.

The pion is then absorbed by the original proton, which changes into a neutron:

$$\pi^- + p \to n \tag{33-6}$$

There are similar processes involving a neutral pion π^0, but then the neutron and proton do not change places.

$$n \to n + \pi^0 \tag{33-7}$$
$$\pi^0 + p \to p \tag{33-8}$$

The value of the neutral pion is that it can explain the nuclear force between two protons, p_1 and p_2,

$$p_1 \to p_1' + \pi^0 \tag{33-9}$$
$$\pi^0 + p_2 \to p_2' \tag{33-10}$$

or between two neutrons, n_1 and n_2,

$$n_1 \to n_1' + \pi^0 \tag{33-11}$$
$$\pi^0 + n_2 \to n_2' \tag{33-12}$$

Since the rest mass energy of a pion is about 140 MeV, a nucleon must have an excess energy greater than this to be able to emit a pion. Energies this large are not normally available inside a nucleus, so we are faced with the question of how the processes we have been discussing can possibly occur. The answer is provided by the uncertainty principle, which allows us to violate the law of conservation of energy by an amount ΔE for a time Δt, if

$$\Delta t \approx \frac{h}{\Delta E} \tag{33-13}$$

If the energy is uncertain by an amount ΔE, we obviously cannot require that it should remain constant to an accuracy better than ΔE.

If m_π is the rest mass of a pion, the extra energy needed by a nucleon to emit a pion is

$$\Delta E \approx m_\pi c^2 \tag{33-14}$$

This extra energy is available only for the very short time

$$\Delta t \approx \frac{h}{m_\pi c^2} \tag{33-15}$$
$$\approx 3 \times 10^{-23} \text{ sec} \tag{33-16}$$

The velocity with which the pion is emitted is certainly less than c, and so the maximum distance it can hope to escape from the nucleon during the time Δt is

$$R_n = c\,\Delta t \tag{33-17}$$

or, $$R_n \approx \frac{h}{m_\pi c} \tag{33-18}$$

A particle which is produced in this way, without enough energy being available to ensure it a permanent existence, is called a **virtual particle.** The process which produces it is called a **virtual process.** The virtual pion exists only by the grace of the uncertainty principle, and it must return to its parent nucleon before its time runs out. The situation can be described more precisely in terms of probability. There is a good chance that the pion will be emitted, travel a distance less than R_n, and then return. There is even a small chance that it will travel much further than R_n, but this is a rare occurrence and becomes less and less likely the further the pion travels.

During its journey the virtual pion may encounter another nucleon and be absorbed by it. In this case an exchange has taken place and we are discussing the process responsible for nuclear forces. The exchange is likely to occur only when the distance between the two nucleons is not much greater than $R_n = 0.89 \times 10^{-12}$ cm. This explains the fact that nuclear forces are short range forces and become very weak when the distance between the two nucleons exceeds a range of the order of 10^{-12} cm.

33-3   The Nature of a Nucleon

This theory provides us with a very interesting picture of an isolated nucleon. It must be imagined as continually emitting a pion and snatching it back again. A proton, or neutron, is surrounded by a cloud of virtual pions, most of which are within 10^{-13} cm of the center, but a few stray out to 10^{-12} cm or further. For an appreciable fraction of the time a proton is split up into a neutron and a positive pion,

$$\text{p} \rightleftarrows \text{n} + \pi^+ \tag{33-20}$$

Similarly, for an appreciable fraction of its existence a neutron consists of a proton and a negative pion,

$$\text{n} \rightleftarrows \text{p} + \pi^- \tag{33-21}$$

This picture helps us to understand the magnetic properties of protons and neutrons. By analogy with the electron (section 29-5), the magnetic moment of a proton would be expected to be one nuclear magneton,

$$\beta_n = \frac{eh}{4\pi m_\text{p} c} \tag{33-22}$$

where m_p is the mass of the proton. The measured value of the proton's magnetic moment is 2.79 β_n. The suggested explanation is that the proton is really a proton for only part of the time. The rest of the time it is a neutron plus a positive pion and this arrangement has a different magnetic moment.

The situation is even more dramatic in the case of the neutron. Since

Neutron

(a) Only part of the time is the neutron
    really a single uncharged particle.

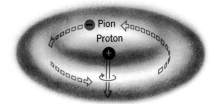

Pion
Proton

(b) For part of the time it is a proton surrounded by a doughnut-shaped rotating
    cloud of negative pions (compare figure 29-5c and the associated discussion of
    clouds of probability).

Figure 33-3   The explanation of the neutron's magnetic moment.

magnetic moments are caused by rotating electric charges, the uncharged
neutron might be expected not to have a magnetic moment at all. In fact,
it is found to have a magnetic moment of 1.91 β_n. The suggested
explanation is that, for part of the time, the neutron consists of a proton
with a doughnut-shaped cloud of negative pions rotating around it
(figure 33-3).

33-4   Agitated Chaos

In place of the old concept of force, mysteriously acting at a distance,
we now have a very different picture of two bodies exerting forces on
one another by throwing particles backward and forward between
themselves. In the case of an electromagnetic force, the particle thrown
backward and forward is the photon. In the case of a nuclear force, it is
the pion. The force on a particle is defined classically as the momentum
transferred to it per second, and this momentum is now seen to come
from the photons or pions which the particle absorbs or emits.

There is a similar change in our concept of the nature of a field.
Previously we have considered a field to be a peculiar property of a
point in space which enables it to exert a force on a particle placed there.
For example, the electric field **E** at a point is such that an electron
placed at that point experiences a force $-e\mathbf{E}$. We can now form a much
more vivid picture of what is happening. The electron placed at the
point experiences an electromagnetic force because it absorbs some of
the real and virtual photons already present in the vicinity of the point.

The field **E** at a point can be considered to be caused by the presence
of charged particles at distant points. Each of these charged particles is
continually emitting real and virtual photons. A real photon corresponds
to an electromagnetic wave radiated outward, never to return. A virtual

Inklings

photon, on other hand, acquires its energy because the Uncertainty Principle allows a temporary violation of the law of conservation of energy, but it must eventually return to the charge that emitted it. The electromagnetic field surrounding a charged particle may therefore be visualized as a cloud of photons increasing in density as the particle is approached. (The reader should compare section 28-1, where the oscillating electric vector of an electromagnetic wave was related to the density of photons in space.)

We thus obtain a very peculiar picture of apparently empty space. A small volume of space contains virtual photons coming from nearby charges and virtual pions coming from nearby nucleons. It also contains other virtual particles related to other types of interaction and their corresponding fundamental processes. However, there is even more. If we look at the small volume for a short time Δt, the Uncertainty Principle may allow it to have an extra energy $\Delta E \sim h/\Delta t$. If Δt is very small, this energy may be large enough to create an electron-positron pair or even a proton-antiproton pair or some other allowed combination of particles. The pair must, of course, annihilate again before its time runs out. Space is therefore in a state of agitated chaos, full of virtual particles and virtual pairs, which are continually being created and almost immediately annihilated.

Potential energy can also be viewed in a more revealing light. Consider an isolated charged particle at rest surrounded by its cloud of virtual photons and concentrate on a "snapshot" taken at a fixed instant of time. Each virtual photon has an energy $h\nu$, and the sum of the quantities $h\nu$ for all the photons present may be looked upon as the potential energy residing in the electric field. An identical argument can be applied to a second charged particle far away from the first. Now bring the two particles together until they are a distance R apart. The electric field is then different, and the distribution of virtual photons is different. The sum of the quantities $h\nu$ is not the same as when the two charged particles were far apart, and the difference is nothing more than the quantity q_1q_2/R, which was previously called mutual electrostatic potential energy. Thus, even the rather nebulous concept of potential energy finds a ready explanation in terms of virtual particles and fundamental processes.

33-5   Types of Interaction

There are very many fundamental processes and consequently many different ways in which fundamental particles can interact. Fortunately, all interactions can conveniently be divided into the four types shown in table 33-1. The classification is based partly upon the nature of the fundamental processes and partly upon the strength of the interaction.

The strength of an interaction depends upon the probability that the associated fundamental process will occur. Consider two nucleons interacting by exchanging pions (figure 33-2). The strength of their interaction is proportional to the frequency with which the pion is

Table 33-1   The four types of interaction

Type of interaction	Relative strength (Coupling constants)	Characteristic time
Strong interactions	1	10^{-23} sec
Electromagnetic interactions	10^{-2}	10^{-21} sec
Weak interactions	10^{-14}	10^{-9} sec
Gravitational interactions	10^{-40}	10^{17} sec (3×10^9 years)

thrown backward and forward. In this case the pion is exchanged about 10^{23} times every second! If the emission of a pion by a nucleon were a very improbable event, occurring say once every second, then nuclear forces would be very much weaker, by a factor of about 10^{23}.

The probability that a fundamental process will occur depends upon many factors, such as the excess energy made available for kinetic energy of the products. Also, some interactions such as nuclear forces are confined to short distances, whereas others, such as electromagnetic forces, can extend over great distances. Nevertheless, when the theory has been formulated in a suitable mathematical form, it is found that, in addition to these factors, the strength of the interaction is proportional to a quantity called the **coupling constant,** which depends only upon the type of interaction. The relative values of the coupling constants are shown in table 33-1.

The strongest interactions are the ones similar to nuclear forces which involve exchange of mesons. They are called **strong interactions.** Next come the **electromagnetic interactions,** which involve the exchange of photons. They are weaker than the strong interactions by a factor of about 100. There is an interesting difference between strong and electromagnetic interactions. Strong interactions have a range somewhat smaller than 10^{-12} cm, beyond which they become extremely weak. This range is related to the rest mass of a pion according to equation 33-18. Since the rest mass of a photon is zero, the corresponding range for electromagnetic forces is infinite. Thus, although electromagnetic forces fall off as $1/R^2$, they have no tendency to decrease very rapidly after a certain distance and are still effective at large distances.

The third category contains the **weak interactions,** with a coupling constant one hundred trillion (10^{14}) times smaller than that of the strong interactions. A fundamental process associated with a *strong* interaction has a typical half-life of 10^{-23} sec. The half-life of a process associated with a *weak* interaction is typically of the order of 10^{-9} sec, longer by the enormous factor 10^{14}. Although the obvious initial reaction to these lifetimes is that they are both unimaginably short, their ratio is the same as the ratio of one hour to the age of the earth.

Two typical weak interactions are β-decay

Inklings

$$n \rightarrow p + e^- + \bar{\nu} \tag{33-23}$$

and the decay of a pion

$$\pi^+ \to \mu^+ + \nu_\mu \tag{33-24}$$

The fourth and final category of interactions contains the very, very weak **gravitational interactions,** weaker than the strong interactions by the stupendous factor of 10^{-40}. It is by no means certain that gravitational forces can be reconciled with quantum mechanics in its present form. In the general theory of relativity gravitational forces are caused by the curvature of space-time (section 25-5). If gravitational interactions are to be included in the present scheme, they must be assumed to be due to the passing backward and forward between two gravitating masses of an undiscovered fundamental particle, which has been called a **graviton.** Since gravitational forces are not short range, but obey the inverse square law like electromagnetic forces, the graviton must have zero rest mass and travel with the speed of light. It is probably a boson, like the pion which is responsible for nuclear forces, or the photon which is responsible for electromagnetic forces. Whereas the pion has a spin of zero and the photon has a spin of 1 unit, the graviton is expected to have a spin of 2 units. In section 32-2 we saw how difficult it is to detect a neutrino. Detection of a graviton is likely to be even worse by a factor of about 10^{26}, which is the ratio of the coupling constants for weak interactions and gravitational interactions. Notice that, if the characteristic time of a fundamental process involving the emission or absorption of gravitons is inversely proportional to the coupling constant, it has the enormously long value of 3×10^9 years, which is comparable with the age of the universe. This is almost certainly not a coincidence. The discussion in section 13-3 revealed that there is an intimate relationship between gravitation and cosmology.

33-6 Strangeness

There seem to be two kinds of weak interaction, which have no obvious connection with one another, but which nevertheless have the same order of magnitude of coupling constant. First there are the processes involving neutrinos, two of which have already been quoted in equations 33-23 and 33-24. Secondly there are processes involving what have come to be called the **strange particles.** These are the kaons, K, and the hyperons, Λ, Σ, Ξ and Ω. Consider, for example, the decay mode of the lambda-particle,

$$\Lambda^0 \to p + \pi^- \tag{33-25}$$

Since this involves the emission of a pion, we might expect it to be a strong interaction with a very short half-life of about 10^{-23} sec. In fact it has a much longer half-life of 2.5×10^{-10} sec and is therefore a weak interaction.

To explain this behavior of the strange particles, it has been found necessary to assign to each of them an integral quantum number, S, which is called its **strangeness.** Table 33-2 includes these strangeness

Table 33-2   Quantum numbers of the metastable hadrons.

To obtain the quantum numbers of the antiparticles, leave I unaltered and merely change the sign of B, I_3, Q, Y, S.

Baryon number B	Particle	Isotopic spin I	Third component of Isotopic spin I_3	Charge number Q	Hyper-charge Y	Strange-ness S
0 Mesons	π^-	1	-1	-1	0	0
	π^0		0	0		
	π^+		$+1$	$+1$		
	K^0	$\frac{1}{2}$	$-\frac{1}{2}$	0	$+1$	$+1$
	K^+		$+\frac{1}{2}$	$+1$		
	η^0	0	0	0	0	0
1 Baryons	n	$\frac{1}{2}$	$-\frac{1}{2}$	0	$+1$	0
	p		$+\frac{1}{2}$	$+1$		
	Λ^0	0	0	0	0	-1
	Σ^-	1	-1	-1	0	-1
	Σ^0		0	0		
	Σ^+		$+1$	$+1$		
	Ξ^-	$\frac{1}{2}$	$-\frac{1}{2}$	-1	-1	-2
	Ξ^0		$+\frac{1}{2}$	0		
	Ω^-	0	0	-1	-2	-3

numbers. This table shows only particles, but the strangeness number of an antiparticle is the same as that of the particle with the sign reversed. For example, the strangeness number of $\overline{K^0}$ is -1. In this, as in most other respects, particle and antiparticle are opposites.

Any process with a half-life of the order of 10^{-23} sec, which therefore gives rise to a strong interaction, is found to conserve strangeness. This is trivial for processes which do not involve strange particles, such as

$$n \rightarrow p + \pi^- \tag{33-26}$$

since a strangeness of 0 is assigned to each particle involved. Consider, however, a process such as

$$\pi^- + p \rightarrow K^0 + \Lambda^0 \tag{33-27}$$

There are no strange particles on the left hand side and so the initial strangeness is zero. The strangeness of K^0 is $+1$ and the strangeness of Λ^0 is -1, giving a total strangeness of 0 on the right hand side also. Strangeness is therefore conserved and this is a strong interaction which proceeds rapidly. On the other hand, the decay of Λ^0,

$$\Lambda^0 \rightarrow p + \pi^- \tag{33-28}$$

does not conserve strangeness. The strangeness of Λ^0 is -1 and both

Inklings

products have a strangeness of 0. This is the reason why this decay is a weak interaction with a comparatively long half-life of 2.5×10^{-10} sec. All processes which do not conserve strangeness are found to be weak interactions which proceed slowly. The reader should consider how this argument applies to the other decay processes discussed in section 32-8.

33-7 Isotopic Spin

The strangeness number S is more than a number chosen deliberately in order to make the idea about conservation of strangeness work. S can be related to other quantum numbers which have a more obvious significance. Two of these, the charge quantum number Q and the baryon number B are easily understood. The isotopic spin quantum number I is a more subtle concept, but it seems to be a valuable clue to some basic aspects of fundamental particles.

The **charge quantum number** Q is $+1$ for a particle with charge $+e$, 0 for a neutral particle, and -1 for a particle with charge $-e$. Amongst the many very short-lived fundamental particles recently discovered, there is some evidence for the existence of particles with $Q = +2$ or -2 and charges $+2e$ and $-2e$. If we wish, we may talk about the charge quantum number of a nucleus and it should be obvious that it is the same thing as its atomic number ($Q = Z$).

The **baryon number** B is the quantity which was introduced in section 32-8 when we were discussing conservation of baryons. For baryons such as p, n, and Σ^+, the baryon number $B = +1$. For their antiparticles $\bar{p}$, $\bar{n}$, and $\bar{\Sigma}^-$, $B = -1$. For the graviton, the photon, all leptons, and all mesons, $B = 0$. The baryon number of a nucleus is clearly the same thing as its mass number A.

The **isotopic spin quantum number** I is related to the fact that the fundamental particles occur in small groups. The members of a small group all have approximately the same mass and similar properties, but they have different electric charges (different values of Q). The nucleons n and p form such a group and, as far as nuclear forces are concerned, a neutron inside a nucleus is exactly equivalent to a proton, the difference becoming apparent only when we consider electrostatic potential energy. Other similar groups are the three sigmas Σ^-, Σ^0, and Σ^+ and the two Xis Ξ^- and Ξ^0. A small group of this kind is called a **charge multiplet**.

The isotopic spin quantum number I is the same for all the members of such a charge multiplet, and its definition is that the number of members of the group is $2I + 1$. For example, there are 2 nucleons, n and p, so the isotopic spin of a nucleon is $\frac{1}{2}$ because $2 \times \frac{1}{2} + 1 = 2$. Since there are 3 sigmas, Σ^-, Σ^0, and Σ^+, the isotopic spin of a sigma is 1. Notice that we do not include the antiparticles when we count the members of the group. If we were to do so, there would be 4 nucleons, $\bar{p}$, $\bar{n}$, n, and p, and 6 sigmas, $\bar{\Sigma}^-$, $\bar{\Sigma}^0$, $\bar{\Sigma}^+$, Σ^-, Σ^0, and Σ^+.

Turning to the mesons, there is only one η^0 and so its isotopic spin $I = 0$. There are 2 kaons, K^0 and K^+, and so $I = \frac{1}{2}$ for all kaons. However, the pions must be approached with special care. An obvious approach might be to say that $I = \frac{1}{2}$ for pions because there are 2 pions, π^0 and π^-, and π^- is the antiparticle of π^+, whereas π^0 is its own antiparticle. This would be incorrect and would invalidate the theoretical usefulness of I. The procedure that works is to take $I = 1$ and say that there are 3 pions, π^-, π^0, and π^+. There are then 3 antiparticles, but the antiparticle of π^- is identical in all respects with π^+, the antiparticle of π^0 is π^0 itself, and the antiparticle of π^+ is identical in all respects with π^-. A comparable case is the sigmas, which also have $I = 1$, but Σ^+ has *strangeness* $S = -1$ and is not identical with $\bar{\Sigma}^+$, which has $S = +1$. This distinction is not relevant to pions which have $S = 0$.

The reason for the peculiar definition of I, utilizing the expression $2I + 1$, is that there is a strong mathematical analogy between isotopic spin and the spin quantum number J associated with angular momentum. Referring back to section 29-3, if a particle has angular momentum $Jh/2\pi$, the angular momentum vector may have $2J + 1$ orientations in space. If the component of its angular momentum along a preferred direction is $m_J h/2\pi$, then m_J may have one of the $2J + 1$ values $-J, (-J + 1)$ $\ldots (J - 1), J$. If $J = \frac{3}{2}$, then m_J may be $-\frac{3}{2}, -\frac{1}{2}, +\frac{1}{2}$, or $+\frac{3}{2}$. If $J = 2$, then m_J may be $-2, -1, 0, +1$, or $+2$.

Similarly, if a charge multiplet has isotopic spin I, then there are $2I + 1$ different particles with different charges. A quantity I_3, called the **third component of isotopic spin**, is analogous to m_J inasmuch as I_3 may have any one of the $2I + 1$ values $-I, (-I + 1) \ldots (I - 1), +I$. Each value of I_3 corresponds to one of the $2I + 1$ different particles. For example, pions have $I = 1$ and the 3 values of I_3 are -1 (π^-), 0 (π^0) and $+1$ (π^+). The nucleons, n and p, have $I = \frac{1}{2}$ and the 2 values of I_3 are $-\frac{1}{2}$ (for the neutron) and $+\frac{1}{2}$ (for the proton).

As we range through the $2I + 1$ values of I_3, we also range through the $2I + 1$ values of Q, but I_3 and Q are not quite the same thing. Figure 33-4 is an attempt to demonstrate that the values of Q are shifted relative to I_3 and the shift is different for different charge multiplets. A new quantum number, the **hypercharge** Y, is defined as twice this shift

$$Y = 2(Q - I_3) \tag{33-29}$$

The strangeness quantum number S is then defined as

$$S = Y - B \tag{33-30}$$

where B is, of course, the baryon number. The reader should study figure 33-4 carefully to see how Y and S are determined in various cases. A little thought will convince you that all members of a charge multiplet have the same Y and the same S.

Table 33-2 lists the values of B, I, I_3, Q, Y, and S for all the metastable hadrons (baryons plus mesons). These quantum numbers are not all independent. A hadron is completely specified if we know B, J, I, I_3, S, and a quantity P called parity, which is concerned with the question of left-handedness and right-handedness (to be discussed in a later sec-

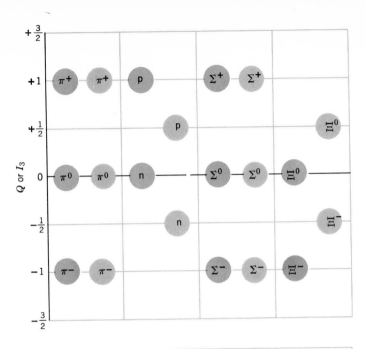

Particles	Pions	Nucleons	Sigmas	Xis
Shift $= (Q - I_3)$	0	$+\frac{1}{2}$	0	$-\frac{1}{2}$
Hypercharge $Y = 2(Q - I_3)$	0	$+1$	0	-1
Baryon number, B	0	$+1$	$+1$	$+1$
Strangeness, $S = Y - B$	0	0	-1	-2

Figure 33-4 The shift of the Q values relative to the I_3 values determines the hypercharge Y and the strangeness S. Green circles give the charge number Q, and black circles give the third component of isotopic spin I_3.

tion). In terms of the above quantum numbers, Y may be deduced from equation 33-30 and then Q may be deduced from equation 33-29.

Table 33-2 is restricted to particles. An antiparticle has the same values of J and I as its particle, but the values of B, I_3, S, Y, and Q are opposite in sign. For example, the hypercharge Y is -1 for $\overline{K^0}$ and $+2$ for $\overline{\Omega^+}$.

33-8 A Little More Orderliness

With the quantum numbers as a guide, considerable success has been achieved in sorting out the fundamental particles into larger groups. The members of a large group all have the same baryon number B, the same spin quantum number J, and the same parity P. Within a large

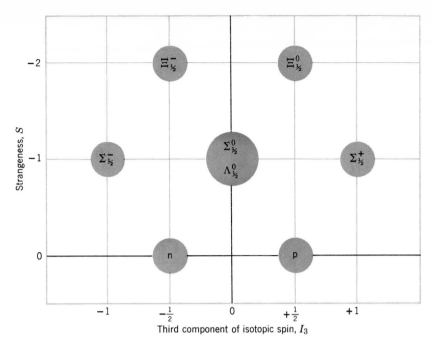

Figure 33-5 The octet of metastable baryons with spin $J = \frac{1}{2}$. $\Sigma_{\frac{1}{2}}^{0}$ and $\Lambda_{\frac{1}{2}}^{0}$ both have $I_3 = 0$ and $S = -1$, but $\Sigma_{\frac{1}{2}}^{0}$ has $I = 1$, whereas $\Lambda_{\frac{1}{2}}^{0}$ has $I = 0$, so they are different particles.

group there are smaller sub-groups, in each of which the particles all have the same I and the same S, but different values of I_3. (These are the charge multiplets of the previous section.)

This is probably best understood by examining figures 33-5 and 33-6. Figure 33-5 shows an octet of metastable baryons with spin $J = \frac{1}{2}$. Figure 33-6 shows a decuplet of baryons with spin $\frac{3}{2}$. All of the particles in this decuplet, with the exception of Ω^-, are short-lived with half-lives of the order of 10^{-23} sec and were therefore not included in table 32-1. For example, the $\Sigma_{1/2}^{+}$ of the octet has a spin of $\frac{1}{2}$ and a half-life of 0.8×10^{-10} sec, so it was included in table 32-1. On the other hand, the $\Sigma_{3/2}^{+}$ of the decuplet, with a spin of $\frac{3}{2}$ and a half-life of the order of 10^{-23} sec, is a kind of "excited state" of $\Sigma_{1/2}^{+}$. (The subscript obviously refers to the spin quantum number J.)

The Ω^- at the tip of the pyramid of the decuplet is metastable for special reasons. It is famous because it had not been discovered when this type of grouping was first attempted, and so there was a vacancy at the tip of the pyramid. The theory was able to predict the properties of the missing particle and, when it was discovered (see figure 32-15), the predictions were all verified. The prediction of its mass was possible because the various horizontal charge multiplets of figure 33-6 are separated by equal intervals of mass, as can be seen from the figure.

The reader will no doubt agree that this is all very complicated and somewhat puzzling, but that something of significance is beginning to emerge.

Inklings

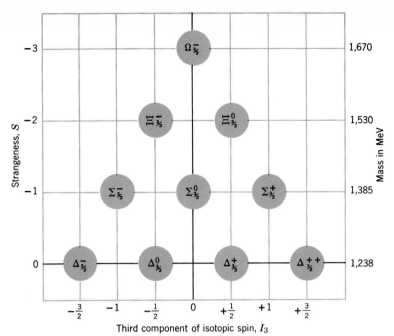

Figure 33-6 The decuplet of baryons with spin $J = \frac{3}{2}$.

QUESTIONS

1. Section 33-4 contains the sentence, "Consider an isolated charged particle at rest surrounded by its cloud of virtual photons, and concentrate on a snapshot taken at a fixed instant of time." Although this conveys the right idea, it is really nonsense because the requirement of "a fixed instant of time" makes the energy infinitely uncertain. How would you reformulate the discussion to avoid this difficulty?

2. Consider whether strangeness is conserved in the processes of equations 31, 44, 47, 56, and 58 of Chapter 32.

3. Draw a diagram similar to figure 33-4 for the particles K, η, Λ, and Ω.

PROBLEMS

A

1. A cosmologist suggests that gravitational forces have a "range" of about 10^{10} light years, which is the size of the observable universe. What is the order of magnitude of the corresponding rest mass of the graviton?

B

2. What are the charge quantum numbers and baryon numbers of (a) an α-particle, (b) a neutral helium atom, (c) a singly ionized helium atom?

3. Find an approximate value of the baryon number of the earth.

C

4. $\bar{Q}$ is defined as the average charge of a charge multiplet. Find a general expression for $\bar{Q}$ in terms of S and B.

5. Give the values of all the quantum numbers (except P) for the particle $\Delta_{3/2}^{++}$ included in figure 33-6. What are the quantum numbers of its antiparticle?

6. Remembering the analogy with angular momentum, can you guess how to obtain the isotopic spin of the deuteron?

Inklings

34

Symmetry

34-1 Symmetry and Conservation Laws

When the diversity of the fundamental particles is properly understood, it is possible that our description of the universe will be reduced to a few basic concepts. Meanwhile, to act as a guide to this future revelation, we have at our disposal certain sweeping generalizations, which seem to transcend the details of any particular theory. Some of these are conservation laws, like the law of conservation of energy. Others are called **symmetry principles,** like the postulate of special relativity that the basic laws of physics are independent of the velocity of the observer. In some cases a conservation law can be seen to be a mathematical consequence of a symmetry principle.

In common usage the word symmetry implies a "balance" between two sides of an object. Referring to figure 34-1, the shape shown in part *a* is immediately recognized as symmetrical, and the shape of part *b* as unsymmetrical. In physics the word has a more precise, but more general meaning. Something has a particular type of symmetry if a particular operation can be performed upon it and still leave it unchanged in some essential respect. This is best illustrated by examples. The symmetry of the shape in figure 34-1*a* is associated with the rather complicated operation illustrated in part *c* of the same figure. This is the operation of reflection about the central line *AB*. If *P* is any point on the figure, draw a line from *P* perpendicular to *AB* and move *P* to the point *P'* an equal distance on the other side of *AB*. Since the figure is symmetrical, *P'* coincides with a point that was originally on the figure. The point which was originally at *P'* is moved by this operation over to *P*. The net result is merely to interchange the two sides of the figure and, since the two sides are identical, the figure remains unaltered. However, in the case of the unsymmetrical shape of part *b*, the reflection operation produces the shape of part *d*, which is clearly different since it is the other way round.

There are some simpler symmetry operations, which nevertheless have a profound significance in physics. One of them is the operation of moving an object bodily through space. In figure 34-2 the triangle has been translated from position 1 to position 2, without rotation, but its

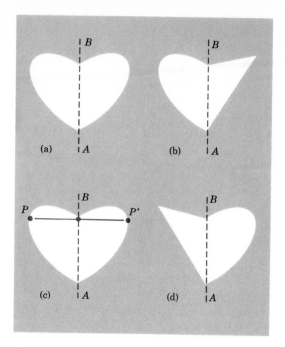

Figure 34-1 Illustrating the common meaning of symmetry.

size and shape remain unaltered. We say that the triangle is symmetrical under the operation of translation in space. In the realm of geometry this sounds rather trival, but when applied to real physical situations it can be far from trivial. Suppose that a triangle is made by looping a rubber band round two nails and then hanging a weight from the midpoint of its lower section (figure 34-3). If this device is moved to the moon it does not preserve its shape. On the moon the weight is not so heavy and therefore it does not pull the band down so far.

This last device seems to have been specially designed to cause trouble, but the effect is present to a lesser degree with all real physical objects. Suppose that the triangle of figure 34-2 is cut out of cardboard.

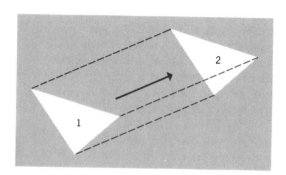

Figure 34-2 Translation in space has not changed the shape of the triangle.

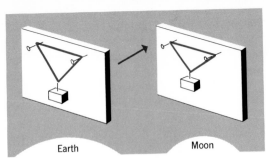

Figure 34-3 The triangle is made by wrapping a rubber band around two nails and hanging a weight from the midpoint of its lower section. If this device is taken to the moon, the triangle does not preserve its shape.

Earth   Moon

Like the rubber band, this cardboard triangle sags a little under the strain of its own weight and the bottom vertex is a little lower than it would be in the absence of gravity. Since the earth's gravitational field is not constant but varies from place to place, when the triangle is translated to a new position in this field the amount of sag changes a little and so the triangle changes its shape, just like the rubber band.

After a thorough searching analysis of this kind, is there anything that does remain unaltered when a physical system is moved from one place to another? We believe that the basic laws of physics, which explain the behavior of the system, are independent of its location in space. The gravitational force acting on the weight which distorts the rubber band is given by Newton's law of universal gravitation,

$$F = \frac{GM_1M_2}{R^2} \tag{34-1}$$

M_1 is either the mass of the earth or the mass of the moon and R is either the radius of the earth or the radius of the moon. Some of the quantities in the equation do have values which vary from place to place. The important thing is that the *form* of the equation is always the same. The force always obeys an inverse square law and is not, say, inverse square on the earth but inverse cube on the moon. Moreover, the fundamental constant G is the same everywhere and does not vary in an arbitrary fashion from place to place with no apparent reason. Similarly, the extent to which the rubber band sags depends upon its elasticity. This depends upon the fact that, when the rubber is stretched, its atoms move further apart and then exert forces on one another in an attempt to restore the initial situation. Interatomic forces are electromagnetic and are explained by Maxwell's equations and the laws of quantum mechanics. Maxwell's equations and the laws of quantum mechanics have the same form on the moon as on the earth.

We therefore say that "*The form of the basic laws of physics is symmetrical under the operation of translation in space.*" If we then look at the present form of the basic laws and consider the question of whether momentum is conserved or not, we discover an interesting fact. The law of conservation of momentum is a mathematical consequence of the fact that the basic laws have this property of assuming the same form at all points in space. The conservation law is a consequence of the symmetry principle.

There is reason to believe that the symmetry principle is more fundamental than the detailed form of the conservation law. The law of conservation of momentum first turned up in Newtonian mechanics, but it had to be modified in the special theory of relativity by allowing the mass to vary with velocity (equation 24-35). In the general theory of relativity the basic equation of gravitation is not as simple as equation 34-1. Moreover, it is not true in the general theory that Maxwell's equations are the same on the moon as on the earth. Electromagnetic phenomena are influenced slightly by the strength of the gravitational field. Nevertheless, the general theory is able to formulate basic equations which have the property of assuming the same form at all points in space. These equations almost certainly lead to a law similar to the law of conservation of momentum, although the exact form of this law is not yet quite certain.

It is clear that, however hard pressed, the physicist will always try to trim his basic laws to make them symmetrical. The point at issue is whether he will always be successful, or whether he will eventually have to give up in despair. So far he has been quite successful.

34-2 Compilation of Symmetry Principles and Conservation Laws

In this section and the next, we shall collect together all the known symmetry principles and conservation laws. The subject is still controversial and it is not always certain how the conservation law follows from the symmetry principle and whether the connection is logically rigorous. It is therefore convenient in some cases to place the emphasis on the symmetry principle and in other cases on the conservation law. Some symmetry principles and conservation laws seem to be applicable only under restricted conditions, but nevertheless are very interesting and important. We shall discuss them in the next section.

A. SYMMETRY UNDER TRANSLATION IN SPACE. The form of the basic laws of physics is the same at all points in space. This leads to the LAW OF CONSERVATION OF LINEAR MOMENTUM.

B. SYMMETRY UNDER ROTATION IN SPACE. The basic laws of physics describing a system apply in the same form if the system is rotated through a fixed angle. The laws of physics have the same form in all directions. This leads to the LAW OF CONSERVATION OF ANGULAR MOMENTUM.

C. SYMMETRY UNDER TRANSLATION IN TIME. The form of the basic laws of physics does not change with time. Once the really basic laws have been found, they apply equally well to events that occurred a billion years ago and events that will occur a billion years in the future. This leads to the LAW OF CONSERVATION OF ENERGY.

D. SYMMETRY UNDER REVERSAL OF TIME. Imagine any physical process which can take place and therefore, of course, obeys the basic laws. Imagine this process running backward in time, like a movie which is passing through the projector in the wrong direction. Then this reversed process also obeys the basic laws of physics and is therefore a possible process. In section 12-7 we quoted the example of a body dropped from rest at a height h and striking the ground with a velocity v ($= \sqrt{2gh}$). Corresponding to this there is a possible time reversed motion in which the body is thrown upward with a velocity v and comes to rest at a height h.

This particular symmetry principle is not an easy one to accept. Suppose that I drop a plate on the floor and it shatters to smithereens. Can I maintain that there is another possible motion in which the fragments fly together and reform an unblemished plate? Yes! There is nothing in the laws of physics to prevent this from happening. However, it would be necessary to throw the fragments together with *exactly* the right velocities so that they hit one another in *exactly* the right places. This is quite impossible to achieve in practice and would certainly never occur of its own accord. The breaking of the plate is a possible occurrence and a reasonably probable occurrence. The reconstitution of the plate by the time reversed process is possible, but highly improbable. Symmetry under time reversal says that a time reversed process *can* occur, but it gives no guarantee that it *will* occur. This has been discussed at greater length in Chapter 12, sections 7, 8, and 9.

Applied to fundamental processes, time reversal says that any process that can go in one direction can also go in the opposite direction. For example, a neutron with sufficient energy can disintegrate into a proton and a negative pion.

$$n \rightarrow p + \pi^- \tag{34-2}$$

Conversely, a negative pion can combine with a proton to form a neutron

$$\pi^- + p \rightarrow n \tag{34-3}$$

β-decay is the disintegration of a neutron into a proton, an electron and an antineutrino.

$$n \rightarrow p + e^- + \bar{\nu} \tag{34-4}$$

Presumably an antineutrino, an electron and a proton can come together and form a neutron

$$\bar{\nu} + e^- + p \rightarrow n \tag{34-5}$$

The reason this reaction has not been observed is associated with the very low probability that all three particles will be in the same place at the same time.

E. RELATIVISTIC SYMMETRY. The basic laws of physics have the same form for all observers, independently of their motion. In the special theory of relativity the basic laws have the same form in all inertial frames and therefore do not depend upon the velocity of the observer. In the general theory, which is less well established, the basic laws are assumed to have the same form for all observers, however complicated their motion.

F. SYMMETRY UNDER INTERCHANGE OF SIMILAR PARTICLES. Fundamental particles do not have individual identities and the interchange of two similar particles makes no difference to a physical process. This was explained in more detail in section 29-6, where it was called the **INDISTINGUISHABILITY OF SIMILAR PARTICLES.**

G. THE LAW OF CONSERVATION OF ELECTRIC CHARGE. If negative charge is considered to cancel an equal amount of positive charge, there is no known physical process which changes the net amount of electric charge. In a fundamental process, the sum of the charge quantum numbers Q is the same on both sides of the equation. This conservation law is believed to be related to certain symmetry properties of the quantum mechanical wave function ψ, but the situation is not understood well enough to be easily explicable in simple terms.

H. THE LAW OF CONSERVATION OF LEPTONS. If an antiparticle is considered to cancel its corresponding particle, there is no known physical process which changes the net number of leptons (section 32-6). There may be a similar **LAW OF CONSERVATION OF μ-LIKE LEPTONS.** It is not yet known if there is a symmetry principle underlying these laws.

I. THE LAW OF CONSERVATION OF BARYONS. If an antiparticle is considered to cancel its corresponding particle, there is no known physical process which changes the net number of baryons (section 32-8). In a fundamental process, the sum of the baryon numbers B is the same on both sides of the equation. No symmetry principle underlying this law has yet been discovered, either. It is interesting that there are conservation laws for fermions but not for bosons. The net number of photons, pions, kaons, etas, or gravitons is not conserved.

34-3   Restricted Symmetries

The symmetries to be discussed in this section are sometimes called "broken symmetries" for the following reason. While they can be usefully applied to a wide range of physical phenomena, they break down when applied to other phenomena. This may be because nature is constructed according to a scheme of partial or imperfect symmetry. It is more likely, however, that we do not yet understand the situation properly and that we have not yet formulated the symmetry principles in such a way that they are universally applicable.

J. CHARGE INDEPENDENCE. The nuclear force between two nucleons is independent of whether they are neutrons or protons. This suggests the symmetry principle that the strength of a strong interaction is unaltered by changing the electric charge on the particles. That is, for example, by changing neutral n into positive p or Σ^+ into Σ^-. Now the concept of isotopic spin I was introduced to cope with the fact that the particles in a charge multiplet are identical as far as strong interactions are concerned, but have different charges. There is a strong mathematical analogy between isotopic spin and angular momentum, and there is, in fact, a LAW OF CONSERVATION OF ISOTOPIC SPIN analogous to the law of conservation of angular momentum.

However, although n and p are equivalent as far as nuclear forces are concerned, their electromagnetic effects are very different because one is charged and the other is not. Consequently, it is found that isotopic spin is conserved in fundamental processes related to strong interactions, but not in processes related to electromagnetic and weak interactions.

K. THE LAW OF CONSERVATION OF STRANGENESS. The very reason for introducing strangeness (section 33-6) was that it is conserved in strong interactions, but not in weak interactions.

L. PARTICLE-ANTIPARTICLE SYMMETRY. For every particle there is a corresponding antiparticle, which seems to be its exact opposite. For every fundamental process which is known to occur, there is another possible process obtained by changing every particle into its antiparticle. We have already given several examples of this. As a further example, corresponding to ordinary β-decay,

$$n \rightarrow p + e^- + \bar{\nu} \tag{34-6}$$

there is a process of "anti-β-decay," in which an antineutron changes into an antiproton,

$$\bar{n} \rightarrow \bar{p} + e^+ + \nu \tag{34-7}$$

Changing a particle into its antiparticle is sometimes called **charge conjugation.** As we shall see shortly, the indiscriminate change of particles into antiparticles sometimes leads to trouble, so this is classified as a broken symmetry.

M. MIRROR SYMMETRY. This symmetry principle states that, for every known physical process there is another possible process which is identical with the mirror image of the first. Consider a clock, which is a complicated mechanism containing toothed wheels, spindles, and levers obeying the laws of mechanics. Suppose that it is powered by a main-spring, the springiness of which can be ultimately traced to electromagnetic forces between its atoms, so that the laws of electromagnetism are also involved. The image of this clock in a mirror has every appearance of being a possible mechanism which would work and obey the laws of physics (figure 34-4). It is true that the numbers on its dial are I, Ƨ, Ɛ, and so on, and its hands go round counterclockwise. Nevertheless, there seems to be no reason why we should not construct in the real world a real copy of the image. It would look rather odd, but we would expect it to work, to obey the laws of physics and keep the same time as an ordinary clock.

To see that this is not all trivial and obvious, let us consider carefully the difference between an object and its mirror image. The symmetry

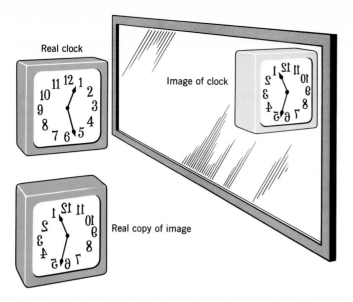

Real clock

Image of clock

Real copy of image

Figure 34-4   A real clock, its image in a mirror, and a real copy of the image.

operation which produces a mirror image is very similar to the one associated with the popular notion of symmetry. From any point P on the object draw a line perpendicular to the surface of the mirror and move P to P' an equal distance on the other side of the mirror (figure 34-5). This procedure changes a right hand glove into a left hand glove. Try shaking hands with yourself in a mirror and you will discover that you can shake hands with the image of your left hand, but not with the image of your right hand. Now a left hand glove is an essentially different thing from a right hand glove. You cannot put a left hand glove

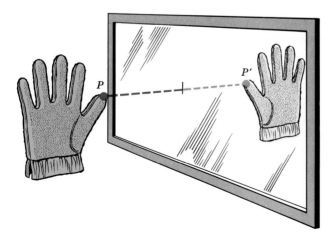

Symmetry

Figure 34-5   The image of a right hand glove is a left hand glove, which is an essentially different thing.

on your right hand. There is no way of turning a left hand glove around in space so that it falls exactly on top of a right hand glove. You cannot step into a mirror and coincide with your image, because you would have to turn around, and your right hand would then be on top of the image of your left hand. Moreover, your real heart would be on your left side, while the image heart would be on the right side of your image. The image of an object is an essentially different thing from the object itself.

The mainspring of a clock relies upon the laws of electromagnetism. In formulating these laws we made use of the right hand (sections 16-2 and 16-3). Since reflection changes a right hand into a left hand, we must consider carefully what happens to the laws of electromagnetism for a mirror image. Consider a current in a long straight wire exerting a magnetic force on a positive charge $+q$ moving in the same direction as the current (figure 34-6). Right hand rule number 1 tells us that the magnetic field lines go around the current in the manner shown. Right hand rule number 2 then tells us that this magnetic field exerts a magnetic force on the moving charge which deflects it toward the current. Now look at the mirror image. The magnetic field lines do not go the right way round the current and the image therefore violates right hand rule number 1. If there is a real physicist using his right hand to figure out the direction of the magnetic field, then the image physicist is using his left hand, and this is why he gets it the wrong way round.

However, the magnetic field is just a mathematical convenience which we use at an intermediate stage in the calculation. What matters from a

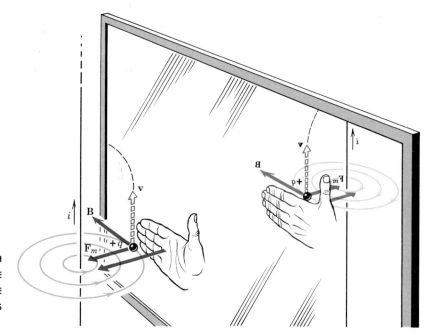

Figure 34-6 In a mirror, electromagnetism has two *left* hand rules instead of two right hand rules.

physical point of view is the behavior of the moving charge and we can see that the image charge is behaving itself and deflecting *toward* the current. The reason is that the image physicist must use his left hand *twice* in order to discover the direction of the force on the charge. You can easily convince yourself that this gives the same result as using the right hand twice. If we concentrate on the things that really matter, we find that mirror images do obey the laws of electromagnetism. In fact there is nothing in classical physics to make us doubt mirror symmetry.

In modern physics mirror symmetry would imply that the mirror image of any known fundamental process is also a possible process consistent with the basic laws. This seems to be true for strong interactions, electromagnetic interactions and gravitational interactions. However, in 1956 C. N. Yang and T. D. Lee suggested that some puzzling features of the behavior of kaons might be resolved if weak interactions violated mirror symmetry. Within two years several experiments had been performed to confirm this suggestion. Mirror symmetry can be expressed mathematically in terms of a quantity called **parity,** *P*, and there is a corresponding **Law of Conservation of Parity.** Weak interactions do *not* conserve parity, although all other types of interaction do.

One of the weak interactions is β-decay and the first experiment which established a violation of mirror symmetry involved the β-decay of cobalt 60 nuclei. With the cobalt nucleus spinning in the direction shown in figure 34-7a, it was found that the electron is *always* emitted downward and the antineutrino is *always* emitted upward. (More precisely, the electron is emitted in all directions with varying degrees of probability, but it is most likely to be emitted directly downward and it is never emitted directly upward. The antineutrino is most likely to be emitted directly upward and it is never emitted directly downward.) The angular momentum of the cobalt nucleus decreases from 5 to 4 units of $h/2\pi$. The electron and antineutrino each have a spin of $\frac{1}{2}$ a unit and so, in order to conserve angular momentum, they must both be spinning in the same direction as the cobalt nucleus, as shown. The mirror image of this process is shown in part *b* of the same figure. Notice how reflection changes the direction of spin. The point *P* reflects into the point *P'* and both are moving into the paper.

The important difference between the real process and its mirror image is this. Let us always look at the cobalt nucleus in the direction in which it appears to spin clockwise, that is, upward for the real process and downward for the image process. Then, in the real process the electron is emitted toward us and this is what *always* occurs. In the image process the electron is emitted away from us, and this *never* occurs. So clearly the mirror image of a known process is not a possible process and mirror symmetry is violated.

The explanation is probably to be found in the nature of the antineutrino. There is good reason to believe that, if you look at an antineutrino in such a direction that it is moving away from you, then it is always rotating clockwise and never counterclockwise. Notice that this is so for the real β-decay process, but not for the image process, and so the ob-

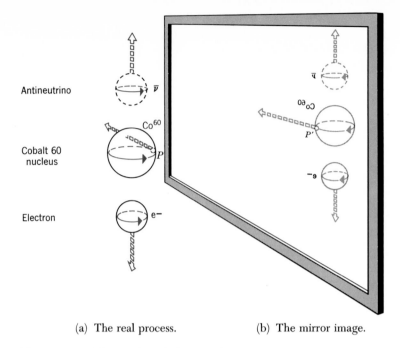

(a) The real process.   (b) The mirror image.

Figure 34-7   β-decay of a cobalt 60 nucleus and its mirror image.

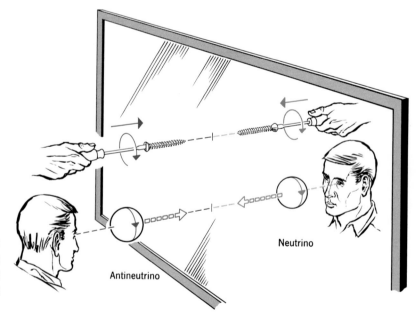

Figure 34-8   The analogy between an antineutrino and a right-handed screw and between a neutrino and a left-handed screw.

ject produced in the image process could not possibly be an antineutrino. In this respect an antineutrino is like a right-handed screw, the ordinary kind of screw. To make the screw move away from you as you screw it in, you must turn it clockwise. The other kind of screw, a left-handed screw, is an essentially different object which cannot be made to coincide with a right-handed screw however much it is turned around in space. Figure 34-8 shows that a left-handed screw is the mirror image of a right-handed screw. A right hand is shown turning the right-handed screw clockwise and driving it forward. The mirror image of this is a left hand turning a left-handed screw *counter*clockwise to drive it forward.

The same figure shows an antineutrino moving toward the mirror and rotating clockwise from the point of view of an observer looking toward the mirror. The mirror image of this also moves toward the mirror, but rotates *counter*clockwise from the point of view of an observer behind the mirror looking toward the mirror. Such an entity is believed to be a neutrino. *The neutrino and antineutrino are mirror images of one another. The neutrino is the analogue of a left-handed screw and the antineutrino is the analogue of a right-handed screw.*

This brings us back to the question of particle-antiparticle symmetry. Figure 34-9a is the β-decay of cobalt 60 again, and part c of the same figure is the result of changing every particle into its antiparticle. The antinucleus is composed of antiprotons and antineutrons and it emits a positron and a neutrino. If everything else remains unaltered, including the directions of the spins, the process in c is not possible. The particle emitted upward is a right-handed screw and cannot possibly be a neutrino. By this and many similar examples, we are forced to conclude that all weak interactions violate the principle of particle-antiparticle symmetry.

Now consider the process of figure 34-9d, which is obtained from the observed process a by changing particles into antiparticles and also reflecting in a mirror. This is a possible process, because the particle emitted upward is a left-handed screw, as a neutrino should be. Therefore, although each symmetry principle is violated separately, the process is symmetrical under the double operation of mirror reflection plus particle-antiparticle interchange.

Similar considerations apply to weak interactions involving μ-neutrinos (see question 34-11). These can be explained similarly by assuming that the μ-neutrino is a left-handed screw and the anti-μ-neutrino is a right-handed screw. The other type of weak interaction, involving strange particles rather than neutrinos, also violates both symmetry principles. In figure 34-7a, the basic experimental fact is that the electron is always emitted downward. In weak interactions involving strange particles there is a similar lopsidedness in the direction of emission of the products. This is more of a mystery because it has not yet received the same type of elegant explanation that can be applied when neutrinos are involved. Again, though, the processes are found to be symmetrical under a combined operation of mirror reflection plus particle-antiparticle interchange.

The end result of this discussion restores our faith in the elegance of

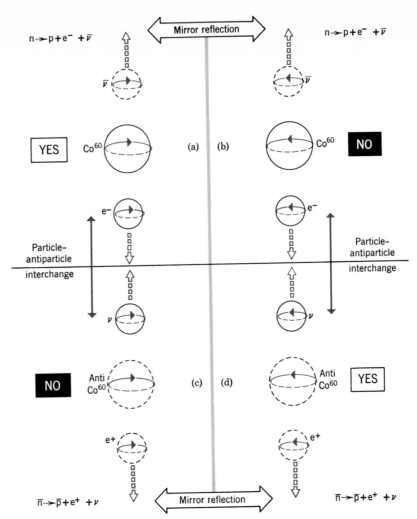

Figure 34-9 The β-decay of cobalt 60 reflected in a mirror and subjected to particle-antiparticle interchange.

nature. Figure 34-9 has a very satisfying symmetry. The expression "broken symmetry" seems unnecessarily harsh when applied to this situation. We have no reason to doubt that, the more we come to understand the working of this universe, the more we shall be impressed by the beauty of its design.

34-4   C, P, and T

The first edition of this book ended with the ringing phrases of the last paragraph of the previous section. The events of the intervening years prompt the addition of the qualifying phrase—"but we must never presume that we have penetrated nature's mysteries far enough to reveal its true beauty." The development that points to this moral is the discovery of weak interactions which appear not to be symmetrical under the combined operation of mirror reflection plus particle-antiparticle interchange.

It has become common practice to denote the operation of particle-antiparticle interchange by the letter C (for Charge conjugation) and mirror reflection by the letter P (for Parity). The double operation of particle-antiparticle interchange followed by mirror reflection is then denoted by CP. The β-decay of cobalt 60 is said to violate C symmetry and P symmetry, but to preserve CP symmetry.

The experiment that first provided evidence for a violation of CP symmetry was concerned with the decay of the neutral kaon, K^0, via weak interactions. Consider the equation

$$K^0 = \tfrac{1}{2}(K^0 + \overline{K}^0) + \tfrac{1}{2}(K^0 - \overline{K}^0) \tag{34-8}$$

If a particle is something like a little hard sphere, this equation has no obvious meaning. However, if we emphasize the wavelike character of a kaon, there is a very real sense in which the wave representing K^0 can be looked upon as a superposition of two waves representing the two "particles":

$$K_1^0 = K^0 + \overline{K}^0 \tag{34-9}$$

$$\text{and} \quad K_2^0 = K^0 - \overline{K}^0 \tag{34-10}$$

K_1^0 rapidly decays into two pions, as it is allowed to do by CP symmetry:

$$K_1^0 \rightarrow \pi^+ + \pi^- \tag{34-11}$$

Consequently, about 10^{-7} sec after the formation of the original K^0, K_1^0 has almost completely disappeared, leaving K_2^0 to be studied in isolation.

K_2^0 is allowed by CP symmetry to decay slowly into three pions:

$$K_2^0 \rightarrow \pi^+ + \pi^0 + \pi^- \tag{34-12}$$

However, CP symmetry completely forbids a decay into *two* pions:

$$K_2^0 \rightarrow \pi^+ + \pi^- \tag{34-13}$$

Nevertheless, about one K_2^0 in every thousand does decay into two pions, violating CP symmetry.

The fundamental process that violates CP symmetry may therefore be written

$$(K^0 - \overline{K}^0) \rightarrow \pi^+ + \pi^- \tag{34-14}$$

If we start with this process, change particles into antiparticles and then reflect in a mirror, the resulting process is *not* found in nature. To see how subtle this is, notice that changing particles into antiparticles makes very little difference to equation 34-14, apart from the sign on the left hand side. Moreover, K and π both have zero spin, so reflection in a mirror has none of the obvious physical significance characteristic of the situations illustrated in figures 34-7 and 34-8.

Although subtle, the violation is very disturbing from a theoretical point of view. If we accept the validity of quantum mechanics and the special theory of relativity, it appears to be possible to prove rigorously that any fundamental process ought to be symmetrical under the triple operation *CPT*. Here *T* represents the operation of time reversal, making each particle retrace its motion as time runs backward. If we start with any possible fundamental process, change particles into antiparticles, reflect in a mirror, and then run the process backward, we ought to end up with another possible process.

Now consider a process such as the one in equation 34-14 which violates *CP* symmetry. If we apply the double operation *CP* to this process, the result is an impossible process. If we then time reverse to complete the triple operation *CPT*, the final result must be a possible process. The third operation *T* has therefore changed an impossible process into a possible one. Going through the series of operations in reverse, we see that time reversal *T* is capable of turning a possible process into one that is impossible.

Time reversal symmetry therefore appears to have broken down. If this is true, there is a very fundamental way in which the laws of nature recognize the forward direction of time. This is very different from the arguments presented in Chapter 12 and section 34-2D, where we said that all time reversed processes are possible, but some are very unlikely. There is now a suggestion that some time reversed processes are categorically forbidden.

Thus, whereas the first edition ended on the hopeful note that we are beginning to understand fundamental processes and particles, the second edition ends on a different, more challenging note. There are still many more wonders to be discovered and to be understood.

QUESTIONS

1. Now that you have come to the end of the book, how would you answer Question 1 of Chapter 14?
2. Which of the following quantities is *always* conserved? (*a*) Momentum, (*b*) angular momentum, (*c*) total rest mass, (*d*) charge, (*e*) number of baryons, (*f*) parity, (*g*) force, (*h*) isotopic spin.
3. Discuss the β-decay of the neutron from the point of view of each of the conservation laws and each of the symmetry principles.
4. A newly discovered particle, the iota, decays into an electron, e^-, and an antineutrino, $\bar{\nu}$. Which of the following possibilities can be excluded and why?
   (*a*) It is a lepton. (*b*) It is a baryon. (*c*) It is a meson. (*d*) It is a photon. (*e*) It is a boson. (*f*) It is a fermion. (*g*) Its velocity is exactly equal to the velocity of light.
5. Which conservation law (or laws) does each of the following violate?

   (*a*) $\mu^- \rightarrow e^- + e^+ + \nu$
   (*b*) $\eta^0 \rightarrow e^+ + \bar{\nu}$
   (*c*) $p \rightarrow e^+ + \nu$
   (*d*) $p + \bar{n} \rightarrow n + \bar{p}$

6. Discuss the following statement, "Of course the basic laws of physics obey the symmetry principles, because if a law did not we would deny that it was truly basic."

7. Sometimes the decay of a particle A into two other particles B and C, $(A \rightarrow B + C)$, is allowed by all the known laws of physics, except that the rest mass of A is smaller than the sum of the rest masses of B and C by an amount m. Is the process possible if A is moving alone through otherwise empty space with a kinetic energy greater than mc^2? (Hint: the answer is obvious once you have related the question to the relevant symmetry principle.)

8. You have established radio contact with a friendly alien space ship approaching the solar system. During the course of an exchange of knowledge, you find it necessary to explain to them which is your right hand. *Precisely* what would you say under the following circumstances? (*a*) You can modify your radio transmission by swinging it through an angle in space or by changing the direction of the oscillating electric vector. (*b*) You cannot modify your radio transmission but you know that the visitors are able to see the constellations. (*c*) Neither of the above is possible. The only possibility is to talk about the nature of physical phenomena. (*d*) The only possibility is to talk about physical phenomena, but you do not know whether the visitors are made of matter or antimatter.

9. A current from a storage battery passes through a coil of wire and the resulting magnetic field deflects a magnetic compass needle. The poles of the battery are marked $+$ and $-$, but there is no indication which is the north pole of the compass needle. Can you decide from the direction in which the compass needle deflects whether you are looking at the real apparatus or its reflection in a mirror? Suppose that all the protons, neutrons and electrons in the apparatus were suddenly changed into antiprotons, antineutrons and positrons. Would the compass needle still deflect in the same direction? Do not answer this question by applying known symmetry principles. Convince yourself that the symmetry principles are valid in this case.

10. Would the process of figure 34-7*b* be a permissible physical process if we conceded that the particle emitted upward is a neutrino and not an antineutrino?

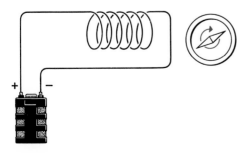

Question 34-9

11. Assuming that the μ-neutrino is a left-handed screw and the anti-μ-neutrino is a right-handed screw, analyze the decay of the pion

$$\pi^+ \rightarrow \mu^+ + \nu_\mu$$

   (a) If the muon and μ-neutrino move directly away from one another and therefore have no *orbital* angular momentum, use the law of conservation of angular momentum to decide whether the muon is spinning like a right-handed or left-handed screw. (b) Draw a diagram similar to figure 34-9 and decide which of the various processes can occur.

12. Without actually going there, is there any way of deciding whether a distant star is made of matter or antimatter? Our knowledge of the star may be obtained from the light emitted by its surface and also from the neutrinos emitted as a result of processes such as the ones summarized in equation 31-40.

13. Discuss the effect upon physical processes of (a) reflection twice in two mirrors at right angles to one another, (b) reflection three times in three mirrors mutually at right angles (like two walls and the floor of a room all meeting in a corner).

14. Write an essay on "Energy." Include an enumeration of the various forms of energy and detailed discussions of processes which transform one kind of energy into another kind. Say which are the basic kinds of energy and explain how some forms of energy, such as heat and chemical energy, can be analyzed in terms of these basic forms.

15. In various places throughout this book we have had to admit our ignorance on some important issues. Make a list of what you consider to be the outstanding unsolved problems in physics. Can you guess how some of them might be resolved? If you had been a nineteenth century physicist, do you think you could have guessed the nature of quantum mechanics?

Mathematical Appendix

The student with little previous experience of mathematics should study this appendix carefully before proceeding beyond Chapter 1. However, he may omit sections A7, A8, and C4 until he encounters a need for them.

A / ALGEBRA

A1   Fundamentals

Algebra is arithmetic in which letters are used to represent numbers. In arithmetic, we would say

$$3 \times 5 = 15 \tag{1}$$

A similar statement in algebra is

$$ab = c \tag{2}$$

which means that, when the number a is multiplied by the number b, the result is the number c. If a represents the number 3, and b represents the number 5, then c must represent the number 15, and equations 1 and 2 are different ways of saying the same thing. However, it is possible for a and b to represent other numbers, for example,

$$a = 2\tfrac{1}{2} \quad \text{and} \quad b = 4\tfrac{1}{3} \tag{3}$$

Once a and b have been chosen, there is no freedom of choice for c, because

$$c = ab \tag{4}$$

that is, $\quad c = (2\tfrac{1}{2}) \times (4\tfrac{1}{3}) \tag{5}$

or $\quad c = 10\tfrac{5}{6} \tag{6}$

Equation 2 is therefore a very general equation in which a and b may represent any two numbers whatsoever, but c is restricted to a particular value which is the product of a and b.

The **symbols** such as a, b, and c which represent numbers may be combined in various ways to produce various **algebraic expressions.** For example, the expression $a(b + c)$ means that the number represented by b is added to the number represented by c and the resulting sum is multiplied by the number represented by a. The expression $[a + (b + c)d](a + c)$ means that (1) b is added to c; (2) the sum is then multiplied by d; (3) a is then added; (4) the result of (3) is then multiplied by the sum of a and c. The expression $(ab)/(b + c)$ means that a is multiplied by b and the result is divided by the sum of b and c.

Example

Evaluate the expression

$$p = \frac{(r + s)[rs + t(3 + s)]}{5st(r + t)}$$

when $r = 2$, $s = 5$ and $t = \frac{1}{2}$.

Inserting the given values of r, s, and t,

$$p = \frac{(2 + 5)[(2 \times 5) + \frac{1}{2}(3 + 5)]}{5 \times 5 \times \frac{1}{2}[2 + \frac{1}{2}]}$$

$$= \frac{7[10 + (\frac{1}{2} \times 8)]}{\dfrac{25}{2} \times [2\frac{1}{2}]}$$

$$= \frac{7[10 + 4]}{\dfrac{25}{2} \times \dfrac{5}{2}}$$

$$= \frac{7 \times 14}{\dfrac{125}{4}}$$

Multiplying numerator and denominator by 4,

$$p = \frac{7 \times 14 \times 4}{125}$$

$$= \frac{98 \times 4}{125}$$

$$= \frac{392}{125}$$

$$= 3\frac{17}{125}$$

It is preferable to express this in decimal form

$$p = 3.136$$

PROBLEMS

Express the meaning of the following expressions in words.

1. $(a + b)(c + d)$

2. $\left(a + \dfrac{b}{c}\right)(b + c)$

3. $\dfrac{a}{c}\left[2a + bc\left(3 + \dfrac{1}{a}\right)\right]$

4. $\dfrac{(x + y + z)(xy + yz + zx)}{\left(\dfrac{1}{x} + \dfrac{2}{y} + \dfrac{3}{z}\right)}$

Evaluate

5. $a(b + c) + \tfrac{1}{2}b(c + 2a)$ when $a = 2$, $b = 3$, $c = 4$

6. $(r + s)\left(\dfrac{t}{r} + \dfrac{s}{t}\right)$ when $r = \tfrac{1}{2}$, $s = 5$, $t = 3\tfrac{1}{2}$

7. $\dfrac{(5ar + b)\left(\dfrac{1}{a} + \dfrac{2}{b}\right)}{(ab + 2br + 3ra)}$ when $r = 1$, $a = 2$, $b = 3$

8. $\dfrac{(x + 3.2y)(1.6xy + 0.3z)}{yz(x + 1.2yz)}$ when $x = 0.25$, $y = 1.13$, $z = 2.32$

A2 Negative Numbers

The expression $a - b$ means that the number b is *subtracted* from the number a. It is clearly different from the expression $a + b$, which is the result of *adding* b to a. However, the idea that b can represent any number may be extended to allow b to represent both positive and negative numbers. If b is a negative number, the addition of b to a is really a subtraction. For example, if

$$a = +7 \quad \text{and} \quad b = -3 \tag{7}$$

then

$$a + b = +7 + (-3) \tag{8}$$
$$= +7 - 3 \tag{9}$$
$$= +4 \tag{10}$$

The subtraction of a negative number is equivalent to addition. In the above case

$$a - b = +7 - (-3) \tag{11}$$
$$= +7 + 3 \tag{12}$$

Algebra

To ensure that the normal procedures of arithmetic shall give consistent results when applied to negative numbers, it is necessary to adopt the following rules for multiplication. **The multiplication of two positive numbers produces a positive number.**

For example $\quad (+9) \times (+7) = +63$ $\qquad\qquad\qquad$ (13)

The multiplication of a positive number and a negative number produces a negative number.

For example $\quad (+1.2) \times (-2.5) = -3$ $\qquad\qquad\qquad$ (14)

and $\qquad\qquad (-\frac{1}{2}) \times (+3.6) = -1.8$ $\qquad\qquad\qquad$ (15)

The multiplication of a negative number by another negative number produces a positive number.

For example, $\quad (-5) \times (-0.32) = +1.6$ $\qquad\qquad\qquad$ (16)

The value of a positive or negative number, regardless of its sign, is called its **numerical value** or sometimes its **absolute magnitude**. The numbers $+3$ and -3 both have a numerical value of 3. It is important to distinguish clearly between a negative number and its numerical value. The statement "The electric charge at P has a negative value of $-q$" implies that the symbol q represents the numerical value of the charge and q is therefore a positive number. If, at a later stage, one is told that the charge is actually -2.5 statcoulombs, the number which must be inserted for q in any formula is $+2.5$ and not -2.5.

Example 1

$\qquad$ Evaluate $p = \dfrac{(a - b)(ab + c)}{(cb - 2a)}$

$\qquad$ when $\quad a = +2, b = -\frac{1}{2}, c = -3$

Inserting the given values of a, b, and c

$$p = \frac{[+2 - (-\frac{1}{2})][(+2)(-\frac{1}{2}) + (-3)]}{[(-3)(-\frac{1}{2}) - 2(+2)]}$$

$$= \frac{[+2 + \frac{1}{2}][-1 - 3]}{[+1\frac{1}{2} - 4]}$$

$$= \frac{(2\frac{1}{2})(-4)}{-(2\frac{1}{2})}$$

The $2\frac{1}{2}$ in the numerator cancels the $2\frac{1}{2}$ in the denominator.

$$p = \frac{-4}{-1}$$

Multiplying numerator and denominator by -1

$$p = \frac{+4}{+1}$$

$$p = +4$$

Example 2

When an electric charge of $+q_1$ statcoulomb is at a distance of R cm from another electric charge of $+q_2$ statcoulomb, their mutual electrostatic potential energy in ergs is

$$\Phi_e = \frac{q_1 q_2}{R}$$

Evaluate this energy when $q_1 = +1.5$ statcoulomb, $q_2 = -0.2$ statcoulomb, and $R = 0.1$ cm.

The inexperienced reader need not be disconcerted by an algebraic expression like this. The greek letter capital phi, Φ, is merely a symbol representing a number, just like a or x or r. The subscript e is used merely to remind us that we are talking about electrostatics and to avoid confusion with a similar quantity, Φ_G, which is the mutual *gravitational* potential energy of two bodies. The symbol q is commonly used for an electric charge, but in this problem there are two charges with different numerical values and it is necessary to distinguish between them. This is done by the subscripts 1 and 2. Therefore, we merely insert the numerical values into the formula in the usual way.

$$\Phi_e = \frac{(+1.5) \times (-0.2)}{0.1}$$

$$= \frac{-0.3}{0.1}$$

$$\Phi_e = -3.0 \text{ erg}$$

Notice that, since one charge is positive and the other negative, the mutual electrostatic potential energy comes out negative.

PROBLEMS

9. Express in words the meaning of

$$\frac{(a - b)(ab + c)}{(cb - 2a)}$$

Evaluate

10. $a - b + ab$

when $a = -0.1$, $b = +20$

11. $\dfrac{(xy + y + z)(x - 1)}{xyz}$

when $x = +2$, $y = -3$, $z = -\frac{1}{2}$

12. $\dfrac{(\alpha\beta + \beta + \gamma)(\alpha - 1)}{\alpha\beta\gamma}$

when $\alpha = +2$, $\beta = -3$, $\gamma = -\frac{1}{2}$

Algebra

13. $$\frac{(\theta - \phi - \eta)\left(\theta\phi - \dfrac{\eta}{\theta}\right)}{(\eta\theta - \phi)}$$

when $\theta = +1$, $\phi = -1$, $\eta = -2$

14. What is the mutual electrostatic potential energy of two charges of value -2.5 statcoulomb and -15.0 statcoulomb, when the distance between them is 5 cm.

15. Write down a formula for the mutual electrostatic potential energy of charges $+q_1$ statcoulomb and $-q_2$ statcoulomb at a distance R cm apart.

A3 Exponents

The expression a^n means that the number a is multiplied by itself n times. n is called the **exponent** of a. For example "a squared" is

$$a^2 = a \times a \tag{17}$$

"a cubed" is

$$a^3 = a \times a \times a \tag{18}$$

"a to the fifth power" is

$$a^5 = a \times a \times a \times a \times a \tag{19}$$

The expression

$$a^n \times a^m$$

clearly means that a is multiplied by itself n times and then by a number which is the result of multiplying a by itself m times. This is the same as multiplying a by itself $n + m$ times, and so

$$a^n \times a^m = a^{n+m} \tag{20}$$

For example

$$2^2 \times 2^3 = 2^{2+3} = 2^5 \tag{21}$$

because $\quad (2 \times 2) \times (2 \times 2 \times 2) = (2 \times 2 \times 2 \times 2 \times 2) \tag{22}$

Similarly

$$\frac{a^n}{a^m} = a^{n-m} \tag{23}$$

because it means that a is multiplied by itself n times and then divided by a m times, which is equivalent to multiplying a by itself $n - m$ times.

For example

$$\frac{2^6}{2^2} = 2^{6-2} = 2^4 \tag{24}$$

because $\dfrac{(2 \times 2 \times 2 \times 2 \times 2 \times 2)}{(2 \times 2)} = (2 \times 2 \times 2 \times 2)$ (25)

We can therefore establish the convention that

$$a^{-m} = \frac{1}{a^m} \tag{26}$$

So that $\dfrac{a^n}{a^m} = a^n \times a^{-m}$ (27)

$$= a^{n+(-m)} \tag{28}$$

$$= a^{n-m} \tag{29}$$

If n is equal to m, then

$$\frac{a^n}{a^n} = a^{n-n} = a^0 \tag{30}$$

but it is also obviously equal to 1, so

$$a^0 = 1 \tag{31}$$

whatever the value of a.

Since a^n means a multiplied by itself n times, the expression $(a^n)^m$ means a^n multiplied by itself m times. This is equivalent to multiplying a by itself nm times

$$(a^n)^m = a^{nm} \tag{32}$$

For example

$$(5^2)^3 = (5 \times 5)^3 \tag{33}$$

$$= (5 \times 5) \times (5 \times 5) \times (5 \times 5) \tag{34}$$

$$= (5 \times 5 \times 5 \times 5 \times 5 \times 5) \tag{35}$$

$$= 5^6 \tag{36}$$

or $(5^2)^3 = 5^{2 \times 3}$ (37)

Fractional exponents can be introduced by assuming the general validity of equation 32, even when n and m are not integers. Then

$$(a^{1/n})^n = a^{n/n} \tag{38}$$

$$= a^1 \tag{39}$$

or

$$(a^{1/n})^n = a \tag{40}$$

$a^{1/n}$ is the number which, when multiplied by itself n times, produces a. It is therefore the "n^{th} root" of a. For example, the square root of a is $a^{1/2}$ and

$$(a^{1/2})^2 = a \tag{41}$$

More generally

$$(a^{1/n})^m = a^{m/n} \tag{42}$$

For example

$$4^{3/2} = (4^{1/2})^3 \tag{43}$$
$$= 2^3 \tag{44}$$
$$= 8 \tag{45}$$

Example 1

Evaluate $\dfrac{3^5 3^2}{3^3}$.

$$\frac{3^5 3^2}{3^3} = \frac{3^{5+2}}{3^3}$$
$$= \frac{3^7}{3^3}$$
$$= 3^{7-3}$$
$$= 3^4$$
$$= 3 \times 3 \times 3 \times 3$$
$$= 9 \times 9$$
$$= 81$$

Example 2

Evaluate $(x^2 + y^2)^{3/2}$ when $x = 3$, $y = 4$.
$$(3^2 + 4^2)^{3/2} = (9 + 16)^{3/2}$$
$$= (25)^{3/2}$$
$$= (25^{1/2})^3$$
$$= 5^3$$
$$= 125$$

Example 3

Evaluate $9^{-3/2}$.

$$9^{-3/2} = \frac{1}{9^{3/2}}$$

$$= \frac{1}{(9^{1/2})^3}$$

$$= \frac{1}{3^3}$$

$$= \frac{1}{3 \times 3 \times 3}$$

$$= \frac{1}{27}$$

PROBLEMS

Evaluate

16. 2^5 **17.** $3^2 \times 3^3$ **18.** $3^2 \times 2^3$

19. $\dfrac{3^5}{3^7}$ **20.** $\left(\dfrac{1}{5}\right)^{-2}$ **21.** $(\tfrac{1}{4})^{-1/2}$

22. $(2^3)^2$ **23.** $(4^3)^{1/2}$ **24.** $4^{-3/2}$

25. $\dfrac{a^3 + b^3}{a^2 + b^2}$ when $a = 2, b = 3$

26. $(a^2 + b^2)^{1/2}$ when $a = 2, b = 3$

27. $(a^{1/2} + ab + b^{1/2})^{1/2}$ when $a = 4, b = 9$

A4   Powers of Ten

Numbers such as 25,310,000 and 0.00000537 are cumbersome, and calculations are made much easier if such numbers are expressed in terms of powers of ten.

For example

$$25{,}310{,}000 = 2.531 \times 10{,}000{,}000 \tag{46}$$

$$= 2.531 \times 10^7 \tag{47}$$

$$\text{Also } 0.00000537 = \frac{5.37}{1{,}000{,}000} \tag{48}$$

$$= 5.37 \times 10^{-6} \tag{49}$$

In each case the result is a number between 1 and 10 multiplied by the appropriate power of ten.

A simple way to find the exponent of ten is the method of shifting the decimal point. If the number is greater than ten, place the point of your pen or pencil on the decimal point and shift it to the left one figure at a time. Each shift is equivalent to dividing by ten, and must therefore be compensated by multiplying by one power of ten. Continue until there is only one figure to the left of the decimal point, meanwhile counting the total number of shifts, which is then the required exponent of ten. If the number is less than 1, again place the point of your pen or pencil on the decimal point and shift it to the *right*, one figure at a time. Each shift is equivalent to multiplying by ten, and must be compensated by dividing by ten, or multiplying by 10^{-1}. Continue until the decimal point is to the right of the first figure which is not zero. The required exponent of ten is then minus the number of shifts.

Algebra

753

Example 1

Express 25,310,000 in powers of ten. Introducing a decimal point, the number is

$$25310000.0$$

After 1 shift	2531000.00×10^1
After 2 shifts	253100.000×10^2
After 3 shifts	25310.0000×10^3
After 4 shifts	2531.00000×10^4
After 5 shifts	253.100000×10^5
After 6 shifts	25.3100000×10^6
After 7 shifts	2.53100000×10^7

Since the decimal point then has only one figure to the left of it, this is the desired answer. Unless the five zeros following the 1 have special significance (see section A5), the answer would normally be written $25,310,000 = 2.531 \times 10^7$

Example 2

Express 0.00000537 in powers of ten.

After 1 shift	$00.0000537 \times 10^{-1}$
After 2 shifts	$000.000537 \times 10^{-2}$
After 3 shifts	$0000.00537 \times 10^{-3}$
After 4 shifts	$00000.0537 \times 10^{-4}$
After 5 shifts	$000000.537 \times 10^{-5}$
After 6 shifts	$0000005.37 \times 10^{-6}$

The six initial zeros are clearly redundant and so $0.00000537 = 5.37 \times 10^{-6}$

Example 3

Evaluate $\sqrt{2 \times 10^9}$

$$\sqrt{2 \times 10^9} = (2 \times 10^9)^{1/2}$$
$$= 2^{1/2} \times 10^{9/2}$$
$$= 1.414 \times 10^{4\frac{1}{2}}$$

However, this is *not* what we want, because fractional powers of ten do not have the same simple significance as integral powers of ten. The correct procedure is:

$$(2 \times 10^9)^{1/2} = (20 \times 10^8)^{1/2}$$
$$= 20^{1/2} \times 10^{8/2}$$
$$= \sqrt{20} \times 10^4$$
$$= 4.47 \times 10^4$$

MATHEMATICAL
APPENDIX

Example 4

Evaluate $v = \sqrt{\dfrac{GM_e}{R}}$ when $G = 6.67 \times 10^{-11}$, $M_e = 6.0 \times 10^{24}$

754   and $R = 7 \times 10^6$.

$$v = \left(\frac{6.67 \times 10^{-11} \times 6.0 \times 10^{24}}{7 \times 10^6} \right)^{1/2}$$

$$= \left(\frac{6.67 \times 6.0 \times 10^{24-11}}{7 \times 10^6} \right)^{1/2}$$

$$= \left(\frac{40.02 \times 10^{13}}{7 \times 10^6} \right)^{1/2}$$

$$= \left(\frac{40.02}{7} \times 10^{13-6} \right)^{1/2}$$

$$= (5.72 \times 10^7)^{1/2}$$

$$= (57.2 \times 10^6)^{1/2}$$

$$= \sqrt{57.2} \times 10^{6/2}$$

$$= 7.56 \times 10^3$$

PROBLEMS

28. If you really understand powers of ten, you should be able to tell immediately, just by glancing at the figures, which is the greater of each pair of numbers

(i) 10^9 or 10^3 (ii) 10^3 or 10^{-9} (iii) 10^{-4} or 10^{-6}

(iv) 4×10^6 or 2×10^6 (v) 4×10^{-6} or 2×10^{-6}

(vi) 2×10^{-3} or 4×10^{-4}

29. Express in powers of ten

(i) 53,200   (ii) 216,532,800   (iii) 232

(iv) 0.0013   (v) 0.00000027   (vi) 0.000956

30. Evaluate

(i) $(9 \times 10^8)^{1/2}$ (ii) $\sqrt{1.6 \times 10^7}$ (iii) $\sqrt{0.000064}$

(iv) $(2.7 \times 10^4)^{1/3}$ (v) $(0.000125)^{1/3}$

31. Evaluate $\dfrac{(3.4 \times 10^5) \times (5 \times 10^{-11})}{(2 \times 10^3)^2}$

32. Evaluate $\dfrac{(2 \times 10^5)^2 (3 \times 10^2)^3}{(5 \times 10^{-6})^2}$

33. Evaluate $T = 2\pi \sqrt{\dfrac{R^3}{GM_e}}$

when $R = 1.6 \times 10^9$ $G = 6.67 \times 10^{-8}$

and $M_e = 8.5 \times 10^{31}$

34. Evaluate $\lambda = \sqrt{\dfrac{\eta T}{\pi \rho}}$

when $\eta = 2 \times 10^{-5}$, $T = 30$

and $\rho = 0.145$

Algebra

A5   Significant Figures

The numerical value of a physical quantity is usually not known with perfect accuracy, the only exceptions being a few quantities which are *defined* to have exact values. For example, the temperature of the ice-point is defined to be 273.15°K exactly (Chapter 12). However, a quantity such as the velocity of light has to be measured, and there is always a limit to the accuracy of the measurement. The currently accepted value for the velocity of light is 2.99793×10^{10} cm/sec, but the uncertainty is 0.00001×10^{10} cm/sec. This means that the exact value is not known, but it lies somewhere between $(2.99793 - 0.00001) \times 10^{10} = 2.99792 \times 10^{10}$ and $(2.99793 + 0.00001) \times 10^{10} = 2.99794 \times 10^{10}$.

Careful attention to experimental errors is important in exact scientific work, but it is not essential in an elementary text such as this one. It is sufficient to use the concept of **significant figures** and a few simple rules associated with this concept. The quoted value of the velocity of light, 2.99793×10^{10} cm/sec, has six significant figures. The first five, 2, 9, 9, 7, and 9 are certainly correct. The sixth, 3, is uncertain, since it might really be 2 or 4. It would therefore be ridiculous to quote a value of 2.997934×10^{10}, since the seventh figure, 4, would have no significance at all.

To count the number of significant figures, start with the first figure on the left which is not zero and end with the first figure on the right which is uncertain. Ambiguity is avoided if all numbers are expressed in powers of ten, as explained above, and if only one uncertain figure is included. If a number is written as 0.001567, with the 7 uncertain, then it has four significant figures, and this is more obvious when it is expressed as 1.567×10^{-3}. The number 3.201×10^5 has four significant figures, since there is no reason why the zero should not be counted. The number 3.210×10^5 has four significant figures if it was written this way with the deliberate intent that the zero should imply that the exact value lies between 3.209 and 3.211×10^5.

In a calculation involving the multiplication and division of several quantities, the result cannot be more accurate than the least accurate of the numbers used. If some of the quantities are quoted with more significant figures than the least accurate, a good working rule is as follows. For the accurate quantities retain one more significant figure than for the quantity with the smallest number of significant figures. Quote the result with one more significant figure than the least accurate of the quantities which went into the calculation. The last figure in the result is then often meaningless, except in a case such as

$$\frac{1.00}{0.99} = 1.01 \tag{50}$$

Here the numerator, with three significant figures, really lies between 0.99 and 1.01 and has an accuracy of 1%. The denominator, although it has only two significant figures, lies between 0.98 and 1.00 and therefore also has an accuracy of 1%. If the result were expressed as 1.0, with only two significant figures, this would imply that it could lie anywhere between 0.9 and 1.1, with an accuracy of 10%. Clearly, the accuracy is nearer to 1%. This last example illustrates that the concept of significant figures is only a crude substitute for a proper treatment of errors.

The reader of this book should not worry unduly about significant figures and very little attention is paid to them in the text. As far as the problems are concerned, it may be assumed that two or sometimes three significant figures are usually adequate for the answer. The important thing is to avoid obvious absurdities, such as the statement that a circle with a diameter of 0.010 cm has a circumference of 0.0314159 cm.

PROBLEMS

35. How many significant figures are there in
    (i) 2.678×10^2 (ii) 0.010 (iii) 0.005608
    (iv) 268.5×10^5 (v) 6.000×10^9 (vi) 0.61×10^{-3}

36. Evaluate the following and express the answer with an appropriate number of significant figures

(i) $\dfrac{(2.36 \times 10^9) \times (8.97482 \times 10^{-4})}{(6.534 \times 10^{11})}$

(ii) $\dfrac{\pi \times (0.007295)^2 \times 101}{(99)^2}$

(iii) $5.62 - 0.01578$ (Think!)

A6 Algebraic Procedures

The procedure for multiplying brackets is to take the terms in the first bracket one by one and multiply each into all the terms in the second bracket one by one. This is best understood by a careful study of the following examples:

$$(a + b)(c + d) = a(c + d) + b(c + d) \tag{51}$$
$$= ac + ad + bc + bd \tag{52}$$
$$(a + b + c)(d + e + f) = a(d + e + f) + b(d + e + f)$$
$$+ c(d + e + f) \tag{53}$$
$$= ad + ae + af + bd + be + bf$$
$$+ cd + ce + cf \tag{54}$$

The following special cases are particularly important.

First, $(a + b)(a - b) = a(a - b) + b(a - b)$ (55)
$$= a^2 - ab + ba - b^2 \tag{56}$$

Since ab and ba mean the same thing, they cancel one another, and so

$$(a + b)(a - b) = a^2 - b^2 \tag{57}$$

Secondly, $(a + b)^2 = (a + b)(a + b)$ (58)
$$= a(a + b) + b(a + b) \tag{59}$$
$$= a^2 + ab + ba + b^2 \tag{60}$$

Algebra

$$(a + b)^2 = a^2 + 2ab + b^2 \tag{61}$$

Thirdly, $\quad (a - b)^2 = (a - b)(a - b) \tag{62}$
$$= a(a - b) - b(a - b) \tag{63}$$
$$= a^2 - ab - ba + b^2 \tag{64}$$

$$(a - b)^2 = a^2 - 2ab + b^2 \tag{65}$$

An equation remains valid if the same quantity is added to both sides or subtracted from both sides. To illustrate this, suppose that we are told that

$$2x - 3 = x + 7 \tag{66}$$

and we wish to find the value of x. Subtract x from both sides of the equation, which then becomes

$$2x - x - 3 = x - x + 7 \tag{67}$$
or $\quad x - 3 = 7 \tag{68}$

Now add 3 to both sides of the equation

$$x - 3 + 3 = 7 + 3 \tag{69}$$
or $\quad x = 10 \tag{70}$

which is the desired result.

An equation remains valid if both sides are multiplied by the same quantity or divided by the same quantity. To simplify the equation

$$\frac{x + y}{x - y} = \frac{x - y}{x + y} \tag{71}$$

let us multiply both sides by $(x - y)$. Then

$$\frac{(x + y)(x - y)}{(x - y)} = \frac{(x - y)(x - y)}{(x + y)} \tag{72}$$

On the left hand side, multiplying by $(x - y)$ and then dividing by $(x - y)$ leaves us where we started, and so the $(x - y)$ in the numerator cancels the $(x - y)$ in the denominator. Therefore,

$$(x + y) = \frac{(x - y)^2}{(x + y)} \tag{73}$$

Similarly, multiplying both sides by $(x + y)$

$$(x + y)^2 = (x - y)^2 \tag{74}$$

Using equations 61 and 65,

$$x^2 + 2xy + y^2 = x^2 - 2xy + y^2 \tag{75}$$

Subtracting $x^2 + y^2$ from both sides

$$2xy = -2xy \tag{76}$$

Adding $2xy$ to both sides

$$4xy = 0 \tag{77}$$

Dividing both sides by 4

$$xy = \frac{0}{4} \tag{78}$$

or $xy = 0$ (79)

A little thought will convince the reader that this is possible only if

either $x = 0$ (80)

or $y = 0$ (81)

Fractions are added by arranging for them all to have a common denominator. Consider the expression

$$p = \frac{1}{a - b} - \frac{1}{(a + b)} \tag{82}$$

Since a quantity is unchanged by multiplying it by a second quantity and simultaneously dividing by this second quantity,

$$p = \frac{1}{(a - b)}\frac{(a + b)}{(a + b)} - \frac{1}{(a + b)}\frac{(a - b)}{(a - b)} \tag{83}$$

Both fractions now have the same denominator $(a + b)(a - b)$, so

$$p = \frac{(a + b) - (a - b)}{(a + b)(a - b)} \tag{84}$$

$$= \frac{a + b - a + b}{(a^2 - b^2)} \tag{85}$$

or $p = \dfrac{2b}{(a^2 - b^2)}$ (86)

Example 1

Find the value of α which satisfies the equation

$$(\alpha + 2)(2\alpha + 5) = 2(\alpha - 1)^2$$

The equation may be written

$$\alpha(2\alpha + 5) + 2(2\alpha + 5) = 2(\alpha^2 - 2\alpha + 1)$$

or $2\alpha^2 + 5\alpha + 4\alpha + 10 = 2\alpha^2 - 4\alpha + 2$

Subtracting $2\alpha^2$ from each side and adding the terms 5α and 4α on the left side

$$9\alpha + 10 = -4\alpha + 2$$

Algebra

759

Adding 4α to each side

$13\alpha + 10 = +2$

Subtracting 10 from each side

$13\alpha = -8$

Dividing both sides by 13

$$\alpha = -\frac{8}{13}$$

or $\quad \alpha = -0.615$

PROBLEMS

37. Evaluate $(3 + 5)(2 + 8)$ by first performing the additions and then the multiplication. Then evaluate it by the rule for multiplying brackets and check that you obtain the same answer.

38. Multiply the brackets in the following expressions
  (i) $(\alpha - \beta)(\alpha^2 + \beta^2)$
  (ii) $(a + b)(a^2 - ab + b^2)$
  (iii) $(a - b)(a^2 + ab + b^2)$
  (iv) $(x + y + z)(xy + yz + zx)$

39. Find x, given that
  (i) $3x + 4 = 2x - 1$
  (ii) $2(x + 5) = 3(x - 2)$
  (iii) $(x - 1)^2 + 4 = (x + 1)^2$
  (iv) $\dfrac{x - 1}{x + 5} = \dfrac{x + 3}{x - 2}$

40. Express the following with a common denominator

  (i) $\dfrac{1}{a} + \dfrac{1}{b} + \dfrac{1}{c}$

  (ii) $\dfrac{x}{yz} + \dfrac{y}{zx} + \dfrac{z}{xy}$

  (iii) $\dfrac{x}{(x + a)} - \dfrac{y}{(y + b)}$

A7 Approximations

The first part of this section need not be read until the student has reached Chapter 9 of the main text. The second part can be deferred until Chapter 23.

In Chapter 9, we consider the expression

$$\frac{1}{R_e} - \frac{1}{R_e + h}$$

where R_e is the radius of the earth (6.37×10^8 cm) and h is a small height very much less than R_e. We proceed as follows. Using a common denominator, $R_e(R_e + h)$, the expression becomes

$$\frac{(R_e + h) - R_e}{R_e(R_e + h)} = \frac{h}{R_e(R_e + h)}$$

$$= \frac{h}{R_e R_e \left(1 + \dfrac{h}{R_e}\right)}$$

$$= \frac{h}{R_e{}^2 \left(1 + \dfrac{h}{R_e}\right)} \qquad (87)$$

We then state that h/R_e is so small compared with 1 that it can be neglected, and we therefore finally conclude that

$$\frac{1}{R_e} - \frac{1}{R_e + h} \simeq \frac{h}{R_e{}^2} \qquad (88)$$

The symbol $\simeq$ means "approximately equal to."

Consider the justification of this procedure. Suppose $h = 10^5$ cm, the largest value it is likely to have within the context of the discussion (which is related to bodies moving up and down near the surface of the earth. 10^5 cm is about three times the height of the Empire State Building). Then

$$\frac{h}{R_e{}^2} = \frac{10^5}{(6.37 \times 10^8)^2}$$

$$= 2.474 \times 10^{-13} \qquad (89)$$

$$\frac{h}{R_e} = \frac{10^5}{6.37 \times 10^8}$$

$$= 1.570 \times 10^{-4} \qquad (90)$$

$$\frac{1}{1 + \dfrac{h}{R_e}} = \frac{1}{1 + (1.57 \times 10^{-4})}$$

$$= \frac{1}{1.000157}$$

$$= 0.999843 \qquad (91)$$

Therefore,

$$\frac{h}{R_e{}^2} \cdot \frac{1}{\left(1 + \dfrac{h}{R_e}\right)} = (2.474 \times 10^{-13})(0.999843)$$

$$= (2.474 \times 10^{-13})(1 - 0.000157)$$

$$= (2.474 - 0.000389) \times 10^{-13} \qquad (92)$$

Algebra

The point of this tedious calculation is to see how much difference it makes to neglect the factor $1 \big/ \left(1 + \dfrac{h}{R_e}\right)$. The result is given with four significant figures and three figures to the right of the decimal point. The neglected factor influences the fourth figure to the right of the decimal point and can therefore be completely ignored.

More generally, let us suppose that x is very much smaller than 1 and consider the expression $1/(1 + x)$. Put $a = 1$ and $b = x$ in equation 57. Then

$$(1 - x^2) = (1 - x)(1 + x) \tag{93}$$

Dividing both sides by $(1 + x)$,

$$\frac{(1 - x^2)}{(1 + x)} = (1 - x) \tag{94}$$

If x is very small, then x^2 is even smaller and can be neglected. For example, if $x = 0.0001$, then $(1 - x) = 0.9999$, but $(1 - x^2) = 0.99999999$.

So, if x is very much less than 1,

$$\frac{1}{1 + x} \approx 1 - x \tag{95}$$

or

$$\frac{1}{1 - x} \approx 1 + x \tag{96}$$

The expression discussed above might therefore be handled in the following way.

$$\frac{1}{R_e} - \frac{1}{R_e + h} = \frac{1}{R_e} - \frac{1}{R_e\left(1 + \dfrac{h}{R_e}\right)}$$

$$= \frac{1}{R_e}\left[1 - \frac{1}{\left(1 + \dfrac{h}{R_e}\right)}\right]$$

$$\approx \frac{1}{R_e}\left[1 - \left(1 - \frac{h}{R_e}\right)\right]$$

$$\approx \frac{1}{R_e}\left[\frac{h}{R_e}\right]$$

$$\approx \frac{h}{R_e^2} \tag{97}$$

The remainder of this section need not be read until the student has reached Chapter 23.

In the theory of relativity it is often necessary to calculate the value of $\sqrt{1 - \dfrac{v^2}{c^2}}$ when the velocity v is very much smaller than the velocity of light c. Under these circumstances $\dfrac{v^2}{c^2}$ is very much smaller than 1 and $1 - \dfrac{v^2}{c^2}$ is just slightly smaller than 1. $\sqrt{1 - \dfrac{v^2}{c^2}}$ is therefore also just slightly smaller than 1, but it is often not good enough to put it approximately equal to 1, because what matters is the small difference between $\sqrt{1 - \dfrac{v^2}{c^2}}$ and 1. Let us therefore define the number z by the equation

$$\sqrt{1 - \frac{v^2}{c^2}} = 1 - z \tag{98}$$

When v is very much smaller than c, z is very much smaller than 1. If the number on the left hand side of equation 98 is equal to the number on the right hand side, then the square of the number on the left hand side must be equal to the square of the number on the right hand side.

$$1 - \frac{v^2}{c^2} = (1 - z)^2 \tag{99}$$

The squared bracket on the right hand side can be evaluated using equation 65 if we put $a = 1$ and $b = z$. Then

$$1 - \frac{v^2}{c^2} = 1 - 2z + z^2 \tag{100}$$

or

$$\frac{v^2}{c^2} = 2z - z^2 \tag{101}$$

If z is very much less than 1, z^2 is very much smaller than $2z$. For example, if $z = 0.001$, then $2z = 0.002$ and $z^2 = 0.000001$. We can therefore neglect z^2 in equation 101 and put

$$\frac{v^2}{c^2} \approx 2z \tag{102}$$

or

$$z \approx \frac{1}{2} \frac{v^2}{c^2} \tag{103}$$

Inserting this value of z into equation 98.

When v is very much smaller than c

$$\sqrt{1 - \frac{v^2}{c^2}} \approx 1 - \frac{1}{2} \frac{v^2}{c^2} \tag{104}$$

Algebra

Since $\dfrac{1}{2} \dfrac{v^2}{c^2}$ is very small compared with 1, it follows from equation 95 that

763

$$1 - \frac{1}{2}\frac{v^2}{c^2} \approx \frac{1}{1 + \frac{1}{2}\frac{v^2}{c^2}} \tag{105}$$

It is then immediately obvious from equations 104 and 105 that

$$\frac{1}{\sqrt{1 - \frac{v^2}{c^2}}} \approx 1 + \frac{1}{2}\frac{v^2}{c^2} \tag{106}$$

Throughout the above arguments we could have used x instead of v^2/c^2 and we would then have obtained the following more general results.

If x is very much smaller than 1

$$\sqrt{1 - x} \approx 1 - \tfrac{1}{2}x \tag{107}$$

$$\frac{1}{\sqrt{1 - x}} \approx 1 + \tfrac{1}{2}x \tag{108}$$

Another problem encountered in the theory of relativity is to evaluate $\sqrt{1 - \frac{v^2}{c^2}}$ when v is not very small compared with c, but is almost equal to c, that is, when

$$\frac{v}{c} = 1 - \alpha \tag{109}$$

and α is very much less than 1, so that v is just slightly smaller than c. Then

$$\sqrt{1 - \frac{v^2}{c^2}} = \sqrt{1 - (1 - \alpha)^2} \tag{110}$$

The squared bracket may be evaluated by using equation 65 with $a = 1$, $b = \alpha$.

$$\sqrt{1 - \frac{v^2}{c^2}} = \sqrt{1 - (1 - 2\alpha + \alpha^2)} \tag{111}$$

$$= \sqrt{2\alpha - \alpha^2} \tag{112}$$

If α is very much smaller than 1, α^2 is very much smaller than 2α and can be omitted. Therefore,

$$\sqrt{1 - \frac{v^2}{c^2}} \approx \sqrt{2\alpha} \tag{113}$$

From equation 109

$$\alpha = 1 - \frac{v}{c} \tag{114}$$

Therefore, when v is slightly smaller than c

$$\sqrt{1 - \frac{v^2}{c^2}} \approx \sqrt{2\left(1 - \frac{v}{c}\right)} \qquad (115)$$

PROBLEMS

41. Without actually performing a long division, evaluate the following

(i) $1 - \dfrac{1}{1.00001}$

(ii) $\dfrac{1}{2} - \dfrac{1}{2.000005}$

(iii) $\dfrac{1}{15} - \dfrac{0.2}{3.00009}$

42. Evaluate $\sqrt{1 - \dfrac{v^2}{c^2}}$ when

(i) $v = 0.000100c$ (ii) $\dfrac{v}{c} = 5.00 \times 10^{-4}$

(iii) $\dfrac{v}{c} = 2.38 \times 10^{-2}$

Note: The *difference* between $\sqrt{1 - \dfrac{v^2}{c^2}}$ and 1 should be given with three significant figures.

43. Evaluate $\dfrac{1}{\sqrt{1 - \dfrac{v^2}{c^2}}}$ when

(i) $v = 0.00100c$ (ii) $\dfrac{v}{c} = 3.00 \times 10^{-3}$

(iii) $\dfrac{v}{c} = 5.76 \times 10^{-5}$

Note: The difference between $\dfrac{1}{\sqrt{1 - \dfrac{v^2}{c^2}}}$ and 1 should be given with three significant figures.

44. Evaluate $\sqrt{1 - \dfrac{v^2}{c^2}}$ when

(i) $\dfrac{v}{c} = 0.999990$ (ii) $\dfrac{v}{c} = 0.99934$

(iii) $\dfrac{v}{c} = 0.9999999916$

Algebra

765

A8 Infinity

The first part of this section need not be read until the student has reached Chapter 9. The second part can be deferred until Chapter 24.

"When R becomes infinite, $1/R$ becomes zero." What does this statement mean? The concept of infinity in mathematics is a subtle one and we shall not attempt to discuss it with meticulous detail. From a practical point of view the statement may be taken to mean "If R is made large enough, $1/R$ can be made so small that it can be neglected." It is clear that the larger we make R, the smaller $1/R$ becomes. When R is 100, $1/R$ is 0.01. When R is 1,000,000, $1/R$ is 0.000001. When R is 1,000,000,000, $1/R$ is 0.000000001, and so on. How small $1/R$ has to be before it can be neglected depends upon the particular problem that is being considered. Nevertheless, however small it has to be, it can obviously be made this small by making R large enough.

Another simple way to express this is as follows. The statement "R becomes infinite" means that R becomes larger than any number you can think of, however large you try to make it. The statement "$1/R$ becomes zero" means that $1/R$ becomes smaller than any number you can think of, however small you try to make it. Infinity is sometimes treated as though it were an ordinary number and is represented by the symbol ∞. The preceding discussion can then be summarized by the equation

$$\frac{1}{\infty} = 0 \tag{116}$$

Now consider another statement of a similar kind. "When α becomes zero, $1/\alpha$ becomes infinite," This means that, by making α small enough, we can make $1/\alpha$ as large as we please. For example, if α is 0.01, then $1/\alpha$ is 100. If α is 0.000001, then $1/\alpha$ is 1,000,000. If α is 0.000000001, then $1/\alpha$ is 1,000,000,000, and so on. In simplified symbolic form

$$\frac{1}{0} = \infty \tag{117}$$

The remainder of this section need not be read until the student has reached Chapter 24.

A case of particular interest in the theory of relativity is the formula for the mass m of a body which has a velocity v.

$$m = \frac{m_0}{\sqrt{1 - \dfrac{v^2}{c^2}}} \tag{118}$$

Here m_0 is the mass of the body when it is at rest and c is the velocity of light. If the velocity of the body were equal to the velocity of light ($v = c$), then v/c would be 1, and the quantity inside the square root sign would be $1 - 1 = 0$. We would then have

$$m = \frac{m_0}{0} = \infty \tag{119}$$

The idea of infinite mass in nature seems ridiculous, so what does this really mean? Actually, the velocity of a material body can never be exactly equal to the velocity of light, but must always be less than it. However, v can approach very close to c, so that v/c is only slightly smaller than 1. Then v^2/c^2 would be only slightly smaller than 1 and $1 - v^2/c^2$ would be a very small quantity. $\sqrt{1 - \dfrac{v^2}{c^2}}$ would also be a very small quantity. $m = \dfrac{m_0}{\sqrt{1 - \dfrac{v^2}{c^2}}}$ would therefore be a very large quantity. The situation, then, is that, as v approaches closer and closer to c, m becomes larger and larger and can be made as large as you please by making the difference between v and c small enough.

The quantity 0/0 has no well-defined meaning. It might seem obvious that the zero in the numerator cancels the zero in the denominator, giving 1, since any number divided by itself gives 1. However, consider a quantity such as

$$\frac{2(1 - x)}{(1 - x)} = 2 \tag{120}$$

If we put $x = 1$, then both numerator and denominator become 0 but it would now seem reasonable to continue to take the answer to be 2. On the other hand, $\dfrac{3(1 - x)}{(1 - x)}$ would be most reasonably taken to be 3. In fact it seems that 0/0 might be any number, depending upon the circumstances.

Example

Calculate m when $m_0 = 1$ gm and
(i) $v = 0.999900c$ (ii) $v = 0.9999999900c$

In equation 118, put $m_0 = 1$ and also, since v is only slightly smaller than c, make use of the approximation of equation 115.

$$\sqrt{1 - \frac{v^2}{c^2}} \approx \sqrt{2\left(1 - \frac{v}{c}\right)}$$

Then
$$m = \frac{1}{\sqrt{2\left(1 - \dfrac{v}{c}\right)}}$$

In case (i) $\left(1 - \dfrac{v}{c}\right) = 0.000100$

$$= 1.00 \times 10^{-4}$$

Therefore
$$m = \frac{1}{\sqrt{2 \times 1.00 \times 10^{-4}}}$$

$$= \frac{1}{1.414 \times 10^{-2}}$$

$$= 70.7 \text{ gm}$$

Algebra

767

In case (ii) $\left(1 - \dfrac{v}{c}\right) = 0.0000000100$

$$= 1.00 \times 10^{-8}$$

Therefore $\qquad m = \dfrac{1}{\sqrt{2 \times 1.00 \times 10^{-8}}}$

$$= \dfrac{1}{1.414 \times 10^{-4}}$$

$$= 0.707 \times 10^4$$

$$= 7{,}070 \text{ gm.}$$

B / GEOMETRY

B1   Triangles

The sum of the three angles of a triangle is equal to 180°. For example, in figure 1 the angle ABC (which means the angle at the point B) is 110° and the angle BCA is 25°. The third angle CAB must therefore be $180° - 110° - 25° = 45°$

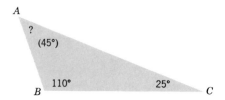

Figure 1   The sum of the three angles of a triangle is 180°.

Two triangles are congruent if they have the same shape and size. It is then possible to lift up one triangle and place it exactly on top of the other. In figure 2, triangles ABC and $A'B'C'$ are congruent. Any angle of one triangle is equal to the corresponding angle of its congruent triangle. For example, angle BAC is equal to angle $B'A'C'$. Any side of one triangle has a length equal to the length of the corresponding side of its congruent triangle. For example, $AB = A'B'$. (The expression AB is frequently used to mean the length of the straight line joining the point A to the point B).

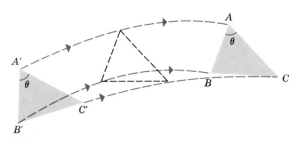

Figure 2   Congruent triangles.

Two triangles are similar if they have the same shape, but not necessarily the same size. The smaller triangle can be placed exactly on top of the larger triangle if it is first expanded uniformly in all directions by the right amount. In figure 3 triangles PQR and $P'Q'R'$ are similar. Any angle of one triangle is equal to the corresponding angle of the other triangle. For example, angle RPQ is equal to angle $R'P'Q'$. The sides are not equal in pairs, but the sides of one triangle bear a constant ratio to the sides of the other triangle. For example

$$\frac{PQ}{P'Q'} = \frac{QR}{Q'R'} = \frac{RP}{R'P'} \tag{121}$$

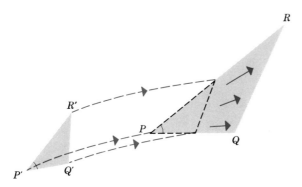

Figure 3   Similar triangles.

An **isosceles triangle** is one in which two of the sides are equal. In figure 4 the isosceles triangle ABC has $AB = AC$. In such an isosceles triangle the angle ABC is equal to the angle ACB.

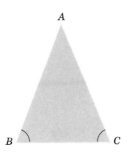

Figure 4   An isosceles triangle.

An **equilateral triangle** is one in which all three sides are equal. In the equilateral triangle of figure 5

$$PQ = QR = RP \tag{122}$$

All three angles of the triangle are then equal to one another. Since the sum of all the three angles must be $180°$, each one must be $60°$ since $60° + 60° + 60° = 180°$.

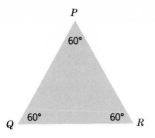

Figure 5 An equilateral triangle.

B2 Right Triangles and Pythagoras' Theorem

A **right angled triangle** (often called a **right triangle**) has one of its angles equal to 90°. The sum of the other two angles must then be 90°. The **hypotenuse** is the side opposite the right angle. In figure 6 it is the angle BAC which is equal to 90°, and the hypotenuse is the side BC. Right angled triangles obey Pythagoras' theorem, which states that: "**The square of the length of the hypotenuse is equal to the sum of the squares of the lengths of the other two sides.**" For triangle ABC of figure 6, this would imply that

$$BC^2 = AB^2 + AC^2 \tag{123}$$

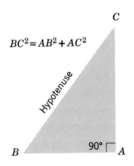

Figure 6 A right triangle.

A famous right triangle has $AB = 3$ units of length and $AC = 4$ units of length. Then, applying Pythagoras' theorem,

$$BC^2 = 3^2 + 4^2 \tag{124}$$
$$= 9 + 16 \tag{125}$$
$$= 25 \tag{126}$$
or $\quad BC = 5$ units of length $\tag{127}$

This is often called a **3-4-5 triangle** (figure 7).

Another important triangle is the **45° right triangle,** which has one angle equal to 90° and the other two each equal to 45° ($90° + 45° + 45° = 180°$). This is shown in figure 8, in which the angle BAC is 90°

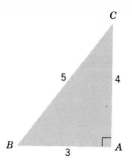

Figure 7 A 3-4-5 triangle.

and the angles ABC and ACB are each $45°$. Since these last two angles are equal, it is an isosceles triangle with

$$AB = AC \tag{128}$$

From Pythagoras' theorem

$$BC^2 = AB^2 + AC^2 \tag{129}$$
$$= 2AB^2 \tag{130}$$

Therefore $BC = \sqrt{2} \times AB \tag{131}$

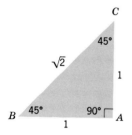

Figure 8 A $45°$ right triangle.

A **$60°$ right triangle** has its angles equal to $90°$, $60°$, and $30°$ ($90° + 60° + 30° = 180°$). It can be looked upon as half of an equilateral triangle (figure 9). Suppose that each side of this equilateral triangle has a length of 2 units.

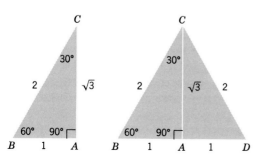

Figure 9 A $60°$ right triangle.

$$BC = CD = BD = 2 \text{ units of length} \tag{132}$$

Then AB is half as long as BD and

$$AB = AD \tag{133}$$

$$= \tfrac{1}{2}BD \tag{134}$$

That is, $AB = 1$ unit of length $\tag{135}$

Applying Pythagoras' theorem

$$BC^2 = AB^2 + AC^2 \tag{136}$$

$$2^2 = 1^2 + AC^2 \tag{137}$$

$$AC^2 = 2^2 - 1^2 \tag{138}$$

$$= 4 - 1 \tag{139}$$

$$= 3 \tag{140}$$

or $AC = \sqrt{3}$ units of length

A **30° right triangle** is clearly identical with a 60° right triangle (figure 10).

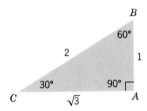

Figure 10 A 30° right triangle.

B3 Circles

Figure 11 illustrates the meaning of various concepts associated with a circle. It should be self-explanatory.

Length of the perimeter of a circle

$$= 2\pi \times (\text{radius}) \tag{141}$$

$$= 2\pi r \tag{142}$$

Area enclosed by a circle

$$= \pi \times (\text{radius})^2 \tag{143}$$

$$= \pi r^2 \tag{144}$$

$$\pi = 3.14159\ldots \tag{145}$$

Unless greater accuracy is obviously needed, π can be taken to three significant figures when working problems in this book.

$$\pi \approx 3.14 \tag{146}$$

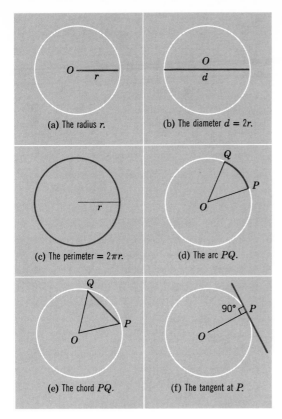

(a) The radius r.

(b) The diameter $d = 2r$.

(c) The perimeter $= 2\pi r$.

(d) The arc PQ.

(e) The chord PQ.

(f) The tangent at P.

Figure 11 Some concepts associated with a circle.

B4 Solid Geometry

Surface area of a sphere
$$= 4\pi \times (\text{radius})^2 \tag{147}$$

Volume of a sphere
$$= \frac{4\pi}{3} \times (\text{radius})^3 \tag{148}$$

Volume of a cylinder
$$= (\text{Area of base}) \times (\text{height}) \tag{149}$$

C / TRIGONOMETRY

C1 Meaning of Sine, Cosine, and Tangent

To define the trigonometric functions for the angle POQ (figure 12), suppose that it is part of a right angled triangle OPQ with the angle PQO equal to $90°$. The hypotenuse of this triangle is clearly OP. The

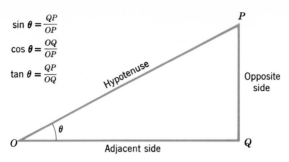

$$\sin \theta = \frac{QP}{OP}$$

$$\cos \theta = \frac{OQ}{OP}$$

$$\tan \theta = \frac{QP}{OQ}$$

Figure 12   The right triangle used to define the trigonometric functions.

side QP is called the **opposite side,** because it is opposite the angle POQ which interests us. The side OQ is called the **adjacent side,** because it is adjacent to the angle POQ. For convenience, let us denote the value of the angle POQ by θ.

The sine of the angle POQ is abbreviated to $\sin \theta$ and is defined as

$$\sin \theta = \frac{\text{Length of opposite side}}{\text{Length of hypotenuse}} \tag{150}$$

$$= \frac{QP}{OP} \tag{151}$$

The cosine of the angle POQ is abbreviated to $\cos \theta$ and is defined as

$$\cos \theta = \frac{\text{Length of adjacent side}}{\text{Length of hypotenuse}} \tag{152}$$

$$= \frac{OQ}{OP} \tag{153}$$

The tangent of the angle POQ is abbreviated to $\tan \theta$ and is defined as

$$\tan \theta = \frac{\text{Length of opposite side}}{\text{Length of adjacent side}} \tag{154}$$

$$= \frac{QP}{OQ} \tag{155}$$

The hypotenuse of a right triangle is always longer than either the opposite side or the adjacent side. It follows that $\sin \theta$ and $\cos \theta$ must always be less than 1. On the other hand, there is no restriction on the relative lengths of the opposite and adjacent sides, and so $\tan \theta$ may have any value between 0 and ∞.

Suppose that the numerical value of $\sin \theta$ is α.

$$\sin \theta = \alpha \qquad (156)$$

Sometimes the value of α is known and we wish to deduce the value of the angle θ. The expression, "the angle whose sine has the value α" is often abbreviated to $\sin^{-1} \alpha$ or sometimes arcsin α

$$\theta = \sin^{-1} \alpha = \arcsin \alpha \qquad (157)$$

Similarly if $\quad \cos \theta = \beta \qquad (158)$

$$\theta = \cos^{-1} \beta = \arccos \beta \qquad (159)$$

Also if $\quad \tan \theta = \gamma \qquad (160)$

$$\theta = \tan^{-1} \gamma = \arctan \gamma \qquad (161)$$

In algebra the exponent -1 is used to denote a reciprocal. $\left(x^{-1} = \dfrac{1}{x}. \right)$

It is important to realize that $\sin^{-1} \alpha$ does *not* mean $\dfrac{1}{\sin \alpha}$.

C2   Simple Relations Between Trigonometric Functions

Using equations 151 and 153

$$\frac{\sin \theta}{\cos \theta} = \frac{(QP/OP)}{(OQ/OP)} \qquad (162)$$

Multiply the numerator and the denominator by OP.
Then,

$$\frac{\sin \theta}{\cos \theta} = \frac{QP}{OQ} \qquad (163)$$

According to equation 155, this is the same as $\tan \theta$. Therefore,

$$\tan \theta = \frac{\sin \theta}{\cos \theta} \qquad (164)$$

Applying Pythagoras' theorem to the right triangle OPQ of figure 12

$$QP^2 + OQ^2 = OP^2 \qquad (165)$$

Dividing both sides by OP^2

$$\frac{QP^2}{OP^2} + \frac{OQ^2}{OP^2} = \frac{OP^2}{OP^2} \qquad (166)$$

or

$$\left(\frac{QP}{OP} \right)^2 + \left(\frac{OQ}{OP} \right)^2 = 1 \qquad (167)$$

From equations 151 and 153, it is seen that the quantities inside the brackets are $\sin \theta$ and $\cos \theta$ respectively. Therefore,

$$(\sin \theta)^2 + (\cos \theta)^2 = 1 \qquad (168)$$

Trigonometry

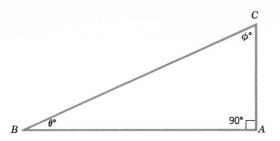

Figure 13 Trigonometric functions of complementary angles.

This is usually written

$$\sin^2 \theta + \cos^2 \theta = 1 \tag{169}$$

In the right triangle of figure 13, angle BAC is $90°$. Denote the value of angle ABC by $\theta°$ and the value of angle BCA by $\phi°$. Since the sum of the three angles of a triangle is $180°$

$$\theta + \phi + 90° = 180° \tag{170}$$

from which we obtain

$$\phi = 90° - \theta \tag{171}$$

The hypotenuse is BC and the side opposite to θ is AC. Therefore

$$\sin \theta = \frac{AC}{BC} \tag{172}$$

However, as far as the angle ϕ is concerned, AC is the adjacent side, and so

$$\cos \phi = \frac{AC}{BC} \tag{173}$$

Comparing equations 172 and 173, it is obvious that

$$\cos \phi = \sin \theta \tag{174}$$

That is,

$$\cos (90° - \theta) = \sin \theta \tag{175}$$

Similarly, BA is the side adjacent to θ and

$$\cos \theta = \frac{BA}{BC} \tag{176}$$

But BA is opposite to ϕ and so

$$\sin \phi = \frac{BA}{BC} \tag{177}$$

Figure 14 The angle *POQ* approaches 0°.

Comparing equations 176 and 177

$$\sin \phi = \cos \theta \tag{178}$$

That is,

$$\sin (90° - \theta) = \cos \theta \tag{179}$$

C3 Some Special Angles

In figure 14 the angle *POQ* is very small, almost zero. With *O* fixed and the length of *OP* constant, imagine that *OP* is slowly rotated clockwise until the angle *POQ* becomes zero. As this happens, the length of the opposite side *QP* becomes smaller and eventually becomes zero. Also, *P* and *Q* eventually coincide, and the adjacent side *OQ* becomes equal to the hypotenuse *OP*. Since $\sin \theta = QP/OP$, it eventually becomes $0/OP = 0$. Since $\cos \theta = OQ/OP$, it becomes 1. Since $\tan \theta = PQ/OQ$, it becomes 0. Therefore,

$$\sin 0° = 0 \tag{180}$$
$$\cos 0° = 1 \tag{181}$$
$$\tan 0° = 0 \tag{182}$$

In figure 15 the angle *POQ* is almost equal to 90°. With *O* fixed and the length *OP* constant, imagine that *OP* is slowly rotated counterclockwise until the angle *POQ* becomes 90°. As this happens, *Q* eventually coincides with *O*, and the opposite side *QP* becomes equal to the hypotenuse *OP*. The length of the adjacent side *OQ* becomes zero. Since $\sin \theta = QP/OP$, it becomes 1. Since $\cos \theta = OQ/OP$, it becomes 0.

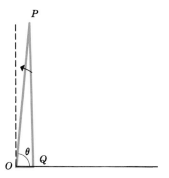

Trigonometry

Figure 15 The angle *POQ* approaches 90°.

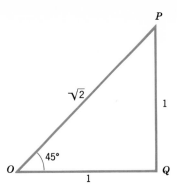

Figure 16 The trigonometric functions of 45°.

Since $\tan \theta = QP/OQ$, it becomes larger and larger as OQ becomes smaller and smaller, and eventually becomes infinite. Therefore,

$$\sin 90° = 1 \tag{183}$$
$$\cos 90° = 0 \tag{184}$$
$$\tan 90° = \infty \tag{185}$$

Notice that equations 183 and 184 could have been obtained from equations 180 and 181 by making use of the fact that $\sin \theta = \cos (90° - \theta)$.

When $\theta = 45°$ (figure 16), the triangle OPQ is a 45° right triangle (compare figure 8). The opposite side QP and the adjacent side OQ are then equal, and both may be made 1 unit of length. The hypotenuse OP then has a length of $\sqrt{2}$ units. Therefore,

$$\sin 45° = \frac{QP}{OP} = \frac{1}{\sqrt{2}} \tag{186}$$

$$\cos 45° = \frac{OQ}{OP} = \frac{1}{\sqrt{2}} \tag{187}$$

$$\tan 45° = \frac{QP}{OQ} = \frac{1}{1} \tag{188}$$

or

$$\sin 45° = \frac{1}{\sqrt{2}} = 0.707 \tag{189}$$

$$\cos 45° = \frac{1}{\sqrt{2}} = 0.707 \tag{190}$$

$$\tan 45° = 1.000 \tag{191}$$

When $\theta = 60°$ (figure 17), the triangle OPQ is a 60° right triangle (compare figure 9). If the adjacent side OQ is taken to be 1 unit of

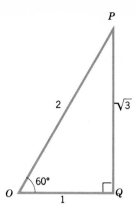

Figure 17 The trigonometric functions of 60°.

length, then the opposite side QP is $\sqrt{3}$ units and the hypotenuse OP is 2 units.

Therefore,

$$\sin 60° = \frac{QP}{OP} = \frac{\sqrt{3}}{2} \tag{192}$$

$$\cos 60° = \frac{OQ}{OP} = \frac{1}{2} \tag{193}$$

$$\tan 60° = \frac{QP}{OQ} = \frac{\sqrt{3}}{1} \tag{194}$$

or

$$\sin 60° = \frac{\sqrt{3}}{2} = 0.866 \tag{195}$$

$$\cos 60° = \frac{1}{2} = 0.500 \tag{196}$$

$$\tan 60° = \frac{\sqrt{3}}{1} = 1.732 \tag{197}$$

When $\theta = 30°$, we use a 30° right triangle, which is just a 60° right triangle turned round (figure 18). Then

$$\sin 30° = \frac{QP}{OP} = \frac{1}{2} \tag{198}$$

$$\cos 30° = \frac{OQ}{OP} = \frac{\sqrt{3}}{2} \tag{199}$$

$$\tan 30° = \frac{QP}{OQ} = \frac{1}{\sqrt{3}} \tag{200}$$

or

Trigonometry

779

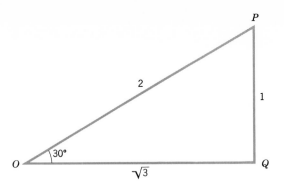

Figure 18 The trigonometric functions of $30°$.

$$\sin 30° = \frac{1}{2} = 0.500 \tag{201}$$

$$\cos 30° = \frac{\sqrt{3}}{2} = 0.866 \tag{202}$$

$$\tan 30° = \frac{1}{\sqrt{3}} = 0.577 \tag{203}$$

Notice that, since $30° + 60° = 90°$, equations 175 and 179 are obeyed.

$$\sin 60° = \cos 30° = \frac{\sqrt{3}}{2} \tag{204}$$

$$\cos 60° = \sin 30° = \tfrac{1}{2} \tag{205}$$

C4 Angles Greater than 90°

When θ is greater than $90°$, its trigonometric functions must be defined more carefully. Suppose that the point P moves on a circle with its center at O (figure 19). Let $P_0'OP_0$ be a diameter of this circle which, for convenience, is taken to be horizontal in figure 19. Suppose that OP initially coincides with OP_0, and then rotates counterclockwise through an angle θ, which can be greater than $90°$. Through P draw the line PQ, which is perpendicular to $P_0'OP_0$ and meets $P_0'OP_0$ at the point Q. The trigonometric functions are then defined as follows:

$$\sin \theta = \frac{QP}{OP} \tag{206}$$

$$\cos \theta = \frac{OQ}{OP} \tag{207}$$

$$\tan \theta = \frac{QP}{OQ} \tag{208}$$

When θ is less than $90°$, the triangle OPQ is identical with the one shown in figure 12, and the definitions just given are the same as those given earlier in equations 151, 153, and 155. However, in general it is

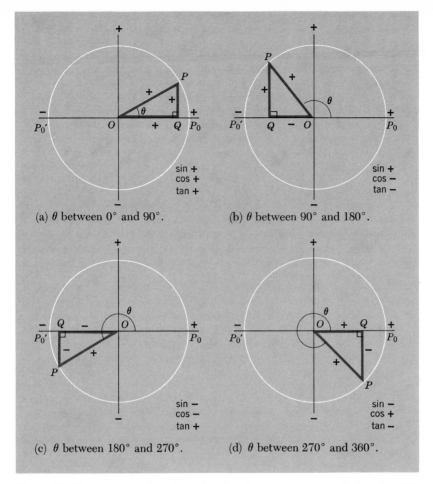

(a) θ between $0°$ and $90°$.

sin +
cos +
tan +

(b) θ between $90°$ and $180°$.

sin +
cos −
tan −

(c) θ between $180°$ and $270°$.

sin −
cos −
tan +

(d) θ between $270°$ and $360°$.

sin −
cos +
tan −

Figure 19   Trigonometric functions of angles that are not necessarily less than $90°$.

necessary to consider carefully the sign of the various lengths involved. The hypotenuse OP is always taken to be positive. The length QP is taken to be positive if P lies above Q, but negative if P lies below Q. The length OQ is taken to be positive if Q is to the right of O, but negative if Q is to the left of O. This is all clearly indicated in figure 19.

If θ lies between $0°$ and $90°$, all the lengths are positive, and therefore all the trigonometric functions are positive.

If θ lies between $90°$ and $180°$, OP is positive, QP is positive, but OQ is negative. Therefore, $\sin \theta = (QP)/(OP)$ is a positive quantity divided by a positive quantity, and so $\sin \theta$ is positive. However, $\cos \theta = (OQ)/(OP)$ is a negative quantity divided by a positive quantity, and so $\cos \theta$ is negative. Also, $\tan \theta = (QP)/(OQ)$ is a positive quantity divided by a negative quantity, and so $\tan \theta$ is negative.

When θ lies between $180°$ and $270°$, OP is positive, QP is negative and OQ is negative. It follows that $\sin \theta$ is negative, $\cos \theta$ is negative, but $\tan \theta$ is positive.

When θ lies between $270°$ and $360°$, OP is positive, QP is negative

Trigonometry

781

and OQ is positive. It follows that $\sin \theta$ is negative, $\cos \theta$ is positive and $\tan \theta$ is negative.

Ignoring the question of sign, the relative magnitudes of OP, QP and OQ can be found by considering the triangle OPQ in any of the four cases shown in figure 19. All that is necessary is to deduce the value of the angle POQ, which lies between 0 and 90°. For example, inspection of figure 19b shows that, when θ lies between 90° and 180°, the angle POQ is $(180° - \theta)$. Therefore,

$$\frac{\text{Absolute magnitude of length } OQ}{\text{Absolute magnitude of length } OP} = \cos (180° - \theta) \qquad (209)$$

However, when we wish to find $\cos \theta$, OQ is taken to be negative and OP to be positive, giving a negative value to $\cos \theta$. It follows that

$$\cos \theta = -\cos (180° - \theta) \qquad (210)$$

Many similar relations can be found. The important ones are quoted in problems 45, 46, and 47.

PROBLEMS

45. If θ lies between 90° and 180°, prove that,
$$\sin \theta = \sin (180° - \theta)$$
$$\cos \theta = -\cos (180° - \theta)$$
$$\tan \theta = -\tan (180° - \theta)$$

46. If θ lies between 180° and 270°, prove that
$$\sin \theta = -\sin (\theta - 180°)$$
$$\cos \theta = -\cos (\theta - 180°)$$
$$\tan \theta = \tan (\theta - 180°)$$

47. If θ lies between 270° and 360°, prove that
$$\sin \theta = -\sin (360° - \theta)$$
$$\cos \theta = \cos (360° - \theta)$$
$$\tan \theta = -\tan (360° - \theta)$$

48. Find the sine, cosine, and tangent of
    (i) 180°   (ii) 270°   (iii) 360°
    (iv) 135°   (v) 120°   (vi) 150°
    (vii) 225°

D / ANSWERS TO PROBLEMS IN MATHEMATICAL APPENDIX

5. 26
6. $\frac{649}{14} = 46\frac{5}{14} = 46.357$
7. $\frac{91}{108} = 0.843$
8. 0.498
10. -22.1

11. $-\frac{19}{6} = -3\frac{1}{6} = -3.167$

12. -3.167

13. -4

14. $+7.5$ erg

15. $\Phi_e = -\dfrac{q_1 q_2}{R}$

16. 32

17. 243

18. 72

19. $\frac{1}{9} = 0.111$

20. 25

21. 2

22. 64

23. 8

24. $\frac{1}{8} = 0.125$

25. $\frac{35}{13} = 2\frac{9}{13} = 2.692$

26. $\sqrt{13} = 3.606$

27. $\sqrt{41} = 6.403$

28. (i) 10^9 (ii) 10^3 (iii) 10^{-4} (iv) 4×10^6
   (v) 4×10^{-6} (vi) 2×10^{-3}

29. (i) 5.32×10^4 (ii) 2.165328×10^8 (iii) 2.32×10^2
   (iv) 1.3×10^{-3} (v) 2.7×10^{-7} (vi) 9.56×10^{-4}

30. (i) 3×10^4 (ii) 4×10^3 (iii) 8×10^{-3}
   (iv) 30   (v) 5×10^{-2}

31. 4.25×10^{-12}

32. 4.32×10^{28}

33. 1.69×10^2

34. 3.63×10^{-2}

35. (i) 4   (ii) 2   (iii) 4   (iv) 4
   (v) 4   (vi) 2

36. (i) 3.242×10^{-6}
   (ii) 1.73×10^{-6}
   (iii) 5.60

37. 80

38. (i) $\alpha^3 + \alpha\beta^2 - \alpha^2\beta - \beta^3$
   (ii) $a^3 + b^3$
   (iii) $a^3 - b^3$
   (iv) $x^2y + zx^2 + xy^2 + y^2z + yz^2 + z^2x + 3xyz.$

39. (i) $x = -5$
   (ii) $x = +16$
   (iii) $x = +1$
   (iv) $x = -\frac{13}{11} = -1.182$

40. (i) $\dfrac{bc + ca + ab}{abc}$

   (ii) $\dfrac{x^2 + y^2 + z^2}{xyz}$

   (iii) $\dfrac{(bx - ay)}{(x + a)(y + b)}$

Mathematical
Appendix:
Answers

41. (i) $0.00001 = 10^{-5}$
    (ii) 1.25×10^{-6}
    (iii) 2×10^{-6}

42. (i) 0.99999999500
    (ii) 0.999999875
    (iii) 0.999717

43. (i) 1.000000500
    (ii) 1.00000450
    (iii) 1.00000000167

44. (i) 4.47×10^{-3}
    (ii) 3.63×10^{-2}
    (iii) 1.30×10^{-4}

48. (i) $\sin 180° = 0,\quad \cos 180° = -1,\quad \tan 180° = 0$
    (ii) $\sin 270° = -1,\quad \cos 270° = 0,\quad \tan 270° = -\infty \text{ or } +\infty$
    (iii) $\sin 360° = 0,\quad \cos 360° = +1,\quad \tan 360° = 0$
    (iv) $\sin 135° = +0.707,\quad \cos 135° = -0.707,\quad \tan 135° = -1$
    (v) $\sin 120° = +0.866,\quad \cos 120° = -0.5,\quad \tan 120° = -1.732$
    (iv) $\sin 150° = +0.5,\quad \cos 150° = -0.866,\quad \tan 150° = -0.577$
    (vii) $\sin 225° = -0.707,\quad \cos 225° = -0.707,\quad \tan 225° = +1$

MATHEMATICAL

APPENDIX

SI
Units

SI is an abbreviation of the French "Le Système International d'Unités," which translates as "The International System of Units." It is basically an MKS system with the meter for the unit of length, the kilogram for the unit of mass, and the second for the unit of time. In addition, however, the unit of electric current, the ampere, is specially defined in the following way.

"The **ampere** is that constant current which, if maintained in two straight parallel conductors of infinite length, of negligible cross section, and placed 1 meter apart in vacuum, would produce between these conductors a force equal to 2×10^{-7} newton per meter of length."

The motivation behind this is to incorporate into the MKS system the practical units of electricity, the ampere, the volt, and the ohm. These practical units are so firmly embedded in electrical engineering practice that it would be difficult to abandon them. More is involved than a mere change of units. SI units are associated with a special form of the fundamental equations of electricity and magnetism. These equations are significantly different from the equations given in this text for use with CGS units. Moreover, the SI formulation of electromagnetic theory uses two new constants.

The Permittivity Constant

$$\epsilon_0 = 8.85418 \times 10^{-12} \text{ coulomb}^2 \text{ newton}^{-1} \text{ meter}^{-2} \tag{1}$$

The Permeability Constant

$$\mu_0 = 4\pi \times 10^{-7} \text{ meter kilogram coulomb}^{-2} \tag{2}$$

The permeability constant, μ_0, is defined to have exactly the value quoted. The value of the permittivity constant, ϵ_0, is then determined by the measured speed of light, c, and the relationship

$$\epsilon_0 \mu_0 = \frac{1}{c^2} \tag{3}$$

When SI units are used, the equations of electricity and magnetism take the following form.

Coulomb's Law

$$F_e = \frac{1}{4\pi\epsilon_0} \frac{q_1 q_2}{R^2} \tag{4}$$

The unit of charge, q, is the **coulomb.** The force F_e is, of course, measured in newtons and the distance R in meters.

Mutual Electrostatic Potential Energy

$$\Phi_e = \frac{1}{4\pi\epsilon_0} \frac{q_1 q_2}{R} \tag{5}$$

Φ_e is an energy measured in joules.

Electric Field

$$E = \frac{F_e}{q} \tag{6}$$

The units of E are newtons per coulomb or volts per meter.

Electric Field at a Distance R from a Charge q

$$E = \frac{1}{4\pi\epsilon_0} \frac{q}{R^2} \tag{7}$$

Electric Potential at a Distance R from a Charge q

$$\phi_e = \frac{1}{4\pi\epsilon_0} \frac{q}{R} \tag{8}$$

The unit of potential is the **volt.**

Electric Current

$$i = \frac{q}{t} \tag{9}$$

The unit of electric current is the **ampere,** which is 1 coulomb per second. Strictly speaking, the ampere is a base unit with a basic definition, as given earlier in this appendix. It would therefore be more correct to define the coulomb as the quantity of electricity transported in 1 second by 1 ampere.

Ohm's Law

$$i = \frac{V}{R} \tag{10}$$

With the current i measured in amperes and the potential difference V measured in volts, the electric resistance R is in ohms.

Joule Heat

$$\frac{H}{t} = iV \tag{11}$$

H/t is an energy dissipated per second and its units are joules per second or **watts.**

The Magnetic Field of a Moving Charge

$$B = \frac{\mu_0}{4\pi} \frac{q_1 v_1 \sin\theta}{R^2} \tag{12}$$

The quantity that has been called the "magnetic field" throughout this text (for heuristic reasons) is more properly called the "magnetic flux density." Its unit is the **tesla.**

Magnetic Force on a Moving Charge

$$F_m = q_2 v_2 B \sin\phi \tag{13}$$

Compare this with equation 16-6 and notice the absence of the speed of light, c. This is one of the ways in which the SI formalism differs sharply from the CGS formalism. One consequence is that, whereas in the CGS formalism the ratio E/B is dimensionless, in the SI formalism the ratio E/B has the dimensions of a velocity.

Magnetic Field of a Segment of Current (Biot-Savart Law)

$$B = \frac{\mu_0}{4\pi} \frac{i_1 l_1 \sin\theta}{R^2} \tag{14}$$

Magnetic Field of a Current in a Long Straight Wire

$$B = \frac{\mu_0 i_1}{2\pi r} \tag{15}$$

Magnetic Force on a Segment of Current

$$F_m = i_2 l_2 B \sin\phi \tag{16}$$

Force between Currents in Parallel Long Straight Wires

$$F_m = \frac{\mu_0}{2\pi} \frac{i_1 i_2 l_2}{r} \tag{17}$$

This is the equation used in the definition of the ampere. If i_1 and i_2 are both 1 ampere and the distance r between the wires is 1 meter, the force F_m on 1 meter of either wire ($l_2 = 1$ meter) is $\mu_0/2\pi = 2 \times 10^{-7}$ newton.   SI Units

Magnetic Moment of a Current Loop

$$M = i_2 \mathcal{A} \text{ ampere meter}^2 \tag{18}$$

Torque on a Current Loop in a Magnetic Field

$$\tau = MB \sin \theta \tag{19}$$

Magnetic Flux

$$N = AB \tag{20}$$

The unit of magnetic flux is the **weber.** Frequently, the units of magnetic field (magnetic flux density) are quoted as webers per square meter rather than teslas.

Faraday's Law of Electromagnetic Induction

Work done by the induced field as it drives unit charge once around a loop

$$= \frac{(N_2 - N_1)}{(t_2 - t_1)} \tag{21}$$

Compare this with equation 18-7 and notice that c is again absent in the SI formula.

Electromagnetic Waves

$$c = \frac{1}{\sqrt{\epsilon_0 \mu_0}} \tag{22}$$

$$E = cB \tag{23}$$

Energy crossing unit area per second,

$$S = \frac{EB}{\mu_0} \tag{24}$$

Momentum crossing unit area per second,

$$P = \frac{S}{c} \tag{25}$$

$$= \sqrt{\frac{\epsilon_0}{\mu_0}} EB \tag{26}$$

The Hydrogen Atom

The radius of the cloud (the size of the atom),

$$r_n = \frac{\epsilon_0 h^2}{\pi m e^2} n^2 \tag{27}$$

The energy associated with a cloud,

$$\epsilon = -\frac{e^4 m}{8\epsilon_0^2 h^2} \frac{1}{n^2} + \text{a small contribution depending on the other quantum numbers} \tag{28}$$

The Bohr Magneton

$$\beta = \frac{eh}{4\pi m} \tag{29}$$

RULES FOR CONVERTING SI EQUATIONS INTO CGS EQUATIONS

The following procedure converts any one of these SI equations into the corresponding CGS equation.

(1) Replace ϵ_0 by $\dfrac{1}{4\pi}$.

(2) Replace μ_0 by $\dfrac{4\pi}{c^2}$.

(3) Replace B by $\dfrac{B}{c}$.

(4) Replace the symbol for a magnetic flux, N, by $\dfrac{N}{c}$.

(5) Replace the symbol for a magnetic moment, M, by cM.

SI Units

Conversion Factors from SI to CGS Units

The SI units are on the left and the CGS units on the right. When an SI unit has a special name, the approved abbreviation of this is in parentheses. The abbreviation in parentheses following a CGS unit is the one commonly used in this text.

Length	1 meter (m)	$= 10^2$ centimeter (cm)
Time	1 second (s)	$= 1$ second (sec)
Mass	1 kilogram (kg)	$= 10^3$ gram (gm)
Density	$1 \, \text{kg/m}^3$	$= 10^3 \, \text{gm/cm}^3$
Force	1 newton (N)	$= 10^5$ dyne
Energy	1 joule (J)	$= 10^7$ erg
Power	1 watt (W)	
	$= 1$ joule/second	$= 10^7$ erg/sec
Pressure	1 pascal (Pa)	
	$= 1$ newton/meter^2	$= 10$ dyne/cm^2
Electric charge	1 coulomb (C)	$= 2.998 \times 10^9$ statcoulomb
Electric current	1 ampere (A)	$= 2.998 \times 10^9$ statampere
Electric potential	1 volt (V)	$= 3.336 \times 10^{-3}$ statvolt
Electric field	1 volt/meter	$= 3.336 \times 10^{-5}$ statvolt/cm
Electric resistance	1 ohm (Ω)	$= 8.897 \times 10^{11}$ statohm
Magnetic field (magnetic flux density)	1 tesla (T)	$= 10^4$ gauss
Magnetic flux	1 weber (Wb)	
	$= 1$ tesla meter^2	$= 10^8$ gauss cm^2
Frequency	1 hertz (Hz)	$= 1$ cycle/second

Prefixes to Express Decimal Multiples and Submultiples of SI Units

Factor	Prefix	Symbol	Factor	Prefix	Symbol
10^{12}	tera	T	10^{-1}	deci	d
10^{9}	giga	G	10^{-2}	centi	c
10^{6}	mega	M	10^{-3}	milli	m
10^{3}	kilo	k	10^{-6}	micro	μ
10^{2}	hecto	h	10^{-9}	nano	n
10^{1}	deka	da	10^{-12}	pico	p
			10^{-15}	femto	f
			10^{-18}	atto	a

For example,

$1\,\text{GHz} = 10^9\,\text{Hz} = 10^9$ cycles per second

$1\,\text{M}\Omega = 10^6$ ohm

$1\,\text{cA} = 10^{-2}$ ampere

$1\,\text{fm} = 10^{-15}$ meter

Although these prefixes are often convenient abbreviations, the beginning student is strongly advised to replace a prefix immediately by the appropriate power of ten.

Tables

The Greek Alphabet

Alpha	A	α
Beta	B	β
Gamma	Γ	γ
Delta	Δ	δ
Epsilon	E	ϵ
Zeta	Z	ζ
Eta	H	η
Theta	Θ	θ, ϑ
Iota	I	ι
Kappa	K	κ
Lambda	Λ	λ
Mu	M	μ
Nu	N	ν
Xi	Ξ	ξ
Omicron	O	o
Pi	Π	π
Rho	P	ρ
Sigma	Σ	σ, s
Tau	T	τ
Upsilon	Υ	υ
Phi	Φ	ϕ, φ
Chi	X	χ
Psi	Ψ	ψ
Omega	Ω	ω

The Elements

Element	Symbol	Atomic number	Mass number°	Element	Symbol	Atomic number	Mass number°
Actinium	Ac	89	227	Iridium	Ir	77	193
Aluminum	Al	13	27	Iron	Fe	26	56
Americium	Am	95	243	Krypton	Kr	36	84
Antimony	Sb	51	121	Lanthanum	La	57	139
Argon	Ar	18	40	Lawrencium	Lw	103	257
Arsenic	As	33	75	Lead	Pb	82	208
Astatine	At	85	210	Lithium	Li	3	7
Barium	Ba	56	138	Lutetium	Lu	71	175
Berkelium	Bk	97	247	Magnesium	Mg	12	24
Beryllium	Be	4	9	Manganese	Mn	25	55
Bismuth	Bi	83	209	Mendeleevium	Md	101	256
Boron	B	5	11	Mercury	Hg	80	202
Bromine	Br	35	79	Molybdenum	Mo	42	98
Cadmium	Cd	48	114	Neodymium	Nd	60	142
Calcium	Ca	20	40	Neon	Ne	10	20
Californium	Cf	98	251	Neptunium	Np	93	237
Carbon	C	6	12	Nickel	Ni	28	58
Cerium	Ce	58	140	Niobium	Nb	41	93
Cesium	Cs	55	133	Nitrogen	N	7	14
Chlorine	Cl	17	35	Nobelium	No	102	254
Chromium	Cr	24	52	Osmium	Os	76	192
Cobalt	Co	27	59	Oxygen	O	8	16
Copper	Cu	29	63	Palladium	Pd	46	106
Curium	Cm	96	248	Phosphorus	P	15	31
Dysprosium	Dy	66	164	Platinum	Pt	78	195
Einsteinium	Es	99	254	Plutonium	Pu	94	244
Erbium	Er	68	166	Polonium	Po	84	209
Europium	Eu	63	153	Potassium	K	19	39
Fermium	Fm	100	253	Praseodymium	Pr	59	141
Fluorine	F	9	19	Promethium	Pm	61	147
Francium	Fr	87	223	Protactinium	Pa	91	231
Gadolinium	Gd	64	158	Radium	Ra	88	226
Gallium	Ga	31	69	Radon	Rn	86	222
Germanium	Ge	32	74	Rhenium	Re	75	187
Gold	Au	79	197	Rhodium	Rh	45	103
Hafnium	Hf	72	180	Rubidium	Rb	37	85
Helium	He	2	4	Ruthenium	Ru	44	102
Holmium	Ho	67	165	Samarium	Sm	62	152
Hydrogen	H	1	1	Scandium	Sc	21	45
Indium	In	49	115	Selenium	Se	34	80
Iodine	I	53	127	Silicon	Si	14	28

° For stable elements the mass number refers to the most abundant isotope. For unstable elements it refers to the isotope with the longest half-life.

The Elements (continued)

Element	Symbol	Atomic number	Mass number°	Element	Symbol	Atomic number	Mass number°
Silver	Ag	47	107	Tin	Sn	50	120
Sodium	Na	11	23	Titanium	Ti	22	48
Strontium	Sr	38	88	Tungsten	W	74	184
Sulfur	S	16	32	Uranium	U	92	238
Tantalum	Ta	73	181	Vanadium	V	23	51
Technetium	Tc	43	97	Xenon	Xe	54	132
Tellurium	Te	52	130	Ytterbium	Yb	70	174
Terbium	Tb	65	159	Yttrium	Y	39	89
Thallium	Tl	81	205	Zinc	Zn	30	64
Thorium	Th	90	232	Zirconium	Zr	40	90
Thulium	Tm	69	169				

Quantum Numbers of the Metastable Hadrons.

To obtain the quantum numbers of the antiparticles, leave I unaltered and merely change the sign of B, I_3, Q, Y, S.

Baryon number B	Particle	Isotopic spin I	Third component of Isotopic spin I_3	Charge number Q	Hyper-charge Y	Strange-ness S
0 Mesons	π^-	1	-1	-1	0	0
	π^0		0	0		
	π^+		$+1$	$+1$		
	K^0	$\frac{1}{2}$	$-\frac{1}{2}$	0	$+1$	$+1$
	K^+		$+\frac{1}{2}$	$+1$		
	η^0	0	0	0	0	0
1 Baryons	n	$\frac{1}{2}$	$-\frac{1}{2}$	0	$+1$	0
	p		$+\frac{1}{2}$	$+1$		
	Λ^0	0	0	0	0	-1
	Σ^-	1	-1	-1	0	-1
	Σ^0		0	0		
	Σ^+		$+1$	$+1$		
	Ξ^-	$\frac{1}{2}$	$-\frac{1}{2}$	-1	-1	-2
	Ξ^0		$+\frac{1}{2}$	0		
	Ω^-	0	0	-1	-2	-3

Periodic Table of the Elements

	Group I	Group II	Transition elements										Group III	Group IV	Group V	Group VI	Group VII	Group O
Period 1	H 1																	He 2
Period 2	Li 3	Be 4											B 5	C 6	N 7	O 8	F 9	Ne 10
Period 3	Na 11	Mg 12											Al 13	Si 14	P 15	S 16	Cl 17	Ar 18
Period 4	K 19	Ca 20	Sc 21	Ti 22	V 23	Cr 24	Mn 25	Fe 26	Co 27	Ni 28	Cu 29	Zn 30	Ga 31	Ge 32	As 33	Se 34	Br 35	Kr 36
Period 5	Rb 37	Sr 38	Y 39	Zr 40	Nb 41	Mo 42	Tc 43	Ru 44	Rh 45	Pd 46	Ag 47	Cd 48	In 49	Sn 50	Sb 51	Te 52	I 53	Xe 54
Period 6	Cs 55	Ba 56	° 57–71	Hf 72	Ta 73	W 74	Re 75	Os 76	Ir 77	Pt 78	Au 79	Hg 80	Tl 81	Pb 82	Bi 83	Po 84	At 85	Rn 86
Period 7	Fr 87	Ra 88	† 89–?														?	

° Rare earth metals	La 57	Ce 58	Pr 59	Nd 60	Pm 61	Sm 62	Eu 63	Gd 64	Tb 65	Dy 66	Ho 67	Er 68	Tm 69	Yb 70	Lu 71
† Actinide metals	Ac 89	Th 90	Pa 91	U 92	Np 93	Pu 94	Am 95	Cm 96	Bk 97	Cf 98	Es 99	Fm 100	Md 101	No 102	Lw 103

The Metastable Fundamental Particles

	Name	Particles Charge +e	Particles Charge 0	Particles Charge −e	Antiparticles Charge +e	Antiparticles Charge 0	Antiparticles Charge −e	Rest mass (Units such that rest mass of electron = 1)	Spin. Units of $\frac{h}{2\pi}$
Photon	photon		γ			γ		0	1 (Boson)
Leptons	neutrino, antineutrino		ν			$\bar{\nu}$		0	½ (Fermions)
	μ-neutrino, anti-μ-neutrino		ν_μ			$\bar{\nu}_\mu$		0	
	electron, positron			e^-	e^+			1	
	muons			μ^-	μ^+			207	
Mesons	π-mesons (pions)	π^+	π^0			π^0	π^-	264 273	0 (Bosons)
	K-mesons (Kaons)	K^+	K^0			$\overline{K^0}$	K^-	966 974	
	η-meson (eta)		η^0			η^0		1074	
Baryons	proton, antiproton neutron, antineutron	p	n			$\bar{n}$	$\bar{p}$	1836.1 1838.6	½ (Fermions)
	Lambda		Λ^0			$\overline{\Lambda^0}$		2183	
	Sigma	Σ^+	Σ^0	Σ^-	$\overline{\Sigma}^+$	$\overline{\Sigma}^0$	$\overline{\Sigma}^-$	2328 2334 2343	
	Xi		Ξ^0	Ξ^-	$\overline{\Xi}^+$	$\overline{\Xi}^0$		2573 2586	
	Omega			Ω^-	$\overline{\Omega}^+$			3276	3/2

Logarithms to Base Ten

Numbers	0	1	2	3	4	5	6	7	8	9	Proportional Parts								
											1	2	3	4	5	6	7	8	9
10	0000	0043	0086	0128	0170	0212	0253	0294	0334	0374	4	8	12	17	21	25	29	33	37
11	0414	0453	0492	0531	0569	0607	0645	0682	0719	0755	4	8	11	15	19	23	26	30	34
12	0792	0828	0864	0899	0934	0969	1004	1038	1072	1106	3	7	10	14	17	21	24	28	31
13	1139	1173	1206	1239	1271	1303	1335	1367	1399	1430	3	6	10	13	16	19	23	26	29
14	1461	1492	1523	1553	1584	1614	1644	1673	1703	1732	3	6	9	12	15	18	21	24	27
15	1761	1790	1818	1847	1875	1903	1931	1959	1987	2014	3	6	8	11	14	17	20	22	25
16	2041	2068	2095	2122	2148	2175	2201	2227	2253	2279	3	5	8	11	13	16	18	21	24
17	2304	2330	2335	2380	2405	2430	2455	2480	2504	2529	2	5	7	10	12	15	17	20	22
18	2553	2577	2601	2625	2648	2672	2695	2718	2742	2765	2	5	7	9	12	14	16	19	21
19	2788	2810	2833	2856	2878	2900	2923	2945	2967	2989	2	4	7	9	11	13	16	18	20
20	3010	3032	3054	3075	3096	3118	3139	3160	3181	3201	2	4	6	8	11	13	15	17	19
21	3222	3243	3263	3284	3304	3324	3345	3365	3385	3404	2	4	6	8	10	12	14	16	18
22	3424	3444	3464	3483	3502	3522	3541	3560	3579	3598	2	4	6	8	10	12	14	15	17
23	3617	3636	3655	3674	3692	3711	3729	3747	3766	3784	2	4	6	7	9	11	13	15	17
24	3802	3820	3838	3856	3874	3892	3909	3927	3945	3962	2	4	5	7	9	11	12	14	16
25	3979	3997	4014	4031	4048	4065	4082	4099	4116	4133	2	3	5	7	9	10	12	14	15
26	4150	4166	4183	4200	4216	4232	4249	4265	4281	4298	2	3	5	7	8	10	11	13	15
27	4314	4330	4346	4362	4378	4393	4409	4425	4440	4456	2	3	5	6	8	9	11	13	14
28	4472	4487	4502	4518	4533	4548	4564	4579	4594	4609	2	3	5	6	8	9	11	12	14
29	4624	4639	4654	4669	4683	4698	4713	4728	4742	4757	1	3	4	6	7	9	10	12	13
30	4771	4786	4800	4814	4829	4843	4857	4871	4886	4900	1	3	4	6	7	9	10	11	13
31	4914	4928	4942	4955	4969	4983	4997	5011	5024	5038	1	3	4	6	7	8	10	11	12
32	5051	5065	5079	5092	5105	5119	5132	5145	5159	5172	1	3	4	5	7	8	9	11	12
33	5185	5198	5211	5224	5237	5250	5263	5276	5289	5302	1	3	4	5	6	8	9	10	12
34	5315	5328	5340	5353	5366	5378	5391	5403	5416	5428	1	3	4	5	6	8	9	10	11
35	5441	5453	5465	5478	5490	5502	5514	5527	5539	5551	1	2	4	5	6	7	9	10	11
36	5563	5575	5587	5599	5611	5623	5635	5647	5658	5670	1	2	4	5	6	7	8	10	11
37	5682	5694	5705	5717	5729	5740	5752	5763	5775	5786	1	2	3	5	6	7	8	9	10
38	5798	5809	5821	5832	5843	5855	5866	5877	5888	5899	1	2	3	5	6	7	8	9	10
39	5911	5922	5933	5944	5955	5966	5977	5988	5999	6010	1	2	3	4	5	7	8	9	10
40	6021	6031	6042	6053	6064	6075	6085	6096	6107	6117	1	2	3	4	5	6	8	9	10
41	6128	6138	6149	6160	6170	6180	6191	6201	6212	6222	1	2	3	4	5	6	7	8	9
42	6232	6243	6253	6263	6274	6284	6294	6304	6314	6325	1	2	3	4	5	6	7	8	9
43	6335	6345	6355	6365	6375	6385	6395	6405	6415	6425	1	2	3	4	5	6	7	8	9
44	6435	6444	6454	6464	6474	6484	6493	6503	6513	6522	1	2	3	4	5	6	7	8	9
45	6532	6542	6551	6561	6571	6580	6590	6599	6609	6618	1	2	3	4	5	6	7	8	9
46	6628	6637	6646	6656	6665	6675	6684	6693	6702	6712	1	2	3	4	5	6	7	7	8
47	6721	6730	6739	6749	6758	6767	6776	6785	6794	6803	1	2	3	4	5	5	6	7	8
48	6812	6821	6830	6839	6848	6857	6866	6875	6884	6893	1	2	3	4	4	5	6	7	8
49	6902	6911	6920	6928	6937	6946	6955	6964	6972	6981	1	2	3	4	4	5	6	7	8
50	6990	6998	7007	7016	7024	7033	7042	7050	7059	7067	1	2	3	3	4	5	6	7	8
51	7076	7084	7093	7101	7110	7118	7126	7135	7143	7152	1	2	3	3	4	5	6	7	8
52	7160	7168	7177	7185	7193	7202	7210	7218	7226	7235	1	2	2	3	4	5	6	7	7
53	7243	7251	7259	7267	7275	7284	7292	7300	7308	7316	1	2	2	3	4	5	6	6	7
54	7324	7332	7340	7348	7356	7364	7372	7380	7388	7396	1	2	2	3	4	5	6	6	7

Numbers	0	1	2	3	4	5	6	7	8	9	Proportional Parts								
											1	2	3	4	5	6	7	8	9
55	7404	7412	7419	7427	7435	7443	7451	7459	7466	7474	1	2	2	3	4	5	5	6	7
56	7482	7490	7497	7505	7513	7520	7528	7536	7543	7551	1	2	2	3	4	5	5	6	7
57	7559	7566	7574	7582	7589	7597	7604	7612	7619	7627	1	2	2	3	4	5	5	6	7
58	7634	7642	7649	7657	7664	7672	7679	7686	7694	7701	1	1	2	3	4	4	5	6	7
59	7709	7716	7723	7731	7738	7745	7752	7760	7767	7774	1	1	2	3	4	4	5	6	7
60	7782	7789	7796	7803	7810	7818	7825	7832	7839	7846	1	1	2	3	4	4	5	6	6
61	7853	7860	7868	7875	7882	7889	7896	7903	7910	7917	1	1	2	3	4	4	5	6	6
62	7924	7931	7938	7945	7952	7959	7966	7937	7980	7987	1	1	2	3	3	4	5	6	6
63	7993	8000	8007	8014	8021	8028	8035	8041	8048	8055	1	1	2	3	3	4	5	5	6
64	8062	8069	8075	8082	8089	8096	8102	8109	8116	8122	1	1	2	3	3	4	5	5	6
65	8129	8136	8142	8149	8156	8162	8169	8176	8182	8189	1	1	2	3	3	4	5	5	6
66	8195	8202	8209	8215	8222	8228	8235	8241	8248	8254	1	1	2	3	3	4	5	5	6
67	8261	8267	8274	8280	8287	8293	8299	8306	8312	8319	1	1	2	3	3	4	5	5	6
68	8325	8331	8338	8344	8351	8357	8363	8370	8376	8382	1	1	2	3	3	4	4	5	6
69	8388	8395	8401	8407	8414	8420	8426	8432	8439	8445	1	1	2	2	3	4	4	5	6
70	8451	8457	8463	8470	8476	8482	8488	8494	8500	8506	1	1	2	2	3	4	4	5	6
71	8513	8519	8525	8531	8537	8543	8549	8555	8561	8567	1	1	2	2	3	4	4	5	5
72	8573	8579	8585	8591	8597	8603	8609	8615	8621	8627	1	1	2	2	3	4	4	5	5
73	8633	8639	8645	8651	8657	8663	8669	8675	8681	8686	1	1	2	2	3	4	4	5	5
74	8692	8698	8704	8710	8716	8722	8727	8733	8739	8745	1	1	2	2	3	4	4	5	5
75	8751	8756	8762	8768	8774	8779	8785	8791	8797	8802	1	1	2	2	3	3	4	5	5
76	8808	8814	8820	8825	8831	8837	8842	8848	8854	8859	1	1	2	2	3	3	4	5	5
77	8865	8871	8876	8882	8887	8893	8899	8904	8910	8915	1	1	2	2	3	3	4	4	5
78	8921	8927	8932	8938	8943	8949	8954	8960	8965	8971	1	1	2	2	3	3	4	4	5
79	8976	8982	8987	8993	8998	9004	9009	9015	9020	9025	1	1	2	2	3	3	4	4	5
80	9031	9036	9042	9047	9053	9058	9063	9069	9074	9079	1	1	2	2	3	3	4	4	5
81	9085	9090	9096	9101	9106	9112	9117	9122	9128	9133	1	1	2	2	3	3	4	4	5
82	9138	9143	9149	9154	9159	9165	9170	9175	9180	9186	1	1	2	2	3	3	4	4	5
83	9191	9196	9201	9206	9212	9217	9222	9227	9232	9238	1	1	2	2	3	3	4	4	5
84	9243	9248	9253	9258	9263	9269	9274	9279	9284	9289	1	1	2	2	3	3	4	4	5
85	9294	9299	9304	9309	9315	9320	9325	9330	9335	9340	1	1	2	2	3	3	4	4	5
86	9345	9350	9355	9360	9365	9370	9375	9380	9385	9390	1	1	2	2	3	3	4	4	5
87	9395	9400	9405	9410	9415	9420	9425	9430	9435	9440	0	1	1	2	2	3	3	4	4
88	9445	9450	9455	9460	9465	9469	9474	9479	9484	9489	0	1	1	2	2	3	3	4	4
89	9494	9499	9504	9509	9513	9518	9523	9528	9533	9538	0	1	1	2	2	3	3	4	4
90	9542	9547	9552	9557	9562	9566	9571	9576	9581	9586	0	1	1	2	2	3	3	4	4
91	9590	9595	9600	9605	9609	9614	9619	9624	9628	9633	0	1	1	2	2	3	3	4	4
92	9638	9643	9647	9652	9657	9661	9666	9671	9675	9680	0	1	1	2	2	3	3	4	4
93	9685	9689	9694	9699	9703	9708	9713	9717	9722	9727	0	1	1	2	2	3	3	4	4
94	9731	9736	9741	9745	9750	9754	9759	9763	9768	9773	0	1	1	2	2	3	3	4	4
95	9777	9782	9786	9791	9795	9800	9805	9809	9814	9818	0	1	1	2	2	3	3	4	4
96	9823	9827	9832	9836	9841	9845	9850	9854	9859	9863	0	1	1	2	2	3	3	4	4
97	9868	9872	9877	9881	9886	9890	9894	9899	9903	9908	0	1	1	2	2	3	3	4	4
98	9912	9917	9921	9926	9930	9934	9939	9943	9948	9952	0	1	1	2	2	3	3	4	4
99	9956	9961	9965	9969	9974	9978	9983	9987	9991	9996	0	1	1	2	2	3	3	3	4

Three-Place Table of Sines, Cosines, Tangents, and Cotangents

Angles	Sines	Cosines	Tangents	Cotangents	
0°00′	0.000	1.00	0.000		90°00′
30	0.009	1.00	0.009	115	30
1°00′	0.018	1.00	0.018	57.3	89°00′
30	0.026	1.00	0.026	38.2	30
2°00′	0.035	0.999	0.035	28.6	88°00′
30	0.044	0.999	0.044	22.9	30
3°00′	0.052	0.999	0.052	19.1	87°00′
30	0.061	0.998	0.061	16.4	30
4°00′	0.070	0.998	0.070	14.3	86°00′
30	0.078	0.997	0.079	12.7	30
5°00′	0.087	0.996	0.088	11.4	85°00′
30	0.096	0.995	0.096	10.4	30
6°00′	0.104	0.994	0.105	9.51	84°00′
30	0.113	0.994	0.114	8.78	30
7°00′	0.122	0.992	0.123	8.14	83°00′
30	0.130	0.991	0.132	7.60	30
8°00′	0.139	0.990	0.140	7.12	82°00′
30	0.148	0.989	0.150	6.69	30
9°00′	0.156	0.988	0.158	6.31	81°00′
30	0.165	0.986	0.167	5.98	30
10°00′	0.174	0.985	0.176	5.67	80°00′
30	0.182	0.983	0.185	5.40	30
11°00′	0.191	0.982	0.194	5.14	79°00′
30	0.199	0.980	0.204	4.92	30
12°00′	0.208	0.978	0.213	4.70	78°00′
30	0.216	0.976	0.222	4.51	30
13°00′	0.225	0.974	0.231	4.33	77°00′
30	0.233	0.972	0.240	4.17	30
14°00′	0.242	0.970	0.249	4.01	76°00′
30	0.250	0.968	0.259	3.87	30
15°00′	0.259	0.966	0.268	3.73	75°00′
30	0.267	0.964	0.277	3.61	30
16°00′	0.276	0.961	0.287	3.49	74°00′
30	0.284	0.959	0.296	3.38	30
17°00′	0.292	0.956	0.306	3.27	73°00′
30	0.301	0.954	0.315	3.17	30
18°00′	0.309	0.951	0.325	3.08	72°00′
30	0.317	0.948	0.335	2.99	30
19°00′	0.326	0.946	0.344	2.90	71°00′
30	0.334	0.943	0.354	2.82	30
20°00′	0.342	0.940	0.364	2.75	70°00′
30	0.350	0.937	0.374	2.67	30
21°00′	0.358	0.934	0.384	2.61	69°00′
30	0.366	0.930	0.394	2.54	30
22°00′	0.375	0.927	0.404	2.48	68°00′
30	0.383	0.924	0.414	2.41	30
	Cosines	Sines	Cotangents	Tangents	Angles

TABLES

800

Three-Place Table of Sines, Cosines, Tangents, and Cotangents

Angles	Sines	Cosines	Tangents	Cotangents	
23°00′	0.391	0.920	0.424	2.36	67°00′
30	0.399	0.917	0.435	2.30	30
24°00′	0.407	0.914	0.445	2.25	66°00′
30	0.415	0.910	0.456	2.19	30
25°00′	0.423	0.906	0.466	2.14	65°00′
30	0.430	0.903	0.477	2.10	30
26°00′	0.438	0.899	0.488	2.05	64°00′
30	0.446	0.895	0.499	2.01	30
27°00′	0.454	0.891	0.510	1.96	63°00′
30	0.462	0.887	0.521	1.92	30
28°00′	0.470	0.883	0.532	1.88	62°00′
30	0.477	0.879	0.543	1.84	30
29°00′	0.485	0.875	0.554	1.80	61°00′
30	0.492	0.870	0.566	1.77	30
30°00′	0.500	0.866	0.577	1.73	60°00′
30	0.508	0.862	0.589	1.70	30
31°00′	0.515	0.857	0.601	1.66	59°00′
30	0.522	0.853	0.613	1.63	30
32°00′	0.530	0.848	0.625	1.60	58°00′
30	0.537	0.843	0.637	1.57	30
33°00′	0.545	0.839	0.649	1.54	57°00′
30	0.552	0.834	0.662	1.51	30
34°00′	0.559	0.829	0.674	1.48	56°00′
30	0.566	0.824	0.687	1.46	30
35°00′	0.574	0.819	0.700	1.43	55°00′
30	0.581	0.814	0.713	1.40	30
36°00′	0.588	0.809	0.726	1.38	54°00′
30	0.595	0.804	0.740	1.35	30
37°00′	0.602	0.799	0.754	1.33	53°00′
30	0.609	0.793	0.767	1.30	30
38°00′	0.616	0.788	0.781	1.28	52°00′
30	0.622	0.783	0.795	1.26	30
39°00′	0.629	0.777	0.810	1.23	51°00′
30	0.636	0.772	0.824	1.21	30
40°00′	0.643	0.766	0.839	1.19	50°00′
30	0.649	0.760	0.854	1.17	30
41°00′	0.656	0.755	0.869	1.15	49°00′
30	0.663	0.749	0.885	1.13	30
42°00′	0.669	0.743	0.900	1.11	48°00′
30	0.676	0.737	0.916	1.09	30
43°00′	0.682	0.731	0.932	1.07	47°00′
30	0.688	0.725	0.949	1.05	30
44°00′	0.695	0.719	0.966	1.04	46°00′
30	0.701	0.713	0.983	1.02	30
45°00′	0.707	0.707	1.00	1.00	45°00′
	Cosines	Sines	Cotangents	Tangents	Angles

TABLES

This is not a standard convention. It is introduced with the hope that it might improve the clarity of the diagrams.

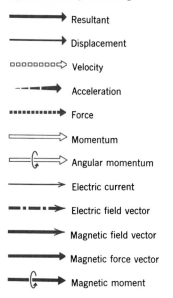

Resultant

Displacement

Velocity

Acceleration

Force

Momentum

Angular momentum

Electric current

Electric field vector

Magnetic field vector

Magnetic force vector

Magnetic moment

FIELD LINES

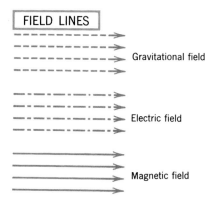

Gravitational field

Electric field

Magnetic field

Answers to Odd-Numbered Problems

Chapter 2

3. (a) 8.05×10^5 cm, (b) 5 cm, (c) 3×10^3 cm, (d) 5×10^3 gm, (e) 5 gm, (f) 4.74×10^7 sec, (g) 3.6×10^2 sec, (h) 4.47×10^4 cm sec^{-1}, (i) 1.39×10^2 cm sec^{-1}, (j) 2×10^{-2} cm sec^{-1}, (k) 1.5×10^3 cm sec^{-1}. **5.** 3.1536×10^7 sec. **7.** 218 sec. **9.** 2×10^9 sec. **11.** 48 min (a) 7.9 hours, (b) 128 days, (c) 14.2 years, (d) 940 centuries. **13.** $\sqrt{14} = 3.74$ cm. **17.** 3.74 cm. **19.** $2c$. **21.** Each story must be at least as high as a man (about 200 cm). The total height must be greater than 2×10^4 cm. **23.** Approximately 1/6000.

Chapter 3 ´

1. (a) 150 m sec^{-1} northeast, (b) 30 m sec^{-1} southwest, (c) 90 m sec^{-1} southwest, (d) 60 m sec^{-1} northeast, (e) 90 m sec^{-1} northeast, (f) 30 m sec^{-1} southwest. **5.** 7.07 miles toward the northwest. **7.** 21.2 knots. **11.** 600 m sec^{-1} horizontally and 1030 m sec^{-1} vertically. **13.** (a) 4.23, $45°$ above the x-axis, (b) 4.23, $45°$ below the x-axis, (c) 5, $53.1°$ above the x-axis, (d) 8.60, $54.5°$ above the x-axis, (e) 7.28, $164°$ above the x-axis, (f) 5.83, $121°$ below the x-axis. **15.** $26.6°$ to the vertical. **17.** $21.8°$ to the horizontal. **19.** 51 mph at $11.3°$ to the tracks. **21.** The boat must be pointed directly toward the opposite bank. 40 sec. 200 m. **23.** 8013 m sec^{-1}. **25.** 2500 m sec^{-1} toward the west, 2500 m sec^{-1} toward the north, 3540 m sec^{-1} vertically upward.

Chapter 4

1. 45 cm. **3.** (a) 2.94×10^3 cm sec^{-1}, (b) 4.41×10^3 cm. **5.** 2.74 sec. **7.** 28 m sec^{-1}. **9.** 0. **11.** 36 mph. **13.** 3.36×10^3 cm sec^{-3}, $1.07 \times$

10^3 cm. **15.** 5 cm sec^{-1}. **17.** 12 sec. **19.** 440 yards. **21.** 37 m, about 15 stories. **23.** 1.26×10^4 m = 12.6 kilometers. **25.** 2.5 m, 0.71 sec, 0.71 sec, 7 m sec^{-1}. **27.** 75 m above the ground. **29.** 450 m. **31.** (a) 90 m, (b) 90 m, (c) 90 m, (d) 80.9 m, (e) 46.6 m, (f) 2.5 m.

35. $\sqrt{\dfrac{2d(a_2 + a_1)}{a_2 a_1}}$

Chapter 5

1. (a) $\dfrac{\pi}{6} = 0.524$, (b) $\dfrac{\pi}{4} = 0.785$, (c) $\dfrac{\pi}{3} = 1.047$, (d) $\dfrac{3\pi}{2} = 4.712$.

3. $\dfrac{10\pi}{3} = 10.45$ radians sec^{-1}. **5.** 3.5 radians sec^{-1}. **7.** 53 cm sec^{-1}, 185 cm sec^{-2}. **9.** 3 cm sec^{-2}. **11.** 0.126 cm sec^{-1}, 1.32×10^{-2} cm sec^{-2}. **13.** 3×10^6 cm sec^{-1}, 2×10^{-7} rad sec^{-1}, 0.6 cm sec^{-2}. **15.** 0.313 rad sec^{-1}. **17.** 9900 miles. **19.** A small body. 2.5×10^8 g. **21.** About a quarter of a second.

Chapter 6

1. 16 dynes. **3.** 0.25 m sec^{-2}. **5.** 19.6 dynes. **7.** 1.67×10^{-12} m sec^{-2}. **9.** 2.04×10^3 gm. **11.** 9.8×10^2 cm sec^{-2}. **13.** (a) 3.3×10^{-3} cm sec^{-2}, (b) 0.6 cm sec^{-2}. **15.** 2.7×10^{31} newtons. **17.** 0.13 sec. **19.** 9.6×10^4 dynes. **21.** 1.5×10^8 dynes. **23.** 8×10^{27} dynes. **25.** On the line joining the centers of the earth and moon, very nearly nine times as far from the earth as from the moon. **27.** (a) 6.67×10^{-10} m^3 kg^{-1} sec^{-2}, (b) 6.67×10^{-8} m^3 kg^{-1} sec^{-2}, (c) 6.00×10^{-10} m^3 kg^{-1} sec^{-2}. **29.** It is not correct. M_2/M_1. **31.** 0.27 cm sec^{-2}.

Chapter 7

1. 9.8 newtons. No. **3.** 19.6 m sec^{-2}. **5.** 2.45 m sec^{-2}. **7.** 8 m sec^{-2}. **9.** 2 gm. **11.** 182.6 days. **13.** 6.7×10^{35} gm. **15.** 371 cm sec^{-2}. **17.** 3. **19.** 245 cm sec^{-2}. **21.** 1.96×10^3 m. **23.** 8.2 m sec^{-2}. **25.** 1.02×10^5 cm sec^{-1}, 27 days. **27.** $F_2 = 775$, $F_3 = 640$, $F_4 = 495$, $F_5 = 340$, $F_6 = 175$, all in newtons. The frictional force on each barge is 25 newtons. **29.** 5%. **31.** $\dfrac{3GM_p M_r}{R^2}$ directly toward the planet.

33. $(1.8 - 0.98 \cos\theta) \times 10^5$ dynes, where θ is the angle between the string and the vertical radius joining the center to the top of the circle. **35.** 1.2×10^{12} years. The age of the universe is about 10^{10} years.

Chapter 8

1. 175 gm cm sec^{-1}. **3.** 72 gm cm sec^{-1}. **5.** $p = mgt$. **7.** (a) 5.01×10^{-18} gm cm sec^{-1}, (b) 5.01×10^{-23} kg m sec^{-1}. **9.** 0.21 m sec^{-1}. **11.** 0.96 m sec^{-1}. **13.** 1.88 kg m sec^{-1}. **15.** $v_A = 2.14$ m sec^{-1}, $v_B =$

2.86 m sec^{-1}. **17.** 7.15 cm sec^{-1} in the direction in which the 2 kg lump was originally moving. **19.** 900 m sec^{-1} in the same direction that the rocket was moving before it exploded. **21.** 1.12. **23.** (a) $m\sqrt{2gh_1}$ upward, (b) $m\sqrt{2gh_2}$ downward, (c) 0.

Chapter 9

1. 24 ergs. **3.** 4.7×10^8 cm sec^{-1}. **5.** 2. **7.** -5.84×10^{37} joules.
9. 400 ergs. **11.** 106 ergs. **13.** 4.9×10^8 joules sec^{-1}. **15.** 4, 1.41.
17. About 5×10^9 joules, 24,800 mph which is the escape velocity. **19.** $\frac{1}{2}$.
21. 1.22×10^3 joules. **23.** None. **25.** 2.68×10^4 ergs. **27.** $2.98 \times$
10^4 cm sec^{-1}, 1.49×10^5 cm sec^{-1}. **29.** $-\dfrac{GM_eM_s}{R}$, decreases.
33. 6.7×10^8 cm.

Chapter 10

1. 692 newton meters. **3.** 2.55×10^{15} gm cm^2 sec^{-1}. **5.** $6.4 \times$
10^4 gm cm^2. **7.** 5×10^4 gm cm^2 sec^{-1}. **9.** 44.6 gm cm^2 sec^{-1}.
11. 0.29 cm sec^{-1}. **13.** 2.56×10^{29} joules. **15.** 2.4×10^{16} gm, 1.24
miles. **17.** 4.1×10^{23} dyne cm. Increase. **21.** 0.500 rad sec^{-1}.
23. (a) ω_0, (b) $\omega_0(I + mR^2)\left[I + m\left\{\left(R + v_r\sqrt{\dfrac{2h}{g}}\right)^2 + \dfrac{2\omega_0{}^2R^2h}{g}\right\}\right]^{-1}$.

Chapter 11

1. (a) 2.01565, (b) 47.984745, (c) 43.989830, (d) 34.00548. **3.** (a) $\frac{1}{12}$,
(b) $\frac{1}{2}$, (c) $\frac{1}{2}$, (d) 68.2. **5.** (a) 6.02×10^{23}, (b) 1.51×10^{23}, (c) 3.34×10^{22}.
7. 1.87×10^3 dyne cm^{-2}. **9.** 2.49×10^4 dyne cm^{-2}. **11.** 2.4×10^5 dyne
cm^{-2}. **13.** 4.35×10^{21}. **15.** 3.52×10^3 cm sec^{-1}. **17.** 4.14×10^{-16} dyne
cm^{-2}. **21.** 59.4 miles. **23.** 5.3×10^{21} gm, 5.5 miles. **25.** $1.3 \times$
$10^{-6} M_m$.

Chapter 12

3. 10^8 ergs. **5.** 6.24×10^9 ergs. **7.** 1.04×10^{10} ergs. **9.** 4.4×10^{13} ergs.
11. 1.25×10^4 erg deg^{-1}. **13.** $\frac{3}{2}R$, $\frac{5}{2}R$. **15.** $V = (N_1 + N_2)\frac{3}{2}kT$, $pV =$
$(N_1 + N_2)kT$, $Q = \frac{3}{2}(N_1 + N_2)k(T_2 - T_1)$, $Q' = \frac{3}{2}(N_1 + N_2)k(T_2 - T_1) +$
$p_1(V_2 - V_1)$.

Chapter 13

1. 5.27×10^{17} statcoulomb. **3.** 8.75 dynes, $+75.5$ ergs. **5.** 10 cm.
7. The 5 gm mass has an acceleration of 7.1 cm sec^{-2}. The 9 gm mass
has an acceleration of 4.0 cm sec^{-2}. **9.** About -5×10^{-11} erg. **11.** 0.36
dyne at 20.6° to the line joining q_3 to q_1. **13.** 6.9 dynes, 24 ergs. **15.** 0.78
dyne at 31° to the x-axis, $+6.58$ ergs. **17.** $(+2.5, 0)$. **19.** A decrease of
2.8 sec. **21.** 5.3×10^{25} cm sec^{-1}. **23.** $-\frac{1}{2}$. **25.** 1.39×10^{-10} erg.
27. 1.2×10^{57}, $n = 1.46 = \frac{3}{2}$ approximately.

Answers to
Odd-Numbered
Problems

Chapter 14

1. -3.33×10^{-6} erg gm^{-1}. 3. 3.98 cm sec^{-2}, -3.98×10^{10} erg gm^{-1}. 5. 1.5×10^{-5} cm sec^{-2}. 7. 24 dynes statcoulomb^{-1}. 9. 4.8×10^6 statvolt cm^{-1}, -4.8×10^{-2} statvolt. 11. -14.4 statvolts. 13. 70 statvolt cm^{-1} from B toward A. 15. 29.4 joule kg^{-1} or 2.94×10^5 erg gm^{-1}. 17. 1.86×10^{-15} statvolt cm^{-1}. 19. 1.32×10^9 cm sec^{-1}. 21. 5.37×10^5. 23. (a) 0, (b) 0.276 statvolt. 25. -2.4×10^{-7} erg. 27. $\dfrac{2qRl}{(R^2 - \frac{1}{4}l^2)^2} \approx \dfrac{2ql}{R^3}$ if $R \gg l$, $\dfrac{ql}{(R^2 - \frac{1}{4}l^2)} \approx \dfrac{q}{R^2}$ if $R \gg l$. 29. Field inside $= 0$, Field outside $= \dfrac{q}{R^2}$. 31. Field inside $= \dfrac{qR}{a^3}$, Field outside $= \dfrac{q}{R^2}$.

Chapter 15

1. 300 statcoulombs. 3. 4.8×10^5 statamp, 1.6×10^{-4} amp. 5. (a) 2.13×10^4 statamp, (b) 7.1×10^{-6} amp. 7. 4.8 ohms, 30 watts. 9. 3.12×10^{15} eV. 11. 3.6×10^5 coulombs, 2.25×10^{24} electrons, 4.32×10^6 joules, about 8800 crates. 13. 0.108 gm. 15. 6.95×10^8 cm sec^{-1}, 1.55×10^7 mph. 17. 5×10^5 volts, 6.9×10^8 cm sec^{-1}.

19. 17.5 electrons cm^{-3}. 21. $\dfrac{R_1 R_2}{(R_1 + R_2)}$. 23. 2.3×10^{-15} sec.

25. 7.4×10^{-2} sec.

Chapter 16

1. 2.8×10^{-7} gauss. 3. (a) 0, (b) 4.5 dynes, (c) 3.2 dynes. 5. 3.75×10^5 statcoulombs. 7. 1.67×10^{-3} gauss vertically upward. 9. 2×10^{-3} dyne vertically upward. 11. 1.03×10^{-9} dyne, 4.4×10^{-5}. 13. 3.47×10^2 gauss, horizontal, toward the southeast. 15. (a) 1.22×10^{-9} dyne, horizontal, toward the northeast, (b) 8.65×10^{-10} dyne, horizontal, toward the west. 17. 1.04 cm. 19. 4.4×10^{-7} gauss. 21. 45.5 cm. 23. 5×10^7 cm sec$^{-1} = 1.1 \times 10^6$ mph. 25. 0.028%. 27. $\dfrac{BAv}{4\pi c}$.

Chapter 17

1. 1.48×10^{-6} gauss. 3. 4.45×10^{-8} gauss. 5. 1.2×10^{-5} dyne, horizontal, toward the east. 7. Zero. 9. 5.0×10^{-5} gauss. 11. (a) 0.100 gauss, (b) 0.167 gauss. 13. 4×10^{-15} dyne toward the northwest. 15. $\dfrac{2\sqrt{2}i}{cr}$ in a plane parallel to both wires, in a direction at $45°$ to either wire, tilting upward toward the right in the diagram. 17. $\dfrac{2i_1 M}{cr(r + a)}$ directly toward i_1. 19. $\dfrac{evr}{2c}$, $\dfrac{e}{2mc}$.

Chapter 18

1. 1.13×10^6 gauss cm². **3.** 1.67×10^{-4} statvolt cm⁻¹. **5.** 1.13×10^5 statvolt cm. **7.** 2.5×10^9 statamps $= 0.84$ amp. **9.** (a) 7.15×10^7 statamp $= 2.38 \times 10^{-2}$ amp, (b) 1.91×10^{-6} statvolt $= 5.7 \times 10^{-2}$ volt, (c) $F_e = 2.3 \times 10^{-16}$ dyne, $F_m = 16 \times 10^{-16}$ dyne, (d) 7.5 dynes, (e) about 2×10^{-3} gauss.

11. $i_l = \dfrac{2ihb\omega}{c^2 rR\left(1 + \dfrac{b^2}{4r^2}\right)}$. **13.** $i_2 = \dfrac{V_0 \alpha r_2}{2c^2 r_1{}^2 \rho^2}$.

Chapter 19

1. 4×10^{-14} sec. **3.** 4.8×10^4 cm sec⁻¹. **5.** 5×10^{10} sec⁻¹. **7.** 6.67×10^5 dyne cm⁻¹. **9.** $x = 3.7 \cos 1.8t$. **11.** 1.7×10^4 cm. **13.** 1.26 sec. **15.** 8.25×10^{13} sec⁻¹. **17.** 0.592 cm sec⁻².

19. $2\pi \sqrt{\dfrac{m}{(k_1 + k_2)}}$. **23.** $2\pi \sqrt{\dfrac{R_e}{g_s}} = 5.1 \times 10^3$ sec.

Chapter 20

1. $0.03 \cos\left(2\pi vt + \dfrac{\pi}{6}\right)$ gauss. **3.** 2.46×10^4 cm. **5.** (a) 5.4×10^5 erg cm⁻² sec⁻¹, (b) 1.8×10^{-5} gm cm⁻¹ sec⁻¹. **7.** 3.3×10^{-5} dyne cm⁻². **9.** 7.96×10^{-4} dyne cm⁻². **11.** 2.16 dyne cm⁻² $= 2.14 \times 10^{-6}$ atmospheres.

Chapter 21

1. 7.5×10^{-2} cm. **3.** 2.46×10^{-2} cm. **5.** 1.82×10^{10} cm sec⁻¹. **7.** 0.183 cm. **9.** $32.2°$. **11.** 4.0×10^{-2} cm. **13.** $\theta_1 = 2 \cos^{-1}\left(\dfrac{n}{2}\right)$, $\sqrt{2} < n < 2$. **15.** $10°$ in the same direction as the slab.

Chapter 22

1. 6.8×10^{-9} sec. **3.** 6.67×10^{-6} sec $= 2.5 \times$ period of light. **5.** $\tan \alpha = \dfrac{v}{c}$. **7.** 1.4×10^{-2} sec, 2.4×10^{21} cm $= 2.6 \times 10^3$ light years.

Chapter 23

1. Approximately 61.3. **3.** 1.66×10^{10} cm sec⁻¹. **5.** 5.2×10^{-6} sec. **7.** 60 cm. **9.** 0.982 c. **11.** 0.9999998 c. **13.** $1\frac{1}{4}$ years $= 456.6$ days.

15. (a) and (b) $2\pi \sqrt{\dfrac{m}{k\left(1 - \dfrac{v^2}{c^2}\right)}}$. **17.** (a) 1.683×10^{12} sec, (b) 0.1% less than (a). **19.** $\dfrac{1}{\sqrt{1 - \dfrac{V_r{}^2}{c^2}}}$ meter. B maintains that the distance between the two brushes is $\sqrt{1 - \dfrac{V_r^2}{c^2}}$ meter, but that they are not parallel to one

another. **21.** 81.65 cm. **23.** (a) 1 meter, (b) $\sqrt{1 - \dfrac{V_r^2}{c^2}}$ meter.

Chapter 24

1. 0.8 c. **3.** 6.7 gm. **5.** 50 cm. **7.** 0.512 MeV. **9.** 4.6×10^{-11} gm.
11. 1.11 gm. **13.** 2.62×10^{10} cm sec^{-1}. **15.** 1.71×10^{10} cm sec^{-1}.
17. 0.94 $c = 2.83 \times 10^{10}$ cm sec^{-1}. **19.** $\dfrac{c}{\sqrt{2}}$. **21.** 2.05×10^{-17} gm
cm sec^{-1}. **23.** 5.8×10^{-5} erg. **25.** 0.7 gm/year. **27.** 0.87 $c = 2.6 \times 10^{10}$ cm sec^{-1}. **29.** 1.96×10^{-6} cm. **31.** 0.966 c. **33.** 4×10^{13} gm, 0.11 mile. **35.** 7.8×10^{-4} u, about 2 parts in 10^4.

Chapter 25

3. 5.003×10^{-5} cm. **5.** 1.6×10^8 cm. **9.** 2.1×10^{-3} sec longer.
11. The rates of the two clocks differ by about 1.5 parts in 10^{39}.
13. $\sqrt{2} = 1.414$. **15.** 3×10^{27} cm $= 3.2 \times 10^9$ light years. **17.** The energy requirement is equivalent to about 450 years of the present power
output. **19.** $R_s = \sqrt{\dfrac{3c^2}{8\pi G\rho}}$, (a) 1.7×10^{13} cm, the distance from the sun
to the earth, (b) 2.6×10^6 cm, Mount Everest. **21.** The clock in the forward direction of the acceleration is fast by about 2.8 parts in 10^{15}.

Chapter 26

1. 3.54×10^{14} sec^{-1}, in the infrared just beyond the red end of the visible
spectrum. **3.** 3.02×10^{12} sec^{-1}. **5.** 1.99×10^{-16} cm. **7.** Its frequency
is 1.5×10^6 sec^{-1}, so it is a radio wave. **9.** (a) 4100 °K, (b) 4.1×10^9 °K,
(c) 4.1×10^{12} °K. **11.** 1.09×10^{15} sec^{-1}. **13.** 1.24×10^{-10} cm.
15. 1.24×10^{20} sec^{-1}. **17.** 1.7×10^{15} sec^{-1}. **19.** 1.55×10^9 volts.
21. 9.17×10^{15} sec^{-1}, No. **23.** About 5000 photons per sec.

Chapter 27

1. 2×10^{-20} gm cm sec^{-1}. **3.** 2.21×10^{-6} cm. **5.** 6.63×10^8 gm cm sec^{-1}. **7.** 7.37×10^{-35} cm. **9.** 3.63×10^{-6} cm. **11.** 6.63×10^{-10}erg $= 4.15 \times 10^2$eV. **13.** 1.006×10^{20}sec^{-1}. **15.** $\dfrac{c}{\sqrt{2}}$. **17.** 1.24×10^{-13} cm. **19.** 3.97×10^5 cm sec^{-1}, 1.33×10^{-13} erg $= 9.05 \times 10^{-2}$ eV.
21. 1.25×10^{-10} cm. **23.** Momentum of electron $=$ momentum of photon $= 6.63 \times 10^{-19}$ gm cm sec^{-1}. Kinetic energy of electron $= 2.4 \times 10^{-10}$ erg $= 1.5 \times 10^2$ eV. Total energy of electron $= 8.2 \times 10^{-7}$ erg $= 5.12 \times 10^5$ eV. Total energy of photon $=$ kinetic energy of photon $= 1.99 \times 10^{-8}$ erg $= 1.24 \times 10^4$ eV. **25.** 1.24×10^9 eV.

Chapter 28

1. 6×10^{-20} gm cm sec^{-1}. **5.** 2×10^{-23} gm cm sec^{-1}. **7.** 2.7×10^{-20} erg. **9.** 0.29 cm. **11.** About 7×10^{8} cm sec^{-1}, No. **13.** About 3×10^{-110} cm sec^{-1}. **15.** Within about 10^{-10} sec. **17.** About 7×10^{-28} gm. The velocity is very nearly equal to c and its uncertainty has a very small value of about $10^{-10}\, c$. **19.** About 10^{17} sec, which is comparable with the age of the universe.

Chapter 29

3. 3.29×10^{15} sec^{-1}. **5.** 4.8×10^{-8} cm. **7.** -3.16×10^{-27} gm cm^2 sec^{-1}. **9.** 2.18×10^{8} cm sec^{-1}, 1.000026. **11.** 2.47×10^{15} sec^{-1}. **13.** 4.85×10^{-3} erg $= 5.9 \times 10^{3}\, m_e c^2$. **15.** 2.55×10^{-11} cm. All energies are multiplied by a factor of 207. The frequency of the photon would be 5.11×10^{17} sec^{-1}. X-rays. **17.** 105,000 °K. **19.** 5.4×10^{-3} cm.

21. $r_n = \dfrac{h^2 n^2}{4\pi^2 e^2 m}$, $E_n = -\dfrac{\pi^2 e^4 m}{h^2 n^2}$.

Chapter 30

1. (a) 92, (b) 92. **3.** 7.22×10^{24}. **5.** (a) 0, (b) 3.01×10^{23}, (c) 3.01×10^{23}, (d) 3.7×10^{23}. **7.** 4.82×10^{24} protons. **9.** 9.31×10^{8} eV. **11.** About 10^{10} cm sec^{-1}. **13.** 4.2×10^{-7} erg. **15.** About 9 gm. **17.** 1.51×10^{-44} gm. **19.** 7.2×10^{12} cm sec^{-1}.

Chapter 31

1. 231.0364 u. **3.** 239.0521 u. **5.** (a) 5×10^{5}, (b) 2.5×10^{5}, (c) 980. **7.** 0.0215 u. **9.** 0.083. **11.** 3.2×10^{7} gm. **13.** 2.6×10^{7} cm sec^{-1}. **15.** $\dfrac{E_\alpha m_\alpha}{m_d} \approx \dfrac{4E_\alpha}{(A-4)}$, where m_d is the mass of the daughter nucleus. **17.** 1.24 MeV. **19.** 28.5 MeV. **21.** 1.04×10^{10} years. **23.** The energy of sunlight is greater by a factor of 5.6×10^{4}. **25.** About half a million years.

Chapter 32

1. 6.6×10^{-14} cm. **3.** 8.2×10^{-11} cm. **5.** 1.19×10^{-2} erg $= 7.46 \times 10^{9}$ eV. **7.** (a) $\eta^\circ \rightarrow 4\pi^\circ$ is possible, $\eta^\circ \rightarrow \pi^+ + \pi^\circ + \pi^\circ + \pi^-$ is borderline, (b) No. **9.** 5.46×10^{-17} gm cm sec^{-1}, $\sqrt{\tfrac{4}{5}}\, c = 2.68 \times 10^{10}$ cm sec^{-1}, 0.635 MeV. **11.** 2.9×10^{-6} erg $= 1.81$ MeV. **13.** (a) 1.68×10^{-4} erg $= 105$ MeV, (b) 5.42×10^{-5} erg $= 33.9$ MeV, (c) 4.05×10^{-4} erg $= 253$ MeV, (d) 5.98×10^{-5} erg $= 37.3$ MeV, (e) 1.23×10^{-4} erg $= 76.7$ MeV, (f) 3.58×10^{-4} erg $= 223$ MeV. **15.** 26.7 MeV. **17.** 2.5×10^{7} cm sec^{-1}.

Chapter 33

1. 2.3×10^{-65} gm. **3.** 3.6×10^{51}. **5.** $B = +1$, $J = \tfrac{3}{2}$, $I = \tfrac{3}{2}$, $I_3 = +\tfrac{3}{2}$, $S = 0$, $Y = +1$, $Q = +2$.

Answers to Odd-Numbered Problems

Index

Index

INDEX

SOME USEFUL CONVERSION FACTORS

1 meter	$= 100$ cm
1 mile	$= 1.61 \times 10^5$ cm
1 mph	$= 44.7$ cm sec^{-1}
1 kg	$= 1000$ gm
1 year	$= 3.16 \times 10^7$ sec approximately
1 light year	$= 9.46 \times 10^{17}$ cm
1 newton	$= 10^5$ dyne
1 joule	$= 10^7$ erg
1 radian	$= 57.30°$
2π radians	$= 1$ revolution
	$= 360°$
1 atmosphere pressure	$= 1.01 \times 10^6$ dyne cm^{-2}
	$= 14.70$ pounds per square inch
	$= 76$ cm of mercury
A pressure of 1 cm of mercury	$= 1.33 \times 10^4$ dyne cm^{-2}
1 electron volt	$= 1.60 \times 10^{-12}$ erg

$$T°K = t°C + 273.15$$

PRACTICAL UNITS OF ELECTRICITY

1 coulomb	$= 3 \times 10^9$ statcoulomb
1 ampere	$= 3 \times 10^9$ statampere
1 volt	$= \dfrac{1}{300}$ statvolt
1 ohm	$= 1.11 \times 10^{-12}$ statohm